Localized Excitations in Solids

W0193009

Localized Excitations in Solids

Edited by R. F. WALLIS
U. S. Naval Research Laboratory
Washington, D. C.

Proceedings of the First International Conference on Localized Excitations in Solids, held at the University of California at Irvine, September 18-22, 1967

Springer Science+Business Media, LLC 1968

Library of Congress Catalog Card Number 68-19187

ISBN 978-1-4899-6218-8 ISBN 978-1-4899-6445-8 (eBook)
DOI 10.1007/978-1-4899-6445-8

© 1968 Springer Science+Business Media New York

Originally published by Plenum Press in 1968.
Softcover reprint of the hardcover 1st edition 1968

All rights reserved

No part of this publication may be reproduced in any form without written permission from the publisher

FOREWORD

The first International Conference on Localized Excitations in Solids was held at the Irvine campus of the University of California on September 18-22, 1967. The Conference was supported by the International Union of Pure and Applied Physics, the U. S. National Science Foundation, the U. S. Office of Naval Research, the U. S. Army Research Office, the University of Southern California, and the University of California, Irvine.

The initial idea for an International Conference on Localized Excitations in Solids was conceived in discussions between Professors Maradudin, Burstein, and Krumhansl. It was felt that such a Conference would render a valuable service in bringing together scientists actively engaged in the study of localized excitations to survey the work to date, describe current research, and point out areas for future exploration. It was also felt that the very rapid increase in research on the subjects of interest made such a Conference particularly desirable at this time. Furthermore, an opportunity would be provided for close interaction between workers in diverse areas of solid state physics such as lattice dynamics, magnetism, and electronic structure.

The localized excitations which formed the subject matter of the Conference are localized phonons, localized magnons, localized plasmons, and localized excitons. These subjects possess the unifying feature that the theory is substantially the same for each. The papers presented covered such topics as the fundamental characteristics of localized excitations, the interactions of localized excitations, and the effects of localized excitations on observable physical properties.

Approximately 120 scientists attended the Conference. Countries represented among the participants were Canada, Denmark, England, France, Germany, Israel, Italy, Japan, New Zealand, Scotland, the Soviet Union, Sweden, and the United States. A particular honor was the presence of Professor I. Waller, one of the founders of the field of lattice dynamics, and Professor E. W. Montroll, one of the first to investigate localized phonons.

The success of the first International Conference on Localized Excitations in Solids was due in large measure to the efforts of Professor A. A. Maradudin of the University of California, Irvine. He provided not only the initiative, but also the insight, effort, and attention to detail that were required to insure that the Conference would be a memorable and valuable experience for the participants. To Professor Maradudin, many thanks for a job well done. Appreciation must also be expressed to Professor Elias Burstein and the other members of the Program Committee for the outstanding program which they arranged. Special thanks are due Mr. Rodney Rose, Business Manager of the Department of Physics, University of California, Irvine, who very capably handled many details associated with the smooth operation of the Conference, the comfortable living arrangements, and the social and recreational events. Thanks are also due Mrs. Sandy Mills and Mrs. Peg Maradudin who were co-chairmen of the ladies program and Mrs. Marty Pelke of the Conference Staff who provided much help in many ways. Finally, deep appreciation must be expressed for the splendid hospitality shown the Conference and its participants by the University of California, Irvine.

R. F. Wallis

CONFERENCE ORGANIZATION

Chairman of the Conference

E. W. Montroll

Conference Secretary

A. A. Maradudin

Planning Committee

A. A. Maradudin
E. Burstein
J. Callaway
A. Heeger
J. J. Hopfield
M. V. Klein
J. A. Krumhansl
R. Orbach
R. O. Pohl
A. J. Sievers
W. G. Spitzer
R. F. Wallis

International Advisory Committee

M. Balkanski
R. Barrie
R. J. Elliott
J. Hori
Yu. M. Kagan
W. Ludwig
G. F. Nardelli
I. Waller

Program Committee

E. Burstein
A. J. Glick
A. Heeger
M. V. Klein
R. G. Wheeler

Publications Committee

R. F. Wallis
D. L. Mills
L. J. Sham
W. G. Spitzer

Session Chairmen

I. Waller
G. F. Nardelli
J. Hori
W. G. Spitzer
Yu. M. Kagan
T. Wolfram
R. Barrie
W. Ludwig
L. Genzel
E. Burstein

CONTENTS

G. THERMAL PROPERTIES OF DEFECT MODES

H. DEFECT MODES ASSOCIATED WITH COLOR CENTERS

PART A. INTRODUCTORY PAPERS

LOCALIZED, GAP, AND RESONANCE MODES

A.A. Maradudin*
University of California
Irvine, California

I. Introduction

It is only seven years since the first experimental discovery of localized vibration modes in crystals,[1] but in those seven years more studies, both theoretical and experimental, of exceptional vibration modes have been carried out than in the first eighteen years following their theoretical discovery by I.M. Lifshitz.[2] It seemed appropriate, therefore, at a Conference devoted to localized excitations in solids to survey the studies of localized vibrations in solids, which have been carried out in the past few years. Since a great deal of the existing work has been surveyed elsewhere recently[3], I will restrict my remarks here largely to the results of very recent investigations.

II. Localized Modes

High frequency localized modes were the first kind of impurity induced exceptional vibration modes to be observed experimentally. They have been observed by various experimental techniques in ionic crystals, in polar and homopolar semi-conductors, and in metals.

In the cases of ionic crystals and polar semi-conductors the most widely studied impurities which give rise to localized vibration modes occupy sites of cubic symmetry. They are charged impurities, and the localized modes to which they give rise therefore have a non-zero dipole moment associated with them which can couple

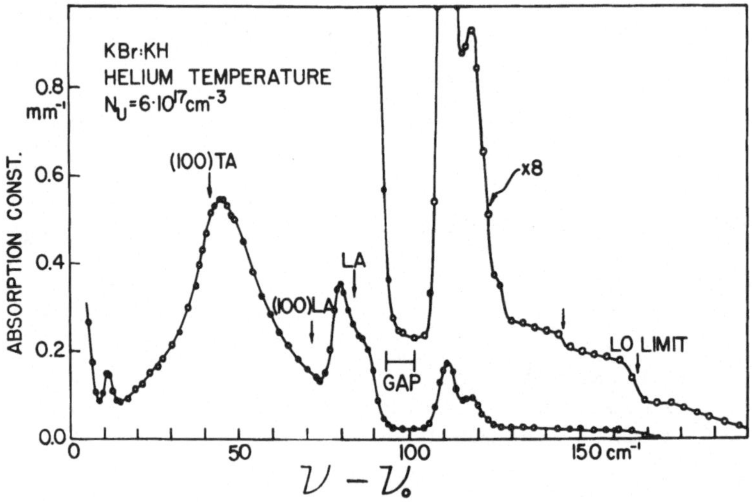

Fig. 1. The high frequency sideband to the localized mode peak due to H^- U-centers in KBr.

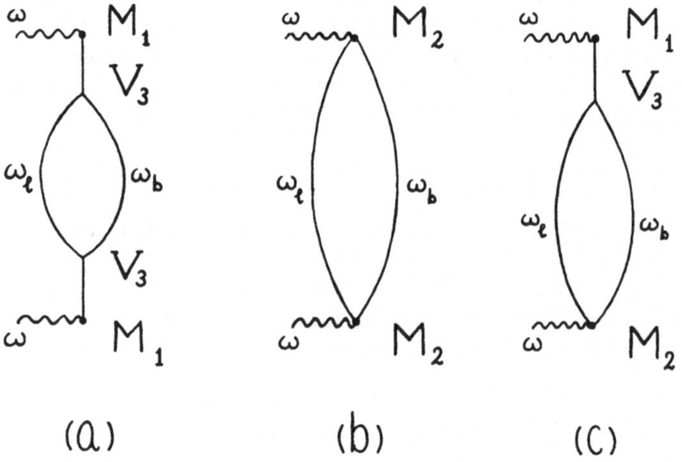

Fig. 2. The three dominant mechanisms which give rise to sidebands to localized mode peaks.

to an external radiation field to extract energy from it. The selection rules governing one phonon absorption processes are that the frequency of the incident light should equal the frequency of the localized mode. There is no wave vector conservation rule which has to be satisfied, because the introduction of impurities into the host crystal destroys the periodicity of the crystal which gives rise to this selection rule in perfect crystals. The selection rules imposed by the symmetry of the impurity site are satisfied in this case and the localized mode is observed as a sharp peak in the absorption spectrum.

All the localized modes observed in ionic crystals by infrared techniques have been introduced by the presence of either hydride or deuteride U-center impurities. Frequencies of these modes have now been measured in many different crystals. Although they are of fundamental importance in devising force constant models to describe the interaction between the U-center and the host crystal such results are otherwise not very exciting. Accordingly, interest has shifted to various properties of U-center localized modes in ionic crystals, such as their response to external forces and properties associated with the anharmonicity of the interatomic forces. Many of these effects will be discussed in considerable detail at this Conference, so that I should like to confine myself to one particular property of localized modes which has consequences for the study of other kinds of exceptional vibration modes as well. I refer to the side bands to the localized mode peak which were first discussed by Fritz.(4) In Fig. 1 are displayed the results for the high frequency side band to the localized mode peak due to hydride U-centers in potassium bromide. Frequency is measured from the center of the localized mode line. These are data of Timusk and Klein.(5) There are three dominant mechanisms which give rise to these side bands. I have tried to illustrate them graphically in Fig. 2. The first is a process in which the incident light interacts with a localized mode through its first order dipole moment whereupon the localized mode breaks up into a second localized mode and a band mode through cubic anharmonic terms in the crystal potential energy. Energy conservation requires that the frequency of the incident light equal the frequency of the localized mode plus or minus the frequency of the band mode. This mechanism was first suggested by Fritz (4) to explain his experimental observations. The second mechanism is one in which the incident light excites two normal modes directly through the crystal second order dipole moment.[6] One of these modes is a localized mode and the other is a band mode. Again, energy conservation requires that the frequency of the incident light equal the frequency of the localized mode plus or minus the frequency of the band mode. The third mechanism for the production of side bands is the interaction between the first two processes.[7] In this case the incident light interacts directly with a localized mode through its first order dipole moment, the localized mode then breaks up into a localized mode and a band mode through the cubic anharmonic terms in the crystal potential energy, and then this process interferes constructively with the process in which the light excites directly a localized mode and a band mode. Interference between one and two phonon processes induced by the anharmonicity of the crystal were first discussed in the context of the scattering of neutrons by anharmonic crystals by Ambegaokar, Baym, and Conway,[8] and numerical calculations of these cross terms by Ambegaokar and Maradudin[9] showed them to be very small in that context. For each mechanism the shape of the sideband which is obtained is a weighted frequency spectrum of the perturbed host crystal, and it is just because they provide a means of studying at least part of this frequency spectrum that the localized mode sidebands are so interesting to study. The consensus at the present time is that the dominant mechanism responsible for the side bands of the localized mode peak is the first that I described, namely that through cubic anharmonic processes, but this is not a unanimous opinion.

The site symmetries of atoms in homopolar crystals of the diamond structure or in polar semiconductors of the zincblende structure is T_d. One of the consequences of this fact is that the localized modes associated with substitutional impurities in such crystals give rise to a first order Raman scattering of light. Just as a vibrating dipole radiates energy, so do the electrons surrounding an atom radiate energy when they are set into vibration by an external electromagnetic field. Because the nucleus which these electrons surround is itself vibrating, the re-radiated energy undergoes a Doppler shift with respect to the frequency of the exciting light, and the shift in the frequency of the re-radiated light is equal to the frequency of the vibrating nucleus. In a localized vibration mode it is only the impurity atoms which are vibrating to a good approximation, and the intensity distribution of the light scattered from a crystal possessing localized modes shows a sharp maximum at a frequency shift equal to the localized mode frequency.

Recently Feldman, Ashkin, and Parker[10] have succeeded in observing the Raman scattering of light by localized vibration modes associated with silicon impurities in germanium. Their experimental results are shown in Fig. 3. The maximum frequency of germanium is 350 cm^{-1}. The peak in the Raman spectrum at 400 cm^{-1} is identified as the localized mode peak. A second peak at 462 cm^{-1} is tentatively identified as the localized mode frequency due to a nearest neighbor pair of silicon impurity atoms. This work represents the first observation of localized vibration modes by Raman spectroscopy, a tool which is bound to increase in usefulness for these kinds of observations as time goes on.

III. Gap Modes

I should now like to discuss briefly a particular kind of localized vibration mode which is characterized by the fact that its frequency, instead of lying above the maximum frequency of the perfect host crystal, falls in the gap in the frequency spectrum of the host crystal between the acoustic and optical branches. Since this mode has a frequency which lies in a stop band for the crystal, the displacements of the atoms vibrating in it also decay exponentially with increasing distance from the impurity site, as in the case of high frequency localized modes. Studies of such localized modes, which I will call gap modes, are constrained by the fact that very few crystals are known which have gaps in their frequency spectra. These are all alkali-halide crystals: sodium iodide, sodium bromide, potassium iodide, potassium bromide, lithium chloride, and cesium fluoride. Of these sodium iodide, sodium bromide, lithium chloride, and cesium fluoride are hygroscopic and consequently unpleasant to work with experimentally. This leaves only potassium iodide and potassium bromide as crystals in which gap modes can be observed readily. Two experimental techniques have been used for this purpose to date. The first is infrared absorption, and the second is what I will call the optical analogue of the Mössbauer effect.

The first experimental observation of a gap mode was in work of Sievers on potassium iodide containing cation or anion impurities. Recently, Sievers and his colleagues[11] have examined with high resolution the gap modes introduced into KI by Cl^- impurities. Their results are shown in Fig. 4. From the phonon dispersion curves of potassium iodide obtained by neutron spectroscopy by Dolling and his colleagues[12] it is

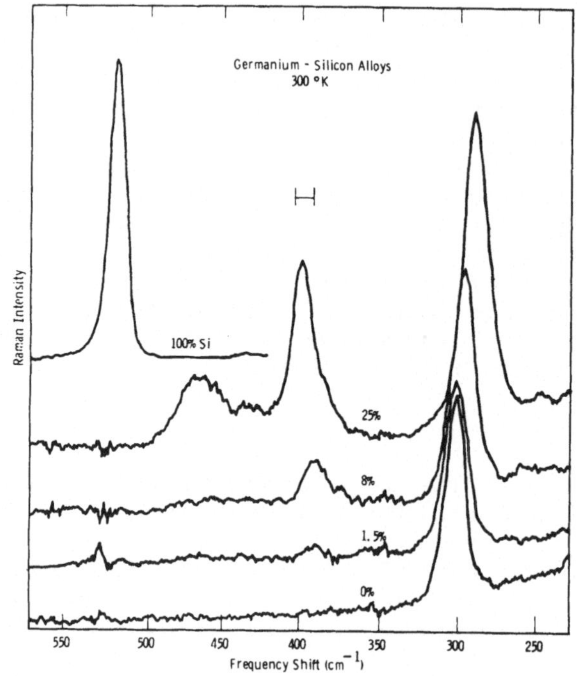

Fig. 3. First order Raman spectrum of Ge containing Si impurities.

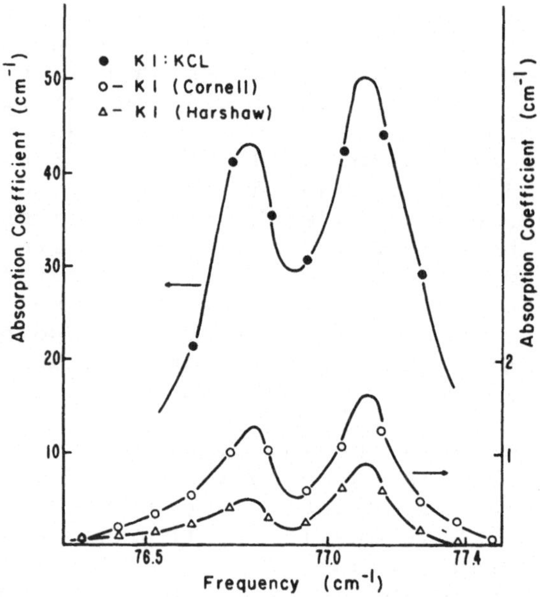

Fig. 4. The IR absorption spectrum of KI containing Cl^- impurities in the I^- sublattice.

known that the frequency spectrum of potassium iodide has a gap which extends from 69.8 cm^{-1} to 95.6 cm^{-1}. The prominent peaks in this figure, at 76.79 cm^{-1} and 77.10 cm^{-1} fall well into the gap in the frequency spectrum, and are identified as the absorption by gap modes pushed into the gap out of the top of the acoustic spectrum by the light chloride impurities. The two prominent peaks are associated with gap modes induced by the two most abundant isotopies of chlorine, chlorine35 and chlorine 37 present in their natural abundance ratio of approximately 3:1. Attempts to fit the two frequencies with the same pair of nearest neighbor central and noncentral impurity force constants by Benedek and Maradudin(13) proved to be unsuccessful. Any pair of force constants which reproduces the experimental frequency for chlorine 35 will not reproduce the experimental frequency for chlorine 37, and vice versa. A similar situation was recently encountered by MacDonald(14) in attempting to fit the frequencies of localized modes due to hydride and deuteride impurities in potassium chloride. These results are puzzling at first glance because one would think that in cases of isotopic substitution the force constants should be the same for the two isotopes. Possible explanations as to why this is not the case are that on the effects of the impurity ion on the surrounding host crystal extend beyond its nearest neighbors, so that a more elaborate force constant model should be employed, and that the larger mean square displacement amplitude of the lighter of the two isotopic species means that it samples the environment of its nearest neighbors more than does the heavier isotope so that through the anharmonic terms in the crystal potential energy it is coupled to the surrounding host crystal by different effective harmonic force constants.

I have mentioned earlier that sidebands to the localized mode peaks associated with U-center impurities in alkali-halide crystals are a weighted frequency spectrum of the host crystal. In particular, if the U-center gives rise to gap modes which, while not infrared active themselves, have the appropriate symmetry, for example A_{1g}, E_g, or F_{2g}, to contribute to the sidebands, they can be observed in the frequency dependence of the sideband. This seems to be the case when U-centers are present in potassium iodide. These centers apparently give rise to a gap mode of A_{1g} symmetry which has been observed in the sideband structure measured by Timusk and Klein(15) and shown in Fig. 5. The peak at 93.5 cm^{-1} in this figure lies in the gap of the frequency spectrum in potassium iodide and is attributed to ab-

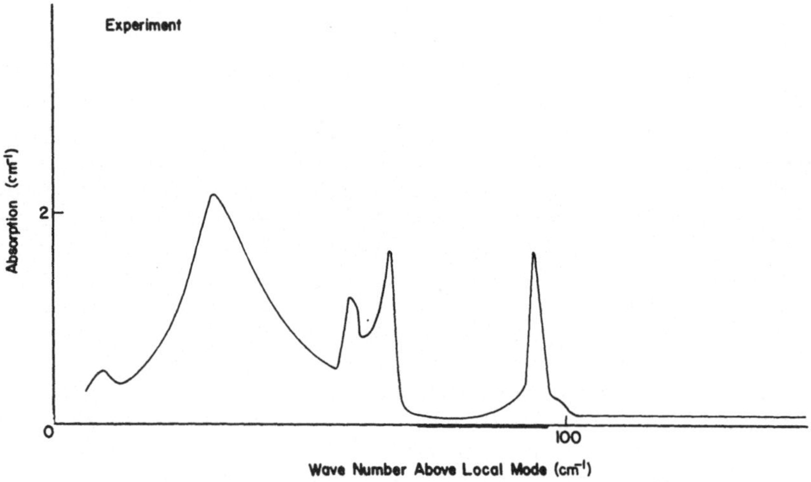

Fig. 5. The high frequency sideband to the localized mode peak due to H^-U-centers in KI.

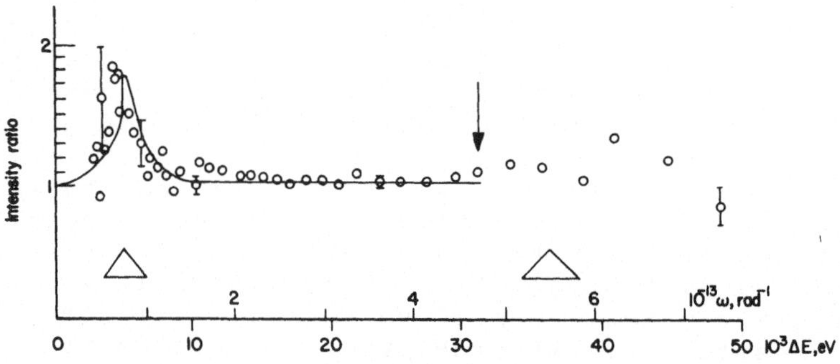

Fig. 6. The ratio of the cross section for neutron scattering by Mg containing 2.8 at % Pb to the cross section for scattering by pure Mg.

sorption by gap modes of A_{1g} symmetry.

IV. Resonance Modes

The last kind of exceptional vibration modes that I would like to discuss are resonance modes. Unlike localized and gap modes resonance modes are not true normal modes of a perturbed harmonic crystal. A crude physical explanation of the nature of these modes can be given as follows. If the impurity ion is very heavy and/or is coupled very weakly to the surrounding host crystal, then its motion in low frequency normal mode vibrations of the perturbed crystal is the following. At very low frequencies, due to infinitesimal translation invariance, it vibrates in phase with its neighbors in the host crystal. However, as the frequency of the normal modes increases because of its heavy mass and/or weak binding to the crystal, it begins to lag more and more behind its neighbors, until a frequency is reached at which it begins to vibrate 180° out of phase with the surrounding lattice in a kind of local optical vibration mode. The frequency at which this occurs is called the frequency of a resonance mode. The mean square vibration amplitude of the impurity atom as a function of its frequency is sharply peaked at the resonance mode frequency, but the mode is not spatially localized in the way that a high frequency localized mode or a gap mode is. In addition, because the frequency of a resonance mode falls in the range where the density of vibrational frequencies of the host crystal is non-zero, it can decay into the continuum of band modes and acquires a width in this manner.

Resonance modes have been studied experimentally in ionic crystals and in metals. I will confine my remarks here to recent experimental studies of resonance modes in metals, where they have been investigated by neutron spectroscopy, by specific heat measurements, and by measurements of electrical resistivity.

It is well known, I think, that the double differential scattering cross section for the inelastic incoherent scattering of neutrons by one phonon processes by cubic Bravais crystals traces out the frequency spectrum of the crystal through its energy dependence. The situation is not quite this simple in the scattering of neutrons from crystals containing impurities. However, as a first approximation, the energy dependence of the double differential scattering cross section in this case also traces out the frequency spectrum of the crystal. This means that if resonance modes have been introduced into the spectrum of normal mode vibrations of a crystal by the introduction of impurities into it, they give rise to peaks in the frequency spectrum at the frequencies of the resonance modes. These peaks can therefore be observed in the energy dependence of the incoherent one phonon scattering cross section for the crystal. Such experiments have recently been carried out for an alloy of magnesium containing small concentrations of lead by Chernoplekov and Zemlyanov[16] whose experimental results are shown in Fig. 6. This figure shows the ratio of the cross section for neutron scattering by magnesium containing 2.8 atomic per cent lead to the cross section for the scattering by pure magnesium. The solid curve is a theoretical prediction for this ratio obtained on the basis of a Debye spectrum for the host crystal, and the arrow indicates the maximum frequency of pure magnesium. The sharp peak at an energy transfer of about 5 millielectron volts is attributed to scattering by resonance modes induced by the heavy lead impurities. From the hexagonal symmetry of the magnesium host lattice one would expect two resonance modes associated with each lead impurity, but the experimental resolution does not permit such a possible fine structure in the resonance to be resolved.

A second method for observed resonance modes in metals through neutron spectroscopy is to study the double differential cross section for the coherent inelastic scattering of neutrons by one phonon processes. In the case of a perfect crystal the energy and momentum conservation conditions which govern these scattering processes enable one to infer the phonon dispersion curves from the results of such measurements. In the case of a disordered alloy, it is known from theoretical calculations that the frequencies of the quasi-phonons, which are now the elementary excitations of the system, are shifted from their values in the pure host crystal. In addition, the disorder in the crystal leads to an uncertainty in the energies of these quasi-phonons which shows up as a width to the measured neutron peaks, over and above the width due to lattice anharmonicity. It was pointed out by Elliott and Maradudin[17] that if impurities in a crystal give rise to low frequency resonance modes, these modes give rise to a pronounced maximum in the frequency dependence of the widths of the experimentally observed neutron bunches at the frequency of the resonance mode, and at the same time the shift in the phonon frequency from its value in the pure host crystal goes through zero at the frequency of the resonance mode. These predictions were subsequently verified experimentally by several workers. In Fig.7 are shown experimental results of Svensson[18] for an alloy of copper containing 3 atomic per cent gold. The upper curve gives the comparison between the experimental and theoretical results for the frequency shift, while the lower curve gives the comparison between the experimental and theoretical results for the line width of the phonons propagating in the [011] direction in this crystal. The various solid and dashed curves are theoretical results calculated on the basis of either the simple explicit expressions of Elliott and Maradudin for the shift and width, on the basis of the complete expression for the line shape and with or without inclusion of corrections for instrumental resolution. The agreement between theory and experiment is seen to be quite good, both qualitatively and quantitatively.

It is found theoretically that resonance modes give rise to a sharp peak in the frequency spectrum of a disordered crystal near the frequency associated with the resonance mode due to an isolated heavy impurity. Consequently, one would expect that the specific heat of such a crystal should increase anamalously in a temperature range where such resonance modes just begin to be excited. This suggestion was first made theoretically by Lehman and DeWames,[19] and subsequently by Kagan and Iosilevskii.[20] It has recently been verified experimentally by several groups. In Fig.8 are shown the experimental results of Cape, Lehman, Johnston and DeWames [21] for the excess specific heat of magnesium containing 0.5 and 1 atomic percent lead over that of pure magnesium. The solid curves are theoretical predictions for the excess specific heat which were calculated on the basis of a rather realistic model of magnesium, and the assumption that the lead atoms can be approximated by isotopic impurities. The agreement between theory and experiment is gratifyingly good.

The final method by which resonance modes have been studied experimentally in metals is through anomalies in the low temperature electrical resistivity to which they give rise. Because the mean square displacement of the impurity atom in a resonance mode is very large, its cross section for the scattering of electrons is also very large. This suggests that the temperature dependence of the electrical resistivity of an alloy whose spectrum contains resonance modes induced by the impurities, should have anomalously large values in a temperature range at which the resonance modes are being excited. The theory of the scattering of conduction electrons by resonance modes in alloys has been worked out by Kagan and Zhernov.[22] Recently, Panova, Zhernov, and Kutaitsev[23] have measured the electrical resistivity of magnesium-lead and magnesium-silver alloys. Their experimental results are shown in Fig.9, on which is displayed the temperature dependence of the relative change of the resistance for alloys of magnesium and lead. Since the experimental values have been divided by the concentration, and since all of the points seem to fall rather nicely on a universal curve, the peak at about 60°K is attributed to the enhancement of the resistivity caused by the anomously large scattering of the conduction electrons by the resonance modes introduced by the heavy lead impurities. The solid curve is a theoretical calculation of the resistivity based on the Kagan-Zhernov theory.

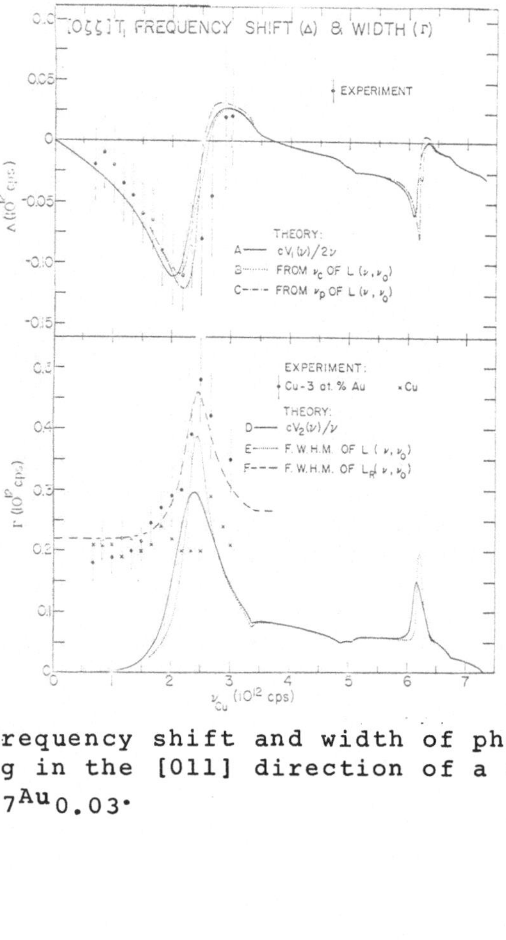

Fig. 7. The frequency shift and width of phonons propagating in the [011] direction of a crystal of $Cu_{0.97}Au_{0.03}$.

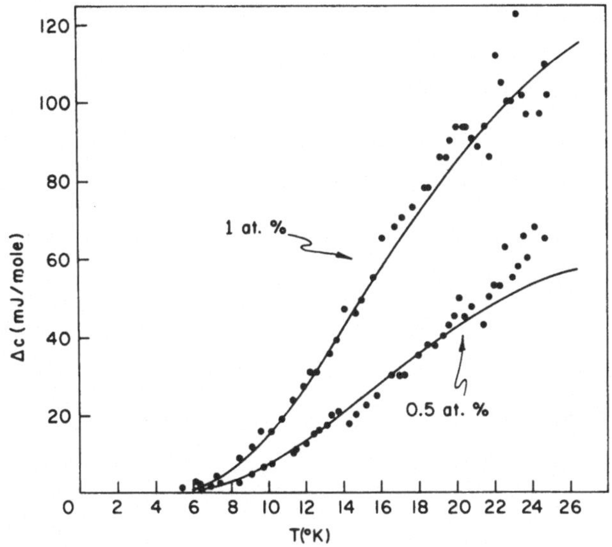

Fig. 8. The excess specific heat of Mg containing Pb impurities over the specific heat of pure Mg.

V. Conclusion

Localized, gap, and resonance modes give rise to a large number of effects in crystals, and a variety of experimental techniques has been used in the study of these various effects. Some of these techniques are now standard methods for studying impurity induced properties in crystals. Others are just beginning to be used, and it is for the future to decide how useful they will be in the study of defect induced properties. At the same time the theory of the dynamical properties of impurity atoms in crystals is also becoming more sophisticated. Difficult problems are being tackled. The challenge presented by a wealth of experimental information against which the prediction of theories can be checked has forced theorists to use more realistic models of impurities and of host crystals. The study of exceptional vibrational modes in crystals has developed so rapidly in the past seven years, that if progress continues at a comparable rate a talk on this subject at a conference on localized excitations seven years hence could undoubtedly be devoted to a comparable number of effects and experimental techniques which are still unthought of today. What an exciting prospect that is!

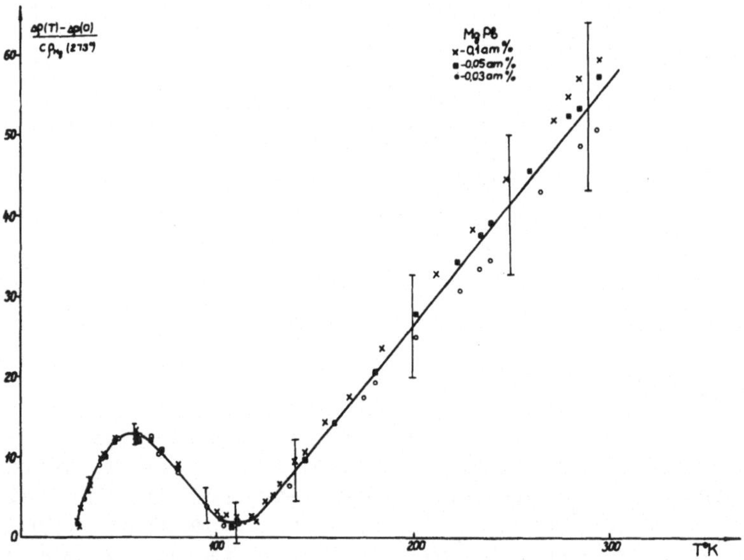

Fig. 9. The electrical resistivity of Mg - Pb alloys.

References

*This research was partially supported by the Air Force Office of Scientific Research, Office of Aerospace Research, United States Air Force, under AFOSR Grant Number 1080-66.

1. G. Schaefer, J. Phys. Chem. Solids 12,233 (1960).

2. I.M. Lifshitz, J. Phys. U.S.S.R. 7,215(1943); 7,249 (1943); 8,89(1944).

3. A.A. Maradudin in Solid State Physics, edited by F. Seitz and D. Turnbull (Academic Press, Inc., New York, 1966) Vol.18,p.273; Vol.19,p.1.

4. B. Fritz, in Lattice Dynamics, edited by R.F.Wallis (Pergamon Press, New York, 1965)p.485.

5. T. Timusk and M.V. Klein, Phys. Rev. 141,664(1966).

6. Nguyen Xuan Xinh, Westinghouse Research Laboratories Scientific Paper 65-9F5-442-P8, Sept.3, 1965; Solid State Communications 4,9 (1966).

7. H. Bilz, D. Strauch, and B. Fritz, J.de Physique, Supplement to vol. 27, May-June 1966, p.C2-3.

8. V. Ambegaokar, J. Conway, and G. Baym, in Lattice Dynamics, edited by R.F. Wallis (Pergamon Press, New York, 1965) p.261.

9. A.A. Maradudin and V. Ambegaokar, Phys. Rev. 135, A1071 (1964).

10. D. Feldman, M. Ashkin, and J.H. Parker,Jr., Phys. Rev. Letters 17, 1209 (1966).

11. I.G. Nolt, R.A. Westwig, R.W. Alexander,Jr., and A.J. Sievers, Phys. Rev. 157,730 (1967).

12. G. Dolling, R.A. Cowley, D.S. Schittenhelm, and I.M. Thorson, Phys. Rev. 147,577 (1966).

13. G. Benedek and A.A. Maradudin (to be published)

14. R.A. MacDonald, Phys. Rev. 150,597 (1966).

15. Reported in T. Gethins, T. Timusk, and E.J. Woll,

Phys. Rev. 157,744 (1967).

16. N.A. Chernoplekov and M.G. Zemlyanov, Zhur. Eksper. i Teor. Fiziki 49,449 (1965). [English translation: Soviet Physics - JETP 22,315 (1966).]

17. R.J. Elliott and A.A. Maradudin, Proc. Intern. Conf. on Inelastic Neutron Scattering From Solids and Liquids, Bombay, 1964 (Internationsal Atomic Energy Agency, Vienna, 1965) p.231.

18. E.C. Svensson, Ph.D. Thesis, McMaster University, 1967 (unpublished).

19. G. Lehman and R.E. DeWames, Phys. Rev. 131,1008 (1963).

20. Yu. M. Kagan and Ya. A. Iosilevskii, Zh. Eksper.i Teor. Fiz. 45,819 (1963) [English translation: Soviet Physics - JETP 18,562 (1964).]

21. J.A. Cape, G. Lehman, W.V. Johnston, and R.E. De Wames, Phys. Rev. Letters 16,892 (1966).

22. Yu. M. Kagan and A.P. Zhernov, Zhur. Eksper. i Teor. Fiziki 50,1107 (1966). [English translation: Soviet Physics - JETP 23,737 (1966).]

23. G. Kh. Panova, A.P. Zernov, and V.I. Kutaitsev, I.V. Kurchatov Atomic Energy Institute Preprint 1299, Moscow (1967).

ASYMPTOTIC DESCRIPTIONS OF DEFECT EXCITATIONS

J. A. Krumhansl*

Laboratory of Atomic and Solid State Physics

Cornell University

I. Introduction

Within the wide variety of elementary excitations of defect solids those referred to as "resonances" or "localized" modes are particularly interesting and occur frequently. Theoretical methods have been developed for handling the defect excitation problem generally; the Green's function approach is particularly useful and has been applied to defect vibrations, localized moments, and localized electronic excitations. The input to these theoretical analyses range from parameterized models to actual experimental lattice dispersion or band data. But frequently certain asymptotic features appear independently of the model details. For example, in the defect vibration case both localized and resonance modes seem to have many of the features of a simple Einstein oscillator. It is the purpose of the present remarks to indicate how this arises. The method has developed from previous work, particularly in collaboration with M. Wagner and with J. A. D. Matthew;[1] it is similar to that of Okazaki et. al.[2] The results are not only pedagogical but also may be used for numerical estimates. For simplicity only defect lattice vibrations in the harmonic approximation are discussed here. The extension to other systems is apparent.[2]

II. General Method

The lattice dynamic Green's function $\underline{\underline{G}}$ in matrix notation obeys the equation

$$(\underline{\underline{V}} - \omega^2 \underline{\underline{1}}) = - \underline{\underline{1}} \qquad (1)$$

where V is the mass normalized potential matrix.

The defect problem may be stated

$$V = V_R + D \tag{2}$$

where V_R denotes some reference configuration and D is a "defect" therein. Given the reference configuration quantities V_R and $G_R = (\omega^2 - V_R)^{-1}$ then in general

$$G = \frac{1}{1 - G_R\, D}\, G_R \tag{3}$$

The usual choice of reference basis is $V_R = V_o$, the potential for the perfect lattice. Then $D = D_o$ contains all deviations from V_o i.e. $D_o = (V - V_o)$. Using the special notation $G_o = P$ equation (3) leads to

$$G = \frac{1}{1 - P\, D_o}\, P \tag{4}$$

This is completely rigorous and has been used in almost all defect vibration theories. However, it does not serve to bring out asymptotic properties most directly, particularly when $D_o \approx -V_R$ over some region; this corresponds to decoupling of the defect from the rest of the crystal.

It may be the case that either because of the nature of the defect or the frequency range in question well defined physical considerations suggest that the total system can be subdivided into a defect subspace and a complementary "crystal" region, whence it is natural to partition the potential matrix in the form

$$V = \begin{bmatrix} L' & a' \\ (a')^T & C' \end{bmatrix} \tag{5}$$

where L', and C' are square matrices in the defect subspace and in the crystal subspace, while a', $(a')^T$ are coupling matrices between L' and C'. This decomposition is exact but not unique, and may be carried out in a variety of ways; the point is that physical knowledge of the asymptotic situation will lead to a good chance of L' to include those few degrees of freedom of the small defect region which participate most strongly in the excitation. It is apparent that a useful reference potential to choose is not the perfect lattice, but rather

$$V_R' = \begin{bmatrix} L' & 0 \\ 0 & C' \end{bmatrix} \tag{6}$$

whence the reference Green's function is

$$\begin{aligned} \underline{\underline{G}}_R' &= (\omega^2 - \underline{\underline{L}}')^{-1} \text{ (on the defect subspace, L')} \\ \underline{\underline{G}}_R' &= (\omega^2 - \underline{\underline{C}}')^{-1} \text{ (on the crystal subspace, C')} \end{aligned} \tag{7}$$

The Green's function $(\omega^2 - \underline{L}')^{-1}$ is that for a small "psuedo-molecule," while $(\omega^2 - \underline{C}')^{-1}$ is that for an excavated crystal i.e. disconnected from the defect subspace.

In this basis the defect matrix is

$$\underline{\underline{D}}' = \begin{bmatrix} 0 & a' \\ (a')^T & 0 \end{bmatrix} \tag{8}$$

Applying (3) one finds

$$\underline{\underline{G}} = [\omega^2 - \underline{\underline{L}}' - \underline{\underline{a}}'(\omega^2 - \underline{\underline{C}}')^{-1} (\underline{\underline{a}}')^T]^{-1} \quad \text{(in L')} \tag{9}$$

$$\underline{\underline{G}} = [\omega^2 - \underline{\underline{C}}' - (\underline{\underline{a}}')^T (\omega^2 - \underline{\underline{L}}')^{-1} (\underline{\underline{a}}')]^{-1} \text{ (in C')} \tag{10}$$

The additional terms have the form of corrections to the reference Green's function. In equation (10) the result is to give phonon scattering, which is observed in thermal conductivity; this will not be considered further here because of space limitations. But optical excitations can couple directly to the defect degrees of freedom so we consider the implications of equation (9) in further detail.

Given a set of symmetry coordinates $\underline{\alpha}$ (meaning row or column vector) for $\underline{\underline{L}}'$ such that

$$\underline{\underline{L}}' \, \underline{\alpha} = \omega_\alpha^2 \, \underline{\alpha} \tag{11}$$

One may construct from these psuedo-molecule normal modes

$$\underline{\underline{G}}_R' = \sum_\alpha \underline{\alpha}(\omega^2 - \omega_\alpha^2)^{-1} \; \underline{\alpha} \; ; \quad \text{(in L')} \tag{12}$$

This defect system would have undamped resonances at the frequencies ω_α. Expanding the matrix inverse in equation (9) as in many body perturbation theory one finds corrections to equation (12), and to leading terms

$$\underline{\underline{G}} \simeq \sum_\alpha \underline{\alpha}(\omega^2 - \omega_\alpha^2 - \Sigma_{\alpha\alpha})^{-1} \underline{\alpha}; \text{ (in L')} \tag{13}$$

where $\Sigma_{\alpha\alpha}$ is the self energy correction i.e. shift and broadening

$$\Sigma_{\alpha\alpha} \simeq \Delta_\alpha + i\Gamma_\alpha = \lim_{\varepsilon \to o} [\underline{\alpha}\ \underline{\underline{a}}'(\omega^2 + i\varepsilon - \underline{\underline{C}}')^{-1} (\underline{\underline{a}}')^T \underline{\alpha}] \qquad (14)$$

Asymptotic behavior is well described by a psuedo molecule (in L') if $\Sigma_{\alpha\alpha}$ is small.

III. An Illustrative Example

We concentrate on one example -- a single point defect in a monatomic simple cubic lattice. This model has been used often in formal model calculations. Only nearest neighbor central force constants are considered with v and v_o, m and m_o being the defect and normal lattice potential constants and masses respectively. The dynamic matrix $\underline{\underline{V}}$ then factors into three cartesian components and its structure for one of these is

$$\left[\begin{array}{c|cccccc} \frac{2v}{m} & -\frac{v}{\sqrt{mm_o}}, & -\frac{v}{\sqrt{mm_o}}, & 0, & 0, & \cdots \\ \hline -\frac{v}{\sqrt{mm_o}} & \frac{v_o}{m_o}+\frac{v}{\sqrt{mm_o}}, & -\frac{v_o}{m_o}, & -\frac{v_o}{m_o}, & 0, & \cdots \\ -\frac{v}{\sqrt{mm_o}} & -\frac{v_o}{m_o}, & \frac{2v_o}{m_o} & \cdots & & \cdots \\ 0 & -\frac{v_o}{m_o} & -\frac{v_o}{m_o} & \frac{2v_o}{m_o} & & \cdots \\ 0 & \vdots & \vdots & \vdots & & \\ \vdots & \vdots & \vdots & \vdots & & \end{array}\right] \qquad (15)$$

The bare molecule Green's function is simply

$$\underline{\underline{G}}_R' = (\omega^2 - \frac{2v}{m})^{-1} \qquad (16)$$

which is the response of a simple Einstein oscillator. From (14) we find the correction shift and width

$$\underline{\underline{G}}'(\text{in L}') \simeq [\omega^2 - \frac{2v}{m} - \frac{4v^2}{mm_o}(\sigma + i\gamma)]^{-1} \qquad (17)$$

Here

$$(\sigma + i\gamma) = \lim_{\varepsilon \to 0} \sum_{\omega_{c'}} \frac{S_{c'} S_{c'}^{*}}{(\omega^2 - \omega_{c'}^{2} + i\varepsilon)} \tag{18}$$

with $S_{c'}$, $S_{c'}^{*}$ being the crystal eigenvectors evaluated on the neighboring site to the defect.

The resonance condition is now

$$\omega_R^{2} \simeq \frac{1}{m}[2v + \frac{4v^2}{m_o}\sigma(\omega_R)] = \frac{v_{eff}}{m} \tag{19}$$

which may be interpreted as a simple Einstein oscillator having a dynamic correction to the direct potential term 2v as the surrounding lattice responds to the motion. At the same time there is the damping term

$$\Gamma = \frac{4v^2}{mm_o} \gamma(\omega) \tag{20}$$

We now consider asymptotic cases in more detail.

Low Frequency Resonances. If $\omega^2 \ll \omega_{c'}^{2} \simeq \mathcal{O}(\omega^2_{Debye})$ then

$$\begin{aligned} \sigma(\omega) &\simeq - \frac{1}{\omega_{c'}^{2}} [1 + \mathcal{O}(\frac{\omega^2}{\omega_{c'}^{2}}) + \cdots] \\ &\simeq - \langle\omega_{c'}^{-2}\rangle + \mathcal{O}(\omega^2/\omega_{c'}^{4}) \end{aligned} \tag{21}$$

while from the density of states n(ω)

$$\gamma(\omega) = \frac{\pi}{2\omega} n(\omega) \tag{22}$$

Thus in leading approximation

$$\omega_R^{2} = \frac{2v}{m}[1 - \frac{2v}{m_o} \langle\omega_{c'}^{-2}\rangle] \tag{23}$$

$$\Gamma = \frac{2v^2}{mm_o} \frac{\pi n(\omega)}{\omega} \tag{24}$$

The heavy mass resonance case (Kagan and Iosilevskii, Brout, and Visscher) corresponds to $v = v_o$, $m \gg m_o$ which in Debye approximation leads to

$$\omega_R^2 = \left(\frac{m_o}{m}\right)\frac{\omega_D^2}{3}\ ,\quad \left(\frac{\Gamma}{\omega_R}\right) = \frac{\pi}{2}\left(\frac{\omega_R}{\omega_D}\right)$$

This is the familiar result, however, (21) indicates how frequency dependent corrections are to be computed.

It is in the case where the low frequency resonance is due to $v \ll v_o$ (e.g. Li in KBr; A. J. Sievers[3]) that the present method is particularly useful. First of all (23) shows that as $(v/v_o) \to 0$, $\omega_R^2 \simeq (2v/m)$.

But much more importantly (24) shows that

$$\Gamma(\omega_R) = \left(\frac{v}{v_o}\right)^2 \Gamma_o(\omega_R) \tag{25}$$

where Γ_o would have yielded the Brout-Visscher damping; <u>the line width is significantly narrower than the mass defect model would have predicted</u>. This point is underlined for two reasons: first, it comes simply and directly out of the present method and is recognized to be a general feature of weak coupling, $v \ll v_o$; second, in the interpretation of low frequency resonance experiments the <u>general</u> usage of the Brout-Visscher damping is clearly wrong.

For reference the results in the Debye approximation are $(v \ll v_o)$

$$\omega_R^2 = \frac{2v}{m}\ ;\quad \frac{\Gamma}{\omega_R} \simeq 3\pi\,\frac{v}{v_o}\left(\frac{\omega_R}{\omega_D}\right) \tag{26}$$

<u>High Frequency Localized Modes</u>. This case has been treated asymptotically[1],[4],[5] using moment expansions of the Green's functions. The same results follow directly here noting that, for $\omega^2 \gg \omega_{c'}^2$, $\sigma(\omega) \simeq \omega^{-2} + \Theta(\omega_{c'}^2/\omega^4) + \cdots$; while $\gamma(\omega) = 0$ outside of the band. Then (19) leads to the resonance condition for $m \ll m_o$

$$\omega_R^2 = \frac{2v}{m} + \frac{2v}{m_o} = \frac{2v}{m}\left(1 + \frac{m}{m_o} + \cdots\right) \tag{27}$$

The first term is again just the Einstein oscillator frequency (for defect potential constant v) with the rest of the crystal held rigid; the second term is the correction due to motion of the neighbors of the defect ion. We remark that this result is still restricted to the nearest neighbor model; the general case has been discussed by Matthew.

IV. Some Selected Numerical Results

The low frequency resonance Li:KBr studied by Sievers and Takeno have been fit using (26). In the case of Li^7 we find (taking $\omega_D \simeq 174°K$)

$$(v/v_o) \simeq 0.0069 \quad (\Gamma/\omega_R) = 0.0075$$

Nearly identical values were found from the laborious calculation of a model Green's function.

The Brout-Visscher $(\Gamma/\omega_R) \simeq 0.18$, showing how seriously the mass defect expression is in error when $(v/v_o) << 1$.

However, the matter is not entirely this simple;[3] $(\Gamma/\omega_R)_{exp} \simeq 0.02$, and the strong temperature dependent of Γ suggests that anharmonic corrections are needed. These would affect the lifetime more significantly than the resonant frequency.

As regards defect isotope effects on the resonance modes the present analysis predicts that for all practical purposes such a system as Li:KBr should show a simple isotope shift e.g. $(\omega_R{}^6/\omega_R{}^7) = (7/6)^{1/2}$ because $(v/v_o) << 1$. The conflicting results of Benedek and Nardelli[6] and Klein[5] are probably in error on this point because whereas long range forces were incorporated in their lattice models only nearest neighbor Green's functions were included in the actual defect calculation; this inconsistency must be corrected. Sievers[3] emphasizes that the isotopic mass ratio does not quite agree with experiment, although the disagreement is really not great (1.08 theoretical vs. 1.105 experimental). We return to this point later, and discuss higher order corrections.

Isotope effects on localized modes have also been examined e.g. U-centers. Equation (27) may be used to estimate either the effect of defect isotopic substitution, or isotopic substitution on the neighbors.

The isotope shift for H^-, D^- interchange has been measured by Mitsuishi and Yoshinaga;[6] it deviates from $\sqrt{2}$, being 1.39 instead in KCl. If in (27) m_o is taken as the nearest neighbor mass m_2 then

$$\frac{\omega_D{}^2}{\omega_H{}^2} \simeq \frac{2}{1}\frac{(1 + 1/m_2)}{(1 + 2/m_2)} \simeq 2(1 - \frac{1}{m_2}),$$

which for KCl gives $(\omega_D/\omega_H) \simeq 1.396$.

A recent experiment by Sievers[7] on U-centers in LiF examines the U-center shift due to interchange of Li^6 and Li^7 i.e. host lattice isotope effect. Then the second term of (27) is affected and

$$\frac{\Delta\omega_R}{\omega_R} \simeq -\frac{1}{2}\left(\frac{m}{m_o}\right)\left(\frac{\Delta m_o}{m_o}\right) \tag{28}$$

Taking again the nearest neighbor mass m_2 for m_o and the average value 6.5 for Li, m = 1, Δm = 1 we find $(\Delta\omega_R/\omega_R) \simeq 10.18 \times 10^{-3}$, whereas Sievers finds an experimental value of 3.8×10^{-3} at low temperatures. The calculated harmonic shift is only in order of magnitude agreement; an extension of the present model to next nearest neighbors reduced the computed value. But as Sievers points out some significant anharmonic effects not included in the present discussion must be considered.

From these examples we conclude that in a variety of asymptotic cases the numerical complications of Green's function evaluation are unnecessary for estimates of resonant frequencies and isotope effects. However, when one goes beyond the dominant terms there would be no way to avoid consideration of both realistic potentials and dynamics in the lattice.

V. Further General Considerations

First we remark on the most immediate generalization of (19), namely, introduction of a realistic σ(ω). Then the resonance condition is found from the graphical construction in Fig. 1 which is self explanatory.

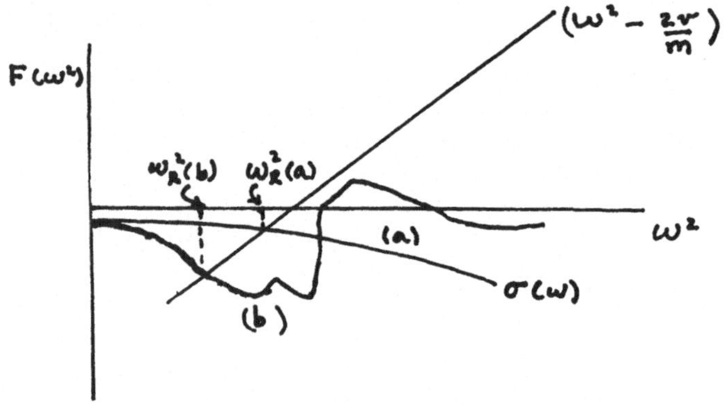

Fig. 1. Graphical Evaluation of Resonance Conditions: (a) Debye Model (b) Real Lattice

In a real lattice the frequency dependence of $\sigma(\omega)$ could strongly modify the details of the resonance condition, particularly if the "cage" around the defect is nearly resonant. For example host isotope effects would not be easily described. This may be the case for the Br^- ions around the small Li^+ in Li:KBr. But in any case it is again apparent from Fig. 1 how as $v^2 \to 0$ the $\sigma(\omega)$ correction to $\omega^2 = (2v/m)$ becomes negligible. For small but finite v a realistic accounting of σ may be necessary to account for finer details of isotope shifts.

Next we comment on long range forces. Suppose the defect is taken to be at site index j = 0 and it now interacts not only with its neighbors but arbitrarily to long range. It is possible to extend (9), by formal perturbation theory, to include these effects. But it is interesting that the lowest order behavior still may be reduced effectively to a simple oscillator asymptotically. Consider the lowest order self energy term (14); written out in matrix components it is

$$\Sigma_o = \sum_{j, j' \neq 0} \left[a_{oj} (\omega^2 - \underline{\underline{C}}')^{-1}_{jj'} a_{j'o} \right] \tag{29}$$

where $(\omega^2 - \underline{\underline{C}}')^{-1}_{jj}$ is the crystal Green's function connecting two sites j, j'. In the very low frequency limit dominated by long wave crystal phonons a first approximation is to set $(\omega^2 - \underline{\underline{C}}')^{-1}_{jj'} = (\omega^2 - \underline{\underline{C}}')^{-1}_{oo}$, effectively neglecting spatial variation of the Green's function. To this limited approximation

$$\Sigma_o \simeq \frac{1}{mm_o} (\omega^2 - \underline{\underline{C}}')^{-1}_{oo} \left| \sum_{j \neq o} v_{oj} \right|^2 \tag{30}$$

By invariance of the energy under infinitesimal uniform translation $(\sum_{j \neq o} v_{oj}) + v_{oo} = 0$ whence again $\Sigma_o \simeq (mm_o)^{-1} \sigma v_{oo}^2$, which is similar in form to the simple nearest neighbor model results. However, it will be recognized that neglecting the spatial variation of the crystal Green's Function is limited to very low frequencies indeed. Otherwise the long range forces must be accounted for by evaluating (29) and higher order self energy corrections in detail.

Lastly, we remark that although the discussion here was explicitly in cartesian coordinate basis the whole development can be set up in the representations of the (point group) symmetry of the defect crystal whenever the defect occupies a symmetry position.

To conclude I would like to acknowledge the stimulus and interest of J. A. D. Matthew, R. J. Elliott, M. Wagner, A. J. Sievers and R. Pohl.

References

* Supported in part by the U. S. Atomic Energy Commission under contract AT(30-1)-3699, Technical Report #NYO-3699-15.

1. J. A. Krumhansl, Proc. Int. Conf. on Lattice Dynamics, Copenhagen, 1963, R. F. Wallis, Ed., Pergamon Press (1965); J. A. D. Matthew, J. Phys. Chem. Solids 26, 2067 (1965); M. Wagner, Phys. Rev. 131, 2520 (1963).

2. M. Okazaki, Y. Toyawawa, M. Inoue, T. Inui, E. Hanamura, J. Phys. Soc. Japan 22, 1337 (1967); 22, 1349 (1967).

3. A. J. Sievers, this conference; also for Li:KBr, see Phys. Rev. 140, A1030 (1965).

4. L. Gunther, J. Phys. Chem. Solids 26, 1695 (1965).

5. M. V. Klein, this conference and previous references.

6. G. Benedek and G. F. Nardelli, Phys. Rev. 155, 1004 (1967).

7. A. J. Sievers, to be published, Solid State Comm.

PART B. DEFECT MODES IN IONIC CRYSTALS

LOCALIZED AND RESONANCE MODES IN IONIC CRYSTALS*

A. J. Sievers

Laboratory of Atomic and Solid State Physics

Cornell University, Ithaca, New York

I. INTRODUCTION

Impurity-activated lattice absorption was first observed in alkali halide crystals by Schäfer[1] in 1958. The infrared absorption occurred at a frequency far above the maximum phonon frequency of the host lattice and was correctly identified as due to a localized lattice mode. Because not only localized modes but other types of lattice impurity modes already had been predicted in a variety of theoretical studies,[2] Schäfer's success signaled that perhaps some of these other impurity modes could also be observed. Here we shall consider the observation of some impurity modes by spectroscopic methods.

Let us begin by considering the results of the theoretical work by Mazer et al.[3] They considered the problem of a diatomic linear chain with a mass defect. An impurity of mass M' replaces either the light mass M_1 or the heavy mass M_2 of the diatomic chain and the nearest neighbor force constants at the impurity are not altered. The local mode solutions for this problem are shown pictorially in Fig. 1. For M' replacing M_2 where $M' \ll M_2$, two impurity modes are generated, a local mode above the top of the optic spectrum and a gap mode between the acoustic and optic branches. Both modes have the same symmetry. In the second case $M' \gg M_2$ and neither local modes or gap modes are predicted. The different behavior for these two cases can readily be understood. At the zone edge of the acoustic branch, the heavy atoms M_2 vibrate π out of phase with each other while the light ions M_1 are at rest. The converse is true at the zone edge of the optic branch, the light atoms M_1 vibrate π out of phase and the heavy atoms M_2 are at rest.

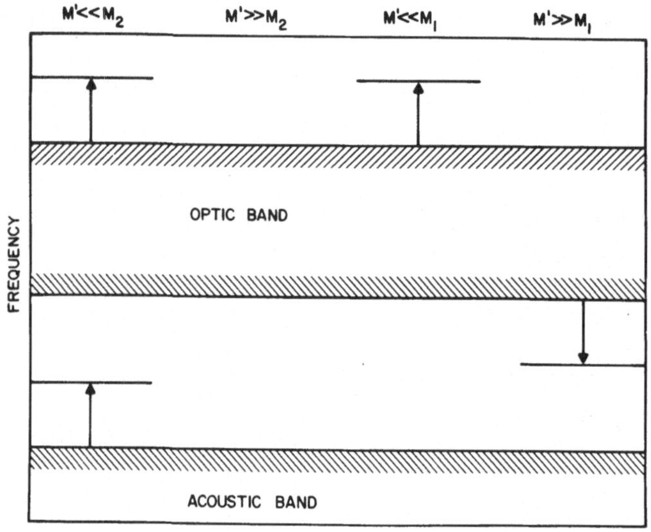

Figure 1: Isotopic impurity in a diatomic linear chain. $M_1 < M_2$.

Let us consider the first case where $M' \ll M_2$. We start by letting $M' = M_2$ and then increase the mass perturbation by reducing M'. Initially the impurity is vibrating out of phase with the other nearest M_2 atoms. As M' is decreased the number of atoms participating in this mode decreases and the mode moves up into the gap. The mode at k = 0 moves out of the optic band because of the mutual repulsion between levels of the same symmetry. For the second case $M' \gg M_2$ and the mode at the zone edge acoustic band is depressed. Because the mode is moving away from the optic band no local mode from the optic band is expected. Two more cases are possible if the impurity M' replaces the lighter atom M_1 in the diatomic chain and are also shown in Fig. 1. These modes can be interpreted in a similar manner as above.

The first case where $M' \ll M_2$ is particularly important because this mass perturbation is appropriate (as a first approximation anyway) to describe the hydride ion in alkali halide crystals. Fig. 1 indicates that not only Schäfer's local mode should be observed but also another localized lattice mode should occur in the frequency gap between the optic and acoustic phonon branches. This "gap" mode should also be infrared active. The amplitude of an impurity ion and its neighbors for a local mode and also for a gap mode have been calculated for a diatomic linear chain by Renk.(4) The atomic amplitudes are represented in Fig. 2. For the local mode the negative ions move in one direction while the positive ions move in the opposite sense. This motion is the same as for the k = 0 mode of the transverse optic branch except that

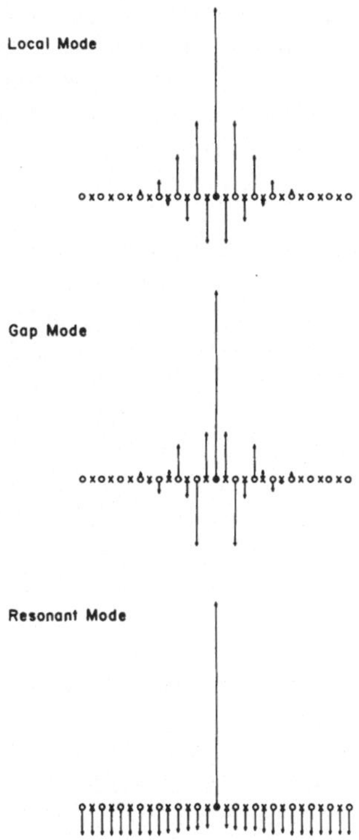

Figure 2: Eigenvector for an infrared active impurity mode in a diatomic linear chain. Local mode and gap mode by Renk (Ref. 4). Resonant mode by Weber (Ref. 5).

the amplitude has a spatial variation. In fact, the maximum amplitude is associated with the impurity ion itself and the amplitude decays exponentially as the distance from the impurity is increased.

An exponential decay of the amplitude with increasing distance from the impurity ion also occurs for gap modes and is shown in Fig. 2. In this mode like ions vibrate π out of phase with each other which is a motion characteristic of normal lattice modes at the edge of the Brillouin zone. Because of the opposing motion of like ions, the oscillating dipole moment associated with this gap mode is smaller than for the local mode.

If the nearest neighbor force constants of the impurity ion are changed, then, in addition to the infrared active modes described above, a number of infrared inactive local modes can occur in which the impurity ion itself is at rest. Force constant perturbations can also produce interesting changes in the amplitudes of the ions in the inband frequency region as was first shown by Szigeti(6) for a monatomic linear chain. Genzel et al.[7] have studied the same problem for a diatomic linear chain lattice and find that resonant lattice modes can occur if weak nearest neighbor force constants are assumed for the impurity ion. Such resonant lattice modes were first predicted for a mass defect in a three dimensional monatomic lattice by Brout and Visscher.[8]

In contrast with the local and gap modes the resonant mode is not spacially localized but extends far into the lattice. The amplitude of the impurity ion and its neighbors for a low frequency resonant mode in a linear chain has been calculated recently by Weber[5] and we show his results in Fig. 2. Most of the ions move in phase as is expected for a low frequency acoustic mode. The relative motion of the impurity ion which is out of phase gives rise to an oscillating electric dipole moment in an ionic crystal. Thus an infrared active mode in the acoustic spectrum is possible. Of course, infrared inactive resonance modes in which the impurity ion does not participate are also possible and some of these should be Raman active.

A theory which describes the infrared lattice absorption coefficient for crystals with mass defects was first given by Wallis and Maradudin and also Takeno et al.[2] Later a more complete description of the optical properties associated with the mass defect problem for a homonuclear crystal was given by Dawber and Elliott.[9] Their calculations showed that the induced absorption coefficient associated with a mass defect can be written as

$$\alpha_I(\omega) = \frac{\pi D}{3\sqrt{\varepsilon}\, c}\left(\frac{\varepsilon+2}{3}\right)^2 \left[m_L{}^2(\omega)\delta(\omega) + m_B{}^2(\omega)g(\omega)\right] \qquad (1)$$

where D is the number of impurities per unit volume, ε is the frequency dependent dielectric constant of the host crystal, c the velocity of light, $(\varepsilon+2/3)^2$ is the local field correction factor to account for the presence of the medium, $m_L(\omega)$ is the dipole moment associated with the local mode vibration, $\delta(\omega)$ is the shape function of the local mode to account for the anharmonicity or randomness in the lattice, $m_B(\omega)$ is the dipole moment associated with the band modes and $g(\omega)$ is the density of modes per unit frequency range per unit volume.

A number of peaks can occur in $\alpha_I(\omega)$. From the first term in Eq. (1), the absorption coefficient has a maximum at the local mode or gap mode frequency. From the second term in Eq. (1) resonant modes can occur in the band mode region associated with peaking of the functions $m_B(\omega)$ and $g(\omega)$ at frequencies where local modes would like to occur but are in a sense lifetime-broadened by the finite density of modes. Also, additional structure can appear in the absorption coefficient where maxima occur in the density of "normal" lattice modes.

II. GAP MODES

A. Monatomic Impurities

To date no impurity besides the U center has been observed to produce an infrared active local mode in alkali halide crystals. Nevertheless, the more or less continuous investigation of the properties of the hydride ion local mode since Schäfer's original work has led to a large body of information on the impurity-lattice interaction. Because the localized lattice mode is covered in detail elsewhere at this meeting, I shall deal with some of the complementary impurity modes here.

In contrast with the experimental situation for high frequency localized modes, lower frequency localized modes have been observed for a variety of impurities at frequencies in the gap region between the optic and acoustic phonon branches of potassium iodide.(10) Let us look at some of the successes and failures of the far infrared studies for this particular alkali halide crystal.

The impurity-induced absorption coefficient for two negative and two positive ion impurities is shown in Fig. 3. For the hydride ion impurity no sharp absorption lines have been observed in the KI gap region(11) which extends from 69.7 to 96.5 cm^{-1}. The band centered at 62 cm^{-1} was identified with the U-center by photochemical conversion.(12) The large absorption coefficient in the gap region probably arises from gap modes associated with a variety of unwanted impurities which, with an instrumental resolution of a few cm^{-1}, appears as a continuous absorption here. Another KI crystal was doped with deuterium ions and a broad peak was observed at 59.4 cm^{-1} but no structure was observed on this weak band.

Both sodium ions and thallium ions produce sharp and broad bands in the acoustic phonon spectrum but no sharp lines in the gap region. The impurity-induced absorption spectrum for sodium centers is shown in Fig. 3. Four concentrations of thallium ions have been studied. The same spectrum has been observed for all crystals and is also shown in Fig. 3. We have determined the band strength by measuring the area under the absorption curves. For at least an order of magnitude in concentration, the total strength varies linearly with impurity concentration.

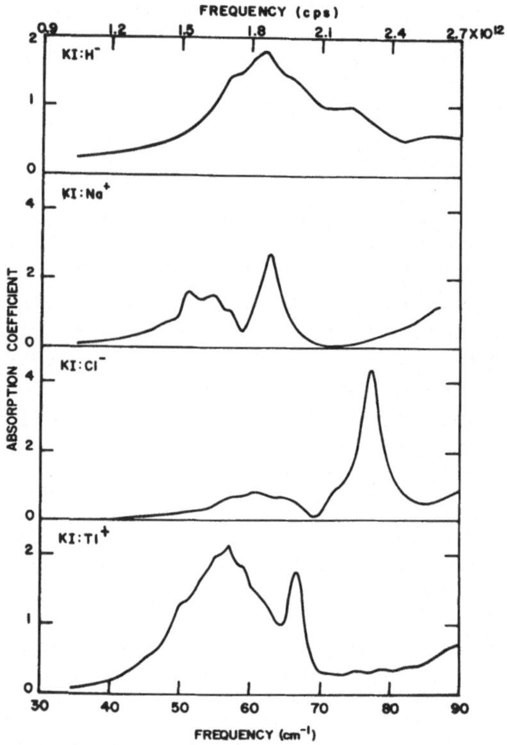

Figure 3: Impurity-induced absorption for some monatomic impurities in KI. Sample temperature is 2°K.

To date the only metal impurity ion which induces an infrared active gap mode in KI is Cs^+. A sharp absorption line is observed at 83.5 cm^{-1} with an additional broad absorption band occurring near the top of the acoustic spectrum. This experimental observation clearly demonstrates that a force constant change and not the impurity mass charge determines the gap mode frequency.

The most complete experimental investigation of gap modes(13,14) has been on the system KI:Cl^-. The observed impurity-induced spectrum shown in Fig. 3 consists of a narrow line at 77 cm^{-1} and a band composed of at least three broad lines. Below 30 cm^{-1}, the impurity-induced absorption is very small. The total integrated absorption strength has been measured for a number of concentrations and varies linearly with the chemically determined impurity concentration. This linear dependence leads us to define an oscillator strength, f, for the absorption line in analogy with the strength defined for the electronic transitions. The oscillator strength f is defined by

$$\int \alpha(\omega)d\omega = \frac{\pi e^2}{\sqrt{\varepsilon_o}\, c^2 M^*} \left[\frac{\varepsilon_o + 2}{3}\right]^2 Nf \;[cm^{-2}] \qquad (2)$$

where ε_o is the dielectric constant of the host crystal at the local mode frequency, e the electronic charge, M^* the effective mass and N the number of impurities per cm^3. For the sharp absorp-line at 77 cm^{-1} we find f = 0.016. (Later we shall find that f $\sim$ 0.1 for resonant modes.)

Because the phonon gap between the acoustic and optic branches in KI extend from 69.7 to 96.5 cm^{-1}, the three broad absorptions centered at 58, 61 and 66 cm^{-1} occur in the acoustic phonon spectrum. We shall see that similar peaks occur at 56, 66 and 67 cm^{-1} in KI:Br^-. In neither case do the absorption peaks seem to identify critical points in the KI density of modes spectrum.(11) However, frequency shifts in the absorption spectrum are to be expected for the two impurities if the absorption lines are to be identified with resonant or incipient resonant modes which occur in the frequency region of the large density of modes. A satisfactory method for investigating the nature of these broad resonant modes has not yet been found and we now turn to the study of the gap mode absorption.

At a higher resolving power of 380 the gap mode at 77 cm^{-1} is found to consist of two lines with center frequencies of 77.10 and 76.79 $\pm$ 0.05 cm^{-1}. The frequency separation is estimated to be 0.31 $\pm$ 0.05 cm^{-1}. The ratio of the line strengths for the doublet components is about 3 to 1 which corresponds nicely to the natural abundance ratio of the two stable chloride isotopes.

The activation of gap modes by bromide impurities in KI also has been studied. An absorption spectrum very similar to that found previously for the chloride doping is observed, namely, three fairly broad absorption bands below the acoustic edge at 56, 66, and 67 cm^{-1}, plus a sharp absorption line at 73.8 cm^{-1}. An additional absorption line at 77 cm^{-1} is attributed to an unwanted Cl^- impurity which a chemical analysis showed to be present in a concentration of 3 x 10^{18} Cl^-/cm^3 in these crystals. The gap mode at 73.8 cm^{-1} should also show an isotope shift; however, a resolving power of 380 is insufficient to resolve the frequency difference.

Let us see what can be concluded from these experimental results. Band modes are observed for all impurities whereas gap modes appear for about one half of the crystal-dopant combinations. At least some of the broad absorption bands in the acoustic spectrum must occur because of the large density of states in this frequency region. Also additional absorptions could arise from resonant modes as has been discussed by Maradudin et al.(13) A comparison of the gap mode frequency with the fractional mass change demonstrates that the experimental results cannot be interpreted with a simple mass defect model. Some success in fitting the gap mode frequencies has been obtained by introducing an additional parameter, the nearest neighbor coupling parameter to the impurity. Nolt et al.(14) have fit the frequencies by using the simple lattice model of Mitani and Takeno.(15) A comparison of the impurity ion diameter(16) with the force constant change gives consistent results in that a large mismatch in diameters corresponds to a large change in coupling constant. Also if the impurity diameter is larger than the host diameter such as Cs^+, then the coupling constant is larger than the host value while if the impurity diameter is less than the host diameter the coupling constant is less than the host value.

The internal consistency of these fits can be checked with the isotope shift data. For Cl^- with $K'/K = 0.38$, the shift in frequency associated with the Cl^- mass change is calculated to be 1.74 cm^{-1} compared to the measured value of 0.31 cm^{-1}. With $K'/K = 0.81$ for the Br^- impurity the model predicts an isotope shift of 0.26 cm^{-1} while the experimental shift is less than 0.15 cm^{-1}. For both impurities the model fails to account for the small isotope shift observed although the calculated shifts are, indeed, much less than the maximum possible shift given by the square root of the isotope mass ratio.

Apparently, the observed shift cannot be obtained within the harmonic approximation. Recently, Benedek and Maradudin(17) have calculated the isotope shift for Cl^- using Hardy's deformation dipole model and also Cochran's shell model for the host crystal. Both models of the host crystal fail to predict the small isotope frequency shift.

B. Molecular Impurities

Two molecular impurities have been studied in KI. They are KI:KCN(18) and KI:KNO_2.(18-20) Absorption lines are found in the gap region and broad bands are located at the edge of the acoustic spectrum. Qualitatively the spectra are similar to that observed for Cl^- in Fig. 3. For the NO_2^- center two sharp lines are found in the gap region at 71.2 and 79.5 cm^{-1} and two broad bands in the acoustic spectrum as shown in Fig. 4a. The nitrite concentration is 1.5×10^{19} NO_2^- ions per cm^3 and the nitrate concentration is 1×10^{18} NO_3^- ions per cm^3. The integrated absorption coefficient for each absorption line has been plotted vs. concentration in Fig. 4B for the three KI:KNO_2 crystals studied. The linear depen-

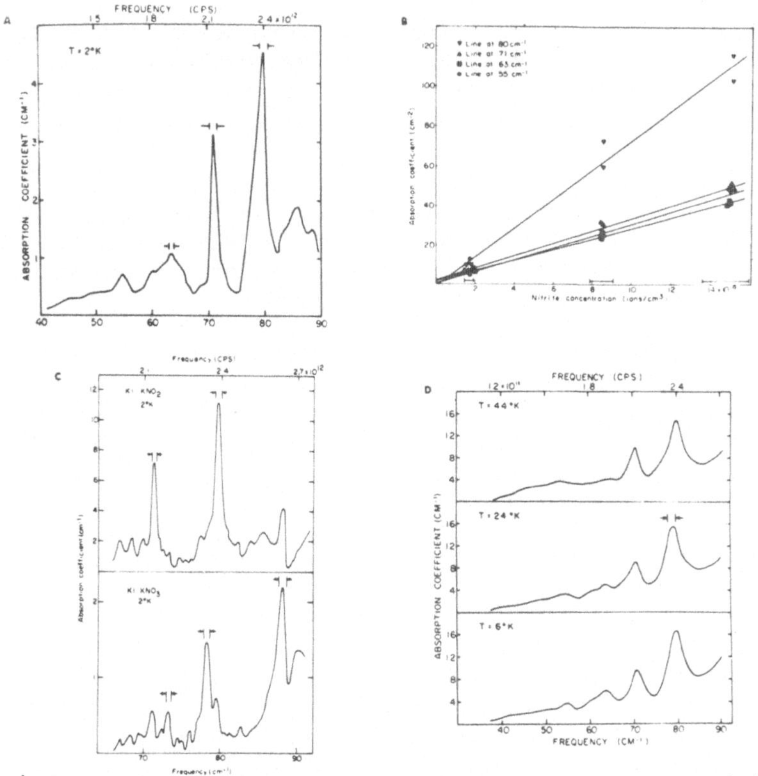

Figure 4: Far infrared absorption spectrum of KI:KNO_2. (A) A medium resolution spectrum of KI:KNO_2; (B) The linear dependence of the absorption strengths with NO_2^- concentration; (C) Top trace high resolution spectrum displaying the two nitrite gap modes. Bottom trace: high resolution spectrum of an air grown KI:KNO_2 crystal; (D) Temperature dependence of an impurity-induced nitrite absorption. After Lytle (Ref. 18).

dence indicates that the four absorption lines are due to the nitrite and not the nitrate center which is also present in all crystals. In the top half of Fig. 4C a measurement at higher resolving power of the sample shown in Fig. 4A is displayed. In the bottom of Fig. 4C the spectrum of an air grown KI:KNO_3 crystal which contains about 5×10^{17} NO_3^- ions per cm^3 and about 7×10^{17} NO_2^- ions per cm^3 is shown. In addition to the three prominent absorption lines at 73.3, 78.4, and 88.2 cm^{-1}, the NO_2^- gap modes are also visible. An unambiguous identification of the three lines with the NO_3^- center is not possible because the absorptions are so very weak. For example, the two higher frequency lines correspond very closely to the frequencies of the OH^- ion absorption lines measured by Renk.[21] Also, the chloride and bromide impurities which are present in all KI crystals,[14] introduce an uncertainty about any absorption which occurs near 77 or 73.8 cm^{-1}.

Finally, in Fig. 4D the temperature dependence of the impurity-induced nitrite absorption is shown at low resolution. The broad bands in the acoustic spectrum wash out as the temperature is increased while the gap modes remain relatively temperature independent.

A number of attempts have been made to explain the dynamics of the nitrite ion in KI. Since all present models are only partially successful only a qualitative description which focuses mainly on the far infrared results is presented here.

Timusk and Staude[22] first reported phonon structure in the NO_2^- electronic absorption at 400 mμ in alkali halide crystals at 4.2°K. For the NO_2^- center, sharp structure was observed in absorption on the high frequency side of the electronic (or electronic plus molecular vibration) transition. It was suggested that the lines which appeared at 63 cm^{-1} and 70 cm^{-1} from the no-phonon line might arise from band absorption and gap mode absorption. The low symmetry of the NO_2^- ion insures that only modes which transform according to the A_1 representation (both infrared active and Raman active) couple with the nondegenerate electronic transitions. A possible interpretation is to identify the two far infrared absorptions at 71.2 and 63 cm^{-1} with the infrared active modes transforming according to the A_1 representation. To do this we must assume the C_{2v} symmetry of the impurity center lifts the three fold degeneracy of the infrared active T_{1u} mode and produces three infrared active non-degenerate modes transforming as the A_1, B_1, and B_2 representations of the C_{2v} point group. From the dielectric constant measurements of Sack and Moriarty[23] and the near-infrared work of Narayanamurti et al.[24] the NO_2^- ion is definitely displaced from the normal ion equilibrium position in KI and also the C_2 axis is probably oriented along the equivalent <110> directions. The near-infrared measurements dictate that the molecule is almost freely rotating about this two fold axis, but not about either of the other two principle axes of the molecule. The effective coupling of the impurity to the lattice will be very different in the plane perpendicular to the C axis as compared with along the C axis. To explain the far infrared results we try the following model. If because of the free rotation an axial potential is used for vibrations within the plane, the B_1 and B_2 type modes would appear degenerate. In this approximation the $B_1 + B_2$ absorption would be twice as strong as the A_1 absorption. A comparison of the experimental oscillator strengths for the two nitrite gap modes gives a ratio of (1.7). Also, the total oscillator strength of these two absorption lines is the same order of magnitude as found for the Cl^- gap mode.[14] The results are consistent although the details of the molecular motion of the NO_2^- center in KI are no doubt complex.

III. RESONANT MODES

A. Isotope Effect

We shall now focus our attention on some resonant modes which have been observed in the low frequency region where the host density of lattice modes increases monatomically with increasing frequency. These modes have the interesting property of being very sensitive to external perturbations. Moreover, in this long wavelength limit the simple Debye spectrum of modes can be used for the host crystal. Hence, some of the simplicity of the dynamics characteristic of the U-center mode is regained in this low frequency region.

The first optically active resonant modes were identified with the substitutional silver ion impurity in alkali halide crystals.[25,26] The far infrared absorption spectra found for silver activated potassium chloride and bromide were quite similar. The spectra are characterized by an absorption line superimposed on the low frequency wing of a broad absorption band. For KCl:AgCl the low frequency absorption line is observed at 38.6 cm^{-1} and is shown in Fig. 5. The frequency corresponds to about one quarter

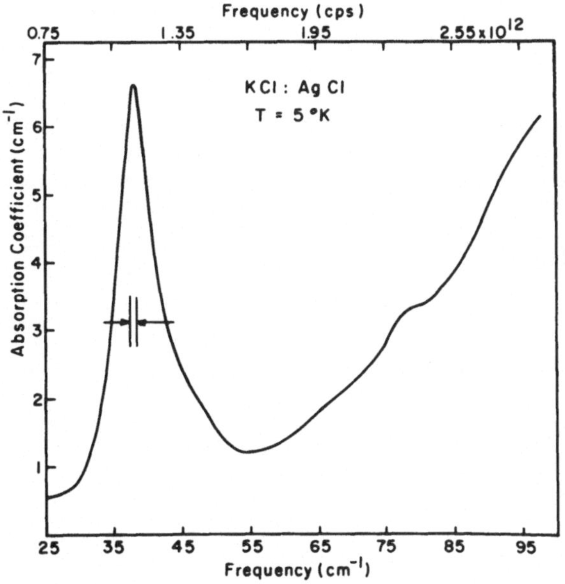

Figure 5: Absorption coefficient for the silver-activated resonant mode in KCl:AgCl.

of the Debye temperature. Only one silver doping has been studied, however, a number of silver dopings have been measured for a KBr host lattice. In this case a strong low frequency absorption is found at 33.5 cm^{-1}. The integrated absorption strength varies linearly with silver concentrations. In addition to this line, broad absorption bands are observed at higher frequencies. These higher frequency absorptions probably arise from the structure in the density of states although no direct one-to-one correlation with the dispersion curves found by Woods et al.(27) has been observed.

Silver concentrations of 10^{18} ions per cm^3 in potassium iodide yield a much sharper absorption line centered at 17.3 cm^{-1} with a full width at one half maximum absorption of 0.45 cm^{-1}. Using isotopically pure silver in the form of Ag metal as a dopant, it has been possible to detect the isotopic frequency shift for the two silver isotopes 107 and 109.(28) The ratio of the frequency of the resonant mode for the light mass to that for the heavy mass is

$$\frac{\omega(107)}{\omega(109)} = 1.008 \pm 0.002$$

This result shows first that the resonant mode is due to silver and second that the frequency ratio varies approximately as the square root of the mass ratio which is

$$\left(\frac{109}{107}\right)^{\frac{1}{2}} = 1.009 \ .$$

Another sharp absorption line associated with a low frequency resonant mode has been observed in $NaCl:Cu^+$ by Weber and Nette.(29) Recently, we have made transmission measurements with $NaCl:Cu^+$ samples at 4.2°K and with an instrumental resolution of 0.09 cm^{-1}. By using isotopically pure $^{63}Cu^+$ and $^{65}Cu^+$ the isotope shift has been resolved.(28) The ratio of the frequencies is

$$\frac{\omega(63)}{\omega(65)} = 1.016 \pm 0.002$$

For this impurity the square root of the mass ratio is

$$\left(\frac{65}{63}\right)^{\frac{1}{2}} = 1.016 \ .$$

Again the frequency ratio is given quite accurately by a simple Einstein oscillator dependence.

The observation of the $^6Li^+$ and $^7Li^+$ resonant mode frequency shift in KBr was first reported[30] in 1965. The resolution of the grating monochromator was barely sufficient to distinguish the experimental frequency shift from a $(mass)^{1/2}$ dependence. Recent measurements show that the ratio of resonant mode frequencies is

$$\frac{\omega(6)}{\omega(7)} = 1.105 \pm 0.004 \ .$$

Now the Einstein oscillator no longer accounts for the isotope shift. The square root of the mass ratio is

$$\left(\frac{7}{6}\right)^{\frac{1}{2}} = 1.08 \ .$$

Originally, to account for the isotope frequency shift in $^6Li^+$ and $^7Li^+$ doped KBr, Sievers and Takeno[30] used a simple lattice model in which long range forces were neglected. In the low frequency limit this model predicted that the resonant mode behaves as an Einstein oscillator. More recently, precise calculations of the isotope shift have been attempted by Benedek and Nardelli[31] using Hardy's deformation dipole model and also by Klein[32] using the shell model to describe the perfect crystal. In both of these calculations long range forces are included. With these harmonic models a much smaller isotope shift has been calculated than was found previously either with the simple lattice model or by experiment. We have looked for anharmonic contributions to the resonant mode frequency which could influence the isotope shift. Two different anharmonic effects have been considered. One contribution to the isotope shift can arise from the anharmonic nature of the potential which describes the resonant mode oscillator. Another contribution can arise from the modulation of this potential by other lattice modes. The first contribution appears to be too small even if a square well potential is assumed for the resonant mode.[28] However, from our static stress measurements on resonant modes[33] the second contribution which arises from the dynamical motion of the lattice is the correct order of magnitude to account for the large experimental isotope shifts. We now turn to these stress experiments.

B. Stress-Induced Frequency Shift

With the application of uniaxial stress along different crystallographic directions, the coupling of the resonant mode excited state to lattice distortions of different symmetries can be distinguished. Consider the electric dipole transition between the two vibrational states ψ which transform according to the A_{1g} and T_{1u} irreducible representations of the O_h point group appropriate to the defect site. The local symmetry of the lattice is lowered by the uniaxial elastic deformation and a splitting of the degenerate T_{1u} state can occur. The lattice distortions which will perturb the resonant mode excited state can be obtained as follows. The matrix elements associated with the stress perturbation H' are for the ground state:

$$\Delta E_g = \langle\psi(A_{1g})|H'|\psi(A_{1g})\rangle \quad (3)$$

and for the excited state

$$\Delta E_e = \langle\psi(T_{1u})|H'|\psi(T_{1u})\rangle \quad (4)$$

For these matrix elements to be invariant under all symmetry operations of the octahedral group O_L, they must transform like the A_{1g} representation. Thus, the stress perturbation must transform according to the representations contained in the direct product of

$$T_{1u} \times T_{1u} = A_{1g} + E_g + T_{2g} \quad (5)$$

and the only modes which interact with the excited state of the resonant mode are the long wavelength acoustic distortions of A_{1g} (spherical), E_g (tetragonal and orthorhombic), and T_{2g} (trigonal) symmetry. Three coupling coefficients A, B, and C then determine the dependence of the resonant mode frequency upon the fully symmetric, the tetragonal, and the trigonal strain components respectively. These coupling coefficients can be estimated from the slopes of the absorption frequency versus applied stress data if the stiffness constants of the crystal are known. Unfortunately, an unsettled question is whether or not the stiffness constants of the host crystal need to be modified to account for a softening of the lattice in the neighborhood of the impurity. For the coefficients which we give below the stiffness constants of the unperturbed crystal have been used.

Table I

Stress coupling coefficients for $KBr:Li^+$ and $KI:Ag^+$.(34)

	$KBr:Li^+$	$KI:Ag^+$
$A(A_{1g})$	830 ± 100	390 ± 90 cm^{-1}/unit strain
$B(E_g)$	360 ± 40	510 ± 70 cm^{-1}/unit strain
$C(T_{2g})$	170 ± 30	- 15 ± 10 cm^{-1}/unit strain

From Table I the resonant mode for $KBr:Li^+$ is most strongly coupled to long wavelength spherically symmetric modes. Also, these coupling coefficients are about the same size as has been found for the U-center mode.[35] Hence, for both the U center local mode and the Li^+ resonant mode similar stress shifts on the order of a few cm^{-1} can be observed. Of course, the shift is much more dramatic for the resonant mode because of the low frequency involved.

Also recorded in Table I are the coefficients for $KI:Ag^+$. The resonant mode is most strongly coupled to tetragonal distortions of the surrounding lattice and progressively less to those of spherical and trigonal symmetry. The presence of appreciable non-central force components are indicated by this result. Such components could arise from some covalent bonding associated with the Ag^+ impurity. Preliminary experimental results indicate that for $NaCl:Cu^+$ the tetragonal coupling coefficient is again largest and all three coefficients are similar to those found for Ag^+. No doubt both of these impurity ions are bound in their respective alkali halide lattice in a similar manner.

For all of these defect systems the resonant mode is very sensitive to the lattice constant. Because the A_{1g} hydrostatic coefficient gives the resonant mode frequency shift per unit volume charge, the dependence of the frequency upon the lattice constant can be written as

$$-\frac{\partial \ln \omega}{\partial \ln a} = \frac{3A}{\omega} \qquad (6)$$

where a is the lattice constant, ω is the resonant frequency and A is the hydrostatic coupling coefficient. For $KBr:Li^+$ this dimensionless coefficient is $3A/\omega = 155$ whereas for the U center in KCl the coefficient is 3.6.

Another method by which the lattice constant dependence of the center frequency can be measured is to observe the center frequency of the infrared-active resonant mode for a number of alkali halide lattices. Such measurements have been carried out for silver doped sodium and potassium salts and these results are shown in Fig. 6. In direct analogy with the Mollwo-Ivey plots used for U centers, the center frequencies are displayed on a log frequency-log lattice constant plot. For five different alkali halide lattices, the center frequencies can be fitted by a straight line in Fig. 6. This line is obtained by setting $3A/\omega = 3$ in Eq. 6. However, $KI:Ag^+$ occurs at a much lower frequency and from the uniaxial stress measurements[34] the coefficient $3A/\omega = 68$. This different slope is shown at the $KI:Ag^+$ point in Fig. 6. Because relaxation effects will be different for the different

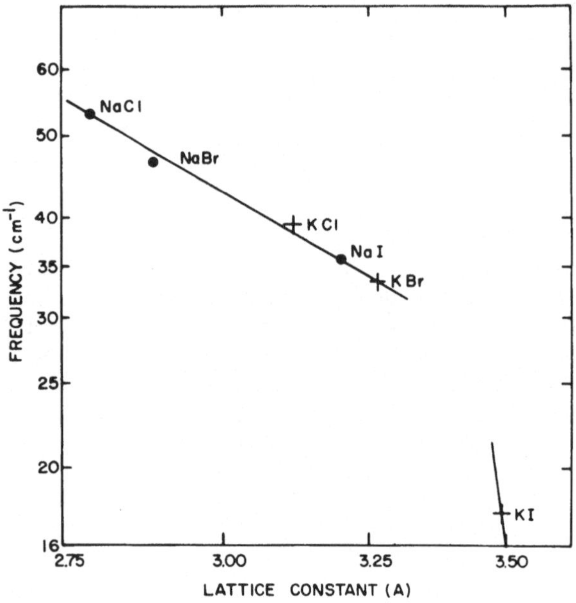

Figure 6: The resonant mode frequency versus host lattice constant for silver doped sodium and potassium halides.

lattices both slopes should not really be given in the same figure. Nevertheless, we must conclude that the Ag^+ resonant mode becomes more sensitive to a lattice constant change as the lattice constant increases. The extreme sensitivity of the resonant mode frequency has also been demonstrated by alloying experiments which are presented elsewhere at this meeting.

C. Temperature Dependent Properties

In contrast with temperature dependent studies on gap modes which are hindered by the strong temperature dependence of the host lattice absorption,[14] temperature dependent studies on resonant modes at moderately low temperatures are relatively straightforward. One of the most striking features which has been observed with some of these low frequency absorptions is the rapid temperature dependence of the line strength.[36] For both KBr:Li^+ and KI:Ag^+ the strength decreases rapidly as the temperature is increased and disappears by about 40°K. For NaCl:Cu^+ the strength decreases much more slowly and the absorption line can still be measured at liquid nitrogen temperatures.[29] This different behavior which has been a puzzle for some time can now be understood from the stress results. Because of the large linear coupling of the resonant mode to long wavelength phonons, the modulation of the resonant mode energy levels by the lattice modes will introduce characteristic properties into the spectrum. The sharp zero phonon lines which have been studied in the optical spectral region are a well known example of this motional effect of the lattice.(37,38) There has been some confusion about the credibility of identifying low frequency transitions with zero phonon lines because the adiabatic approximation is usually envoked in the optical region. However, a number of investigators(32,37,39) have indicated that if the phonon-impurity interaction shifts but does not mix the impurity states then the adiabatic approximation is not required.

A characteristic property of a zero phonon line is that the line strength varies with temperature with a Debye Waller like factor. The magnitude of this exponent is controlled by the square of the coupling constant of the defect mode to long wavelength phonons--a quantity which is readily obtained from stress measurements.(40) For the three resonant modes we have considered here the coupling decreases in the following order $Li^+ > Ag^+ > Cu^+$. Now it is clear that the strength of the lithium resonant mode should vary with temperature much more rapidly than the strength of the copper resonant mode.

IV. SUMMARY

Whereas only the U-center modes have been observed at frequencies above the phonon spectrum of the pure crystal, a large number of gap modes have been found. These gap modes produce extremely sharp absorption lines with widths less than 0.2 cm^{-1}. The magnitude of the isotope frequency shift observed for the Cl^- gap mode in KI has not yet been accounted for satisfactorily. With low symmetry centers such as polyatomic molecules, the complexity of the spectra in the gap region is offset to some degree by the additional spectroscopic measurements which are now possible in the near infrared and optical region.

A number of resonant modes have been studied in alkali halide crystals. Sharp absorption lines have been identified with weak coupling of the impurity ion to the host lattice. A number of isotope shifts have been measured but again agreement with theory is unsatisfactory. By measuring the stress dependences of the absorption line the anharmonic coupling of the resonant mode to long wavelength phonon modes has been determined. Both kinds of measurements indicate that a strong coupling exists. Hence, the qualitative features of these resonant modes are fairly well understood but to predict which impurity-lattice system will lead to a lattice resonant mode is not yet possible.

The key to understanding many of the curious properties of lattice resonant modes has been the static stress measurements. I should like to mention that these pioneering measurements were made by I. G. Nolt.

References

* This research was mainly supported by the U. S. Atomic Energy Commission under contract AT(30-1)-2391, Technical Report No. NYO-2391-62. Additional support was received from the Advanced Research Projects Agency through the use of space and technical facilities of the Materials Science Center at Cornell University.

1. G. Schäfer, J. Phys. Chem. Solids 12, 233 (1960).
2. For a complete list of references on the mass defect problem see: A. A. Maradudin, E. W. Montroll, and G. H. Weiss, Theory of Lattice Dynamics in the Harmonic Approximation, Solid State Physics Suppl. 3, F. Seitz and D. Turnbull, Editors [Academic Press, 1963].
3. P. Mazer, E. W. Montroll, and R. B. Potts, J. Wash. Acad. Sci. 46, 2 (1956).
4. K. F. Renk, Z. Physik 201, 445 (1967).
5. R. Weber, Ph.D. Thesis, Physikalisches Institut der Universität, Freiburg Im Breisgau (1967).
6. B. Szigeti, J. Phys. Chem. Solids 24, 225 (1963).
7. L. Genzel, K. F. Renk, and R. Weber, phys. stat. sol. 12, 639 (1965).
8. R. Brout and W. M. Visscher, Phys. Rev. Letters 9, 54 (1962).
9. P. G. Dawber and R. J. Elliott, Proc. Roy. Soc. 273, 222 (1963); P. G. Dawber and R. J. Elliott, Proc. Phys. Soc. 81, 453 (1963).
10. A. J. Sievers, Low Temperature Physics, J. Daunt, D. Edwards, F. Milford, and M. Yaqub, eds. (Plenum Press, New York), LT9, Part B, p. 1170.
11. G. Dolling, R. A. Cowley, C. Schittenhelm and I. M. Thorson, Phys. Rev. 147, 577 (1966).
12. J. H. Schulman and W. D. Compton, Color Centers in Solids, (MacMillan Company, New York, 1962) p. 49.
13. A. J. Sievers, A. A. Maradudin, and S. S. Jaswal, Phys. Rev. 138, A272 (1965).
14. I. G. Nolt, R. A. Westwig, R. W. Alexander, Jr., and A. J. Sievers, Phys. Rev. 157, 730 (1967).
15. Y. Mitani and S. Takeno, Prog. of Theoret. Phys. (Kyoto) 33, 779 (1965).
16. L. Pauling, Nature of the Chemical Bond, (Cornell University Press, Ithaca, 1945).
17. G. Benedek and A. A. Maradudin, to be published.
18. C. D. Lytle, M.S. Thesis, Cornell University, Ithaca (1965).

19. A. J. Sievers and C. D. Lytle, Physics Letters 14, 271 (1965).
20. K. F. Renk, Physics Letters 14, 281 (1965).
21. K. F. Renk, Physics Letters 20, 137 (1966).
22. T. Timusk and W. Staude, Phys. Rev. Letters 13, 373 (1964).
23. H. S. Sack and M. C. Moriarty, Solid State Comm. 3, 93 (1965).
24. V. Narayanamurti, W. D. Seward, and R. O. Pohl, Phys. Rev. 148, 481 (1966).
25. A. J. Sievers, Phys. Rev. Letters 13, 310 (1964).
26. R. Weber, Physics Letters 12, 311 (1964).
27. A. D. B. Woods, B. N. Brockhouse, R. A. Cowley, and W. Cochran, Phys. Rev. 131, 1025 (1963).
28. A more detailed description of the isotope shifts is in preparation. R. D. Kirby, I. G. Nolt, R. W. Alexander, Jr. and A. J. Sievers, to be published.
29. R. Weber and P. Nette, Phys. Letters 20, 493 (1966).
30. A. J. Sievers and S. Takeno, Phys. Rev. 140, 1030 (1965).
31. G. Benedek and G. F. Nardelli, Phys. Rev. 155, 1004 (1967).
32. M. V. Klein, Physics of Color Centers, W. Beall Fowler, ed. (Academic Press, 1968) Chapter 7.
33. I. G. Nolt and A. J. Sievers, Phys. Rev. Letters 16, 1103 (1966).
34. I. G. Nolt, Ph.D. Thesis, Cornell University, Ithaca (1967).
35. W. Barth and B. Fritz, phys. stat. sol. 19, 515 (1967).
36. S. Takeno and A. J. Sievers, Phys. Rev. Letters 15, 1020 (1965).
37. R. H. Silsbee, Phys. Rev. 128, 1726 (1962); R. H. Silsbee and D. B. Fitchen, Rev. Mod. Phys. 36, 423 (1964).
38. T. Timusk and M. V. Klein, Phys. Rev. 141, 664 (1966).
39. D. E. McCumber, Phys. Rev. 133, A163 (1964).
40. R. W. Alexander, Jr. and A. J. Sievers, to be published.

IN BAND ABSORPTION OF MONOVALENT IMPURITIES IN SODIUM CHLORIDE*

H. F. Macdonald and M. V. Klein

Department of Physics and Materials Research Laboratory

University of Illinois, Urbana, Illinois

INTRODUCTION

In this paper preliminary results of an investigation of the far infrared absorption spectrum of sodium chloride containing monovalent impurities are reported. In recent thermal conductivity measurements on this system Caldwell and Klein (1) observed depressions of the thermal conductivity curves which were asymmetrical with respect to the peak and interpreted their results in terms of scattering of phonons by lattice resonant modes associated with the impurities. A shell model was used to describe the phonons of the perfect lattice and a simple model of the impure lattice was assumed. This involved a change of mass at the defect site together with a change of central force constant to the six nearest neighbors. This model was only moderately successful in predicting the shape of the thermal conductivity curves and the present work was undertaken in order to obtain more detailed information about the resonance modes responsible for the thermal conductivity depressions.

EXPERIMENTAL

Absorption measurements at liquid nitrogen and liquid helium temperatures in the range from 33 to 160 cm^{-1} were made using a Beckman IR 11 spectrometer which was fitted with a beam condenser. Crystals of pure sodium chloride and of sodium chloride doped with silver, bromine, and fluorine were mounted in a cryostat which enabled them to be moved into or out of the radiation beam while at low temperature. The spectrometer was used in the single beam mode and the transmittance of the sample obtained directly as the ratio of the radiation intensity with the sample in the beam to that with the sample out of the beam. Corrections for the reflection losses at the two crystal surfaces were applied and the impurity induced absorption was obtained as the difference in absorption constant between the doped and pure crystals.

The temperature dependence of the absorption of Ag^+ in sodium chloride is shown in Fig. 1. The resonance mode responsible for the thermal conductivity depression at about 35-50° K occurs at 53 cm^{-1} at 7° K and 52 cm^{-1} at 80° K, as found previously by Weber.[2] In the same temperature range its width increases from 12 to 14 cm^{-1}. In addition to this peak, there is a broad absorption towards higher energy, which shows relatively sharp peaks in the region of the acoustic modes of pure sodium chloride. This structure consists of a weak band at 120.5 cm^{-1} and a strong one at 131 cm^{-1}. The main features of the spectrum are essentially tem-

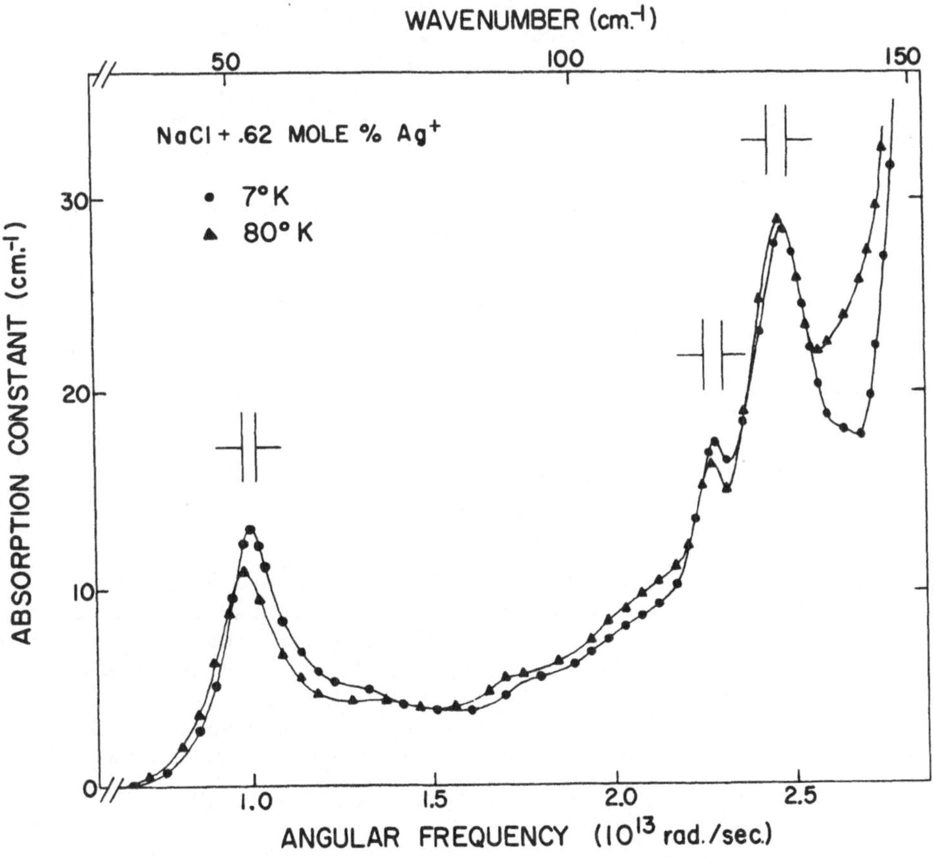

Fig. 1. Temperature dependence of the absorption of NaCl:Ag^+.

perature independent, apart from a slight increase in the absorption with increasing temperature around 140 cm^{-1}. In this region the background absorption of the pure crystal is rising rapidly at nitrogen temperature and the observed increase is just within the limit of experimental error.

For the bromine impurity a weak resonance mode is found at 83 cm^{-1}, corresponding to a thermal conductivity depression at approximately 60° K. Again a broad absorption towards higher energy is observed with a strong peak at 140 cm^{-1}. The spectrum shows little variation with temperature, apart from a slight broadening of the resonance mode peak.

In the case of fluorine the thermal conductivity showed appreciable depression only below 15° K, with a much weaker depression near 40° K. This suggests that a resonant mode absorption might occur around 10 cm^{-1} but this has not been found experimentally.[3] However a sharp line, which was first measured by Sievers,[3] occurs at 60 cm^{-1} at liquid helium temperature together with two bands towards higher energy -- a broad band peaking at 112 cm^{-1} and a narrow one at 144 cm^{-1}. Two weak satellite lines are also found close to the 60 cm^{-1} band at 48.5 and 72.5 cm^{-1}. On warming to liquid nitrogen temperature the 60 cm^{-1} band broadens and shifts to 66 cm^{-1} while the two high energy bands remain almost unchanged.

The peak positions and widths of the high energy structure in $NaCl:Ag^+$, Br^-, and F^- at liquid helium temperature are summarized below:

Crystal	Peak Pos.	Width	Peak Pos.	Width
$NaCl:Ag^+$	120.5 cm^{-1}	3 cm^{-1}	131 cm^{-1}	7.5 cm^{-1}
$NaCl:Br^-$	----------	-------	140	9.0
$NaCl:F^-$	112	18	144	8.5

THEORY

The nearest neighbor central force model used by Caldwell and Klein to fit their thermal conductivity results may also be used to calculate the optical absorption associated with the impurity. The application of this model to the case of Ag^+ in sodium chloride has been described in detail by Klein,[4] and similar calculations have been performed by Benedek and Nardelli.[5] Klein obtained the following result for the impurity induced absorption constant:

$$\alpha(\omega) = \frac{(n_\infty^2+2)^2 4\pi\omega Ne^{*2}\mathcal{N}}{9n(\omega)\ c\ (\omega_o^2-\omega^2)^2} \mathrm{Im}(TO_x,0|\underline{t}(\omega^2-i0^+)|TO_x,0)$$

Here n_∞ is the high frequency refractive index, $n(\omega)$ the refractive index at a frequency ω, N the impurity concentration per unit volume, ω_0 the frequency of the k=0 transverse optical phonon, e* the effective charge, η the number of unit cells in the crystal, and μ the reduced mass of the unit cell in the pure crystal. $|TO_x,0)$ is the k=0 TO phonon eigenvector polarized in the x-direction and $\underline{t}$ is that part of the T-matrix perturbed by the defect. In the case of a nearest neighbor model this contains only those terms involving the defect and its six nearest neighbors. The high frequency refractive index appears in the expression for the absorption constant because of the inclusion in the theory of the contribution to the polarization from the electron distribution, as well as that due to the ionic motion.

From the above expression for the absorption constant it can be seen that the radiation couples directly to a vibrational configuration which consists of a k=0 TO phonon. The resulting frequency spectrum is no longer the delta function at $\omega=\omega_0$ obtained for the pure crystal, but consists in addition of a continuous spectrum, given by the above expression, perhaps with resonances. The resonant modes responsible for the thermal conductivity depressions consist primarily of a relative motion of the defect and two of its nearest neighbors along a line joining them. This motion has the same T_{1u} symmetry as the k=0 TO phonon. In addition, motion of the atoms of the host crystal at lattice mode frequencies may involve relative displacements of the defect and its nearest neighbors which have T_{1u} symmetry. This will give rise to extra absorption features which will reflect, in a modified form, the density of states of the host crystal. It is this type of process which produces the high energy structure observed in the experimental spectra described in the previous section.

DISCUSSION

The optical absorption of Ag^+ in sodium chloride calculated using the theory of the previous section is compared with the experiment results at 7° K in Fig. 2. Using a fractional change in the nearest neighbor central force constant for the silver ion of $\Delta f/f = -0.525$, which gives the resonance mode peak at the experimental position of 53 cm^{-1}, the dashed curve results. Although the shape of the calculated resonance peak fits fairly well, its intensity is about two times greater than that observed. In addition the theory predicts a minimum in the optical absorption in the region of the main high energy band.

The failure of the central force model to account for the high energy structure in the case of the silver impurity indicates that some improvement in the model is required. There is reason to believe that the silver-halide ion bond has important non-central components, since the elastic constant C_{44} is much less than C_{12}

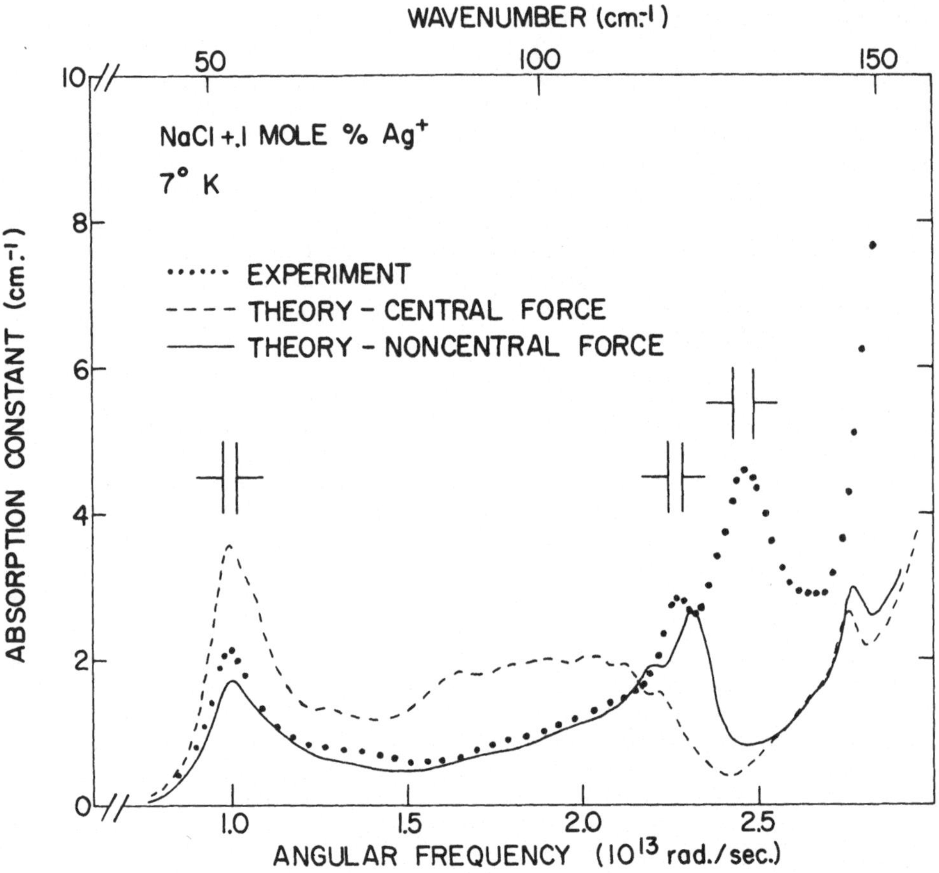

Fig. 2. Comparison of theory and experiment for NaCl:Ag^+.

in the silver halides. Assuming an unchanged central force constant but with a fractional change in the nearest neighbor noncentral force constant of $\Delta f/f = -0.345$ the full line curve shown in Fig. 2 is obtained. The intensity of the resonance peak agrees well with the experiment and some high energy structure also appears in the calculated absorption curve. This has a reduced intensity and is shifted to lower energy by about 6% compared with the experimental result, but its shape shows the correct qualitative features.

The result of a calculation for Br^- in sodium chloride using a fractional change in the nearest neighbor central force constant of $\Delta f/f = -0.130$ is given in Fig. 3. The theoretical curve shows the main features observed experimentally, namely a weak resonance peak together with a stronger peak towards higher energy. The calculated high energy peak again has a reduced intensity and is shifted to lower energy by about 6% compared with the experiment.

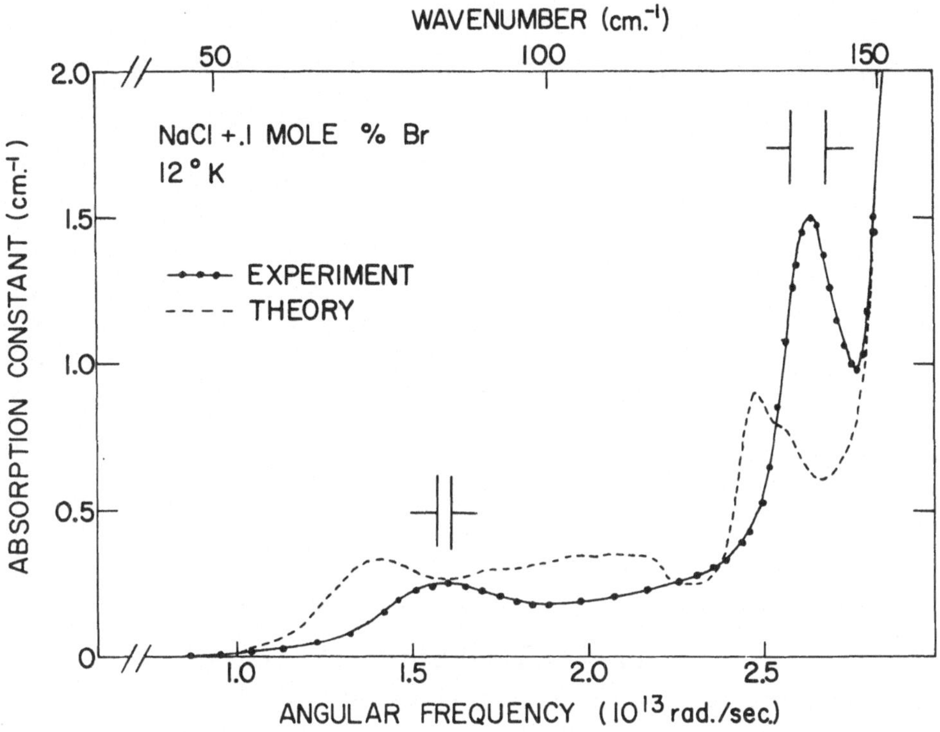

Fig. 3. Comparison of theory and experiment for NaCl:Br^-.

In the case of the fluorine impurity the change in the nearest neighbor central force constant was obtained by fitting the theoretical curve to the strong resonance peak at 60 cm^{-1}. The calculated curve using a fractional force constant change of $\Delta f/f$ = -0.893 is compared with the experiment result at 7° K in Fig. 4. The calculated absorption is much weaker than that observed experimentally and shows no feature corresponding to the broad absorption centered at 112 cm^{-1}. There is a weak peak at 128 cm^{-1} which is about 11% lower in energy than the observed band at 144 cm^{-1}.

For both the fluorine and silver impurities the background absorption at the high energy limit of the experimental curve rises more steeply than predicted by theory. Of the three impurities studied the agreement between theory and experiment is poorest in the case of fluorine and it is clear that considerable improvement in the model is required.

For the silver and bromine impurities the agreement between theory and experiment is reasonable and the relative intensities of the two absorption curves are fairly well reproduced. However, apart from the silver resonance peak, which was fitted to the experimental result, the structure in the calculated curves is displaced to lower energy in both cases. This brings into question

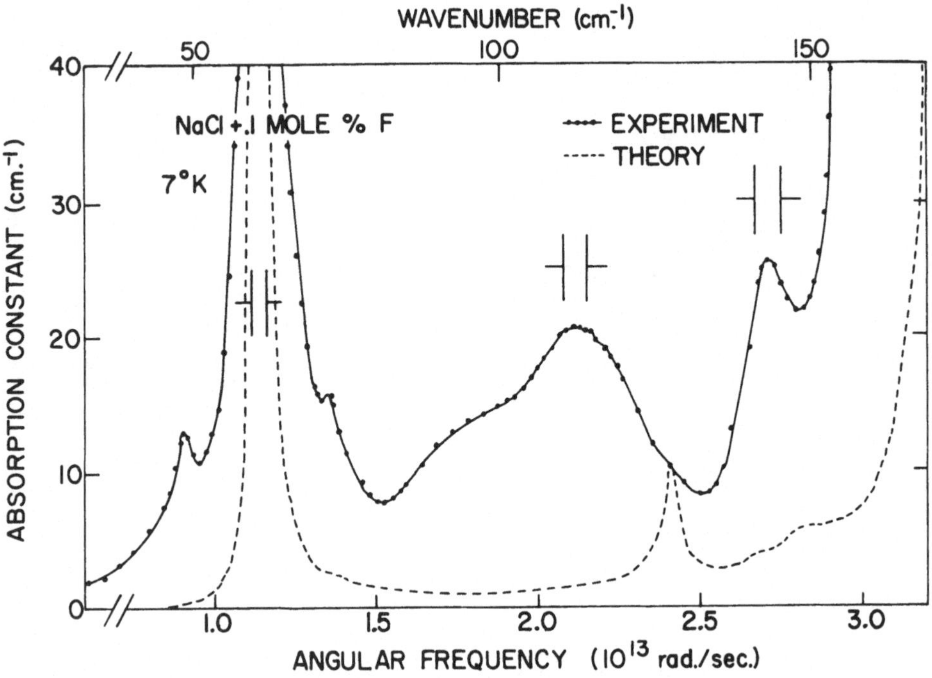

Fig. 4. Comparison of theory and experiment for NaCl:F^-.

the accuracy of the shell model phonons used in this calculation. For example an overall increase of less than 10% in the acoustic phonon energies would not only bring the peak positions of the absorption features into close agreement but would also improve the discrepancy in the intensities due to the factor $(\omega_o{}^2-\omega^2)^2$ in the denominator of the expression for the absorption constant. A comparison of the shell model phonon energies with those obtained by Karo and Hardy [(6)] using a deformable dipole model indicates that a deviation of this order is not unreasonable.

At the present time we conclude that for the silver impurity the inclusion of nearest neighbor non-central forces is necessary to account for the structure observed in the region of the acoustic modes of sodium chloride. For bromine it appears that a pure central force model is adequate. In both cases there is evidence to suggest that a slight modification of the acoustic phonon energies of the host crystal would considerably improve the agreement between theory and experiment.

The authors would like to acknowledge the collaboration with T. P. Martin and J. Gould in much of this work.

REFERENCES

*This work was supported by the Advanced Research Projects Agency under Contract SD-131.

1. R. F. Caldwell and M. V. Klein, Phys. Rev. 158, 851 (1967).

2. R. Weber, Phys. Letters 12, 311 (1964).

3. A. J. Sievers (private communication).

4. M. V. Klein, in Physics of Color Centers, W. B. Fowler, ed., to be published.

5. G. Benedek and G. F. Nardelli, Phys. Rev. 155, 1004 (1967).

6. A. M. Karo and J. R. Hardy, Phys. Rev. 141, 696 (1966).

LITHIUM-ACTIVATED RESONANCES IN ALKALI HALIDE ALLOYS*

B. P. Clayman and A. J. Sievers

Laboratory of Atomic and Solid State Physics

Cornell University, Ithaca, New York

Infrared-active lattice resonant modes have been observed as absorption lines in the far-infrared spectral region for a large number of lattice-defect systems.(1) Many of these absorption lines have been identified with the quasi-localized lattice modes which occur when the lattice impurity has a stable equilibrium position at the normal lattice site.(2) In particular, uniaxial stress experiments on the sharp absorption line centered at 16.2 cm^{-1} in $KBr:Li^+$ indicate that the lithium impurity in this system occupies the normal lattice site.(3) More recent studies of the stress-induced frequency shifts for the lattice resonant modes in $KI:Ag^+$ (17.3 cm^{-1}) and $NaCl:Cu^+$ (23.4 cm^{-1}) show similar characteristics.(4) In all these cases there is strong evidence for a singlet ground state being involved in the I.R. transition because the absorption strengths or integrated intensities in the various polarizations remain constant for frequency splittings which are large compared to the sample temperature of 2^oK. Under these conditions, a degenerate ground state would produce dichroism resulting from thermal population effects.

There are other systems, however, in which the impurity is thought to occupy a position displaced from the normal lattice site. $KCl:Li^+$ is undoubtedly the most thoroughly explored system of this type. The electrocaloric effect,(5) the dielectric constant measurements,(6) and the ultrasonic attenuation results(7) are all consistent with an off-center picture in the sense that they show evidence for a number of closely spaced energy states. More recently, the off-center configuration has been proposed for a number of other lattice-defect systems.(8) Thus, there emerges the present picture of at least two distinct classes of lattice impurity modes, those associated with the impurity at the

Table I

Major Component	Minor Component Concentration		Dopant Concentration	
KBr	KCl	0.0%	LiBr	.009%
		1.57%		.006%
		2.22%		.026%
		3.93%		.022%
		7.80%		.020%
		10.2%		.016%
		14.2%		.014%
KBr	KI	2.1%	LiBr	.005%
		3.4%		.016%
		4.5%		.010%
KBr	NaBr	0.04%	LiBr	.009%
KCl	KBr	0.0%	LiCl	.020%
		1.2%		.017%
		3.2%		.013%
		9.2%		.012%

normal lattice site (Class I) and those with an off-center configuration (Class II). If this picture is correct then a continuous change in some as yet undefined parameter should give rise to a transition between the different classes.

The large hydrostatic coupling coefficients obtained from our stress measurements(3),(4) imply that large frequency shifts occur for small changes in the lattice constant, and that the lattice constant might be an appropriate parameter with which to induce a transition between two classes. From X-ray studies(9) it is known that the average lattice constant of an alloy of two alkali halides has a value intermediate to the lattice constants of the two constituents. This method of varying the lattice constant has been exploited to study F-center(10) and U-center(11) systems. Thus by alloying KI or KCl with $KBr:Li^+$, it is possible to increase or decrease the average lattice constant over an appreciable range in a predictable fashion. We have measured the frequency shift of the lithium-induced resonant mode as a function of this average lattice constant.

The far-infrared transmission spectra were obtained using a Strong-type lamellar interferometer operated in an aperiodic fashion.(12) The radiation was detected by a bolometer operating at about 0.3°K similar to that developed by Drew.(13) All measurements were made with the samples at a temperature of 4.2°K. The instrumental resolution ranged from 0.2 to 0.5 cm^{-1}.

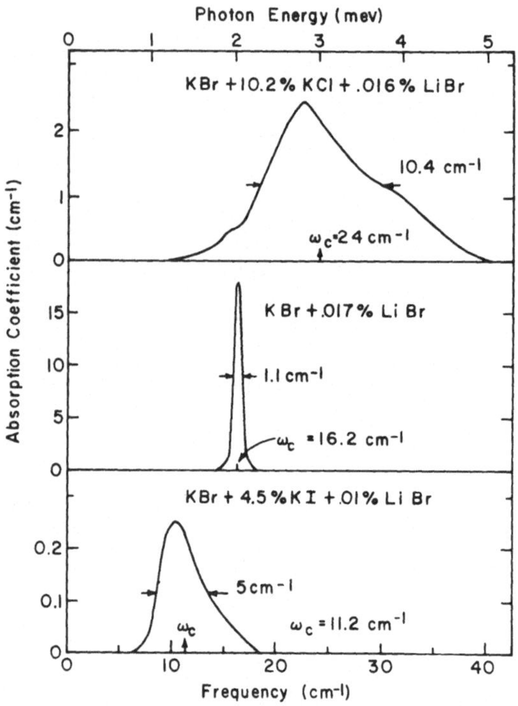

Figure 1: Lithium-induced absorption in two KBr alloys compared with that of KBr:LiBr. Concentrations are expressed in mole percent as determined by chemical analysis. The frequency of the centroid of the absorption line is denoted by ω_c. Note the different absorption coefficient scales for the three cases.

The Li^+-doped alloy crystals examined in this study are listed in Table I. The cation concentrations were determined by gravimetric techniques and the anion concentrations by flame spectrophotometry. All the samples were obtained from single crystal boules grown by the Kryopoulos method.

Typical far-infrared absorption spectra for two of the lithium doped KBr alloys are shown in Fig. 1. In general the center frequency of the absorption shifts to higher frequencies and the line broadens with the addition of KCl or NaBr; it shifts to lower frequencies and broadens upon the addition of KI. For KCl concentrations from zero to 8 mole percent the integrated absorption strength of the lithium activated resonance remains constant. For larger concentrations of KCl the absorption line is quite broad and the strength cannot be determined accurately. On the other hand, when KI is added to KBr:Li^+ the integrated absorption strength of the lithium resonance decreases with

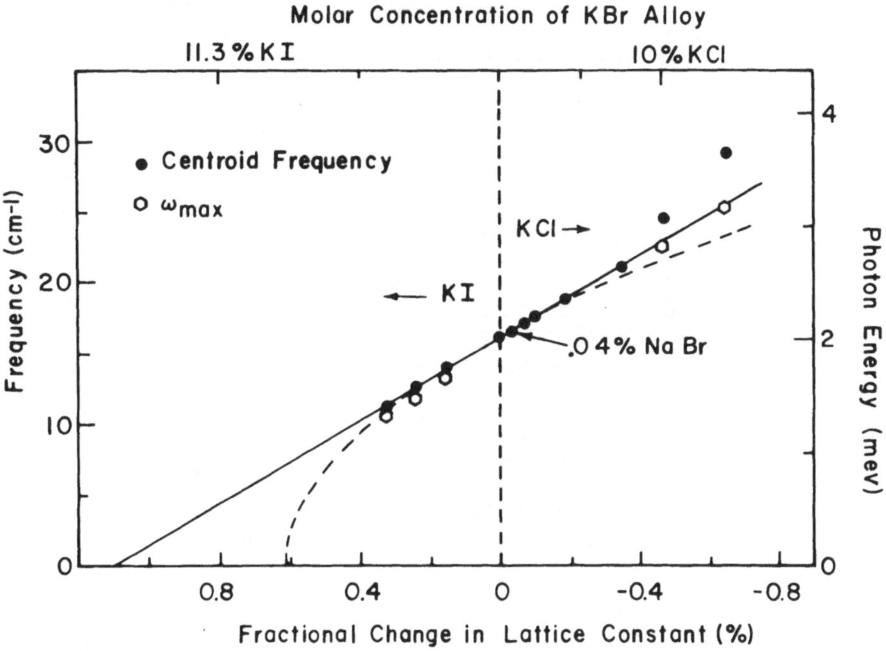

Figure 2: Dependence of absorption frequency upon lattice constant change. Average lattice constant is determined by the Vegard relation; the lattice constant of pure KBr is 3.26 Å at 4.2°K. The molar concentration of the alloy constituent is also shown. The frequency of maximum optical density, ω_{max}, is shown when it differs appreciably from the centroid frequency, ω_c. The linear and quadratic curves are described in the text.

increasing alloy concentration.

The dependence of the centroid frequency of the resonant mode absorption upon the molar alloy concentration and also upon the calculated lattice constant change is shown in Figure 2. The average lattice constant $\bar{a}$ is calculated from the measured alloy concentration by Vegard's relation[14]

$$\bar{a} = a_1 + (a_2 - a_1)x \qquad (1)$$

where a_1 and a_2 are the lattice constants of the two components and x is the molar concentration of component 2.

No resonant mode absorption has been observed for the $KI:Li^+$ system but we have studied the complementary problem of the influence of a lattice constant change on the high frequency

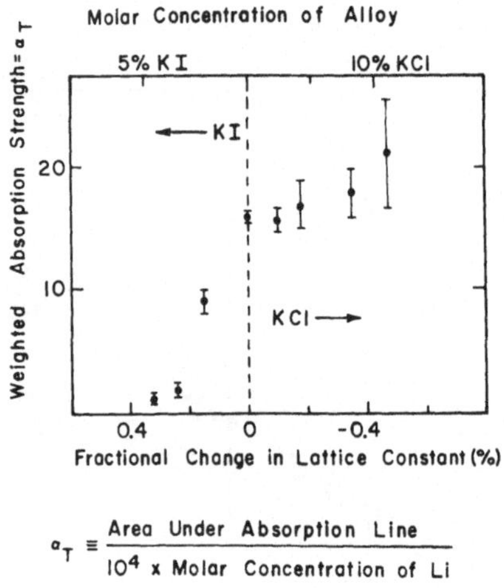

Figure 3: Dependence of weighted absorption strength upon lattice constant change and molar alloy concentration.

($\sim$42 cm^{-1}) absorption in $KCl:Li^+$ by alloying with KBr. For the pure crystal this line has a full width at one half maximum absorption of 14 cm^{-1}. In the alloy systems the lithium induced absorption is increased in width, sharply reduced in strength, and appears to move to lower frequencies with increasing amounts of the minor KBr alloy constituent.

Figure 3 shows the integrated absorption strengths of the Li^+-induced absorption in KBr alloys, weighted by Li^+ concentrations plotted versus molar alloy concentration and lattice constant change. Due to the uncertainty in determining the baseline for such broad lines, the values for the strengths are approximate; but it is clear that the weighted absorption strength is approximately constant for alloys of KCl in KBr. On the other hand, the strength drops sharply, even for low concentrations of KI in KBr.

For the various alloys the observed resonance probably stems from the lithium centers which have unperturbed nearest neighbors. However, if the lithium centers are in a random distribution, the absorption strength should vary as

$$S \sim (1 - x)^6 \qquad (2)$$

where x is the alloy concentration in mole fraction. Because the experimental results indicate that the strength varies much more slowly than Eq. (2) predicts for KCl in KBr and much more rapidly for KI in KBr, the lithium ions are probably not in a random arrangement. We speculate that during crystal growth the lithium ions may prefer to remain in a substitutional site which is near the maximum number of the more polarizable ions of the two component alloy. Unfortunately, it proved impossible to follow the lithium resonance through the entire range of KBr-KCl alloys. For concentrations of KCl in KBr greater than 15%, strong one phonon absorption in the host crystal obscures the lithium-induced line. At the other extreme, even for low concentrations of KBr in KCl the character of the lithium-induced absorption changes dramatically and seems to enhance a broad band absorption rather than appearing as a single resonant mode. Superimposed on this broad background is a weak absorption which we tentatively identify as the remnant of the 42 cm^{-1} line.

Because of this inaccessibility of the intermediate alloy region, we cannot make any definitive statement about the possibility of an instability for an intermediate KCl-KBr alloy or about the connection of the 16.2 cm^{-1} line in KBr with the 42 cm^{-1} line in KCl. However, if this connection exists, it is inconsistent with the model of Bowen et al.[15] They calculated the energy level structure compatible with symmetry considerations for Class I modes and Class II modes in an octahedral environment. The compatibility diagram indicates that the A_{1g} - T_{1u} spacing for the Class I mode (16 cm^{-1} for KBr:LiBr) should become a tunneling splitting (less than 2 cm^{-1} for KCl:LiCl) for the Class II mode.

Our most striking experimental result is the simple manner in which the lithium mode approaches zero frequency as the average lattice constant is increased by alloying KI with KBr as shown in Fig. 2. This trend toward zero frequency indicates that this particular mode is becoming unstable and a transition from Class I to Class II is going to occur. An expansion of the "effective force constant" for this mode in powers of the lattice constant gives the quadratic dependence shown in Fig. 2 with the predicted instability occurring at a lattice constant change of plus 0.61% or an alloy concentration of 8.7% KI. A somewhat better fit to the data is found with a linear relation. The extrapolation of the empirical linear relation in Fig. 2 to zero frequency gives a lattice constant change of plus 1.1% or an alloy concentration of 15.5% KI.[16] Although many ions can participate in this lattice mode, the instability probably represents a spatially localized phenomenon which is related to a change in the nature of the local potential. Because the transition does not depend on the Li^+:Li^+ interaction, it does not represent a cooperative phenomenon as does the lattice instability in Cochran's theory of ferroelectricity.[17]

The authors wish to thank Professor R. H. Silsbee for several helpful discussions and Dr. R. K. Skogerboe for the chemical analyses.

References

* This research was mainly supported by the U. S. Atomic Energy Commission under contract AT(30-1)-2391, NYO-2391-53. Additional support was received from the Advanced Research Projects Agency through the use of space and technical facilities of the Materials Science Center at Cornell University.

1. A. J. Sievers, Phys. Rev. Letters 13, 310 (1964); R. Weber, Phys. Letters 12, 311 (1964); A. J. Sievers and S. Takeno, Phys. Rev. 140, 1030 (1965); R. Weber and P. Nette, Phys. Letters 20, 493 (1966); A. J. Sievers, R. W. Alexander, Jr., and S. Takeno, Solid State Commun. 4, 483 (1966).
2. R. Brout and W. M. Visscher, Phys. Rev. Letters 9, 54 (1962); P. G. Dawber and R. J. Elliott, Proc. Roy. Soc. 273, 222 (1963); G. Benedek and G. F. Nardelli, Phys. Rev. 155, 1004 (1967). A. A. Maradudin in Solid State Physics, Vols. 18 and 19, edited by F. Seitz and D. Turnbull (Academic Press Inc., New York, 1966).
3. I. G. Nolt and A. J. Sievers, Phys. Rev. Letters 16, 1103 (1966).
4. I. G. Nolt and A. J. Sievers, Bull. Am. Phys. Soc. 12, 79 (1967) and to be published.
5. G. Lombardo and R. O. Pohl, Phys. Rev. Letters 15, 291 (1965).
6. H. S. Sack and M. S. Moriarty, Solid State Commun. 3, 936 (1965); A. Lakatos and H. S. Sack, Solid State Commun. 4, 315 (1966).
7. N. E. Byer and H. S. Sack, Phys. Rev. Letters 17, 72 (1966).
8. E. Kraetzig, T. Timusk, and W. Martienssen, Phys. Stat. Sol. 10, 709 (1965); W. Dreybrodt and K. Fussgaenger, ibid., 18, 133 (1966).
9. R. J. Harvighurst, E. Mack, Jr., and F. C. Blake, J. Amer. Chem. Soc. 47, 29 (1925).
10. A. Smakula, N. C. Maynard, and A. Repucci, Phys. Rev. 130, 113 (1963).
11. W. Barth and B. Fritz, Phys. Stat. Sol. 19, 515 (1967).
12. P. L. Richards, J. Opt. Soc. Am. 54, 1474 (1964); I. G. Nolt and A. J. Sievers, to be published.
13. H. D. Drew and A. J. Sievers, Bull. Am. Phys. Soc. 12, 77 (1967).
14. L. Vegard, Z. Physik 5, 17 (1921).

15. S. P. Bowen, M. Gomez, J. A. Krumhansl, and J. A. D. Matthew, Phys. Rev. Letters 16, 1105 (1966); and, also, M. Gomez, S. P. Bowen, and J. A. Krumhansl, Phys. Rev. 153, 1009 (1967).
16. Recent studies by H. K. A. Kahn show that it is not possible to obtain homogeneous solid solutions for KBr containing more than 10 mole % KI with the normal Kyropolous growing method. H. K. A. Kan, Ph.D. thesis, Cornell University (1966), Materials Science Center Report #427.
17. W. Cochran, Advances in Physics 9, 387 (1960); 10, 401-420 (1961).

ON THE TEMPERATURE DEPENDENCE OF WIDTHS OF HIGH FREQUENCY LINES OF CRYSTAL AND LOCAL VIBRATIONS

A.A. Klochikhin, T.F. Maksimova, A.I. Stekhanov
A.F. Ioffe Physico-Technical Institute
Leningrad K-21, U.S.S.R.

The influence of temperature on the width of vibrational lines of the infrared absorption and Raman-scattering spectra of crystals has been studied in a number of experimental works.(1-9)

The most detailed quantitative data on this problem have been recently obtained in the works on infrared absorption in alkali-halide crystals.(1,2) In these works the width of absorption line at the frequency $\omega_t(\vec{K}=0) = \omega_t^o$ caused by the transversal optical vibrations with the wave vector $\vec{K}$ equal to zero has been studied. In the investigated crystals the frequency ω_t^o is somewhat less than the maximum frequency of the eigenfrequency spectrum of lattice vibrations. A strong temperature dependence of the line width has been observed in the temperature range $T \gtrsim \Theta_D \sim \omega_t^o$ where Θ_D is the Debye temperature. At temperatures $T < \Theta_D$ the line widths depended weakly on temperature, remaining close to their values at T=0.

In the theoretical works(10-16) the qualitative explanation of this kind of temperature dependence has been given with the help of the third and fourth order processes of the anharmonic phonon interaction. In the later works(17,18) a satisfactory quantitative agreement with the experimental results has been reached.

The different character of the temperature dependence of the absorption line widths has been observed for the local vibrations of the U=centers (H^- and D^- ions) in alkali-halide crystals.(3-7) In this case the local vibration frequencies ω_{loc} exceed the maximum frequency of the spectrum of the crystal-matrix eigenfrequencies. The absorption line width appears to be strongly dependent on temperature at $T > \omega_{loc}$ both in the region $T > \Theta_D$ and $T > \Theta_D$, in a number of cases almost down to $0^{o}K$.

Such a temperature dependence of the local vibration absorption lines has been explained in the works (5,7,19,20) by the anharmonic interaction of local vibrations with vibrations of the continuous spectrum.

In a number of experimental works (21-22) attention was paid to the strong temperature dependence of OH-vibration line widths in the absorption and scattering spectra of crystals containing the hydrogenic bond. However at present there is no single point of view on the causes of this phenomenon.

In connection with this of great interest is the investigation of light absorption and scattering by inner vibrations of other complex ions and molecules.

Complex ions or molecules are members of the regular lattice of many crystals. Their inner vibrations formoptical modes whose frequencies have a wide range of values. In addition, some of these ions can be introduced into alkali-halide crystals as impurities. Then their inner vibrations with frequencies $\omega_{in} > \omega_{max}$ display themselves in a crystal as local vibrations.

In this work we have undertaken experimental investigations of the influence of temperature on the inner vibration lines of the groups CH, OH, SO_4 in regular c crystals and impurity groups OH,NO_2,NO_3 and SO_4 in alkali-halide crystals. The data obtained are discussed in terms of theoretical representations on the temperature dependence of the high frequency vibration line widths, which were developed in our paper.(23)

I. Experimental Results

We have investigated Raman-scattering spectra of crystals of Rochelle salt ($NaKC_4H_4O_6 \cdot 4H_2O$), barite ($BaSO_4$), sodium bromide with two water molecules ($NaBr\ 2H_2O$), and the infrared absorption spectra of the crystals NaCl, KCl and KBr containing as impurities the ions OH,$SO_4$$NO_2$ and NO_3. The investigations were made

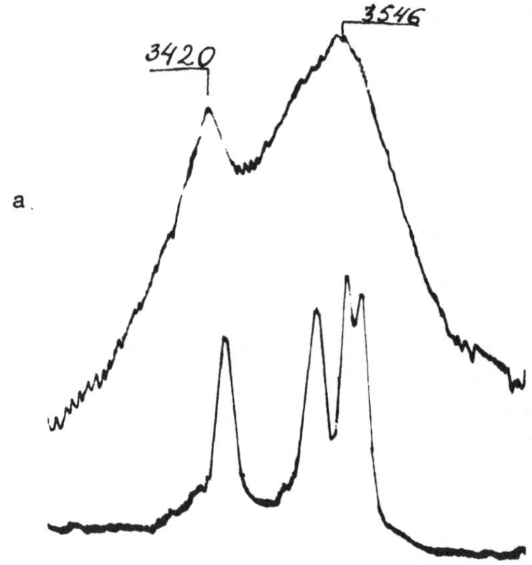

Fig. 1. Microphotograms of the Raman spectrum of the crystal $NaBr \cdot 2H_2O$

T, °K a - 300 b - 77

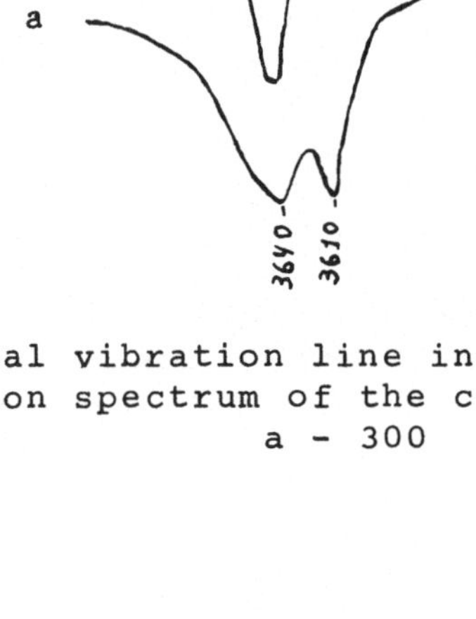

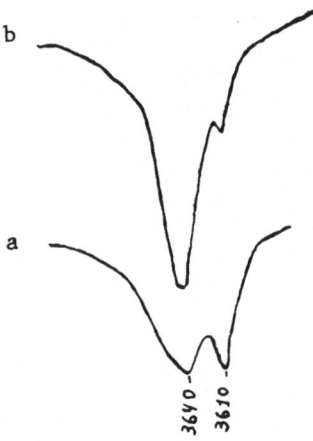

Figure 2. OH-local vibration line in the infrared absorption spectrum of the crystal KCl(OH)
T, °K a - 300 B-100

at room temperature and at liquid nitrogen temperatures.

Raman-scattering of light in the crystals was exited by the 2537Å mercury line. The spectra were detected by means of a Zeiss Q-24 quartz spectrometer with 10Å/mm dispersion in the 3000Å-wave length region. To obtain data on the intensity and width of vibrational lines we have made microphotometric investigations of spectra. The absorption spectra were detected on a Zeiss UR-10 infrared spectrometer.

In studying the scattering spectra we used specially selected big natural barite single crystals. The Rochelle salt and $NaBr \cdot 2H_2O$ single crystals were grown from water solutions. The infrared absorption spectra of the alkali-halide crystals with impurities have been investigated on single crystals grown from the alkali-halide salt melt by the Kiropolus method. Impurities were introduced into the charge in the form of corresponding salts of concentrations in between 0.1-1mol%.

Fig. 1 (a,b) shows microphotograms of Raman-spectra of the crystal $NaBr \cdot 2H_2O$ at room and liquid nitrogen temperatures. At room temperature in the high frequency region one can observe a wide line of OH-vibrations which goes from 3350 to 3550cm^{-1}. In the line one can note two wide maxima near 3420 and 3546cm^{-1}. In lowering the temperature down to 77°K four separate narrow and intense lines with frequencies 3406,3422,3450 and 3534cm^{-1} are observed in the place of the smeared lines.

The Raman-spectrum of the Rochelle salt crystal has been investigated by us in the region of the group CH valence vibration lines. At room temperature two intense lines at 2940 and 2980cm^{-1} are observed here. Cooling down to liquid nitrogen temperature affects the CH line widths but not so strongly as in the case of the OH-vibration line in $NaBr \cdot 2H_2O$. The CH lines get noticeably narrower and shift to high frequencies with decreasing temperature.

Side by side with the investigation of the temperature dependence of CH and OH groups we have studied the temperature behavior of the inner vibration lines of the barite crystal group SO_4. In the frequency range 1000 to 1200cm^{-1} five narrow and intense lines are detected at the frequencies 1083, 1104, 1140, 1145, and 1167cm^{-1}. The crystal temperature decrease from room temperature down to liquid nitrogen temperature affects the spectrum shape slightly. One can note only a small line

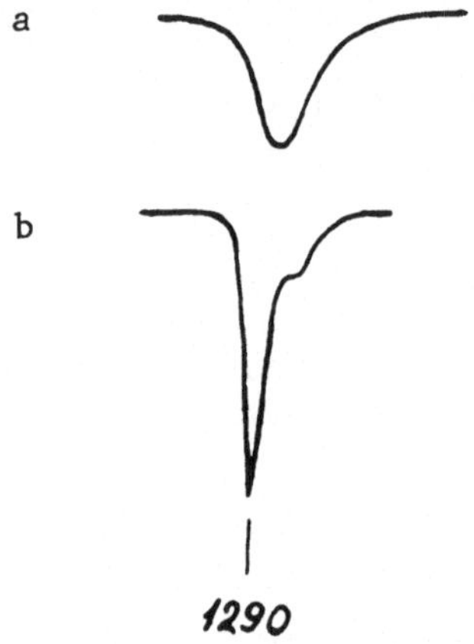

Fig. 3. NO_2 ion local vibration line in the infrared absorption spectrum of the crystal $KCl(NO_2)$
T, °K a - 300 b - 100

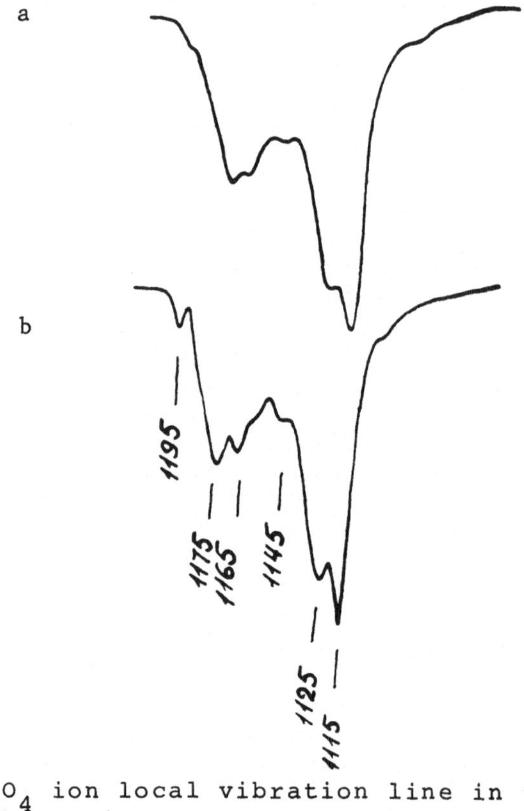

Fig. 4. SO_4 ion local vibration line in the infrared absorption spectrum of the crystal $KCl(SO_4)$
T, °K a - 300 b - 100

narrowing.

Fig. 2 (a,b) shows the infrared absorption spectra of KCl (OH) crystal in the region 3600-3800cm^{-1}. At room temperature an intense line of the OH group local vibration with the frequency 3640cm^{-1}, on which a weaker line at 3610cm^{-1} belonging apparently to the compound frequency is superimposed is detected . Similar lines have been also observed in the crystals NaCl(OH) and KBr(OH). In lowering the temperature down to liquid nitrogen temperature the width of the line at 3640cm^{-1} decreases strongly, its peak intensity increasing. From the peak intensity change we have estimated the change in the halfwidth of the line at 3640^{-1}in the temperature range 100 to 300°K. At room temperature the OH-line halfwidth magnitude is about 70cm^{-1}. In lowering the temperature by a factor of three the halfwidth decreases approximately by a factor of 2.5, its change following a law close to linear.

In the infrared absorption spectra of KCl, NaCl and KBr crystals containing as impurities the molecular ions NO_2 and NO_3 at room temperature we have observed very narrow and intense lines with frequencies of 1285 and 1400cm^{-1} caused by the inner vibration of the NO_2 and NO_3 groups respectively. The investigation of the temperature dependence of these lines in the range 100 to 300°K showed that their width and peak intensity change strongly with temperature. Fig.3(a,b) shows the shape of the 1285cm^{-1} line of a KCl(NO_2) crystal at temperatures of 300 and 100°K. Estimations showed the line halfwidth to be about 20cm^{-1} at room temperature, and decreasing by at least a factor of two in the temperature range used in these investigations.

In the KCl(SO_4) crystal and also in NaCl and KBr crystals containing SO_4 ion impurities a wide line in the region 1100-1200cm^{-1} caused by the three-fold-degenerate vibration of the SO_4 ion appears to be the most intense line in the spectrum. Fig.4(a,b) shows this line in the KCl(SO_4) crystal at room and liquid nitrogen temperatures. As can be seen, it has a complicated structure and contains a minimum of six components with the frequencies 1110,1120,1145,1160,1170,1190cm^{-1}. In lowering the temperature the intensity of the whole line increases slightly, and the split off components get narrow. As a consequence of this the line structure becomes sharper and clearer.

II. Discussion

We now compare our experimental results with theory. As can be seen in all cases investigated the line widths proved to be dependent on temperature in the temperature range of our investigations. The strongest temperature dependence of the line width appeared to be for the OH, CH groups in regular crystals and OH,NO_2 and NO_3 impurity groups in alkali-halide crystals.

As was shown in our work[23] the temperature dependence of the high frequency vibration line width in regular crystals has to take place at temperatures $T \gg \Delta d$, where Δd is the dispersion of the high frequency vibration mode. The existence of the temperature dependence of the line width in our case in the temperature range 77^oK to room temperature shows that the corresponding mode dispersion in any case is less than $50 cm^{-1}$. Thus the assumption of a small value of the inner vibration mode dispersion in comparison with their frequencies made in the theoretical considerations in the work[23] is confirmed. For a more accurate estimate of the dispersion value one has to investigate the temperature behavior of the line widths at lower temperatures.

The strong temperature dependence of the local vibration lines of NO_2 and NO_3 ions is in agreement with the dependence predicted theoretically for the case of individual high frequency local vibrations.

Comparatively weak temperature dependence has been observed for the line width of the SO_4 ion threefold-degenerate vibration, the line being split by the crystal field. The SO_4 ion possesses nine inner vibrational degrees of freedom. Anharmonic interactions are possible between all its vibrations. This is why according to the theoretical results of reference[23] one should expect a weak temperature dependence of the observed line width. Note that the temperature dependences of widths are similar both for the SO_4 ion being in a regular lattice and for the SO_4 ion in alkali-halide crystals.

A comparison of the local vibration line half-widths of SO_4 and OH ions in alkali-halide crystals shows that the OH line width is approximately 5-7 times larger. This value is in good agreement with the estimates of reference[23] obtained without taking into account the differences in the values of the potential energy derivatives. The anharmonic interaction is known to be mainly connected with the atom repulsion potential whose parameters differ slightly for different alkali-halide crystals. It follows from our results that this is apparently valid for the interaction of atoms of different atomic groups also.

The comparison of the OH group line halfwidths in the regular crystal $NaBr \cdot 2H_2O$ and of the impurity OH groups in the KCl crystal shows that the halfwidths in both cases are large. Numerous attempts at a theoretical explanation of the large OH-vibration line widths in crystals and of the temperature dependences of the hydrogen bond line width can be found in the literature. However, in these theories the anharmonic vibration interaction has not been taken into account. Our results show that the decisive role in the explanation of the large value and temperature dependence of the hydrogen bond line width should apparently be given to the anharmonic interaction between vibration modes. The large magnitude of the OH-vibration halfwidth in this case is explained by the large vibration amplitude of the light hydrogen atom. The influence of the hydrogen bond in this case affect the value of the anharmonic constants. As is known from the theory of the hydrogen bond, the hydrogen atom potential function in the hydrogen bridge is essentially anharmonic.(24) Note that in all cases the additional dependence on temperature can be given by the compound frequencies.

REFERENCES

1. G. Heilmann, Z. Physik 152, 368, 1958.
2. M. Hass. Phys. Rev. 117, 1497, 1960.
3. G. Schaefer. J. Phys. Chem. Sol. 12, 233, 1960.
4. D.N. Mirlin, I.I. Reshina. Fiz. Tverdogo Tela, 5, 3352, 1963.
5. W.Hayes, G.D. Jones, R.J. Elliott, C.T. Sennett. Lattice Dynamics. J. Phys. Chem. Sol. Suppl. 1, 475, 1965
6. B. Fritz. Lattice Dynamics. J. Phys. Chem. Sol. Suppl. 1, 485, 1965.
7. M.A. Ivanov, M.A. Krivoglaz, D.I. Mirlin, I.I.Reshina, Fiz. Tverd. Tela B, 192, 1966.
8. A.I. Stekhanov, M.B. Eliashberg, Fiz. Tverd. Tela, 5, 9985, 1963.
9. A.I. Stekhanov, T.I. Maksimova, Fiz. Tverd. Tela.8, 924, 1966.

10. M. Born, M. Blackman, Zs.Phys. 82, 551, 1933.
M. Blackman. Zs. Phys. 86, 421, 1933.
11. A.A. Maradudin, R.F. Wallis, Phys. Rev. 123, 777, 1961.
12. V.S. Vinogradov, Fiz. Tverd. Tela, 4, 712, 1962.
13. J. Neuberger, R.D. Hatcher, J. Chem. Phys. 34,1733, 1961.
14. D.W. Jepsen, R.F. Wallis, Phys. Rev. 125, 1496,1962.
15. L.E. Gurevitch, I.P. Ipatova, J. Exper. Teor. Fiz., 45,231, 1963.
16. V.N. Kashcheev, Fiz. Tverd. Tela, 5, 2339, 1963.
17. R.F. Wallis, I.P. Ipatova, A.A. Maradudin, Fiz. Tverd. Tela. 8, 1064, 1966.
18. I.P. Ipatova, A.A. Maradudin, R.F. Wallis, Phys. Rev. 155, 882. 1967.
19. M.A. Ivanov, L.B. Kvashnina, M.A. Krivoglaz, Fiz. Tverd. Tela, 7, 2047, 1965.
20. I.P. Ipatova, A.A. Klochikhin, J. Exper. Teor. Fiz. 50, 1603, 1966.
21. A.I. Stekhanov, Dokladi Ak.Nauk USSR, 92, 281,1953, 106, 433, 1956. Isvestia Akad. Nauk USSR ser.fiz. 21, 31, 1957, 22, 1109, 1958.
22. A.I. Stekhanov, A.A. Klochikhin, Z.A. Gabrichidze, E.A. Popova, Physical Problems of Spectroscopy. Proceedings of the XIII Meeting. Leningrad. 1960. Izvestia Ak. Nauk U.S.S.R., N. 1962.
23. A.A. Klochikhin, T.I. Maksimova, A.I. Stekhanov, Preprint of Ioffe Phys. Technical Institute, 1967.
24. N.D. Sokolov. Uspekhi Fiz. Nauk. 57, 205, 1955.

THEORETICAL MODELS *

Miles V. Klein

Department of Physics and Materials Research Laboratory

University of Illinois, Urbana, Illinois

INTRODUCTION

In this paper we shall describe some mathematical models used to calculate the properties of localized phonon modes and resonance states. The treatment will cover only those phenomena that can be described by a "single defect" theory. This includes effects appearing to lowest order in the concentration of randomly distributed impurities as well as those that can be treated in a "molecular" field approximation. Both harmonic and anharmonic effects will be mentioned. Many detailed applications of these models will be discussed in other papers of this conference.

Models occur in more than one sense in a subject such as this. There are the mathematical models assumed to describe the perfect host crystal and models for the harmonic perturbation introduced by the impurity. There are also models for the anharmonic coupling assumed at or near the defect site. These go as input into a theoretical calculation. The computer then turns out the numerical results, which one must try to interpret. Sometimes it is hard to give a simple picture that explains a given feature in the output, but in many cases one can introduce parameters such as effective masses which can be evaluated numerically and which behave as do similar parameters in much simpler molecular or mechanical models. Thus models can be valuable in the output side of a calculation.

GENERAL THEORETICAL FRAMEWORK

The fundamental theoretical formalism has been discussed in many recent papers and review articles.[1,2,3] For a harmonic system a classical description can be given that is easily extended to a correct quantum-mechanical description, since in the Heisenberg picture the equations of motion are the classical Newtonian equations of motion. The Green's function matrix described below is closely related to the double-time displacement-displacement Green's function used in quantum statistical mechanics.

We work with the reduced displacement vector $\underline{v} = \underline{M}^{1/2}\underline{u}$, where $\underline{M}$ is the mass matrix for the pure crystal and $\underline{u}$ the vector for the physical displacements from equilibrium. The equation of motion for the normal modes of a harmonic crystal with a single impurity is

$$(\underline{A} - \omega^2\underline{I} + \underline{\gamma})\underline{v} = 0. \tag{1}$$

Here $\underline{A} = \underline{M}^{-1/2}\underline{\Phi}\underline{M}^{-1/2}$ is the dynamical matrix of the perfect crystal, $\underline{I}$ is the unit matrix, and

$$\underline{\gamma}(\omega^2) = \underline{M}^{-1/2}\ \Delta\underline{\Phi}\underline{M}^{-1/2}\ - \Delta\underline{M}\underline{M}^{-1}$$

represents the perturbation or defect matrix due to $\Delta\underline{\Phi}$ the change in coupling constant matrix and $\Delta\underline{M}$ the change in mass matrix. For the theory to be useful $\underline{\gamma}$ should be nonzero only in a small region of configuration space, i.e., couple to only a few configurations. We call this the "space of the defect."

All the dynamic properties of the perturbed harmonic crystal may be obtained from $\underline{\bar{G}}(z)$, the Green's function matrix for Eq. (1) defined by

$$\underline{\bar{G}} = (\underline{A} - z\underline{I} + \underline{\gamma})^{-1} \quad . \tag{2}$$

We introduce the Green's function $\underline{G}$ for the perfect crystal by

$$\underline{G} = (\underline{A} - z\underline{I})^{-1} \quad , \tag{3}$$

which can be written in terms of the eigenvalues $\omega_k{}^2$ and eigenvectors v_k of the perfect crystal as

$$\underline{G} = \Sigma_k\ v_k\ v_k{}^{\dagger}\ /\ (\omega_k{}^2 - z) \quad . \tag{4}$$

$\underline{\bar{G}}$ may be written in terms of $\underline{G}$ as

$$\underline{\bar{G}} = \underline{G} - \underline{G}\ \underline{t}\ \underline{G} \quad , \tag{5}$$

where the t-matrix in Eq. (5) is localized to the same space of the defect as $\underline{\gamma}$ and is given by

$$\underline{t} = \underline{\gamma}(1 + \underline{g\gamma})^{-1} = (1 + \underline{\gamma g})^{-1}\underline{\gamma} \quad . \tag{6}$$

Here $\underline{g}$ represents the part of $\underline{G}$ in the localized defect space. The main task then is to determine $\underline{g}$ and $\underline{t}$ and then to solve (6).

IRREDUCIBLE CONFIGURATIONS

To perform the matrix inversion in Eq. (6) it is almost essential to use basis coordinates or configurations in the defect space that belong to the irreducible representations of the point symmetry group at the impurity site. Equation (6) then reduces to block form with each block belonging to a different irreducible representation. If a given representation occurs just once, we call it "uncoupled." Its block then has a dimension r equal to the dimension of Γ, and each matrix in the block is a multiple of the r dimensional unit matrix. Then (6) becomes a scalar equation of the form

$$t(\Gamma) = \gamma(\Gamma)[1 + g(\Gamma)\gamma(\Gamma)]^{-1} \quad . \tag{7}$$

If the representation Γ occurs m times in the space of the defect, then Eq. (7) becomes a m by m matrix equation describing these "coupled configurations" having the same symmetry. In what follows all matrix equations will refer to such an irreducible subspace of the space of the defect.

THE IRREDUCIBLE CONFIGURATIONS OF THE XY_6 DEFECT MODEL

As an illustration we shall discuss the irreducible configurations for an impurity with octahedral symmetry. Consider the XY_6 molecule shown in Fig. 1.[4] In the <u>mass defect approximation</u> the impurity mass alone is changed. The space of the defect consists of the 3 Cartesean coordinates of the displacement of the impurity. These already belong to representation $T_{1u}(\Gamma_{15})$ of the octahedral group. In this case the diagonal (and only nonzero) element of the matrix $\underline{\gamma} = \underline{\gamma}_m$ is $-\Delta M\omega^2/M_o$ where M_o is the mass of the host atom at the impurity site.

When we add central force constant changes between the impurity and its nearest neighbors, 6 more degrees of freedom are added to the space of the defect, corresponding to the motion of the neighbors along the bond directions. The new irreducible coordinates are the following: An A_{1g} (Γ_1) "breathing" coordinate, two doubly degenerate E_g (Γ_{12}) "tetragonal" coordinates, and three triply degenerate T_{1u} coordinates. One of the latter is shown in the figure labelled $|T_{1u};2)$. It will be coupled by $\underline{\gamma}$ and $\underline{g}$ matrices to the impurity coordinate $|T_{1u};1)$, since they both have the same symmetry. Equation (7) will then be a 2 by 2 matrix equation.

There are other possible basis configurations in the T_{1u} subspace which are linear combinations of the ones mentioned above. One such basis is labelled by primes in Fig. 1. A change in central force constant (Δf_c) couples only to $|T_{1u};2')$ and its two permutations. The diagonal (and only) matrix element of $\underline{\gamma} = \underline{\gamma}_c$

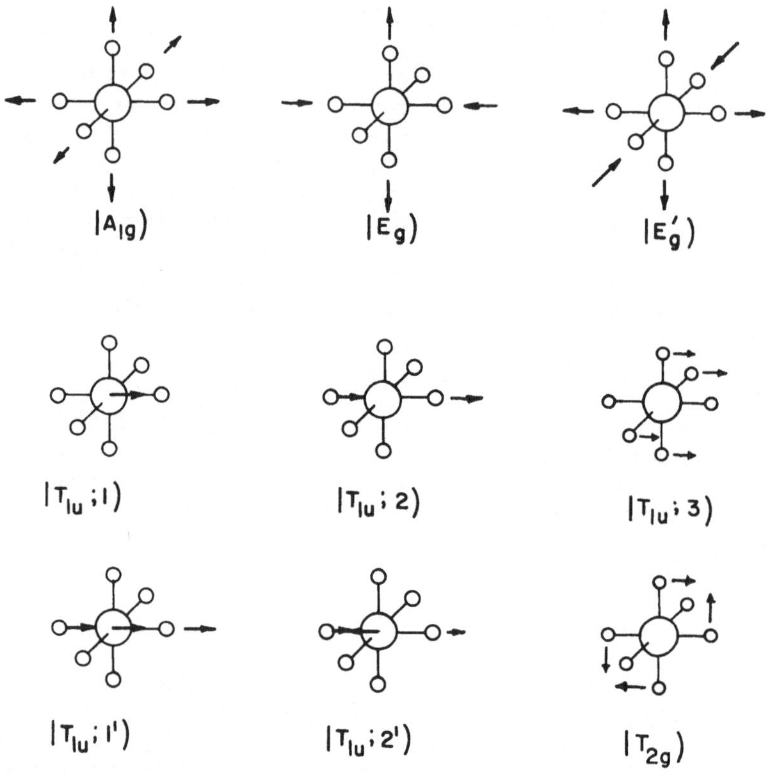

Fig. 1. Some irreducible configurations of the XY_6 octahedral molecule.

is $2\Delta f_c/M_r$, with $M_r = 2M_oM_n/(M_o + 2M_n)$ and M_n = mass of a neighbor atom.

The T_{1u} configurations are the only ones that will contribute to the first order defect-induced infrared absorption. The A_{1g} and E_g modes contribute to the first order defect-induced Raman scattering.

If the non-central force constant between the impurity and its nearest neighbors is changed, 12 new degrees of freedom associated with bond bending are added. They belong to the following triply degenerate irreducible representations: T_{1g}, T_{2u}, T_{2g}, and T_{1u}. The first two coordinates are both infrared and Raman inactive. One of the Raman active T_{2g} configurations is shown in Fig. 1. One of the new T_{1u} configurations is also shown. It will mix with the other two coordinates of the same symmetry, making Eq. (5) a 3 by 3 matrix equation.

HARMONIC MODELS FOR THE HOST LATTICE

One model applied extensively to describe the harmonic host lattice is the nearest neighbor force constant model.[5] It yields simple analytical expressions for the phonon dispersion curves. There are also simplifying relationships among Green's function matrix elements that are not true for more complicated systems. The Green's functions needed can be computed from one dimensional integrals involving Bessel functions. The resulting expressions show very accurately how the singularities in the frequency spectrum affect the computed results. The disadvantage is that these singularities are not very realistic since the model does not reproduce important qualitative features of the real phonon dispersion curves. In addition the neglect of all long range forces means that such models will not predict the behavior for low frequency resonance states revealed by other models that include such forces.[2] We shall have more to say about these resonances later.

In ionic crystals the simplest model that includes the long range Coulomb interaction is the rigid-ion model.[6] It takes into account the electrostatic interaction between dipoles produced when the ions are displaced from equilibrium, but it neglects the dipoles produced by deformation of the electron distribution when the ions are displaced. The dispersion curves calculated with such a model do not agree very well with experiment.

Much better agreement is obtained with Cochran's shell model[7] or Hardy's deformation dipole model.[8] The former model is based on a picture of the ions proposed by Dick and Overhauser, in which each of the ions consists of a small heavy core coupled elastically and isotropically to a light rigid shell, which approximates the behavior of the rare-gas-shell electron cloud. Short range repulsive forces act between shells only. The deformation dipole model is based on an idea of Born and Huang that dipoles must appear when the ions deform while undergoing a relative displacement. These dipoles are the equivalent of the shell-core relative displacement. In both models the dipolar coordinates can be eliminated from the equations of motion to obtain the dynamic matrix. Except for one term the resulting matrices are the same.[9] Because of this the resulting dispersion curves are not identical, even if the same input parameters are used.[10]

A recent modification of the shell model is the "breathing shell model" of Schröder.[11] It adds a symmetric inward displacement of the shells to the degrees of freedom of the ordinary shell model, without introducing new force or charge parameters. This modification seems to improve the fit to the neutron-determined dispersion curves.

A final "model" that may be useful in phonon problems is not really a model at all but a phenomenological method of fitting experimental data with a minimum number of parameters. It is the old Fourier expansion technique that has been recently revived and applied successfully to electron energy band problems by G. Dresselhaus and M. S. Dresselhaus.[12] In a recent unpublished paper they have applied the method to phonons in silicon and germanium. The method makes extensive use of group theory. It can reveal when redundant parameters have been introduced and makes most efficient use of the parameters that remain.

HARMONIC PERTURBATIONS

I have already discussed some aspects of simple harmonic perturbations such as the mass defect approximation and the nearest neighbor-central-force-constant-change model. In order to obtain better agreement between calculations and experiments theorists are now using more complex defect models, including non-central force constant changes and going out to farther shells of neighbors.

The most sophisticated harmonic perturbation that I know of is provided by the shell model treatment of the defect. One then has a consistent treatment of both host crystal and impurity. This allows one to account explicitly for changes in shell-core force constants and shell charges. The method has been developed by Page and Strauch, from whom we shall learn more details. The essential ingredient in obtaining solutions with this model is an extension of the Green's function method to include shell coordinates.

The kind of harmonic model assumed for the perturbed crystal will affect the expression used for the infrared absorption coefficient. In the simplest case, treated by Dawber and Elliott,[13] a charged, rigid impurity in an uncharged, unpolarizable host lattice was assumed. The total electric field acting on the impurity in a case like this is simply the applied field. Two simple modifications can be added to make the Dawber-Elliott picture more realistic (apart from adding force constant changes). An effective charge $e^* \neq e$ can be introduced to take some of the deformation properties of the impurity atom into account, and the host lattice can be given a high frequency index of refraction $n_\infty \neq 1$. In a cubic field the local field at the impurity will be the applied field plus the Lorentz local field correction. These two features change the simple result for the absorption constant by a factor

$$\left[\frac{e^*(n_\infty^2 + 2)}{3\,e}\right]^2 \quad . \qquad (8)$$

If the atoms surrounding the impurity have deformation dipoles, then each will contribute to the dipole moment of the perturbed crystal due to the distortion of its charge cloud as it participates in the localized mode or resonance state. This will be true even for a high frequency localized mode, where essentially the only core to move is that of the impurity, since the short range repulsive forces will polarize the impurity and its near neighbors. Whereas the impurity polarization is included in e* in (8), only part of the polarization of the neighbors is included, namely that induced by the long-range Coulomb interaction. Leigh and Szigeti have examined this problem of how the distortion of the host lattice atoms contributes to the effective charge associated with a localized excitation from a quite general point of view, as they will report to this conference.

Consider now an ionic crystal containing a monovalent impurity treated in the harmonic approximation. In a rigid ion treatment of both impurity and host lattice there is no need to make explicit local field corrections, since the Coulomb interaction between ions is already taken into account. Nevertheless a new feature emerges due to the dispersive effects of the k = 0 transverse optic phonon. The electric field couples to a configuration proportional to the unperturbed displacement vector |TO,k=0) for this phonon. The absorption coefficient is proportional to the imaginary part of

$$(TO,k=0|\underline{\bar{G}}|TO,k=0) = (TO,k=0|\underline{G}|TO,k=0) - (\omega_o^2 - \omega^2)^{-2}\,(TO,k=0|\underline{t}|TO,k=0). \qquad (9)$$

Here we have used Eq. (5). ω_o is the frequency of the optical phonon. The second term gives the contribution of the impurity to the absorption.

The resulting expression for the absorption neglects the contribution of the electronic polarization to the dipole moment of the crystal. A simple way to take this into account is to add the factor (8). We at Illinois have done this in most of our calculations, for we feel that the correction using n_∞ is better than none at all. As mentioned above, this procedure neglects the deformation dipoles produced on the ions surrounding the impurity.

Clearly, what is needed is a consistent expression for the absorption constant using both host and defect shell models or an equivalent picture, but as far as I know, this has not yet been attempted.

ANHARMONIC PERTURBATIONS

Some theoretical calculations of anharmonic impurity properties have used the very simplest anharmonic perturbations with a few parameters, sometimes only one. For calculations with pure anharmonic ionic crystals one often obtains the cubic and quartic coupling constants from the same Born-Meyer potential that gives the approximately correct elastic constants.(14) For NaCl the measured third and fourth order elastic constants are more or less consistent with this sort of calculation.(15) For other types of crystal, such as valence crystals, partially ionic crystals, silver halides, more caution may have to be exercised.

Caution is probably also in order when one wants to obtain anharmonic coefficients at an impurity by differentiating an assumed force law. It is hard enough to estimate the perturbed harmonic force constants because of uncertainties about the static displacements and the force laws. In crystals with mixed bonding, this can be even more of a problem.

Apart from quantitative estimates the form of the anharmonic potential can always be determined from symmetry arguments.

THE LOCALIZED PHONON EXCITATIONS

Both localized modes and resonance states (or pseudo-localized modes) will be referred to here as localized excitations. A narrow resonance has a formal similarity to a localized mode and can be treated on essentially the same footing. We work now with coupled irreducible representations of the same symmetry. We split $\underline{g}$ into real and imaginary parts:

$$\underline{g}(z) = \underline{R}(\omega^2) - \mathcal{I}(\omega^2) \quad \text{with } z = \omega^2 - i0^+ . \qquad (10)$$

$\mathcal{I}$ is zero for ω^2 outside the allowed band of phonon frequencies. The condition for a resonance state or localized mode is

$$\mathrm{Re}\,\{\det(1 + \underline{g}(\omega_o{}^2)\underline{\gamma}(\omega_o{}^2)\,)\} = 0 \qquad (11)$$

When $\mathcal{I}(\omega_o{}^2)$ is small this is equivalent to

$$\mu_o = -1 \qquad (12)$$

where μ_o is an eigenvalue of $\underline{R\gamma}$. The corresponding eigenvector $\underline{v}_o$ then represents the linear combination of configurations in representation Γ that give the localized excitation. If the representation is uncoupled, $\underline{v}_o$ is determined by symmetry alone.

The atoms outside the defect space participate in the excitation with an amplitude proportional to the matrix element of $\underline{G}$ between the configuration $\underline{v}_o$ and all configurations of the same symmetry outside the space. The reduced displacement of the atoms in the space of the defect is given by

$$\underline{v}(t) = \frac{q(t)\ \underline{v}_o}{\sqrt{(\tilde{v}_o\ \underline{\gamma}\ v_o)\ (d\mu_o/d\omega^2)_{\omega_o^2}}} \quad . \tag{13}$$

Here q(t) is the dynamical coordinate of the localized mode or the approximate dynamical coordinate of the resonance state. It behaves like all such harmonic oscillator coordinates, e.g., its mean square expectation value in a state where n quanta are excited is

$$\langle q^2 \rangle = \hbar\ (2n + 1)/(2\omega_o) \quad . \tag{14}$$

If we have an uncoupled irreducible representation, Eq. (13) becomes a scalar equation

$$v(t) = q(t)\ /\ \sqrt{\gamma(d\mu_o/d\omega^2)_{\omega_o^2}} \tag{13'}$$

where γ = diagonal matrix element of the defect matrix in that representation.

An uncoupled configuration will have a well-defined mass M_e associated with it. For instance for Fig. 1 we would have

$$M_e = M_n \text{ for } A_{1g},\ E_g,\ |T_{1u};2)$$
$$M_e = M_o \text{ for } |T_{1u};1)$$
$$M_e = 2M_n + M_o \text{ for } |T_{1u};1')$$
$$M_e = M_r = 2M_oM_n/(2M_n + M_o) \text{ for } |T_{1u};2')$$

The real physical displacement corresponding to (13) is then

$$u(t) = M_e^{-1/2}v(t) = [M_e\gamma\mu_o']^{-1/2}q(t) \ . \tag{15}$$

For an ordinary simple harmonic oscillator we would have instead of (15) simply

$$u(t) = (M^*)^{-1/2}q(t) \quad . \tag{16}$$

Equations (15) and (16) will have the same form if we define an effective mass by

$$M^* = M_e\gamma(d\mu_o/d\omega^2)_{\omega_o^2} \quad . \tag{17}$$

In the case of two or more coupled configurations there is in general no counterpart of M_e or M^* unless there is a configuration $\underline{v}_o$ for which the diagonal matrix element of $\underline{\gamma}$ is much larger than all other matrix elements. This is true for $|T_{1u};1)$ at high frequency if there is a finite mass change and also true for $|T_{1u};2')$ at low frequency if only central force constant changes and mass changes are considered. The concept of an effect mass is somewhat limited, but when it is applicable, it can be very useful.

There are three general types of localized phonon excitations: (1) high frequency localized modes, (2) localized gap modes, and

(3) low frequency resonance states. We shall now discuss a simple physical model which reveals important qualitative features of the first and last of these excitations with respect to effective masses and effective force constants.

HIGH FREQUENCY LOCALIZED MODE

We consider a high frequency localized mode of the U-center type such that the frequency is much higher than any unperturbed phonon frequency. If one then makes a high frequency expansion of the Green's function matrix (16) the localized mode condition of Eq. (11) or (12) can be rearranged to give in the central force constant model

$$M_o'\omega_o^2 = \Phi + 2\Delta f \quad . \tag{18}$$

Here M_o' is the impurity mass, Δf the change in central force constant (the 2 comes from the 2 neighbors), and Φ is given by

$$\Phi = \left(\frac{\partial^2 V}{\partial u_x^2}\right)_o \quad . \tag{19}$$

$\Phi/2$ is the restoring sping constant in the perfect crystal when the atom at the origin is displaced and all other atoms held at rest. It is clearly the appropriate force constant parameter for a high frequency localized mode since the inertia of all atoms but the impurity effectively keeps them at rest. We then have a simple Einstein oscillator with

$$u(t) \approx (M_o')^{-1/2} q(t) \text{ and } M^* \approx M_o' \ .$$

The relevant configuration is of course $|T_{1u};1)$ in Fig. 1.

There is a simple mechanical model shown in Fig. 2 which reveals the features of the localized mode. M_1 represents the light impurity; the two nearest neighbors have mass M_2. At high frequency the neighbors are at rest, and the restoring spring constant acting upon M_1 is $2(f + K_1)$. Thus $(f + K_1)$ is the effective high frequency force constant K_{hi}. We may make the following identifications with the lattice case:

$$M_1 \longleftrightarrow M_o'$$

$$\Phi + 2\Delta f \longleftrightarrow 2K_1 + 2f = 2K_{hi} \quad . \tag{20}$$

LOW FREQUENCY RESONANCE STATE

Resonances occuring at very low frequency are dominated by soft force constants, since the perturbation = $\Delta M\omega^2/M_o$ becomes relatively unimportant. We take the central force constant model and ask: What happens as the force constant change Δf is made more and more negative? The answer is that the $|T_{1u};2')$ configuration is the first to become unstable, i.e., to have a resonance at zero frequency. This occurs when $\Delta f = -f_o$ with

$$f_o = (M_r/2) \ (2'|\underline{g}(0)|2')^{-1} \quad . \tag{21}$$

When we make a unit relative displacement in the $|T_{1u};2')$ con-
figuration and allow all the other atoms in the crystal to follow
without constraint, then $2f_o$ will be the restoring force. Now at
low frequencies there are no intertial forces to prevent the atoms
from following; thus f_o is the proper low frequency effective force
constant for the unperturbed cyrstal. The definitions of $2f_o$ and Φ
are similar, except that a constraint is imposed in the case of Φ.
When the constraint is removed to define $2f_o$ the elastic energy per
unit displacement is lowered. Hence we must have

$$\Phi > 2f_o \quad . \tag{22}$$

For a stable but soft system, we expect the net low frequency force
constant to be

$$f' = f_o + \Delta f \quad , \tag{23}$$

whereas at high frequency Eq. (18) says it must be

$$f_{hi} = \Phi/2 + \Delta f > f' \quad . \tag{24}$$

The model of Fig. 2 shows the same instability at zero frequency for the motion of M_1 relative to the center of mass of its two
neighbors. The net low frequency force constant for this relative
motion turns out to be

$$K' = f + K_o \quad , \tag{25}$$

where $K_o = \dfrac{K_1 K_2}{K_1 + K_2} = K_1 - \dfrac{K_1^2}{K_1 + K_2} \quad .$

Thus $\quad K' = f + K_1 - K_1^2/(K_1 + K_2) < f + K_1 = K_{hi} \quad . \tag{26}$

Note that in order to obtain a softer effective force constant
at low frequencies than at high frequencies, we must have K_1 non-
zero, i.e., we need nonzero "long range" forces. This important
qualitative behavior is not revelaed by lattice models using only
nearest neighbor forces. With long range forces, one can have an
impurity system that is on the verge of being unstable at low frequencies, but which possesses a positive (but small) effective

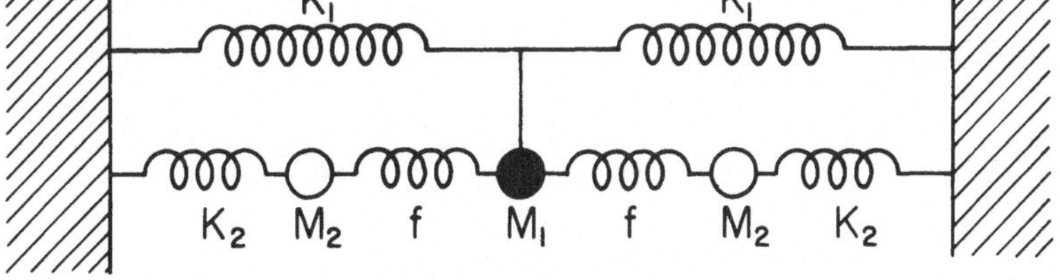

Fig. 2. Simple mechanical model.

force constant at high frequencies. If the impurity mass is small enough, such a system can even possess a high frequency localized mode!

To obtain a nearly unstable system the explicit central force constant f in Fig. 2 must be negative. It is not hard to imagine how this could happen for a small impurity in an ionic crystal.

Let us return now to the impurity in the crystal and suppose that

$$f' = f_o + \Delta f \qquad (27)$$

is positive but small. There will than be a resonance at a small but finite frequency. The resonance consition can be calculated to first order in ω_o^2 with the result expressed in the form

$$\omega_o^2 = 2f'/M^* \qquad (28)$$

Equation (17) for the effective mass reduces to

$$M^* = M_r\, A\, (M_o' + \delta M_o/M_o) \qquad (29)$$

with A and δM_o given in terms of T_{1u} matrix elements of $\underline{R}$ by

$$\delta M_o/M_o = \frac{1}{(2'|R|1)^2} \quad \frac{d(2'|R|2')}{d\omega^2} \; -1,$$

$$A = (2'|R|1)^2/(2'|R|2')^2 \qquad .$$

Some computed results for A and δM_o are shown in Table I.

An important feature of Eq. (29) is that the effective mass correction in δM_o is additive. This means that the resonance may occur even in the limit $M_o' \to 0$. The reason for this can be seen best in the simple mechanical model to be discussed shortly.

If we rewrite Eq. (18) using (23) in the form

$$M_o'\omega_o^2 = 2f' + (\Phi - 2f_o) \qquad , \qquad (30)$$

then the constant frequency curves in a plot of f' versus M_o' will be straight lines that do not go through the origin because of the term $(\Phi - 2f_o)$. A family of straight lines is also obtained from (28) and (29). Again they miss the origin because of the nonzero δM_o. The result is shown schematically in Fig. 3. The dot shows that a very light impurity with soft force constants can have both

Host	Impurity site	M_o (a.m.u.)	A	δM_o (a.m.u.)
NaCl	+	23	0.59	16
NaCl	-	35.5	1.05	4
KBr	+	39.1	0.67	19
KBr	-	79.9	1.31	5

Table I. Computed Results (Shell model Green's functions used)

a high frequency localized mode and a low frequency resonance. It has been suggested that such an effect may occur for rocking motions of the OH^- ion in alkali halides.[17] The H part of the ion would do most of the moving. Having lost part of its electron cloud to the oxygen, it might act like a very soft U-center. Here the localized mode would correspond to the librator and the low frequency resonance to the mysterious "32 wavenumber band."[18]

Behavior similar to Fig. 3 can be seen in the detailed f' versus M' plots of Benedek and Nardelli.[2] They were the first to point out that these curves do not go through the origin and that the reason lay with the long range forces.

For the system of Fig. 2 the approximate low frequency equations analagous to (28) and (29) are

$$\omega_o^2 = 2K'/M^* \quad , \tag{31}$$

$$M^* = (M_1K_1^2 + 2M_2K_1^2)/(K_1 + K_2)^2 \quad . \tag{32}$$

Note that as $M_1 \to 0$, we have $M^* \to 2M_2K_1^2/(K_1 + K_2)^2$. (33)

This mechanical model has all the important qualitative features of the real impurity system. Note that the long range forces have two effects: (1) a lowering of the effective force constant at low frequencies and (2) and additive pseudo-inertial effect represented

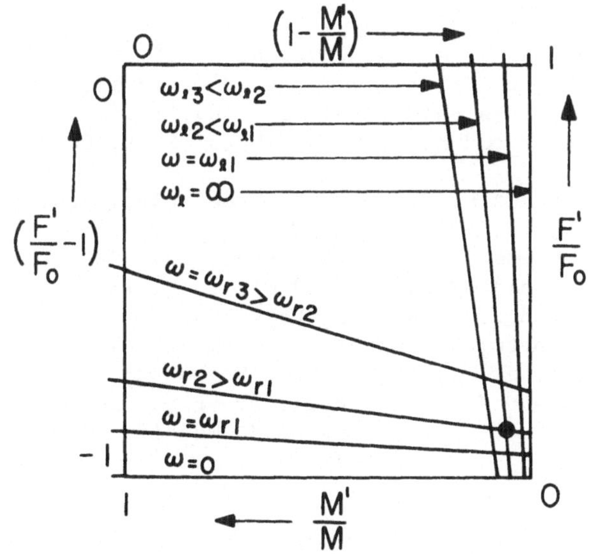

Fig. 3. Approximate constant frequency lines for high frequency localized modes and low frequency resonances.

by δM_o or by (33). The displacement of the two neighbors is resisted not only by the true inertia of M_1 but also by K_1.

REFERENCES

*This work was supported in part by the Advanced Research Projects Agency under Contract SD-131.

1. A. A. Maradudin, Rept. Progr. Phys. 28, 331 (1965); A. A. Maradudin in Solid State Physics, F. Seitz and D. Turnbull, eds. (Academic Press, New York, 1966), Vols. 18 and 19.
2. G. Benedek and G. F. Nardelli, Phys. Rev. 155, 1004 (1967).
3. M. V. Klein in Physics of Color Centers, W. B. Fowler, ed. to be published.
4. J. H. Van Vleck, J. Chem. Phys. 7, 72 (1939).
5. H. B. Rosenstock and G. F. Newell, J. Chem. Phys. 21, 1607 (1953). See also A. J. Sievers and S. Takeno, Phys. Rev. 140, A1030 (1965) and Dr. Takeno's paper at this conference.
6. E. W. Kellerman, Phil. Trans. Roy. Soc. (London) 238, 513 (1940).
7. W. Cochran, Phys. Rev. Letters 2, 495 (1959); Proc. Roy. Soc. (London) A253, 260 (1959); A. D. B. Woods, W. Cochran, and B. N. Brockhouse, Phys. Rev. 119, 980 (1960).
8. J. R. Hardy, Phil. Mag. 4, 1278 (1959); J. R. Hardy and A. M. Karo, Phil. Mag. 5, 859 (1960); J. R. Hardy, Phil. Mag. 7, 315 (1962).
9. R. A. Cowley, W. Cochran, B. N. Brockhouse, and A. D. B. Woods, Phys. Rev. 131, 1030 (1963).
10. A. M. Karo and J. R. Hardy, Phys. Rev. 141, 696 (1966); R. F. Caldwell and M. V. Klein, Phys. Rev. 158, 851 (1967).
11. U. Schröder, Solid State Comm. 4, 347 (1966); V. Nüsslein and U. Schröder, Phys. Stat. Sol. 21, 309 (1967).
12. G. Dresselhaus and M. S. Dresselhaus, Phys. Rev. 160, 649 (1967).
13. P. G. Dawber and R. J. Elliott, Proc. Phys. Soc. (London) 81, 453 (1963).
14. E. R. Cowley and R. A. Cowley, Proc. Roy. Soc. (London) A287, 259 (1965); ibid A292, 209 (1966); I. P. Ipatova, A. A. Maradudin, and R. F. Wallis, Phys. Rev. 155, 882 (1967).
15. Karl D. Schwartz, J. Acoust. Soc. Am. 41, 1083 (1967).
16. J. A. D. Matthew, J. Phys. Chem. Solids 26, 2067 (1965).
17. M. V. Klein and J. Gould, Bull. Am. Phys. Soc. 12, 380 (1967).
18. C. K. Chau, M. V. Klein, and B. M. Wedding, Phys. Rev. Letters 17, 521 (1966).

A THEORY OF IMPURITY-INDUCED INFRARED LATTICE ABSORPTION OF ALKALI HALIDE CRYSTALS

Shozo Takeno

Research Institute for Fundamental Physics

Yukawa Hall, Kyoto University, Kyoto, Japan

1. INTRODUCTION

A number of experimental and theoretical studies have been made on the physical properties of phonon impurity modes in solids, such as localized modes and resonant modes.[1] In recent years, a considerable amount of experimental data using the impurity-induced infrared absorption have been accumulated for various mixed alkali halide crystals.[2-3] In almost all theoretical works, only the change in mass accompanying the introduction of impurities into crystals has been taken into account.[1] Although the change in force constants has been taken into account in some cases, the results have been expressed in numerical terms rather than in analytical terms.[4-5] With such numerical results, it is not always easy to understand the general aspect of physical situation.

It is the purpose of this paper to develop a theory of the impurity-induced infrared absorption of mixed alkali halide crystals and to give as analytical and general results as possible for the properties of infrared-active localized modes and resonant modes. For this purpose a specific but exactly soluble crystal and impurity models are used, which, when comparing theory with experiment, enables us to cover a series of mixed alkali halide crystals.

2. FORMULATION AND SPECIFIC CRYSTAL AND IMPURITY MODELS

We introduce the Fourier transform $G_{\alpha\beta}(mn,\omega)$ of the retarded double-time Green's function $G_{\alpha\beta}(mn,t)$ defined by[6]

$$G_{\alpha\beta}(mn,t)=-i\theta(t)\left\langle [u_\alpha(m,t),u_\beta(n,0)]\right\rangle, \qquad (2.1)$$

where $\theta(t)$ is Heaviside's step function, $u_\alpha(n,t)$ is the Heisenberg operator of the α component $u_\alpha(n)$ of the displacement vector u(n) of an atom at site n in a crystal and the angular bracket means a canonical ensemble. The Green's functions $G_{\alpha\beta}(mn,\omega)$ obeys the equation

$$G_{\alpha\beta}(mn,\omega)=(\hbar/2\pi)g_{\alpha\beta}(mn,\omega)+\sum_{ij\gamma\delta} g_{\alpha\gamma}(mi,\omega)V_{\gamma\delta}(ij)G_{\delta\beta}(jn,\omega), \qquad (2.2)$$

where $g_{\alpha\beta}(mn,\omega)$ is the phonon Green's functions in the case of the pure crystal, the V's describe perturbations due to the introduction of impurities into the crystal, and the sum extends over all lattice sites perturbed by the impurities.

As an application, we consider a NaCl-type crystal lattice containing Ni substitutional ion impurities in which each ion interacts only with its nearest neighbours with equal central and noncentral force constants. The position of an ion in the lattice is expressed by a set of three integers $n=(n_1,n_2,n_3)$. Let host lattice site with $n_1+n_2+n_3$=even (even lattice site) and odd (odd lattice site) be occupied by ions with masses M_1 and M_2, respectively. Let us suppose that all the impurities in the lattice are identical and occupy some of even lattice sites. The mass of the impurity is denoted by M '. Since each component of motion is independent of one-another, we limit our discussion to the x-component of motion, the x-axis being taken to be in the direction of one of the crystal axes. The x-component of the displacement vector of the n ion is denoted by $u_x(n)\equiv u(n)=u(n_1,n_2,n_3)$. Let K and K' be the force constants for a host ion not directly coupled with the impurity and for the impurity, respectively. It is convenient to introduce the number r, such that r=1 and 2 for even and odd lattice sites, respectively. Then in terms of $\omega_r^2=6K/M$ and the perturbation parameters

$$\lambda=(M_1'-M_1)/M_1, \qquad \mu=(K'-K)/K, \qquad (2.3)$$

the time-independent equation of motion satisfied by u(n) for any ion, host ion or impurity ion, can be written in the form

$$M_r\Big(\omega^2-\omega_r^2\sum_{p=1}^{3}\sum_{\delta=\pm}L_p^\delta\Big)u(n)$$
$$=M_1\sum_i\Big[\Big\{\mu\omega_1^2\sum_{p=1}^{3}\sum_{\delta=\pm}L_p^\delta-\lambda\omega^2\Big\}\Delta(ni)-\mu\omega_1^2\sum_{p=1}^{3}\sum_{\delta=\pm}L_p^\delta\Delta_p^\delta(ni)\Big]u(i), \qquad (2.4)$$

where the L's are defined by

$$L_1^\pm u(n)=(1/6)\{u(n_1,n_2,n_3)-u(n_1\pm1,n_2,n_3)\}, \qquad (2.5)$$
$$L_2^\pm u(n)=(1/6)\{u(n_1,n_2,n_3)-u(n_1,n_2\pm1,n_3)\},$$

$$L_3^{\pm} u(n) = (1/6)\{u(n_1,n_2,n_3) - u(n_1,n_2,n_3 \pm 1)\}$$

and $\Delta(ni) = \Delta(n_1,i_1)\Delta(n_2,i_2)\Delta(n_3,i_3)$, $\Delta_1^{\pm}(ni) = \Delta(n_1,i_1 \pm 1)\Delta(n_2,i_2)\Delta(n_3,i_3)$, $\Delta_2^{\pm}(ni) = \Delta(n_1,i_1)\Delta(n_2,i_2 \pm 1)\Delta(n_3,i_3)$, $\Delta_3^{\pm}(ni) = \Delta(n_1,i_1)\Delta(n_2,i_2)\Delta(n_3,i_3 \pm 1)$, the Δ's being Kronecker's delta. In Eq. (2.4) $i=(i_1,i_2,i_3)$ denotes the positions of the impurities and the sum extends over all positions of the impurities.

Corresponding to Eq.(2.4), Eq.(2.2) takes the form

$$G(mn,\omega) = (h/2\pi)\, g(mn,\omega) +$$

$$M_1 \sum_i \left[\left\{ \mu\omega_1^2 \sum_{\rho=1}^{3} \sum_{\delta=\pm} L_\rho^\delta - \lambda\omega^2 \right\} g(mi,\omega) G(in,\omega) - \mu\omega_1^2 \sum_{\rho=1}^{3} \sum_{\delta=\pm} L_\rho^\delta g(mi,\omega) G_\rho^\delta(in,\omega) \right], \tag{2.6}$$

where $G(mn,\omega) \equiv G_{xx}(mn,\omega)$, $g(mn,\omega) \equiv g_{xx}(mn,\omega)$, $G_1^{\pm}(in,\omega) = G(i_1 \pm 1, i_2, i_3; n; \omega)$ $G_2^{\pm}(in,\omega) = G(i_1, i_2 \pm 1, i_3; n; \omega)$, $G_3^{\pm}(in,\omega) = G(i_1, i_2, i_3 \pm 1; n; \omega)$. The L's in the above equation operate only on the argument i in $g(ni,\omega)$.

Within the framework of harmonic approximation, almost all vibrational properties of the crystal are contained in solutions of Eq.(2.6), which depend on the perturbation parameters λ and μ and the Green's functions characterizing the vibration of the pure crystal. The Green's functions $g(mn,\omega)$ can be evaluated from solutions of Eq.(2.4) in which the right-hand side is set equal to zero. A detailed discussion on this problem has been given elsewhere[8], so we present here only the result of calculations. Let $g_0(\omega)$ and $g_1(\omega)$ be the diagonal parts of the g's for even and odd lattice sites, respectively. The relationship between $g_0(\omega)$ and $g_1(\omega)$ is

$$g_1(\omega) = [M_1(\omega^2 - \omega_1^2)/M_2(\omega^2 - \omega_2^2)]\, g_0(\omega). \tag{2.7}$$

An expression for $g_0(\omega)$ with infinitesimal positive or negative imaginary part is given by

$$\lim_{\epsilon \to 0+} g_0(\omega \pm i\epsilon) = g_0'(\omega) \mp i \,\mathrm{sgn}\,\omega\, g_0''(\omega), \tag{2.8}$$

where

$$g_0'(\omega) = \begin{cases} B_1(\omega)\int_0^\infty \exp[-B_2(\omega)t]\, I_0^3(t)\,dt & \text{for } \omega > \omega_M \\ B_1'(\omega)\int_0^\infty \exp[-B_2'(\omega)t]\, J_0^3(t)\,dt & \text{for } \omega_- < \omega < \omega_+ \\ B_1(\omega)\int_0^\infty \sin[B_2(\omega)t]\, J_0^3(t)\,dt & \text{otherwise,} \end{cases} \tag{2.9}$$

$$g_0''(\omega) = \begin{cases} B_1(\omega)\int_0^\infty \cos[B_2(\omega)t]\, J_0^3(t)\,dt & \text{for } 0 < \omega < \omega_- \text{ or } \omega_+ < \omega < \omega_M \\ 0 & \text{otherwise,} \end{cases} \tag{2.10}$$

in which $B_1(\omega) = (3/M_1\omega_1^2)[M_2(\omega^2 - \omega_2^2)/M_1(\omega^2 - \omega_1)^2]^{1/2}$, $B_1'(\omega)^2 = -B_2(\omega)^2$, $B_2(\omega) = 3[\{(\omega/\omega_1)^2 - 1\}\{(\omega/\omega_2)^2 - 1\}]^{1/2}$, $B'(\omega)^2 = -B_2(\omega)^2$. In the above equations $\omega_M^2 = 6KM_1M_2/(M_1+M_2)$ is the maximum frequency of the optical frequency band, $\omega_+^2=6K/M_1$ for $M_1<M_2$ or $6K/M_2$ for $M_1>M_2$ is the minimum frequency of the optical band, and $\omega_-^2=6K/M_1$ for $M_1>M_2$ or $6K/M_2$ for $M_1<M_2$ is the maximum frequency of the acoustical frequency band. $J_0(t)$ and $I_0(t)$ are the Bessel functions of real and imaginary arguments with order zero respectively. Numerical values for the integrals in Eqs.(2.9) and (2.10) have been prepared by several authors.[9-11] In the high frequency and the low frequency limits, $(\omega/\omega_M)^2 \gg 1$ and $(\omega/\omega_1)^2 \ll 1$, we can ontain analytical expressions for $g_0(\omega)$ as follows

(i) case $(\omega/\omega_M)^2 \gg 1$

$$g_0'(\omega)=(1/M_1\omega_1^2)\left\{1+(\omega_1/\omega)^2\right\}, \qquad g_0''(\omega)=0 \tag{2.11}$$

(ii) case $(\omega/\omega_1)^2 \ll 1$

$$g_0'(\omega)=-3/2M_1\omega_1^2, \qquad g_0''(\omega)=(3\pi/8)\left\{1+(M_2/M_1)\right\}^{1/2}(\omega/M_1\omega_1^3). \tag{2.12}$$

The above result is useful for studying the analytical properties of localized modes in U centers and of resonant modes appearing near the bottom of the acoustical band.

§3. THE ABSORPTION COEFFICIENT

We take the effective charge e(n) of host ion at site n, which is not coupled directly with the impurities, in the form

$$e(n)=e(n_1,n_2,n_3)=\begin{cases} e & \text{for } n_1+n_2+n_3=\text{even} \\ -e & \text{for } n_1+n_2+n_3=\text{odd}, \end{cases} \tag{3.1}$$

For the i impurity and its nearest neighbours, we put

$$\begin{aligned} &e(i)=e(i_1,i_2,i_3)=e+\Delta e, \\ &e(i_1\pm1,i_2,i_3)=e(i_1,i_2\pm1,i_3)=e(i_1,i_2,i_3\pm1)=-e-(\Delta e/6). \end{aligned} \tag{3.2}$$

Then, the x-component $P_x \equiv P$ of the total dipole moment of the crystal can be divided into two parts

$$P=P_1+P_2, \tag{3.3}$$

where

$$P_1=e(S_1-S_2), \tag{3.4}$$

$$P_2=\sum_i \Delta e\left\{u(i)-(1/6)\sum_{i^*} u(i^*)\right\}, \tag{3.5}$$

where $S_1=\sum^1 u(n)$ and $S_2=\sum^2 u(n)$ are the sums of u(n) over all even and odd lattice sites, respectively, and i^* denotes the positions of the nearest neighbours of the i impurity. Summing up Eq.(2.4) for all even and odd lattice sites, separately, we obtain a set of simultaneous equations for S_1 and S_2. Putting solutions of these equations into Eq.(3.4), we obtain

$$P_1=\frac{\mu-\lambda+(M_1/M_2)\mu(1+\lambda)}{1+\mu}\,\frac{e\,\omega^2}{\omega^2-\omega_M^2}\sum_i u(i). \tag{3.6}$$

P_2 can be evaluated from equation satisfied by u(i) (See Eq.(2.4)) as follows

$$P_2=\epsilon e\,\frac{1+\lambda}{1+\mu}\,\frac{\omega^2}{\omega_1^2}\sum_i u(i), \tag{3.7}$$

where

$$\epsilon=\Delta e/e. \tag{3.8}$$

Combining Eqs.(3.6) and (3.7) with Eq.(3.3), we obtain

$$P=\frac{e\,\omega^2}{1+\mu}\,\beta(\omega)\sum_i u(i), \tag{3.9}$$

where $\beta(\omega)=\{\epsilon(1+\lambda)/\omega_1^2\}+\{[\mu-\lambda+(M_1/M_2)\mu(1+\lambda)]/(\omega^2-\omega_M^2)\}$. We see that besides the factor $\epsilon\,\omega^2\beta(\omega)/(1+\mu)$ the total dipole moment of the crystal has been expressed entirely in terms of the displacements of the impurities.

We now apply the result obtained above to a calculation of the absorption coefficient. We shall suppose that an external electric field is weak, so we confine ourselves to the linear response of the vibration system to the external field. According to the general theory of Kubo,[12] then the absorption coefficient is expressed in terms of the Fourier transform of the retarded double-time Green's functions formed by the total electric dipole moment of the crystal. We shall assume that the external electric field is in the direction of the x-axis. Then, we see from Eq. (3.9) that the absorption coefficient can be expressed entirely in terms of the sum of the Green's functions $G(ij,\omega)$ for all i and j, where j as well as i refers to impurity sites. Thus, we obtain an expression for the absorption coefficient $A(\omega)$ (per volume):

$$A(\omega)=-\frac{8\pi^2e^2\omega^5}{\hbar\,\eta\,c\,V}\,\frac{\beta(\omega)^2}{(1+\mu)^2}\,\mathrm{Im}\Big\{\lim_{\epsilon\to0+}\sum_{ij}G(ij,\omega+i\epsilon)\Big\}, \tag{3.10}$$

where η is the index of refraction of the crystal, c is the light velocity, V is the volume of the sample, and Im{B} means the imaginary part of B.

§4. INFRARED-ACTIVE MODE

If the i impurity is completely isolated from the other impurities, the symmetry of an infrared-active mode is expressed by the equation $G(i_1 \pm 1, i_2, i_3; n; \omega) = G(i_1, i_2 \pm 1, i_3; n; \ ) = G(i_1, i_2, i_3 \pm 1; n; \omega)$. We shall assume that the concentration of the impurities is small, so the symmetry property of the infrared-active mode remains unchanged. Using this, we can reduce Eq.(2.6) for m=i and i^* and n=j(impurity site) as follows

$$
\begin{aligned}
D_{11} G(ij) + D_{12} G(i^* j) &= (\hbar/2\pi) g(ij) + M_1 \omega^2 \sum_{i_1 (\neq i)} g(i i_1) \{(\mu - \lambda) G(i_1 j) - \mu G(i_1^* j)\}, \\
D_{21} G(ij) + D_{22} G(i^* j) &= (\hbar/2\pi) g(i^* j) + M_1 \omega^2 \sum_{i_1 (\neq i^*)} g(i^* i_1) \{(\mu - \lambda) G(i_1 j) - \mu G(i_1^* j)\},
\end{aligned}
\tag{4.1}
$$

where $D_{11} = 1 + \mu - M_1 \omega^2 (\mu - \lambda) g_0(\omega)$, $D_{12} = -\mu \{1 - M_1 \omega^2 g_0(\omega)\}$, $D_{21} = -M_1 \omega^2 (\mu - \lambda) g_{0i}(\omega)$, $D_{22} = 1 + \mu M_1 \omega^2 g_{01}(\omega)$ in which $g_{01}(\omega) \equiv g(ii^*, \omega) = g(i^* i, \omega)$, and we have omitted the argument ω in the G's and the g's in Eq.(4.1).

Here we present only the result of calculations in which the second terms in the right-hand side of Eqs.(4.1) are neglected. The absorption coefficient is obtained by retaining only the diagonal terms for the G's in Eq.(3.10) and solving Eq.(4.1) for G(ii, ω):

$$
A(\omega) = (4\pi e^2 N_i / \eta c V) \omega^5 \beta(\omega)^2 g_0''(\omega) / \{[\mathrm{Re}\, D(\omega^2)]^2 + [\mathrm{Im}\, D(\omega^2)]^2\}, \tag{4.2}
$$

where

$$
D(\omega^2) = 1 + \mu + \mu(1+\lambda)(\omega/\omega_1)^2 + \{\lambda(1+\mu) - \mu(1+\lambda)(\omega/\omega_1)^2\} M_1 \omega^2 g_0(\omega). \tag{4.3}
$$

In Eq.(4.2) Re $D(\omega^2)$ and Im $D(\omega^2)$ are abbreviations of the real and the imaginary parts of $\lim_{\epsilon \to 0+} D((\omega + i\epsilon)^2)$, respectively. The eigenfrequency ω_0 of the infrared-active mode is obtained from a solution of the equation Re $D(\omega^2) = 0$. In the out-of-band region the absorption coefficient becomes delta-function like while in the in-band region it has a typical resonance form. If the resonance is sufficiently strong, we can approximate the absorption coefficient in the vicinity of $\omega = \omega_0$ by a Lorentzian function with a width to given by

$$
\Gamma_0 / 2 = \mathrm{Im}\, D(\omega_0^2) / |d\, \mathrm{Re}\, D(\omega^2)/d\omega|_{\omega = \omega_0}. \tag{4.4}
$$

Using Eqs.(2.11) and (2.12), we can obtain the analytical properties of a high-frequency localized mode and a low-frequency resonant mode as follows:

$$
\omega_0^2 = -\{\lambda(1+\mu)/(1+\lambda)\} \omega_1^2 \qquad \text{for} \quad \omega_0^2 \gg \omega_m^2 \tag{4.5}
$$

$$
\left.
\begin{aligned}
\omega_0^2 &= \{2(1+\mu)/(3\lambda + \lambda\mu - 2\mu)\} \omega_1^2 \\
\Gamma_0 / 2 &= (3\pi/16) \{1 + (M_2/M_1)\}^{1/2} |\lambda| \omega_0^4 / \omega_1^3
\end{aligned}
\right\} \quad \text{for} \quad \omega_0^2 \ll \omega_1 \tag{4.6}
$$

Putting Eq.(2.3) into Eqs.(4.5) and (4.6), we can easily verify that there exists the relation $\omega_0^2=6K'/K'$ if $M' \ll M$ for the localized mode and if $K' \ll K$ for the resonant mode. Thus we see that in these limitting cases these impurity modes are well described by the Einstein oscillator model.

We have compared the results obtained above with the experimental data of Schaefer[2] and Sievers,[3] where we have used Eq. (4.5) for localized modes in U centers, Eq.(4.3) for gap modes, and Eq.(4.6) for low-frequency resonant modes. Some of the results obtained are shown in Table I.

We have found that for all the samples for U centers the softening of force constants upon introduction of H^- ion impurity takes place. For gap modes and resonant modes, we have estimated the ratio of the ionic radius of an impurity to that of the host ion which the impurity replaces. We see that a large mismatch in ionic radius gives rise to a large change in force constants. Particularly interesting is the case of Li^+ in KBr, in which we can conclude that a remarkable softening of force constants must take place.

A discussion on the other vibrational properties of the infrared-active mode, such as the localization of the modes, the extra specific heat due to the resonant mode has been omitted together with a plot of absorption-coefficient-curves.

Table I The values of K'/K for localized modes in U centers, localized gap modes, and low-frequency resonant modes in mixed alkali halide crystals.

Sample	(cm^{-1})	M'/M	K'/K	ionic radius ratio	type of impurity mode
H^- in NaCl	563	0.0282	0.62		lo.
H^- in KBr	445	0.0128	0.53		lo.
Cl^- in KI	77	0.280	0.38	0.84	ga.
Cs^+ in KI	84	3.40	1.50	1.25	ga.
Li^{+6} in KBr	17.9	0.15	0.005	0.45	re.
Li^+ in NaCl	44.0	0.30	0.026	0.63	re.

lo.:localized mode, ga.:gap modes, re.:resonant modes.

REFERENCES

1) See, for example, A.A.Maradudin, Solid State Physics (F.Seits and D.Turnbull, editors, Academic Press, 1966), Vol.18, 19 and references cited therein.
2) G.Schaefer, J.Phys. Chem. Solids 12, 293 (1960).
3) A.J.Sievers, Lectures on Elementary Excitations and their Interactions in Solids, NATO Advanced Study Institute, Cortina, Italy, 1966.
4) K.Patnaik and J.Mahanty, Phys. Rev. 155, 987 (1967).
5) G.Benedek and G.F.Nardelli, Phys. Rev. 155, 1004 (1967).
6) D.N.Zubarev, Soviet Phys. -Uspekhi 3, 320 (1960).
7) R.J.Elliot and D.W.Taylor, Proc. Phys. Soc. (London), 83, 189 (1964).
8) S.Takeno, Prog. Theor. Phys. 38 (1967), to be published and to appear in Rev. Mod. Phys. .
9) A.A.Maradudin, E.W.Montroll, G.H.Weiss, R.Herman and H.W. Milnes, Academic Royale de Belgique, 14 (1960).
10) Y.Mitani and S.Takeno, Prog. Theor. Phys. 33, 779 (1965).
11) T.Wolfram and J.Callaway, Phys. Rev. 130, 2207 (1963).
12) R.Kubo, J. Phys. Soc. Japan, 12, 570 (1957).
13) S.Takeno, Prog. Theor. Phys. 28, 33 (1962); A.J.Sievers and S.Takeno, Phys. Rev. 140, A1030 (1965).

THE TEMPERATURE DEPENDENCE OF THE FUNDAMENTAL LATTICE VIBRATION ABSORPTION BY LOCALIZED MODES

I.P. Ipatova and A.V. Subashiev
A.F. Ioffe Physico-Technical Institute
Leningrad K-21, U.S.S.R.

A.A. Maradudin*
University of California
Irvine, California

Since the pioneering work of Schaefer[1] on the infrared lattice vibration absorption by localized modes associated with U-centers in alkali-halide crystals, controversy has surrounded the temperature dependence of the area under the localized mode peaks in the absorption spectra of imperfect crystals. Schaefer reported that the area under the localized mode peak decreases with increasing temperature, and the same result was obtained by several subsequent investigators.[2,3] However, Mirlin and Reshina[4], and subsequently other workers,[5,6] presented experimental evidence that the area under the localized mode peak is essentially independent of temperature.

In this paper we present somewhat nonstandard calculations of the localized mode contribution to the absorption coefficient of a crystal containing substitutional impurities at sites of O_h symmetry.

Our starting point is the Hamiltonian describing the atomic vibrations of a crystal consisting of N unit cells each of which contains r ions, in which is found an impurity ion at a site of O_h symmetry which gives rise to a triply degenerate localized mode:

$$H = E_z + H_B + H_L + H_{LB}. \qquad (1)$$

In this expression E_z is the zero-point energy, H_B is

the Hamiltonian of the band modes,

$$H_B = \sum_p^{3rN-3} \hbar\omega_p b_p^+ b_p, \qquad (2)$$

H_L is the Hamiltonian of the localized modes,

$$H_L = \hbar\omega_o \sum_{s=1}^{3} b_s^+ b_s, \qquad (3)$$

and H_{LB} is the interaction Hamiltonian which couples the localized and band modes,

$$H_{LB} = 3 \sum_{ps_1s_2} V_{ps_1s_2} A_p A_{s_1} A_{s_2}. \qquad (4)$$

In these expressions b_m^+ and b_m are creation and destruction operators for phonons in the mth normal mode of the perturbed crystal, and $A_m = b_m + b_m^+ = A_m^+$ is a phonon field operator. Here, and in all that follows we use the convention of labeling the 3rN-3 band modes by an index p, the 3 localized modes by an index s, and denote by m a mode which can be either a band mode or a localized mode. In Eq.(2) ω_p is the frequency of the pth band mode, while ω_o in Eq.(3) is the frequency of the triply degenerate localized mode. We have neglected the anharmonic contributions to H_B and H_L because they serve primarily to renormalize the frequencies $\{\omega_p\}$ and ω_o, and do not affect the coupling between the localized and band modes which is our primary concern here. The choice of the interaction Hamiltonian H_{LB} calls for more comment. We have omitted all anharmonic terms in H_{LB} of fourth and higher order. The neglected terms contribute small corrections to the results we obtain by retaining only the cubic anharmonic terms. We have also omitted the cubic anharmonic terms which couple a single localized mode to a pair of band modes, because these terms are responsible for broadening the localized mode absorption line and for shifting the position of its center but do not affect the temperature dependence of the strength of the absorption line.

A group theoretic analysis of the coefficient $V_{ps_1s_2}$ yields the result that with s_1 and s_2 both labeling localized modes, which transform according to the Γ_4^- irreducible representation of the point group O_h the index p can run only over modes which transform according to the irreducible representations Γ_1^+, Γ_3^+, and Γ_5^+ of the group O_h. However, if the Leibfried [(7)] approximation to the cubic anharmonic force constants is made, the contribution to H_{LB} from band modes which transform according to the Γ_5^+ irreducible representation is identically zero. With this approximation H_{LB} can be written in the form

$$H_{LB} = 3 \sum_{a}^{N(\Gamma_1^+)} C(\Gamma_1^+ a) A(\Gamma_1^+ a) \left[A_1^2 + A_2^2 + A_3^2 \right] +$$

$$+3 \sum_{a}^{N(\Gamma_3^+)} C(\Gamma_3^+ a) \{ A(\Gamma_3^+ a1) \left[2A_3^2 - A_1^2 - A_2^2 \right] + 3^{\frac{1}{2}} A(\Gamma_3^+ a2) \left[A_1^2 - A_2^2 \right] \}, (5)$$

where $N(\Gamma)$ is the number of times the irreducible representation Γ occurs in the reduction of the mechanical representation of the group of the impurity site.[8] The coefficient $C(\Gamma a)$ is the single anharmonic coupling coefficient associated with the a$^{\underline{th}}$ band mode which transforms according to the irreducible representation Γ. $A(\Gamma_1^+ a)$ is the phonon field operator for the a$^{\underline{th}}$ band mode which transforms according to the one-dimensional irreducible representation Γ_1^+, while $A(\Gamma_3^+ a1)$ and $A(\Gamma_3^+ a2)$ are phonon field operators for the ath band mode which transform according to the first and second column of the two-dimensional irreducible representation Γ_3^+. A_1, A_2, A_3 are the phonon field operators for the three, degenerate, localized modes.

We now assume that we can solve for the eigenstates and eigen energies of the localized mode Hamiltonian $H_L + H_{LB}$:

$$(H_L + H_{LB}) |n\rangle = E_n |n\rangle. \qquad (6)$$

Both $|n\rangle$ and E_n depend parametrically on the band mode operators $\{A_p\}$ because of the dependence of H_{LB} on these operators. An eigenstate of the total Hamiltonian $H_B + H_L + H_{LB}$ will now be written in the product form $|n\rangle|N(n)\rangle$ and is a solution of the equation

$$(H_B + H_L + H_{LB}) |n\rangle|N(n)\rangle = \mathcal{E}_N^{(n)} |n\rangle|N(n)\rangle \ . (7)$$

If, for the localized mode states $\{|n\rangle\}$ which are of primary importance to us it is the case that to a sufficiently good approximation

$$\langle n'|H_B|n\rangle = \delta_{nn'} H_B \ , \qquad (8)$$

the band mode eigenstates $\{|N(n)\rangle\}$ are solutions of the equations

$$H_n |N(n)\rangle = \mathcal{E}_N^{(n)} |N(n)\rangle \ . \qquad (9)$$

where

$$H_n = H_B + E_n \ . \qquad (10)$$

We have solved Eq. (6) for $|n\rangle$ and E_n by perturbation theory for the ground state of the localized mode and for its three lowest excited states, which in the absence of H_{LB} are degenerate. The energies $\{E_n\}$ were calculated only to the first order of perturbation theory and are therefore linear functions of the $\{A_p\}$. The terms quadratic in the $\{A_p\}$ which are obtained in the second order of perturbation theory are proportional to the square of the cubic anharmonic force constants, and have the effect of renormalizing the band mode frequencies $\{\omega_p\}$. We have already neglected anharmonic shifts of the band mode frequencies in writing H_B, so that it would be inconsistent to retain the second order perturbation to E_n at this stage of the calculation, and we have not done so. We find that if we approximate $|n\rangle$ by the corresponding eigenstate $|n)$ of H_L Eq. (8) is satisfied with an error which is proportional to the square of the cubic anharmonic force constants and to a product of four band mode creation and destruction operators. Such anharmonic corrections to H_B have already been neglected at earlier stages of this calculation, and it is consistent to do so here.

With the preceding approximations the imaginary part of the dielectric response function which describes the processes in which light of frequency ω is absorbed by the crystal with the excitation of the localized mode from its ground state to its first excited states is given by

$$\varepsilon^{(2)}_{\mu\nu}(\omega) = \frac{2\pi}{\hbar}\frac{n_d}{V}\sum_n M^{(n)}_{\mu} M^{(n)}_{\nu} \int_{-\infty}^{\infty} dt e^{-i\omega t} \langle e^{-i\frac{t}{\hbar}H_o} e^{i\frac{t}{\hbar}H_n}\rangle_o . \tag{11}$$

In this expression n_d is the number of impurity ions, V is the volume of the crystal, the sum on n is over the three first excited states of the localized mode, the angular brackets $\langle\cdots\rangle_o$ denote an average with respect to the canonical ensemble described by the Hamiltonian $H_o=H_B+E_o$ and

$$M^{(n)}_{\mu} = \delta_{\mu n}\, e(\hbar/2M'\omega_o)^{\frac{1}{2}} \quad n=1,2,3 \tag{12}$$

where e is the dipole moment effective charge of the impurity ion, and M' is its mass.

The expression for the absorption coefficient which follows from Eq. (11) is

$$K(\omega) = \pi\frac{n_d}{V}\frac{\omega}{\eta c}\frac{e^2}{M'\omega_o} e^{-2M} \int_{-\infty}^{\infty} dt e^{-i(\omega-\omega_o)t+g(t)} \tag{13}$$

where $\eta(\omega)$ is the real part of the refractive index of the crystal, c is the speed of light, and

$$2M(T) = \frac{36}{\hbar^2} \sum_a^{N(\Gamma_1^+)} c^2(\Gamma_1^+ a) \frac{2n(\omega(\Gamma_1^+ a)+1}{\omega^2(\Gamma_1^+ a)} +$$

$$+ \frac{144}{\hbar^2} \sum_a^{N(\Gamma_3^+)} c^2(\Gamma_3^+ a) \frac{2n(\omega'(\Gamma_3^+ a))+1}{\omega^2(\Gamma_3^+ a)} \tag{14}$$

$$\equiv \frac{12}{\hbar^2} \sum_{ps_1 s_2} V^2_{ps_1 s_2} \frac{2n_p+1}{\omega_p^2} \tag{15}$$

$$g(t) = \frac{12}{\hbar^2} \sum_{ps_1 s_2} \frac{V^2_{ps_1 s_2}}{\omega_p^2} \left[(n_p+1)e^{-i\omega_p t} + n_p e^{i\omega_p t}\right] \tag{16}$$

The integrated absorption under the localized mode peak is therefore given by

$$\alpha = \int_{-\infty}^{\infty} K(\omega)_{\ell.m.} d\omega = 2\pi^2 \frac{n_d}{V} \frac{e^2}{M'\eta c} e^{-2M} \tag{17}$$

Contact between the result of the preceding adiabatic treatment of infrared lattice vibration absorption by localized modes and the result of a more conventional approach via a many-body perturbation theoretic evaluation of the Kubo formula for the imaginary part of the dielectric response function can be made as follows.

The localized mode contribution to the imaginary part of the dielectric response function of a crystal can be written as[9]

$$K(\omega) = \frac{n_d}{V} \frac{4\pi e^2 \omega}{M'\eta c} \frac{2\omega_o \Gamma_o(\omega)}{\left[\omega^2 - \Omega_o^2(\omega)\right]^2 + 4\omega_o^2 \Gamma_o^2(\omega)} \tag{18}$$

where the functions $\Omega_m(\omega)$ and $\Gamma_m(\omega)$ are related to the diagonal element $P_{mm}(z) = P_m(z)$ of the proper self energy of the phonon m by

$$-\frac{1}{\beta\hbar} P_m(\omega \pm io) = \Delta_m(\omega) \mp i\Gamma_m(\omega) \tag{19}$$

$$\Omega_m^2(\omega) = \omega_m^2 + 2\omega_m \Delta_m(\omega). \tag{20}$$

Our use of the subscript o in labeling Γ and Ω in Eq. (18) is meant to emphasize the fact that for localized modes $\Gamma_s(\omega)$ and $\Omega_s(\omega)$ are independent of the mode index s.

We now focus our attention on the shape of the absorption line in the vicinity of its center, that is for ω in the neighborhood of the frequency ω_c which is defined by

$$\omega_c^2 = \Omega_o^2(\omega_c). \tag{21}$$

Assuming that $\Omega_o(\omega)$ and $\Gamma_o(\omega)$ are small and slowly varying functions of ω for $\omega \approx \omega_c$ we find that

$$K(\omega) \approx \frac{n_d}{V} \frac{2\pi e^2}{M'\eta c} Z_o \frac{\gamma_o}{(\omega-\omega_c)^2+\gamma_o^2} \tag{22}$$

where

$$\gamma_o = \frac{\omega_o \Gamma_o(\omega_c)}{\omega_c[1-\Omega'(\omega_c)]} \tag{23}$$

and

$$Z_o^{-1} = 1-\Omega_o'(\omega_c) = 1-\frac{\omega_o}{\omega_c}\Delta'_o(\omega_c) \cong 1-\Delta'_o(\omega_o). \tag{24}$$

The integrated absorption under the localized mode peak is therefore

$$\alpha = \int_{-\infty}^{\infty} K(\omega)\,d\omega = 2\pi^2 \frac{n_d}{V} \frac{e^2}{M'\eta c} Zo. \tag{25}$$

The temperature dependence of the integrated absorption must therefore be contained in the localized mode quasi-particle strength Z_o.

In the lowest order of perturbation theory the expression for $\Delta_o'(\omega)$ can be written in the form

$$\Delta_o'(\omega) = -\frac{18}{\hbar^2} \sum_{p_1 p_2} V_{sp_1p_2} \left\{ (n_{p_1}+n_{p_2}+1)\left[\frac{1}{(\omega-\omega_{p_1}-\omega_{p_2})^2} - \frac{1}{(\omega+\omega_{p_1}+\omega_{p_2})^2}\right] + 2(n_{p_1}-n_{p_2}) \frac{1}{(\omega+\omega_{p_1}-\omega_{p_2})^2} \right\}$$
$$-\frac{36}{\hbar^2} \sum_{ps_1} V^2_{pss_1} \left\{ (n_p+n_o+1)\left[\frac{1}{(\omega-\omega_p-\omega_o)^2} - \frac{1}{(\omega+\omega_p+\omega_o)^2}\right] + (n_p-n_o)\left[\frac{1}{(\omega+\omega_p-\omega_o)^2} - \frac{1}{(\omega-\omega_p+\omega_o)^2}\right]\right\}. \tag{26}$$

We are interested in this expression evaluated at $\omega = \omega_o$ in the limit that $(\omega_o/\omega_L) = \lambda >> 1$. In this limit the second term dominates the first, and we obtain

$$\Delta_o'(\omega_o) \approx -\frac{36}{\hbar^2} \sum_{ps_1s_2} V^2_{pss_1} \frac{2n_p+1}{\omega_p^2} . \tag{27}$$

If we recall that $\Delta_s(\omega)$ is independent of s, we can rewrite Eq.(27) as

$$\Delta_o'(\omega_o) \approx -\frac{12}{\hbar^2} \sum_{ps_1s_2} V^2_{ps_1s_2} \frac{2n_p+1}{\omega_p^2} = -2M(T). \quad (28)$$

Combining Eqs.(24) and (28) we see that the localized mode quasi-particle strength Z_o is given by

$$Z_o = \left[1 + 2M(T) + \cdots\right]^{-1} \cong e^{-2M(T)} \quad (29)$$

In view of the results of the adiabatic treatment of the integrated absorption by localized modes we conjecture that if only the dominant terms in the large parameter λ are retained in each order of the perturbation theoretic expansion for $\Delta_o'(\omega)$, the resulting expansion for Z_o sums rigorously to $\exp(-2M(T))$. We have not yet succeeded in proving this conjecture. In fact, a straightforward perturbation theoretic calculation of Z_o^{-1} is complicated by the fact that in the fourth order of perturbation theory the proper self energy $P_s(z)$ acquires a pole at $z \approx \omega_o$, which may signify the breakdown of the adiabatic treatment, and in any case necessitates some rearrangement of the perturbation series for the proper self energy to render it well behaved for $z \approx \omega_o$.

A simple, approximate, calculation of 2M(T) as a function of temperature was carried out for the case of H^- U-centers in KCl on the basis of the following approximations: (a) neglect of the anharmonicity of the Coulomb interaction [10]; (b) the Leibfried approximation for the short range cubic anharmonic force constants [7]; and the assumption that the band modes are unaffected by the presence of the impurity ion [11]. The short range repulsive potential used in our calculations was the Born-Mayer potential determined by Jaswal [12]. In the limit of low temperatures our result for 2M(T) is

$$2M(T) = 0.176\left[1+6.6(T/101)^2\right] \quad (30)$$

with T given in $^\circ$K. For the same system Mitra and Singh [3] obtained the experimental result that

$$2M(T) = 0.052\left[1+6.6(T/200)^2\right]. \quad (31)$$

The theoretical result for 2M(T) is larger than the experimental result by a factor of about 3, and increases more rapidly with increasing temperature. However, the order of magnitude agreement between the theoretical and experimental results, particularly in view of the approximations made in obtaining the former, suggests that the assumptions underlying our calculation are correct, and that the agreement between theory and experiment may be improved by a calculation free from the approximations made in obtaining Eq.(30). Such a calculation is now underway, and the results will be reported elsewhere.

Acknowledgment

The authors would like to thank Mr. B.I. Bennett for carrying out the numerical evaluation of 2M(T).

References

*This research was partially supported by the Air Force Office of Scientific Research, Office of Aerospace Research, United States Air Force, under AFOSR Grant Number 1080-66.

(1) G. Schaefer, J. Phys. Chem. Solids. 12,233 (1960).
(2) S. Takeno and A.J. Sievers, Phys. Rev. Letters 15, 1020 (1965).
(3) S.S. Mitra and R.S. Singh, Phys. Rev. Letters 16, 694 (1966).
(4) D.N. Mirlin and I.I. Reshina, Fiz. Tverd. Tela 6, 945 (1964); [English translation: Soviet Physics-Solid State 6,728 (1964).]
(5) L. Kleinman, M.H.L. Pryce, and W.G. Spitzer, Phys. Rev. Letters 17,304 (1966).
(6) B. Fritz, U. Gross, and D. Bäuerle, phys. stat.sol. 11,231 (1965).
(7) G. Leibfried, in Handbuch der Physik, second edition, vol. 7, part 1 (Springer, Berlin, 1955)p.104
(8) G. Ya. Liubarskii, The Application of Group Theory in Physics (Pergamon Press, Inc., New York, 1960).
(9) See, for example, Nguyen Xuan Xinh, Westinghouse Research Laboratories Scientific Paper 65-9F5-442-P8, September 3, 1965; Phys. Rev. (to appear).
(10) I.P. Ipatova, A.A. Maradudin, and R.F. Wallis, Phys. Rev. 155,882 (1967).
(11) R.J. Elliott, W. Hayes, G.D. Jones, H.F. Macdonald, and C.T. Sennett, Proc. Roy. Soc. (London) A289,1 (1965).
(12) S. Jaswal, Phys. Rev. 140, A687 (1965).

Low-lying Resonant Modes in Anharmonic Crystals*

Giorgio Benedek

Istituto di Fisica dell'Università;Gruppo Nazionale di Struttura della Materia,Milan,Italy

I. INTRODUCTION

With respect to the anharmonic properties,the U-center localized mode has received till now much more attention ([1]) than the low-lying resonant modes induced by weakly bound impurities. Nevertheless,some recent works on the stress effect ([2,3]),the sidebands ([4]),and the temperature dependence ([5]) of the resonant infrared (IR) absorption give evidence that the resonant-mode properties are influenced by the local anharmonicity to a much larger extent than the local modes. This statement seems to be supported also by the fact that the harmonic theory predictions for the isotope shifts disagree with experiments; such disagreement seems indeed to be larger for resonant-modes ([5-7]) than for gap or local modes([8]). Here,our purpose is to investigate the above properties on the basis of a simple anharmonic theory and to discuss briefly the KBr:Li isotope shift and the off-center ion problems,where anharmonicity is found to play an important role.

II. THEORY

We consider a substitutional defect in a lattice of NaCl structure,affecting locally the mass as well as the harmonic,third and fourth order anharmonic force constants. Then we neglect the host lattice anharmonicity, i.e. we assume the third and fourth order anharmonic force constant matrices $\Phi^{(3)}$ and $\Phi^{(4)}$, respectively, to have non-zero elements only within a small region around the defect site. This assumption is not so simply justified as for the case of local modes (9), which display a strong localization in space with exponential law. However, the following arguments can be used. The contribution of the lattice anharmonicity to the stress-induced frequency shift of the resonant modes is much smaller than the contribution of the local anharmonicity; indeed in the Grüneisen approximation we can express the cubic anharmonic coefficient A by (2,3)

$$A/\Omega = \tilde{\gamma}R + \gamma(1-R), \tag{1}$$

where Ω is the resonant frequency, and

$$\tilde{\gamma} = -\frac{1}{2}\frac{\partial \ln \tilde{f}^*}{\partial \ln V}, \qquad R = \frac{\partial \ln \Omega^2}{\partial \ln \tilde{f}^*} \tag{2}$$

$\tilde{f}^*$ is the nearest-neighbor (n.n.)-defect effective force constant, V the crystal volume and γ the host lattice average mode gamma (3); for spacially localized modes R is close to unity, while $\tilde{\gamma}$ is found from both theory and experiment to be at least one order of magnitude larger than γ. This fact could supply a further "a posteriori" justification of the above assumption. In our problem we follow the thermodynamic Grenn function method and the approximations given previously, e.g., by Cowley (10) for the host lattice. Using the notations given in Ref.6, the simplest approach is to add to the frequency-dependent harmonic perturbation $\Lambda(\Omega^2)$ of the dynamical matrix, including the changes in mass $-\varepsilon=\Delta M_{\pm}/M_{\pm}$ and n.n. central force constant $\lambda=(\tilde{f}^*-f^*)/M_{\pm}$, the matrices $B(\Omega)$ and D representing the self-energies coming from the third and fourth order anharmonic interactions, respectively, represented by the diagrams

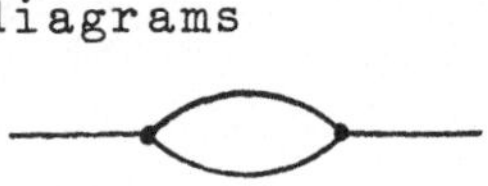 and .

As $B(\Omega)$ and D are assumed to be completely defined in the 21-dimensional subspace of our harmonic perturbation $\Lambda(\Omega^2)$, we express them in the representation of the symmetry coordinates used in Ref.6, which transform according to the irreducible representations (irr.rep.) Γ of the point group O_h. Let $n(\Gamma)$ denote the number of times the irr. rep. Γ is contained in our 21-dimensional system. For the optic-active modes, which transform according to the irr.rep. Γ_{15}, $B(\Omega)$ and D are three-dimensional matrices, as $n(\Gamma_{15}) = 3$. They are found to be

$$B_{jj'}^{(\Gamma_{15})}(\Omega) = -18\hbar\Sigma_{\Gamma ss'}\Sigma_{\Gamma' rr'}(\Gamma_{15}j|V^{(3)}|\Gamma s;\Gamma' r)$$

$$x(\Gamma s';\Gamma' r'|V^{(3)}|\Gamma_{15}\ j') \qquad (3a)$$

$$x\iint d\omega d\omega' \rho_{ss'}^{(\Gamma)}(\omega^2)\rho_{rr'}^{(\Gamma')}(\omega'^2)R(\omega,\omega',\Omega)$$

$$D_{jj'}^{(\Gamma_{15})} = 12\hbar\Sigma_{\Gamma ss'}(\Gamma_{15}j;\Gamma s|V^{(4)}|\Gamma s';\Gamma_{15}j') \qquad (3b)$$

$$x\int d\omega(2n(\omega)+1)\rho_{ss'}^{(\Gamma)}(\omega^2) \ ,$$

where Γ and Γ' run over all the irr. rep.; s and s' and r and r' run from 1 to $n(\Gamma)$ and $n(\Gamma')$, respectively. In Eqs.(3), $n(\omega) = (\exp(\hbar\omega/kT)-1)^{-1}$ is the occupation number, $\rho_{ss'}^{(\Gamma)}(\omega^2)$ denotes the perturbed projected density of states for the squared frequencies for the irr. rep. Γ, (11) and is a $n(\Gamma)$-dimensional matrix, and

$$R(\omega,\omega';\Omega) = \frac{n(\omega)+n(\omega')+1}{\omega+\omega'+\Omega+io^+} + \frac{n(\omega)+n(\omega')+1}{\omega+\omega'-\Omega-io^+} \qquad (4)$$

$$+ \frac{n(\omega)-n(\omega')}{\omega'-\omega+\Omega+io^+} + \frac{n(\omega)-n(\omega')}{\omega'-\omega-\Omega-io^+}$$

is the temperature-dependent propagator for Stokes and anti-Stokes two-phonon states. Taking into account Eq.(4) the double integral of Eq.(3a) can be easily reduced to a single integral whose imaginary part is a convolution and real part its Hilbert transform. This reduction is advantageous in the numerical calculation. It is well known that the third order term $B(\Omega)$:(i) produces a finite shift of the Γ_{15} resonance frequency in the complex plane, the imaginary part being the additional inverse life time due to the anharmonic interaction (1); (ii) gives rise to sidebands of the main IR absorption peak, which correspond to the excitation of two-phonon states. A previous analysis (4) of the spectra of lithium and silver doped KBr has revealed that in the low-frequency region these states can be at least as important as the one-phonon states.

In the lattice displacement representation (l=Bravais index, κ =index of the ion within the unit cell, α= Cartesian index) $V^{(3)}$ and $V^{(4)}$ are related to the anharmonic force constant matrices through the host lattice masses M_κ by the equations

$$V^{(3)}_{\alpha\alpha'\alpha''}\begin{pmatrix} l l' l'' \\ \kappa\kappa'\kappa'' \end{pmatrix} = (M_\kappa M_{\kappa'} M_{\kappa''})^{-\frac{1}{2}} \Phi^{(3)}_{\alpha\alpha'\alpha''}\begin{pmatrix} l l' l'' \\ \kappa\kappa'\kappa'' \end{pmatrix} / 3! \tag{5a}$$

$$V^{(4)}_{\alpha\alpha'\alpha''\alpha'''}\begin{pmatrix} l l' l'' l''' \\ \kappa\kappa'\kappa''\kappa''' \end{pmatrix} = (M_\kappa M_{\kappa'} M_{\kappa''} M_{\kappa'''})^{-\frac{1}{2}} \Phi^{(4)}_{\alpha\alpha'\alpha''\alpha'''}\begin{pmatrix} l l' l'' l''' \\ \kappa\kappa'\kappa''\kappa''' \end{pmatrix} / 4! \tag{5b}$$

Within our approximation, the non-vanishing matrix elements of $\Phi^{(3)}$ are

$$\left.\begin{matrix} \Phi^{(3)}_{xxx}(001) = b_{11} = \phi''' & ; \; \Phi^{(3)}_{yyx}(001) = b_{12} \\ \Phi^{(3)}_{\alpha\beta\gamma}(1'1''1) = -\Phi^{(3)}_{\alpha\beta\gamma}(1'1''-1); & \Phi^{(3)}_{yxy}(001) = b_{44} \end{matrix}\right\} = \frac{\phi''}{r} - \frac{\phi'}{r^2} \tag{6}$$

and those which can be obtained from symmetry and translational invariance conditions.Here,0 denotes the defect site,and ±1 the n.n. ions in the x direction. In Eqs.(6) the derivatives with respect the interionic distance r of a hypotetical defect-n.n. ion effective potential ϕ have been introduced. In the symmetry-coordinate representation we have

$$(\Gamma_{15x}1|\Phi^{(3)}|\Gamma_{15x}1;\Gamma_1) = 2(b_{11}+2b_{12})/\sqrt{6}$$

$$(\Gamma_{15x}1|\Phi^{(3)}|\Gamma_{15x}1;\Gamma_{12t}) = 2(b_{11}-b_{12})/\sqrt{3}$$

$$(\Gamma_{15x}1|\Phi^{(3)}|\Gamma_{15y}2;\Gamma_1) = \sqrt{2}(\Gamma_{15x}1|\Phi^{(3)}|\Gamma_{15x}2;\Gamma_{12t}) =$$
$$= -\sqrt{2}(\Gamma_{15x}2|\Phi^{(3)}|\Gamma_{15x}2;\Gamma_1) = -2(\Gamma_{15x}2|\Phi^{(3)}|\Gamma_{15x}2;\Gamma_{12t}) = \tag{7}$$
$$= -b_{11}/\sqrt{3}$$

$$(\Gamma_{15x}1|\Phi^{(3)}|\Gamma_{15y}1;\Gamma'_{25z}) = 2b_{44} ,$$

where the index t means "tetragonal and the cartesian orientations for the irr.reps. Γ_{15} and Γ'_{25} have been specified. The matrix elements which are not reported in Eqs.(7) are negligible for our purposes, or zero. The elements of $L^{(3)}$ can be readily obtained from Eqs.(7) and (5a). As concerns $\Phi^{(4)}$,only those elements which involve four times the defect site, or three times the defect site and once a n.n. site, are considered. This is consistent with the assumption of strong spacial localization of the resonant mode. Our elements are then

$$\Phi^{(4)}_{xxxx}(0001) = d_{11} = \phi''''$$

$$\Phi^{(4)}_{xxyy}(0001) = d_{12} = \frac{\phi'''}{r} - \frac{2\phi''}{r^2} + \frac{2\phi'}{r^3} = \frac{1}{r}(b_{11}-2b_{12})$$

$$\left.\begin{array}{l}\Phi^{(4)}_{yyyy}(0001) = d_{22}\\ 3\Phi^{(4)}_{yyzz}(0001) = 3d_{23}\end{array}\right\} = \frac{3\phi''}{r^2} - \frac{3\phi'}{r^3} = \frac{3b_{12}}{r} \qquad (8)$$

$$\Phi^{(4)}_{xxxx}(0000) = -2d_{11}-4d_{22} \quad ; \quad \Phi^{(4)}_{xxyy}(0000) = 2d_{23}-4d_{12},$$

and those obtained from symmetry operations. Usually d_{11} >> $-d_{12}$>> d_{22}; then,by setting $d_{22}=3d_{23}=0$, we consider only the elements

$$(\Gamma_{15x}1;\Gamma|\Phi^{(4)}|\Gamma';\Gamma_{15x}1) = -2d_{11}\delta_{\Gamma;\Gamma_{15x}1}(\delta_{\Gamma';\Gamma_{15x}1}- -\sqrt{2}\delta_{\Gamma';\Gamma_{15x}2})-4d_{12}\delta_{\Gamma;\Gamma_{15y}1}(\delta_{\Gamma';\Gamma_{15y}1}-\sqrt{2}\delta_{\Gamma';\Gamma_{15y}2})$$

$$(\Gamma_{15x}2;\Gamma|\Phi^{(4)}|\Gamma';\Gamma_{15x}1) = \sqrt{2}(d_{11}\delta_{\Gamma;\Gamma_{15x}1}\delta_{\Gamma';\Gamma_{15x}1}+ +2d_{12}\delta_{\Gamma;\Gamma_{15y}1}\delta_{\Gamma';\Gamma_{15y}1}) \qquad (9)$$

From our choice of the anharmonic coefficients(Eqs. 7 and 9) it appears that the projected densities here involved are those belonging to Γ_1,Γ_{12} and Γ'_{25} (these are 1x1 matrices and are given in Ref.11) and the following elements belonging to the irr.rep. Γ_{15}

$$\rho^{(\Gamma_{15})}_{11}(\omega^2) = \frac{1}{\pi}\mathrm{Im}\{(G_1(z)+\lambda\chi Q(z))/\mathcal{D}(z)\} \qquad (10)$$

$$\rho_{12}^{(\Gamma_{15})}(\omega^2) = \frac{\sqrt{2}}{\pi}\mathrm{Im}\{(G_2^{\pm}(z)+\lambda\chi^{\frac{1}{2}}Q(z))/\mathscr{D}(z)\}$$

$$\rho_{22}^{(\Gamma_{15})}(\omega^2) = \frac{2}{\pi}\mathrm{Im}\{(G_3^{\pm}(z)+(\lambda+\varepsilon\omega^2)Q(z))/\mathscr{D}(z)\}$$

where $\chi = M_{\pm}/M_{\mp}$ is the ratio of the substituted-ion mass to the n.n. ion mass; the Green functions $G_{\mu}^{\pm}(z)$ and the harmonic Γ_{15} resonant denominator $\mathscr{D}(z)$ are defined in Ref 6; $Q(z)=G_1^{\pm}(z)G_3^{\pm}(z)-G_2^{\pm}(z)^2$ and $z=\omega^2+i0^+$. At this point the new perturbation matrix $\Lambda_a=\Lambda(\Omega^2)+B(\Omega)+D$ is completely specified and the resonance conditions for Γ_{15} modes, as well as the IR absorption coefficient, can be easily investigated following Ref.6. In order to obtain the correct temperature dependence of the resonance frequency we should consider the effect of thermal expansion. The local contribution of the thermal expansion can be easily included in the present theory by replacing the harmonic force constant change λ with

$$\lambda + \frac{2r}{M_{\pm}} \frac{b_{11}+2b_{12}}{1+\lambda/f_1^*} \varepsilon_{xx}^T \qquad (11)$$

where ε_{xx}^T is the host-lattice thermal strain at the temperature T, and f_1^* is the Γ_1 effective force constant for the host lattice [12].

III.NUMERICAL RESULTS FOR KBr:Li

From experiment it appears that the system KBr:^{6}Li exhibits a low resonant frequency at 17.81 cm^{-1} [5].The experimental isotope frequency shift for replacing ^{7}Li with ^{6}Li is $\Delta=(10.2\pm0.5)\%$; this value disagrees strongly with the value $\Delta=1.5\%$ of the harmonic theory [6,7]. Beyond, the IR absorption strength and the life-time (i. e., the inverse half-width) of this resonance decrease very rapidly, if compared with the behavior of local modes, when the absolute temperature increases. In order to attempt a quantitative interpretation, of these data by means of the present theory, we have derived the values of b_{11} and b_{12} from Nolt and Sievers' experimental stress coefficients $A/(\bar{c}_{11}+2\bar{c}_{12})$ and $B/(\bar{c}_{11}-\bar{c}_{12})$ by means of the relations [4]

$$A = (r\Omega/6\tilde{f}^*)(b_{11}+2b_{12}) \quad ; \quad B = (r\Omega/12\tilde{f}^*)(b_{11}-b_{12}). \qquad (12)$$

the values of the local elastic constants $\bar{c}_{ij}$ have been calculated elsewhere on the basis of the harmonic theory ([12]); note that the anharmonic corrections are found to have a small influence on the c's. We have used A=9.05 and $B=5.88 \times 10^{13} sec^{-1}$. The b's cannot be derived from Eqs.(12) because $\tilde{f}^*$ is not known:however,the fitting of the present theory on the experimental resonance frequency will give simultaneously $\tilde{f}^*$ and the b's. Furthermore, the coefficient d_{11} can be obtained if the spacial behavior of the potential function is given. We assumed a Born-Mayer like potential for Li ion, in view of the fact that the quantity $d\ln\phi/d\ln r=1+b_{11}/b_{12}=2(A+B)/(A-2B)$, which expresses the spacial behavior of ϕ, is equal to -11.7+3.0 . In the calculations of the Green functions entering our problem we used the eigenfrequencies and polarization vectors of the host lattice calculated at 4096 points in the BZ according to the Hardy-Karo model. The fitting of the resonance frequency gives $\tilde{f}^*/f^*=0.0167$ (harmonic theory:$\tilde{f}^*/f^*=0.015$) and $\Delta=7.0\%$. On the other hand it is possible to fit exactly the experimental isotope shift with $\tilde{f}^*/f^*=0.0175$,but the resonance frequency is slightly lowered to 15.5 cm^{-1}. To investigate the T-dependence of the strength factor f(T) and the squared half-width $\Gamma(T)$ (both defined in Ref.6), we have used the phenomenological relations ([5])

$$f(T) = e^{-(T/\theta_1)^2} f(0) \quad , \quad \Gamma(T) = \Gamma(0)\coth(\theta_2/T) \qquad (13)$$

Note that both f(0) and $\Gamma(0)$ contain already the T=0°K anharmonic contributions. The present theoretical results compared with the experimental ones are

$$\begin{aligned} \theta_1 &= 32.5°K & (\theta_1)_{exptl} &= 27.9°K \\ \theta_2 &= 4.15°K & (\theta_2)_{exptl} &= 8.0°K \\ \Gamma(0) &= 4.79 \times 10^{23} sec^{-2} & (\Gamma(0))_{exptl} &= 5.0 \times 10^{23} sec^{-2} \end{aligned} \qquad (14)$$

For comparison the harmonic theory ([8]) gives $\Gamma(0)=2.07 \times 10^{23} sec^{-2}$.The agreement with the experimental data is satisfactory except for the temperature dependence of the complex frequency shift: this is probably due to the fact that the perturbative series of the anharmonic terms converges quite slowly and should be truncated at a higher order than that was assumed in the present approach. However,the essential anharmonic character of resonant modes is apparent; it is of particular interest to note that third order anharmonicity contributes a remarkable lowering of the Γ_{15} resonant frequency with respect to the harmonic prediction, making the defect to approach the instability in its lattice site. It is possible that anharmonicity plays an important role in the discussion of the off center configurations of small ions, which would therefore assume a dynamical meaning. This situation seems to be enhanced for silver impurities, for which calculations are in progress.

The author wishes to thank Prof.G.F.Nardelli for his continued advice, and Prof.M.V.Klein for preprints of his recent papers.

REFERENCES

(*) Work supportedby EOAR under Grant N.67-08 with the European Office of Aerospace Research, U.S.Air Force.
(1) For an extended bibliography on this subject see A.A. Maradudin, Solid State Physics, edited by F.Seitz and D.Turnbull (Academic Press Inc., N.Y.1966), Vol 19.
(2) I.G.Nolt and A.J.Sievers, Phys.Rev.Letters 16, 1103(1966)
(3) G.Benedek and G.F.Nardelli, Phys.Rev.Letters, 16, 517, (1966)
(4) G.Benedek and G.F.Nardelli, Phys.Rev.Letters, 17, 1136, (1966)
(5) S.Takeno and A.J.Sievers, Phys.Rev.Letters, 15, 1020, (1965); A.J.Sievers, in Lectures on Elementary Excitations and Their Interactions in Solids-Cortina 1966.
(6) G.Benedek and G.F.Nardelli, Phys.Rev., 155, 1004(1967)
(7) M.V.Klein, in Physics of Color Centers, W.Beall Fowler editor; Cap.7.(to be published)
(8) Giorgio Benedek and A.A.Maradudin, to be published; MacDonald, Phys.Rev.150, 597(1966).
(9) W.M.Vissher, Phys.Rev., 134, A965(1964).
(10) R.A.Cowley, in Phonons-Scottish Universities'Summer School, 1965; R.W.H.Stevenson, editor(Oliver and Boyd, 1966), pag 170.
(11) G.Benedek and G.F.Nardelli, Phys.Rev., 154, 872(1967).
(12) Giorgio Benedek and G.F.Nardelli, to be published.

A METHOD FOR DETERMINING STRICT BOUNDS ON THE FREQUENCIES OF LOCALIZED PHONONS

P. Dean

National Physical Laboratory

Teddington, England

1. INTRODUCTION

This paper presents a new method for determining the frequencies of localized modes due to impurities or defects in harmonic solids. The method consists of finding strict and close upper and lower bounds for the local-mode frequencies by means of a simple matrix calculation; it is not a Green's function method, and therefore does not depend upon the availability of tabulated data. A full account of the method, with proofs and references to basic theorems, is to be published elsewhere(1).

The physical basis of the technique is similar to one which has been used by Litzman and Celý(2) and others(3) in the past ten years; it depends upon the properties of a fairly small "defect region" which comprises the defect itself plus some surrounding atoms, all the other atoms in the solid being regarded as infinitely massive. It is well known that such a system can simulate fairly accurately those modes of vibration of the total solid which are highly localized at the defect, and that one can obtain lower bounds for frequencies above the continuous spectrum by a simple eigenvalue calculation. In this paper we show how one can overcome the main limitations, i.e. the vagueness in accuracy, of such calculations as they have been used hitherto. We give formulae for strict and close upper as well as lower bounds; these formulae depend only upon the properties of the defect region and a knowledge of the band edges of the pure host material.

The method of this paper has the advantage that it can be used in situations where the Green's function method is quite impracticable: it can be used, for example, for computing local modes in amorphous materials. However, in its present form the method cannot be used for the study of in-band resonances.

2. BASIS OF THE METHOD

The time-independent equations of motion for a rigid-ion lattice dynamical model can be written in matrix form as

$$(M - \omega^2 I)u = 0 \tag{1}$$

where M is a symmetric matrix whose order is the number of degrees of freedom in the system, I is the unit matrix of the same order, and u is the column vector whose elements are the mass-modified components of atomic amplitude (cf. Dean[4]).

We regard the total system represented by equation (1) as containing two regions: (i) a "defect region" $\mathfrak{D}$ which consists of the lattice defect or impurity atoms together with some surrounding atoms, and (ii) the "host structure" $\mathcal{H}$ containing all the remaining atoms in the system. (The precise specification of $\mathfrak{D}$ and $\mathcal{H}$ will vary from case to case and depends both upon the type of system under study and the accuracy required in the computation of the local mode frequencies.) If the rows and corresponding columns of the matrix M are suitably arranged, equation (1) may be written in the partitioned form

$$\left[\begin{array}{c|c} D-\omega^2 I_D & C \\ \hline C^T & H-\omega^2 I_H \end{array}\right] \left[\begin{array}{c} u_D \\ \hline u_H \end{array}\right] = 0, \tag{2}$$

where D and H are matrices which refer to atoms in $\mathfrak{D}$ and $\mathcal{H}$ respectively, and C (and its transpose C^T) represent interactions between the two regions. The column vectors u_D and u_H in (2) contain the mass-modified amplitudes of atoms in $\mathfrak{D}$ and $\mathcal{H}$, and I_D and I_H are unit matrices whose orders are those of D and H respectively. The matrix D has a precise physical significance: it is the dynamical matrix for the region $\mathfrak{D}$ when the atoms in $\mathcal{H}$ which interact directly with atoms in $\mathfrak{D}$ are confined to their equilibrium positions. (Thus D, for example, represents the dynamical matrix for $\mathfrak{D}$ if the atoms in $\mathcal{H}$ are regarded as being of infinitely heavy mass.) A similar meaning, with the roles of $\mathfrak{D}$ and $\mathcal{H}$ reversed, can be attributed to the matrix H.

It is clear from (2) that, for modes which are highly localized within the region $\mathfrak{D}$, the equation

$$(D - \omega^2 I_D) u_D = 0 \tag{3}$$

is approximately valid. This is the essential feature basic to the technique of Litzman and Cely[2] and other authors[3]. A further point utilized by these authors is that the largest eigenvalues of D represent lower bounds for the squared frequencies of local modes which lie above the continuous spectrum of M. This follows by applying Rayleigh's theorems[5] to the total system, allowing the masses of atoms in $\mathcal{H}$ to change from their true values to infinitely heavy masses. Our contribution, in this paper, is to invoke theorems due to Kato[6] and Temple[7], as described in detail by Dean[1]. Our results indicate that it is a simple matter, on solving the eigenvalue problem for D, to calculate both upper and lower strict bounds for local mode frequencies; these results apply to gap modes as well as to modes whose frequencies lie above the continuous branches of the **spectrum.**

3. STRICT BOUNDS FOR LOCAL-MODE FREQUENCIES

We consider separately various cases of physical interest. Each of these cases is treated in more detail in reference (1), where proofs and references to theorems are given.

3.1 Single Out-of-Band Frequency

Consider the determination of strict bounds for a single frequency above the continuous spectrum. The formulae we give apply whether this frequency is degenerate or not. The degenerate case occurs, for example, if the defect consists of an impurity atom in a crystal of cubic symmetry.

Let σ denote the largest eigenvalue of D, and ω_L the maximum frequency (top of the continuum) of the total unperturbed system (i.e. the total system with no defect). Bounds for λ, the squared frequency of the local mode of the total perturbed system, are given by

$$\sigma \leq \lambda \leq \sigma + \frac{\varepsilon^2}{\sigma - \omega_L^2} , \tag{4}$$

where

$$\varepsilon^2 = (C^T v, C^T v)/(v, v) \tag{5}$$

and v is the eigenvector of D associated with the eigenvalue σ. (If σ is a degenerate eigenvalue v can be any corresponding eigenvector.) The bracket symbol on the right-hand side of (5) denotes the scalar product of two vectors.

Clearly, the condition $\sigma > \omega_L^2$ must be valid if (4) is to be used. In practice, this condition is usually valid; if not it can be made so by increasing the size of the defect region $\mathfrak{D}$. The evaluation of ε^2 in (5) presents no difficulty. The interaction matrix C^T appropriate to any given $\mathfrak{D}$ is easily written down, and the vector $C^T v$ then computed.

3.2 More than One Out-of-Band Frequency

This section deals with the case of a lattice defect or a group of impurity atoms which induce more than one distinct frequencies above the continuous region of the spectrum. Each of the out-of-band frequencies may be either degenerate or non-degenerate.

There are several formulae for obtaining strict bounds. It is largely a question of judgment which is the most efficient in any particular case. If there is any doubt, each formula should be used and a set of closest bounds obtained by comparing results.

Let $\sigma_1 \geqslant \sigma_2 \geqslant \cdots \geqslant \sigma_s$ be the s largest eigenvalues of D, and $\lambda_1 \geqslant \lambda_2 \geqslant \cdots \geqslant \lambda_s$ the out-of-band eigenvalues (squared frequencies) of M. One set of bounds for the λ's is given by

$$\sigma_j \leqslant \lambda_j \leqslant \sigma_j + \sum_{i=j}^{s} \frac{\varepsilon_i^2}{\sigma_i - \omega_L^2} \quad (j = 1, 2, \ldots s) \qquad (6)$$

where, again, ω_L is the maximum frequency of the total unperturbed system. The number ε_i^2 is given by

$$\varepsilon_i^2 = (C^T v_i, C^T v_i)/(v_i, v_i), \qquad (7)$$

where v_i is the eigenvector of D corresponding to σ_i (or, if σ_i is degenerate, any eigenvector of D corresponding to σ_i).

An alternative formula for strict bounds on the local mode squared frequencies is

$$\sigma_j \leqslant \lambda_j \leqslant \sigma_j + \frac{\varepsilon_j^2}{\sigma_j - \nu_j} \quad (j = 1, 2, \ldots s), \qquad (8)$$

where ν_j is an upper bound for the nearest eigenvalue of M smaller than λ_j. Clearly the use of this formula entails finding bounds for the eigenvalues in the order λ_s, $\lambda_{s-1}, \ldots \lambda_1$, and it depends upon the inequalities $\nu_j < \sigma_j$ being satisfied. Formula (8) will be most useful for cases in which there are just a few well-separated (degenerate or non-degenerate) out-of-band frequencies.

If the total defect system has point group symmetry it may be possible to use the formula

$$\sigma_j \leq \lambda_j \leq \sigma_j + \frac{\varepsilon_j^2}{\sigma_j - \omega_L^2} \tag{9}$$

for each out-of-band eigenvalue separately. This will be possible if the out-of-band modes belong to different irreducible representations, and an essential requirement is that the defect region $\mathfrak{D}$ be chosen to maintain the point group symmetry of the system. If there exist other out-of-band modes belonging to the same irreducible representation as (say) j then, in finding strict bounds for λ_j, a slightly modified version of the formula (9) may be used. In this formula a number ξ_j is substituted for ω_L^2, where ξ_j is an upper bound for the nearest eigenvalue of M smaller than λ_j belonging to a mode of the same irreducible representation. Of course, it may be simpler not to utilize the symmetry properties of the lattice and to rely on either (6) or (8). Whichever formula is used the closeness of bounds can always be improved by increasing the size of the defect region.

3.3 Gap Modes

In this section we give formulae for the determination of strict bounds on the frequencies of local modes which exist in the gaps between bands of the frequency spectrum.

In the case of a single (degenerate or non-degenerate) local-mode squared frequency λ, strict bounds are given by

$$\sigma - \frac{\varepsilon^2}{\beta - \sigma} \leq \lambda \leq \sigma + \frac{\varepsilon^2}{\sigma - \alpha} . \tag{10}$$

Here σ is an eigenvalue of D which satisfies $\alpha < \sigma < \beta$, where α and β are the limits of the band gap. There may sometimes be a possible ambiguity in choosing σ if the defect region $\mathfrak{D}$ is small, but this ambiguity can always be removed by increasing the size of $\mathfrak{D}$. Clearly, a satisfactory σ will have been chosen if the limits computed from (10) lie within the range α to β. The number ε^2 in (10) is given by equation (5) where, now, v is the vector of D (or, in the case of a degenerate σ, any vector of D) corresponding to the eigenvalue σ.

In the case of s frequencies $\lambda(1) \geq \lambda(2) \geq \dots \geq \lambda(s)$ in the band gap, strict limits may be found from the formula

$$\sigma(j) - \sum_{i=1}^{j} \frac{\varepsilon(i)^2}{\beta - \sigma(i)} \leq \lambda(j) \leq \sigma(j) + \sum_{i=j}^{s} \frac{\varepsilon(i)^2}{\sigma(i) - \alpha} \tag{11}$$

where the eigenvalues $\sigma(1), \sigma(2), \dots \sigma(s)$ of D satisfy $\beta > \sigma(1) \geq \sigma(2) \geq \dots \geq \sigma(s) > \alpha$, and the defect region $\mathfrak{D}$ is not too small. The numbers $\varepsilon(i)^2$, in (11), are defined by

$$\varepsilon(i)^2 = (C^T v(i), C^T v(i))/(v(i), v(i)), \tag{12}$$

where $v(i)$ is the eigenvector of D corresponding to $\sigma(i)$. An alternative formula for strict bounds of $\lambda(j)$ is

$$\sigma(j) - \frac{\varepsilon(j)^2}{\mu(j)-\sigma(j)} \leqslant \lambda(j) \leqslant \sigma(j) + \frac{\varepsilon(j)^2}{\sigma(j)-\nu(j)}, \tag{13}$$

where $\nu(j)$ is an upper bound for the nearest eigenvalue smaller than $\lambda(j)$, and $\mu(j)$ is a lower bound for the nearest eigenvalue greater than $\lambda(j)$. Clearly, the conditions $\mu(j) > \sigma(j) > \nu(j)$ must be satisfied if (13) is to be utilized.

If the total system has point group symmetry it may be possible to improve on the bounds given by (13) by extending $\nu(j)$ to lower, and $\mu(j)$ to higher values by neglecting modes which do not belong to the same irreducible representation as j (as in the case for local modes above the continuous spectrum, discussed in section 3.2). If the symmetry property is utilized it is necessary for the defect region $\mathfrak{D}$ to be chosen to maintain the point group symmetry of the total system.

As a general guide, the accuracy one can achieve for gap modes will be poorer than that for modes whose frequencies lie above the continuum if defect regions of similar sizes are used. However, in all cases one can attain any required accuracy simply by increasing the size of the region $\mathfrak{D}$.

4. GENERAL REMARKS

Experience on a number of numerical examples(1) indicates that the arithmetic mean of the lower and upper bounds computed by the methods of section 3 provides a value for the local mode frequency even more accurate than that suggested by the interval between the bounds. Thus, for the case of the single out-of-band frequency (cf. equation (4)), the formula

$$\lambda \triangleq \sigma + \tfrac{1}{2}\varepsilon^2(\sigma - \omega_L^2)^{-1} \tag{14}$$

usually represents a very good approximation indeed, and the error in λ is likely to be appreciably less than the maximum possible error of

$$\delta\lambda \equiv \tfrac{1}{2}\varepsilon^2(\sigma - \omega_L^2)^{-1}. \tag{15}$$

As an example of the accuracy of formula (4) consider the simple cubic lattice with equal nearest-neighbour central and non-central forces. Let the impurity consist of a single substitutional defect atom whose mass is one quarter of the masses of the other atoms. A defect region comprising the defect atom and its six nearest neighbours yields the values $\lambda\omega_L^{-2} = 2.1056$, $\delta\lambda\omega_L^{-2} = 0.0017$. When the defect region consists of the impurity atom and its 26 nearest neighbours the values $\lambda\omega_L^{-2} = 2.1063$, $\delta\lambda\omega_L^{-2} = 0.0002$ are obtained. These values compare with the result $\lambda\omega_L^{-2} = 2.107 \pm 0.0025$ computed by the Green's function method using linear interpolation between elements in the table provided by Maradudin _et al_[8].

The method of this paper is most useful in the study of modes whose frequencies lie well away from band edges. It becomes relatively less favourable to use in cases where defect-mode frequencies lie close to band edges. Thus, it gives a good accuracy (comparable to the accuracy of the Green's function method) for a defect region of just seven atoms in the example quoted, where the mass ratio is $\frac{1}{4}$; however, if the mass ratio were closer to unity a larger defect region would be required to achieve this accuracy.

The method has considerable potential in cases where the Green's function method cannot in practice be used. Thus it may be used to study defect modes in, for example, amorphous materials. The reason is that the presence of short range structural order usually enables one to construct a local defect region $\mathfrak{D}$ surrounding any defect or impurity of interest. To apply the method requires only a knowledge of the way in which $\mathfrak{D}$ is connected (via interatomic forces) to the outside region $\mathcal{H}$, and a knowledge of the maximum vibrational frequency ω_L of the total unperturbed system. The interactions between $\mathfrak{D}$ and $\mathcal{H}$ are really part of the definition of $\mathfrak{D}$ itself, and can be assessed from short range order considerations; ω_L (or an upper bound for ω_L — which can equally be used in the formulae of this paper) can be found from elementary considerations, using Frobenius' theorem[9]. On the other hand, the Green's function method requires, in essence, a knowledge of the total unperturbed dynamical system.

Perhaps the most obvious advantage of the method is that it does not depend, for its use, upon the availability of tabulated data. Thus it can be applied readily to any force-constant model of a rigid-ion harmonic solid. The method can be extended to polarizable ions, but this extension will not be considered in this paper.

ACKNOWLEDGMENTS

I am grateful to Dr. J. H. Wilkinson for helpful advice, and to Dr. R. J. Bell for useful discussions. The work described in this paper was carried out at the National Physical Laboratory.

REFERENCES

[1] P. Dean, to be published.
[2] O. Litzman and J. Celý, Czech. J. Phys. B 11, 320 (1961).
[3] W. Ludwig, in The Theory of Crystal Defects, edited by B. Gruber (Academia, Prague, 1966) pp. 57 - 165; N. Krishnamurthy and T. M. Haridasan, Phys. Letters 21, 372 (1966); P. L. Land and B. Goodman, J. Phys. Chem. Solids 28, 113 (1967).
[4] P. Dean, J. Inst. Maths Applics. 3, 98 (1967).
[5] cf. A. A. Maradudin, E. W. Montroll and G. H. Weiss, Theory of Lattice Dynamics in the Harmonic Approximation (Academic Press, New York, 1963).
[6] T. Kato, J. Phys. Soc. Japan 4, 334 (1949).
[7] G. Temple, Proc. Roy. Soc. A 211, 204 (1952).
[8] A. A. Maradudin, E. W. Montroll, G. H. Weiss, R. Herman and H. W. Milnes, Mem. Acad. Roy. Belg. 14, No. 1709 (1960).
[9] cf. E. Bodewig, Matrix Calculus (North Holland Publ. Co., Amsterdam, 1956).

ASPECTS OF LOCAL EXCITATION VIBRATIONAL INTERACTION IN ALKALI HALIDE CRYSTALS ACTIVATED BY O_2^- AND S_2^- MOLECULES

K. Rebane, L. Rebane, O. Sild

Institute of Physics and Astronomy

Estonian S.S.R. Academy of Sciences, U.S.S.R.

Crystals activated by simple impurity molecules appear to be suitable objects for investigating some aspects of vibronic and vibrational interactions in the impurity center. The fact is that the important problem for the local dynamics of the crystal - the problem of existence and properties of the local vibration - is solved trivially here: the intramolecular vibration slightly modified by the crystal shows itself to be the local vibration.

The luminescence and absorption spectra of these crystals often have a fine and rich vibrational structure. The part of the vibrational structure directly due to the local vibration can easily be interpreted. The detailed structure can serve as a source of information about the interaction of electrons and the local vibration with crystal vibrations, about the details of the local dynamics in the luminescence center.

Alkali halide crystals activated by O_2^- and S_2^- molecules have been chosen as the representatives of systems having the above mentioned properties.

The O_2^- spectrum in alkali halide crystals has been studied in papers ([1,2]), the S_2^- spectrum in ([3,4]). According to the data on the paramagnetic resonance ([5,6]) O_2^- and S_2^- molecules replace anions in alkali halide crystals . The axes of the molecules are pointed at six equivalent directions [110] of the crystal lattice.

In our previous papers the luminescence spectra of O_2^- were measured at 4,2°K in KCl, KBr, NaCl and NaBr [7,8]. The temperature dependence of luminescence spectra of KCl-O_2^- and KBr-O_2^- has been studied more in detail [9]. The excitation spectra, the temperature quenching of luminescence [10], the effect of pressure upon luminescence spectra [11] and the life-time of the excited electronic state [8,11] have also been studied (for details see the results [9]). Luminescence spectra of S_2^- in KCl, KBr and RbBr will be published [12]. Methods of preparing samples and methods of the experiment were described in [7,9].

In the present paper attention is given to the following three problems: (1) the vibronic structure due to the local vibration, (2) the fine structure of vibronic sub-bands, (3) the unusual dependence of the half-width of sub-bands on their number in the vibronic series.

The luminescence spectra of crystals activated by isoelectronic O_2^- and S_2^- molecules have much in common. They consist of a series of distinct vibronic sub-bands (6 - 9 sub-bands have been registered). At helium temperature each sub-band resolves into one narrow and intensive peak (the main maximum) and a number of more or less distinct maxima above the continuous background. According to the theory this kind of spectrum corresponds to the case of strong interaction of the electronic transition with one local vibration and an essentially weaker interaction with the rest of the vibrations.

By frequencies of main maxima the frequency and the anharmonicity of the local vibration of the center in the ground electronic state have been determined (and the corresponding potential curve has been plotted [13,14]). The results are presented in Table 1. These quantities proved to depend little on the host crystal in the case of O_2^- centers; for S_2^- centers the frequency of the local vibration consecutively decreases in the series of KCl, KBr and RbBr crystals. The potential curves for KBr-O_2^- and KBr-S_2^- in the excited electronic state were calculated on the basis of the intensities of sub-bands measured at 90°K.

Table 1.

Center	Host crystal	Ground electronic state		Excited electronic state	Stokes losses upon radiation	
		Vibrational frequency Ω_g, cm^{-1}	Anharmonicity, cm^{-1}	Vibrational frequency Ω_e, cm^{-1}	P, eV	Number of quanta $P:\hbar\Omega_g$
O_2^-	KCl KBr NaCl NaBr	1150±5	18±5	830	1,28	10
S_2^-	KCl KBr RbBr	630 619 575	5,0 5,5 4,0	400	0,64	9

Differences between frequencies $\Delta\omega$ of the main maximum and the maxima of the fine structure for each vibronic sub-band n in the spectra of KBr–O_2^- (a) and KBr–S_2^- (b) crystals are presented in Fig.1. For comparison the upper part of the figure brings out the frequencies of KBr crystal vibrations corresponding to singular points of the Brillouin zone which have been

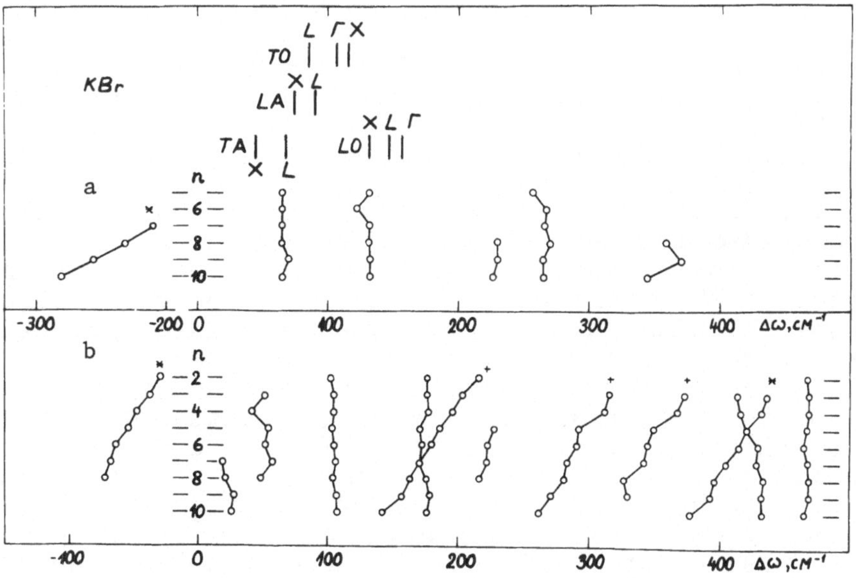

Fig. 1. Frequency differences $\Delta\omega$ for vibronic sub-bands n.

obtained from the theoretical papers by Hardy and Karo ([15]) (T and L - transversal and longitudinal branches; O and A - optical and acoustical branches; X, L, Γ - the singular points ⟨100⟩, ⟨111⟩ and ⟨000⟩).

The characteristic feature of KBr-O_2^- and of all the rest of O_2^- spectra studied by us is the clear recurrence of frequencies of the fine structure (in Fig. 1 the points corresponding to different n are arranged vertically). The series marked with the asterisk appears to be the exceptional one. That shifting series of lines is interpreted as the main maximum of sub-bands belonging to isotopic $O^{16}-O^{18}$ centers ([16]). Differences between frequencies $\Delta\omega$ in the O_2^- spectra in various host crystals are essentially different.

The fine structure of vibronic sub-bands in the spectrum of KBr-S_2^- contains two groups of maxima: the frequencies of the first group have no noticeable dependence on n and are the analogs to the frequencies of the fine structure in O_2^- spectra; the frequencies of the second group consecutively decrease at the increase of n. The spectra of KCl-S_2^- and RbBr-S_2^- crystals have the analogous character: series of shifted maxima can be noticed in them as well.

Two series of shifting lines in the spectrum of KBr-S_2^- (marked with the asterisk) may be connected with the isotopic $S^{32}-S^{34}$ molecule. The values of isotopic shifts of these lines are in good agreement with the theoretical values calculated on the basis of the potential curves. It is difficult to identify the other shifting maxima in the S_2^- spectrum (marked with the cross) with isotopic recurrences if one considers the lack of correspondence between their frequencies and the values of expected isotopic shifts; there is also a contradiction between the relatively high intensity of these maxima and the possible concentration of the isotopes. Therefore we should like to connect the rest of the shifting maxima with somewhat changed potential curves. The shifting lines can be obtained if we take into account that they belong to the transitions leading to the potential curve with the frequency of the local vibration decreased by 9 cm^{-1}. The different potential curve may appear as a result of the splitting of the ground electronic state of S_2^- ($^2\Pi_g \rightarrow B_{2g} + B_{3g}$) or belong to the slightly different ("deformed") S_2^- centers. Thus, on the basis of the data in Table 1 we come to the following conclusion concerning the characteristic frequencies of the fine vibrational structure of vibronic sub-bands: (a) they are different in case of different host crystals; (b) they are different for O_2^- and S_2^- centers in one and the same host crystal; (c) no regular coincidence between them and the frequencies in the singular points of the Brillouin zone can be noticed. Therefore one has to confess that in this case it is difficult to interpret the frequencies of the fine structure as singularities in the spectrum of the host crystal. The observed approximate coincidence of two frequencies in the spectrum of O_2^- with TA, L and LO, X vibrations of the KBr crystal has to be considered occasional (the decisive argument here is point (b)).

In the fine structure the frequencies of pseudolocal vibrations are observed. In order to explain the two characteristic frequencies 65 and 130 cm^{-1} in the spectrum of KBr-O_2^- we propose the hypothesis that they are in essence the translational and the librational vibrations of the impurity molecule. The identification of the "optical" frequency with the librational one is reasonable, as the estimation of the possible frequency of the librational vibration on the basis of the data about the potential barrier ([17]) gives the value up to 200 cm^{-1}. The "acoustical" frequency can be connected with the translational vibration of the molecule, for O_2^- is lighter than the Br ion; as a result, high acoustical frequencies form the corresponding packet which represents a pseudolocal vibration.

Thus, we have neglected the role of a nearly free rotational motion when interpreting low-temperature spectra in the initial (excited) electronic state as well as in the final electronic state. This may be justified if (in addition to the sufficiently high barrier) the Stokes losses upon the rotational (librational) degrees are small. The circumstance of the existence of only 1 - 2 lines in the vibrational series of the fine structure is found to be the proof of the smallness of the Stokes losses upon all kind of ion shifts in the impurity center except the intramolecular vibration. Such a value of the Stokes losses upon the librational vibration can be understood if we suppose that their equilibrium positions (determined by the symmetry of the center) remain the same (along [110]) in the excited electronic state. If the equilibrium positions of the librational vibration differ from those in the ground electronic state, the Stokes losses upon the rotational degree of freedom are appreciable and the high quasirotational levels of the final electronic state make an essential contribution to the structure of the spectrum. The theoretical description of the structure of the vibronic-rotational spectrum in case of appreciable Stokes losses upon the rotation and the potential barriers of considerable height between the equivalent orientations of the molecule requires the analysis which takes account of the great anharmonicity of vibrations and the high transparency of the barriers.

In the paper ([18]) a rather remarkable experimental fact was established: the half-widths of vibronic sub-bands in the luminescence spectrum of KBr-O_2^- produced by the generation of different numbers of vibrational quanta belonging to the intramolecular vibration of O_2^- d e c r e a s e as the number of the vibrational level n increases. The decrease in half-widths amounts to ca 45 per cent with the increase of n from 6 to 10 (T = 90°K).

Considering the effect as caused by vibrations one has in principle two possibilities for its explanation: (1)mixing the local vibration with crystal vibrations; (2) anharmonic interaction between local and crystal vibrations.

The numerical estimation of the magnitude of the effect leads to the conclusion that the first possibility may yield a total change in the half-width up to 10 per cent only through the whole series, i.e. 1 - 2 per cent if n is changed by unity. This is evidently insufficient for the explaining of the experimental data in the case of KBr-O_2^- where a change of 10 - 15 per cent in the half-width is observed as n changes by unity.

The anharmonic interaction may explain the sign and the size of the observed effect if in the expansion of the potential energy $V(x,y)$(x is the coordinate of the local vibration, y - the complex of coordinates of the crystal vibrations) the coefficient D has the order of magnitude $D \approx -0.1\hbar\omega_k$ in the member Dx^2y where ω_k is a certain effective frequency of crystal vibrations.

If the mechanism proposed by us as an explanation is correct then the normal behavior - the increase of half-widths of sub-bands with the number of the vibrational level in the series - should be observed in the absorption spectrum of $KBr-O_2^-$.

No dependence of half-widths of vibrational sub-bands upon n has been observed in the luminescence spectrum of S_2^-.

1. J. Rolfe, F. Lipsett, W. King, Phys.Rev. 123, 447 (1961).
2. J. Rolfe, J. Chem. Phys. 40, 1664 (1964).
3. J. H. Schulman, R.D. Kirk, Solid State Comm. 2, 105 (1964).
4. R.D. Kirk, J.H. Schulman, H.B. Rosenstock, Solid State Comm. 3, 235 (1965).
5. W. Känzig, M.H. Cohen, Phys. Rev. Letters 3, 509 (1959).
6. J.R. Morton, J. Chem. Phys. 43, 3418 (1965).
7. K. Rebane, L. Rebane, Izv. AN ESSR, ser. fiz.-mat. i tehn. nauk 14, 309 (1965).
8. K.K. Rebane, L.A. Rebane, O.I. Sild, Internat.Conf. on Luminescence, Preprints, Budapest 2, C.5, 115 (1966).
9. L.A. Rebane, Trudy IFA AN ESSR N37 (1967).
10. K.K. Rebane, L. Rebane, R. Avarmaa, Izv. AN ESSR, ser. fiz.-mat. i tehn. nauk 16, 118 (1967).
11. A.I. Laisaar, A. Niilisk, Trudy IFA AN ESSR N37 (1967).
12. L. Rebane, Izv. AN ESSR, ser. fiz.-mat. nauk 16 N4 (1967).
13. A. Sh. Yerenchinov, O.I. Sild, Trudy IFA AN ESSR N37 (1967).
14. R.A. Avarmaa, Trudy IFA AN ESSR N37 (1967).
15. J.R. Hardy, A.M. Karo, Phys. Rev. 129, 2024 (1963).
16. L. Rebane, Izv. AN ESSR, ser. fiz.-mat. i tehn. nauk 15, 301 (1966).
17. R. Pirc, B. Žekš, P. Gosar, J. Phys. Chem. Solids 27, 1219 (1966).
18. L.A. Rebane, T.I. Saar, Izv. AN ESSR, ser. fiz.-mat. i tehn. nauk 15, 297 (1966).

ON THE LOCALIZED VIBRATIONS OF NON-BRIDGING OXYGEN ATOMS IN VITREOUS SILICA

R. J. Bell and P. Dean

National Physical Laboratory

Teddington, England

1. INTRODUCTION

A structure sensitive band, in the region 900 - 950 cm^{-1} of the infrared spectrum of vitreous silica, has for some time been thought to be associated with the presence of non-bridging oxygen atoms. In this paper we present supporting theoretical evidence for this idea: we show that the band is probably due to highly localized vibrations of the non-bridging oxygen atoms.

2. THE STRUCTURE OF VITREOUS SILICA

At the present time there are several conflicting theories of the structure of vitreous silica[1-4]. Each of these theories accepts that the basic structural unit is the SiO_4 tetrahedron, with the silicon atom at its centre and four oxygen atoms at the vertices, and it is generally agreed that the tetrahedral units are linked together via the common (or "bridging") oxygen atoms. However, the theories differ in proposing the types of geometrical configuration which occur in the structure. The most generally accepted theory, the so-called "random network" theory due originally to Zachariasen[1], states that the SiO_4 tetrahedral units form a connected three-dimensional random network, the mechanism of disorder being a randomness in the relative orientation of adjacent tetrahedra. The Si-O-Si angles can take any values in a range from approximately 120° to 180°; there is some recent evidence that the mean angle is about 150°. In a completely connected structure of this kind each oxygen atom connects two silicons. It is thought, however, that in real samples of vitreous silica a small proportion of the oxygen atoms may constitute so-called "non-bridging" atoms which are connected to one silicon atom only.

The other theories of the structure of vitreous silica include the vitron theory, proposed by Tilton[2] and developed recently by Robinson[3], and the crystallite theory of the Russian school[4]. It may be that the structure of vitreous silica (which is certainly not unique, but varies slightly with the method of preparation) contains features proposed by more than one of the theories mentioned here. Thus the true structure may lie somewhere between a connected random network and a crystallite system, i.e. it may be a connected network with small regions almost crystalline in character and other regions essentially of a random structure.

The theoretical results presented in this paper are based upon the random network picture, but they would almost certainly remain valid if based upon either of the other two theories mentioned here. Thus the conclusions of this paper are basically independent of the precise model chosen for the structure.

3. MODELS AND SPECTRA

In this work (which is part of a larger programme of work on the structure and spectra of glasses) we constructed a number of physical models of typical atomic arrangements in vitreous silica. The models contain several hundred polystyrene spheres (to represent silicon and oxygen atoms) connected together by steel rods. Figure 1 shows a section of one such model, built to resemble part of a random network structure. The measured positions of atoms in such models are used as a basis for calculations of the atomic vibrational properties of the glass. There seems no way at present to generate in a reasonable time a set of atomic coordinates consistent with, say, the random network theory by means of computer program. It is for this reason that in our work on glasses we build models which are consistent with experimental data on structure, and then use the measured atomic coordinates as a basis for theoretical calculations.

In figures 2 and 3 we depict X-ray and neutron radial distribution functions for the model of figure 1. These histograms represent quite well the form of curves obtained from X-ray and neutron diffraction experiments on real vitreous silica. The mean Si-O-Si angle for the model structure of figure 1 is 140°. This is rather less than some recent estimates for this mean angle in the real glass, which place it at about 145°. It is therefore not surprising that the density of the model is rather larger (at 2.7 gm/cc) than the experimentally determined density of

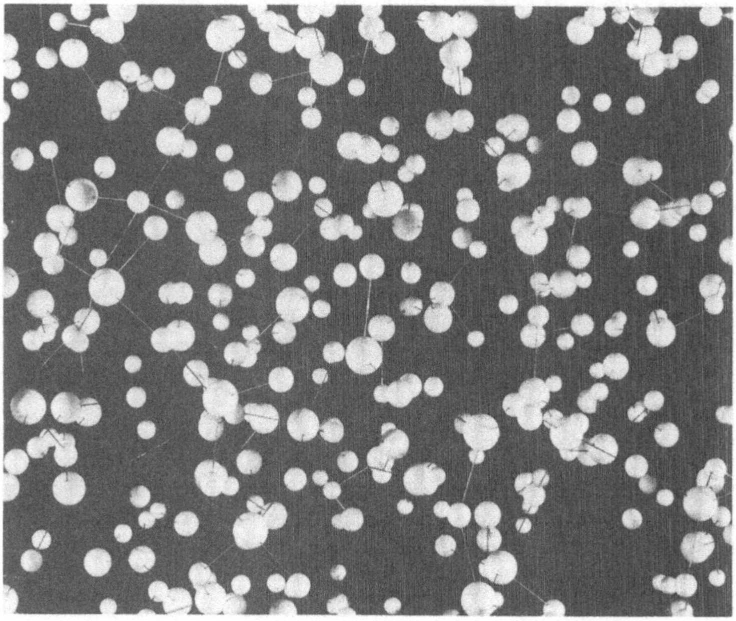

Figure 1. Part of a model of a sample arrangement of atoms in vitreous silica. The large spheres represent silicon atoms, and the small spheres oxygen atoms. The spheres are connected together by steel rods.

vitreous silica of 2.2 gm/cc; however, this does not affect the conclusions of this paper.

In figures 4 and 5 we depict vibrational spectra for our model based upon an interatomic force field of nearest-neighbour central (γ) and non-central (γ') forces. In the computation of these spectra the ratio of force constants γ'/γ was taken as 3/17, which is close to the best value for this force field representation of vitreous silica(5). The spectra of figures 4 and 5 correspond to the cases of fixed and free-end boundary conditions, respectively. In imposing the fixed end condition, we restrict the surface oxygen atoms of the model, each of which is connected to only one silicon atom, to remain stationary, i.e. each surface oxygen atom is considered as infinitely heavy; this restriction is removed and the surface oxygen atoms are allowed to vibrate when the free-end boundary condition is imposed. The spectra are depicted as histograms, since the model (of figure 1) on which the computations are based contains only a finite number of degrees of freedom: 816 for the fixed end and 1002 for the free end boundary condition. The spectra were computed by a method derived from a numerical technique used originally by

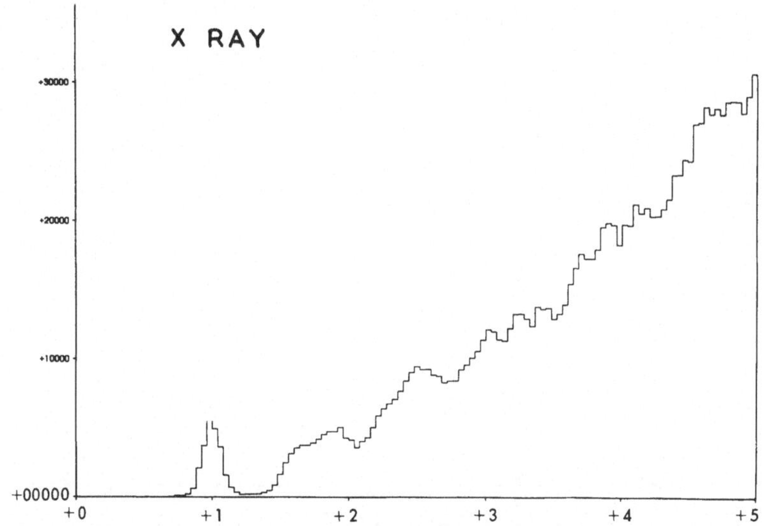

Figure 2. The calculated radial distribution function, with X-ray form factors included, for the model referred to in the text. The horizontal scale is in units of the Si-O bond length; the vertical scale is arbitrary.

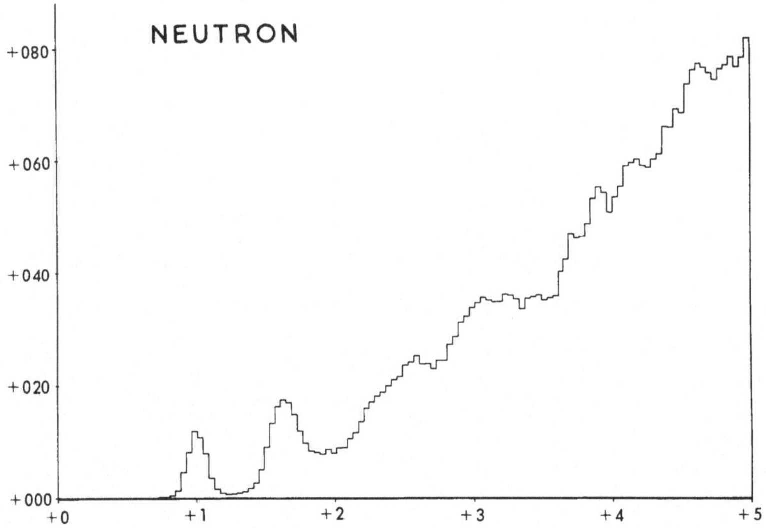

Figure 3. The calculated radial distribution function, with neutron form factors included, for the model referred to in the text. The horizontal scale is in units of the Si-O bond length; the vertical scale is arbitrary.

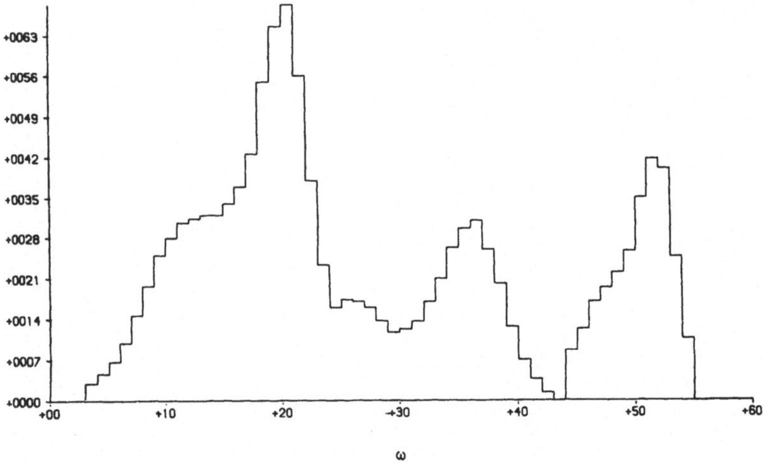

Figure 4. The computed vibrational frequency spectrum for the model described in the text with the fixed-end boundary condition imposed. On the horizontal scale each unit is approximately 22 cm^{-1}; thus the point at $\omega = 50$ corresponds to a frequency of about 1100 cm^{-1}. The units on the vertical scale refer to the number of modes in each interval of the histogram.

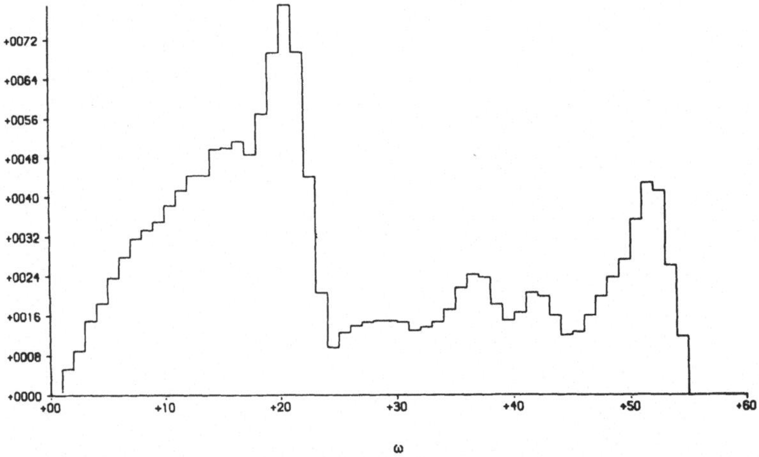

Figure 5. The computed vibrational frequency spectrum for the model described in the text with the free-end boundary condition imposed. The scales are as described for figure 4.

Dean and Bacon[6] in work on the spectra of disordered two-component lattices. Details of the method as applied to glasses will be published.

The positions of the main peaks at about 450 cm^{-1}, 800 cm^{-1} and 1100 cm^{-1} in the computed vibrational spectra of figures 4 and 5 fit fairly accurately the positions of the main absorption and scattered bands in the infrared and Raman spectra of vitreous silica (cf., for example, the composite spectra provided by Simon[7]). A study of the atomic displacement eigenvectors in our work has enabled us to interpret some features. We find that the high frequency band at about 1100 cm^{-1} is due to modes which are essentially Si-O-Si bond stretching vibrations, although these modes are not, in general, localized to just one Si-O-Si group of atoms. The band at about 800 cm^{-1} is due to modes in which oxygen atoms vibrate in the plane of the triagle formed with their neighbouring silicon atoms, the motion being perpendicular to the Si-Si line; again, these modes are not, in general, highly localized to just a few atoms. The band at about 450 cm^{-1} has, as yet, resisted a simple interpretation; the few modes we have studied from this region of the spectrum extend spatially throughout the whole model and yet have no obvious wavelike properties which identify them as simple elastic waves. Our results here for the bands at 1100 cm^{-1} and 800 cm^{-1} are supported by a consideration of the relative intensities of these bands in the infrared and Raman spectra of vitreous silica. However, we shall not discuss these details in this paper.

4. MODES DUE TO NON-BRIDGING OXYGEN ATOMS

The major difference between the spectra of figures 4 and 5 is in the region at about 900 cm^{-1}. The band gap for the model with fixed ends is replaced in the spectrum for the free-end model by a new and fairly intense band. A study of atomic displacement eigenvectors for modes in the region of the new band shows them to be due to Si-O stretching vibrations involving surface (i.e. non-bridging) oxygen atoms, and to be highly localized in character. In a typical mode at about 900 cm^{-1} one finds that, perhaps, 99% of the energy of the mode is localized in the vibrations of about three, fairly close, non-bridging oxygen atoms. The number of modes in the band at about 900 cm^{-1} is probably equal to the total number of surface oxygen atoms in the model with free ends. An exact identification of numbers cannot be made (except, possibly, by computing displacement eigenvectors for all the modes in the region from about 800 cm^{-1} to 1000 cm^{-1}), as the new band merges into the existing bands on either side.

Simon[8] has already tentatively assigned the presence of a band at about 950 cm^{-1} in the infrared spectrum of vitreous silica as due to non-bridging oxygen atoms. This band is known to be structure sensitive, i.e. its intensity varies from sample to sample of the glass and appears (as we discuss later) to depend upon the history of the particular specimen. The appearance of the band at about 900 cm^{-1} in our computed spectra, and the evidence that it is due to localized motions involving surface non-bridging oxygen atoms, lends support to Simon's conjecture. This support does depend, though, on the assumption that an interior non-bridging oxygen atom would produce a localized Si-O stretching mode similar to those at the surface of our model. This seems likely, since the local force field environment on the silicon-oxygen pair is identical in both cases out to at least several nearest-neighbour distances.

An example of the structure-sensitive nature of the band in the region 900-950 cm^{-1} of the spectrum of vitreous silica is afforded by fast neutron irradiation studies(8). If a sample of pure vitreous silica, showing little evidence of a band in this region of the infrared spectrum is subjected to a heavy dosage (of the order of 10^{20} neutrons per cm^2) of fast neutrons a well-defined vibrational band appears at about 920 cm^{-1}. It has been suggested(8) that an effect of fast neutron bombardment is the melting, and subsequent structure rearrangements, of small regions in the glass due to the generation of high temperature waves (thermal spikes) and high pressure waves. The molten regions, which may reach temperatures as high as 10^4 °K, are subsequently rapidly quenched by thermal conduction, leaving behind regions whose structure is different from the original atomic arrangement of the glass. It seems probable that the rearranged regions contain a number of non-bridging oxygen atoms which are responsible for the emergence of the new band in the infrared spectrum. Fast neutron-irradiated quartz shows a similar behaviour to vitreous silica in the emergence of a new spectral band in the region 900-950 cm^{-1}; the mechanism of formation of this band is probably similar to that for irradiated vitreous silica.

The intensity of the band in the region 900-950 cm^{-1} of the infrared spectrum of vitreous silica (and neutron-irradiated quartz) is probably proportional to the number of non-bridging oxygen atoms in the structure. Thus it is likely that the intensity of this band can be used as a gauge of the connectivity of the atomic network in any sample. It is interesting that a band emerges in this region of the spectrum if alkali oxide is added to vitreous silica. Again, this is probably due to the formation of non-bridging oxygen atoms - for the connectivity of the atomic structure of vitreous silica is believed to decrease with increasing alkali oxide content.

5. GENERAL REMARKS

One general problem associated with the study of spatially localized atomic vibrations in glasses is that the conventional method for the study of localized defect modes in solids, the Green's function method, cannot be used. This is because the Green's function matrix itself cannot, in practice, be derived or computed for an amorphous atomic structure. To obtain information on localized modes alternative methods must be used, and one such method is the kind of approach on which the work of this paper is based, i.e. the direct computation of the vibrational properties of reasonably large sample structures. Another approach is one described by Dean(9), an outline of which appears as a paper in this volume.

The study of localized modes in amorphous materials is potentially a field of considerable interest. It seems likely that such modes, perhaps induced by the presence of foreign atoms in the glass, can be used to probe local structure and force fields via the infrared spectrum.

ACKNOWLEDGMENTS

The spectra of figures 4 and 5 were computed by programs written by Mr. N. F. Bird. The work described in this paper was carried out at the National Physical Laboratory.

REFERENCES

[1] W. H. Zachariasen, J. Amer. Chem. Soc. 54, 3841 (1932); J. Chem. Phys. 3, 162 (1935).

[2] L. W. Tilton, N. B. S. J. Res. 59, 139 (1957).

[3] H. A. Robinson, J. Phys. Chem. Solids 26, 209 (1965).

[4] E. A. Porai-Koshits, in The Structure of Glass, edited by E. A. Porai-Koshits (Consultants Bureau, New York, 1966), Vol. 6, p. 3.

[5] W. A. Weyl and E. C. Marboe, The Constitution of Glasses (Interscience, New York, 1962), Vol. 1, p. 288.

[6] P. Dean and M. D. Bacon, Proc. Roy. Soc. A 283, 64 (1965).

[7] I. Simon, in Modern Aspects of the Vitreous State, edited by J. D. Mackenzie (Butterworths, London, 1960), Vol. 1, Chapter 6.

[8] I. Simon, J. Amer. Ceramic Soc. 40, 150 (1957).

[9] P. Dean, to be published.

LOCAL LATTICE INSTABILITY FOR IMPURITY CENTERS

N. Kristoffel, G. Zavt

Institute of Physics and Astronomy

Estonian S.S.R. Academy of Sciences, U.S.S.R.

1. INTRODUCTION

The asymmetric distortion of the lattice near impurity centers can occur if the electronic spectrum of the impurity contains orbitally degenerated or closely situated levels (respectively the Jahn-Teller or the Pseudo-Jahn-Teller effect)(1,2). Such a mechanism of lowering symmetry is possible for the perfect lattice as well (3,4). In particular, the Pseudo-Jahn-Teller effect can give rise to the ferroelectric-type second-order phase transitions (5-7).

For the perfect crystal, however, the "pure vibrational"mechanism of instability caused only by the nature of the forces acting in the lattice is also possible (8). In the present paper we shall show that such a type of instability leading to the asymmetric local distortion of the lattice takes place also in case of impurity centers. To sum up, it comes to the following.

The condition for the stability of the crystal lattice in the harmonic approximation is the positive definition of all squared normal mode frequencies. If the insertion of the impurity in the crystal changes its force constants in such a way that the squared frequency of some mode becomes negative, then that means the instability of the lattice around the impurity.Such a mode with a negative squared frequency is a localized one (displacements fall rapidly with distance) that makes for the localization of the effect in a small region. If the anharmonicity of vibrations is taken into account, one obtains the lattice distortion (change of equilibrium positions) instead of instability. It will be shown that the distortion is asymmetric because the appearance of instability is more probable for non-totally symmetric modes (F_{1u} in case of the substitutional impurity in $NaCl$-type lattices). This mechanism is interesting, especially as it yields asymmetric distortion in cases when the Jahn-Teller effect does not work, e.g. for the ground electronic state of the impurity.

On the other hand, some phenomena have been observed recently which are apparently caused by the mechanism under discussion. We mean the Li^+-ion in KCl which is displaced from the lattice site [9,10]. Dienes et al. [11] have made an attempt to calculate the asymmetric situation of Li^+ by using complicated expressions for the potential energy of the lattice. It seems that the present theory offers a more natural explanation of this case.

2. LOCAL LATTICE INSTABILITY

Let $\mathcal{D}'$ be the dynamical matrix of the lattice with a point imperfection, ω_j^2 are its eigenvalues, $w_\alpha(^s_l;j)$ - eigenvectors ($j=1,\cdots 3pN$) where $l=1,\cdots N$ numbers the primitive cells. $s=1,\cdots p$ distinguishes different atoms in a cell and $\alpha = x,y,z$. The lattice contains an impurity at the site $l=0$, $s=s_0$ whose mass is m'. Let us consider within classical limits the space correlation function for atomic displacements $\langle x_\alpha(^s_l)\, x_\beta(^{s'}_{l'})\rangle$ where brackets denote thermal averaging. It is easy to show (see [12]) that

$$\langle x_\alpha(^s_l)\, x_\beta(^{s'}_{l'})\rangle = \frac{kT}{(m_s m_{s'})^{1/2}}\, \mathcal{G}_{\alpha\beta}(^{ss'}_{ll'};0) . \qquad (1)$$

Here $\mathcal{G}_{\alpha\beta}(^{ss'}_{ll'};0)$ is the value at point $\omega^2=0$ of the Green function for the imperfect lattice:

$$\mathcal{G}_{\alpha\beta}(^{ss'}_{ll'};\omega^2) = \left(\frac{m_s m_{s'}}{M_s(l)M_{s'}(l)}\right)^{1/2} \sum_j \frac{w_\alpha(^s_l;j)\, w^*_\beta(^{s'}_{l'};j)}{\omega_j^2 - \omega^2} \qquad (2)$$

where $M_s(l)=\begin{cases} m' & l=0, s=s_0 \\ m_s & \text{in other cases.}\end{cases}$

The Green function (2) satisfies the Dyson equation

$$\mathcal{G} = G - GV\mathcal{G} \qquad (3)$$

with

$$G = (\mathcal{D}-\omega^2)^{-1} , \qquad (4)$$

$$V_{\alpha\beta}\binom{ss'}{ll'} = -\left(\frac{m'}{m_{s_0}} - 1\right)\omega^2 \delta_{l0}\delta_{l'0}\delta_{ss_0}\delta_{s's_0}\delta_{\alpha\beta} + \Delta\mathcal{D}_{\alpha\beta}\binom{ss'}{ll'} \qquad (5)$$

where $\mathcal{D}$ is the dynamical matrix of the ideal lattice and $\Delta\mathcal{D} = \mathcal{D}' - \mathcal{D}$.

Having distinguished the defect region A in the lattice (the region where the change of force constants takes place) one can divide the matrices in Eq. (3) into four submatrices and the Dyson equation into four equations respectively. The solution of one of them belonging to region A is

$$\mathcal{G}^{AA} = (I + G^{AA}V)^{-1} G^{AA} . \qquad (6)$$

We shall now use the method described in ([12]). By introducing the eigenvalues $\Omega(\omega^2)$ and the orthonormalized right (ξ) and left (η) eigenvectors of the matrix $-G^{AA}V$ and by classifying them according to the irreducible representations of the site group created by the impurity center, one can rewrite Eq.(6) in the following form

$$\mathcal{G}_{\alpha\beta}\binom{ss'}{ll'};\omega^2) = \sum_{\mu\kappa i}\sum_{l''s''\gamma} \frac{\xi_\alpha^{\mu\kappa i}\binom{s}{l}\eta_\gamma^{\mu\kappa i}\binom{s''}{l''}G_{\gamma\beta}\left(\binom{s''s'}{l''l'};\omega^2\right)}{1 - \Omega_{\mu\kappa}(\omega^2)} \qquad (6a)$$

$$(l,s),(l',s') \in A$$

where μ is the type of irreducible representations, κ distinguishes the repeated representations and i - the lines of the degenerated representation.

Finally we introduce symmetrized combinations of displacements in space A

$$\varphi_{\mu\kappa i} = \sum_{ls\alpha} \xi_\alpha^{\mu\kappa i}\binom{s}{l} x_\alpha\binom{s}{l} \qquad (7)$$

and calculate the function $\langle|\varphi|^2\rangle$. Using (1) and (6a) one obtains

$$\langle|\varphi_{\mu\kappa i}|^2\rangle = \kappa T \frac{C_{\mu\kappa i}(0)}{1 - \Omega_{\mu\kappa}(0)} \qquad (8)$$

where

$$C_{\mu\kappa i}(\omega^2) = \sum_{ls\alpha}\sum_{l's'\alpha'} \eta_\alpha^{\mu\kappa i}\binom{s}{l} G_{\alpha\beta}\left(\binom{ss'}{ll'};\omega^2\right) . \qquad (9)$$

Since in the absence of the perturbation $\Omega_{\mu\kappa} = 0$, the quantities $C_{\mu\kappa i}$ represent the thermal average of the squared linear combination formed by the displacements of the ideal lattice and therefore $C_{\mu\kappa i} > 0$. Hence the condition $1 - \Omega_{\mu\kappa}(0) > 0$ must be fulfilled for the impure lattice for all μ and κ. Now we shall show that the opposite condition

$$1 - \mathcal{R}_{\mu\kappa}(0) < 0 \qquad (10)$$

is the condition for the local lattice instability. Indeed, the equation determining the frequencies of localized modes is $1 - \mathcal{R}_{\mu\kappa}(\omega^2) = 0$. Instability appears if the equation is satisfied at $\omega^2 < 0$. The quantities $\mathcal{R}(\omega^2)$ diminish evenly if ω^2 moves away from the lower boundary of the phonon spectrum ($\omega = 0$) in the negative direction (this is a consequence of the properties of the Green functions (4)). Hence, if (10) is satisfied, $\mathcal{R}(\omega^2)$ becomes inevitably equal to unity at some $\omega^2 < 0$ which means the appearance of the localized mode with imaginary frequency. Thus, the inequality (10) is the necessary and sufficient condition for creating instability over the given (μ, κ) mode. Since $\mathcal{R}_{\mu\kappa}(0)$ is the function of the parameters describing the change of force constants, (10) shows how great this change must be in order to make the lattice instable. Such an instability is localized in a small region near the impurity because any effect of imperfection decreases rapidly with distance if ω^2 lies outside the phonon bands.

3. CONDITIONS FOR THE APPEARANCE AND THE NATURE OF INSTABILITY

Using the results of our earlier paper ([12]), we have investigated the conditions for the appearance of instability in the case of a KCl lattice containing a monovalent impurity at the cation site. It is assumed that only short-range repulsive forces between the impurity and its nearest neighbors are changed. Thus the matrix $\Delta\mathcal{D}$ in Eq. (5) contains only two parameters which we denote by

$$P_1 = \frac{\mathcal{D}'_{xx}(100)}{\mathcal{D}_{xx}(100)}, \qquad P_2 = \frac{\mathcal{D}'_{xx}(010)}{\mathcal{D}_{xx}(010)}$$

where $\mathcal{D}$ and $\mathcal{D}'$ are assumed to include both the repulsive and Coulomb forces.

The normal modes of vibrations of the impure lattice are classified according to the following irreducible representations of the O_h group: A_{1g}, E_g, F_{2g}, F_{1g}, F_{2u}, $3F_{1u}$. It has been found that for the appearance of instability over the former five types of vibrations (excluding $3F_{1u}$) constants P_1 or P_2 must be negative and unreally great (e.g. for A_{1g} mode P_1 must be less than -1.65). Hence the only real possibility is connected with F_{1u} modes. There are three sets of F_{1u} vibrations and thus one obtains a third-order algebraic equation with respect to $\mathfrak{R}(F_{1u})$. By solving this equation we have found that only one type of F_{1u} vibrations can produce instability. Vector ξ_z for one of these vibrations (we call it quasioptical and denote by $F_{1u}^{(1)}$; two other vibrations correspond to the preferred x and y -axes) has the form shown in Fig.1. The values of δ_1/δ and δ_2/δ (see Fig.1) depend upon ω^2 but for all $\omega^2<0$ they are considerably smaller than unity.

The result of the investigation of conditions under which the inequality (10) is fulfilled for $F_{1u}^{(1)}$ vibrations is presented in Fig.2. The lattice becomes instable over these modes for all P_1 and P_2 lying below the plotted curve . The curve itself is in conformity with Eq. $\mathfrak{R}(F_{1u}^{(1)})=1$, which in accordance with (8) corresponds to infinitely great mean-square displacements: the force affecting this combination of displacements disappears. It follows from Fig.2 that the conditions for the appearance of instability are most favorable if P_1 and P_2 are small and if at least one of them is negative. Such a situation can be realized if short-range repulsive forces between the impurity(which is considered to be situated from the beginning in the lattice site) and its nearest neighbors become very small. Under such conditions the Coulomb interactions make a predominant contribution to the above-mentioned force constants, that corresponds to $P_1<0$ and small absolute values of P_1 and P_2 .

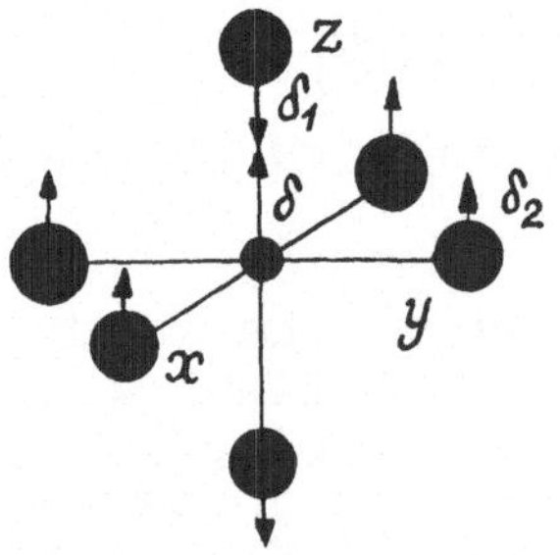

Fig.1. The quasioptical F_{1u}-vibration in the defect region.

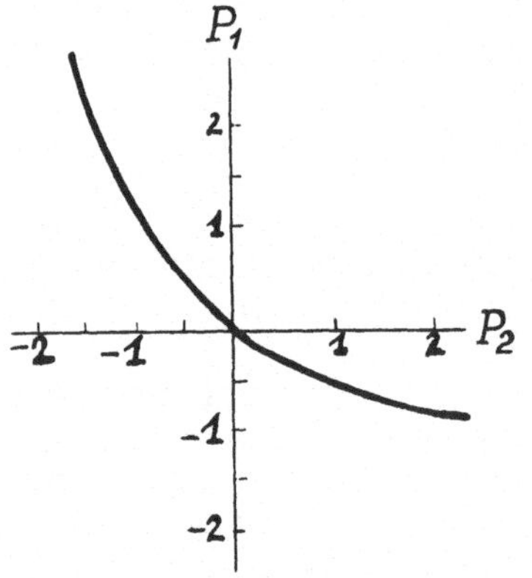

Fig.2. The instability region (below curve)for impurity parameters.

One can illustrate the nature of the mechanism of instability under discussion by considering the typical ionic interaction potential for the one-dimensional lattice. Let R_0 be the point of the minimum of the potential, R_1 - the point of inflexion; usually $R_1 \sim 1{,}3R_0$. Instability appears if distance R' between the impurity which is situated in the lattice site and its nearest neighbors is found to be greater than R_1. In such a case the force acting on the combination of displacements of type (8) is not restoring. The Coulomb forces make a contribution of order 10 per cent to force constants for distance R_0. Thus, for the points $R' > R_0$, the short-range forces must be ~ 5 per cent of the value in a perfect lattice. The examined mechanism apparently explains the reason for instability in the case of the Li^+ impurity in a KCl matrix ([9,10]). Indeed, it was found in the quantum-mechanical calculation ([13]) that the part of the interaction energy between Li^+ and the nearest Cl^- corresponding to the overlapping of their electronic clouds is practically equal to zero at distances of order R_0.

4. STABILIZATION OF IMPURITY CENTER IN LOW-SYMMETRY CONFIGURATIONS

For determining the real configuration of the lattice which is instable in the harmonic approximation it is necessary to take into consideration anharmonic terms in the potential energy. Now we shall consider in a rather approximate way the possible low-symmetry configurations of the cubic lattice which is assumed to be instable over quasioptical F_{1u} vibrations. Since the amplitude of these vibrations falls very rapidly with increasing distance from the impurity, one can restrict the lattice relaxation to the defect region (impurity + its nearest neighbors), assuming the rest of the ions to be fixed in their sites. Then the normal coordinates of $F_{1u}^{(1)}$ vibrations (we denote them by $y_i, i=1,2,3$) have the form shown in Fig.1. Let us single out the part of the potential energy of the impure lattice depending on the coordinates of $F_{1u}^{(1)}$ vibrations, which with an accuracy up to fourth-order anharmonic terms is

$$\mathcal{U} = -\frac{\varkappa}{2}(y_1^2+y_2^2+y_3^2) + A(y_1^4+y_2^4+y_3^4) + B(y_1^2y_2^2+y_1^2y_3^2+y_2^2y_3^2)\,. \quad (11)$$

Here $\varkappa = |\omega^2(F_{1u}^{(1)})|$ is the root of Eq. $1 = \mathcal{R}_{F_{1u}}(\omega^2)$ in the region of instability. Generally speaking, the expression (11) must additionally include the third-order terms of the form $y_i^2 y_\mu$, where μ belongs to A_{1g}, E_g or F_{2g} vibrations. These terms may seem small in comparison with the usual third-order anharmonic terms. This is connected with the fact that the predominant part of the displacement in $F_{1u}^{(1)}$ vibrations falls on the impurity itself while at least one of the vibrations constituting the third-order terms must be even (i.e. of a type in which the impurity itself is at rest). The investigation shows that the potential (11) can possess minima leading to the low-symmetry distortion of the impurity center in case of the following configurations of the ions (y_{i0}).

(1) Eight equivalent configurations of the type $|y_{10}|=|y_{20}| =|y_{30}|$ where the center is distorted along the third-order axes (the impurity itself is displaced in that direction). The symmetry is reduced from O_h to C_{3v}. In this case

$$y_{i0} = \pm\frac{1}{2}\left(\frac{\varkappa}{A+B}\right)^{1/2} \tag{12}$$

$$\mathcal{U}_{111} \equiv \mathcal{U}(y_{i0}) = -\frac{3\varkappa^2}{16(A+B)} . \tag{13}$$

(2) Six equivalent configurations of the type $y_{10} \neq 0$, $y_{20}=y_{30}=0$ where the center is distorted along the fourth-order axes (symmetry C_{4v}). Now

$$y_{i0} = \pm\frac{1}{2}\left(\frac{\varkappa}{A}\right)^{1/2} \tag{14}$$

$$\mathcal{U}_{100} = -\varkappa^2/16A . \tag{15}$$

(3) Twelve equivalent configurations of the type $|y_{10}| = |y_{20}| \neq 0, y_{30}=0$ where the center is distorted along the second-order axes with symmetry C_{2v}. Now

$$y_{i0} = \pm\frac{1}{2}\left(\frac{\varkappa}{2A+B}\right)^{1/2} \tag{16}$$

$$\mathcal{U}_{110} = -\frac{\varkappa^2}{4(2A+B)} . \tag{17}$$

We shall now expand $\mathcal{U}$ around the new extrema up to quadratic terms and diagonalize the obtained quadratic forms. We have found that $\mathcal{U}_{111}$ may be the minimum if $A>0$, $B<0$ and $A>|B|$; $A>0$, $B>0$ and $2A>B$; $A<0$, $B>0$ and $B>|2A|$, $\mathcal{U}_{100}$ is the minimum at $A>0$, $B>0$ and $B>2A$ while $\mathcal{U}_{110}$ may be the minimum only if $B=2A$. In all cases quantity $\mathcal{U}(y_{i0})$ yields a gain in energy (due to the lattice distortion).

Thus, depending on concrete conditions, the impurity center can become stable in low-symmetry configurations of a definite type with destroyed central symmetry and the corresponding static electric momentum is created. The characteristic feature of the centers under consideration seems to be $A>0$, $B<0$, $A>|B|$, where symmetry is lowered to C_{3v}. For KCl-Li it is found ([10]) that Li^+ is displaced namely along the directions of the type ⟨111⟩. Quantity $\mathcal{U}_{111}$ has been estimated at an order of some hundreths eV; that is a typical value for the asymmetrical distortion of impurity centers. For that reason the effects of instability are essential only at relatively low temperatures. An increase of temperature favors the transitions of the center between the equivalent minima of $\mathcal{U}$. The barrier for the transition from the definite $\mathcal{U}_{111}$ to the other minimum on the opposite side of the same axis is found to be higher than for the transition to the neighboring minimum. For that reason the "rotational" mechanism is preferable to the "vibrational" one in the relaxation of the dipole momentum (see also([14,15])).

1. U. Öpik, M. Pryce, Proc. Roy. Soc.(London) A238, 423 (1957).
2. N. Kristoffel, Opt. and Spectr. 9, 615 (1960).
3. J. Birman, Phys. Rev. 125,1959 (1962);127, 1093 (1962).
4. N.N. Kristoffel, Fiz. Tverdogo Tela 6, 3266 (1964).
5. G. Shukla, K. Sinha, J. Phys. Chem. Solids 27, 1837 (1966).
6. I.B. Bersuker, Phys. Lett. 20, 589 (1966).
7. N. Kristoffel, P. Konsin, phys. stat. sol. 21, K39 (1967).
8. M. Born, K. Huang, Dynamical Theory of Crystal Lattices (Moscow) (1958).
9. G. Lombardo, R. Pohl, Phys. Rev. Lett. 15,291(1965).
10. N. Byer, H. Sack, Phys. Rev. Lett. 17, 72 (1966).
11. G. Dienes, R. Hatcher, R. Smoluchowski, W. Wilson, Phys. Rev. Lett. 16, 25(1966).
12. G.S. Zavt, N.N. Kristoffel, Fiz. Tverdogo Tela 8, 2271 (1966).
13. A.K. Godkalns, Trudy Inst. Fiz. i Astron. Akad. Nauk E.S.S.R. N20, 148 (1963).
14. S. Bowen, M. Gomez, J. Krumhansl, J. Matthew, Phys. Rev. Lett. 16, 1105 (1966).
15. M. Baur, W. Salzman, Phys. Rev. 151, 710 (1966).

PART C. DEFECT MODES IN SEMICONDUCTORS

INFRARED ABSORPTION STUDIES OF LOCALIZED VIBRATIONAL MODES IN SEMICONDUCTORS

W. Hayes

Clarendon Laboratory

University of Oxford

INTRODUCTION

There has been considerable general interest recently in the lattice dynamics of crystals containing defects. The most detailed experimental information concerning the dynamics of imperfect lattices has been obtained by infrared spectroscopic methods. The introduction of point defects into ionic crystals may activate two types of infrared vibrational absorption. (a) Localized mode absorption arising from the vibrations of light impurities. This type of absorption occurs outside the regions of band mode vibration and gives rise to sharp absorption lines. The amplitude of a localized vibrational mode is large near the defect and dies away rapidly with distance from the defect. (b) Resonant mode absorption which may be activated by all impurities. This type of absorption gives rise in general to broader peaks within the region of the band modes of the host crystal. The amplitude of resonance modes is enhanced near the defect but these modes may be transmitted through the lattice and closely resemble unperturbed lattice modes at large distances from the defect. For a recent review of work on local and resonance modes in ionic crystals see, for example, Maradudin.(1)

We shall confine our attention here to the results of experimental investigations of localized vibrations of light impurities in semiconductors at frequencies higher than the highest band mode vibrations. For the investigation of these impurities by infrared methods large concentrations ($\sim 10^{18}$/cc) of impurities are required. If these impurities are electrically active the associated free carrier absorption makes infrared investigations impracticable. The problem has been dealt with in silicon by counter doping.(2)

Although counter doping is not usually fully effective it is possible, in silicon, to produce the additional necessary compensation by electron irradiation. In the III-V and II-VI compounds work has, in general, been confined to impurities which do not generate free carriers. The system GaAs:Li is rather unusual;(3) although Li is electrically active in GaAs it is self-compensating and may also be readily used to closely compensate either donor or acceptor doped GaAs. The investigation of local modes in homopolar crystals such as silicon is in some respects easier than in III-V and II-VI compounds because of the intense restrahl absorption in the latter compounds.

We shall divide our description of experimental investigations into two parts. In Part I we shall describe the vibrations of isolated point defects and in Part II the vibrations of paired defects.

We have added an Appendix which contains some remarks on the experimental comparison of anharmonicity in local modes and band modes.

EXPERIMENTAL RESULTS

I. Isolated Point Defects

Recent work on the localized vibrations of Be ions in CdTe(4) is illustrative of many of the properties of localized vibrational modes in semiconductors and will be described first. We expect Be to dissolve in Cd sites and since the valence configuration of Be matches that of Cd no charge carrier problems arise. When considering the localized vibrations of Be we shall first, as an approximation, consider the host lattice to be static and fit the localized mode spectrum at a fixed temperature to the Hamiltonian(5)

$$H = \frac{1}{2m}(p^2 + m^2\Omega^2 r^2) + B\,xyz + C_1(x^4 + y^4 + z^4) + C_2(x^2y^2 + y^2z^2 + z^2x^2) + \ldots \quad (1)$$

in which the potential has been expanded as a power series in the displacement of the Be ion. The expansion of the potential reflects the T_d point symmetry of a Cd site. The term in B is a consequence of the lack of inversion symmetry associated with T_d. The energy levels of this Hamiltonian up to the third harmonic ($\nu = 3$) are given in Figure 1. Electric dipole transitions are allowed in T_d from the singly degenerate ground state ($\nu = 0$) to only the Γ_5 triply degenerate excited states. It is apparent that the fundamental, one second harmonic and two third harmonics are symmetry allowed and all four transitions of Be in CdTe have been observed. The positions of these transitions at 4°K are given in Figure 1.

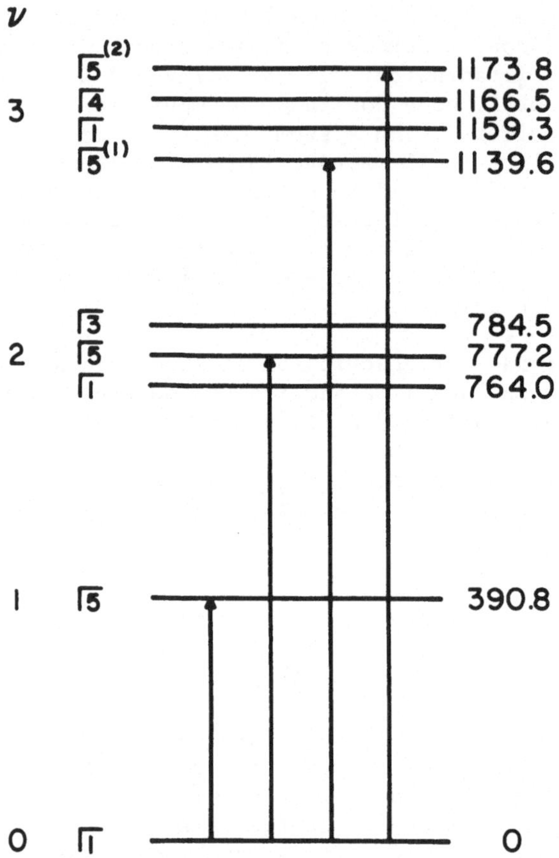

Figure 1. Energy level diagram for localized vibrational modes of Be in CdTe. The observed transitions (cm^{-1}) are indicated. The positions of the remaining levels have been calculated using (1) and (2) (text).

These provide sufficient information to determine the parameters of eq. (1) and we find

$$\begin{aligned} \Omega/C &= 399.1 \text{ cm}^{-1} \\ /B/ &= 2.30\times10^{13} \text{ ergs/cm}^3 \\ C_1 &= -2.97\times10^{20} \text{ ergs/cm}^4 \\ C_2 &= 4.47\times10^{19} \text{ ergs/cm}^4 \end{aligned} \qquad (2)$$

Using (2) we can calculate the positions of the remaining levels in Figure 1 and these are indicated.

The higher harmonics are observable because of anharmonicity. For example, the term in B in (1) gives a contribution to the intensity of the second harmonic through the matrix element

$$\langle 000/x/011\rangle = \frac{\langle 000/Bxyz/111\rangle\langle 111/x/011\rangle}{E_o - E_3} + \frac{\langle 000/x/100\rangle\langle 100/Bxyz/011\rangle}{E_2 - E_1} \qquad (3)$$

Using (3) and (2) we calculate the ratio of the oscillator strengths of the fundamental and second harmonic to be

$$\frac{f_\omega}{f_{2\omega}} = 130$$

which may be compared with the observed ratio of 50. This discrepancy arises from our neglect so far of coupling of the local oscillator to the lattice. The Hamiltonian (1) should also include terms of the form[5]

$$H_3 = \sum_{\underset{\sim}{k}} \sum_{xyz} \Phi(\underset{\sim}{k})\,(b_x + b_x^+)^2\,[a(\underset{\sim}{k}) + a^+(\underset{\sim}{k})] \qquad (4)$$

$$H_4 = \sum_{\underset{\sim}{k}\underset{\sim}{k}'} \sum_{xyz} \Delta(\underset{\sim}{k},\underset{\sim}{k}')\,(b_x + b_x^+)^2\,[a(\underset{\sim}{k}) + a^+(\underset{\sim}{k})][a(\underset{\sim}{k}') + a^+(\underset{\sim}{k}')]; \qquad (5)$$

these expressions have been written in terms of the creation and destruction operators b^+ and b which refer to local modes and the operators a^+ and a which refer to band modes. The parameters Φ and Δ have been discussed in some detail.[5]

It is possible in CdTe, which has a degree of ionicity, to excite the second harmonic of the local mode, for example, by virtual excitation with light of a restrahl phonon which then decays through H_3 (4) into two local modes. The effective dipole matrix element contains terms of the form[5]

$$\frac{\langle n_R, 0/\mu/n_R + 1, 0\rangle\langle n_R + 1, 0/H_3/n_R, 2\rangle}{\hbar(\omega_R - 2\,\Omega)} \qquad (6)$$

where the states $/n_R, \nu)$ represent the excitation of restrahl and local phonons. We also expect a contribution to the higher harmonic intensities through the electron-local mode interaction. This may be described as the virtual excitation of an exciton which then decays into two local modes and the mechanism may be formally described in a manner similar to (6). Preliminary approximate calculations using the results of stress experiments on excitons in CdTe[6] suggest that the contribution of this mechanism to the intensity of the Be second harmonic is comparable with that of (3). It seems likely that admixture of electronic states into the local mode states gives a greater contribution to the intensity of higher harmonics than the admixture of restrahl states (see also ref. (5)).

Experimentally it is found that the anharmonic coupling to the lattice gives a measurable change of the peak position of the localized vibrational lines with temperature. For example, the position of the peak position of the second harmonic of Be in CdTe fits the expression

$$\omega = \omega_o - \beta T \qquad (7)$$

in the range 77 to 300°K with $\beta = 0.041\ cm^{-1}\ °K^{-1}$. This linear temperature dependence may be accounted for[5] using (5). Other effects arising from anharmonicity are changes of width of the localized vibrational lines with temperature and the observation of sidebands on the localized vibrational lines due to simultaneous excitation of localized and lattice band modes.[5] These effects are at present under investigation in the case of CdTe:Be.[4]

Preliminary measurements on localized modes of Li in CdS and ZnSe have recently been reported.[8]

Localized vibrations of B in Si have been extensively investigated.[2] Boron is an acceptor in Si and was partly compensated by the addition of P donors; final compensation was achieved by electron irradiation. Boron has two naturally occurring isotopes, ^{10}B and ^{11}B, with relative abundances of 18.4 and 81.6%. The fundamental and second harmonics of the two isolated isotopes have been found[2] (Table I) and the intensity ratio is 50. The ratio of the energies (Table I) is 1.04 which agrees with the value of 1.04 calculated for a high degree of localization.

Table I. Localized modes of isolated impurities in silicon and germanium.

Impurity	Crystal	Frequency cm^{-1} Fund.	2nd Harm.	Temp. °K	Ref.
^{10}B	Si	644	1284	290	2
^{11}B	Si	620	1238	290	2
^{12}C	Si	611	1217	77	11
^{13}C	Si	590	1175	77	11
^{14}C	Si	573	-	77	11
Si^{+}	Ge	389	-	300	13

+ The position of the local mode of silicon varies with silicon concentration; the value quoted is for 1% Si.

The theory of Dawber and Elliott,[9][10] which assumes no change in force constants for the impurity ion, may be used to calculate the local mode frequency of B in Si. The local mode frequency $\omega_L = \sqrt{z}$ is obtained from the expression[9][10]

$$1 + \epsilon Z \int \frac{\nu(\mu) d\mu}{\mu - Z} = 0 \qquad (8)$$

where $\epsilon = \frac{M-M'}{M}$ is the mass defect; $\nu(\mu)$ is the density of unperturbed band modes per unit range of $\mu = \omega^2$ normalized so that

$$\int_0^{\mu_m} \nu(\mu) d\mu = 1$$

where $\sqrt{\mu_m} = \omega_m$ is the maximum frequency of the unperturbed lattice. Using (8) and a phonon spectrum of Si calculated by F.A. Johnson values of 660 and 686 cm^{-1} have been calculated[2] for the local modes of ^{11}B and ^{10}B. These values are a few per cent higher than the measured values (Table I) suggesting a small reduction (~ 10%) of the local force constants.

Local modes of carbon isotopes in Si have been measured by Newman and Willis[11] (Table I). Calculation of the local mode frequencies[11] again suggests a reduction of the local force constants of 10%. Since C has the same valence structure as Si the C atom does not create free carriers. The coupling of the local mode to the light arises in this case presumably from transfer of charge between the carbon atom and its silicon neighbors.[12] The fundamental of the local mode of Si in Ge has been measured by Raman methods[13] (Table I).

The III-V compounds have received considerable attention recently. Local modes of Al and P in GaSb were reported by Hayes[14] (Table II). The Al impurities replace Ga ions and the P impurities replace Sb ions and since the impurities have the same valence structures as the ions they replace no free carrier problems are created by the impurities. The fundamental and second harmonic of Al in InSb have been measured[15][16] (Table II). The fundamental and second harmonic of both Al and P in GaAs have also been measured[17][18] (Table II); in addition, measurements of the two third harmonics of P in GaAs have been reported[16] (Table II) although some doubt has been expressed about the measured position of the weaker third harmonic.[16] The calculation of the frequencies of these local modes has been hampered by lack of detailed knowledge of the phonon spectra of the host crystals. However, approximate calculations suggest that the changes in force constants for the impurities in III-V compounds are only a few percent (see e.g., ref. 15).

Localized modes of isolated interstitial Li^+ (donor) ions and substitutional Li^- (acceptor) ions in GaAs have been measured[13] (Table II). A theoretical investigation of the vibrations of isolated interstitial Li^+ ions in Si has been carried out;[19][20] there is, however, no firm experimental evidence for the spectrum of this entity (see Part II).

Table II. Localized vibrations of isolated impurities in III-V compounds at 77°K

Impurity	Crystal	Frequency cm^{-1} Fund.	2nd Harm.	3rd Harm.	Ref.
Al_{Ga}	GaSb	316.7	-	-	14
P_{Sb}	GaSb	324.0	-	-	14
Al_{In}	InSb	295.7	590.6	-	15,16
P_{As}	GaAs	355.4	709.7	1058 1092	16,17,18
Al_{Ga}	GaAs	362	722	-	17
$^{7}Li^{+}_{int.}$	GaAs	380.0	-	-	3
$^{6}Li^{+}_{int.}$	GaAs	407.2	-	-	3
$^{7}Li^{-}_{subst.}$	GaAs	422.2	-	-	3
$^{6}Li^{-}_{subst.}$	GaAs	451.4	-	-	3
Si_{Ga}	GaAs	384	-	-	21
Si_{As}	GaAs	399	-	-	21

Silicon dissolves in GaAs as both a donor (replacing Ga) and an acceptor (replacing As) and the local modes of the two entities have been observed(21) (Table II). The Si doped crystals were compensated with Li or Cu and the Li impurity gives rise to additional complex spectra (see Part II).

II. Paired Defects

When defects are incorporated into crystals in sufficiently high concentration complexing of the defects may be expected to occur. When compensation of impurities in semiconductors is achieved by counter doping, electrostatic effects will enhance the pairing of impurities of opposite charge. Lithium dissolves in many semiconductors as a highly mobile interstitial ion and pairs readily with impurity ions of opposite charge.(3)(22) Elastic deformation may also effect the pairing of defects.

A recent discussion of the theory of the vibrations of paired impurities has been given by Elliott and Pfeuty.(23) In particular, they consider two impurities of mass defect ϵ_1 and ϵ_2 on neighboring substitutional sites in the Si lattice. The site symmetry is C_{3V} and the mechanical representation for displacement vectors of the two defect atoms is

$$2\Gamma_1 + 2\Gamma_3 \tag{9}$$

corresponding to two singlet vibrations along the bond axis (Γ_1) and two doublet vibrations perpendicular to the bond axis (Γ_3). In effect, the two threefold degenerate representations (Γ_5; see Part I) of the isolated impurities have been split into two singlets and two doublets and transitions to all four states from the Γ_1 ground state are observable.

Localized mode frequencies for $^{10}B(\epsilon_1 = 0.644)$ paired in silicon with impurities of varying ϵ_2 are plotted in Figure 2. The force constants for the pairs are assumed to be the same as for lattice ions. It is apparent that if ϵ_2 is sufficiently large to produce a local mode (> 0.15) there are four local modes corresponding to (9). The splitting of the local modes is largest when $\epsilon_2 = \epsilon_1$.

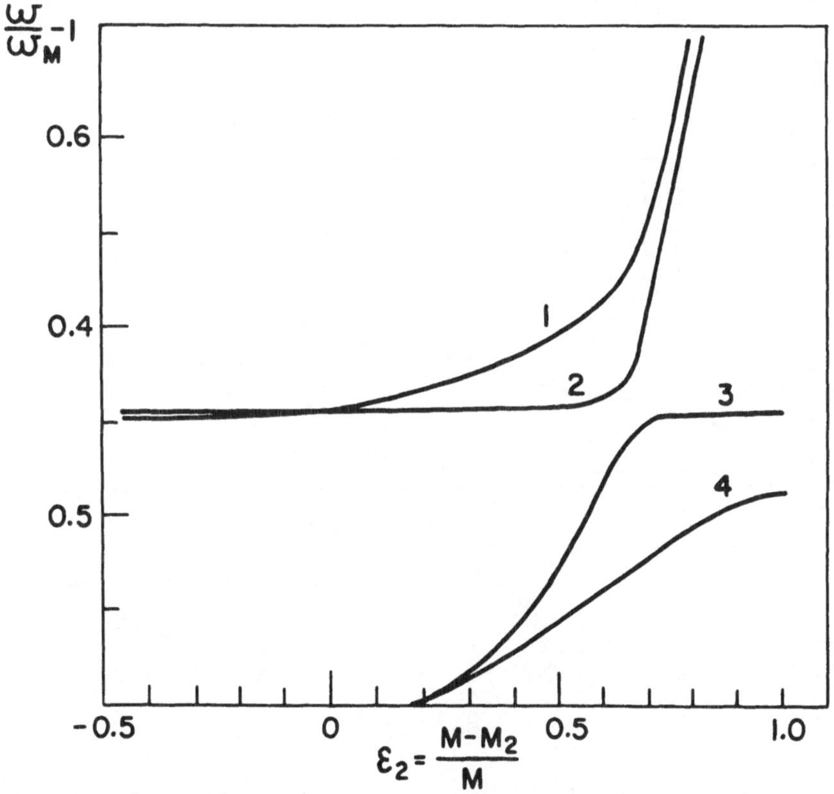

Figure 2. Calculated localized mode frequencies, in terms of the maximum lattice frequency ω_m, for $^{10}B(\epsilon_1 = 0.644)$ paired in silicon with impurities of varying ϵ_2. The curves 1 and 4 represent transitions to Γ_1 states and the curves 2 and 3 to Γ_3 states (see (9), text) (after Elliott and Pfeuty[23]).

The intensities of local modes for $^{11}B(\epsilon_1 = 0.608)$ paired in silicon with impurities of varying ϵ_2 are given in Figure 3; it is assumed that there are no force constant changes and that the impurity charges are of opposite sign. It is apparent that for $\epsilon_1 = \epsilon_2$ the low frequency modes corresponding to oscillation in phase have zero intensity; for impurities with the same charge and mass defect (e.g., boron pairs) the high frequency modes, corresponding to oscillation out of phase, would have zero intensity. When the defect has inversion symmetry, as far as the vibrations are concerned ($\epsilon_1 = \epsilon_2$), the point group is D_{3d} and the vibrational representations are

$$\Gamma_1^+ + \Gamma_1^- + \Gamma_3^+ + \Gamma_3^- \qquad (10)$$

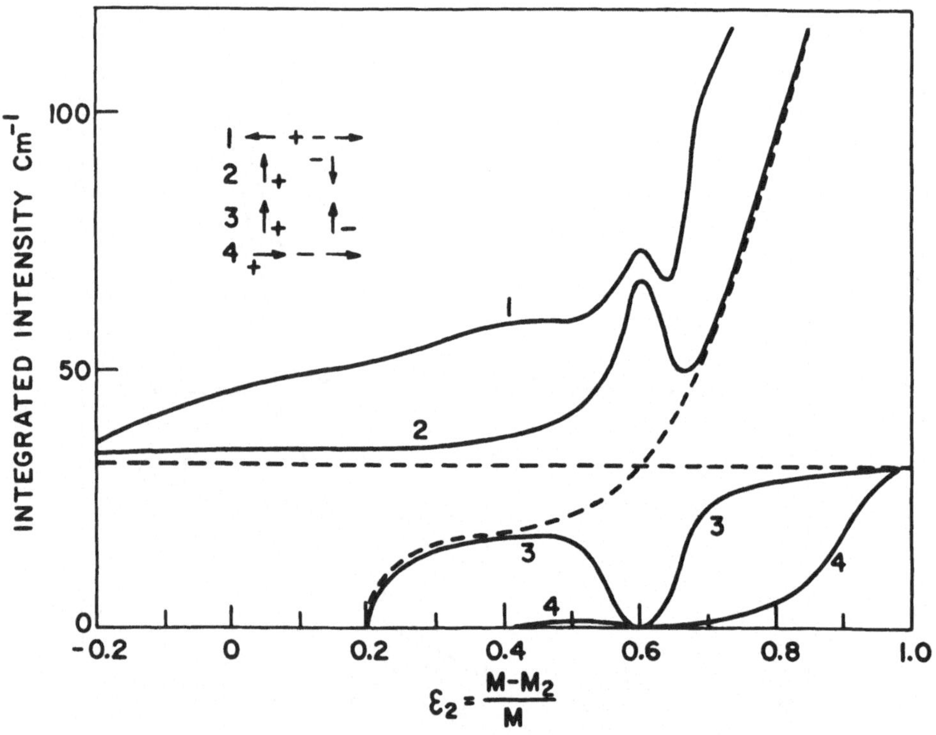

Figure 3. Calculated intensities of localized modes for $^{11}B(\epsilon_1 = 0.608)$ paired in silicon with oppositely charged impurities of varying ϵ_2. It has been assumed that the Γ_3 modes (curves 2 and 3) are singly degenerate for convenience in the comparison of intensities. Dashed curves show intensities of isolated impurities (after Elliott and Pfeuty[23]).

The parity restriction on electric dipole transitions now reduces the number of observable transitions from four to two. When the charges on the impurities are unlike only the high frequency Γ_i^+ modes are optically active (Figure 3) and when the charges are the same only the low frequency, Γ_i^-, modes are optically active.

Figure 3 shows that for $\epsilon_2 > \epsilon_1$ the high frequency modes are determined predominantly by the light mass motion and for small ϵ_2 the low frequency modes disappear into the band and the high frequency modes are determined predominantly by the ^{11}B motion. It is also apparent from Figure 3 that the transverse modes couple strongly to each other only near $\epsilon_2 = \epsilon_1$ whereas the longitudinal modes couple over a greater range.

Newman and Smith[(24)] have recently reported the observation of localized vibrational lines in silicon at 570.0, 560.0 and 552 cm^{-1} which they have attributed to ^{10}B - ^{10}B, ^{10}B - ^{11}B and ^{11}B - ^{11}B pairs. The intensity ratios of the lines are in good agreement with those calculated from isotope abundances. These lines represent the low frequency singlet (curve 4) of Figure 2; for boron pairs of equal mass the excited state involved is the optically active Γ_1^- state. The frequencies calculated[(23)] for the ^{10}B and ^{11}B pairs are 590 and 572 cm^{-1} suggesting a small change in force constants. The low frequency doublets for the pairs have not been found; this may be due to the fact that the weak lines (curve 3, Figure 2) are too close to the intense fundamentals of the unpaired impurities (Table I). Additional weak lines have been observed in boron doped silicon compensated with P and it has been suggested[(2)(23)] that there may be due to complexes involving B and P.

In mixed crystals of Si and Ge a Raman active line has been found varying from 389 to 402 cm^{-1} in crystals whose silicon concentration varies from 1 to 33%; this has been allocated to the local mode of Si (Table I). In the high concentration samples two additional weak Raman lines appear[(13)] at 476 and ~ 448 cm^{-1} with an intensity ratio of 2:1; these presumably arise from the Raman active Γ_3^+ and Γ_1^+ modes of nearest neighbor Si pairs.

The splitting of a localized mode of substitutional Li^- in GaAs (Table I) due to a nearby compensated heavy donor (Te^+) has been observed by Hayes.[(3)]

Lithium diffuses interstitially in Si and forms nearest neighbor pairs with substitutional B.[(25)] When Si:B crystals are compensated with Li, local mode spectra arising from pairing of the various isotopes are observed. If we assume that the substitutional-interstitial B-Li pair has C_{3v} symmetry we conclude that two Γ_1 and two Γ_3 modes should be observable for each pair of isotopes (see Figures 2 and 3). This system has been investigated in a number of laboratories.[26][27][28] A list of the measured local mode lines is given in Table III.

Table III. Energies of localized vibrational lines of B-Li pairs in silicon at 290°K.

Isotopes	Energy (cm^{-1})	Transition
^{10}B - ^{6}Li	683	$\Gamma_1 \rightarrow \Gamma_3$
^{10}B - ^{7}Li	681	$\rightarrow \Gamma_3$
^{11}B - ^{6}Li	657	$\rightarrow \Gamma_3$
^{11}B - ^{7}Li	655	$\rightarrow \Gamma_3$
^{10}B - ^{6}Li	584	$\rightarrow \Gamma_1$
^{10}B - ^{7}Li	584	$\rightarrow \Gamma_1$
^{11}B - ^{6}Li	564	$\rightarrow \Gamma_1$
^{11}B - ^{7}Li	564	$\rightarrow \Gamma_1$
$^{10,11}B$ - ^{6}Li	534	?
$^{10,11}B$ - ^{7}Li	522	?

It is apparent that the lines from 564 cm^{-1} to higher energies are dominated by motion of the B since changing the Li isotopes does not alter the energy appreciably. The Γ_3 lines (assigned by their greater intensity) have a higher frequency than the Γ_1 lines indicating that the force constants affecting motion perpendicular to the pair axis are greater than parallel to the axis.

The lines at 534 and 522 cm^{-1} (Table III) are determined predominantly by motion of Li since changing the B isotopes does not affect their position. We expect to see an additional line for each Li isotope but this has not been observed. The detailed interpretation of the lines at 534 and 522 cm^{-1} is therefore uncertain (see also Part I). Possible explanations of the failure to observe the additional Li lines have been considered by Balkanski and Nazarewicz.[28] A knowledge of the degeneracy associated with these lines would be helpful and this may be obtainable by application of stress.[7]

The binding energy of Li-B pairs in Si has been investigated by Spitzer and Waldner[29] by measuring intensity changes with temperature in the local mode spectrum and they obtain a value of 0.39 ev. Infrared absorption bands have been observed at overtone frequencies of the local modes of Li-B in Si and at summation frequencies of the two strongest modes with lattice phonons.[30]

Local modes arising from paired Li and Si ions in GaAs have also been investigated.[21]

Oxygen impurity in silicon[31] gives rise to infrared absorption[32] with some of the characteristics of a local mode. It has been assumed that the oxygen dissolves in interstitial sites, bonding primarily with two silicons, and the infrared spectrum has been assigned to the vibrations of the entity Si_2O. Absorptions are observed at 1205, 1130 ($\sim 9\ \mu$) and 515 cm^{-1}. The 9 μ band is the most prominent feature of the spectrum and is assumed to arise from the asymmetric stretching mode of Si_2O. This band shows sharp fine structure at 4°K and attempts have been made to explain this structure on a librational model.[32][33] Interaction of C and Li with oxygen in silicon has also been investigated by infrared methods.[11][27]

In conclusion, although it is evident that local modes are of interest in themselves, it seems worth emphasizing that the local mode spectrum gives detailed information concerning the impurity configuration in crystals.

This paper was written during the tenure of a summer appointment at the RCA Laboratories, Princeton, New Jersey. I am particularly indebted to Dr. Z.J. Kiss for hospitality. I would like to thank Mr. A. Spray for permission to quote some of his unpublished measurements on CdTe:Be.

APPENDIX

Anharmonic contributions to the Debye-Waller factor of CaF_2 have been observed in neutron scattering experiments at elevated temperatures;[34] these are evident through a temperature dependence of the structure factor for some odd index reflections. A rigorous treatment of the effect of anharmonicity requires an average of anharmonic perturbations over all lattice modes.[35] A less rigorous approach has been used by Dawson, Hurley and Maslen[36] who assume that the vibrations of each ion in the crystal are governed by a single particle potential

$$V(r) = V_o + 1/2\ A\ r^2 + Bxyz + \dots \qquad (11)$$

The terms in the expansion of V(r) are determined by the site symmetry. In CaF_2, the cations have O_h symmetry and, for these, only even terms occur in the expansion of V(r); the anion sites have T_d symmetry and both even and odd terms occur. If we neglect quartic and higher terms in the expansion of V(r), effects of anharmonicity on the structure factor are determined only by the fluorines. Values of B for F^- ions obtained from neutron scattering data for CaF_2[36] and BaF_2[37] are given in Table IV.

The potential (11) has also been used for the analysis of localized vibrational modes of H^- ions on F^- sites in CaF_2[5] and BaF_2[7] (see also eq. (1)).

Table IV. Comparison of the cubic anharmonic parameter B for F^- and H^- ions in CaF_2 and BaF_2.

Crystal	$B(ergs/cm^3)x10^{-12}$	
	F^-	H^-
CaF_2	- 6·4	(-) 7·87
BaF_2	- 3·5	(-) 3·98

and the magnitude of B obtained from these experiments is also given in Table IV. It is apparent that the magnitude of the cubic anharmonicity of the H^- ion is remarkably close to that of the F^- ion it replaces. Cubic anharmonicity should also effect the neutron scattering of semiconductors and a comparison of B for lattice ions and impurity ions (see eq. (2)) would be of interest.

REFERENCES

1. A.A. Maradudin, Solid State Phys. 19, 1 (1966).
2. J.F. Angress, A.R. Goodwin, and S.D. Smith, Proc. Roy. Soc. A287, 64 (1965).
3. W. Hayes, Phys. Rev. 138, A1227 (1965).
4. W. Hayes and A. Spray, to be published.
5. R.J. Elliott, W. Hayes, G.D. Jones, H.F. Macdonald and C.T. Sennett, Proc. Roy. Soc. A289, 1 (1965).
6. D.G. Thomas, J. Appl. Phys. 32, 2298 (1961).
7. W. Hayes and H.F. Macdonald, Proc. Roy. Soc. A297, 503 (1967).
8. A. Mitsuishi, A. Manabe, H. Yoshinaga, S. Ibuki and H. Komiya, Proc. Internat. Conf. Phys. Semicond., Kyoto (1966), p. 72.
9. P.G. Dawber and R.J. Elliott, Proc. Roy. Soc. A273, 222 (1963).
10. P.G. Dawber and R.J. Elliott, Proc. Phys. Soc. 81, 453 (1963).
11. R.C. Newman and J.B. Willis, J. Phys. Chem. Solids 26, 373 (1965).
12. M. Lax and E. Burstein, Phys. Rev. 97, 39 (1955).
13. D.W. Feldman, M. Ashkin and J.H. Parker, Jr., Phys. Rev. Lett. 17, 1209 (1966).
14. W. Hayes, Phys. Rev. Lett. 13, 275 (1964).
15. A.R. Goodwin and S.D. Smith, Phys. Lett. 17, 203 (1965).
16. S.D. Smith, R.E.V. Chaddock and A.R. Goodwin, Proc. Internat. Conf. Phys. Semicond., Kyoto (1966), p. 67.
17. O.G. Lorimor, W.G. Spitzer and M. Waldner, J. Appl. Phys. 37, 2509 (1966).
18. W.G. Spitzer, J. Phys. Chem. Solids 28, 33 (1967).
19. L. Bellomonte and M.H.L. Pryce, Proc. Phys. Soc. 89, 967 (1966).
20. L. Bellomonte and M.H.L. Pryce, Proc. Phys. Soc. 89, 973 (1966).

21. O.G. Lorimor and W.G. Spitzer, J. Appl. Phys. 37, 3687 (1966).
22. H. Reiss, C.S. Fuller and F.J. Morin, Bell System. Tech. J. 35, 535 (1956).
23. R.J. Elliott and P. Pfeuty, J. Phys. Chem. Solids, to be published.
24. R.C. Newman and R.S. Smith, Phys. Lett. 24A, 671 (1967).
25. E.M. Pell, J. Appl. Phys. 31, 1675 (1960)
26. W.G. Spitzer and M. Waldner, J. Appl. Phys. 36, 2450 (1965).
27. K.M. Chrenko, R.S. McDonald and E.M. Pell, Phys. Rev. 138, A1775 (1965).
28. M. Balkanski and W. Nazarewicz, J. Phys. Chem. Solids 27, 671 (1966).
29. W.G. Spitzer and M. Waldner, Phys. Rev. Lett. 14, 223 (1965).
30. M. Waldner, M.A. Hiller and W.G. Spitzer, Phys. Rev. 140, A172 (1965).
31. W. Kaiser, P.H. Peck and C.F. Lange, Phys. Rev. 101, 1264 (1956).
32. H.J. Hrostowski and B.J. Alder, J. Chem. Phys. 33, 980 (1960).
33. B. Pajot, J. Phys. Chem. Solids 28, 73 (1966).
34. B.T.M. Willis, Acta Cryst. 18, 75 (1965).
35. A.A. Maradudin, P.A. Flinn, Phys. Rev. 129, 2529 (1963).
36. B. Dawson, A. Hurley and V.W. Maslen, Proc. Roy. Soc. A298, 289 (1967).
37. B.T.M. Willis, private communication.

INFRARED ABSORPTION DUE TO LOCALIZED VIBRATIONS IN CdSe CONTAINING S AND CdTe CONTAINING Li

M. Balkanski, R. Beserman, and L. K. Vodopianov

Laboratoire de Physique des Solids, Faculté des Sciences

Paris

I. INTRODUCTION

Localized modes have been observed by means of infrared spectroscopy in homopolar semiconductors like Si containing light impurities,(1) in intermetallic semiconductors like GaAs with Li and P(2)(3) and InSb with Al(4), as well as in ionic crystals.(5) The frequencies of the localized modes due to substitutional atoms and their coupling with the radiation field can be deduced on the basis of simple theory(6) when no changes in force constant are involved.

Some semiconductor crystals of the II-VI group such as ZnTe and CdTe doped with Li and Al(7) have been investigated recently.

We are presenting here optical spectra for CdSe containing S and CdTe containing Se and Li. The frequencies of the absorption peaks attributed to localized modes are compared with that calculated for a mass defect in wurtzite CdS(8) subject to appropriate normal modes frequency scaling.

II. THEORY

The perfect lattice vibrational frequencies ω are roots of the secular equation

$$\mathrm{Det}\left|M\omega^2 - \varphi\right| = 0 \qquad (1)$$

where φ is the potential energy matrix for the perfect crystal formed by atoms with mass M. The introduction of an imperfection into the crystal brings a perturbation C to the potential energy. The matrix representing the potential energy of vibration of the crystal with impurities is $\varphi + C$. The secular equation for the perturbed crystal is then

$$\mathrm{Det}\,|M\omega^2 - \varphi - C| = 0 \quad . \tag{2}$$

It is possible to define a matrix G whose coefficients are the Green's functions of the crystal:

$$G(M\omega^2 - \varphi - C) = I - GC \quad . \tag{3}$$

The localized modes, due to presence of impurities are then solutions of the secular equation

$$\mathrm{Det}\,|I - G(\omega^2)C(\omega^2)| = 0 \quad . \tag{4}$$

G is the Green's function matrix whose elements can be calculated from the eigenvectors $\sigma_{\alpha\kappa}(j,\vec{q})$ and the eigenvalues $\omega^2(j,\vec{q})$ of the perfect crystal. The perturbation matrix $C = \Delta M\omega^2 I$ is the product of the 3 × 3 unit matrix I and the mass difference $\Delta M = M - M'$, M' being the impurity mass. Solutions of Eq. (4) have been obtained for CdS with substitutional mass defect.[8] In this wurtzite crystal the site symmetry is C_{3v} and one obtains two localized modes Γ_1 singlet and Γ_3 doubly degenerate. When the impurity replaces Cd atoms, the localized mode frequencies are solutions of the following equations: $\epsilon_1\omega^2 G_1 = 1$ for the Γ_1 mode representing the vibration of the impurity atoms along the c axis, and $\epsilon_1\omega^2 G_2 = 1$ for the Γ_3 mode - vibration perpendicular to c axis. Substitution of the S atoms leads to $\epsilon_3\omega^2 G_3 = 1$ for the Γ_3 mode and $\epsilon_3\omega^2 G_4 = 1$ for the Γ_1 mode. Here $\epsilon = (M - M_i)/M$. The position of the localized modes is obtained directly from the plot $\omega^2 G(\omega^2)$ as a function of frequency in taking $\omega^2 g(\omega^2) = 1/\epsilon$. The results shown in Figure 1 concern CdS

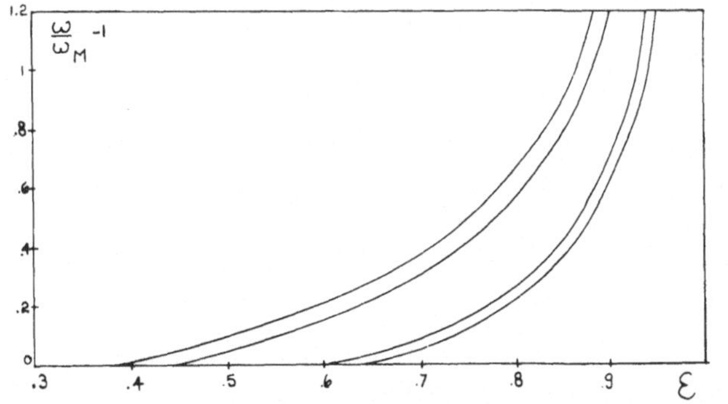

Fig. 1. Localized mode frequencies versus the mass parameter ϵ. Curves 1 and 3 give the Γ_1 mode for substitution of Cd and S respectively; curves 2 and 4 give Γ_3 mode for the same substitutions.

when Cd or S are replaced by a substitutional impurity. In CdS with ω = 308 cm^{-1} a localized mode appears only for $\epsilon > 0.36$ for the substitution of S and $\epsilon > 0.58$ for Cd, respectively, for $M < 20.5$ and $M < 47$. The splitting between the Γ_1 and Γ_3 modes increases when the impurity becomes lighter.

The model on which this calculation is based being simple, one is justified to extend the results to other II-VI compounds when no considerable changes in force constants are expected when going from one compound to another.

In the cubic crystals of this group, like CdTe for example, the site symmetry is T_d and therefore one triply degenerate mode Γ_5 is to be expected.

III. EXPERIMENTAL RESULTS

A. S in CdSe

Absorption spectra for S doped CdSe are given in Figure 2. Two bands are observed at 260 cm^{-1} and 270 cm^{-1}, the second having a peak absorption coefficient twice as large as the first. The frequency position of this peak corresponds to the reflectivity maximum observed for CdSe-S alloys(9,10) at low concentration of S.

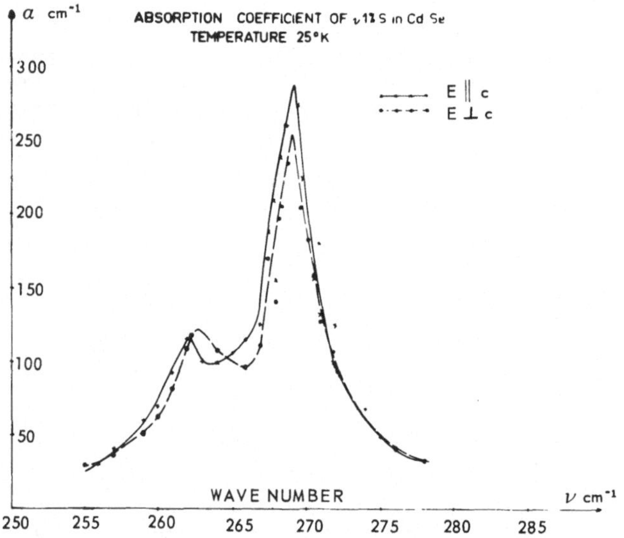

Fig. 2. Absorption spectrum of S doped CdSe at 25°K for two different polarizations of light $E\parallel c$ and $E_\perp c$.

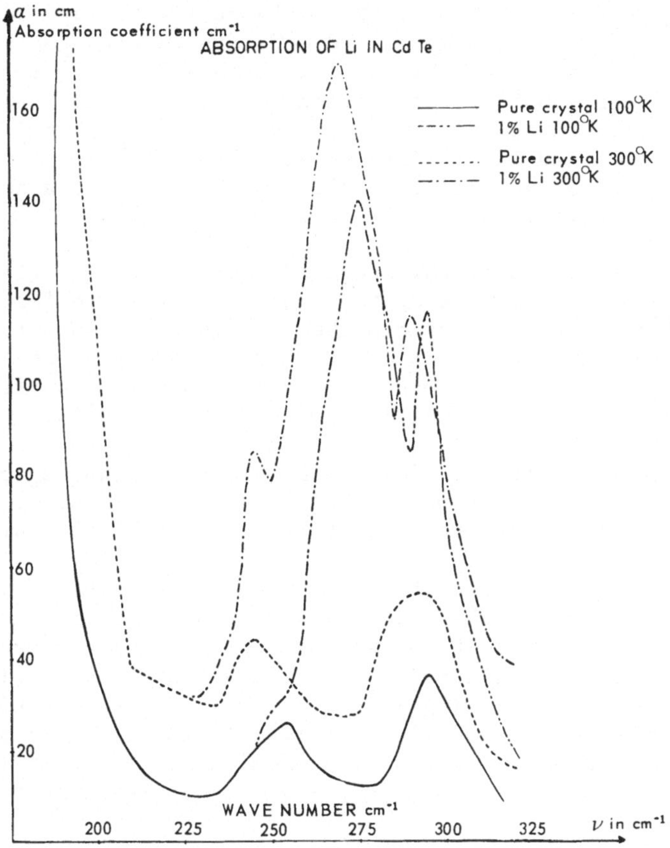

Fig. 3. Absorption spectra for Li doped CdTe at room temperature and at 100°K compared to the spectra of pure crystals.

B. Se in CdTe

Substitution of Te by Se in the CdTe crystals leads to a localized mode at approximately 170 cm^{-1}.

Li doped CdTe has also been investigated and a broad band was observed in the region of 270 cm^{-1}. The experimental results at two different temperatures are shown in Figure 3.

IV. DISCUSSION

In CdS localized modes due to Li^7 at 457 cm^{-1} and to Li^6 at 474 cm^{-1} have been reported. The calculated(8) values for substitutional Li^7 are 615 cm^{-1} for the Γ_1 and 665 cm^{-1} for the Γ_3. This very large discrepancy would indicate that Li could not be treated as a substitutional mass defect, without change in force constants.

If we admit that the density of states for CdS and CdSe crystals are similar, the localized mode frequencies for S in CdSe can be easily deduced from Figure 1 with ω_M = 210 cm^{-1} for CdSe and $\epsilon = (M_{Se}-M_S)/M_{Se} = 0.595$. One obtains $\omega(\Gamma_1)$ = 244 cm^{-1} and $\omega(\Gamma_3)$ = 256 cm^{-1}. From Figure 2 we have the experimental values: $\omega(\Gamma_1)$ = 260 cm^{-1} and $\omega(\Gamma_3)$ = 270 cm^{-1}. The difference between the calculated and observed values could be attributed to a decrease of the first neighbors force constants when S is substituted for Se.

The line width is of the order of 4 cm^{-1} and the relative line width $\Delta\omega/\omega = 0.015$ which is of the order of magnitude for relative broadening(11) when $\Delta\omega$ describes the broadening of an optical spectral line due to anharmonic interaction involving the vibrational energy of the local mode. A temperature line shift is observed of the order of 3 cm^{-1} between 25°K and 300°K.

When Se is substituted for Te in CdTe the frequency of the localized mode observed experimentally is about 170 cm^{-1}. No calculations are available presently to compare with this number.

Li in CdTe leads to a broad band at 270 cm^{-1} whose width is of the order of 35 cm^{-1}. The relative band width $\Delta\omega/\omega$ is here 0.130 which is by a factor of 10 larger than what would be expected in the case of single substitutional impurity vibrations. Li having the possibility of occupying substitutional and interstitial sites and having two isotopes should give rise to four localized modes. The observed band shown in Figure 3 should then be split into four peaks. This system is presently being further investigated.

REFERENCES

1. S. D. Smith and J. F. Angress, Phys. Rev. Letters 6, 131 (1963); M. Balkanski and W. Nazarewitz, J. Phys. Chem. Solids 25, 437 (1964).
2. W. Hayes, Phys. Rev. Letters 138, A1227 (1965).
3. S. D. Smith, R. E. V. Chaddock, A. R. Goodwin, Interntl. Conf. on Semiconductor Physics, Kyoto (1966), p. 67.
4. A. R. Goodwin, S. D. Smith, Phys. Letters 17, 206 (1966).
5. A. J. Sievers and S. Takeno, Phys. Rev. 140, A1030 (1965).
6. P. G. Dawber and R. J. Elliott, Proc. Roy. Soc. A273, 222 (1963).
7. A. Matsuishi, A. Manabe, H. Yoshinaga, S. Ibuki, and H. Komiya, Interntl. Conf. on Semiconductor Physics, Kyoto (1966), p. 72.
8. P. Pfeuty, J. L. Birman, M. A. Nusimovici, and M. Balkanski, International Conference on Localized Excitations in Solids, Calif. (1967).
9. M. Balkanski, R. Beserman and J. B. Besson, Solid State Comm. 4, 201 (1966).
10. H. W. Verleur and A. S. Barker, Jr., Phys. Rev. 155, 750 (1967).
11. P. G. Klemens, Phys. Rev. 122, 443 (1961).

INFRA-RED ABSORPTION BY A CHARGED IMPURITY AND BY ITS NEIGHBOURS

R. S. Leigh and B. Szigeti

University of Reading, England

In Sections 1 and 2 we show that the internal field effects in valency crystals are such that the "apparent charge" of an impurity atom (i.e. its vibrational absorption) is determined almost entirely by short-range effects irrespective of its true charge. In Section 3 we estimate some of the apparent charges from the observed local mode absorption of B^- in silicon.

1. INTRODUCTORY ARGUMENT

In a recent paper[1], which will be referred to as I, we discussed the vibrational absorption induced by charged impurity atoms in valency crystals, such as B^- or P^+ impurity in silicon. We showed that due to the electrostatic field of the impurity, the displacements of uncharged host atoms can result in a dipole moment. In addition, polarization by short-range forces, charge transfer and other short-range effects are also important when nearest neighbours of the impurity are displaced.[1][2]

In I we defined an "apparent charge" tensor η_n for the nth atom by

$$\eta_{n\xi;x} = \frac{\partial M_x}{\partial u_{n\xi}} \qquad \xi = x, y \text{ or } z \tag{1}$$

where $u_{n\xi}$ is the x, y or z displacement of the nth atom and M_x is the x-component of the dipole moment of the whole specimen. The present notation for the η is slightly different from the notation in I. For a cubic crystal structure the x, y and z directions are along the cubic axes. The specimen has the shape of a slab with the x-direction lengthwise, i.e. perpendicular to the short edges. It is important to specify the shape of the specimen as $\partial M_x/\partial u_{n\xi}$ depends on shape. If m_n is the mass of the nth atom, the integrated vibrational absorption is proportional to

$$\sum_n \frac{1}{m_n}\left[\left(\eta_{nx;x}\right)^2 + \left(\eta_{ny;x}\right)^2 + \left(\eta_{nz;x}\right)^2\right] \qquad (2)$$

independently of the force constants. The absorption at any particular frequency is a general second order expression in the η and of course also depends on the force constants.

In what follows, we shall restrict ourselves to the diamond structure. Assuming that for low impurity concentration the absorption due to the various impurities superposes linearly, we consider the case when the specimen contains a single impurity atom with charge e. For this case, the apparent charges η_n have been calculated in I from macroscopic relations, taking only the electrostatic interaction into account. It is thought that if all the actual charge is on the impurity then the η_n so obtained represent good approximations from second neighbours of the impurity outward, but are quite unreliable for its nearest neighbours with which it shares its valency electrons, partly because the electrostatic interaction cannot be represented by the macroscopic equations and also because short-range effects are very important in this case. Our aim now is to obtain information concerning the apparent charges of the impurity and its nearest neighbours.

The vector $\vec{r}_n$ denotes the position of the nth atom with respect to the impurity. The suffix 0 refers to the impurity and the suffix 1 to that one of its nearest neighbours for which the x, y and z components of $\vec{r}_n$ all have the same sign. Since the impurity is in tetrahedral symmetry, $\eta_{oy;x} = \eta_{oz;x} = 0$, and for brevity we shall denote $\eta_{ox;x}$ by η_o. From symmetry, $\eta_{nx;x}$ is the same for the four nearest neighbours and hence equal to $\eta_{1x;x}$. Also from symmetry, if n denotes any one of the four nearest neighbours, then $|\eta_{ny;x}| = |\eta_{nz;x}| = |\eta_{1y;x}|$, but in half of the cases the sign is opposite to the sign of $\eta_{1y;x}$.

In another recent paper[3], which will be referred to as II, we considered the diagonal elements $\eta_{nx;x}$. While the absorption is a sum of second-order terms in the η and converges fairly rapidly with increasing r_n, it was pointed out that $\eta_{nx;x}$ itself decreases only as $1/r_n^3$ and hence distant parts of the crystal may contribute to the linear sum $\Sigma_n \eta_{nx;x}$.

Rigid displacement of the whole specimen by x' in the x-direction obviously results in $M_x = ex'$. Hence from (1)

$$\Sigma_{(crystal)} \eta_{nx;x} = e \tag{3}$$

where the sum goes over all the atoms in the crystal. Further, let A denote a sphere with the impurity at its centre and with radius small compared with the thickness of the specimen but large compared with interatomic distance, so that the macroscopic electrostatic relations hold exactly outside A. It was shown in II that the following relation then holds exactly, independently of any assumptions:-

$$\Sigma_A \eta_{nx;x} = Se \tag{4a}$$

where

$$S = 1 + \tfrac{1}{3}\epsilon_0 \left(p_{11} + 2p_{44}\right) \tag{4b}$$

The sum goes over all atoms inside A. ϵ_0 is the dielectric constant and p_{11} and p_{44} the photoelastic constants of the pure material. Table 1 lists S for valency crystals, reproduced from II. These were calculated from measured values of ϵ_0, p_{11} and p_{44}.

In view of the definition of the η, eq. (4a) may be interpreted to mean this. In a dielectric we consider a macroscopic sphere embedded in its own medium and having a charge e at its centre. If this sphere is displaced rigidly by the small amount X relative to the rest of the specimen, the resulting dipole moment is eSX. According to Table 1, in valency crystals this is negative and its magnitude is much less than eX. This may appear surprising, as it used to be thought that the dipole moment for such a displacement should be $(\epsilon_0 + 2)eX/3$, i.e. positive and much larger than eX.

As a matter of fact, if the effective field were the Clausius-Mosotti field and the polarizability per atom were a constant, then

Table 1. Values of S.

diamond	-0.38
silicon	-0.16
germanium	-0.2

the values of p_{11} and p_{44} would be such that S would be equal to $(\epsilon_0 + 2)/3$ (cf. the numerical values calculated by Mueller[4] from these assumptions). But the actual values of S show that in valency crystals the Clausius-Mosotti assumptions are far from true: the polarizability per atom changes in a displacement and this acts against the "Clausius-Mosotti effects" just mentioned and slightly outweighs them. This is in close analogy to the behaviour of $\partial\epsilon/\partial(-v)$, the change of dielectric constant when the volume is compressed. Again, the Clausius-Mosotti theorem predicts that this quantity should be large and positive, because of the increase of density, while in valency crystals the actual values are small and negative: obviously the decrease of polarizability per atom in compression again slightly outweighs the increase of density.

Actually the measured values of p_{11} and p_{44} show a very large scatter, but in every case one finds $S^2 \ll 1$.

2. CONCLUSIONS CONCERNING INTERNAL FIELD AND SHORT-RANGE EFFECTS

We now consider the case when all the actual charge e is on the foreign atom. If the macroscopic equations used in I to calculate the η were valid even for nearest neighbours, then according to eq. (4) in II the contributions of the host atoms to the left-hand side of the present eq. (4a) would cancel. Eq. (4a) would then reduce to

$$\eta_0 = Se \qquad (5)$$

This is obvious, because if the macroscopic equations were valid for nearest neighbours then we could choose A equal to the volume occupied by the impurity. As stated above, when the Clausius-Mosotti theorem holds S is equal to $(\epsilon_0 + 2)/3$. When it does not hold, we see that the factor S still represents the internal field effects, at least in the macroscopic sense.

Although in valency crystals η_0 would thus be very small, nevertheless, according to Table 2 in I, the macroscopically calculated η would still give a fairly large over-all absorption, due mainly to the off-diagonal η of the host atoms. But in silicon eq. (5) would give $\eta_0^2 = 0.16^2 e^2$, and this is far too small to explain the local mode absorption[5][6]. Neglecting the absorption by nearest neighbours, Angress et al.[5] concluded from the observed local mode absorption of B- in silicon that $\eta_0^2 \approx e^2$. A more detailed analysis of the local mode is given in Section 3, but in any case, since the displacements of the neighbours in the local modes are small, it is obvious that this estimate for η_0^2 must be approximately correct. η_0^2 must therefore be very much larger than the value given by eq. (5). Instead of assuming that all the static charge is on the foreign atom we may assume that some of the static charge has spread over to the nearest neighbours, but this would obviously not help to reduce the discrepancy between eq. (5) and the actual value of η_0.

It is therefore clear that η_0 is due mainly to short-range effects which are not included in (5). Among the host atoms, of course mainly the nearest neighbours are affected by short-range effects. Therefore, in accordance with the arguments which lead to eq. (10) in II, eq. (4a) does not reduce to (5) but to

$$\eta_0 + 4\eta_{1x;x} = Se \qquad (6)$$

The "short-range effects" responsible for the difference between eqs. (5) and (6) include not only charge transfer and deformation by short-range forces, but also the deviation of the electrostatic effects from their values given by the macroscopic equations. Electrostatic effects due to the impurity and the host atoms having different polarizabilities also come under this category. We group these short-range electrostatic effects together with the other short-range effects because (a) in valency crystals all these effects are inextricably mixed up with each other; and (b) their contributions to η_0 may equally well have the opposite sign to e as the same sign.

In view of what has been said, we conclude that in valency crystals the internal field effects are such that the apparent charge η_0 of a charged impurity is determined almost entirely by short-range forces. The charge of course has an effect on these short-range forces, but this effect may equally well decrease η_0 as increase it. This conclusion is borne out by the fact that the local mode absorption for carbon impurity in silicon[7] is about five times as big as for B^- in silicon, and hence, neglecting nearest neighbours in the local mode, η_0 for carbon is approximately 2.3 times as big as η_0 for boron.

It is clear from the derivation of our various results in I that these conclusions concerning the internal field effects do not apply to pure crystals. Nevertheless they must have a significance for pure crystals as S is determined by the properties of the pure crystal alone. The relative strengths of the local mode absorption of carbon and boron in silicon are in close analogy to the fact that the vibrational absorption of the non-ionic SiC is larger[8] than that of the weakly ionic GaP and of other III - V compounds[9][10]. In Burstein's view[11] the absorption of these substances is determined essentially by short-range effects, and he considers that the static charge on the ions is very small. But in the case of B^- impurity in silicon, the static charge on the boron need not be small: the internal field effects are such that the static charge has little effect on η_0.

On the other hand, the static charge of the impurity does affect the η of the atoms near it. According to I, a charged impurity has a long-range effect on the apparent charges of the atoms in its neighbourhood, while uncharged impurities do not exert this effect.

3. LOCAL MODE ABSORPTION AND APPARENT CHARGES

3.1 Method of Calculation

On our assumptions, the short-range interactions affect mainly η_0, $\eta_{1x;x}$ and $\eta_{1y;x}$. But the observed local mode absorption and eq. (6) only represent two relations for these three unknowns. If we assume that $\eta_{1x;x}$ and $\eta_{1y;x}$ are, in a first approximation, determined by central two-body forces between the impurity and its nearest neighbours, and that in equilibrium the valency electrons are shared equally between the impurity and its nearest neighbours, we get

$$\eta_{1y;x} = \eta_{1x;x} \tag{7}$$

because the bond is in the (1,1,1) direction. (7) is obviously based on very inexact assumptions and we are only justified to conclude that $\eta_{1x;x}$ and $\eta_{1y;x}$ are of similar magnitude.

Using the measured local mode absorption of B^- in silicon and eqs. (6) and (7), an estimate can be obtained for the magnitudes of η_0 and the η_1. According to the calculations of J. Slater (private communication), the x-displacements for that local mode of the ^{10}B isotope where the boron vibrates in the x-direction are given by $(U_{0x;L})^2 = 0.8$ and $U_{1x;L} = -0.08\, U_{0x;L}\sqrt{m_{Si}/m_B}$. Here L labels the local mode, the m denote the atomic masses and the U-matrix was defined in I. In the same mode, the nearest neighbours also vibrate in the y and z directions; the amplitude of these components can be calculated from the normalization of the U-matrix if one neglects the motion of the atoms beyond first neighbours. According to Goodwin[6], for a total concentration of 5×10^{19} boron/cc, the integrated local mode absorption of ^{10}B at liquid nitrogen temperature is 0.0143 eV cm^{-1}. From these data, making use of symmetry considerations and of eq. (2.1a) in I and of eqs. (6) and (7), we can calculate η_0 and the η_1 .

3.2 Results for η_0 and the η_1 , and Comments

As the absorption is quadratic in the η, the procedure just described yields two sets of solutions, namely

(a) $\eta_0 = +0.81e; \quad \eta_{1x;x} = \eta_{1y;x} = -0.24e;$

(b) or $\eta_0 = -0.87e; \quad \eta_{1x;x} = \eta_{1y;x} = +0.18e.$

As η_0 is determined by short-range effects it may easily have opposite sign to e and therefore at present we cannot choose between the solutions (a) and (b). As far as the magnitudes are concerned, the discrepancy between the two sets is not larger than the uncertainty of the theoretical procedure and of the experimental data. Thus, while at present we do not know the sign of the effect, either set gives a reasonable estimate of the magnitudes.

For either set of η we find from eq. (2.1a) in I that the nearest neighbours make an appreciable contribution to the local mode absorption. For set (a) η_0^2 is responsible only for 60% of the local mode absorption, the rest being due to the η_1 and to cross-terms between η_0 and the η_1 . In the case of set (b), where η_0 and e have opposite signs, η_0^2 is responsible for 70%.

Table 2 in I lists the vibrational contributions of the foreign atoms, of nearest neighbours and of further neighbours to the integrated absorption. That table is based essentially on long range effects only, and the figures in the first two rows should therefore be modified in the light of our new estimates of η_0 and the η_1 . The new results do not alter significantly the over-all picture presented by that table: the figures still indicate that even if all the static charge is on the foreign atoms, the vibrations of the host atoms contribute rather more than half of the total absorption. The relative contribution of the host atoms to the in-band absorption is even more, because most of the boron vibration is in the local mode. In the case of a heavier impurity, such as Sb^+, almost all the absorption would be due to the vibration of the silicon atoms.

If the impurity shares its charge e with its nearest neighbours then its relative contribution to the total absorption may be even less. In that case short-range effects may be important also for second neighbours so that these may then have to be included in eq. (6).

The large absorption by the vibration of the host atoms may of course have important effects on the shape of the in-band absorption. Work is in progress at present with the aim of obtaining further information on the η from a detailed study of the experimental spectrum. It is already clear that important features of the spectrum which could not be explained on the assumption that only the impurities absorb can be explained on the basis of the present ideas. A reliable determination of the η, however, may be hampered by the lack of knowledge of the force constants. In view of what has been said here and in I, further information on the apparent charges η would be of considerable interest in connection with bonding and related electronic properties.

References

1. R. S. Leigh and B. Szigeti, Proc. Roy. Soc. A 301, 211 (1967) (referred to as I).
2. B. Szigeti, J. Phys. Chem. Solids, 24, 225 (1963).
3. R. S. Leigh and B. Szigeti, Phys. Rev. Letters, 19, 566 (1967) (referred to as II).
4. H. Mueller, Phys. Rev. 47, 947 (1935).
5. J. F. Angress, A. R. Goodwin and S. D. Smith, Proc. Roy. Soc. A 287, 64 (1965).
6. A. R. Goodwin, Ph.D. Thesis, University of Reading (1966).
7. R. C. Newman and J. B. Willis, J. Phys. Chem. Solids, 26, 373 (1965).
8. W. G. Spitzer, D. A. Kleinman and C. J. Frosch, Phys. Rev. 113, 133 (1959).
9. D. A. Kleinman and W. G. Spitzer, Phys. Rev. 118, 110 (1960).
10. M. Hass and B. W. Henvis, J. Phys. Chem. Solids 23, 1099 (1962).
11. E. Burstein, Proceedings of the Copenhagen Conference on Lattice Dynamics, J. Phys. Chem. Solids, Suppl. I, p. 315 (1965).

Note. In II after eq. (10) we said that η_o has presumably the same sign as e. It is clear from what has been said in the present paper that we now think η_o may equally well have opposite sign to e.

THEORY OF FIRST-ORDER RAMAN SCATTERING BY CRYSTALS OF THE DIAMOND STRUCTURE CONTAINING SUBSTITUTIONAL IMPURITIES

Nguyen Xuan Xinh

Aerospace Engineering Sciences Department

University of Colorado, Boulder, Colorado

1. INTRODUCTION

The Raman effect that we consider here is the inelastic scattering of light by the lattice vibrations due to the fluctuations in the crystal electronic polarizability induced by the lattice vibrations. There is conservation of energy, i.e. in a first order scattering process, the frequency ω_s of the scattered radiation differs from the frequency ω_i of the incident radiation by the frequency of one phonon. In perfect crystals, there is also conservation of momentum due to the translational symmetry of the crystal. In actual experiments, the Bragg condition is generally satisfied and since the magnitude of the wave vector of the light in the crystal is very small compared to the maximum magnitude of the phonon wave vector, only the phonons of essentially zero wave vector $\underset{\sim}{k}\simeq 0$ are involved in first order scattering. Actually, a good approximation in the description of Raman scattering consists in replacing the small wave vector of light by zero and we shall do so in what follows. In a crystal of the diamond structure, no atom is at a center of inversion symmetry for the lattice, therefore the perfect crystal exhibits a first order Raman spectrum which in the harmonic approximation consists of discrete δ-function lines centered at frequencies shifted from the frequency of the incident light by that of the $\underset{\sim}{k}=0$ optical phonons. If impurities are present in the crystal, its translational symmetry is destroyed, therefore the selection rule $\underset{\sim}{k}=0$ is relaxed and the incident light can interact with all the phonons of the crystal. The first order spectrum is no longer a line spectrum but becomes continuous and its structure would reflect the singularities of the frequency spectrum of the perfect host crystal and resonant and/or localized modes if they occur. We report in this paper a calculation of the first order Raman spectra of crystals of the diamond structure containing randomly distributed substitutional impurities, using the harmonic approximation for the crystal potential and the mass defect approximation for the impurity atom. Numerical calculations are carried out for a Si crystal containing 1% Ge and C^{12}, respectively. The theoretical results will be discussed in connection with Feldman and co-workers' experimental data on Raman scattering by mixed crystals of the Ge:Si system.

2. THEORY

We are concerned only with the case where the impurity concentration is sufficiently low that the mutual interaction between the impurity atoms through their vibrational fields is small and can be neglected. Assuming that the frequency of the incident radiation is not close to any electronic transition frequency of the crystal, the same phenomenological approach as described in (1) is used here. It is based on Born and Huang's semiclassical theory of Raman scattering (2). Our starting point is therefore the following formula for the intensity per unit solid angle of the scattered radiation(1,2):

$$I(\omega) = \frac{n_d \omega_i^4}{2\pi c^3} \sum_{\substack{\alpha\beta \\ \gamma\delta}} n_\alpha n_\beta i_{\alpha\gamma,\beta\delta}(\omega) E_\gamma^- E_\delta^+ \quad , \tag{1}$$

where $\omega = \omega_s - \omega_i$, n_d is the total number of impurity atoms in the crystal, c is the speed of light, $\underset{\sim}{n}$ is a unit polarization vector of the scattered radiation, and $\underset{\sim}{E}^+$ and $\underset{\sim}{E}^- = (E^+)^*$ are the amplitudes of the positive and negative frequency components of the electric field of the incident radiation, respectively. The functions $i_{\alpha\gamma,\beta\delta}(\omega)$ have the following expression(1):

$$i_{\alpha\gamma,\beta\delta}(\omega) = \frac{1}{2\pi} \int_{-\infty}^{\infty} dt\, e^{-i\omega t} < P_{\beta\delta}(t) P^*_{\alpha\gamma}(0) > \quad , \tag{2}$$

where $P_{\alpha\gamma}(t)$ is the electronic polarizability tensor operator in the Heisenberg representation and where the angular brackets denote a thermodynamic average over the canonical ensemble described by the vibrational part of the crystal Hamiltonian. The fluctuations in the electronic polarizability of the crystal due to the lattice vibrations are described by expanding its components in Taylor's series in powers of the ionic displacements $\{\underset{\sim}{u}(\ell\kappa)\}$ as

$$P_{\alpha\beta} = P^{(0)}_{\alpha\beta} + \sum_{\ell\kappa\mu} P_{\alpha\beta,\mu}(\ell\kappa)\, u_\mu(\ell\kappa)$$

$$+ \frac{1}{2} \sum_{\substack{\ell\kappa\mu \\ \ell'\kappa'\nu}} P_{\alpha\beta,\mu\nu}(\ell\kappa,\ell'\kappa')u_\mu(\ell\kappa)\, u_\nu(\ell'\kappa') + \ldots \qquad (3)$$

where the indices α,β,μ denote the Cartesian coordinates and the index-pair $(\ell\kappa)$ labels the equilibrium lattice site of the κth kind of atom ($\kappa = 1,2$) in the ℓth unit cell. We assume that the host crystal satisfies the cyclic boundary condition. The main contribution to the first order Raman scattering arises from the first-order terms in the expansion Eq.(3). We shall simply replace, in what follows, $P_{\alpha\beta}$ by these terms.

Let substitutional impurities be introduced at random into the host crystal. We describe the disordered crystal thus obtained by making the following approximations. The impurity is considered as a simple mass defect, i.e. we neglect all force constant changes. Let p denote the concentration in impurities ($0<p<1$) and M' and M the mass of the impurity atom and that of the host atom it replaces respectively. The random distribution of impurities is described by postulating that the mass $M_{\ell\kappa}$ of the atom occupying the site $(\ell\kappa)$ is equal to M' with the probability p and M with the probability $(1-p)$. Actually, the polarizability coefficients $\{P_{\alpha\beta,\mu}(\ell\kappa)\}$ have different values according to whether they are associated with a host atom or with an impurity atom. We make the approximation that they have the same values in both cases and consider them as parameters to be determined by experiment. It results from these approximations that the lattice of the disordered crystal still has the same symmetry as the perfect host crystal and it is straightforward to show that the tensor $P_{\alpha\beta,\mu}(0\kappa)$ has only one independent element whose value we denote by a, that the only non-vanishing components are $a = P_{xy,z}(01) = P_{xz,y}(01) = P_{yx,z}(01) = P_{yz,x}(01) = P_{zx,y}(01) = P_{zy,x}(01)$ and that $P_{\alpha\beta,\mu}(02) = -P_{\alpha\beta,\mu}(01)$. Consequently the tensor $i_{\alpha\gamma,\beta\delta}(\omega)$ has only one independent component that can be represented by $i_{xy,xy}(\omega)$ whose expression is

$$i_{xy,xy}(\omega) = \frac{1}{2\pi} \sum_{\substack{\ell\kappa\mu \\ \ell'\kappa'\nu}} P_{xy,\mu}(0\kappa)\, P_{xy,\nu}(0\kappa')$$

$$\times \int_{-\infty}^{\infty} dt\, e^{-i\omega t} < u_\mu(\ell\kappa;t)\, u_\nu(\ell'\kappa';0) > . \qquad (4)$$

The ionic displacement can be expanded in terms of the phonon creation and destruction operators $b^+_{\underset{\sim}{k}j}$ and $b_{\underset{\sim}{k}j}$ according to[3]:

$$u_\alpha(\ell\kappa) = \left(\frac{\hbar}{2NM}\right)^{1/2} \sum_{\underset{\sim}{k}j} \frac{e_\alpha(\kappa|\underset{\sim}{k}j)}{(\omega_j(\underset{\sim}{k}))^{1/2}} \exp\left[2\pi i \underset{\sim}{k}\cdot \underset{\sim}{x}(\ell)\right] A_{\underset{\sim}{k}j} \quad , \qquad (5)$$

where $A_{\underset{\sim}{k}j} = b_{\underset{\sim}{k}j} + b^+_{-\underset{\sim}{k}j} = A^+_{-\underset{\sim}{k}j}$, and where N is the total number of unit cells in the crystal. $\omega_j(\underset{\sim}{k})$ and $\underset{\sim}{e}(\kappa|\underset{\sim}{k}j)$ are respectively the eigenfrequency and eigenvector of the vibrational mode of wave vector $\underset{\sim}{k}$ and branch index j of the perfect host crystal. $\underset{\sim}{x}(\ell)$ is the usual cell equilibrium position vector. By substituting Eq. (5) into Eq. (4) and successively carrying out the summations over ℓ, ℓ', μ, ν, $\underset{\sim}{k}$ and $\underset{\sim}{k}'$, we obtain:

$$i_{xy,xy}(\omega) = \frac{\hbar N}{4\pi M} \sum_{jj'} \frac{C_{xy}(j)\, C_{xy}(j')}{(\omega_j(\underset{\sim}{0})\, \omega_{j'}(\underset{\sim}{0}))^{1/2}} \int_{-\infty}^{\infty} dt\, e^{-i\omega t} \langle A_{\underset{\sim}{0}j}(t) A_{\underset{\sim}{0}j'}(0)\rangle , \qquad (6)$$

where we have defined the coefficient $C_{xy}(j)$ by

$$C_{xy}(j) = \sum_{\kappa} P_{xy,z}(0\kappa)\, e_z(\kappa|\underset{\sim}{0}j) = a[e_z(1|\underset{\sim}{0}j) - e_z(2|\underset{\sim}{0}j)]. \qquad (7)$$

If j refers to an acoustic mode, one has[4]: $(M\ )^{-1/2} e_\alpha(1|\underset{\sim}{0}j) = (M\ )^{-1/2} e_\alpha(2|\underset{\sim}{0}j)$ and the corresponding coefficient $C_{xy}(j)$ vanishes. Therefore, the acoustic modes do not contribute to the Raman scattering in the present approximation. For a diamond lattice, the eigenvectors of the $\underset{\sim}{k} = 0$ optical modes ($j = 1,2,3$) are triply degenerate and can be chosen to be[5]

$$e_\alpha(1|\underset{\sim}{0}j) = \frac{1}{\sqrt{2}}\, \delta_{\alpha j} \quad , \quad e_\alpha(2|\underset{\sim}{0}j) = -\frac{1}{\sqrt{2}}\, \delta_{\alpha j} \quad . \qquad (8)$$

Taking account of Eqs. (7) and (8), Eq. (6) reduces to the term for which $j = j' = 3$,

$$i_{xy,xy}(\omega) = \frac{\hbar N a^2}{2\pi M\, \omega_3(\underset{\sim}{0})} \int_{-\infty}^{\infty} dt\, e^{-i\omega t} \langle A_{\underset{\sim}{0}3}(t)\, A_{\underset{\sim}{0}3}(0)\rangle \ . \qquad (9)$$

In fact, it can be shown that the correlation function $\langle A_{\underset{\sim}{0}j}(t)\, A_{\underset{\sim}{0}j}(0)\rangle$ associated with an optical mode depends upon the index j only through the frequency $\omega_j(\underset{\sim}{0})$ and does not contain the corresponding eigenvector in its expression. Because the $\underset{\sim}{k} = 0$

optical modes in a crystal of the diamond structure are triply degenerate, we can therefore rewrite Eq.(9) as

$$i_{xy,xy}(\omega) = \frac{\hbar N a^2}{2\pi M\, \omega_j(\underset{\sim}{Q})} \int_{-\infty}^{\infty} dt\, e^{-i\omega t} \langle A_{\underset{\sim}{Q}j}(t)\, A_{\underset{\sim}{Q}j}(0)\rangle \quad , \qquad (10)$$

where j refers to any one of the three optical branches. This equation is valid for any configuration of the impurities over the lattice sites. However, the actual configuration is not known. Therefore, we assume an equal probability for every configuration of the impurity atoms over the lattice sites of the crystal that is compatible with the postulated occupation probabilities and subsequently take the average of the correlation function $\langle A_{\underset{\sim}{Q}j}(t)\, A_{\underset{\sim}{Q}j}(0)\rangle$ over all these configurations. This means that we replace Eq. (10) by

$$i_{xy,xy}(\omega) = \frac{\hbar N a^2}{2\pi M\, \omega_j(\underset{\sim}{Q})} \int_{-\infty}^{\infty} dt\, e^{-i\omega t} \langle A_{\underset{\sim}{Q}j}(t)\, A_{\underset{\sim}{Q}j}(0)\rangle_A \, , \qquad (11)$$

where the subscript A denotes the configuration average.

The Fourier transform of the correlation function in Eq. (11) can be calculated by an extension of a method used by Maradudin for evaluating the infrared lattice absorption spectrum of disordered ionic crystals[3]. Using a phonon proper self energy calculated to the first order in the impurity concentration p and taking account of the transformation properties of the vibration eigenvectors under the operations of the symmetry group of the diamond lattice, we finally obtain the following result for the case where $|\omega| < \omega_L$, ω_L being the maximum vibration frequency of the perfect host crystal:

$$i_{xy,xy}(\omega) = a^2\, \frac{2\hbar N}{\pi M}\, n(\omega)\, \mathrm{sgn}\, \omega$$

$$\times \frac{\varepsilon p \omega^2\, \beta(\omega)}{[\omega_j^2(\underset{\sim}{Q}) - \omega^2 + \varepsilon p \omega_j^2(\underset{\sim}{Q})\, \alpha(\omega)]^2 + \varepsilon^2 p^2 \omega_j^4(\underset{\sim}{Q})\, \beta^2(\omega)}$$

$$= a^2\, \frac{2\hbar N}{\pi M}\, \mathrm{sgn}\, \omega\, F(\omega) \quad , \quad |\omega|<\omega_L \quad , \qquad (12a)$$

where $n(\omega)$ is the Bose thermal distribution function, $\varepsilon=1-(M'/M)$ and where the functions $\alpha(\omega)$ and $\beta(\omega)$ are defined by

$$\alpha(\omega) - i\beta(\omega) = \left[1 - \frac{\varepsilon\omega^2}{6N} \sum_{\underset{\sim}{k}j} \frac{1}{(\omega^2 - \omega_j^2(\underset{\sim}{k}))_P} + i\pi\, \frac{\varepsilon\omega^2}{6N} \sum_{\underset{\sim}{k}j} \delta(\omega^2 - \omega_j^2(\underset{\sim}{k}))\right]^{-1}$$

(12b)

In the case where an isolated lighter impurity atom gives rise to a localized vibration mode, the latter makes an additional contribution to the intensity of the first-order Raman scattering that is found to have the expression:

$$i_{xy,xy}(\omega) = a^2 \frac{2\hbar N}{M} n(\omega) \operatorname{sgn} \omega$$

$$\times \frac{\varepsilon p\, \omega_r^2 A^{-1}(\omega_r)}{\omega_r^2 - \omega_j^2(\underset{\sim}{Q})} \left. \frac{\delta(\omega^2 - \omega_r^2)}{1 + \varepsilon p\, \omega_j^2(\underset{\sim}{Q}) \dfrac{\frac{d}{d\omega^2} A(\omega)}{A^2(\omega)}} \right|_{\omega = \omega_r}, \quad |\omega| > \omega_L \,, \qquad (13a)$$

where

$$A(\omega) = 1 - \frac{\varepsilon\omega^2}{6N} \sum_{\underset{\sim}{k}j} \frac{1}{(\omega^2 - \omega_j^2(\underset{\sim}{k}))_P} \,, \qquad (13b)$$

P denoting the Cauchy principal value and where ω_r denotes the solution of the equation

$$f(\omega^2) \equiv \omega^2 - \omega_j^2(\underset{\sim}{Q}) - \varepsilon p\, \omega_j^2(\underset{\sim}{Q})\, A^{-1}(\omega) = 0 \,. \qquad (13c)$$

The frequency ω_r generally lies a little above the localized mode frequency ω_o which is the solution of $A(\omega) = 0$.

3. NUMERICAL CALCULATIONS

For illustrative purposes, we have applied the preceding results to the case of silicon crystals containing 1% carbon and germanium respectively, using a frequency spectrum of Si computed by Dolling(6) from experimental dispersion curves measured at 296°K by neutron spectroscopy(7). Let us introduce the dimensionless frequency $x = \omega/\omega_L$. The frequencies of the localized and resonance modes due to isolated C^{12} and Ge impurities, treated as mass defects, in Si have been computed by Maradudin(8). The light C^{12} impurity gives rise to a resonance mode at a frequency $x = 0.945$ (500.6 cm^{-1}) and to a localized mode whose frequency is $x_o = 1.195$ (633 cm^{-1}) which is only about 4% greater than the experimental value (607 cm^{-1}) obtained by Newman and Willis.(9) This good agreement between the theoretical and experimental values of the localized mode frequency suggests that the substitutional C^{12} impurity in silicon is fairly well described by a simple mass defect model whose use in the present calculation is thus justified. On the other hand, the heavy Ge impurity gives rise to resonance modes at $x = 0.855$ (452.9 cm^{-1}) and at $x = 0.925$ (490 cm^{-1}). It also gives rise to "near resonance modes" at $x = 0.22(116.5\ \mathrm{cm}^{-1})$, $0.63\ (333.7\ \mathrm{cm}^{-1})$ and $0.72\ (381.4\ \mathrm{cm}^{-1})$.

The intensity of first-order Raman scattering is, according to Eq. (12a), proportional to $F(\omega)$. The solid curves in Fig. 1 represent the calculated first-order Raman spectra $F(x)$. According to Eqs. (12a,b), the numerator of $F(\omega)$ has the expression

$$\varepsilon p\omega^2\beta(\omega) = \frac{\dfrac{\pi p\varepsilon^2\omega^4}{6N}\sum_{\underset{\sim}{k}j}\delta(\omega^2-\omega_j^2(\underset{\sim}{k}))}{\left[1-\dfrac{\varepsilon\omega^2}{6N}\sum_{\underset{\sim}{k}j}\dfrac{1}{(\omega^2-\omega_j^2(\underset{\sim}{k}))_P}\right]^2+\left[\dfrac{\pi\varepsilon\omega^2}{6N}\sum_{\underset{\sim}{k}j}\delta(\omega^2-\omega_j^2(\underset{\sim}{k}))\right]^2} \qquad (14)$$

The numerator of the expression in the right hand side of this equation can also be written as $(\pi/2)\varepsilon^2 pn(x)x^3g(x)$, where $g(x)$ is the frequency spectrum of the perfect host crystal. The dashed curves in Fig. 1 represent the frequency spectrum $g(x)$ weighted by the factor $(\pi/2)\varepsilon^2 pn(x)x^3$.

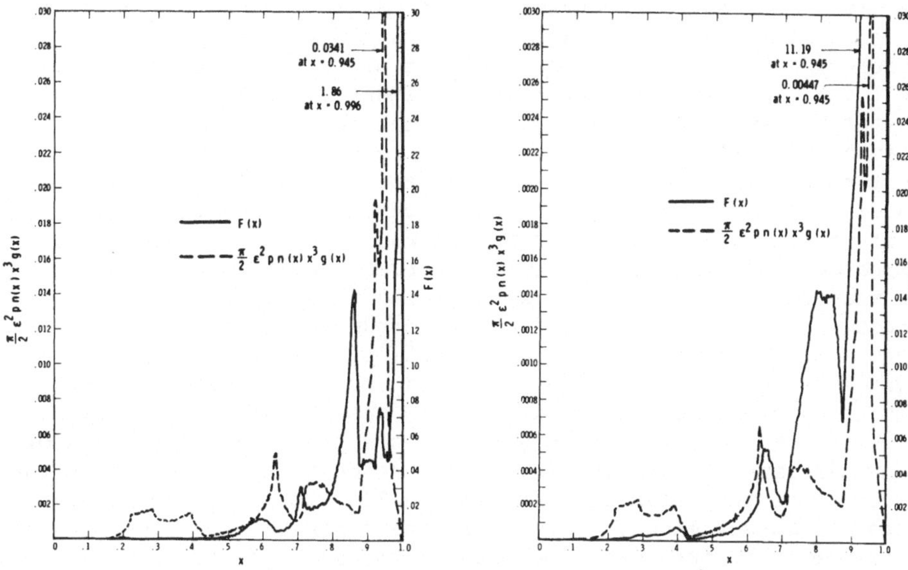

a b

Fig. 1 - Calculated first order Raman spectra of Si containing random substitutional impurities (solid curves) compared to the frequency spectrum $g(x)$ of the perfect host crystal weighted by a factor $(\pi/2)\varepsilon^2 pn(x)x$ (dashed curves). The Si crystal contains: a) 1% germanium, b) 1% carbon 12.

In the case of Ge impurities, the resonance modes show up in the first-order Raman spectrum Fig. 1a as sharp peaks at x = 0.86 and 0.94 while the bumps in the Raman spectrum at x = 0.2, 0.59, and 0.71 reflect the near resonance modes. The strong peak at x = 0.996 corresponds to the δ-function line at x = 1 in the Raman spectrum of the perfect crystal which becomes finite, broadened and is shifted to a lower frequency because of the presence of a finite concentration of impurities. On the other hand, in the case of C^{12} impurities, as mentioned before, the only resonance mode occurs at x = 0.945, i.e. at the immediate vicinity of x = 1. In the first-order Raman spectrum displayed in Fig. 1b, the peak arising from this resonance mode lumps together with and is indistinguishable from the peak coming from the original δ-function line at x = 1 in the Raman spectrum of the perfect silicon crystal. The remainder of the Raman spectrum of the C^{12}-doped crystal reproduces fairly faithfully the (appropriately weighted) frequency spectrum of the perfect host crystal. In the frequency range $|\omega| > \omega_L$, the Raman spectrum would present a simple peak at the frequency of the localized modes.

4. DISCUSSION

The preceding results show that in the harmonic approximation the Raman spectrum of a crystal of the diamond structure containing substitutional impurities are no longer a line spectrum. Because the $\underset{\sim}{k} = 0$ optical modes of the perfect crystal with which light interacts are no longer exact eigenstates of the disordered crystal but are superpositions of these, the light can in fact interact with all the modes of the perfect crystal. The original first-order line spectrum of the perfect crystal is therefore replaced by a continuous one in the presence of impurities. The function $\beta(\omega)$ being the distribution function of the squares of the normal mode frequencies of the perfect crystal, weighted by the mean square amplitude of an isolated impurity atom, the structure of the continuous Raman spectrum must reflect the singularities of the frequency spectrum of the perfect host crystal as well as resonance modes and localized modes arising from the presence of impurities. This can be seen in the calculated spectra of a silicon crystal with a finite concentration of Ge or C^{12} impurities. In Fig. 1a the resonance modes dominate the structure of the Raman spectrum and in Fig. 1b where the resonance modes are absent except near x = 1, the Raman spectrum reproduces fairly well the frequency spectrum of the perfect host crystal.

At the present time, no experiment on Raman scattering by a Si crystal containing either carbon or germanium substitutional impurities has been reported. It seems that a carbon-doped Si crystal suitable for such an experiment is difficult to obtain. The maximum atomic concentration of C atoms that go substitutionally into Si is[9] of the order of 10^{-4} - 10^{-3} and is probably too small to show up in a Raman scattering that is a very weak effect (the ratio of the intensity of the scattered radiation compared with that of the incident radiation is typically of the order of 10^{-7}).

Raman spectra of mixed crystals of the system Ge:Si has been obtained by Feldman et al.[10]. In crystals with Si atomic concentrations ranging from about 1% to 33%, a line has been found whose Raman frequency ν varies with increasing Si concentration p from 389 to 402 cm^{-1}, according to the relation $\nu = (389 \pm 2) + (0.5 \pm 0.1)\,p$. The same frequency, calculated by these authors from the present theory, for $p \leq 15\%$, using a germanium frequency distribution function obtained by Dolling and Cowley from neutron scattering measurement data, has the value $\nu = 373 + 2.6\,p$. As has been suggested by these authors, the differences between these two expressions may be attributed in part to the neglect in the theory of force constant changes and anharmonicity or to inaccuracies in the experimental phonon data. Also, it should be kept in mind that the present theory is concerned with low impurity concentrations only. Feldman and co-workers' experimental data however are in good qualitative agreement with the predictions of the present theory: the frequency of the Raman active localized modes due to light impurities in finite concentration is higher than that of the localized modes induced by an isolated impurity. Also, the observed behavior of the Raman line arising from the $\underset{\sim}{k} \simeq 0$ optical modes of pure Ge in the mixed Ge:Si system is also in agreement with the theory: its frequency is shifted downward and, as can be seen in the measured spectra, its width is broadened by the presence of an increasing finite concentration of impurities. Let us finally mention that for mixed Ge:Si crystals containing more than about 5% Si, an additional band at about 462 cm^{-1} has been observed by Feldman and co-workers who have tentatively assigned it to the localized modes of nearest neighbor pairs of Si atoms.

Acknowledgment. I am grateful to Professor A. A. Maradudin for many helpful discussions on this work.

This research was supported by the Advanced Research Projects Agency, Directorate of Materials Sciences, and was technically monitored by the Air Force Office of Scientific Research under Contract AF 49(638)-1245. The work was accomplished at the Westinghouse Research Laboratories, Pittsburgh, Pennsylvania 15235 while the author was on leave of absence from Laboratoire de Magnétisme et de Physique du Solide, C.N.R.S., 92-Meudon Bellevue, France.

REFERENCES

1. Nguyen X. Xinh, A. A. Maradudin, and R. A. Coldwell-Horsfall, J. de Physique (Paris), 26, 717 (1965).

2. M. Born and K. Huang, Dynamical Theory of Crystal Lattices (Oxford, Clarendon Press, 1956).

3. A. A. Maradudin, in Astrophysics and the Many Body Problem (W. A. Benjamin, Inc., New York, 1963), p. 109.

4. A. A. Maradudin, E. W. Montroll, and G. H. Weiss, Theory of Lattice Dynamics in the Harmonic Approximation (Academic Press, Inc., New York, 1963).

5. M. Lax, Symmetry Principles in Solid State Physics (to be published).

6. G. Dolling, private communication to Professor A. A. Maradudin.

7. G. Dolling, Inelastic Scattering of Neutrons in Solids and Liquids (International Atomic Energy Agency, Vienna, 1963) vol. II, p. 37.

8. A. A. Maradudin, in Solid State Physics, vol. 18 and 19, edited by F. Seitz and D. Turnbull (Academic Press, New York, 1966).

9. R. C. Newman and J. B. Willis, J. Phys. Chem. of Solids, 26, 373 (1965).

10. D. W. Feldman, M. Ashkin, and James H. Parker, Jr., Phys. Rev. Letters, 17, 1209 (1966).

LOCAL MODE ABSORPTION FROM BORON COMPLEXES IN SILICON

R.C. Newman and R.S. Smith

J.J. Thomson Physical Laboratory,

University of Reading, Berks., U.K.

INTRODUCTION

Infrared absorption due to the localized vibrations of isolated boron atoms[1] and boron-lithium pairs[2] in silicon has been reported in several previous papers. Recently the technique of lithium compensation has been used to study a crystal containing approximately equal concentrations (2×10^{20} atoms cm^{-3}) of boron and phosphorus[3] and new satellite lines were detected and ascribed to boron-phosphorus pairs. Crystals doubly doped with boron and an n-type impurity, which was either phosphorus, arsenic or antimony, and where the final compensation was effected by an electron irradiation have also been the subject of recent investigations[4,5]. Measurements on such crystals will be described and related to the existence of the pairs B-B, B-P, B-As and B-Sb.

EXPERIMENTAL DETAILS

The crystals used in this investigation were as follows:
Crystal A, B/P at 5×10^{19} cm^{-3} grown by the F.Z Technique;
Crystal B, B/As at 5×10^{19} cm^{-3}, pulled crystal;
Crystal C, B/Sb at 10^{19} cm^{-3}, pulled crystal;
Crystal D, B/P at 10^{19} cm^{-3}, pulled crystal.
Crystals C and D were grown in the same apparatus under identical conditions.

Polished samples about 2mm in thickness were irradiated with 1.5MeV electrons on both sides to a dose of about 5×10^{18} electrons cm^{-2}. Spectra were recorded on a Grubb-Parsons grating spectrometer flushed with dry air or nitrogen and with the samples mounted on the cold finger of a standard liquid nitrogen cell. Differential spectra were also obtained by mounting a similarly cooled undoped (or doped) sample in the reference beam of the spectrometer.

THEORY

Pairing of a boron atom with another substitutional impurity will lower the boron site symmetry from T_d to C_{3v} and the triply degenerate local mode of the boron will be split into a singlet mode (A_1) along the pair axis and a doubly degenerate mode (E) in the perpendicular direction. For a heavy impurity, other perturbed modes will lie in the continuum[6] . Hence boron-donor pairs should give rise to four infrared active local mode frequencies since there are two boron isotopes with a relative abundance of 4.3:1 for B^{11} and B^{10} respectively. It follows that the strength of the absorption from B^{11}-donor pairs should be about four times greater than that from B^{10}-donor pairs. Since the E modes are double degenerate, they might be expected to be stronger than the A_1 modes, although radical changes in the relative dipole moments could lead to the opposite situation.

For pairs of boron atoms, calculations[6] have shown that all the normal modes lie above the Raman energy. If both atoms have the same mass the site symmetry is D_{3d}, while for B^{11}-B^{10} pairs the symmetry is again reduced to C_{3v}. The displacement vectors for the four modes in D_{3d} symmetry are shown in Table 1; $^1\Gamma^-$ means a singlet mode of odd parity etc. Only the modes of odd parity should be infrared active. For B^{10}-B^{11} pairs, all four modes should be active, although absorption due to modes (3) and (4) would be very weak. Observation of mode (3) would give the most information about the force constant between the two boron atoms.

Table 1
Vibrational modes for B-B pairs

Mode	Displacements		Type
(1)	←	←	$^1\Gamma^-$
(2)	↑	↑	$^2\Gamma^-$
(3)	←	→	$^1\Gamma^+$
(4)	↑	↓	$^2\Gamma^+$

RESULTS

The absorption in crystals A, B and C near the isolated boron fundamentals at 623 (B^{11}) and 646 cm^{-1} (B^{10}) after subtraction of the lattice absorption is shown in Figs. 1-3. Satellite lines are in evidence in every case; the strength of these satellites appeared to be independent of the electron irradiation dose, for doses greater than 5×10^{18} cm^{-2} which was required to make the specimens transparent. No similar satellite structure was found in crystal D except for a line at 608.5 cm^{-1} which was also present in crystals A and B and can be attributed to the presence of carbon (C^{12})[7]. Values of the absorption co-efficients, in units of cm^{-1}, of other lines related to boron pairs are given in Table 2; the energies of all these lines were independent of the type of donor present.

Other lines at 733 and 760 cm^{-1} were present in all samples, including crystal D, and appear to be due to a centre involving a single boron atom, the configuration of which is still uncertain. However, it has been shown[4] that the relative strengths of the lines P_1, P_2 and P_3 are very close to those to be expected for the relative abundances of B^{11}-B^{11}, B^{11}-B^{10} and B^{10}-B^{10} pairs respectively, formed from naturally occurring boron. Calculations, assuming that the B-Si force constant was equal to the Si-Si force constant, gave the energies of the ${}^{1}\Gamma^{-}$ modes of B^{11}-B^{11} and B^{10}-B^{10} pairs as 572 and 590 cm^{-1}. The small discrepancy between experiment and theory can be removed if the B-Si force constant is reduced, as is also necessary for isolated boron[1]. It is therefore reasonable to identify the lines P_1, P_2 and P_3 with the ${}^{1}\Gamma^{-}$ type of mode of a pair of nearest neighbour boron atoms.

The calculated position[6] of the ${}^{2}\Gamma^{-}$ mode of B^{11}-B^{11} pairs is at 637 cm^{-1} which after reduction by the factor 553/572, to

Table 2
Absorption data for satellite lines due to boron

crystal \ Line energy	P_1 552.5 cm^{-1}	P_2 560.0 cm^{-1}	P_3 570.0 cm^{-1}	P_4 703.5 cm^{-1}
A	5.96	2.77	0.30	.06
B	1.86	0.80	0.10	see text
C	0.62	0.29	*	*
D	*	*	*	*

* not detected.

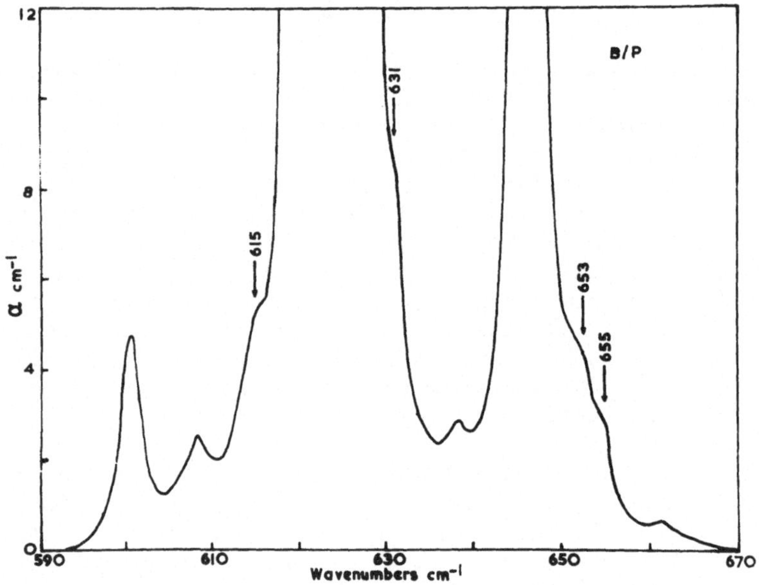

Fig. 1. Local mode absorption in silicon doped with B/P

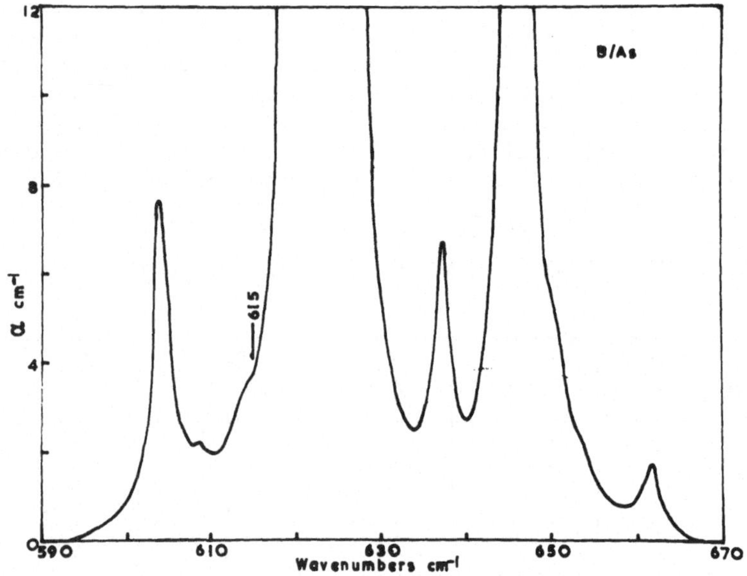

Fig. 2. Local mode absorption in silicon doped with B/As

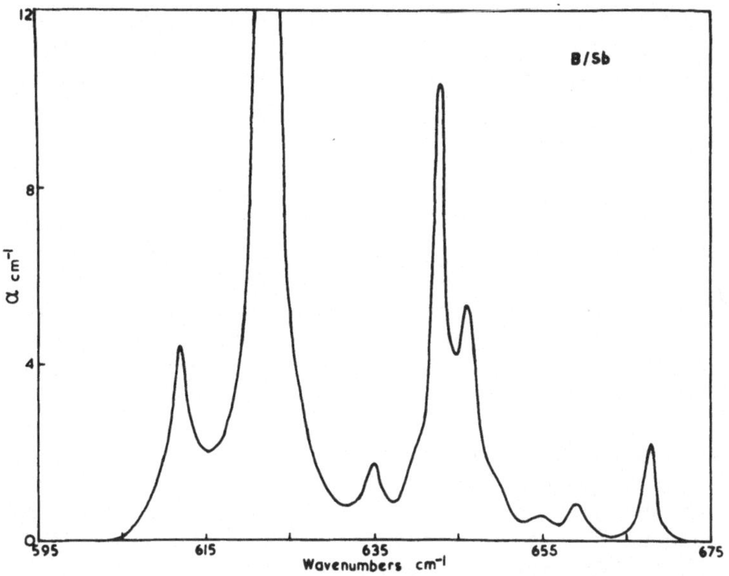

Fig. 3. Local mode absorption in silicon doped with B/Sb

allow for force constant changes, becomes 616 cm^{-1}. There is a shoulder on the low energy side of the isolated B^{11} line which is common to samples A and B, Figs 1 and 2. A differential spectrum was obtained with crystal A in the sample beam and crystal B in the reference beam of the spectrometer and this shoulder was then resolved, because of the partial cancellation of the isolated B^{11} absorption, and its position was located at 615 cm^{-1}. A relatively crude estimate of the peak absorption co-efficient of this band gave 2.4 cm^{-1} for crystal A and 0.8 cm^{-1} for crystal B in qualitative agreement with the relative strengths of the line P_1 in the two samples. On the basis of these correlations, it is proposed that the line at 615 cm^{-1} is due to the $^2\Gamma^-$ mode of B^{11}-B^{11} pairs. The strength of this line is then estimated to be about 0.3 cm^{-1} in sample C at which level it would not be detectable. The position of the corresponding band for B^{11}-B^{10} pairs falls in the region of the B^{11} fundamental where it would not be observable.

It is further tentatively suggested that the line P_4 which had a half width of about 3 cm^{-1} is the $^1\Gamma^{(+)}$ mode of B^{11}-B^{10} pairs. The strength of this band in sample A was close to the limit of detection and although there was an indication of structure at this position in sample B the presence of a line could not be established with certainty. The line is not thought to be due to an irradiation damage centre as it has not been observed in a variety of other irradiated crystals. The calculated energy[6] for this mode is 735 cm^{-1}, which after a correction for a reduction in the Si-B force constants becomes 711 cm^{-1} ; this value is within 7 cm^{-1} of the observed position of line P_4 and assuming the correctness of our interpretation would indicate that the B-B force constant is not very different from that between Si-Si.

The remaining strong satellite lines shown in Figs 1-3 depend on the n-type impurity which is present and it is concluded that they arise from boron donor pairs. The lines from samples A and B have been analysed previously[5] and a summary of the assignments made is given in Table 3. At the expected position of the A_1 mode for B-P pairs, two lines of almost equal intensity at 653 and 655 cm^{-1} were resolved in a differential spectrum between samples A and B. Subsequently an irradiated sample of crystal A has been annealed up to 300°C and then reirradiated and further examined. This led to a significant increase in the strength of the B^{11}-P pair lines and also the line at 653 cm^{-1}, while the line at 655 cm^{-1} was unchanged. A small shoulder and a line respectively were also observed at 655 cm^{-1} in samples B and C and it is therefore concluded that this line does not arise from B-donor pairs. The positions of the B-P pair lines then agree closely with those found in samples compensated with lithium[3] see Table 3 (small discrepancies are thought to be due to calibration errors). The annealing treatment of sample A also led to an increase in the absorption in the band mode region[1] near 20.5 μm; absorption in this region has been predicted[6] for B-P pairs but further measurements would be required to establish a definite correlation. Weak lines coincident with those ascribed to B-As pairs were also seen in sample A and are thought to be due to the presence of a small concentration of arsenic in this crystal; these lines were not observed in samples compensated with lithium[3].

Table 3
Positions of B-P and B-As vibrational satellites

pairs	B^{11} -donor E mode	B^{11} -donor A_1 mode	B^{10} -donor E mode	B^{10} -donor A_1 mode
B-P	600.6cm^{-1}	631 cm^{-1}	622 cm^{-1} *	653 cm^{-1}
B-P†	599.7cm^{-1}	629 cm^{-1}	622.9cm^{-1}	-
B-As	604.1cm^{-1}	637.4cm^{-1}	625 cm^{-1} *	661.8 cm^{-1}

* estimated position - line obscured by isolated B^{11} line at 623 cm^{-1}. † results from ref (3).

The remarkable aspect of the spectrum from sample C Fig.3 is that it indicates that about half of the boron atoms have formed pairs. Antimony appears to be the only impurity present in a sufficiently high concentration to account for this pairing. Inspection of Fig. 3 shows that the lines at 612 and 643 cm^{-1} arise from B^{11} complexes while those at 635 and 668 cm^{-1} involve B^{10}. Spectra from other specimens of crystal C showed somewhat less pairing with a proportionate reduction in the absorption coefficients of all the satellite lines. This spectrum is also different from that of crystal B since the higher energy satellites from each isotope are now the stronger lines. If these lines are ascribed to B-Sb pairs, then these observations imply either (a) that the higher energy satellite now corresponds to the E-modes of vibration or (b) that the dipole moment associated with vibrations along the pair axis (A_1 mode) for B-Sb is increased very considerably compared with that for B-As pairs.

DISCUSSION

The observed splittings between the A_1 and E modes for all three types of boron donor pairs are greater than estimated on the basis of changes only in the mass of the perturbing atom and indicate that the force constants must be changed for these centres. It is interesting to note that the actual splittings are almost independent of the type of donor atom in the pair but that the centre of gravity of the lines shifts to higher energies for the

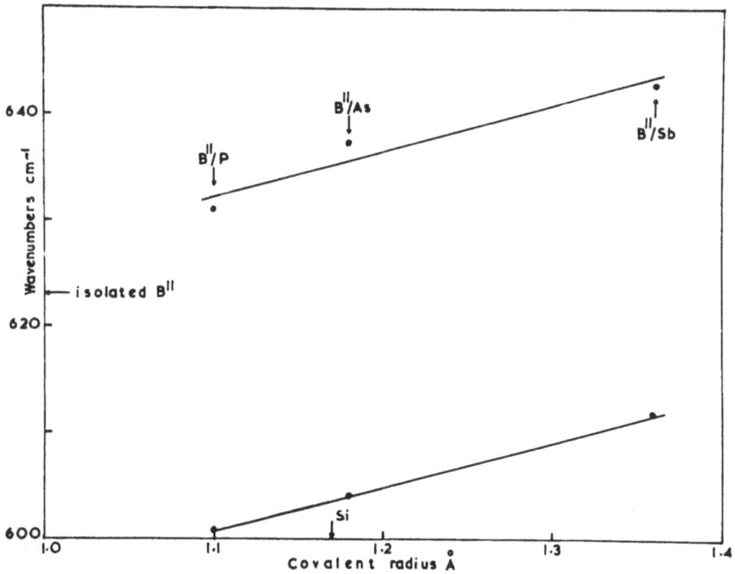

Fig. 4. Positions of B^{11}-donor local mode lines versus donor size

sequence phosphorus, arsenic and antimony as shown in Fig. 4. It is suggested that this may be due to the increasing covalent radii of the donor impurities which may account for increases in the force constant. The elastic strain energy around the large antimony atoms would also be relieved if they paired with the small boron atoms and this could account for the much higher degree of pairing in the B/Sb samples than in the B/P crystals.

ACKNOWLEDGEMENTS

One of us (R.S.S.) would like to thank the S.R.C. for a research studentship. Thanks are also due to M.Dixon of the A.E.I. Central Research Laboratory, Rugby for growing crystals C and D used in this investigation.

REFERENCES

(1) J.F. Angress, A.R. Goodwin and S.D. Smith, Proc.Roy.Soc. A287, 64, (1965).
(2) M. Balkanski and W. Nazarewicz, J.Phys.Chem.Solids, 27, 671, (1966).
(3) V.Tsvetov, W. Allred and W.G. Spitzer, Appl. Phys. Letters, 10, 326, (1967).
(4) R.C. Newman and R.S. Smith, Phys. Letters, 24A, 671, (1967).
(5) R.C. Newman and R.S. Smith, Solid State Comm. to be published.
(6) R.J.Elliott and P. Pfeuty, J.Phys. Chem. Solids, 28, 1789, (1967).
(7) R.C.Newman and J.B. Willis, J.Phys.Chem.Solids, 26, 373, (1965).

LOCALIZED VIBRATIONAL MODES OF DEFECT PAIRS IN SILICON

V. Tsvetov*, W. Allred†, and W. G. Spitzer†

Department of Materials Science

University of Southern California

INTRODUCTION

There have been many recent studies(1) of the effect of impurities, both substitutional and interstitial, on the vibrational modes of crystals. The conditions for introducing high frequency localized vibrational modes have been discussed in the literature, and experimental observation of infrared absorption bands associated with localized modes have been reported for a number of systems. Several cases have involved impurities, with concentrations ranging between 10^{16} and 10^{20} cm^{-3}, in semiconductor crystals. In many cases the impurity used is an electrical dopant, and the resulting absorption from the large free carrier concentration must be reduced by the introduction of an electrically compensating impurity. This is the reason for some cases involving pairs of impurities.

Of interest to the present study is the experimental work done with B and B-Li impurities in Si. Measurements(2) place the triply degenerate localized mode for isolated substitutional B near 624^{-1} cm^{-1} at liquid nitrogen temperature for ^{11}B and $647cm^{-1}$ for ^{10}B. Both frequencies are larger than the highest unperturbed silicon phonon frequency of $\sim 518cm^{-1}$. When interstitial Li compensates B some rather striking effects have been reported(3,4). Most of the triply degenerate B band is split, indicating doubly and singly degenerated modes separated by $90cm^{-1}$ with the doubly

---

* Participant of the Educational and Scientific Exchange Program USSR-USA.

† The research reported in this paper was sponsored in part by a research grant from the National Aeronautics and Space Administration Grant NGR-05-018-083.

degenerate modes at higher frequency. In addition a band is observed slightly above the top of the unperturbed spectrum. The frequency of this mode is dependent upon the Li isotope but insensitive to the B isotope. A qualitative explanation has been offered in which these bands are related to the formation of ion pairs, B_{Si}-Li_i, where B_{Si} is B on a Si site and Li_i is interstitial Li. The point group symmetry at the B and Li sites is reduced from tetrahedral (T_d) to axial (C_{3v}) by the pairing.

This work is an experimental study of Si doped heavily with B and largely compensated with a substitutional donor impurity, i.e. P, As, or Sb. The compensation is completed by diffusing Li. New absorption bands have been observed which can be attributed to B-P, B-As, and B-Sb nearest neighbor pairs. In some cases, notably B-P, the strength of some of the absorption bands depends upon the time and temperature of Li diffusion. This latter effect is interpreted in terms of impurity precipitation.

EXPERIMENTAL METHOD

Silicon ingots were pulled from a melt of high purity Si and the desired impurities. The B and Sb were introduced in elemental form while P and As were added as $Ca_3(PO_4)_2$ and $Ca_3(AsO_4)_2$. In most cases B and the donor impurity were added to the melt simultaneously with concentrations adjusted to make the crystal p-type but, where possible, nearly compensated. The B concentration [B] was always $\gtrsim 10^{20}cm^{-3}$ while the donor concentrations varied from near $10^{20}cm^{-3}$ for [P], $\sim 5\times10^{19}cm^{-3}$ for [As], and $\sim 2 - 3\times10^{19}cm^{-3}$ for [Sb]. The samples were single crystal or polycrystalline with crystallities of several mms. in dimension. The B used was enriched ^{11}B, i.e. 98% ^{11}B and 2% ^{10}B. Four point probe measurements of the resistivity ρ gave values consistent with the doping levels. The silicon was diffused with natural Li, i.e. ~93% ^7Li and 7% ^6Li. The resulting compensated samples had $\rho \gtrsim 100\Omega$ cm. The times, temperatures, and conditions for Li diffusion have already been discussed[4,5]. Spectral half widths were always less than $1.0cm^{-1}$ wave number and generally near $0.5cm^{-1}$. All optical measurements were made at liquid nitrogen temperature.

EXPERIMENTAL RESULTS

A. Localized Mode Frequencies

In Fig. 1 results are shown for three samples, each doped with ^{11}B, a substitutional donor, and Li. The term donor will be used only for the substitutional donor impurity and not for the diffused Li. The bands near 523, 536, 566, 586, 622, and 656 agree closely with bands reported for samples not containing the donor. These absorption bands are related to the presence of B

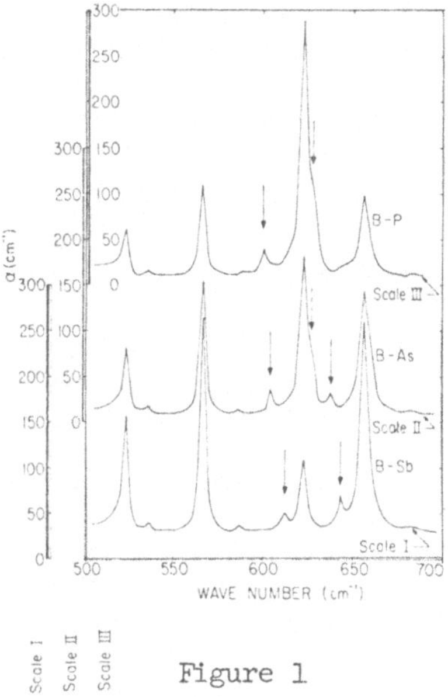

Figure 1

and Li. New bands[6] have been observed for each donor, and their positions in Figure 1 are indicated by the arrows. There are two bands in each case with a possible third one in the B-As sample. The frequencies depend upon the donor employed. In a B-P crystal the new bands have very nearly the same isotope shift as isolated B band. Measurements at 5°K gave nearly the same results as those obtained at nitrogen temperature. The line widths and absorption peak positions as a function of temperature for the new band are, within experimental accuracy, the same as for the isolated B line. Several B-P doped ingots have been grown and the new bands are present in all cases. An ingot was doped with B and elemental Ca, and Li compensated. The concentration of Ca introduced was the same as that by the $Ca_3(PO_4)_2$ and only B and B-Li bands were observed.

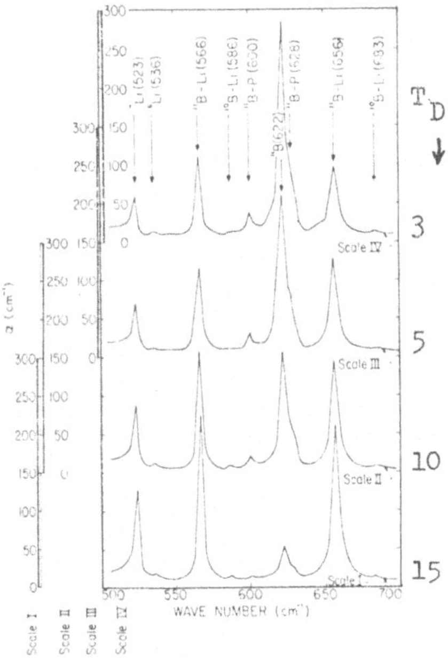

Figure 2

B. Dependence of Absorption Band Intensities on Li Diffusion Conditions

In B-P doped samples the relative intensities of several of the local mode absorption bands are a function of the time and temperature of Li diffusion. The change is indicated in Fig. 2 for a set of four B-P doped samples. The samples were adjacent to one another in the ingot and all Li diffused at 800°C. The diffusion time ranged from 3 to 15 hours. There is a decrease in strength with increase in the diffusion time for both the $622 cm^{-1}$ with a concomitant increase in the bands near 656, 566, and $523 cm^{-1}$. The relationship between these changes is demonstrated in Fig. 3 where

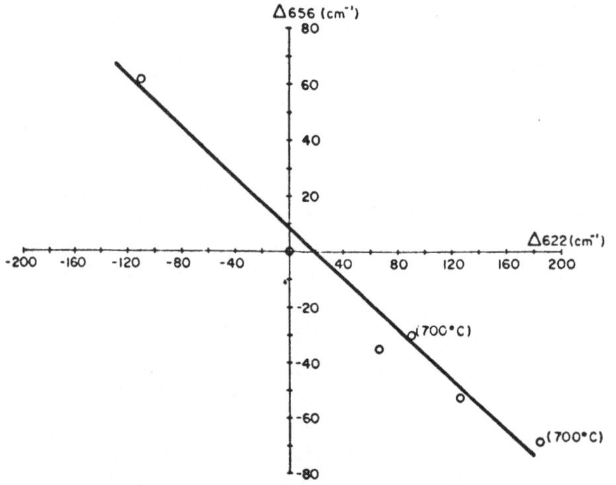

Figure 3

the change in the 622 cm^{-1} band peak height is plotted against that of the 656cm^{-1} band. The changes are measured with respect to the second sample from the bottom in Fig. 2 where both bands are nearly the same. Included in Fig. 3 are two samples labeled (700°C). One was diffused for 8 hours (point with Δ622=186 cm^{-1}) and the other for 96 hours at that temperature. Within the scatter of the data, the points indicate a linear relation given by $\Delta 622 = -(0.45)\,\Delta 656$. The same result is obtained if the 566cm^{-1} band is used in place of the 656cm^{-1}.

No such effect has been observed in the B-As doped samples. The B-Sb samples show a small decrease in band strength for all bands with longer times of diffusion at 800°C as well as an increase in the apparent background absorption.

DISCUSSION OF RESULTS

A. Localized Mode Frequencies

Table I lists the local mode absorption bands observed in the present measurements. Frequencies are the position of the peak absorption coefficient. The new bands have been labeled as ^{11}B-P, ^{11}B-As, and ^{11}B-Sb pair bands. The reasons for this assignment are similar to those used to establish the identification of the B-Li pair bands[4]. The nearly full B isotopic shift, the proximity to the isolated B line, and the shift in frequency with change in donor species indicate the identification of the vibrational modes as ones which primarily involve B motion but the point group symmetry at the B site has been lowered from T_d to C_{3v} by the nearest neighbor substitutional donor. The triply degenerate isolated ^{11}B mode should be split into two bands with one twice the strength of the other. If the donor were a 2nd nearest neighbor, the symmetry becomes C_{2v} and three bands could result but with reduced splittings. In the B-P and B-Sb cases only two bands are observed. In the B-As case there is a third band near 627cm^{-1}. We tentatively ascribe this latter band to an unresolved second nearest neighbor interaction where the remaining structure is

TABLE I

Liquid Nitrogen Temperature Frequencies of Absorption Bands

Band	^{11}B-^{7}Li	^{10}B-^{7}Li	^{11}B-P	^{11}B-As	^{11}B-Sb
B-Li Pair	656.1±0.6	683	---	---	---
Bands	566.1±0.6	586	---	---	---
B Band	622.2±0.5	647	---	---	---
Li Band	523.3±0.5	523.3±0.5	---	---	---
B-Donor	---	---	~628	636.7±0.4	642.7±0.3
Bands	---	---	600.1±0.5	603.7±0.3	611.9±0.3
	---	---	---	~627	---

under the large 622 isolated ^{11}B line.

The relative strengths of the B-(Donor) bands are difficult to determine as they are weak and close to the strong isolated B and B-Li bands. An estimate is made by assuming the background in the region of B-D bands to be a constant plus the tails of the nearby major bands. The absorption tails are estimated by assuming at each local mode can be represented by a collection of non-interacting harmonic oscillators of resonance frequency ν_o. This model gives an absorption coefficient

$$\alpha = \frac{C}{n} \left\{ \frac{\nu^2}{(1-\nu^2)^2 + \nu^2\delta^2} \right\} ,$$

where ν and δ are the frequency and damping constant normalized by the resonance frequency, n the refractive index, and C a constant. It is assumed that n is independent of frequency. Figure 4 shows the curve fitting for a B-P sample. The only major band close enough to the B-P pair bands to have any effect is the isolated ^{11}B band at 622cm^{-1}. C/n was adjusted to give α_{peak} = 250cm^{-1} with δ of 0.008 (5.0cm^{-1}). Measurements over a more extended frequency range indicate that for this sample, $\alpha_{background}$ = 14cm^{-1}. Subtraction of

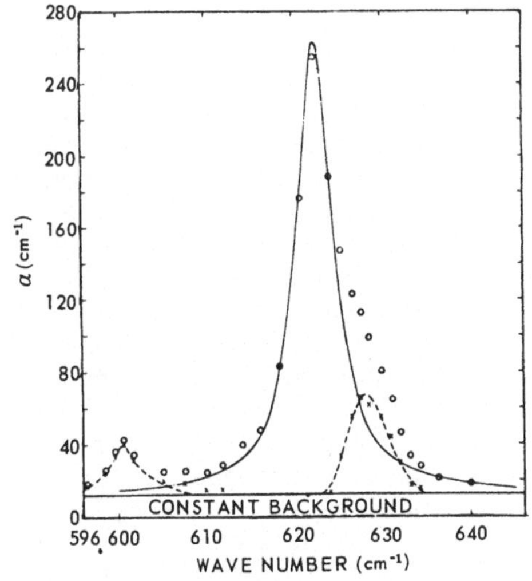

Figure 4

TABLE II

Sample	$\int \alpha d\nu$ (cm^{-2})	
	High frequency pair band	Low frequency pair band
B-P	297	133
B-P	200	132
B-As	65	110
B-Sb	120	90

$\alpha_{background}$ and the calculated curve from the measured data points give the dashed curves in Fig.4.

The results for estimates of $\int \alpha d\nu$ for the pair bands in several samples are given in Table II. While the results are not conclusive, they suggest that the high frequency band is the doubly degenerate one for B-P and B-Sb and the low frequency one for B-As. Comparison of the total integrated absorption of the B-Donor pair bands to that for isolated B gives $\Sigma_{pair} \int \alpha d\nu = 0.1$ to $0.2 \int_B \alpha d\nu$ for both B-P and B-As and 0.4 to $0.5 \int_B \alpha d\nu$ for B-Sb. The difference could arise either from a somewhat larger pairing energy for the B-Sb case or from the equilibrium configuration for B-Sb being characteristic of a somewhat lower temperature. The latter case would imply a larger diffusion constant for Sb than for As and P which is not in agreement with published data.[7]

The qualitative conclusions given here are in general agreement with the recent theory of Elliott and Pfeuty.[8] They calculate the effect of pairs of defects on the lattice modes by using Green's function methods. Of particular interest are their calculations for the frequencies of the localized modes of defect pairs in silicon where one of the defects is B_{Si} and the other a substitutional mass M_2. Their results for no change in force constants are reproduced here in Figure 5. The solid lines are for ^{11}B, the dashed for ^{10}B, ω_M the maximum unperturbed Si phonon frequency, and M the Si mass. For the cases studied here the mass defect $\varepsilon_2 < 0$ and only two modes should be observed, those labeled modes 1 and 2 on Fig. 5. The theory indicates that in both modes B motion dominates. The fact that the observed splitting of modes 1 and 2 is much larger than that given in Fig. 5 is evidence that a change in force constant is necessary. The increase in frequency of the pair bands as one goes from P to As to Sb is in the same order as the tetrahedral covalent radii which are

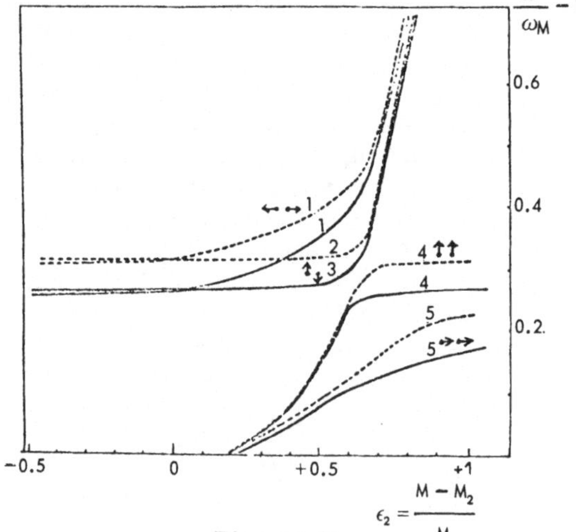

Figure 5

1.10, 1.18, and 1.36Å resepectively.[9]

B. Impurity Precipitation Effects

It was observed that Δ 622 = -(0.45) Δ 656 = -(0.45) Δ 566. With the half widths of 7.0cm^{-1} for the 656cm^{-1} band, 4.5cm^{-1} for 566cm^{-1} band and 5.0cm^{-1} for the 622cm^{-1} band it is observed that the change in $\int \alpha d\nu$ for the isolated B band is equal and opposite in sign to the change in total $\int \alpha d\nu$ for the B-Li pair bands. This result indicates that the total substitutional [B] does not change, and the total absorption strength per center is the same for isolated B and B-Li pairs. This result, the assumption that the isolated B concentration equals the isolated substitutional P concentration, and the data for B-Li doped samples may be used to express the strength of the bands in terms of the concentration of the different defects. In these estimates the B-D pairs were neglected since their absorption bands indicate the concentrations to be relatively small.

The results of Fig. 2 are consistent with a model involving P precipitation or any other process which removes P as an electrically active dopant. From the strength of the 622cm^{-1} band, the substitutional [P] is found to decrease from 9x10^{19}cm^{-3} to ~1.4x10^{19}cm^{-3} after 15 hrs. of Li diffusion at 800°C. It is known[10] that the solid solubility of a donor may be considerably enhanced by a large concentration of acceptors. Annealing of B-P doped samples for 17 hrs. at 740°C had no effect on either the resistivity or on the local mode strengths obtained after Li diffusion. During Li diffusion, Li compensates the excess [B] which leads to a decrease in the solubility of P to a value near that in Si without acceptors. Processes of precipitation of P in Si and in Ge-Si alloys have been reported[11,12] for concentrations similar to those used in the present experiments. At 800°C the time constant for achieving the equilibrium [P] is ~ 10 hrs. while at 700°C the time constant $\gtrsim$ 1000 hrs. These time constants are only approximate as the data are not sufficient to establish the kinetics of the process. Unfortunately, the present authors have been unable to find data on the P solid solubility in Si at 800°C.

The B-As doped samples with [As] of 5x10^{19}cm^{-3} did not show any decrease in [As] with Li diffusion at 800°C indicating that the As solid solubility is above this concentration.

In the B-Sb case the [B] ~ 1.9x10^{20}cm^{-3} and [Sb] ~ 2.5x10^{19} cm^{-3}. Although changes are small compared to those observed in the B-P case, comparison of samples Li diffused at 800°C indicates a decrease of ~ 20% in the isolated Sb concentration as a result of long term diffusion. The Sb concentration is close to the published[13] solid solubility data at 800°C which indicates ~ 2x10^{19}cm^{-3}. In addition a decrease is observed in the total substitutional B concentration to ~ $1.4 \times 10^{20} cm^{-3}$. This concentration is close to the [B] in the B-P doped samples where no B precipitation was observed.

REFERENCES

(1) For a discussion of this topic see A.A. Maradudin, Solid State Physics (Edited by F. Seitz and D. Turnbull, Academic Press, New York, 1966) Vols. 18, 273.

(2) J.F. Angress, A.R. Goodwin, and S.D. Smith, Proc. Roy. Soc. (London) 287A, 64 (1965).

(3) M. Balkanski and W.A. Nazarewicz, J. Phys. Chem. Solids 27, 671 (1966).

(4) M. Waldner, M.A. Hiller, and W.G. Spitzer, Phys. Rev. 140, A172 (1965); also W.G. Spitzer and M. Waldner, J. Appl. Phys. 36, 2450 (1965).

(5) E.M. Pell, J. Phys. Chem. Solids 3, 77 (1957).

(6) V. Tsvetov, W. Allred, and W.G. Spitzer, App. Phys. Letters, 10, p. 326 (1967).

(7) H. Reiss and C.S. Fuller, Semiconductors (Edited by B. Hannay, Reinhold Publishing Co., New York, 1959) p 244.

(8) R. Elliott and P. Pfeuty, J. Phys. Chem. Solids (to be published).

(9) L. Pauling, The Nature of the Chemical Bond (Cornell Univ. Press, Ithaca, New York, 1960) p 246.

(10) H. Reiss, C.S. Fuller, and F.J. Morins, Bell System Technical Journal 35, 535 (1956).

(11) M.L. Joshi and S. Dash, IBM Journal of Research and Development 10, 446 (1966).

(12) L. Ekstrom and J.P. Dismukes, J. Phys. Chem. Solids 27, 857 (1966).

(13) F.A. Trumbore, Bell System Technical Journal 39, 205 (1960).

Note: It has become apparent at this conference that the assignment of the observed frequencies to the different modes on the basis of the relative strengths of the absorption bands is a questionable procedure. Therefore the assignments are not well established as they are based on $\int \alpha d\nu$ values.

THEORY OF VIBRATIONS OF PAIRS OF DEFECTS IN SILICON

P. M. Pfeuty

Department of Theoretical Physics, Oxford University

Laboratoire de Physique des Solides, Faculte des Sciences, Paris

1. INTRODUCTION

The theoretical[(1)(2)] and experimental study[(3)(4)] of the vibrations of defects in silicon has been developed in the last years. Localised modes have been observed by infrared optical absorption when a light charged boron impurity is introduced. To remove the free carrier absorption a lithium[(4)] or phosphorus[(3)] donnor impurity has to be added. Some of the absorption bands can be explained as due to isolated boron impurities and agree with a simple mass defect model[(2)]. The other bands appearing with sufficiently high impurity concentrations[(5)(6)(7)] can only be explained with a model of pair defects. Two different pair defect configurations have been identified, a pair of substitutional impurities and a pair of one substitutional and one interstitial impurity. The theory of these pair defects have been recently made[(8)] using a Green's function technique and is extended to calculate the frequency and the intensity of the observed localized mode absorption bands.

2. THEORY

We will only consider the one defect problem. If s is the number of sites perturbed by the defect, the localised mode frequencies are solutions of the secular equation

2.1 $$\mathrm{Det}\,(1 - \underline{g}\underline{C}) = 0$$

The 3s x 3s matrices $\underline{g}$ and $\underline{C}$ are respectively the Green's function and the perturbation matrix associated with the defect subspace.

The optical absorption coefficient is given by:

2.2 $$\alpha(\omega) = \frac{4\pi\omega\Lambda p}{nc} \sum_{i,i'} e_i e_{i'} G^I_{ii'}(\omega)$$

2.3 $G^I_{ii'} = \lim_{\psi \to o} \mathrm{Im}\, G_{ii'}(\omega + i\psi)$ where $G_{ii'}$ is an element of the imperfect crystal Green's function matrix

$$\underline{G} = (g\underline{C} - 1)^{-1}\, g$$

p is the defect concentration, n the refractive index and Λ the local field correction.

To treat the defects considered here and because of the symmetry of the problem only six independent Green's functions are used.

2.4
$$g_1 = g_{1x\,1x}(0,0) \qquad g_4 = g_{1x\,1x}(0,\ell)$$
$$g_2 = g_{1x\,2x}(0,0) \qquad g_5 = g_{1z\,1z}(0,\ell)$$
$$g_3 = g_{1x\,2y}(0,0) \qquad g_6 = g_{1x\,1y}(0,\ell)$$

where $x_\ell = \frac{1}{2}\, a(110)$

The real part and the imaginary part of these functions have been yet computed(8) using the perfect crystal dynamical matrix eigenfrequencies for silicon.

3. CALCULATIONS

3.1 The Two Nearest Neighbour Substitutional Impurities

The two impurities in the same cell are atoms of mass defect $\varepsilon_1 = \frac{M-M_1}{M}$ and $\varepsilon_2 = \frac{M-M_2}{M}$ and we first suppose that these are the only two perturbed sites. The defect belongs to the point symmetry C_{3v}. The mechanical representation for the defect is

3.1 $$2\Gamma_1 + 2\Gamma_3$$

There are two Γ_1 and two Γ_3 localised modes whose frequencies are solutions of 2.1 and are given by

3.2 $$M\omega^2 = \frac{(\varepsilon_1+\varepsilon_2)[g_1-\phi(g_1^2-g_u^2)]}{2\varepsilon_1\varepsilon_2(g_1^2-g_u^2)} \pm \frac{\{(\varepsilon_1+\varepsilon_2)^2[g_1-\phi(g_1^2-g_u^2)]^2-4\varepsilon_1\varepsilon_2(g_1^2-g_u^2)[1-2\phi(g_1-g_u)]\}^{1/2}}{2\varepsilon_1\varepsilon_2(g_1^2-g_u^2)}$$

where $g_u = g_2 + 2g_3$ and $\phi = \phi\Gamma_1$ for the Γ_1 modes

$g_u = g_2 - g_3$ and $\phi = \phi\Gamma_3$ for the Γ_3 modes

If the atoms are the same $\varepsilon_1 = \varepsilon_2 = \varepsilon$ the point group is D_{3d} and the vibrational representatives are

3.3 $$\Gamma_1^+ + \Gamma_1^- + \Gamma_3^+ + \Gamma_3^-$$

the solutions are now

3.4 $$\varepsilon M\omega^2 = \frac{1}{g_1 \mp g_u} - \phi(1 \pm 1) \quad \text{the signs correspond to } \Gamma_i^\pm.$$

We see that the Γ_i^- modes where the impurity atoms are vibrating in the same way don't depend on ϕ.

The optical absorption from 2.2 is proportional to

$G^I = G_{11}^I \pm 2G_{12}^I + G_{22}^I$ where G is given in (8). The first sign corresponds to like charges and the other to unlike charges.

3.2 The Interstitial Defect

We consider an interstitial atom of mass M' on the tetrahedral site of the diamond lattice. The defect is limited to the interstitial 1 at (0,0,0) and the four first neighbours

$i = 2, 3, 4, 5$ at t_i where

$\underline{t}_2 = {}^a/4(-1,1,1)$ $\underline{t}_3 = {}^a/4(1,-1,1)$, $\underline{t}_4 = {}^a/4(1,1,-1)$, $\underline{t}_5 = {}^a/4(1,1,1)$

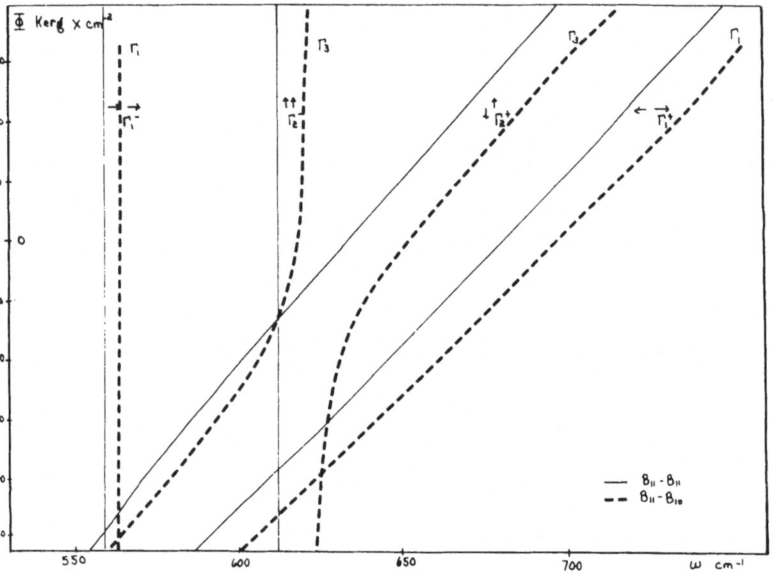

Fig. 1

Effect of the change of force constant ϕ($\phi\Gamma_1$ and $\phi\Gamma_3$ for the Γ_1 and Γ_3 modes) on the frequency of the boron-boron pair localised modes ε_{B10} = 0.595, ε_{B11} = 0.558.

Table 1

Impurity Center	Mode Symmetry	Frequency cm^{-1}			
		Experiments	Calculated (1)	Calculated (2)	Calculated (3)
$B_{11} - B_{11}$	Γ_1^-	552, 3	576	558, 8	558
$B_{11} - B_{11}$	Γ_3^-	615	641	612, 5	
$B_{10} - B_{10}$	Γ_1^-	570	590	571, 6	570
$B_{10} - B_{10}$	Γ_3^-		664	635	
$B_{10} - B_{11}$	Γ_1	560	581	564, 5	
B_{10}	Γ_5	645, 8	680	645, 8	646
B_{11}	Γ_5	622, 8	654	622, 8	623

(1) from eq. 3.2 $\varepsilon\ B_{11}$ = 0.607, $\varepsilon\ B_{10}$ = 0.643

(2) from eq. 3.2 $\varepsilon\ B_{11}$ = 0.558, $\varepsilon\ B_{10}$ = 0.595

(3) model including the change of force constants δ

Comparison between the observed and calculated values of the localised mode frequencies for B-B pairs.

3.5 The representations are $\Gamma_1 + \Gamma_3 + \Gamma_4 + 3\Gamma_5$ only Γ_5 is infrared active. Using symmetry coordinates of Γ_5 type, equation 2.1 is reduced to a 3 dimensional secular equation. The coupling of the interstitial with the neighbours is written

3.6 $$V = \frac{\beta}{2} \sum_{i=2-5} (\underline{r}_1 - \underline{r}_i)^2$$

After manipulation (8) equation 2.1 becomes

3.7 $$\frac{4\beta}{M'\omega^2} = [1-\beta(g_1+2g_4+g_5)] - \frac{2\beta^2 g_6^2}{1-\beta[g_1-g_5-g_6]} = H$$

If only the interstitial is charged the optical absorption is given by the imaginary part of

3.8 $$G = \frac{H}{(4\beta-M'\omega^2 H)}$$

3.3 The Substitutional-Interstitial Defect

This defect consists of an interstitial on a tetrahedral site and a substitutional impurity on one of the neighbouring sites chosen as site 5 of the last problem. The symmetry is now reduced to C_{3v} and the vibrational representations are

3.9 $$4\Gamma_1 + \Gamma_2 + 5\Gamma_3$$

both Γ_1 and Γ_3 are infrared active. The coupling can now be written

$$2V = \beta\sum_{i=2-5} (\underline{r}_1-\underline{r}_i)^2 + \frac{16\alpha}{a^2}[(\underline{r}_1-\underline{r}_5)\cdot\underline{t}_5]^2 + \sum_{i=2-4} \frac{4\gamma}{a^2}[(\underline{r}_5-\underline{r}_i)\cdot(\underline{t}_5-\underline{t}_i)]^2$$

Using symmetry coordinates of Γ_1 and Γ_3 type, equation 2.1 is reduced to a four dimension and a five dimension secular equation. The matrix expressions are given in (8). A six atom model has been considered including the change of force constant ϕ' between the substitutional atom and his first neighbour along $\underline{t}_5$.

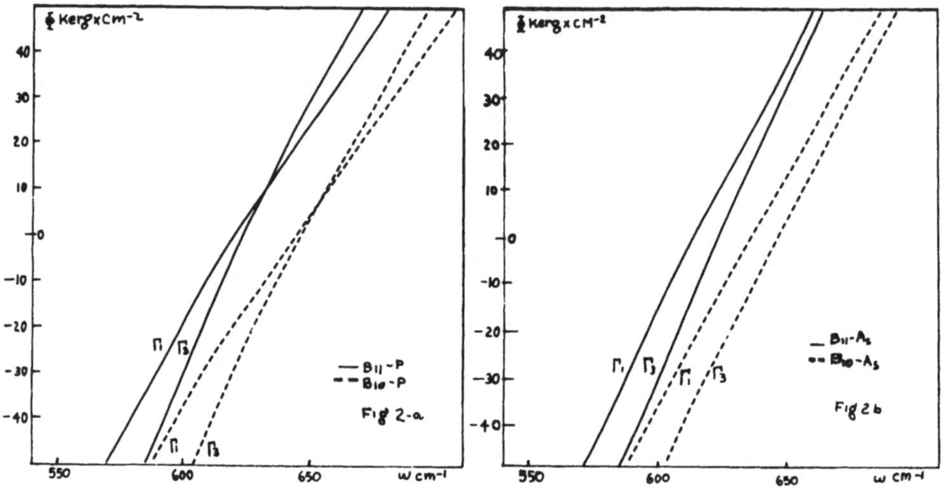

Fig. 2

Effect of the change of force constant ϕ($\phi\Gamma_1$ and $\phi\Gamma_3$ for the Γ_1 and Γ_3 modes) on the frequency of the boron-donnor pair localised modes $\varepsilon_{B10} = 0.595$, $\varepsilon_{B11} = 0.558$. Fig. 2a corresponds to B-P pairs and Fig. 2b to B-As pairs.

Table 2

Impurity Center	Mode Symmetry	Frequency cm^{-1} Experiments	Calculated	ϕ kergs x cm^{-2} $\phi(\Gamma_1)$	$\phi(\Gamma_3)$	Intensity Ratio Experiments	Calculated
B_{11} - P	Γ_1	631	631	8,6			
B_{10} - P	Γ_1	655	654,4	8,6			
B_{11} - P	Γ_3	600,6	600,7		-28,75		1,07
B_{10} - P	Γ_3	622	621		-28,75		1,3
B_{11} - As	Γ_1	637,4	637	25,8			
B_{10} - As	Γ_1	661,8	662	25,8			
B_{11} - As	Γ_3	604,1	604,6		-23		1,7
B_{10} - As	Γ_3	625	626		-23	1,3	1,74

Comparison between the observed and calculated values of the localised mode frequencies for B-P and B-As pairs.

4. DISCUSSION

4.1 The Two Nearest Neighbour Substitutional Impurities

Recently in boron doped silicon compensated with either phosphorous or arsenic localised absorption bands have been observed and ascribed to the vibrations of B-B[5], B-As, and B-P pairs[6][7] in nearest neighbour sites. We will first consider the B-B pairs. For the pairs B_{10}-B_{10} and B_{11}-B_{11} some of the infrared active Γ_i^- modes have been observed. For the B_{10}-B_{11} pairs all the modes are active, but the high frequency modes correspond to a low absorption intensity and have not yet been observed. The calculated Γ_i^- mode frequencies from equations 3.2 and 3.4 are higher than the experimental values (see Table 1). As for the single defect[2], this is due to a weakening of the next neighbour force constants. To calculate this effect we consider a 6 atoms defect model including the two identical impurities and the six silicon first neighbours. The symmetry remains D_{3d} and the vibrational representations are $3\Gamma_1^+ + \Gamma_2^+ + 4\Gamma_3^+ + 3\Gamma_1^- + \Gamma_2^- + 4\Gamma_3^-$. The Γ_1^- modes frequency is a solution of a three dimensional secular equation which depends on the Green's functions $g_1g_2g_3g_4g_5g_6$ and on the change of force constants $\Delta\Phi_{1x2x}(0,0) = \Delta\Phi_{1x2y}(0,0) = \delta$. We obtain a good fit with $\delta = -21,2$ kergs x cm^{-2} (the silicon force constant = 55,6 kergs x cm^{-2}). A similar calculation has been done with a five atoms model of symmetry T_D for the single impurity and we got $\delta = -6,5$ kergs x cm^{-2}. When two impurities are put together the force constants coupling the impurities with the surrounding silicon atoms are reduced. This may be due to a transfer of electrons from the boron silicon bonds to the boron boron bonds. We used also an "effective mass defect" model. The new values $\varepsilon B_{10} = 0.595$ and $\varepsilon B_{11} = 0.558$ are deduced from the single impurity localised modes and used in equations 3.2 and 3.4 to calculate the Γ_i^- frequencies. These calculated values agree with the experiments. All these results are summarized in Table 1. The effect of ϕ is shown in Fig. 1 where the frequency of the modes is plotted versus $\phi\Gamma_1$ and $\phi\Gamma_3$ for B_{11} - B_{11} and B_{10} - B_{11} pairs. ϕ acts only on the Γ^+ Raman active modes and actually such Raman scattering

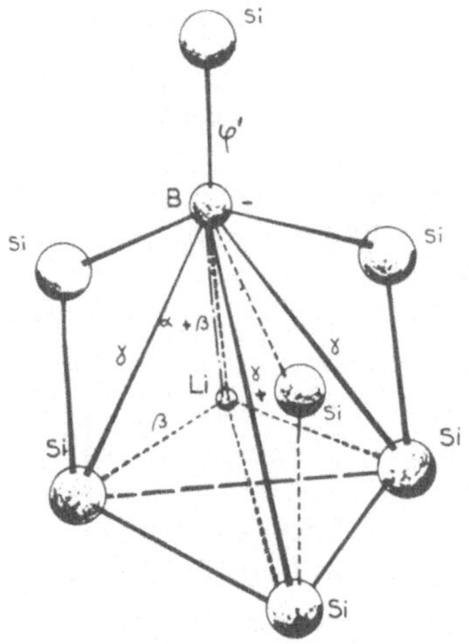

Fig. 3

Structure of the boron lithium defect showing the different force constant parameters.

Table 3 Comparison between the observed and calculated values of the localised mode frequencies for B-Li pairs.

Isotope	Observed Frequency cm^{-1}	Degeneracy	Calculated frequency A	B	C
B_{10}	681	2		707	678
B_{11}	653	2		679 (Li_7) 683 (Li_6)	651 (Li_7) 655 (Li_6)
B_{10}	584	1		609	584
B_{11}	564	1		590	565.5
Li_6	534	2 or 3	533.5	533	533
Li_7	522	2 or 3	522	522	522

The models calculated are

A interstitial Li $\beta = 22$

B B-Li pair 5 atoms model $\beta = 23$ $\alpha = -60$

C B-Li pair 6 atoms model $\beta = 23$ $\alpha+\phi' = -58$ $\gamma = -7$

(all force constants in units $Kergs/cm^2$)

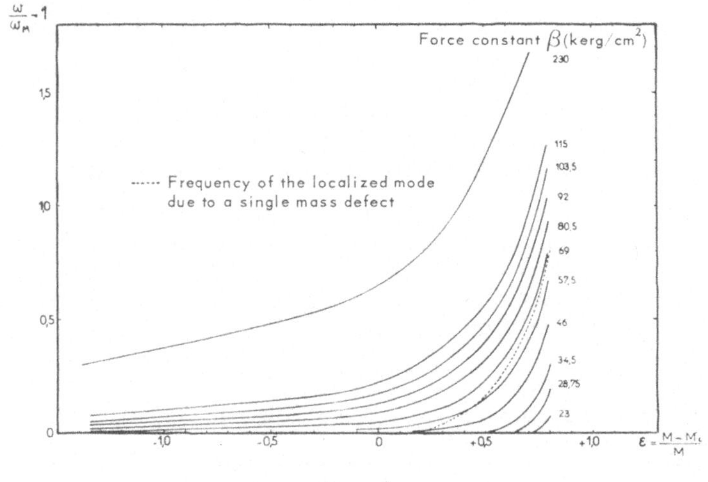

Fig. 4

Frequency $\frac{\omega}{\omega_M}$ - 1 of the triply degenerate localised mode due to an interstitial impurity of mass M' coupled to the first neighbours with the force constant β.

experiments in imperfect crystals have only been made in germanium doped with silicon(9).

For B-P and B-As pairs, the heavy impurity changes the force constants around the boron atom and the triplet associated with isolated boron is split into a Γ_1 and a Γ_3 mode. The frequency of these modes deduced from equation 3.2 with $\varepsilon B_{11} = 0.558$ and $\varepsilon B_{10} = 0.595$ is plotted versus $\phi\Gamma_1$ and $\phi\Gamma_3$ in Fig. 2a and 2b for B-P and B-As pairs. $\phi\Gamma_1$ and $\phi\Gamma_3$ are deduced from the comparison with the experimental results. All the results are summarized in Table 2. $\phi\Gamma_1$ is positive and increases when we replace phosphorous by arsenic. $\phi\Gamma_3$ is negative and does not depend on the donnor impurity. $\phi\Gamma_1$ and $\phi\Gamma_3$ are related to the Born type change of force constants δ and δ' respectively for the B-Si bond and for the pair bond. $\phi\Gamma_1 = \delta + 3\delta'$, $\phi\Gamma_3 = 4\delta$. From these relations we conclude that when a donnor approaches boron the boron silicon force constant decreases ($\delta = -5{,}7$ kergs x cm^{-2}), this effect being independent on the nature of the donnor, and the force constant along the pair bond increases with a greater magnitude for arsenic ($\delta' = 10{,}5$ kergs x cm^{-2}) than for phosphorous ($\delta' = 5{,}3$ kergs cm^{-2}). It would be useful to have results for the pair boron antimony. All these considerations suppose that the lines have been identified from intensity measurements. For the B_{10}-As pair the measured Γ_3 to Γ_1 intensity ratio is only equal to 1,3. This result cannot be explained if we suppose that point charges are located on the defect atoms. The calculated values obtained with this model don't agree with the experiments (see Table 2).

4.2 Boron Lithium Pairs in Silicon

Boron and lithium introduced in silicon form substitutional interstitial pairs shown in Fig. 3. From the experimental results (4) some lines are only due to lithium vibrations. At first we will suppose that these lines are due to isolated lithium vibrations. The frequencies of interstitial localised modes as given by 3.7 are plotted in Fig. 4 against the interstitial mass M' for various values of the force constant β. From the comparison with the observed frequencies for Li_6 and Li_7 we get $\beta = 22$ kergs x cm^{-2}. Recently(10) the force constants β_1 and β_2 coupling the lithium interstitial respectively with the first and second neighbours silicon atoms have been calculated

$\beta_1 = 6$ kergs x cm^{-2} $\beta_2 = 17$ kergs x cm^{-2}. Our results are in good agreement if we suppose that β includes both β_1 and β_2.

If we now consider the boron lithium pair, the observed B_{10} and B_{11} lines are each split into two. Those at higher frequency are almost twice as intense as the lower ones and are interpreted as corresponding to the doublet vibrations. The lithium lines are fitted with $\beta = 23$ kergs x cm^{-2} $\alpha = \gamma = 0$ but the boron lines are too high and not split. These lines are shifted to low frequency by weakening the next nearest neighbour force constant by γ and then split by weakening the B-L_i force constant by α and the boron silicon force constant along the bond by ϕ'. The results are summarized in Table 3.

Acknowledgments

I am grateful to Dr. R. J. Elliott for his guidance during the course of this work and I thank Dr. R. C. Newman for making data available to me before publication.

References

1. R. J. Elliott, "Phonons" Aberdeen Summer School Lectures, 1965 (Oliver and Boyd).
2. R. J. Elliott and P. G. Dawber, Proc. Phys. Soc. 81, 521 (1963)
3. J. F. Angress, A. R. Goodwin and S. D. Smith, Proc. Roy. Soc. A, 287, 64 (1965).
4. M. Balkanski and W. Nazarewicz, J. Phys. Chem. Solids 27, 671 (1966).

   M. Waldner, M. A. Hiller and W. G. Spitzer, Phys. Rev. 140A, 172 (1965).
5. R. C. Newman and R. S. Smith, Phys. Lett. 24A, 671 (1967).
6. R. C. Newman and R. S. Smith, to be published.
7. Y. V. Tsvetov, W. Alfred, W. G. Spitzer, to be published.
8. R. J. Elliott and P. Pfeuty, J. Phys. Chem. Solids, to be published.
9. D. W. Feldman, M. Ashkin, James H. Parker, Phys. Rev. Lett. 17, 24, 1209 (1966).
10. L. Bellomonte and M. H. L. Pryce, Proc. Phys. Soc. 89, 967, 973 (1966).

CALCULATION OF THE OPTICAL ABSORPTION DUE TO LOCALIZED MODE-LATTICE VIBRATION COMBINATION BANDS IN BORON-DOPED SILICON

L. Bellomonte and M.H.L. Pryce

University of Southern California

INTRODUCTION

The absorption spectrum associated with localized modes in boron-doped silicon, compensated by a variety of donors to reduce free-carrier absorption, has been extensively studied[1]. In boron-doped, lithium - compensated, silicon, Waldner, Hiller and Spitzer[2] have observed weak absorption at higher frequencies, which they ascribe to overtones of the localized modes and to combination tones of a localized boron mode with silicon lattice modes. The latter are of course spread over a wide range, but they show some structure, which these authors tentatively correlate with critical points in the lattice spectrum of silicon. In particular, there are peaks which are interpreted as combinations of localized boron modes with frequencies at $517cm^{-1}$ and $377cm^{-1}$. The association with the boron localized modes is confirmed by isotopic substitution. The frequency at $517cm^{-1}$ is close to the Raman frequency of silicon ($518cm^{-1}$). That at $377cm^{-1}$ is close to the longitudinal acoustic frequency at the L-point ($377cm^{-1}$) and to the longitudinal acoustic frequency at the W-point ($371cm^{-1}$). In addition to these identifiable peaks there is a peak (e-band) which appears to be a combination of a localized boron frequency with a frequency at $233cm^{-1}$, which does not correspond to any known feature in the lattice spectrum of silicon.

These facts are somewhat puzzling, for one would not expect combination bands to show any peaking at the Raman frequency, **where** the density of modes is zero, since there is no selection rule indicating particularly strong coupling with phonons of small wave number - in fact, rather the reverse. Nor is there any obvious reason why there should be strong coupling with phonons around the W- or L-points.

We have therefore tried to calculate the theoretically expected shape of the combination band spectrum, in order to throw some light on the interpretation of the observations. The system studied by Waldner et al is too complicated for a first calculation, since the boron is associated with an interstitial lithium in a paired system, and this not only destroys the tetrahedral symmetry of the boron site, but also implies that we should study the dynamics of defect pairs - and it is complicated enough to study an isolated defect quantitatively. We have therefore confined our attention to isolated substitutional boron defects, in the hope that certain well-marked features would be predicted, which we might then expect to find reproduced, in modified form, in the B-Li system. We do find that there is a tendency for peaking at around $370cm^{-1}$ and $480cm^{-1}$, but not at $517cm^{-1}$. One interesting, and unexpected, result of our calculation is the prediction of a rather sharp peak around $225cm^{-1}$, which would fit in rather nicely with the observed e-bands. However, we do not altogether trust this prediction, which is associated with a resonant mode of the boron-silicon system, and depends rather critically on details of the lattice spectrum of silicon. Such resonant behavior should have been predicted in previous calculations of the one-phonon band absorption by Dawber and Elliott(3) and by Maradudin(4). The fact that we do, and they do not, can be traced to small differences in the assumptions made about the density of modes of pure silicon, and we have no confidence in our assumptions being nearer the truth than theirs. However, the apparent coincidence of the calculated resonance with the e-band tempts one to think there may be a real effect.

THE MODEL

We have assumed that the major contribution to the combination bands comes through the agency of a second-order dipole moment, and that anharmonic effects are relatively smaller. We have therefore neglected anharmonicity in our calculations. It may be pointed out, parenthetically, that the observed frequency of the overtones of the localized modes is, to within the experimental accuracy, twice the frequency of the fundamentals, which is certainly consistent with good harmonicity.

In order to simplify the calculations, we have neglected any change in the force constants between boron and silicon from those between silicons in the pure silicon crystal. Physically, this is probably a serious oversimplification. The sole contribution of the substitution of a silicon atom by a boron atom therefore arises from the change of mass.

We have assumed that a certain fraction of the charge (-e) of the defect is localized on, and moves with, the boron atom; and that the remainder is localized on, and moves with, the nearest-neighbor atoms. We have also assumed that this fraction is not constant, but depends linearly on the boron-silicon bond distance. Physically, this corresponds to a flow of charge between the atoms as their distance changes. We have further assumed that the electric fields arising from these charges set up dipoles on the boron and nearest-neighbor silicon atoms; and furthermore we have tried to take into account so-called shell effects and deformation dipoles.

The resulting dipole moment associated with the defect then depends on the displacements of the atoms in a non-linear fashion. We have expanded it up to the second order. The first order moment, of course, gives rise to one-quantum absorption, and so to the localized-mode and the band-mode absorption. The second-order moment is what concerns us for the combination bands.

RESULTS

In order to arrive at quantitative results, we have to pay detailed attention to two aspects of the problem. First we have to make quantitative assumptions concerning the mechanisms giving rise to the second-order dipole moment, such as the polarizabilities of the atoms, and their response to deformations of their electronic clouds. We have had to make some very crude approximations, and in consequence we do not claim much accuracy for absolute values, though we do expect that the general shape of the absorption band should correspond reasonably with our calculations.

Secondly, we have to make dynamical calculations of the displacements of the atoms in the normal modes of the imperfect crystal resulting from boron substitution. These are well known from the work of Dawber and Elliott(3) and Maradudin(4). Actually, we have used somehat different mathematical techniques, because the greater complexity of the second-order calculations makes it expedient to break up the problem according to the irreducible representations of the tetrahedral point symmetry group, under which our model is invariant, but we shall not discuss them here. They will be published at length elsewhere.

The final outcome of our calculation is that the absorption cross-section, σ , per boron atom can be written in the form

$$\sigma(\Omega+\omega) = C\frac{(\Omega+\omega)}{\Omega}\left[\coth\left(\frac{\hbar\Omega}{2kt}\right)+\coth\left(\frac{\hbar\omega}{2kt}\right)\right]\left[\frac{8}{n}\epsilon^2\Omega^4\Sigma_K\frac{1}{(\Omega^2-\tau_K)^2}-\epsilon\right]^{-1}$$

$$\cdot\left[M(z) + \frac{L(z)}{\left[\varphi(z) - \frac{1}{\epsilon z}\right]^2 + \pi^2E^2(z)}\right]$$

where the symbols have the following meaning.

Ω is the (circular) frequency of the localized mode;
ω is the frequency of the lattice mode;
z stands for ω^2;
C is a constant derived from physical constants as follows

$$C = \frac{1}{2}\,\frac{\pi^2\beta^2\hbar e^2}{m^2c\,\sqrt{\epsilon_o}},$$

where β is an effective field factor, estimated to be somewhere between unity and $(\epsilon_o+2\ )^2/9$, ϵ_o is the dielectric constant of silicon (at the frequency $\Omega+\omega$), e is the fundamental charge, m is the electronic mass, c the velocity of light;
E(z) is the normalized density of modes (with respect to dz);
ϵ is the mass-defect parameter $(M_{Si}-M_B)/M_{Si}$;
$\varphi(z)$ is the function defined by the principal value

$$\varphi(z) = \int_P \frac{E(\tau)\,d\tau}{(z-\tau)},$$

which plays a fundamental role in any discussion of the dynamics of the defect problem;

M(z) and L(z) are complicated functions which reflect the general behavior of the density of modes and the movements of the atoms giving rise to the dipole moment.

The first factor in square brackets is the thermal factor, which indicates how the second-order absorption varies with temperature.

The denominator

$$\left[\varphi(z) - \frac{1}{\epsilon_z}\right]^2 + \pi^2E^2(z)$$

has the characteristics of a resonance denominator.

The numerical results of our calculations, plotted for the B^{10} isotope, are shown in Fig.1, in which $\sigma(\Omega+\omega)$ is plotted against ω.

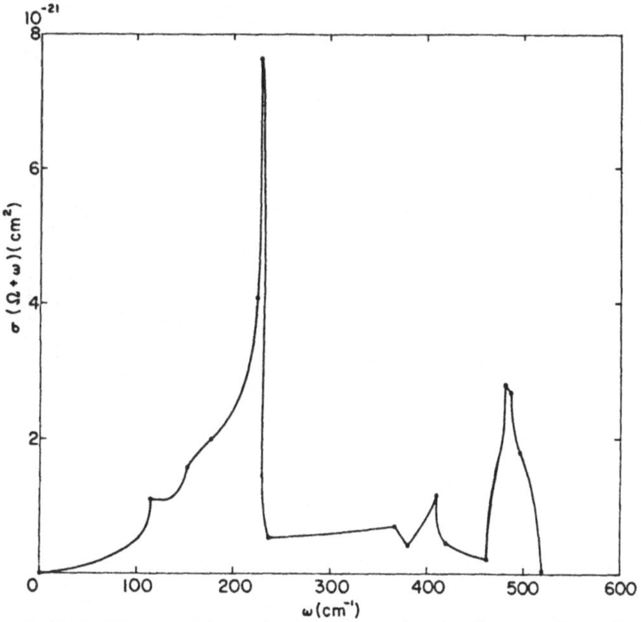

Fig.1 Calculated Absorption Cross Section of Combination Spectrum- for Zero Temperature

DISCUSSION

As already mentioned, the functions M and L reflect, more or less, the behavior of the density of modes E(z). This is plotted (as a function of wave number) in Fig. 2. The resonance denominator depends on $\varphi(z)$, which together with $1/\epsilon z$ is plotted on Fig. 3.

The predicted peak in σ at approximately $\Omega + 225cm^{-1}$ arises from the minimum in the resonance denominator, associated with a small density of modes, E(z), and a near-cancellation of $\varphi(z) - 1/\epsilon z$. The latter can be traced to a sharp maximum in $\varphi(z)$ at the frequency, which in turn can be traced to the sharp drop in E(z) between 219 and $223cm^{-1}$. These frequencies correspond to stationary points of the frequency at points Q and Σ in reciprocal space. Q is on the edge of the Brillouin zone, between L (111) and W((1½o); Σ is inside the Brillouin zone, in the (110) directions, near to the zone boundary at $(\frac{3}{4}\frac{3}{4}0)$, and is a local maximum of a transverse acoutic frequency. The frequency at Σ has been estimated by Dolling[5] by fitting his neutron diffraction data to a shell model calculation, and is presumably fairly reliable. There are no neutron diffraction measurements directly relevant to Q, and we have used the theory of Johnson and Loudon to estimate the frequency there, and to determine the nature of the van Hove singularity.

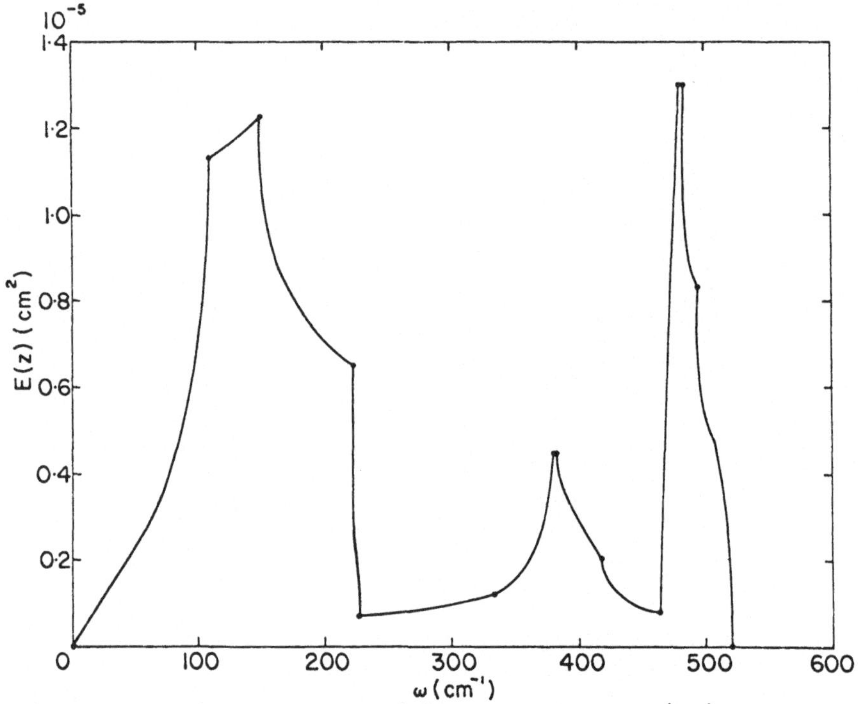

Fig.2 Assumed Density of Modes E(z), vs Frequency (ω).

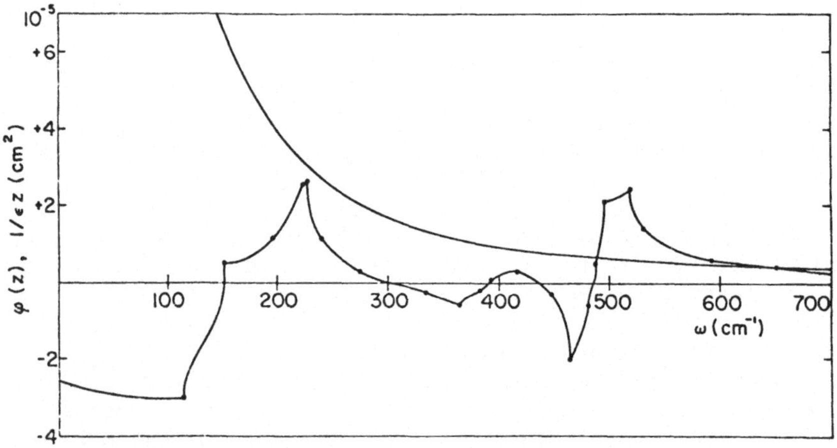

Fig.3 The Functions φ(z) and 1/ ε z, vs Frequency (ω).
ε = 0.643.

The fact that we predict a resonance around 225cm^{-1} in the combination spectrum, whereas Dawber and Elliott, and Maradudin, in their discussion of the one phonon spectrum do not, is mainly due to the fact that their assumed density of modes does not include any strong contribution from Q or Σ , presumably because the simple theories of the silicon spectrum on which they based their numerical work do not predict any maximum frequency at Σ. Dawber and Elliott do show a maximum in the one-phonon spectrum at about $150cm^{-1}$, as associated with the TA critical points at X(100) and at L(111). There is the further point, however, that the numerator associated with $225cm^{-1}$ in the one-phonon spectrum is particularly small, and it would appear from our calculations that the resonance would be more marked in the combination than in the one-phonon spectrum. However, our conclusion concerning the sharp resonance depends so critically on our assumptions concerning the density of modes that we cannot feel any real confidence in its reality.

ACKNOWLEDGEMENTS

This research was supported in part by the Air Force Office of Scientific Research, Office of Aerospace Research, United States Air Force, under grant AFOSR 1115-66, and in part by a grant from the National Aeronautics and Space Administration, NGR - 05-018-083.

One of us (LB) is on leave from the University of Palermo.

REFERENCES

(1) S.D. Smith and J.F. Angress Phys. Letters 6, 131 (1963).
M. Balkanski and W. Nazarewicz (a) J. Phys. Chem. Solids 25, 474, (1964).
(b) J. Phys. Chem. Solids 27, 671, (1966).
W.A. Spitzer and M. Waldner (a) Phys. Rev. Letters 14, 223, (1965).
(b) J. Appl. Phys. 36, 2450, (1965).

(2) M. Waldner, M.A. Hiller, W.G. Spitzer Phys. Rev. 140A, 172, (1965).

(3) P.G. Dawber and R.J. Elliott (a) Proc. Roy. Soc. A273, 222, (1963).
(b) Proc. Phys. Soc. 81, 453, (1963).

(4) A.A. Maradudin in Solid State Physics, Edit. by F. Seitz and D. Turnbull, Vol. 18, p.297, (1966).

(5) G. Dolling in Inelastic Scattering of Neutrons in Solids and Liquids. Vol. II. p.37, I.A.E.A. Vienna (1963).

(6) F.A. Johnson and R. Loudon Proc. Roy. Soc. A281, 274, (1964).

PRELIMINARY CALCULATION OF LOCAL MODES IN CdS*

P. Pfeuty★†, J. L. Birman★,
M. A. Nusimovici†, and M. Balkanski†

★Physics Department, New York University
New York
†Laboratoire de Physique des Solides
Universite de Paris, France

I. Introduction

In ionic crystals, localized modes are well known experimentally[1] and theoretically only for the cubic crystals[2] and in particular for the alkali halides[3]. In a wurtzite crystal a point defect corresponds to the point group C_{3v} and two localized modes are expected corresponding respectively to the representations Γ_1 and Γ_3. Recently, the phonon dispersion curves have been calculated for CdS wurtzite[4]. From the phonon eigenvectors and eigenfrequencies we calculated the local mode frequencies for mass defects in CdS wurtzite.

II. Theory

A Green's function technique is used to study the defect vibrations[5]. In the wurtzite structure CdS the unit cell ℓ contains four atoms k = 1,2,3,4 which are respectively CdI, CdII, SI, SII. We just consider a single mass defect (the force constants are left unchanged). The defect subspace is three dimensional and the localized mode frequency is a solution of the secular equation

$$\text{Det}\ (1-\underline{g}(\omega)\underline{c}(\omega)) = 0 \qquad (1)$$

where $\underline{g}$ and $\underline{c}$ are respectively the Green's function matrix and the perturbation matrix related to the defect subspace.

The matrix $\underline{c}$ is a unit three dimensional matrix multiplied by $(M_k - M')\,\omega^2$ where M' is the impurity mass and M_k the mass of the replaced host ion.

The matrix $\underline{g}$ Green's functions elements are deduced from the phonon eigenvectors $\underset{\sim}{\sigma}(j\underset{\sim}{q})$ and eigenfrequencies $\omega^2(j,\underset{\sim}{q})$

$$g_{\alpha k,\alpha' k'}(\ell,\ell')(\omega^2)=\sum_{j\underset{\sim}{q}}\frac{\sigma_{\alpha k}(j\underset{\sim}{q})\sigma^{+}_{\alpha' k'}(j\underset{\sim}{q})\exp i\underset{\sim}{q}(r\underset{\sim}{\ell}-r\ell)}{N\,M_k^{1/2}M_{k'}^{1/2}\,(\omega^2-\omega^2(j,\underset{\sim}{q}))} \tag{2}$$

From the point group symmetry of the crystal around the defect the Green's functions satisfy the following relations

$$g_{\alpha k,\alpha' k}(0,0)(\omega^2) = \delta_{\alpha\alpha'}\, g_{\alpha k,\alpha k}(0,0)(\omega^2) \quad \text{and}$$

$$g_{xk,xk}(0,0)(\omega^2) = g_{yk,yk}(0,0)(\omega^2) \tag{3}$$

The impurity may replace either Cd or S and four different Green's functions are used in this problem with the following notations

$$\begin{aligned} g_1 &= M_1\, g_{z1\ z1}(0,0)(\omega^2)\\ g_2 &= M_1\, g_{x1\ x1}(0,0)(\omega^2)\\ g_3 &= M_3\, g_{z3\ z3}(0,0)(\omega^2)\\ g_4 &= M_3\, g_{x3\ x3}(0,0)(\omega^2) \end{aligned} \tag{4}$$

where M_1 and M_3 are respectively the mass of Cd and the mass of S.

For many defect vibration problems and especially for band resonance calculations it is convenient to know

$$\text{limit}_{\varphi\to 0}\ g(\omega+i\varphi) = g'(\omega) + \frac{i\pi}{2\omega}\,\nu(\omega) \tag{5}$$

The values of

$$\nu_{\alpha k,\alpha' k'}(\ell,\ell')(\omega) = \frac{1}{N}\sum_{j\underset{\sim}{q}}\sigma_{\alpha k}(j,\underset{\sim}{q})\sigma^{+}_{\alpha' k'}(j\underset{\sim}{q})\exp i\underset{\sim}{q}(r\underset{\sim}{\ell}-r\underset{\sim}{\ell'})\delta(\omega(j\underset{\sim}{q})-\omega) \tag{6}$$

are obtained by using the computed CdS spectrum[4] for 125 values of $\underset{\sim}{q}$ in the reduced Brillouin Zone. The frequency band $(\omega<\omega M)$ was divided into Z = 61 equally spaced regions and the sum in(6)

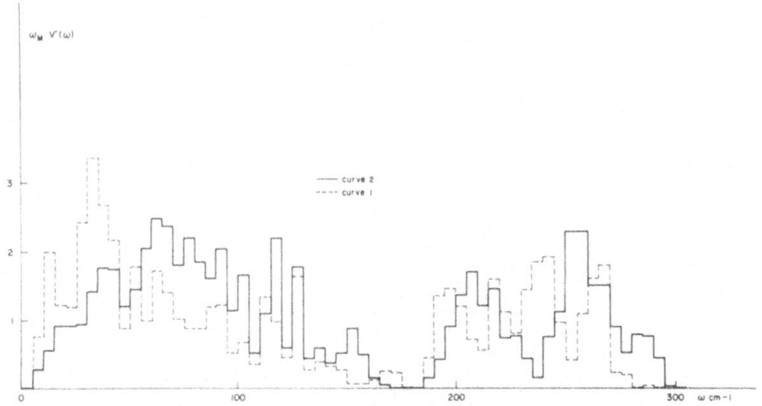

Fig. 1 Imaginary Part of the Green's Functions.

$\omega_M \nu_1(\omega)$ Corresponds to Curve 1 and $\omega_M \nu_2(\omega)$ to Curve 2.

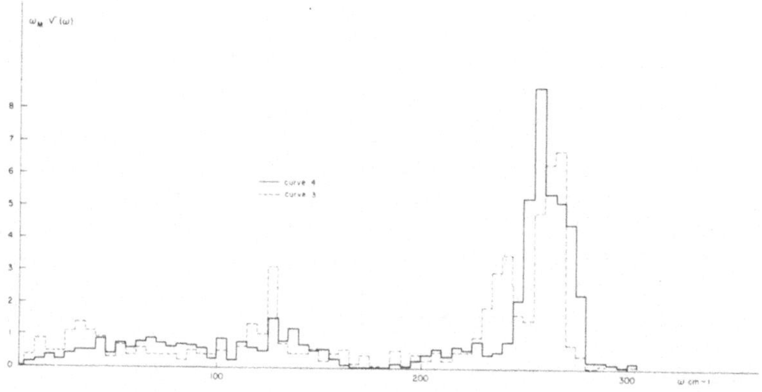

Fig. 2 Imaginary Part of the Green's Functions.

$\omega_M \nu_3(\omega)$ Corresponds to Curve 3 and $\omega_M \nu_4(\omega)$ to Curve 4.

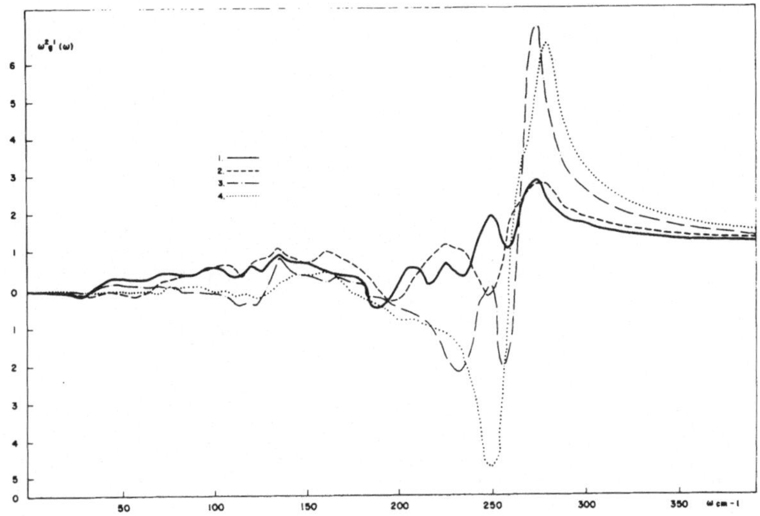

Fig. 3 Real Part of the Green's Functions.
$\omega^2 g_1'(\omega)$ Corresponds to Curve 1, $\omega^2 g_2'(\omega)$ to Curve 2, $\omega^2 g_3'(\omega)$ to Curve 3 and $\omega^2 g_4'(\omega)$ to Curve 4.

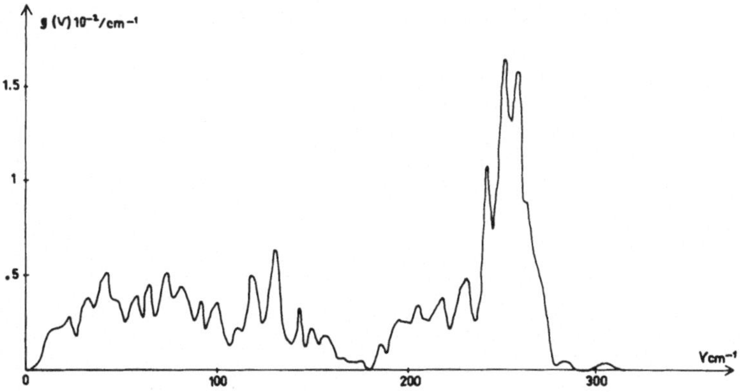

Fig. 4 One Phonon Density of States for CdS.

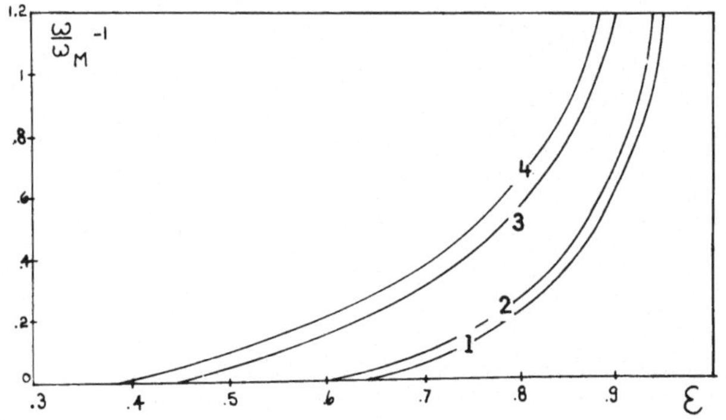

Fig. 5 Localized Mode Frequencies for a Mass Defect

$\frac{\omega}{\omega_M}-1$ is plotted versus ϵ_1 for Curves 1 and 2 Where the Impurity Replaces Cd and Versus ϵ_3 for Curves 3 and 4 Where the Impurity Replaces S. Curves 1 and 3 Correspond to Γ_1 Modes and Curves 2 and 4 to Γ_3 Modes.

carried over $m \frac{\omega M}{Z} < \omega < (m+1) \frac{\omega M}{Z}$ to form a histogram for ν. The values of $g'(\omega)$ are then obtained at the mid points of the histogram sections by using the Kramers Kronig relation

$$g'i\ (\omega) = \int_0^{\omega M} \frac{\nu i\ (\omega')d\omega'}{\omega^2 - \omega'^2}$$

The results for ωM $\nu i(\omega)$ and for $\omega^2 g'i(\omega)$ are plotted in Fig. 1 2 and 3.

The localized mode frequencies from(1)and(3)are solutions of the following equations

$$\epsilon_1\ \omega^2 g_1 = 1 \qquad \text{For the } \Gamma_1 \text{ mode}$$
$$\epsilon_1\ \omega^2 g_2 = 1 \qquad \text{For the } \Gamma_3 \text{ mode} \qquad (8)$$

where the impurity replace Cd and

$$\epsilon_3\ \omega^2 g_3 = 1 \qquad \text{For the } \Gamma_1 \text{ mode}$$
$$\epsilon_3\ \omega^2 g_4 = 1 \qquad \text{For the } \Gamma_3 \text{ mode} \qquad (9)$$

where the impurity replaces S

$$\epsilon_1 = \frac{M_1 - M'}{M_1} \quad \text{and} \quad \epsilon_3 = \frac{M_3 - M'}{M_3}$$

III. Discussion

3.1 The Green's Functions

The imaginary part of the Green's functions $\nu_i\ (\omega)$ may be compared with the one phonon density of states in CdS recently calculated [6] and shown in Fig. 4. The functions $\nu_1(\omega)$ and $\nu_2(\omega)$ related to the heavy Cd ion displacement vectors are enhanced in the low frequency acoustic region and the functions $\nu_3(\omega)$ and $\nu_4(\omega)$ related to the light S ion displacement vectors are enhanced in the high frequency optic region. Similar results had been obtained for KI cubic crystals[5] - the difference between ν_1 ν_2 and ν_3 ν_4 are on the other hand difficult to explain from the phonon dispersion curves[4].

3.2 Localized Modes

The localized mode frequencies obtained from equations (8)and

(9) are plotted in Fig. 5 versus ϵ. If we suppose $\omega M = 308 cm^{-1}$ for CdS a localized mode appears only if $\epsilon_3 > 0.36$ for an impurity replacing S and $\epsilon_1 > 0.58$ for an impurity replacing Cd. If ϵ is large enough, two modes appear out of the band; a singlet Γ_1 and a doublet Γ_3 with a higher frequency than the singlet. The splitting increases when the impurity becomes lighter. Up to now no localized modes have been observed in CdS as due to mass defects. Predictions for different impurities are summarized in Table 1.

TABLE 1

Localized Mode Frequencies For Substitutional Impurities In CdS

Impurity	Replaced Ion	M'	ϵ	$\omega\Gamma_1$ cm^{-1}	$\omega\Gamma_3$ cm^{-1}
Be	Cd	9	0.92	540	580
Mg	Cd	24	0.786	365	380
Ca	Cd	40	0.683	320	330
O	S	16	0.5	324	339
Li	Cd	7	0.94	615	665

For light impurities ($\epsilon \to 1$) all the localized mode frequencies **tend to infinity.** This can be shown from the limit form of the Green's functions when ω is much bigger than $\omega(j\underset{\sim}{q})$

$$g_{\alpha k,\alpha k}(0,0)(\omega^2) \sim \sum_{jq} \frac{|\sigma_{\alpha k}(j\underset{\sim}{q})|^2}{N} \frac{1}{M_k \omega^2}$$

but $\sum_j \sigma_{\alpha k}(j\underset{\sim}{q})\, \sigma^+_{\beta k'}(j\underset{\sim}{q}) = \delta_{\alpha\beta}\, \delta_{kk'}$, then

when $\omega \to \infty$ $\quad g_{\alpha k,\alpha k}(0,0)(\omega^2) \to \dfrac{1}{M_k \omega^2}$ and the solution

for ϵ is $\epsilon = 1$.

3.3 Band Modes

When a heavy mass impurity is introduced a band resonance may appear at $\omega < \omega M$. The response of the lattice is related to the imaginary part of the imperfect crystal Green's functions

$$\mathrm{Im}\, Gi(\omega) = \frac{\dfrac{\pi \nu i(\omega)}{2\omega}}{(1 - \epsilon\omega^2 g' i(\omega))^2 + \left(\dfrac{\pi\epsilon\omega\nu i(\omega)}{2}\right)^2}$$

So a resonance is well defined at the ω which is a solution of $1 - \epsilon\omega^2 g' i(\omega) = 0$ only if $\nu_i(\omega)$ is small. For Se in CdS $\epsilon = -1.5$ and a resonance is expected at $210 cm^{-1}$.

Conclusion.

This preliminary calculation is first to study the defect vibrations in a wurtzite system. The validity of the results depends on the lattice dynamics of the perfect crystal. Both defect model and perfect lattice dynamics have to be refined.

REFERENCES

(1) G. Schaefer, J. Phys. Chem. Solids 12 233 (1960).
(2) R. F. Wallis and A. A. Maradudin, Prog. Theor. Phys. 24 1055 (1960).
(3) S. S. Jaswal and D. J. Montgomery, Phys. Rev. 135A1257 (1966).
(4) M. A. Nusimovici and J. L. Birman, Phys. Rev. 156 925 (1967).
(5) A. A. Maradudin, Rep. on Progress in Physics Vol. XXVIII p331 (1965).
(6) M. A. Nusimovici and J. L. Birman, II-VI Compounds International Conference Brown University 1967 to be published.

* Work supported in part by the U.S. Army Research Office (Durham) under Grant No. DA ARO 31-124 G424 and the Aerospace Research Laboratories Office of Aerospace Research, Wright-Patterson AFB, Dayton, Ohio under Contract No. AF(33)(615)1746. Calculations have been made available through the New York University Computing Center facilities.

PART D. LOCALIZED EXCITONS

ISOELECTRONIC IMPURITIES IN SEMICONDUCTORS

Roger A. Faulkner

Palmer Physical Laboratory, Princeton University, Princeton, New Jersey
Bell Telephone Laboratories, Murray Hill, New Jersey

J. J. Hopfield

Palmer Physical Laboratory, Princeton University, Princeton, New Jersey

INTRODUCTION AND GENERALIZATIONS

The term "Isoelectronic Impurity" is used to denote an impurity center in a crystal arising when an atom from the same column of the periodic table as one of the constituents of the host crystal substitutes for that constituent. Isoelectronic substitutions in certain wideband gap semiconductors have a profound effect on the optical properties of the materials for photon energies in the vicinity of the band gap. In particular, the impurity systems GaP:N, GaP:Bi, and ZnTe:O exhibit sharp absorption lines and CdS:Te exhibits an absorption band lying within the band gaps of the pure crystals.[1,2,3,4] These lines can be attributed to exciton bound states at the impurities.

These systems have been classified by Thomas, Hopfield and Lynch[2] as isoelectronic acceptors and isoelectronic donors on the basis of whether the impurity is attractive for electrons or attractive for holes. For isoelectronic acceptors, the mechanism for binding an exciton is conceived to be as follows: an electron in the conduction band is attracted to the uncharged impurity by a short range potential and is bound to it, the result being a charged center which can subsequently bind a hole. Isoelectronic donors would operate in the inverse sequence, first binding a hole and then an electron.

In one system, GaP:N, in addition to the principal bound state ("A" line), other bound states have been observed in both optical absorption and fluorescence which lie lower in energy than the "A" line and form a sequence of levels converging to the "A" line.[1] These levels have been identified as bound states of excitons to two Nitrogen impurities at various interatomic spacings. They are labelled NN_1,NN_2,NN_3, etc. in increasing energy (diminishing binding energy). There is no apparent regularity in these lines except that they converge to the "A" line.

In GaP:N, the "A" line is actually a doublet, called "A" and "B", arising from the degenerate valence band. The A-B separation is .8 meV. Lying above the A-B doublet by 1.8 meV is a pair of doublets (seen weakly in fluorescence) lying quite close together and which are clearly replicas of the A-B doublet.[5] GaP has a three-valley conduction band and these higher states can be identified as the antisymmetrical pair split off from the symmetrical A-B doublet by intervalley mixing.

GaP is an indirect band gap semiconductor and as such, the optical absorption above the band gap is dominated at low temperatures by the presence of Nitrogen.[6]

A reasonable theory of isoelectronic impurities must be able to correlate all of these experimental findings, the single impurity and the double impurity energy levels, their strengths in optical absorption, and the optical absorption above the band edge.

In almost all of what follows, the impurity system GaP:N will be the only isoelectronic system considered. This one has been most extensively investigated and displays the greatest variety of effects.

Because the interaction of electrons in a crystal with an isoelectronic impurity is probably of short range, of the order of the interatomic spacing, the details of the calculations will depend on the band structure throughout the Brillouin Zone and effective mass theory will be inadequate. Figure 1 shows some of the band structure of GaP as calculated by Bergstresser and Cohen[7] using an empirical pseudopotential approach. The indirect band gap of 2.3 eV is between the points Γ_{15} and X_1. Of particular interest later is the fact that the first conduction band bends over in the center of the zone and comes within .5 eV of the absolute minimum at X_1. There are three inequivalent points X in the zone, giving GaP a three-valley conduction band.

We can argue that we can ignore the interaction of holes in the valence bands with the Nitrogen impurity because we expect a resonance behavior for conduction band electrons due to the existence of a shallow bound state at the impurity, making a large scattering phase shift even though the attraction is rather weak.

For holes, the potential is repulsive and no resonance will exist. The analogy here is obviously to neutron-proton scattering and the existence of the weakly bound deuteron. In the case of the bound state, the short range repulsive core the hole experiences due to the impurity can be ignored for the most part.

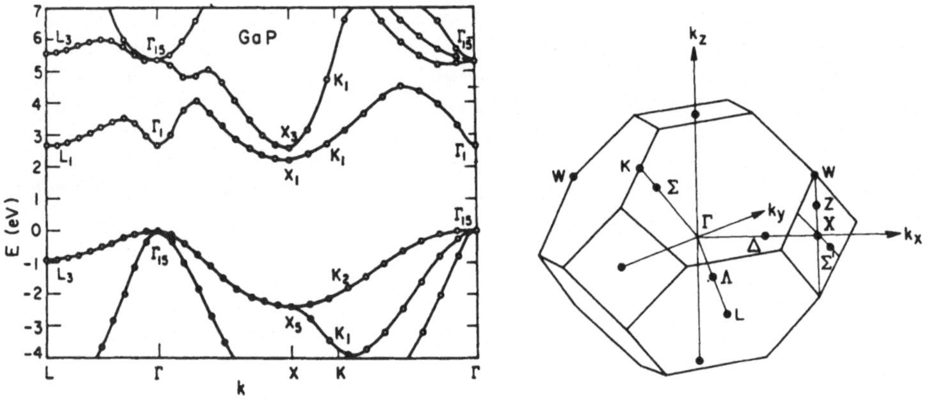

Band structure of GaP.

Marvin L. Cohen† and T. K. Bergstresser

Figure 1

A SIMPLE MODEL

Assume a short-range interaction, $V(\vec{x})$, for electrons with the Nitrogen impurity. This interaction is composed of the bare interaction with the impurity and the interaction with the self-consistent rearrangement of the charge density of the host crystal. Denoting the momentum space wave function for an electron and a hole by $\psi(\vec{k}_1,\vec{k}_2)$, one obtains the Schrodinger equation:

$$[E - \varepsilon_c(\vec{k}_1) - \varepsilon_v(\vec{k}_2)]\psi(\vec{k}_1,\vec{k}_2) = \int_{(B.Z.)} d^3q \; (c\vec{k}_1 \mid V \mid c\vec{q})\psi(\vec{q},\vec{k}_2)$$

$$- \frac{4\pi e^2}{K} \frac{1}{(2\pi)^3} \int d^3q \, \frac{1}{q^2} \, \psi(\vec{k}_1+q,\vec{k}_2+q) \qquad (1)$$

For the sake of simplicity only one conduction band and one

valence band are being considered. $\varepsilon_c(\vec{k})$, the conduction band energy, is measured up from the conduction band minima at points X and $\varepsilon_v(\vec{k})$, the valence band energy, is measured down from the valence band maximum at Γ. In the spirit of the discussion in the first section, the interaction of the hole with the impurity is being ignored.

Consider the matrix element of $V(\vec{x})$ with Bloch Functions appearing in Eq. (1). It may be written in terms of Wannier Functions:

$$(c\vec{k}\,|\,V\,|\,c\vec{q}) = \frac{\Omega}{(2\pi)^3} \sum_{\vec{R}_1,\vec{R}_2} (c\vec{R}_1\,|\,V\,|\,c\vec{R}_2)e^{-i(\vec{k}\cdot\vec{R}_1-\vec{q}\cdot\vec{R}_2)} \qquad (2)$$

where Ω = volume of a primitive cell.

If the Wannier Functions are well localized and if $V(\vec{x})$ has a range of the order of the interatomic spacing, the dominant term in the summation is the $\vec{R}_1 = \vec{R}_2 = 0$ term.

We obtain a simple and almost completely soluble model if we retain only this term from the sum and call it J:

$$(c\vec{k}\,|\,V\,|\,c\vec{q}) \simeq \frac{J\Omega}{(2\pi)^3} \qquad (3)$$

If the impurity were centered about the lattice site at $\vec{R}$ instead of the origin of coordinates, we would have:

$$(c\vec{k}\,|\,V(\vec{x}-\vec{R})\,|\,c\vec{q}) = \frac{J\Omega}{(2\pi)^3}\, e^{-i(\vec{k}-\vec{q})\cdot\vec{R}} \qquad (4)$$

This form will be useful later when states arising from a pair of Nitrogens separated by a lattice vector $\vec{R}$ are considered.

The Schrodinger Equation for the simple model then becomes (with $\hbar = 1$):

$$[E - \varepsilon_c(\vec{k}_1) - \varepsilon_v(\vec{k}_2)]\psi(\vec{k}_1,\vec{k}_2) = \frac{J\Omega}{(2\pi)^3}\int_{(B.Z.)} d^3q\ \psi(\vec{q},\vec{k}_2)$$

$$- \frac{4\pi e^2}{\kappa}\frac{1}{(2\pi)^3}\int d^3q\,\frac{1}{q^2}\,\psi(\vec{k}_1+\vec{q},\vec{k}_2+\vec{q}). \qquad (5)$$

THE BOUND STATE PROBLEM IN THE SIMPLE MODEL

For the bound state problem, let us approximate the wave function $\psi(\vec{k_1},\vec{k_2})$ by a product wave function $\psi_e(\vec{k_1})\psi_h(\vec{k_2})$ and look only at the electron part of Schrodinger's equation first:

$$[E - \varepsilon_c(\vec{k})]\psi(\vec{k}) = \frac{J\Omega}{(2\pi)^3} \int_{(B.Z.)} d^3q \;\; \psi(\vec{q}) \tag{6}$$

Equation (6) can be solved immediately by noting that the right-hand side is independent of $\vec{k}$ and that the multiplier on the left can never be zero for E negative:

$$\psi(\vec{k}) = \frac{N}{[E - \varepsilon_c(\vec{k})]} \tag{7}$$

$$N = \frac{J\Omega}{(2\pi)^3} \int_{(B.Z.)} d^3q \;\; \psi(\vec{q}) \tag{8}$$

Substituting (7) into (8) gives the eigenvalue equation for E. Let us define:

$$f(E) = \frac{\Omega}{(2\pi)^3} \int_{(B.Z.)} d^3q \;\; \frac{1}{[\varepsilon_c(\vec{q}) - E\text{-}i0]} \tag{9}$$

for all E, both positive and negative. This function will find much use later when continuum wave functions are needed.

The eigenvalue equation written in terms of f(E) is:

$$1 + Jf(E) = 0 \tag{10}$$

It becomes quite clear at this point that the effective mass approximation is not adequate in this problem, because if we set $\varepsilon_c(\vec{q}) = q^2/2m^*$ and extend the integration over all space, the integral diverges. However, we can use the relation:

$$\frac{1}{\varepsilon - E} = \frac{1}{\varepsilon} + \frac{E}{\varepsilon(\varepsilon - E)} \tag{11}$$

to obtain the equation:

$$1 + J\left\langle\frac{1}{\varepsilon}\right\rangle = -\frac{J\Omega}{(2\pi)^3} \int_{(B.Z.)} d^3q \;\; \frac{E}{\varepsilon_c(\vec{q})(\varepsilon_c(\vec{q}) - E)} \tag{12}$$

where:

$$\left\langle \frac{1}{\varepsilon} \right\rangle = \frac{\Omega}{(2\pi)^3} \int_{(B.Z.)} d^3q \; \frac{1}{\varepsilon_c(\vec{q})} \tag{13}$$

is the average over the Brillouin Zone of $1/\varepsilon_c(\vec{q})$.

The second integral can be extended over all $\vec{k}$-space and effective masses used if $|E|$ is small.

Remembering that GaP has three valleys in the conduction band and denoting the anisotropic effective masses at the points X by (m_1, m_1, m_2), we obtain:

$$1 + J\left\langle \frac{1}{\varepsilon} \right\rangle = J \frac{3\Omega}{2\pi} m_1 \sqrt{2m_2 |E|} \tag{14}$$

For the bound state to exist, $1 + J\langle 1/\varepsilon \rangle$ must be less than zero. Therefore, there is a critical value of $|J|$ below which no bound state exists. The special combination of J and $\langle 1/\varepsilon \rangle$ which appear in Eq. (14) occurs so often in what follows that it deserves a name:

$$Q \equiv \frac{1 + J\left\langle \frac{1}{\varepsilon} \right\rangle}{J} \tag{15}$$

The energy of the bound state is then given by:

$$\frac{3\Omega}{2\pi} m_1 \sqrt{2m_2 |E|} = Q \tag{16}$$

Or, alternately, since Q is unknown and E is known experimentally, Eq. (16) determines the value of Q.

The normalization constant for the wave function can be determined by requiring:

$$\int_{(B.Z.)} d^3q \; |\psi(\vec{q})|^2 = 1$$

This gives for $\psi(\vec{k})$:

$$\psi(\vec{k}) = \frac{(2m^* |E|)^{\frac{1}{4}}}{\sqrt{3}\, 2\pi m^*} \; \frac{1}{[\varepsilon_c(\vec{k}) + |E|]} \tag{17}$$

where $m^* = (m_1 m_1 m_2)^{1/3}$ is the geometric mean of the effective masses.

In the spirit of the previous discussion, we can now take the hole to have a hydrogenic 1S wave function. We put these two wave functions together to form a trial wave function and vary two parameters to minimize the energy.

Doing this gives us a value for Q, the one important unknown parameter in the theory:

$$Q = .022 \frac{1}{eV} = \frac{1 + J\langle \frac{1}{\varepsilon} \rangle}{J} \tag{18}$$

The bound state energy is very sensitive to the value of J and J is very close to the critical value given by $1 + J\langle 1/\varepsilon \rangle = 0$. This could be only accidental, but it is more likely that it means there is a self-consistent rearrangement of the charge density when the impurity is added which tends to "peg" the interaction strength at the critical value.

The cross section for optical absorption by a single Nitrogen center is proportional to:

$$\left| \int d^3k \ \psi(\vec{k},\vec{k}) \right|^2 = \left| \int d^3k \ \psi_e(\vec{k})\psi_h(\vec{k}) \right|^2$$

Since the hole wave function is concentrated about $\vec{k} = 0$, we set the argument of $\psi_e(\vec{k})$ equal to zero and remove it from the integral, making the cross section proportional to $|\psi_e(0)|^2$, the square of the electron wave function at the center of the zone. Now the significance of the fact that the conduction band bends over in the center, as was mentioned before, becomes clear: the optical absorption cross section is proportional to $1/[\varepsilon_c(\vec{k}=0) + |E|]^2$.

Since no broadening processes are considered here, the absorption line is represented as a delta-function. However, we can calculate the integrated absorption strength and compare that to the area under the experimental absorption curve. Good agreement is found using the following parameters:

m_e^* = electron mass at X = .35

m_h^* = hole mass at Γ = .2

n = index of refraction = 3.45

$$|X| = |\langle c\ \vec{k}=0\ |\ x\ |\ v\ \vec{k}=0\rangle| = \text{optical absorption matrix element} = 2\ \text{Å}$$

OPTICAL ABSORPTION ABOVE THE BAND EDGE

To treat the optical absorption induced by the Nitrogen impurity above the band edge, we shall need the continuum wave functions.

The absorption consists essentially of three parts:

(a) The production of electrons bound to the Nitrogen impurity and free holes.

(b) The production of unbound electrons and free holes.

(c) The production of unbound excitons.

Process (a) is the simplest because we already have the electron part of the wave function, Eq. (17). The hole can be taken to be truly free if we ignore the coulomb interaction. This is a bad approximation near the absorption edge for this channel, but it improves as we go up in energy. Taking the hole to be a free particle gives us the wave function:

$$\psi(\vec{k}_1,\vec{k}_2) = \psi_e(\vec{k}_1)\delta(\vec{k}_2-\vec{k}_o)$$

where $\vec{k}_o$ is the final hole momentum and $\psi_e(\vec{k}_1)$ is given by Eq. (17). Using the effective mass approximation for the hole, we obtain the square root absorption curve shown in Fig. 3:

$$\sigma(\omega) = \frac{16}{n}\left(\frac{e^2}{\hbar c}\right)|X|^2\left(\frac{Eg+\Delta}{\Delta}\right)^2\left(\frac{4\pi}{3}\right)\left(\frac{m_h^*}{m_e^*}\right)^{3/2}\frac{\sqrt{|E_b|}}{\hbar\omega}$$

$$\times\sqrt{\hbar\omega - (Eg - |E_b|)} \qquad (19)$$

$$\Delta = [\varepsilon_c(\vec{k}=0) - \varepsilon_c(\vec{k} = \text{zone edge})] = .5\ \text{eV}$$

$$Eg = \text{energy gap} = 2.3\ \text{eV}$$

$$|E_b| = \text{binding energy of an electron to Nitrogen} = .008\ \text{eV}$$

Process (b) is one step more complicated. Now we will need the continuum electron wave functions.

We can solve Schrodinger's equation for the electron, Eq. (6), for E > 0:

$$\psi_{\vec{k}_o}(\vec{k}) = \delta(\vec{k}-\vec{k}_o) + \frac{1}{[E+i0 - \varepsilon_c(\vec{k})]} \frac{J\Omega}{(2\pi)^3} \int_{(B.Z.)} d^3q \; \psi_{\vec{k}_o}(\vec{q}) \qquad (20)$$

$\vec{k}$ is the argument of ψ and $\vec{k}_o$ is the label of the state represented by ψ.

$\delta(\vec{k}-\vec{k}_o)$ is the unperturbed wave function

$E = \varepsilon_c(\vec{k}_o)$ is the energy of the state

Writing ψ in the form above preserves the normalization:

$$\int d^3k \; \psi^*_{\vec{k}'}(\vec{k})\psi_{\vec{k}''}(\vec{k}) = \delta(\vec{k}'-\vec{k}'') \qquad (21)$$

Integrating (20) over the Brillouin zone yields:

$$\int_{(B.Z.)} d^3k \; \psi_{\vec{k}_o}(\vec{k}) = \frac{1}{[1 + Jf(E)]}$$

where f(E) was defined in Eq. (9).

The electron wave function then becomes:

$$\psi_{\vec{k}_o}(\vec{k}) = \delta(\vec{k}-\vec{k}_o) + \frac{J\Omega}{(2\pi)^3} \frac{1}{[1 + Jf(E)]} \frac{1}{[E+i0 - \varepsilon_c(k)]} \qquad (22)$$

Setting the hole wave function equal to a delta function as before, integrating and performing the necessary sums over the final states give:

$$\sigma(\omega) = \frac{32}{3}\left(\frac{e^2}{\hbar c}\right)\frac{|X|^2}{n}\left(\frac{Eg+\Delta}{\Delta}\right)^2\left(\frac{m_h^*}{m_e^*}\right)^{3/2}\frac{1}{\hbar\omega}$$

$$\times \int_0^{(\hbar\omega-Eg)} dE \, \frac{\sqrt{E}\sqrt{\hbar\omega-Eg-E}}{[\varphi(E) + E]} \qquad (23)$$

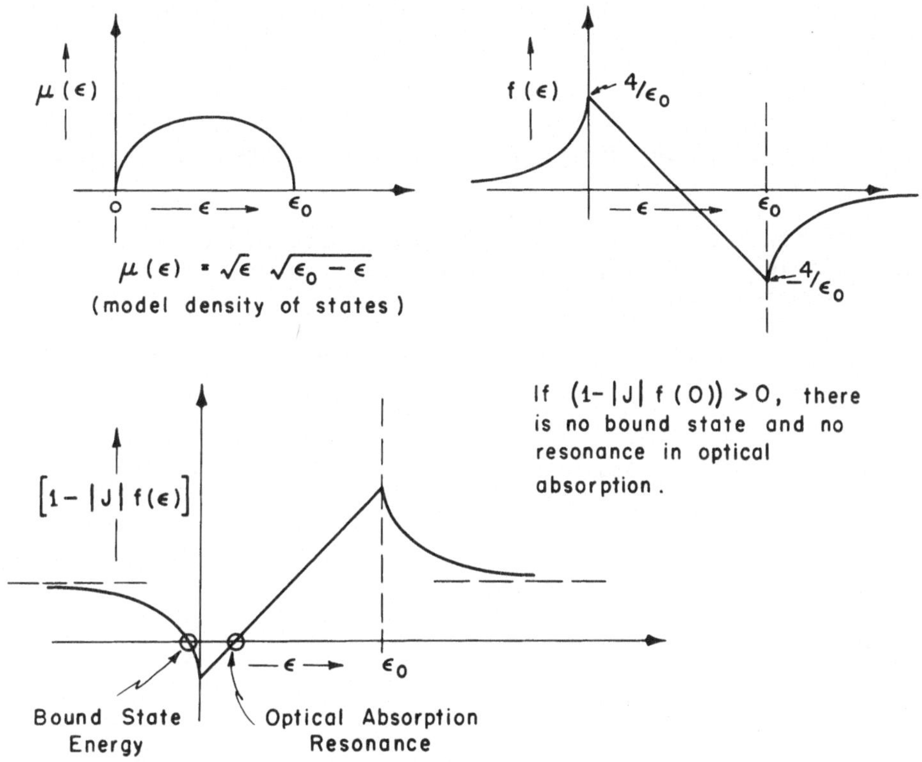

Figure 2

where

$$\varphi(E) = \frac{[1 + J\,\mathrm{Re}(f(E))]^2}{36\pi^4(2m_e^*)^3\left[\dfrac{J\Omega}{(2\pi)^3}\right]^2}$$

We now need to know Re(f(E)) for $E > 0$. The trick used before for $E < 0$ does not work for this case, so we must do something else. If we knew the density of states for the entire conduction band, f(E) could be calculated. In the absence of this knowledge, and in order to keep the calculations simple, a model density of states is used as shown in Fig. 2. Re(f(E)) can be quickly calculated from this density of states, and is also shown in Fig. 6.

There is a point at which $1 + J\,\mathrm{Re}(f(E)) = 0$ which means that $\varphi(E)$ goes to zero also, leading to a resonance in the integral of Eq. (23).

Using the model density of states, the cross section for optical absorption for this process can be calculated and is shown in Fig. 3.

The final process, the production of unbound excitons, is the most complicated of all, but it can be calculated in much the same way as process (b).

Now we must consider Schrodinger's equation for both the electron and the hole:

$$[E - \varepsilon_c(\vec{k}_1) - \varepsilon_v(\vec{k}_2) + V_{COUL}]\psi(\vec{k}_1,\vec{k}_2) = \frac{J\Omega}{(2\pi)^3}\int_{(B.Z.)} d^3q\;\psi(\vec{q},\vec{k}_2) \tag{24}$$

This equation can be solved:

$$\psi(\vec{k}_1,\vec{k}_2) = \psi_o(\vec{k}_1,\vec{k}_2) + \frac{1}{[E+i0 - \varepsilon_c(\vec{k}_1) - \varepsilon_v(\vec{k}_2)]}\frac{J\Omega}{(2\pi)^3}\int_{(B.Z.)} d^3q\;\psi(\vec{q},\vec{k}_2) + \text{Power Series in } V_{COUL} \tag{25}$$

where $\psi_o(\vec{k}_1,\vec{k}_2)$ is the wave function of the unperturbed exciton.

If we ignore all coulomb interactions except that necessary for $\psi_o(\vec{k}_1,\vec{k}_2)$, that is, if we drop the power series in V_{COUL} from Eq. (25), we obtain a solvable equation. This approximation is also bad near the absorption edge, but it improves for higher energies.

Doing this and integrating the resulting equation on the first argument gives:

$$\int_{(B.Z.)} d^3q\;\psi(\vec{q},\vec{k}_2) = \frac{1}{1 + Jf(E - \varepsilon_v(\vec{k}_2))}\int_{(B.Z.)} d^3q\;\psi_o(\vec{q},\vec{k}_2) \tag{26}$$

again exhibiting the resonance denominator. This gives us the wave function:

$$\psi(\vec{K}_1,\vec{K}_2) = \psi_o(\vec{K}_1,\vec{K}_2)$$

$$+ \frac{1}{[E+i0 - \varepsilon_c(\vec{K}_1) - \varepsilon_v(\vec{K}_i)]} \frac{1}{[1 + Jf(E - \varepsilon_v(\vec{K}_2))]}$$

$$\times \frac{J\Omega}{(2\pi)^3} \int_{(B.Z.)} d^3q \ \psi_o(\vec{q},\vec{K}_2) \qquad (27)$$

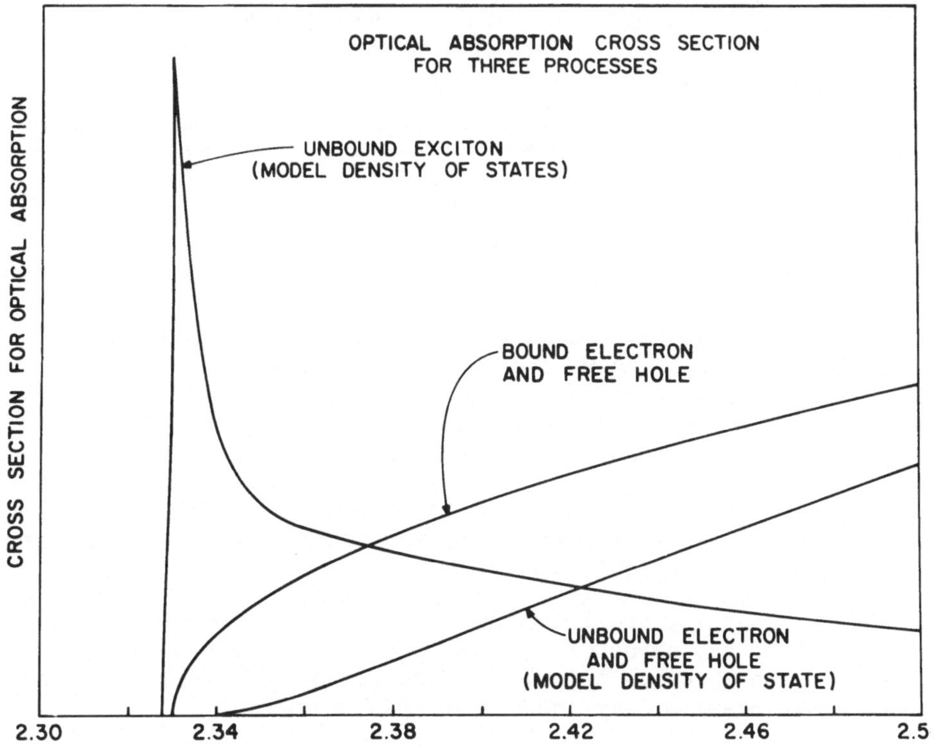

Figure 3. Theoretical curves for three optical absorption processes in GaP:N.

With some labor $\int d^3k \ \psi(\vec{K},\vec{K})$ can be calculated from this expression, using again the model density of states for $f(E-\varepsilon)$. The result is shown in Fig. 3.

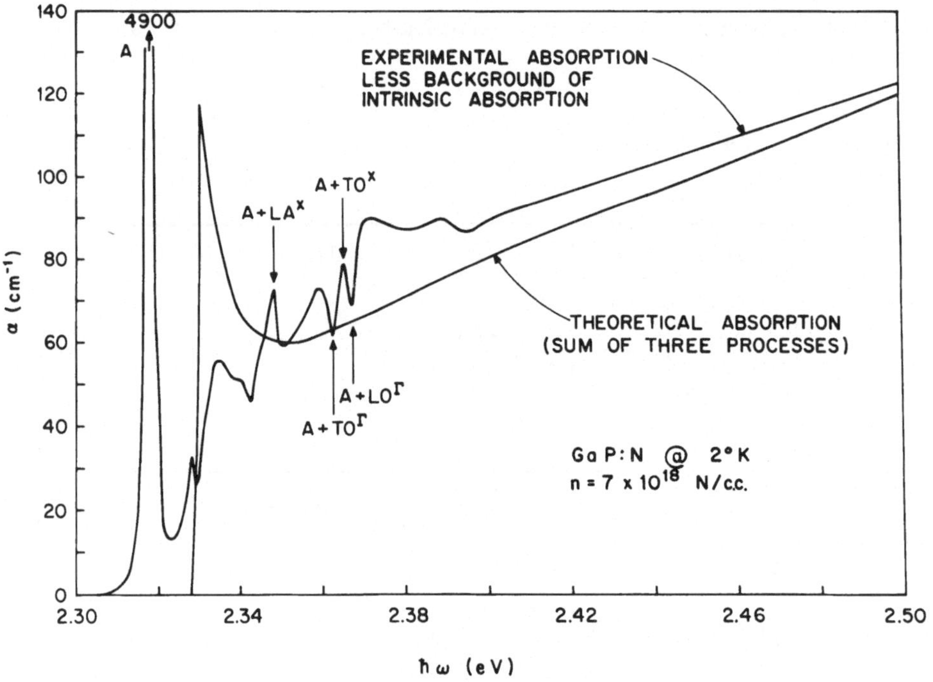

Figure 4

The sharp peak in this curve occurs at the threshold for the production of bound electrons and free holes and is a resonance effect due to this cross channel square root threshold. If we were to consider more terms in the coulomb interaction, this peak would become much less pronounced and not as sharp.

Figure 4 shows the sum of the three processes considered superimposed on the experimental curve which has had the background intrinsic absorption subtracted out.

The first thing one is struck by on comparing the two curves is the lack of any structure on the theoretical curve such as is present on the experimental curve. The reason, of course, is that phonons have been totally ignored in this treatment.

Several resonance effects are present in this absorption curve. Hopfield, et al has discussed these effects in a recent paper.[6] Of particular note are the two positions marked $A + TO^{\Gamma}$ and $A + LO^{\Gamma}$ where ordinarily phonon replicas of the principal absorption line ("A" line) would be expected. Instead, due to cross channel interference, valleys occur instead of peaks.

The second thing one is struck by is the sharp peak in the theoretical curve which is much less pronounced in the experimental curve. As was discussed previously, a better treatment of the coulomb interaction should wash the theoretical peak out, bringing it more into line with experiment.

With these exceptions noted, the theoretical curve is seen nonetheless to account very well for almost all of the absorption induced by the Nitrogen impurity well above the band edge.

CROSS SECTION FOR ELECTRON-IMPURITY SCATTERING

From the wave function of an electron interacting with the Nitrogen impurity, Eq. (22), the scattering cross section for Bloch electrons in the conduction band can be calculated:

$$\sigma(E) = \frac{4\pi/3}{2m_e^* E + \left[\frac{2\pi(1 + J\,\mathrm{Re}(f(E)))}{2m_e^* J\Omega}\right]^2} \tag{28}$$

where E is the energy of the electron relative to the minimum of the conduction band.

Evaluating this expression for zero energy, we obtain $\sigma(0) = 10{,}000\ (\text{Å})^2$.

We can do the same thing for the hole except that now $(1 + J\langle 1/\varepsilon\rangle)$ is no longer near zero as it was for the electron but is of order unity. This reduces the scattering cross section by four orders of magnitude, making it approximately $1\,(\text{Å})^2$. This can be considered an a posteriori justification for ignoring the hole-impurity interaction in the beginning.

DOUBLE IMPURITY BOUND STATES

Let us now return to the bound state problem, now considering the problem of binding an electron to a pair of impurity atoms separated by a lattice vector, $\vec{R}$. We let one atom be at the origin of coordinates and the other be centered about the lattice site at $\vec{R}$. Then, according to Eq. (4), the potential matrix element should be:

$$(c\vec{k}\,|\,V_{II}\,|\,c\vec{q}) \simeq \frac{J\Omega}{(2\pi)^3}\left[1 + e^{-i(\vec{k}-\vec{q})\cdot\vec{R}}\right] \tag{29}$$

and Schrodinger's equation for the electron becomes:

$$[E - \varepsilon_c(\vec{k})]\psi(\vec{k}) = \frac{J\Omega}{(2\pi)^3} \int_{(B.Z.)} d^3q \; (1 + e^{-i(\vec{k}-\vec{q})\cdot\vec{R}})\psi(\vec{q}) \tag{30}$$

This equation can be solved almost as easily as before if we write:

$$\psi(\vec{k}) = \frac{A + B\, e^{-i\vec{k}\cdot\vec{R}}}{(\varepsilon_c(\vec{k}) - E)}, \qquad E < 0$$

From this form we obtain two equations for the two unknowns, A and B. Requiring the determinant of the coefficients to vanish gives the eigenvalue equation:

$$[1 + Jf(E)]^2 = [Jf(E,\vec{R})]^2 \tag{31}$$

where we have a new function defined as:

$$f(E,\vec{R}) = \frac{\Omega}{(2\pi)^3} \int_{(B.Z.)} d^3q \; \frac{e^{i\vec{q}\cdot\vec{R}}}{[\varepsilon_c(\vec{q}) - E]} \tag{32}$$

To do a good job of evaluating $f(E,\vec{R})$, knowledge of the band structure throughout the Brillouin Zone is again necessary. However, a good insight into the nature of the wave function interference effects in this problem can be gained by simply using the effective mass approximation and extending the integral over all $\vec{k}$-space. The oscillatory complex exponential causes the integral to converge.

First recalling that there are three valleys in the GaP conduction and using isotropic effective masses in each, we obtain for $f(E,\vec{R})$:

$$f(E,\vec{R}) = m^* \frac{\Omega}{2\pi} \frac{e^{-\sqrt{2m^*|E|}R}}{R} \left\{ e^{i\frac{1}{2}\vec{G}_1\cdot\vec{R}} + e^{i\frac{1}{2}\vec{G}_2\cdot\vec{R}} + e^{i\frac{1}{2}\vec{G}_3\cdot\vec{R}} \right\} \tag{34}$$

where the $\vec{G}$'s are reciprocal lattice vectors in the (100), (010) and (001) directions, respectively. $\frac{1}{2}\vec{G}$ is the position of one of the conduction band minima in the Brillouin Zone. Because $e^{i\vec{G}\cdot\vec{R}} = 1$ for all $\vec{G}$ and all $\vec{R}$, $e^{i\frac{1}{2}\vec{G}\cdot\vec{R}}$ can equal +1 or -1. Therefore, the factor in brackets in Eq. (34) can be ±1 or ±3. A factor of ±3 would lead to a larger binding energy than a factor of ±1.

If we also use the fact that the masses in each valley are anisotropic, $m^* = (m_1, m_1, m_2)$, we get further effects:

$$f(E,\vec{R}) = m_1 \frac{\Omega}{(2\pi)} \frac{1}{R} \{w_1 + w_2 + w_3\} \tag{35}$$

where, e.g.:

$$w_1 = e^{i\frac{1}{2}\vec{G}_1 \cdot \vec{R}} \frac{\exp\left\{-\sqrt{2m_2 |E|}\, R \sqrt{\cos^2\theta_1 + \frac{m_1}{m_2}\sin^2\theta_1}\right\}}{\sqrt{\cos^2\theta_1 + \frac{m_1}{m_2}\sin^2\theta_1}}$$

θ_1 = angle between $\vec{R}$ and $\vec{G}_1$.

With $m_1 = .26$ m and $m_2 = 1.49$ m, this function becomes quite sensitive to the orientation of $\vec{R}$ in addition to its ordinary dependence of the magnitude of $\vec{R}$.

This scheme, due partly to the crudeness of the evaluation of $f(E,\vec{R})$ does not yield bound state energies in agreement with experiment. Computer calculations have been done which improve the situation somewhat as we shall see shortly.

COMPUTER CALCULATIONS OF THE BOUND STATE PROBLEM

The simple model cannot be considered an adequate theory of isoelectronic impurities, despite its success in certain areas. Most notable among its failures is its inability to predict the energies of any bound states or, given the energy of the single Nitrogen bound state, to predict the energies of the double Nitrogen lines correctly. Furthermore, the excited states of the single Nitrogen level do not appear, even incorrectly.

Therefore, a more ambitious program was undertaken - to calculate the bound states from more or less first principles.

To do this requires a number of items of information. First, the band structure and Bloch Functions of the host crystal must be known throughout the Brillouin Zone. This can be generated from the empirical pseudopotential form factors give for GaP and many other semiconductors by Bergstresser and Cohen.[7] Secondly, a pseudopotential representing the difference between the impurity and the atom it replaces is needed. The atomic structure calculations of Herman and Skillman[8] provide the core electron wave functions necessary for this task. Figure 5 shows the pseudopotential

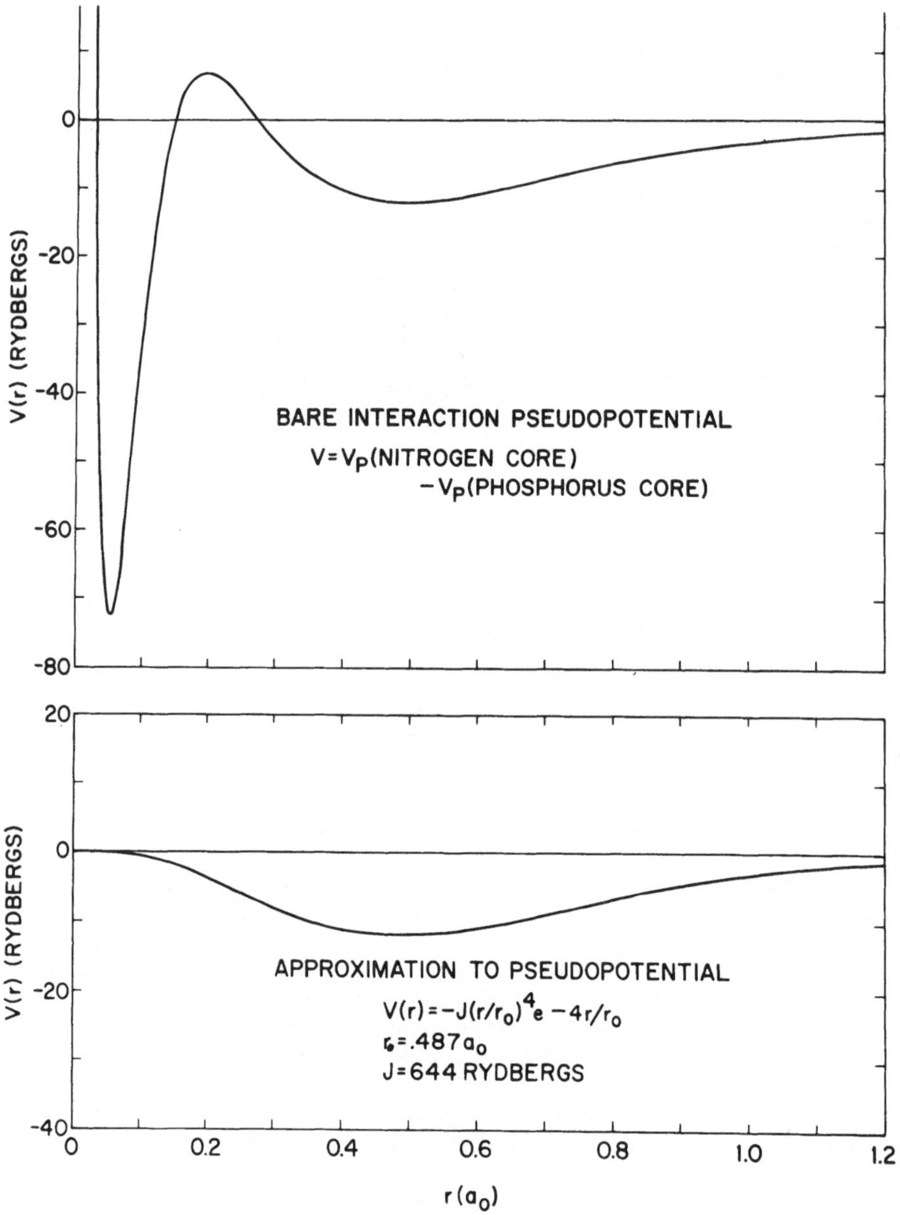

Figure 5. Pseudopotential for Nitrogen substituting for Phosphorous in GaP from the core wave functions of Herman and Skillman.

for a smooth S-state calculated from these core functions along with the simple approximation actually used in the calculation. The approximation ignores the innermost structure of the pseudopotential because that region does not contain sufficient volume to appreciably affect the wave functions. The greatest strength of the pseudopotential comes from the broader well farther from the origin.

The technique is now simple in theory, but more complicated in practice.

We write Schrodinger's equation:

$$(E-H_o)\,|\psi\rangle = V\,|\psi\rangle$$

where H_o is the crystal Hamiltonian and V is the potential of Fig. 5.

For an E which lies outside the spectrum of H_o, we can write:

$$(1-GV)\,|\psi\rangle = 0 \;; \qquad G = \frac{1}{(E-H_o)}$$

Since we have a short range potential, the appropriate set of basis functions to use is the set of Wannier functions. Expressing the matrix (1-GV) in terms of these functions, the condition that a bound state exist is that the determinant of (1-GV) vanishes. The matrix considered is finite because $(n\vec{R}\,|\,V\,|\,n\vec{R}')$, the matrix element of V with Wannier functions, becomes negligible for sufficiently large $\vec{R}$ or $\vec{R}'$.

The problem one faces now is how to generate Wannier functions from Bloch functions. The relevant formula is straightforward enough:

$$w_n(\vec{x}-\vec{R}) = \sqrt{\frac{\Omega}{(2\pi)^3}} \int_{(B.Z.)} d^3k \; e^{-i\vec{k}\cdot\vec{R}}\psi_n(\vec{k},\vec{x})$$

However, numerical calculations of $\psi_n(\vec{k},\vec{x})$ do not specify its phase relative to Bloch functions with different values of $\vec{K}$. An even greater problem is a useful definition of an energy band. Do you require, e.g., band #4 to lie lower in energy than band #5 everywhere in the Brillouin Zone or do you allow them to be inverted in certain regions?

Joseph Callaway and A. James Hughes have investigated these problems in their recent paper.[9] They treated the case of the neutral vacancy in silicon. Since silicon contains a center of inversion, they were able to work with real matrices in the generation of their Bloch functions and so the choice of phase reduced to a choice of sign. Their method of defining energy bands is based on considerations of continuity of symmetry through the Brillouin Zone rather than strict continuity of energy. Their method yields Wannier functions which are well localized and which possess the symmetries of the one-dimensional representations of the point group.

GaP has no center of inversion, and the numerical Bloch functions have complex components. Also, the energy bands do not cross so neatly in GaP as in silicon. Whether for these reasons or because of a lack of fortitude on the part of the investigator, simpler phase and band assignments were chosen for this problem. The Bloch functions were taken to be real and positive at $\vec{x} = 0$ and band assignments were made straightforwardly according to increasing energy. The Wannier functions obtained in this way were not as well localized as those of Callaway and Hughes, so more lattice sites were needed for the matrix (1-GV). A total of 19 were used as opposed to Callaway's 10. To keep the numerical labor within reason, only the first two conduction bands were considered. This resulted in 38×38 matrices for the single Nitrogen problem and as large as 76×76 for the double Nitrogen problem.

It was found that the potential used gave a bound state well into the forbidden gap, approximately one eV. A reduction factor, λ, was used to reduce the strength of the potential to fit the experimentally observed single Nitrogen level in hope that the same parameter would give reasonable values for the double Nitrogen levels. The energy of the single Nitrogen state was found to be extremely sensitive to the value of this parameter, giving -.021 eV at $\lambda = .48$ and -.080 eV at $\lambda = .50$. A similar situation existed in the simple model in its sensitivity to the potential strength, J. Also in the computer calculations, the excited states in the single Nitrogen problem never appeared. This is a property of the short range of the potential chosen for the calculation. A longer range potential would reduce the intervalley matrix elements responsible for splitting off the excited states and bring them below the band edge. This, coupled with the sensitivity of the energy level to the potential strength suggest that the self-consistency effects, which would be of longer range than the present bare potential, are of prime importance in this problem.

Using a multiplying factor of .472, which puts the single Nitrogen level at .0078 eV, the double Nitrogen levels come out as shown in Fig. 6. They are arranged into two groups, allowed and forbidden, according to whether $\psi(\vec{k}=0) \neq 0$ or $\psi(\vec{k}=0) = 0$, respectively. The leftmost column contains the experimental levels

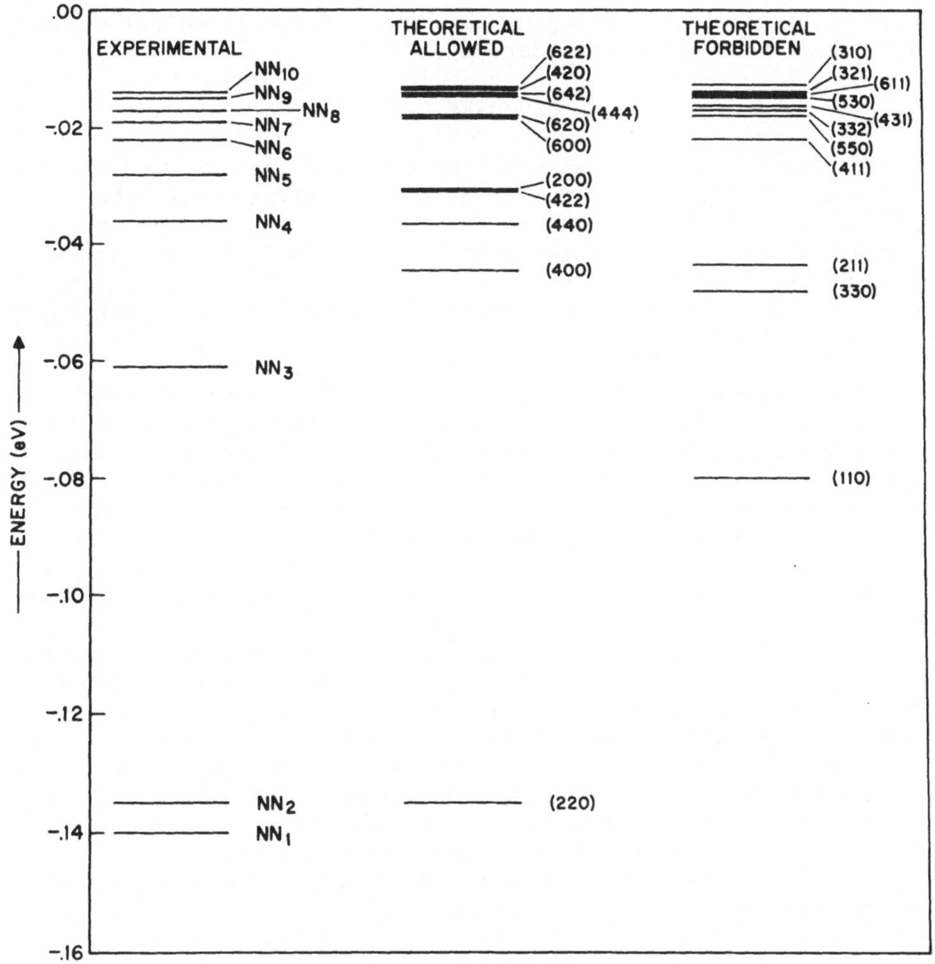

Figure 6. Energy levels of the double Nitrogen states. Only the electron binding energy is presented here and the experimental levels have been raised by 13 meV.

for comparison. The numbers in parentheses represent the lattice vector separating the two Nitrogen impurities.

Although these energy levels obviously do not agree with experiment, the range of energy is correct and the average spacing of levels is correct. Changing the range of the potential would shift the levels about considerably.

SUMMARY

In summary, we can say that one can go a long way toward understanding the nature of isoelectronic impurities using the simple model presented here. The most appealing point about this model is that it correlates the position of the principal bound state with its oscillator strength in optical absorption and with the absorption strength for frequencies above the band edge using only one essential parameter.

The sensitivity of the bound state energy to the strength of the potential in both the simple model and the computer calculations indicates a need for screening effects of the host crystal to be calculated. A longer range potential would both reduce the sensitivity of the bound state energy and bring the excited states mentioned before down below the band edge.

With a better understanding of isoelectronic traps will also come a better understanding of the related problem of central cell corrections for donor and acceptor states. Isoelectronic systems offer an ideal situation for the study of such effects, being by their nature purely central cell corrections with no extraneous features to complicate the picture.

Understanding the nature of the single neutral impurity problem would open the way for an understanding of more complicated systems such as nearest neighbor donor-acceptor states. With vision, one can imagine a whole chemistry of complexes within the vacuum represented by the perfect crystal.

REFERENCES

1. D. G. Thomas and J. J. Hopfield, Phys. Rev. 150, 680 (1966).
2. D. G. Thomas, J. J. Hopfield, and R. T. Lynch, Phys. Rev. Letters 17, 312 (1966).
3. J. D. Cuthbert and D. C. Thomas, Phys. Rev. 154, 763 (1967).
4. A. C. Aten, J. H. Haanstra, and H. de Vrier, Philips Research Reports 20, 395 (1965).
5. P. J. Dean, private communication.
6. J. J. Hopfield, P. J. Dean, and D. G. Thomas, Phys. Rev. 158, 748 (1967).
7. M. L. Cohen and T. K. Bergstresser, Phys. Rev. 141, 789 (1966).
8. Frank Herman and Sherwood Skillman, Atomic Structure Calculations, Prentice-Hall, Englewood Cliffs, N. J.
9. Joseph Callaway and A. James Hughes, Phys. Rev. 156, 860 (1967).

AN ACCOUNT OF BOUND EXCITONS IN SEMICONDUCTORS

D. G. Thomas

Bell Telephone Laboratories, Incorporated

Murray Hill, New Jersey

I. INTRODUCTION

The term "bound exciton," used in connection with semiconductors, is generally taken to describe a state in which a hole and an electron is localized at some defect or impurity in the crystal. Often such states are shallow, that is, a free exciton is bound to the impurity by only a few millielectron volts, but it is quite logical to include much deeper states, for there is no difference in kind between deep and shallow states. In addition, one often thinks of bound excitons as being attached to point defects in the solid. However, it is becoming increasingly clear that pairs of nearest neighbor impurities play important roles in semiconductor physics, and there seems to be no reason for excluding excitons bound to such centers from the present classification. Having gone this far, one does not want to exclude excitons bound to next nearest neighbor pairs of defects, and so one ends up including effects due to electrons bound to donors interacting with holes bound at acceptors, even though the donors and acceptors are far apart.

There is interest in such effects for several reasons. These bound states, particularly at low temperatures, often provide the mechanism by which radiative recombination of holes and electrons takes place. These holes and electrons can be generated by band gap radiation, (photoluminescence), or by forward biasing a p, n junction, (electroluminescence), or by electron bombardment, (cathodoluminescence). It often happens that the emission arising from the decay of bound excitons exhibits a spectrum which contains sharp lines. These lines can often be interpreted, and so information is obtained about the nature of the center involved. This information can be relevant to what chemical impurities are involved and how they are disposed in the lattice. Since the concentrations of the centers are often very low it is difficult to determine their properties by conventional means, and the spectroscopic technique of examining bound excitons is becoming useful for semiconductor materials control.

Recently there have appeared several reviews of the subject.[1,2,3] This article will describe a few of the earlier experimental results to illustrate the type of phenomena encountered, and will summarize recent results obtained over the last year or two.

II. EXCITONS BOUND TO DONORS AND ACCEPTORS

In 1958, Lampert[4] suggested that excitons could bind to neutral and ionized donors or acceptors to give a variety of complexes analogous to the hydrogen molecule or the hydrogen molecule ion, as illustrated in Fig. 1. The criteria for the formation of such complexes has been the subject of several theoretical papers.[5,6,7]

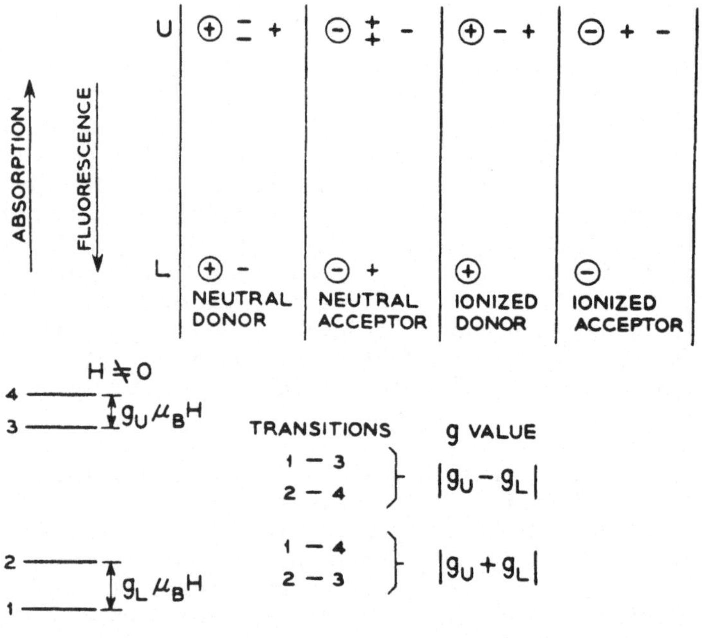

Fig. 1. A schematic representation of the binding of excitons to neutral and ionized donors and acceptors. The energy levels and magnetic splittings are indicated for a degeneracy of two for both holes and electrons as is the case for CdS.

A. Bound Excitons in Silicon

Perhaps the first clear experimental identification of such states was provided by Haynes[8] in 1960 who examined the low temperature photoluminescent spectra from silicon crystals. One such spectrum is shown in Fig. 2. The sharp line at 1.149 eV, (a "zero

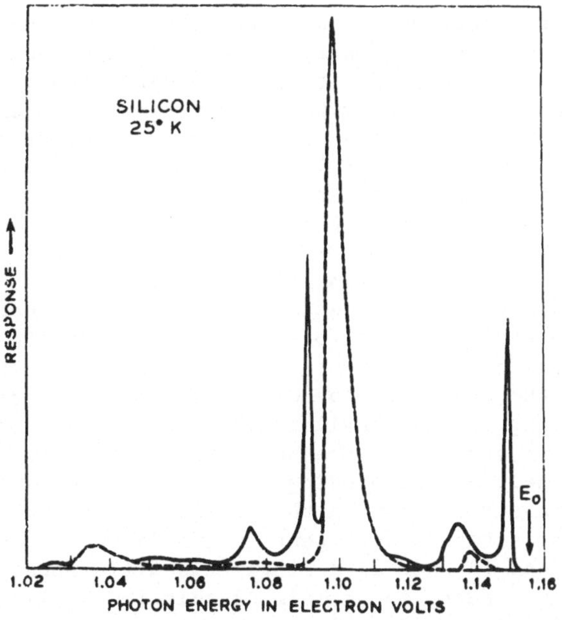

Fig. 2. The photoluminescence of silicon as measured by Haynes at 25°K. The dotted curve is the spectrum from a pure crystal; there is no emission at the intrinsic exciton energy E_0 but rather at the exciton energy less momentum conserving phonon energies. The solid curve is the spectrum from a crystal containing 8×10^{16} arsenic atoms/cc. There is now a zero phonon bound exciton line at 1.149 eV as well as its momentum conserving replicas.

phonon" line), is due to the decay, without phonon cooperation, of an exciton bound to a neutral arsenic donor atom. At 1.091 eV there is a phonon replica of the zero phonon line, corresponding to the state decaying with the emission of a phonon of energy 0.058 eV. This phonon energy is that of a transverse optical phonon which conserves momentum - that is, it has a wave vector which is equal to the wave vector difference between the maximum of the valence band and the minimum of the conduction band in silicon. The existence of this difference is, of course, a result of the fact that silicon has an indirect band gap. Also shown in Fig. 2 is the emission from a pure crystal of silicon. This spectrum is dominated by a peak at 1.10 eV which corresponds to an intrinsic exciton, (energy E_0), decaying with the emission of the same momentum conserving phonon. Notice that no emission occurs at the zero phonon exciton energy, because this is now a highly forbidden transition. It is the presence of the impurity state, which provides coupling to the lattice, that allows the momentum conserving selection rule to be relaxed, and so produces the zero phonon line which is somewhat analogous to a Mössbauer line. A zero phonon line can often be recognized as such by its extreme narrowness, and can usually be proved to be such by the observation of an emission and absorption line at the same frequency. In the present case where the transitions have a low oscillator strength the absorption is not strong, but it has been recently observed by Dean, et al.[9] to be, as expected, a sharp line at 1.149 eV.

For this type of rather weakly bound transition to donors and acceptors the phonon wings correspond to those which conserve momentum; as the binding energy of an exciton to such a center increases so does the strength of the zero phonon component relative to that of the phonon sidebands. This happens because an increase of binding energy localizes the wavefunction at the center and causes greater overlap of the hole and electron wavefunction in the Brillouin zone. As the binding energy increases still further, the difference in strain between the ground and excited states increases and one enters a different regime in which the whole spectrum of phonons is used to relieve this difference of strain as the transition occurs; an example of this is the I_1 transition in CdS described below. (In this case a quantitative account has been given of the shape of a part of the phonon wing.[42] This account has emphasized the fact that there will be the strongest interaction with acoustic phonons whose wavelengths are comparable to the dimensions of the center in question.) These remarks, however, are qualitative and there are exceptions. Thus at isoelectronic traps, described below, quite shallow states such as nitrogen in gallium phosphide, (which is an indirect gap semiconductor), exhibit strong phonon wings in which there is no emphasis on the momentum conserving components. In general, an account of the phonon sidebands of bound excitons in semiconductors and an interpretation of the considerable structure which they exhibit, remains as a field to be explored.

Haynes studied similar effects for several donors and acceptors in silicon and found that the excitons were bound to the centers with an energy of about 0.1 of the binding energy of the particular donor or acceptor. Thus the exciton is bound to the arsenic donor in Fig. 2 with an energy of 0.0065 eV, and an arsenic donor has a binding energy of about 0.055 eV.

It is interesting to note that even at low temperatures the efficiency with which these silicon crystals emit light is very low and consequently the spectra are not easy to observe. The reason for this is that these bound excitons decay predominantly by a non-radiative Auger mechanism which is described in Section IV-B.

B. Bound Excitons in CdS

One of the great advantages of silicon is that crystals can be obtained which contain essentially only one known impurity. In many other materials, particularly the wider band gap semiconductors, this is not so and one has to deduce what one can from the spectra themselves. A case in point is provided by CdS, a direct band gap semiconductor. At low temperatures, a large number of sharp lines can be seen at energies below those of the intrinsic excitons. Some of these lines are shown in Fig. 3, together with absorption lines due to the intrinsic excitons as well as the bound excitons.

The interpretation of these lines has been helped by the study of their Zeeman effect.[10] The idea is quite simple. Referring to Fig. 1, if the exciton is bound to a neutral donor or acceptor then two of the particles in the excited state, for instance the two electrons in the case of the donor, will be paired off to produce a bonding state, and will therefore have no magnetic moment. In this case, therefore, the ground state will split in a magnetic field with a g value corresponding to that of an electron, while the excited state will split with a g value characteristic of a hole. For an acceptor the hole will be in the ground state and an electron in the upper. If, therefore, one could tell whether a hole or electron is in the ground state one could at least say whether one was dealing with a donor or an acceptor. In CdS the g value of the electron, g_e, is almost isotropic, whereas that of the hole is of the form,

$$g_h = g_{h0} \cos \theta,$$

where θ is the angle between the hexagonal c axis of the crystal and the magnetic field, and g_{h0} is a parameter which has different values depending upon the state of binding of the hole. The splittings of the lines enable one to determine the values of g_e and g_{h0}. To determine whether a hole or electron is in the ground state, it is convenient to look at the absorption spectrum at very low temperatures with $C \perp H$, when $g_h = 0$. If there is a splitting in the ground state it will become apparent because there will be a preferential population of the lower level of the ground state,

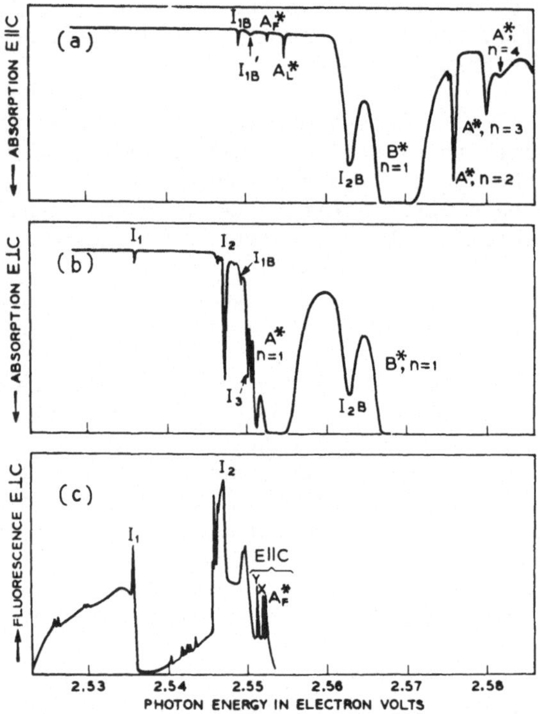

Fig. 3. The absorption and emission spectra of a typical as grown crystal of CdS. The lines indicated by an asterisk correspond to intrinsic exciton states. The others correspond to zero phonon exciton transitions. The absorption spectra show the results for light with E∥C and E⊥C; the emission spectrum is dominantly polarized with E⊥C.

which will be observed as a change of intensity with temperature of the absorption lines. If such thermalization effects are seen with C⊥H then an electron is in the ground state, if not then a hole is there. In this way a prominent line called I_1 in Fig. 3 has been shown to be due to an exciton decaying at a neutral acceptor, and several lines collectively labelled I_2 are due to excitons bound to donors. Halsted and Aven[11] have pointed out that for the II-VI compounds the ratios of the energies with which excitons are bound to donors to the donor binding energies are nearer 0.2 than 0.1 as found for silicon; for acceptors, however, the ratio is again near 0.1. In the present case excitons are bound to donors by about 7×10^{-3} eV and donor binding energies are estimated to be 24 - 30×10^{-3} eV.

While such magnetic analyses have worked out in the case of CdS, they evidently have not in the case of CdSe.[12] Here for instance, despite the fact that the hole when part of a free exciton has a g value of the form $g_{h0} \cos \theta$, all g values in the bound exciton lines are isotropic. It is perhaps surprising that although these states are only bound by 0.03 eV or less the hole and electron characteristics seem to be quite different from those of the free particles. It may, however, be pointed out that in CdSe a 0.03 eV binding energy is large in the sense that the exciton binding energy is 0.015 eV, and donor binding energies are probably comparable to this. Hence, the picture of an exciton being bound to a donor with an energy small compared with the donor binding energy is perhaps inappropriate.

In the above discussion of the magnetic moments of the complexes formed by the binding of excitons to neutral donors or acceptors it has been implied that the two similar particles in the excited state must pair off and themselves produce zero angular momentum. This is generally true for s like electrons and is true for holes in CdS which, because of the hexagonal crystal field, have only two-fold degeneracy. It is not true for four-fold degenerate holes such as occur in most cubic semiconductors where bound states at neutral acceptors are possible in which the two holes can, even in zero magnetic field, combine in different ways to give a variety of energy levels, which will result in complex Zeeman patterns. Thus donors and acceptors in CdS will give simple line patterns, but complications are expected for instance with acceptors in silicon, or cubic ZnSe or cubic SiC. Dean, et al.[9] have observed such complications in the zero field pattern of the absorption lines of excitons bound to acceptors in silicon. It is clear that despite the complications, thermalization measurements and observations of magnetic splittings will still enable generic classifications of the lines to be made.

Such Zeeman data do not reveal the chemical identity of the centers. In principle, the intentional addition of likely impurities can lead to chemical identification, although such experiments require care, as the doping levels at which sharp lines can be seen are rather low. Some recent work[13] has tentatively identified certain of the I_2 lines in CdS with Cl and Al donors. Of greater interest is the I_1 line, for it is a lone shallow acceptor and is clearly connected with the so-called green edge emission, which is believed to arise at low temperatures from donor-acceptor pair recombination, the acceptor also being responsible for I_1. Annealing experiments show that I_1 is associated with what is probably a cadmium vacancy which is expected to be a double acceptor, yet the Zeeman results indicate a single acceptor. Now it is often observed that I_1 appears as two lines[14] separated by 2×10^{-4} eV; these lines behave identically in a magnetic field and yet they appear to originate from different centers, for their relative intensities can, despite reports to the contrary,[14] vary enormously. It has been suggested, therefore, on the basis of certain doping experiments,[13] that the lower energy member of the doublet arises from a Cd vacancy with a Cl, (or Br or I), donor next to it on a sulphur site, and the upper member from a Cd vacancy with an Al donor on the next nearest Cd site. In this way a single acceptor is formed which is closely analogous to the A center seen in ZnS. Schneider[15] and others using paramagnetic resonance methods have shown that this center consists of a zinc vacancy with a Cl or Al donor near it, and Shinoya[16] and others, particularly by the use of polarization measurements of the emission, have shown that hole-electron recombination at this center causes the self-activated luminescence near 4600 Å. This energy is roughly 1.3 eV below that of the band gap of ZnS, whereas the I_1 line is only about 0.046 eV below the band gap of CdS. The binding energy is thur much larger for the A center and so one can understand that in this case the difference between the Al and Cl centers is about 0.12 eV, again much larger than the suggested value for CdS. For ZnS the Al band lies at a lower energy than the Cl band. Because of strong phonon coupling, no sharp line transitions can be seen for ZnS. A convenient summary of this elegant work with ZnS can be found in Ref. 3 on pages 303 and 445.

If further work confirms that the suggested interpretation of the I_1 line in CdS is correct it is interesting to note that the anisotropy of the bound hole is still controlled by the symmetry of the lattice rather than that of the center itself which will be, for such an associated pair of defects, quite different from that of the lattice.

C. Excitons Bound to an Ionized Center in CdS

A certain line called I_3 can be seen in absorption in CdS[10] and is included in Fig. 3. Unlike the lines considered so far, this line does not have a linear Zeeman pattern. Its behavior in a magnetic field is shown in Fig. 4. It is clear that at zero magnetic field there are really two transitions, but that the lower energy one is forbidden and in certain orientations can be introduced by a magnetic field. The absence of any thermalization effects in absorption shows that the splitting is in the excited state of the center. This line corresponds to an exciton bound to an ionized donor as schematically shown in Fig. 1. Because there can be no pairing of particles in the excited state the moments of the hole and electron can interact in a type of jj coupling scheme,

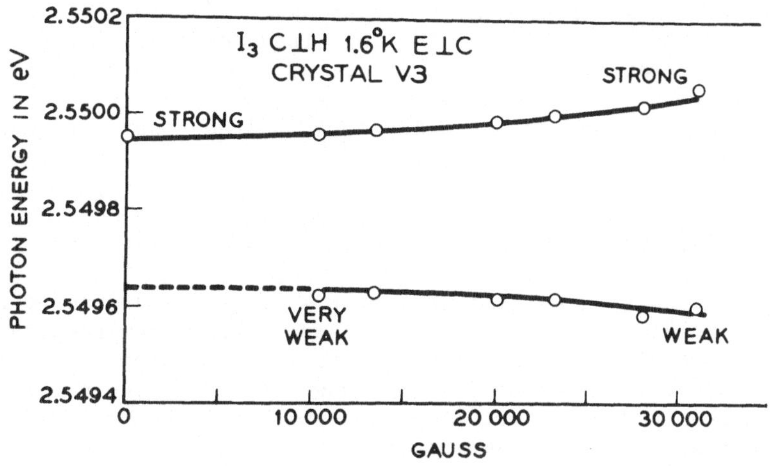

Fig. 4. The Zeeman pattern for E⊥C of the line I_3 in CdS which corresponds to an exciton bound to an ionized donor. A nonlinear splitting is seen which reveals a zero field splitting due to jj coupling between the hole and electron.

to give in this case a $|J_z| = 1$ and a $|J_z| = 2$ state. Transitions from the J=0 ground state are allowed to the $|J_z| = 1$ state but forbidden to the $|J_z| = 2$ state.

This type of zero field splitting is frequently encountered in excitons bound to isoelectronic traps, (see below), which in common with ionized donors or acceptors, have no holes or electrons in the ground state. Excitons bound to ionized donors or acceptors might be expected to be important luminescent centers, for they cannot undergo the type of nonradiative Auger recombination described later. However, the I_3 line in CdS is not strong in fluorescence, although its fluorescence is observed in some crystals.[14] The reason for this is probably that because the state is so shallow, (an exciton is bound to it by only 3.8×10^{-3} eV), it will have a large radius and consequently will be able to tunnel rapidly to lower energy states. So far there do not seem to be any definite reports of deep centers of the type represented by I_3. The reason for this may be that an electron, say, bound tightly to a positively charged center will be a highly localized neutral complex which, unlike a less localized complex, will not have sufficient polarizability to result in a bound state with a polarizable hole. The attractive forces in this case of course have to be of a Van der Waals nature. At an isoelectronic trap, (see Section III), the situation is different in that an electron, say, is bound to a neutral center to form a negative complex which can then bind a hole by normal coulomb forces. Thus deep ionized donors or acceptors probably do not bind excitons.

D. Non-Equilibrium Distributions of Holes and Electrons

It may be asked how is it that neutral donors and acceptors can be present simultaneously in a semiconductor at low temperatures, for normally that species present in lower concentration will be completely compensated. The answer is that band gap radiation will make free holes and electrons which will neutralize compensated centers. At low temperatures the centers will remain neutral, or will gradually ionize by slow radiative recombination of the separated holes and electrons, (i.e., donor acceptor pair recombination). By observing absorption lines it has been possible to detect in CdS the change of population brought about by illumination.[10] In these experiments conducted at helium temperatures band gap light increases the intensities of the I_1 and I_2 lines associated with neutral acceptors and donors respectively, while infrared light, which ionizes these centers, decreases their intensities. The reverse is true for the I_3 line which arises from an ionized donor.

E. Excitons Bound to Donors and Acceptors in Other Materials

As expected excitons bound to donors and acceptors have been detected in emission in germanium. Benoit à la Guillaume and others have observed the effects at photon energies near 0.74 eV. A great deal of careful work has been carried out by Choyke, Hamilton and Patrick on excitons bound to neutral and ionized nitrogen donors in silicon carbide. Several interesting effects are observed in this system. There is first the fact that SiC can exist in many polytypes ranging from pure cubic to pure hexagonal crystals, with several well defined intermediate forms. In many of the polytypes there are inequivalent nitrogen sites each of which give rise to separate bound exciton spectra. The spectra possess sharp phonon satellites which correspond in energy to "momentum conserving" lattice vibrations. By observation of the energies of excitons bound to donors and to ionized donors, a nitrogen donor binding energy of 0.1 eV has been determined in cubic SiC. In addition, in the cubic crystal as the temperature is increased from 4.2°K additional lines are seen which have been related to a 5×10^{-3} eV spin orbit splitting of the hole state and a 2×10^{-3} eV valley orbit splitting of the donor state. In diamond

Dean and coworkers have described bound exciton states. The band gap of diamond is large, (5.48 eV). A prominent donor is nitrogen which has a very large binding energy of about 4 eV. It is reported that lines near 5.26 eV are due to an exciton trapped at a neutral nitrogen donor adjacent to which is an ionized aluminum acceptor. Related to this is the so-called N3 system at 3 eV which corresponds to an exciton bound to the same complex but without an electron in the ground state. Finally an ionized isolated nitrogen donor is stated to bind an exciton which radiates near 1.4 eV. If true, this latter effect is in conflict with the suggestion made above that deep ionized centers cannot bind both a hole and an electron.

A convenient summary of and references to the work discussed in this section can be found in Ref. 1.

III. EXCITONS BOUND TO ISOELECTRONIC TRAPS

A. Isolated Nitrogen Traps in GaP

So far cases have been discussed in which excitons have been bound to centers which in their ground states possess a hole or electron, or if not are charged. Thus binding may be thought of as being due to the type of forces which lead to the existence of the hydrogen molecule and the hydrogen molecule ion. This picture is, of course, a great oversimplification, for the binding energies of excitons to centers of the same type vary as the chemical nature of the centers change; (recall for example Hayne's Rule relating exciton binding energies to the respective donor binding energies). Crudely, these departures from the effective mass approximation have their origin in the short range potentials associated with the presence of a foreign atom in the crystal. They are often referred to as "chemical shifts" or "central cell corrections." The importance of these forces is rather dramatically demonstrated by the existence of so-called "isoelectronic traps."[17] These consist of an atom replacing a host atom from the same column of the Periodic Table, which is able to bind a hole and electron. Often such substitutions can be made without a bound state being produced. An example of this is the addition of arsenic to GaP which can be readily achieved, and which simply results in an alloy system with continuous properties between GaP and GaAs. If, however, nitrogen is added to GaP, which is only possible in comparatively small concentrations, a new phenomenon arises in which discrete states are formed in the forbidden band.[18]

Figure 5 shows the fluorescence from hole electron recombination occurring at an isolated nitrogen atom in GaP. There are two sharp lines A and B separated by 0.8×10^{-3} eV, 11×10^{-3} eV below the intrinsic exciton energy. These lines occur at the same energies

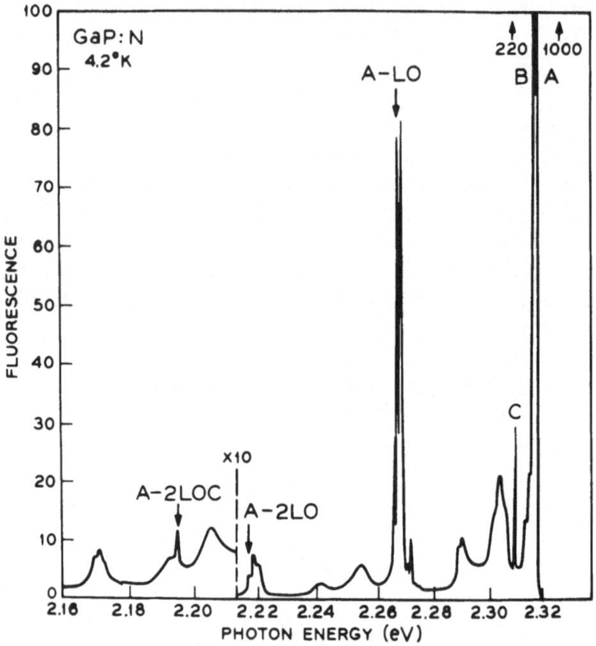

Fig. 5. The photoluminescence spectrum due to isolated nitrogen in GaP. The A and B lines are the J = 1 and J = 2 zero phonon lines. The line marked A-2LOC is the double local mode line which moves to higher energies when N^{15} is substituted for N^{14}.

in absorption. They are therefore zero phonon lines, and Zeeman experiments show that the A line is a triplet and the B line a quintet, as expected for jj combination of a two-fold degenerate electron and a four-fold degenerate hole, (GaP is cubic). There are several ways in which the identification with nitrogen can be made, but perhaps the most convincing is the effect that isotopic substitution has on certain vibrational side bands. Figure 5 shows that the side bands are dominated by lattice modes, but that there is a line labelled A-2 LOC which falls at an energy 0.122 eV below that of A. Now an LO lattice phonon in GaP has an energy of 50.4×10^{-3} eV, and a TO phonon an energy of 45.4×10^{-3} eV. If nitrogen substitutes for phosphorus a crude calculation indicates that a local mode should be present with an energy of 60×10^{-3} eV. Since this lies outside the range of the allowed lattice vibrations a sharp line is expected. The line at 0.122 eV below A is quite sharp and corresponds to approximately twice the expected local mode frequency. (In fact the single local mode can be seen weakly off the B, J=2, line; the experiment is performed at 1.6°K when the B state is preferentially populated. This selection rule forbidding the excitation of the single "breathing" local mode by collapse of the J=1 state is a result of the fact that it is the symmetrical electron state which is tightly bound to the center, and which therefore provides the main interaction with the lattice.) If N^{15} is substituted for the normal N^{14} the double local mode line moves to higher energies, (i.e., has a lower vibrational energy), by 3.4×10^{-3} eV which proves that nitrogen is involved in the center, and is in reasonable agreement with the value expected, (4×10^{-3} eV), on the basis of the simple model. There is also a small shift, (about 0.7×10^{-4} eV), of the zero phonon line to higher energies when the heavier isotope is introduced.

B. The NN Lines in GaP

At higher nitrogen concentrations new lines can be seen in both absorption and emission.[18] These are also zero phonon lines possessing phonon sidebands, and are shown in Fig. 6. These lines

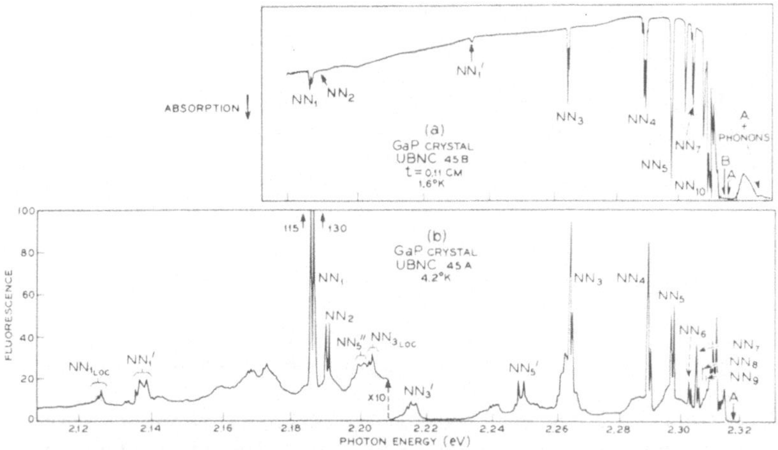

Fig. 6. The absorption and emission spectra of a GaP crystal containing a high concentration of nitrogen. The NN lines correspond to excitons bound to pairs of nitrogen atoms at various possible separations. Single local mode replicas can be seen from these lines.

are labelled NN_1, NN_2, etc. and correspond to excitons bound to pairs of nitrogen atoms. This can be shown by comparing the absorption produced by the NN lines with that produced by the isolated nitrogen atoms, that is, the A line or its vibrational sidebands. If, as is believed to be the case, there is a random distribution of nitrogen atoms among the phosphorus lattice sites then the pair absorption should vary as the square of the A line absorption. This is shown to be the case in Fig. 7, in which the log of absorp-

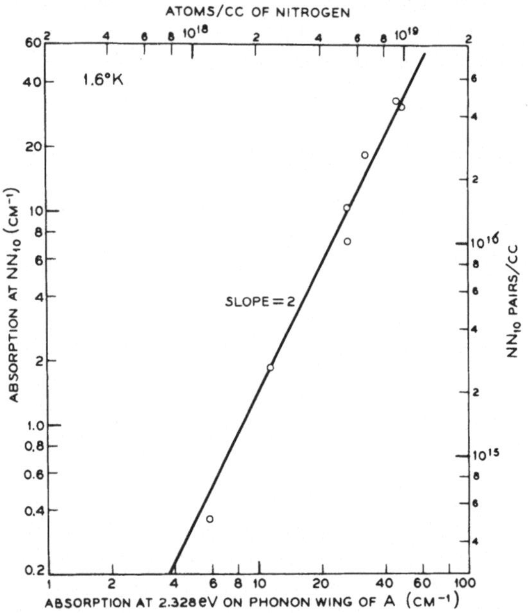

Fig. 7. The absorption at one of the NN lines, NN_{10}, in GaP as a function of the absorption produced by the A line, (due to isolated nitrogen). The concentrations indicated are obtained from this optical data.

tion of NN_{10} is plotted against the log of absorption of a replica of the A line. A line of slope two is obtained as expected. With certain assumptions it is possible from this data to determine the concentrations of both isolated and paired nitrogen centers. A random distribution must be assumed, and also that the oscillator strengths of all the transitions are approximately the same. With a knowledge of the lattice geometry the concentrations may be deduced, and these are shown in Fig. 7.

There are several sets of NN lines which converge at higher energies onto the A line. Their energy separations do not vary smoothly, however, as can be seen from Fig. 6. It is clear that these separate sets of lines correspond to pairs of nitrogen atoms with different internuclear separations. Circumstantial evidence has been presented[18] which indicates that the lowest energy line, NN_1, corresponds to nearest neighbor pairs, NN_2 to next nearest neighbors and so on. Qualitatively one may say that if one nitrogen atom binds an electron, two close together will bind it more tightly. As the atoms become further apart the binding becomes more and more like that to an isolated atom. Hence the energy generally will increase as the separation increases. It is interesting to notice that this energy shift is in the opposite sense to that shown by donor-acceptor pair spectra where coulombic forces are at work between oppositely charged particles.

Qualitatively one expects nitrogen to be more attractive to electrons than is phosphorus because of the more exposed nuclear charge of nitrogen. Thus the electron is to some degree localized at the central nitrogen atom and the hole attracted into a coulombic orbit. There is, therefore, a resemblance to an acceptor, and the state has been termed an "isoelectronic acceptor."[19] Experimental evidence that this is so comes from the absence of a single local mode vibrational sideband off the A line referred to above. In addition, the large ratio, (105:1),[20] of the oscillator strengths of the A to B transitions, and the simple nature of the phonon wings provides additional evidence for the localization of the electron as discussed in connection with GaP:Bi in Section III-C. A more quantitative discussion of the nature of the binding and of the complications arising from the presence of two nitrogen atoms is given in the accompanying article by Faulkner and Hopfield.[21]

C. Bismuth in GaP

If nitrogen replacing phosphorus in GaP produces a center attractive to electrons, a heavier element isoelectronic with phosphorus might be expected to produce a center attractive to holes; an electron would then be bound coulombically and an "isoelectronic donor" would result. Arsenic does not provide an attractive potential sufficiently deep to overcome the kinetic energy of a bound hole and no bound state exists. The situation with antimony is not clear. However bismuth forms a well defined bound state with a depth of 0.1 eV.[22] At low temperatures this center, although present at a maximum concentration of only about 10^{18} atoms/cc is a remarkably efficient fluorescent center. The emission spectrum is dominated by phonon sidebands which display much structure. Figure 8 shows the zero phonon A line and the

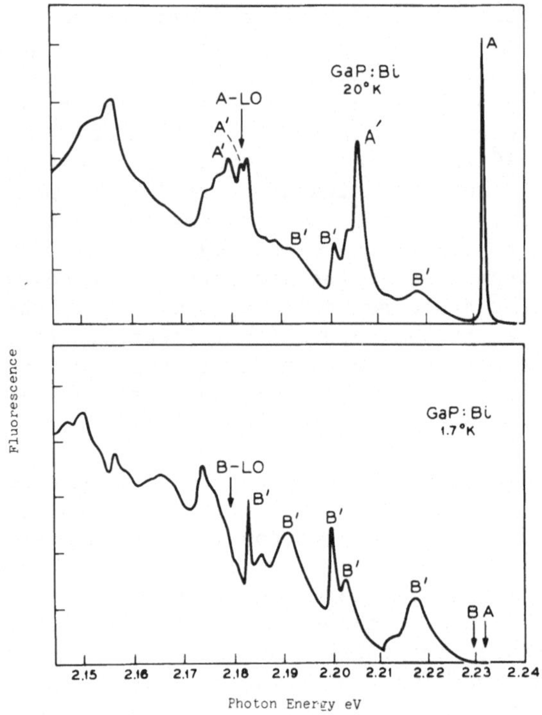

Fig. 8. A section of the photoluminescent spectrum of the isoelectronic trap bismuth in GaP. The top figure relates to 20°K and shows transitions derived from both the A and B states, while the bottom shows 1.7°K data and shows transitions only from the B state.

close by phonon wings at 20°K, and the spectrum as it appears at 1.7°K.

In many respects the spectrum is different from that of GaP:N. The A zero phonon line is clear, but the B line, (separated from A by 2.7 meV), is extremely weak. In addition the phonon sidebands are different for the A and B transitions, and the ratio of oscillator strengths of A to B is only about 5. In GaP:N the A and B zero phonon lines are both strong with respect to the phonon sidebands, these bands are very similar for both transitions, and as mentioned above the ratio of oscillator strengths is 105. The bismuth results can be qualitatively understood[19] by supposing that a hole is now the localized particle. It is the localized particle which will interact with the lattice vibrations. The symmetry of the hole, lower than that of the electron, can interact with lattice vibrations which are capable of mixing the A and B states. As a result the normally forbidden J=2 state will decay, via the A state, using phonons which cause this mixing. The phonon wings can, therefore, be different for the A and B states, the B zero phonon line will be weak, and the B transition will not be highly forbidden. The small difference of oscillator strength between the A and B state is the reason for the prominence of the B transition phonon wings seen in the 20°K section of Fig. 8. At this temperature the B state is not particularly favored in population and for GaP:N there would be negligible emission from the B state.

D. Another Interpretation of Transitions at Isoelectronic Traps

The above interpretation of the nature of the transitions at isoelectronic traps in GaP has been questioned.[23] It is contended that momentum conserving transitions are dominant in the phonon wings as might be expected for an indirect gap material, particularly for the weakly bound state at the isolated nitrogen atom. For the bismuth transition the J=1 state is supposed to decay predominantly via the close by Γ_{1c} band as an intermediate state using a momentum conserving LA phonon with electron scattering. This phonon does not cause conversion of the J=1 to J=2 state and so is not used in the decay of the J=2 state. Instead, a more remote intermediate state is used, the X_{5v} valence band which involves hole scattering, and allows conversion of the J=2 state into the J=1. All of the momentum conserving phonons can now be involved and so different phonon sidebands are seen for the J=2 transition.

However, in the opinion of the author this is not an accurate representation of the state of affairs. There is a qualitative difference between phonon sidebands showing momentum conserving phonons and the sidebands seen, for instance, off the isolated nitrogen atom. In the former case[24] very sharp precisely defined phonon replicas are seen, whereas for GaP:N the whole range of phonons can be seen in the phonon sidebands. For GaP:Bi it is simply not true to say that particular phonons are used, for again it is clear that a large range are employed, and where there are peaks these do not precisely agree with momentum conserving phonons. In addition, the optical oscillator strengths of the isoelectronic traps can be much greater than that of the coulombic traps. Thus the strength of the transition at Bi in GaP is about 300 times that at the donor S in GaP. The point is that isoelectronic traps are quite different from weakly bound coulombic states. In the former case there is a strong but short range potential which may cause only weak binding, yet introduces a major perturbation into the wavefunction which greatly increases the probability of the transition and can involve lattice vibrations covering a wide range of wave vectors. At the coulombic centers there is a long range potential, for which weak binding implies small wavefunction distortions. As a result, only weak indirect transitions are possible which do involve momentum conserving phonons.

E. Other Isoelectronic Traps

Certain other isoelectronic traps have been identified. Oxygen substituting for tellurium in ZnTe apparently gives a trap, attractive to electrons, with a depth of 0.4 eV.[19] The identification of this center with oxygen has depended on the addition of oxygen producing the center in higher concentration. Other circumstantial evidence supports the conclusion and it is believed to be correct. However, it has been noticed that the concentration of the centers, as judged by the absorption spectrum, does vary depending upon the thermal history of the sample. It appears that there can be some form of association of the oxygen atom with a crystal imperfection which can lead to quenching of the optical transition. It would be desirable to observe a local mode due to the presence of the oxygen, but in the fluorescent spectrum such has not been seen, perhaps because the high intensity of the normal multiphonon emission processes masks the local mode emission.

It would seem very reasonable to suppose that oxygen in CdS or ZnSe would give centers similar to that in ZnTe, but to date there does not appear to be conclusive evidence that such centers have been observed, although there are reports of oxygen affecting the luminescent properties of CdS.[25]

Tellurium substituting for sulphur in CdS forms an isoelectronic trap with a depth of about 0.25 eV[26] which is the source of powerful fluorescence at 6000 Å at low temperatures. At concentrations of Te exceeding about 10^{19} atoms/cc a lower energy fluorescence band near 7300 Å appears, corresponding to a trap depth of roughly 0.6 eV.[27] Unfortunately neither the 6000 Å nor the 7300 Å band has any structure so that a detailed analysis is not possible. It appears probable, however, that the 6000 Å band is associated with a hole tightly bound to an isolated tellurium atom with a loosely bound electron, and that the 7300 Å band is similar except that there are two tellurium atoms on nearest sulphur sites. Just as for nitrogen, two isoelectronic traps close together are expected to provide a deeper trap than the isolated atom. It is believed that for separations between the tellurium atoms greater than the minimum one, the hole because of its small radius does not feel the influence of the second tellurium atom. The electron is not attracted to the tellurium and so, to within the rather low precision to which the band position can be defined, the binding energy is the same as that to a single tellurium atom. It is of interest that the two atom center is of sufficient depth that it can cause efficient fluorescence at room temperature.

So far there have been no reports of isoelectronic traps at cation sites. Be, Mg, Ca and Ba have been added to ZnTe without finding any bound states that could be ascribed to a simple isoelectronic center.[28] Mg has been added to CdTe with similar results.[29]

IV. THE DECAY OF BOUND EXCITONS

It is usually the case that at low temperatures free holes and electrons are very rapidly trapped at centers, providing there are a sufficient number of suitable centers present. Thus, one might expect that after pulse excitation making free holes and electrons, there would be a simple exponential decay of the emitted light. In fact, one frequently finds that the decay is far from exponential and that it can vary from one crystal to another. One explanation for this is that the holes and electrons can be captured separately at distant acceptors and donors. The recombination lifetime will depend on the donor-acceptor separation, and as a result of there being many separations a complex time decay can result.[30] However, recombination of an exciton bound to a well defined crystal defect can lead to a simple time decay at isoelectronic traps. Anomalies arise, however, for decay at neutral donors or acceptors. These two topics will now be briefly discussed.

A. Time Decay of Excitons Bound to Isoelectronic Traps

There are two states for an exciton bound to an isoelectronic trap - the J=1 A state, and the J=2 B state. Each will have a characteristic decay time, τ_A and τ_B, and it may be assumed that thermal equilibrium exists between the A and B states, since the phonon processes connecting A and B occur much more rapidly than does photon emission. Under these circumstances the observed time decay constant, τ', is a function of temperature and is expected to be given by,

$$\tau' = \tau_B\{1 + \frac{3}{5}\exp(-\Delta E/kT)\}/\{1 + \tau_B/\tau_A \times \frac{3}{5}\exp(-\Delta E/kT)\} \qquad (1)$$

where ΔE is the energy difference between the A and B states and the factor 3/5 is the ratio of the degeneracies of the A and B states.

Since ΔE is of the order of only a few millivolts, it is often necessary to maintain the crystal at very low temperatures, and to measure short lifetimes pulse excitation has to be used which has a very abrupt cutoff. One way to do this is to use a pulsed electron beam from a Van de Graaf generator with a fall time of less than 3×10^{-9} sec. The beam is directed through a thin wall in a Dewar vessel onto the crystal which can be immersed in helium or in a suitable cooling gas.[20]

At low temperatures exponential decays are observed, and the τ' values can be fitted to Eq. 1. An example is given in Fig. 9 for the decay of an exciton bound to a nitrogen atom in GaP. Below

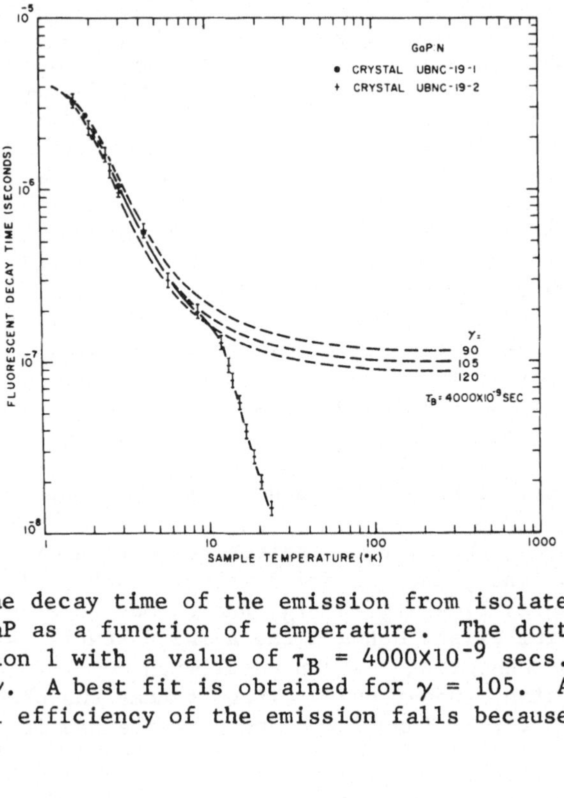

Fig. 9. The decay time of the emission from isolated nitrogen atoms in GaP as a function of temperature. The dotted lines represent equation 1 with a value of $\tau_B = 4000\times10^{-9}$ secs. and various values of γ. A best fit is obtained for $\gamma = 105$. Above about 10°K the overall efficiency of the emission falls because of thermal ionization.

10°K the experimental points fit well Eq. 1, (represented by dotted lines for various values of $\gamma = \tau_A/\tau_B$, with $\tau_B = 4\times10^{-6}$ sec). The best fit gives $\tau_A = 38\times10^{-9}$ sec and $\tau_B = 4\times10^{-6}$ sec. The ratio γ can also be determined from a measurement of the relative intensities of the A and B lines and their phonon wings at a particular temperature. The method is not as accurate as the lifetime measurements, but satisfactory agreement between the two results is obtained. Above 10°K the decay is no longer exponential and the fluorescence quenches, because at this temperature the bound exciton begins to dissociate at an appreciable rate. Rather similar measurements have been made on the systems GaP:Bi and ZnTe:O.

An important consequence of being able to determine the lifetime of the optical decay process, (which is equivalent to the "oscillator strength" of the transitions), is that if there is also a measurement of the absorption produced by the transition in a particular crystal, it is possible to determine the concentration of the center in that crystal. In the examples GaP:N and GaP:Bi the concentrations determined in this way agree within a factor of 5 or less with determinations made by independent means.[20] For the case of ZnTe:O this is the only way at present of determining the concentration.

Table I summarizes some of the results concerning the decay of excitons bound to isoelectronic traps.

TABLE I

Some Characteristics of Excitons Bound to Isoelectronic Traps[20]

Host and Trap	GaP:N	GaP:Bi	ZnTe:O
Photon energy of A line, eV, 4.2°K	2.3171	2.2315	1.9855
$\Delta E = E_A - E_B$	0.8×10^{-3}	2.7×10^{-3}	1.5×10^{-3}
τ_A sec	38×10^{-9}	900×10^{-9}	9×10^{-9}
τ_B sec	4000×10^{-9}	4300×10^{-9}	320×10^{-9}
Oscillator Strength of A Transition	1×10^{-1}	5×10^{-3}	7×10^{-1}

B. Auger Decay

Time measurements have been extended to the decay of excitons bound to neutral donors and acceptors,[31] a particular example being the so-called C line in GaP which is due to the decay of an exciton bound to a neutral S atom. In this case because of the pairing of the two equivalent electrons in the complex,

$$\textcircled{S^+} = + \quad ,$$

there is only one level, and so no complications arise from A and

B levels. The C line was found at 1.6 and 4.2°K to have a decay time of 21 ± 4×10^{-9} sec.

Now the radiative lifetime, τ_R, may be determined indirectly using the formula,

$$\tau_R^{-1} = \frac{8\pi n^2 g_e}{\lambda^2 g_u} \int \alpha d\nu \qquad (2)$$

where n is the refractive index, g_e and g_u the degeneracies of the lower and upper states, N the number of centers/cc, λ is the vacuum wavelength, $\alpha d\nu$ is the total optical absorption produced by the center, with α and dν in cm^{-1}. It is necessary to measure the concentration of the center and the absorption it produces in a particular crystal. The concentration of the sulphur donor can be determined from Hall effect measurements and the absorption is measured directly. The value of τ_R so obtained is 11×10^{-6} sec, which is roughly 500 times longer than the observed value. This discrepancy is far beyond experimental error. Similar results have been obtained using the donor arsenic in silicon; the discrepancy here amounts to a factor of nearly 10,000. The results are summarized in Table II, which also indicates approximate values for the overall quantum efficiencies that have been observed for the photoluminescence of the transitions. This figure, of course, depends upon the particular crystal used.

TABLE II

Comparison of Radiative and Observed Lifetimes of Excitons Bound to Neutral Donors

1.6°K	τ_R	τ_{OBS}	τ_R/τ_{OBS}	Efficiency
S in GaP	11 μs	21 ns	500	1/700
As in Si	750 μs	80 ns	9400	1/4000

The only way to account for these discrepancies appears to be to suppose that the bound exciton can decay nonradiatively much more rapidly than it can radiatively. The low radiative efficiencies that are always observed for these transitions, as illustrated in Table II, are consistent with this suggestion. This process can occur by the Auger emission of the second electron in the case of a donor, and the second hole in the case of an acceptor. This particle will be ejected into the appropriate band with a kinetic energy roughly equal to the hole-electron annihilation energy, and the energy will then be dissipated by phonon emission as the particle cascades to the band minimum. (The direct recombination of a hole and electron with the emission of phonons is a very unlikely event because it would involve the simultaneous emission of a large number of phonons.)

So far no direct proof of the correctness of this picture has been provided - for instance, the detection in some way of the hot particle - but it is supported by strong circumstantial evidence. Thus, while the fluorescent efficiencies of these transitions is always observed to be low, at isoelectronic traps or at donor-acceptor pairs the efficiencies, at low temperatures, are often very high, approaching 100 percent. In both these cases there is no third particle which can be ejected carrying off the recombination energy. As a result the hole and electron are constrained to combine radiatively. Another example is provided by CdS. Here the well known green "edge emission" is known to be due to donor-acceptor pair emission[32] and in suitable crystals can lead to high fluorescent efficiencies. There are also present, from the same impurities as give the pair emission, bound exciton lines[10] which have recently been found to decay with a time constant less than 4 ns.[27] It is not possible to determine the optical lifetime of these excitons bound to neutral donors and acceptors from absorption measurements, because the concentrations are not known. However, the optical lifetime is not expected to be radically different from, say, the decay time of an exciton bound to the isoelectronic trap Te, which is known to be 300 ns. Thus, roughly it appears that the Auger process is at least 100 times more efficient than the radiative one. In agreement with this is the observation that when bound excitons alone are responsible for the emission the fluorescent efficiency is low. (Such crystals can conveniently be produced by annealing in a Cd atmosphere, which removes the acceptor line I_1 and leaves only the donor atoms.) The high efficiency of the tellurium isoelectronic trap in CdS where no Auger effect is possible, also supports the proposed model.

The theory of Auger emission as applied to bound excitons is not very far advanced[31,33] but such results as there are are certainly in line with the contention that the Auger process is more probable than the optical decay.

The results described so far all relate to experiments carried out at very low temperatures where the concentration of free holes or electrons is very low, and where a bound complex does not thermally ionize. In fact, the quenching of luminescence always becomes more important as the temperature rises. It is certain that in GaP, the C line as such is not responsible for the high temperature nonradiative recombination, for its luminescence quenches also. The same centers may, however, still be responsible for nonradiative recombination. Thus, a neutral donor could react with two holes, or with a hole and an electron or with an exciton, to annihilate a hole electron pair and eject a hot particle by an Auger process. In addition, one can imagine two particles bound at a center interacting with one free particle, again leading to Auger emission. Such processes have been discussed in the literature,[34-36] and the set of experiments described here may be said to lend general support to the importance of such transitions.

A point perhaps to be made is this. That while Auger processes have long been considered to be important in narrow band gap semiconductors such as tellurium,[37] where at reasonable temperatures there may be appreciable concentrations of free holes and electrons so that triple collisions have a fair probability of occurrence, this has not generally been supposed to be the case in wider band gap materials since appreciable concentrations of both holes and electrons do not coexist. However, Auger recombination between one or more trapped particles and free particles would appear to be an important nonradiative recombination mechanism. It may account for increased quenching at higher temperatures where greater concentrations of free holes and electrons are present, and also for the "concentration quenching" which is produced by the presence of large quantities of electrically active impurities - although not by large concentrations of isoelectronic traps since these do not supply free carriers.

C. Two Electron Decay

If crystals of GaP are prepared which show in fluorescence the C line, but very little else, especially no donor acceptor pair emission, then some weak but very interesting lines can be seen at energies below the C line.[38] These lines are seen in Fig. 10

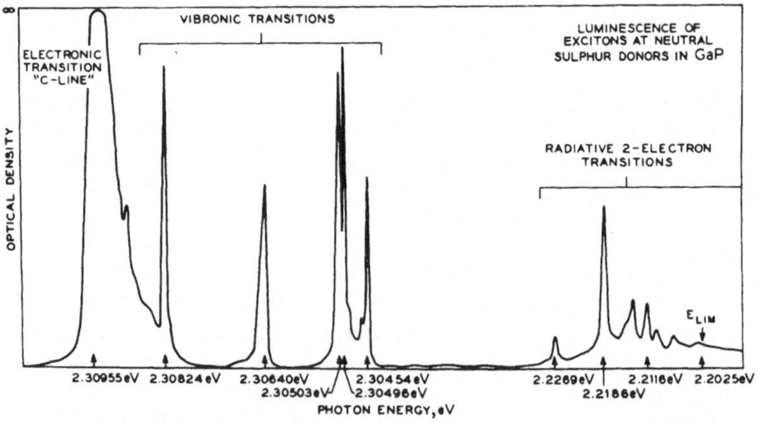

Fig. 10. The photoluminescent spectrum of the "C line" in GaP taken from a photographic plate. The C line itself is overexposed but is in fact a sharp line. The vibronic transitions mostly represent momentum conserving phonon wings. The two electron transitions are described in the text.

between 2.23 and 2.20 eV and are labelled as radiative two electron transitions. The lines labelled as vibronic transitions at higher energies are the expected lines associated with the decay of the C line with the simultaneous emission of momentum conserving lattice phonons. The two electron lines decay at the same rate as the C line, (~ 25 ns), and have a constant intensity ratio to the C line from one crystal to another. They therefore represent yet another path through which the C line may decay.

It was noticed that the pattern of intensities and energies of these lines showed a remarkable resemblance to certain lines seen in the infrared absorption spectrum of donors in semiconductors. This is illustrated in Fig. 11, which shows at the bottom an enlarged view of the two electron spectrum of sulphur, (and in the inset the similar spectrum derived from the selenium donor), and at the top the infrared absorption spectrum of the phosphorus donor in silicon.[39] With the exception of the inverse energy direction, the general similarity of the spectra is apparent

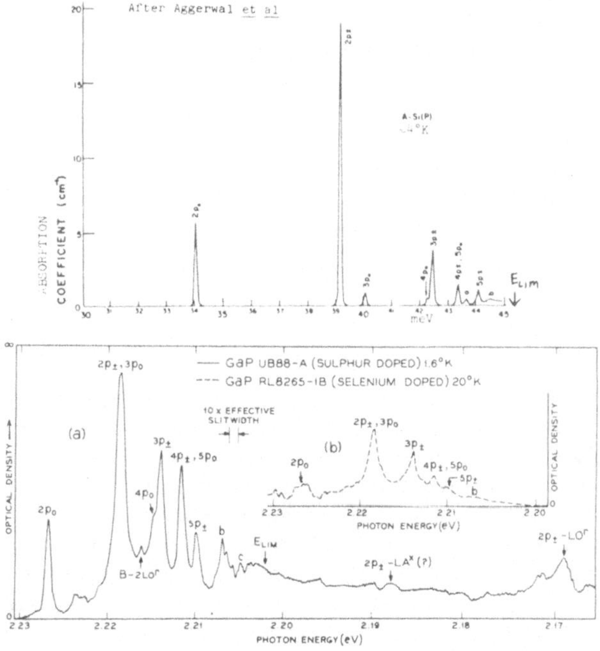

Fig. 11. A comparison of the two-electron spectrum in sulphur doped GaP, (lower), with that of the donor P in silicon as obtained in the infrared, (upper). The similarity will be noted. E_{LIM} marks the series limit of the spectra. The inset in the lower section shows the two-electron spectrum of selenium in GaP. In accordance with independent observations it happens to fall at the same energies as the sulphur spectrum.

These observations can be explained by supposing that the bound exciton can decay giving some of the energy to the second electron which is left in an excited state of the donor, or is even ejected into a continuum state. The residual energy appears as the photon. Thus, lines are seen which are separated from the zero phonon line by energies equal to the excitation energies of the donor. The spectra are called two electron spectra since two electrons participate in the transitions. The reason for the similarity to the infrared spectrum is apparent, and the lines in the lower section of Fig. 11 have been labelled on this basis. Extrapolation of these lines to an ionization limit, which is done again by analogy with the silicon absorption lines, leads to an accurate value for the donor binding energy. For sulphur this is found to be 0.107 ± 0.002 eV. The differences between donor binding energies are known quite accurately from an analysis of different donor acceptor pair spectra.[40] Tellurium has a binding energy 0.015 eV less than that of sulphur. Two electron spectra have been seen for the donor tellurium and a binding energy of 0.093 ± 2 eV is obtained, which is in excellent agreement with the expected value of 0.092 ± 2 eV. There can be no doubt, therefore, that the interpretation given is the correct one. Rather similar effects have been seen for impurities in silicon by P. J. Dean and Others.[41]

This type of transition is of value because it enables donor and acceptor excitation spectra to be examined in small crystals at low concentrations, in a region of the spectrum that is more accessible than the infrared region. In addition, it provides a link between the simple radiative one electron transition and the completely nonradiative Auger transition.

V. CONCLUSIONS

It is clear that many impurities in semiconductors can bind a hole and electron and in favorable cases radiative, or at least partially radiative, recombination can occur, giving rise to sharp spectral lines which lend themselves to detailed investigation. Such investigation can give information concerning both the nature of the impurities and the recombination mechanisms. Excitons may be bound to neutral donors and acceptors, and sometimes to these ionized centers, by forces which are analogous to those existing in the hydrogen molecule or the hydrogen molecule ion. Binding may also occur to nearest neighbor donor acceptor pairs, and especially to widely separated pairs, although the latter situation can hardly be described as a bound exciton. In addition to normal coulombic binding there are usually present short range "central cell" forces which are hard to understand quantitatively but which have an important effect on binding energies. In the so-called isoelectronic traps they are the essential binding force.

While the sharp zero phonon lines are very useful for investigations such as the Zeeman effect, much of the optical strength of the transitions lies in the phonon sidebands. These, too, can be of help in understanding the nature of the transitions. Thus, if they show local mode lines, isotopic substitution leads to predictable effects, and isoelectronic "donors" or "acceptors" have characteristically different vibrational sidebands.

The decay at low temperatures of excitons bound to donor-acceptor pairs or to isoelectronic traps is quite simple and can lead to highly efficient emission. If, however, there is a third mobile particle present such as exists for excitons bound to neutral donors or acceptors, then other decay mechanisms are possible in which energy is transferred to the third particle. It seems that nonradiative Auger decay is by far the most probable of the decay processes, and this draws attention to the probability that this is very likely responsible for nonradiative decay at higher temperatures where, however, not all particles are bound.

This type of semiconductor spectroscopy should continue to be of help in unraveling the properties of impurities, and how they interact with holes and electrons and with light.

VI. ACKNOWLEDGEMENTS

The author wishes to thank J. J. Hopfield and P. J. Dean for many helpful conversations and ideas.

REFERENCES

1. P. J. Dean in Luminescence of Inorganic Solids, Edited by P. Goldberg, Academic Press, New York, 1966, p. 119.
2. S. Shionoya, loc. cit., p. 205.
3. R. E. Halsted, Physics and Chemistry of II-VI Compounds, Edited by M. Aven and J. S. Prener, John Wiley and Sons, New York, 1967, p. 385.
4. M. A. Lampert, Phys. Rev. Letters 1, 450 (1958).
5. J. J. Hopfield, Proc. Int. Conf. on Semiconductors, Paris, 1964, Dunod, p. 725.
6. R. R. Sharma and S. Rodriguez, Phys. Rev. 153, 823 (1967).
7. M. Suffczynski, W. Gorzkowski, and R. Kowalczyk, Phys. Letters 24A, 453 (1967).
8. J. R. Haynes, Phys. Rev. Letters 4, 361 (1960).
9. P. J. Dean, W. F. Flood, and G. Kaminsky, Phys. Rev., to be published.
10. D. G. Thomas and J. J. Hopfield, Phys. Rev. 128, 2135 (1962).
11. R. E. Halsted and M. Aven, Phys. Rev. Letters 14, 64 (1965).
12. D. C. Reynolds, C. W. Litton, and T. C. Collins, Phys. Rev. 156, 881 (1967).

13. D. G. Thomas, R. Dingle and J. D. Cuthbert, Proc. Int. Conf. on II-VI Semiconductors, Sept. 1967, W. A. Benjamin, to be published.
14. D. C. Reynolds and C. W. Litton, Phys. Rev. 132, 1023 (1963).
15. J. Schneider, A. Räuber, W. Dischler, T. L. Estle, and W. C. Holton, J. Chem. Phys. 42, 1839 (1965).
16. T. Koda and S. Shionoya, Phys. Rev. 136, A541 (1964).
17. D. G. Thomas, Proc. Int. Conf. on Physics of Semiconductors Kyoto, J. Phys. Soc. Japan 21, (Supplement), 265 (1966).
18. D. G. Thomas and J. J. Hopfield, Phys. Rev. 150, 680 (1966).
19. J. J. Hopfield, D. G. Thomas, and R. T. Lynch, Phys. Rev. Letters 17, 312 (1966).
20. J. D. Cuthbert and D. G. Thomas, Phys. Rev. 154, 763 (1967).
21. R. A. Faulkner and J. J. Hopfield, this volume.
22. F. A. Trumbore, M. Gershenzon, and D. G. Thomas, Appl. Phys. Letters 9, 4 (1966).
23. L. Patrick, Phys. Rev. Letters 18, 45 (1967).
24. P. J. Dean, Phys. Rev. 157, 655 (1967).
25. C. Z. van Doorn, Solid State Comm. 3, 355 (1965).
C. Z. van Doorn and G. Koch, Solid State Comm. 4, 345 (1966).
26. A. C. Aten, J. H. Haanstra, and H. de Vries, Philips Research Reports 20, 395 (1965).
27. J. D. Cuthbert and D. G. Thomas, to be published.
28. J. L. Merz and R. T. Lynch, Proc. Int. Conf. on II-VI Semiconductors, Sept. 1967, W. A. Benjamin, to be published.
29. K. Itoh, loc. cit.
30. D. G. Thomas, J. J. Hopfield, and W. M. Augustyniak, Phys. Rev. 140, A202 (1965).
31. D. F. Nelson, J. D. Cuthbert, P. J. Dean, and D. G. Thomas, Phys. Rev. Letters 17, 1262 (1966).
32. K. Colbow, Phys. Rev. 141, 742 (1966).
33. Z. Khas, Czech. J. Phys. B8, 568 (1965).
34. L. Bess, Phys. Rev. 105, 469 (1957).
35. P. T. Landsberg, C. Rhys-Roberts, and P. Lal, Proc. Phys. Soc. (London) 84, 915 (1964).
36. M. K. Sheinkman, Fiz. Tverd. Tela 7, 28 (1965), [Translation: Soviet Phys.-Solid State 7, 18 (1965)]. E. I. Tolpygo, K. B. Tolpygo, and M. K. Sheinkman, Fiz. Tverd. Tela 7, 1790 (1965), [Translation: Soviet Phys.-Solid State 7, 1442 (1965)].
37. See for instance, J. Blakemore, Semiconductor Statistics, Pergamon Press, New York 1962, p. 214.
38. P. J. Dean, J. D. Cuthbert, D. G. Thomas, and R. T. Lynch, Phys. Rev. Letters 18, 122 (1967).
39. R. L. Aggarwal, P. Fisher, V. Mourzine, and A. K. Ramdas, Phys. Rev. 138, A882 (1965).
40. F. A. Trumbore and D. G. Thomas, Phys. Rev. 137, A1030 (1965).
41. P. J. Dean, J. R. Haynes, and W. F. Flood, Phys. Rev., to be published.
42. J. J. Hopfield, Proc. Int. Conf. on Semiconductors, Exeter 1962, Institute of Physics and Physical Society, p. 75.

A LOCALIZED EXCITON BOUND TO CADMIUM AND OXYGEN IN GALLIUM PHOSPHIDE

C. H. Henry, P. J. Dean, D. G. Thomas, and J. J. Hopfield*

Bell Telephone Laboratories, Inc., Murray Hill, N.J.

and *Princeton University, Princeton, N.J.

INTRODUCTION

In the past several years there has been an intensive effort to understand the origin of the red luminescence in gallium phosphide. Practical considerations have played an important part in stimulating this effort. Carefully constructed gallium phosphide diodes emit red light with efficiences of greater than one per cent at room temperature.[1] Further improvement of these devices would be facilitated by a detailed understanding of the red luminescence. Gershenzon, Trumbore,[2,3] and Nelson[4] have attributed the red luminescence to recombinations between electrons trapped on O donors and holes trapped on distant Cd (or Zn) acceptors. We have found that part of the low temperature red luminescence is not due to the recombination of distant donor-acceptor pairs, but to the decay of a bound exciton. In this paper we present a detailed study of the optical properties of this exciton.

The exciton luminescence was discovered independently by Morgan, Welber and Bhargava (MWB),[5] who showed conclusively by means of an O isotope experiment that the site binding the exciton is composed, in part, of O. They proposed that the exciton was bound to a Cd-O nearest neighbor acceptor donor pair. The Zeeman experiment we report here shows that the exciton site has a [111] symmetry axis and that the site has no unpaired spin. We are also able to verify by isotope experiments that the exciton site is composed of Cd as well as O. Henry, Dean and Cuthbert have recently shown that the exciton site is neutral.[6] All this evidence is consistent with the model for the exciton site proposed by MWB. Henry, Dean and Cuthbert[6] have further clarified the nature of the red luminescence by showing that in addition to the exciton band, there is a nearby red pair band arising from the recombination of an electron trapped on the above-mentioned exciton site with holes trapped on distant Cd (or Zn) acceptors and not from the recombination of a conventional donor acceptor pair.

EMISSION AND ABSORPTION SPECTRUM

The bound exciton emission spectrum at 20.4°K is shown in Fig. 1. The red pair luminescence was made negligible by slowly cooling the samples after growth. This treatment reduced the amount of free Cd present in the sample.[6] The sample was grown from Ga solution containing 10% Cd. The emission spectrum consists of a no phonon line A followed by a number of narrow phonon replicas L of a 7.0 meV resonant vibration. Two high energy phonons O_1 and O_2 are also discernible. It is probable that much of vibronic sideband is due to a broad spectrum of phonon frequencies which do not give rise to structure. The absorption spectrum (found by luminescence excitation) is shown in Fig. 2. Peak absorption coefficients up to 1.5 cm^{-1} were found. There is little mirror symmetry between absorption and emission although the presence of a low energy phonon L and a high energy phonon O are still discernible. These phonons are slightly shifted in frequency between absorption and emission. The vibronic sideband is broader in the absorption spectrum than in the emission spectrum and the no phonon line A is weaker relative to the rest of the band in absorption than in emission. The direct absorption spectrum is similar to the luminescence excitation spectrum except

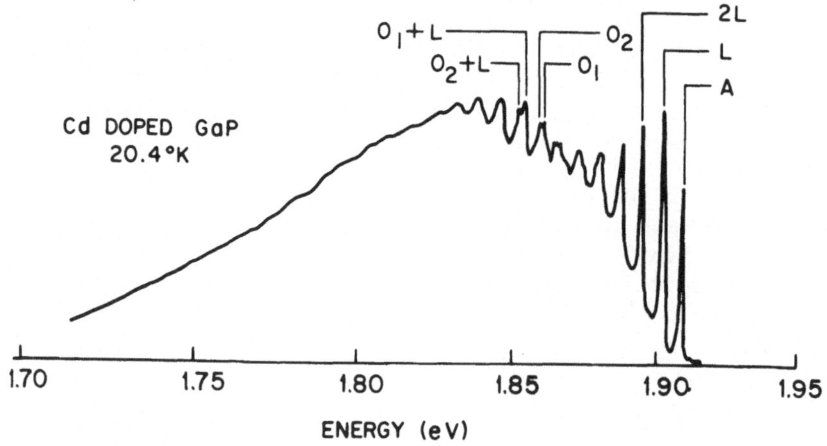

Fig. 1 - Emission spectrum of the Cd-O exciton. The sample was excited with 4880 Å light of an argon ion laser. (Labels O_1 and O_2 stand for optical phonons and not for oxygen.)

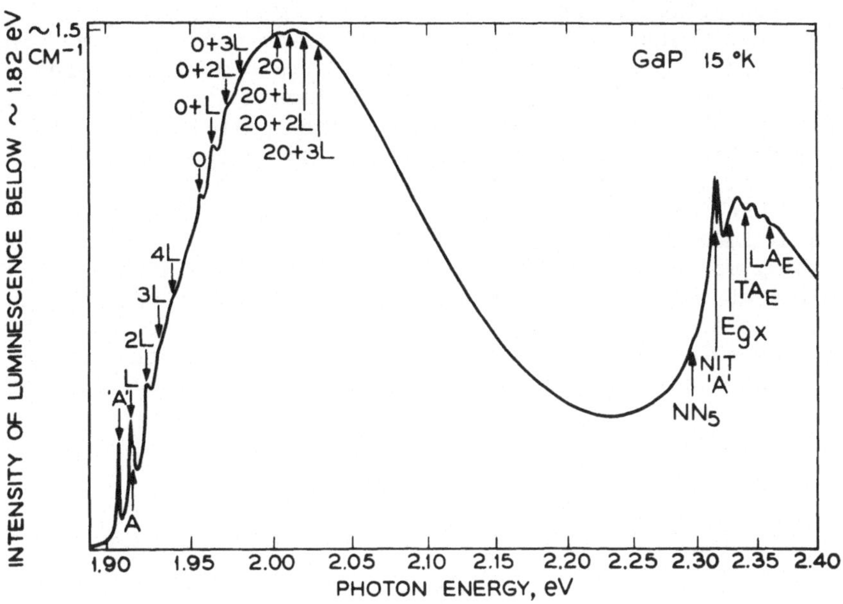

Fig. 2 - Absorption spectrum of the Cd-O exciton measured by luminescence excitation.

that the direct absorption at the N exciton and at higher energies is much larger than in the excitation spectrum. The peak absorption coefficient at line N is ~1000 times that of the Cd-O band for this crystal. Surface recombination becomes important at these very high absorption coefficients, causing the sharp dip in the center of the N exciton line in Fig. 2. The light directly absorbed by the bound exciton ($h\nu \sim 2.0$ eV) is much more efficient in producing bound exciton luminescence than the light absorbed above the band gap ($h\nu \gtrsim 2.30$ eV). This is because radiation at energies greater than the band gap produces green luminescence much more efficiently than red luminescence at low temperature.

The energy of the Cd-O A line is 1.9083 ±.0002 eV. Dean and Thomas have measured the threshold for the production of free excitons and found it to be 2.329 ±.002 eV. Thus the exciton is bound by 421 ±2 meV. As mentioned above, we believe the exciton to be bound to Cd and O. The large exciton binding energy results from the tight binding of the electron by the O donor.

ZEEMAN EXPERIMENT

If the crystal is cooled from 20°K to 1.5°K, a new no

phonon transition B appears in emission, lying 2.3 meV lower in energy than transition A. Transition B is forbidden and can only be observed in the presence of elastic strain or in a magnetic field as shown in Fig. 3(a,b). Transitions A and B both split anisotropically in a magnetic field. The splittings of both lines appear to have the same dependence upon the direction of the magnetic field in the crystal but the B line splittings are much larger than those of the A line. In a field of 31 KG the B line had a maximum splitting of 0.86 meV and the A line had a maximum splitting of .20 meV. The A line splitting could not be clearly resolved at all angles. The angular dependences of the B line splittings are shown in Fig. 3(d). The data were fitted by the solid theoretical curve assuming the center has a 111 symmetry axis, that the excited states can be split into 2 levels, that the

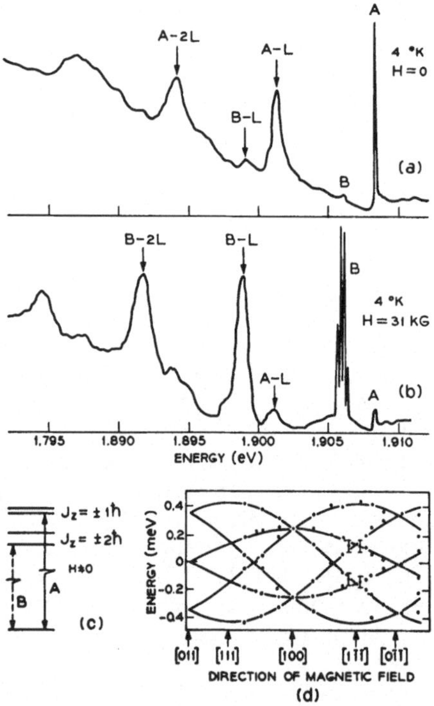

Fig. 3 - (a) (Cd-O) exciton luminescence at 4°K with H=0.
(b) Cd-O exciton luminescence at 4°K with H≠0. The 3 sets of components of the B line arise from different orientations of the center with respect to H. The third and outer-most set cannot be seen because, **for the orientation** giving rise to this component, $(H_\perp)^2$ is quite small.
(c) Level diagram.
(d) The splittings of transition B vs. direction of H as H was rotated in a plane perpendicular to the [011] **direction.**

ground state has no unpaired spin, and that the levels split apart in proportion to the component of H that is parallel to the symmetry axis. The constant of proportionality was found by fitting the splitting for H along the [100] direction.

The exciton states are made up of products of electron states, for electrons that are mainly near the X_1 minimum of the conduction band, and hole states, for holes at the Γ_{15} maximum of the valence band.[7] The low C_{3v} symmetry ([111] symmetry axis) of the exciton site removes the 4-fold degeneracy of the hole states. The four hole states will be split into two Kramers doublets $J_z = \pm 3/2\hbar$ and $J_z = \pm 1/2\hbar$, where J_z is the angular momentum along with the [111] symmetry axis. The electron states will have $J_z = \pm 1/2\hbar$. We can account for what we observed by assuming that the four lowest states of the exciton are made from the $J_z = \pm 3/2\hbar$ hole states and the $J_z = \pm 1/2\hbar$ electron states. These states form four exciton states; two with $J_z = \pm 1\hbar$ and two with $J_z = \pm 2\hbar$. The decay of the former pair of exciton states is allowed and gives rise to transition A, while the decay of the latter two states is forbidden and gives rise to transition B. If the free hole and electron states split in a magnetic field as $E(\text{hole}) = \pm 3/2\ g_h \mu_\beta H_z$ and $E(\text{electron}) = \pm 1/2\ g_e \mu_\beta H$, then the exciton states will split with

$$E_A = \pm(3/2\ g_h - 1/2\ g_e)\mu_\beta H_z$$

$$E_B = \pm(3/2\ g_h + 1/2\ g_e)\mu_\beta H_z.$$

Both transtions split in proportion to the component of H along the symmetry axis H_z and thus have the same pattern but different magnitudes of the splittings. From the measured magnitudes of the splittings, we find $g_h = 0.98 \pm .05$, $g_e = 1.82 \pm 0.2$. These values agree with the free electron and hole g values previously measured by Thomas, Gershenzon and Hopfield.[8]

A component of the magnetic field perpendicular to the [111] syymetry axis will not split the exciton states, but it will mix them. It is this mixing that allows transition B to be observed when a magnetic field is present. The ratio of the A and B transition rates is given by

$$W_B/W_A = [g_e \mu_\beta H_\perp/(E_A - E_B)]^2.$$

The components of the B transition in Fig. (3d) which are split by nearly the maximum amount, have a very small value of $(H_\perp)^2$. For this reason these components were very difficult to observe.

PHONON STRUCTURE

The shapes of the phonon side-bands bound in emission and absorption are not understood. It is quite probable that most of the phonon broadening arises from a smooth spectrum of phonon frequencies which do not lead to structure. Nevertheless, we felt it would be useful to measure carefully the energies of the phonons that do appear as structure. These energies are tabulated in Table I. The vibrations observed in emission appear to be of two types: vibrations which produce simple replicas off both the A line and the B line and vibrations which mix the excited levels and which show up at 1.6°K as replicas of the B line even when the B transition itself cannot be observed. A group theoretical analysis using the exciton wave functions discussed above shows that the former type of vibrations have symmetry Γ_1 and the latter type of vibrations have symmetry Γ_3 for group C_{3v}.[9] Table I lists the phonon energy, symmetry type and the symbol, if any, used to indicate the phonons in Figs. 1 and 2, and the relative strength of the replica.

TABLE I: RESOLVABLE ONE-PHONON REPLICAS

Emission			Absorption		
Energy (meV)	Type	Strength	Energy (meV)	Type	Strength
7.0 ±.05	Γ_1(L)	Strong	7.3 ±.1	(L)	Moderate
47.3 ±.2	$\Gamma_1(O_1)$	Moderate	8.6 ±.1		Moderate
49.7 ±.2	$\Gamma_1(O_2)$	Moderate	16.4 ±.2		Moderate
40.9 ±.2	Γ_3	Weak	37.2 ±.4		Weak
45.2 ±.2	Γ_3	Weak	48.6 ±.2*	(O)	Moderate
46.3 ±.2	Γ_3	Weak			
49.9 ±.2	Γ_3	Moderate			
50.4 ±.2	Γ_3	Moderate			

*The shape of this replica suggests the presence of two unresolved components separated by ca. 1 meV.

It is possible that some of the structure seen at 7.3 meV, 8.6 meV, and 16.4 meV in Fig. 2 may be higher excited electronic states of the exciton and not vibrational structure. These measurements differ somewhat from those of MWB. In particular, MWB label the replica at 49.7 meV as the longitudinal optical phonon and show the replica at 37.2 meV as moderately strong while we find it barely observable, although the other structure in our spectra is significantly sharper than shown by MWB, and label it as weak. All the replicas of type Γ_3 were less than .3 meV wide. The Γ_1 replicas were about .7 meV wide.

ISOTOPE SHIFTS

Cadmium

Crystals were grown doped with isotopes Cd^{110} and Cd^{114}. As shown in Fig. 4, a Cd isotope shift of the 7.0 meV phonon replica was detected. The shift was measured to be 1.36 ±.20 per cent. If Cd were vibrating alone, the frequency would be proportional to $(M_{Cd})^{-1/2}$ and the expected isotope shift would be 1.81 per cent. Thus the frequency change we observed was about 3/4 as large as the change expected when Cd is vibrating alone. This shows that this in-band resonant vibration is highly localized on the Cd atom.

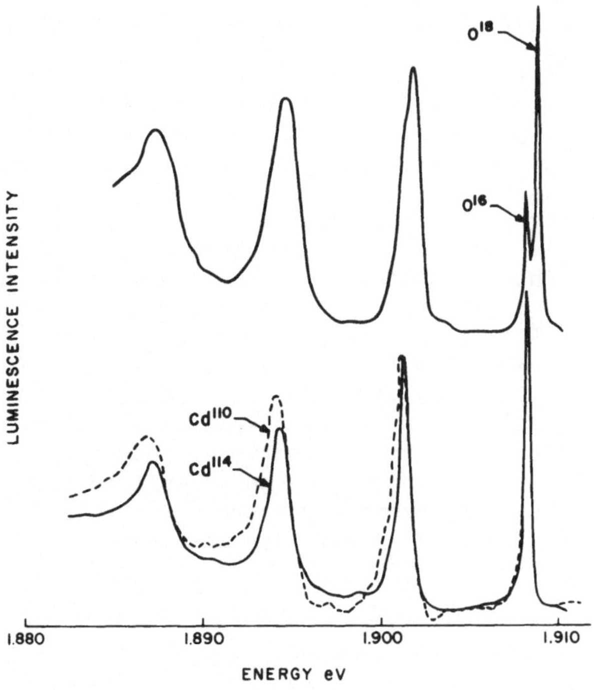

Fig. 4 - The upper figure shows the O isotope shift of the A line. The lower figure shows the Cd isotope shift of the 7.0 meV phonon replica.

By our classification this vibration is of type Γ_1 (of C_{3v}). The Γ_1 vibrations do not lower the symmetry of the center. Thus the Cd atom is probably vibrating along the symmetry axis. Neither the phonons labeled O_1 and O_2 in Fig. 1 nor the no phonon line A were observed to shift.

Oxygen

MWB reported an isotope shift of the no phonon line A when samples were prepared doped with O^{18}. The abundant isotope O^{16} is present in the samples without being intentionally added. We have repeated their experiment and verified their result. We felt that a verification of this important experiment was necessary because the no phonon line A reported by MWB was .75 meV wide, making detection of an isotope shift difficult. As shown in Fig. 4, there is a shift of .65 meV in the A line. No large shifts of the other phonon frequencies L, O_1, and O_2 could be detected. The detection of smaller shifts was impossible because we have not yet produced a sample containing O^{18} but no O^{16}.

Isotope shifts of the no phonon line, only slightly less than the shifts reported here, have been observed for excitons bound to pairs of nitrogen atoms in gallium phosphide.[10] Such shifts are expected whenever the bonds to the neighboring atoms are affected by the electronic state which is undergoing transition. These band changes produce differences between the vibronic frequency spectra in the ground and excited electronic states. MWB have offered a quantitative explanation of this isotope shift, based on an understanding of the vibrational spectrum of the O atom. While we believe that their general explanation of why a shift should occur is correct, we seriously question their detailed calculation, which is based upon phonon assignments which had to be guessed. For example, they took the mass for vibration labeled O_1 in Fig. 1 to be the mass of the O atom. If this were correct, this vibrational energy should change by almost 3 meV when O^{16} is replaced by O^{18}. We observed no isotope change in the energy of this vibration within ±.4 meV. We believe that the no phonon shift is related to isotopic changes in the frequency of broad, relatively high energy, phonon replicas which cannot be resolved in the bound exciton spectra. For this reason, a quantitative interpretation of this remarkable no-phonon shift is lacking at present.

Acknowledgments

We wish to thank C. J. Frosch, R. T. Lynch and F. A. Trumbore for sample preparation, E. I. Gordon for the argon ion laser tube and T. N. Morgan for sending us a preprint of the MWB paper.

References

1. R. A. Logan, H. G. White and F. A. Trumbore, Appl. Phys. Letters 10, 206 (1967).
2. M. Gershenzon, F. A. Trumbore, R. M. Mikulyak and M. Kowalchik, J. Appl. Phys. 36, 1528 (1965).
3. M. Gershenzon, F. A. Trumbore, R. M. Mikulyak and M. Kowalchik, J. Appl. Phys. 37, 483 (1966).
4. D. F. Nelson and K. F. Rodgers, Phys. Rev. 140, A1667 (1967).
5. T. N. Morgan, B. Welber and R. N. Bhargava (to be published), abbreviated MWB in text.
6. C. H. Henry, P. J. Dean, J. D. Cuthbert (to be published).
7. M. L. Cohen and T. K. Bergstresser, Phys. Rev. 141, 789 (1966).
8. D. G. Thomas, M. Gershenzon and J. J. Hopfield, Phys. Rev. 131, 2397 (1963).
9. G. F. Koster, J. Dimmock, R. G. Wheeler, H. Statz, Properties of the Thirty-Two Point Groups, M.I.T. Press (Cambridge, 1963).
10. D. G. Thomas and J. J. Hopfield, Phys. Rev. 150, 680 (1966).

NEW RADIATIVE RECOMBINATION PROCESSES INVOLVING NEUTRAL DONORS AND ACCEPTORS IN SILICON AND GERMANIUM

P. J. DEAN, J. R. HAYNES* AND W. F. FLOOD

BELL TELEPHONE LABORATORIES, INC., MURRAY HILL, N. J.

I. INTRODUCTION

Radiative transitions due to the recombination of excitons localized at neutral donor and acceptor impurities were first recognized by Haynes in the low temperature near band-gap luminescence of lightly doped silicon.(1) Shortly afterwards, similar transitions were identified in doped germanium(2) and subsequently in many other materials. Both no-phonon and phonon-assisted bound exciton recombinations were seen. The identified phonons are those which conserve momentum (M. C. phonons) in the indirect inter-band transitions, as is shown from a comparison between the intrinsic absorption and luminescence spectra. The impurity center remains in its ground electronic state during these transitions. Additional, weaker, impurity-induced luminescence lines were observed in the early work, particularly in germanium by Benoit a la Guillaume and Parodi(2) (hereafter B.P.). The transitions responsible for these bands, which were inadequately explained in the early work, are the subject of the present paper. It will be shown that they all involve recombinations in which the impurity center is left in an excited state. This type of transition has recently been identified in GaP, and is referred to as a "two-electron" transition.(3)

*Deceased.

II. EXPERIMENTAL

The methods used to detect the inefficient photoluminescence of excitons localized at neutral impurity centers in silicon have been described elsewhere.[4] For the later experiments the Dewar was modified so that photoluminescence could be measured with the crystals directly immersed in the liquid refrigerant. The temperature of the refrigerant was adjusted by pumping on the vapor. The low quantum efficiency is due to competitive Auger recombinations within these bound exciton complexes.[5] Auger recombinations reduce the luminescence efficiency for bound exciton recombinations in arsenic-doped silicon to only 0.01 percent. The "two-electron" transitions discussed in the present paper represent only a few percent of the total radiative recombinations.

III. PARTIALLY RADIATIVE "TWO-ELECTRON" TRANSITIONS

The energy level diagram for arsenic donors in silicon shown in Fig. 1 includes the effect of valley-orbit interaction, which splits the six-fold degenerate s-like donor envelope states into the sub-levels with irreducible representations A_1, E and T_1 of the tetrahedral point group of the donor site.[6] In addition, the axial symmetry of the ellipsoidal conduction band valleys lifts the degeneracy of p-like hydrogenic envelope states with different magnetic quantum number, m. The energy levels have been established by infrared absorption, which gives the $1s(A_1) \rightarrow np$ energy separations[7] and, at higher temperatures, the $1s(T_1) \rightarrow np$ and $1s(E) \rightarrow np$ separations.[8][9] Unlike these infrared photo-excitation transitions, the donor excitations observed in the "two-electron" luminescence spectra do not involve a change in parity.

The transitions As_{2s}, As_{H^-} and As_{VO} in Fig. 1 have been observed in the various two electron luminescence recombinations respectively discussed in Section IV A→C. Transition As_{VO} has also recently been reported in Raman scattering from phosphorus donors in silicon.[10]

IV. RESULTS AND DISCUSSION - SILICON

A. "Two-Electron" Transitions from Bound Exciton States

The principal components As^{O}, As^{TO} and B^{O}, B^{TO} in the low temperature luminescence spectra in Fig. 2 are due to the recombination of excitons bound to neutral arsenic and boron impurities, respectively without and with the emission of a

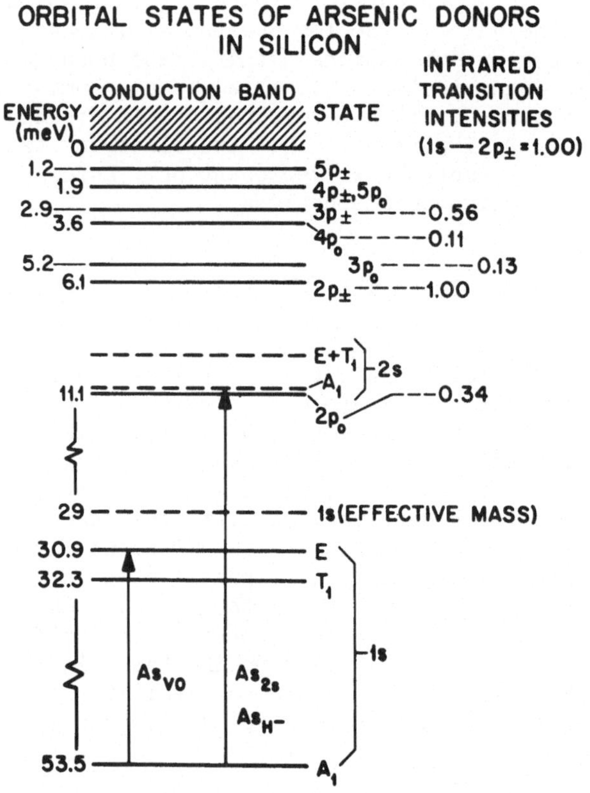

Fig. 1 - Energy level diagram of orbital states of shallow donors in silicon. The indicated optical transitions are those observed in the "two-electron" recombinations.

57.8 meV transverse optical (TO)MC phonon. Additional phonon replicas of the bound exciton transitions in Fig. 2 involve emission of the 18.5 meV transverse acoustical (TA) phonon and, jointly, of the TO MC phonon and the 64.5 meV zone center optical (O^{Γ}) phonon. Corresponding intrinsic phonon-assisted recombinations of free excitons occur, and are labelled I^{TA}, I^{TO} and I^{TO+O}.

The extrinsic line near 1.05 eV in Fig. 2a is due to TO phonon-assisted bound exciton "two-electron" transitions in which the arsenic donor is left in a high excited state. The three upper arrows denote the positions and relative intensities of the three principal components for this transition predicted from the data in Fig. 1, assuming that 1s → np type donor excitations occur. The disagreement between experiment and these predicted components is well outside the limits of experimental error. The lower arrow represents the calculated position for the $1s(A_1) \rightarrow 2s(A_1)$ transition, which has an uncertainty of the order of the calculated splitting between the $2s(A_1)$ and $2s(E+T_1)$ states ($\sim \pm$ 1meV). The agreement with experiment indicates that this is the predominant donor excitation in the "two-electron" transition. The relative intensities of excitations to higher s-like donor states are not known precisely, but are expected to be low.(11) The corresponding "two-electron" transitions could not be resolved in the present work. Similar results have been obtained for phosphorus and antimony donors.(12)

The position of the "two-electron" transition near 1.06 eV associated with boron acceptors (Fig. 2b) is also consistent with 1s → 2s rather than the infrared active 1s → 2p hole excitations.

B. Free Carrier "Two-Electron" Recombinations Involving H^--like Impurity States

The intensities of components As^{TO} and As^{TO}_{2s} decrease <u>together</u> with increasing temperature, as would be expected from the interpretation in Section IVA. A new component appears displaced nearly 10 meV to higher energies ($As^{TO}_{H^-}$ in Fig. 3). This component has a Maxwell-Boltzmann (M.B.) energy profile, broadened in Fig. 3 by the spectral resolution, consistent with transitions from an excited state containing one free particle.(13) The threshold energies of the M.B. components fitted to $As^{TO}_{H^-}$ in Fig. 2a are ~ 6 meV larger than expected for TO phonon-assisted recombinations of free holes at neutral arsenic donors if E_D = 53.5 meV,(7) however. Energy discrepancies of similar magnitude were observed in the corresponding components for phosphorus and antimony donors and for boron acceptors.

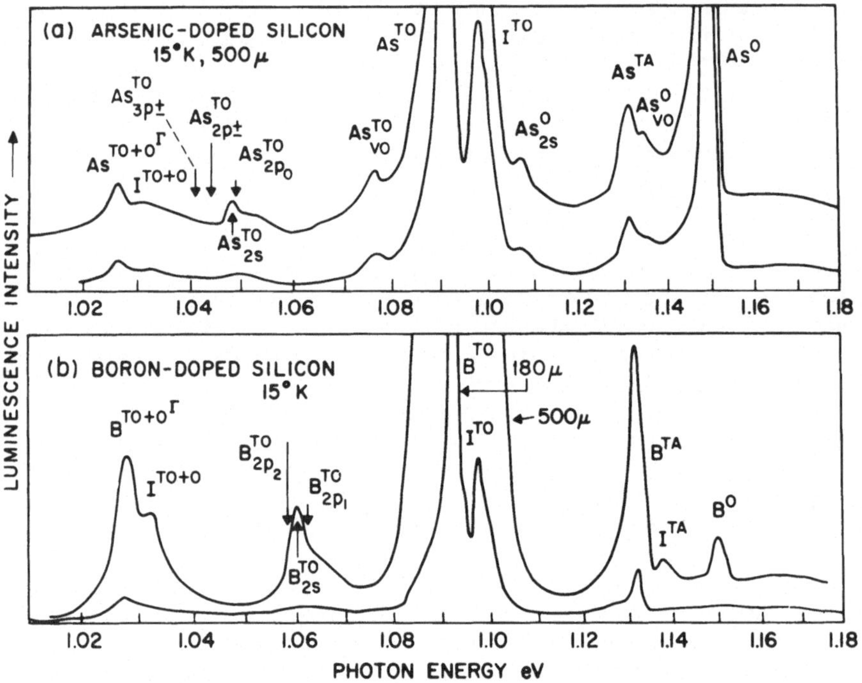

Fig. 2 - Low temperature photoluminescence spectra (a) from a silicon crystal containing 2×10^{17} cm^{-3} arsenic atoms and (b) 6×10^{16} cm^{-3} boron atoms. Components I are intrinsic, As and B are extrinsic. (See text and Fig. 1.) The linewidths of the principal extrinsic components are slit-limited, and the slitwidths (500μ, etc.) are indicated. The ordinate is nearly proportional to the number of photons per unit energy interval.

A "two-electron" transition is not possible in the recombination of a free carrier at a neutral impurity center. The relative intensities of the $\underset{H}{TO}$ components vary as the square of the excitation intensity, however, suggesting that the electronic excited state involves <u>two</u> excess carriers. These experimental results can be understood if an H^--like bound complex is formed by the capture of a second electron (or hole) at the neutral donor (or acceptor) before radiative capture of the free hole (or electron) occurs. We assume that the donor electron (or acceptor hole) is raised to the 2s level, of ionization energy E_s, during the two-electron recombination of the second electron (or hole) with the free carrier.

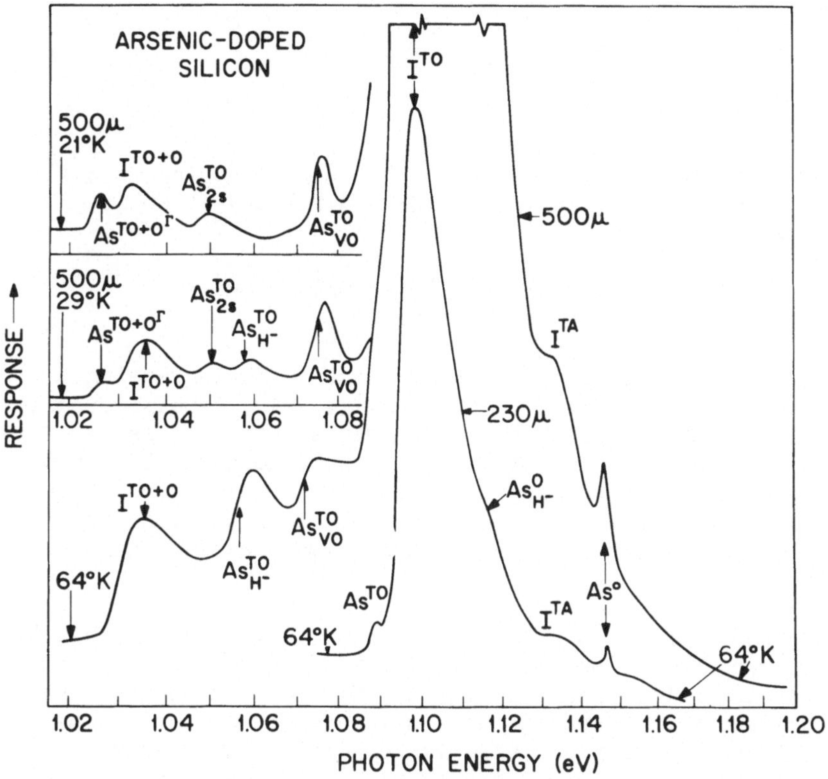

Fig. 3 - Higher temperature photoluminescence spectra from a crystal containing $2x10^{17}$ cm^{-3} arsenic atoms. Note the appearance of component $As_{H^-}^{TO}$ and its growth relative to As_{2s}^{TO}, but not relative to As_{VO}^{TO}, with increasing temperature.

The transition energy hν, is then

$$h\nu = E_g - E_D + E_s - E_{H^-} \quad (1)$$

where E_{H^-} is the ionization energy of the second electron. Experimentally E_{H^-} is ~4 meV for donors and acceptors in silicon,(14) whereas analogy with the H^- ion suggests that E_{H^-} ~2 meV.

This estimate of E_{H^-} is consistent with the temperature dependence of the ratio, R, of the free exciton (Section IVC) and $_{H^-}^{TO}$ "two-electron" component intensities according to the relationship

$$R_{T_1}/R_{T_2} = \exp\ [(E_x - E_{H^-})/kT_1 - (E_x - E_{H^-})/kT_2] \qquad (2)$$

Measurements between 29 °K and 64 °K give $R_{29}/R_{64} = 3 \pm 1$, so that $E_x - E_{H^-} = 5 \pm 1$ meV, i.e. $E_{H^-} = 3 \pm 1$ meV if E_x is 8 meV.[13]

C. Free Exciton "Two-Electron" Recombinations at Neutral Donors

The extrinsic component As_{VO}^{TO} persists with increasing temperatures to 80°K (Fig. 3) but, unlike $As_{H^-}^{TO}$, is also prominent at low temperatures (Fig. 2a). The suggests that it is associated with the decay of free excitons, consistent with the shape of this line at higher temperatures. There is evidence from the line-shape that the interaction cross-section for this transition is substantially reduced at higher exciton kinetic energies.[12]

The position of As_{VO}^{TO}, 22.5 meV below I^{TO}, suggests that the free excitons recombine in the vicinity of arsenic donors causing the donor excitation As_{VO} in Fig. 1. Similar components have been observed for phosphorus and antimony donors in silicon.[12] The relative intensity of these valley-orbit "two-electron" transitions increases considerably with decrease in the $1s(A_1) \rightarrow 1s(E)$ energy separation.

Corresponding transitions to the $1s(T_1)$ valley-orbit state are insignificant in the "two-electron" spectra, in agreement with the prediction that such transitions can occur only through admixture of higher lying states.[10] Free exciton "two-electron" transitions to the higher orbital states shown in Fig. 1 are also negligible, as they are in neutral-impurity Raman scattering.[10][11]

There are no low-lying energy states of this type for acceptors in silicon, since the valence band maxima occur at $\underline{k} = 0$. "Two-electron" transitions of comparably low displacement energy have not been observed for acceptors (Fig. 2b). In particular the ~ 23 meV Raman-active transition induced by boron[10] has not been detected in the luminescence spectra.

V. RESULTS AND DISCUSSION - GERMANIUM

All of the "two-electron" transitions discussed for silicon in Section IV are present in the luminescence spectra of phosphorus-doped germanium, as indicated by the notation in Fig. 4. The temperature dependence of the various components is consistent with these assignments. Thus component P^{O}_{VO} is strong when the intrinsic free exciton component I^{LA} is strong (at 14°K), rather than when the bound exciton component P^{O} is strong (at 4.2°K or below).(15) Component $P^{O}_{H^-}$ is only present at the higher temperatures and is broad, consistent with the recombination of free carriers.

B.P. noticed that the energy of $P^{O}_{H^-}$ is ~1.5 meV greater than predicted for recombination of free carriers at neutral phosphorus donors, assuming that E_x is ~3 meV. Similar discrepancies were observed for four other donors and acceptors in germanium.(2) According to Section IVB (Eq. (1)), this energy difference is $(E_s-E_{H^-})$. Since E_s ~3 meV, E_{H^-} is ~ 1.5 meV, i.e. about 40 percent of the value observed in silicon.

The position of the $\binom{O}{VO}$ components for phosphorus, arsenic and antimony donors in germanium is consistent with the accurately known $1s(A_1) \rightarrow 1s(T_1)$ valley-orbit transition energies. This energy is 4.23 ± 0.02 meV for phosphorus donors. These components do not occur in the acceptor luminescence spectra and were unexplained by B.P. Only the highest energy component, labelled C by B.P., are now attributed to the resonance decay of localized excitons. Only one such component is in fact predicted on the j-j coupling scheme for the electrons and holes in these complexes. The ratio of the localization energy to the impurity ionization energy is ~ 0.1 for these bound exciton complexes. As observed for silicon(1) this ratio is slightly but consistently lower for acceptors compared with donors.

The strong broad band just below 0.710 eV, which increases in relative intensity between 4.2°K (Fig. 4a) and 2°K (Fig. 4b) was also unexplained by B.P. We believe that it is due to the phonon-assisted recombination of the free excitonic molecule, which should be prominent in relatively pure crystals at very low temperatures.(16)

VI. SUMMARY

Evidence has been presented for "two-electron" luminescence transitions in silicon which involve the decay either of a free exciton or a localized exciton, leaving a nearby neutral impurity in an excited state. In addition,

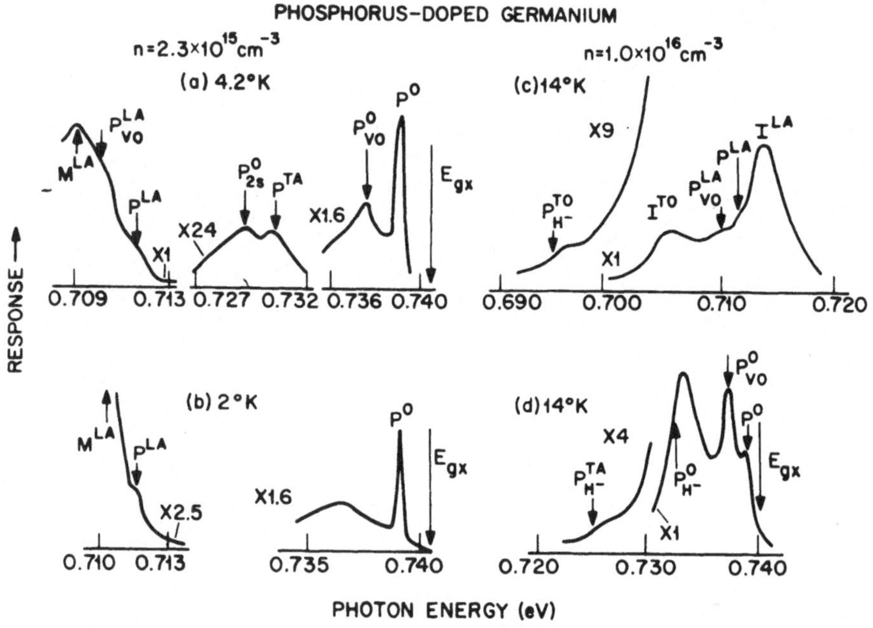

Fig. 4 - Photoluminescence spectra from phosphorus-doped germanium after Benoit a la Guillaume and Parodi (Ref. 2). The detailed interpretation of the extrinsic components in terms of "two-electron" transitions is similar to the silicon spectra (Figs. 2-3). Some previously unidentified components are thereby accounted for (eg P^0_{VO}), and a more satisfactory interpretation offered for the positions of others (eg $P^0_{H^-}$). The spectral resolution is comparable with the width of the narrowest lines.

"two-electron" transitions also predominate in the recombination of free carriers at these impurity centers. This is possible since H^--like states are formed at the neutral centers before the radiative recombination of the minority carriers occurs. These newly-identified "two-electron" transitions can be recognized in previously published photoluminescence spectra of doped germanium,(2) and all of the components in these spectra have now been identified. The theoretical stability of H^--like donor or acceptor complexes in semiconductors was pointed out by Lampert in 1958,(17) but the present work provides the first experimental evidence for their stability. The formation of stable, bound, H^--like states influences the scattering of free carriers by neutral impurities, as recently discussed by Honig.[18]

REFERENCES

(1) J. R. Haynes, Phys. Rev. Letters 4, 361 (1960).
(2) C. Benoit a la Guillaume and O. Parodi, Proceedings of the International Conference on Semiconductor Physics, Prague, 1960 (Czechoslovakian Academy of Sciences, Prague, 1961) p. 426.
(3) P. J. Dean, J. D. Cuthbert, D. G. Thomas and R. T. Lynch, Phys. Rev. Letters 18, 122 (1967).
(4) J. R. Haynes, Methods of Experimental Physics, edited by K. Lark Horovitz and V. A. Johnson (Academic Press, N.Y., 1959 6, part B) p. 322.
(5) D. F. Nelson, J. D. Cuthbert, P. J. Dean and D. G. Thomas, Phys. Rev. Letters 17, 1262 (1966).
(6) W. Kohn, Solid State Physics, edited by F. Seitz and D. Turnbull (Academic Press, N.Y., 5, 1957) p. 257.
(7) J. W. Bichard and J. C. Giles, Canadian J. Phys. 40 1480 (1962).
(8) R. L. Aggarwal, Solid State Commun. 2, 1963 (1964); R. L. Aggarwal and A. K. Ramdas, Phys. Rev. 140, A1246, (1965).
(9) F. P. Ottensmeyer, J. C. Giles and J. W. Bichard, Canadian J. Phys. 42, 1826 (1964).
(10) G. B. Wright and A. Mooradian, Phys. Rev. Letters 18, 608 (1967).
(11) From the parity viewpoint, the "two-electron" transitions in silicon are analogous to the electronic Raman scattering of the emitted radiation by neutral impurity centers. (R. J. Elliott, and R. Loudon, Physics Letters 3, 189, 1963). The Raman scattering efficiency is zero in the effective mass approximation unless the initial and final states are derived from the same envelope state of the impurity. This can explain the prominence of the $1s(A_1) \rightarrow 1s(E)$ transition both in the Raman scattering (Ref. 10) and in the free exciton "two-electron" recombinations (Section IVC). More detailed considerations are evidently necessary to account for the dominance of the $1s \rightarrow 2s$ impurity excitation in the bound exciton "two-electron" spectra, however.
(12) P. J. Dean, J. R. Haynes and W. F. Flood, Phys. Rev. September, 1967.
(13) J. R. Haynes, Proceedings of the International Conference on the Physics of Semiconductors, Prague, 1960 (Academic Press, N.Y., 1960) p. 423.

(14) In the calculation of E_{H^-} from Eq. (1), E_g was obtained from the low temperature exciton energy gap, E_{gx}, using the relationship $E_g = E_{gx} + E_x$. The internal binding energy of the free exciton, E_x, was taken to be ~ 8 meV (Ref. 13).
(15) Component P^0_{VO} is noticeably narrower at 14°K than at 4.2°K. A similar effect has been noted in silicon (Ref. 12). This effect may be due to the increasing relative strength at the lower temperatures of bound-exciton valley-orbit "two-electron" transitions and of the influence of an unresolved splitting in the free exciton transition.
(16) J. R. Haynes, Phys. Rev. Letters 17, 860 (1966).
(17) M. A. Lampert, Phys. Rev. Letters 1, 450 (1958).
(18) A. Honig, Phys. Rev. Letters 17, 186 (1966).

MULTI-PHONON STRUCTURE IN THE EXCITON ABSORPTION SPECTRA OF CuCl

K.S. SONG

Institut de Physique, Strasbourg, France *

INTRODUCTION

An oscillatory structure has been observed recently by Ringeissen (1)(2)(3) in the exciton absorption spectra of CuCl at the liquid helium temperature. A series of equally spaced peaks, separated by about 220 cm^{-1}, are observed superposing on the contiuum absorption. The first of these peaks is separated by the same energy from the n = 2 exciton line of the sharp exciton series. A similar, but of much weaker intensity, structure is also observed beginning from the n = 3 line. Ringeissen has interpreted these peaks as excitonic transitions with simultaneous emission of longitudinal optical phonons whose energy is known to be about 220 cm^{-1} (2). This kind of structure has been observed in photo-current spectra of several semi-conductors and in particular in CuCl (4). Of all these CuCl is the unique known case in which such a structure is observed in absorption.

According to our recent band structure calculation of CuCl (5)(6)(7), the highest valence band is of 3d of Cu origine which is very flat. The calculated hole mass of the spin-orbit split band at Γ_7 is found to be 13 m, while the electron mass at the bottom of the conduction band is 0,30 m. Under these condition we may apply the criteria obtained by Dykman and Pekar (8) for self-trapped excitons in ionic crystals.If the excitons are self-

* Present Address : Institute for Solid State Physics, Tokyo University, Tokyo, Japan.

trapped in CuCl, then the observed oscillatory structure can be interpreted as due to multi-phonon processes resulting from the strong coupling between the self-trapped excitons and the optical phonon field.
The transition probability of such multi-phonon processes can then be calculated through the formalism developped by Huang and Rhys [9] for F centres.
We tried in a previous paper [10] to calculate the transition probability through a perturbation treatment in which the exciton-optical phonon interaction is considered as perturbation. Though one can explain the observed structure qualitatively, this treatment was found inadequate for this problem.

In chapter 2 we will solve formally the wave equation of the system composed of electronic part and the lattice part following the method of Pekar's polaron theory. In chapter 3 we will calculate the transition probability with emission of longitudinal optical phonons using the results given by Huang and Rhys. Finally, in chapter 4, we will apply the results to CuCl and discuss the experimental results in comparison with our theory.

CHAPTER 2 - SELF-TRAPPED EXCITON

The experimental results indicate that the coupling between excitons and the polarization field is very strong (see absorption spectrum of Ref.(3)). On the other hand the coupling constant of polarons α is very large for the hole : $\alpha_h \simeq 10$ and $\alpha_e \simeq 2$ for electron. This shows that the interaction between the exciton and the polarization field can not be treated as a perturbation.

The total Hamiltonian of the system can be written as follows :

$$\begin{aligned} H &= H_{ex} + H_{\ell} + H_{int} \\ &= -\frac{\hbar^2}{2m_e}\Delta_e - \frac{\hbar^2}{2m_h}\Delta_h - \frac{e^2}{\varepsilon_\infty |r_e - r_h|} + \frac{1}{2}\sum_w \hbar\omega\left(q_w^2 - \frac{\partial^2}{\partial q_w^2}\right) \\ &\quad + e\int \frac{P(r')\cdot(r_e - r')}{|r_e - r'|^3}\, d\tau' - e\int \frac{P(r')\cdot(r_h - r')}{|r_h - r'|^3}\, d\tau' \qquad (1) \end{aligned}$$

where ε_∞ = optical dielectric constant

$\hbar\omega$ = longitudinal optical phonon energy
q_w = normal coordinates of mode w for the unperturbed lattice
P(r) = macroscopic polarization field

Following Pekar's polaron theory [11] we look for solutions of above Hamiltonian Ψ minimizing the energy of the total system :

$$\delta\bar{H} = \int \Psi^* H \, \Psi \, d\tau dq = 0 \quad \text{with } dq = \prod_w dq_w \qquad (2)$$

under the normalization condition

$$\int |\Psi|^2 \, d\tau dq = 1 \qquad (3)$$

As a first approximation Ψ is put under following form:

$$\Psi(r_e, r_h, q) = \psi(r_e, r_h)\Phi(q) \qquad (4)$$

in which ψ and Φ are respectively wave functions of the electronic system (exciton) and the lattice system, each of them being separately normalized.
Considering ψ and Φ as two variational parametres, we can eliminate the lattice coordinates from the Hamiltonian by putting $\partial\bar{H}/\partial\Phi = 0$ and obtain the following results [7] :

$$\begin{aligned}\bar{H} = {} & \frac{\hbar^2}{2m_e} \int |\nabla_e \, \psi(r_e, r_h)|^2 \, d\tau_e d\tau_h + \frac{\hbar^2}{2m_h} \int |\nabla_h \psi(r_e, r_h)|^2 \\ & d\tau_e d\tau_h - \frac{e^2}{\varepsilon_\infty} \int |\psi(r_e, r_h)|^2 \frac{1}{|r_e - r_h|} \, d\tau_e d\tau_h \\ & + \sum_w \hbar\omega \, (n_w + 1/2) - \frac{1}{8\pi}\left(\frac{1}{\varepsilon_\infty} - \frac{1}{\varepsilon_0}\right) \int |D_e(\psi, r') \\ & - D_h(\psi, r')|^2 \, d\tau' \end{aligned} \qquad (5)$$

where $D_e(\psi, r')$ and $D_h(\psi, r')$ are functionals of ψ, the exciton wave function, defined as follows :

$$D_e(\psi, r') = e \int |\psi(r_e, r_h)|^2 \frac{(r_e - r')}{|r_e - r'|^3} \, d\tau_e \, d\tau_h$$

$$D_h(\psi, r') = e \int |\psi(r_e, r_h)|^2 \frac{(r_h - r')}{|r_h - r'|^3} \, d\tau_e \, d\tau_h \qquad (6)$$

The wave function of the lattice system is expressed in terms of a new set of normal coordinates q'_w corresponding to the excited state of the crystal :

$$\Phi(q') = \prod_w \varphi_w(q_w - q_{wo}) \qquad (7)$$

with

$$q_{wo} = \left(\frac{1}{4\pi\hbar\omega\bar{\varepsilon}}\right)^{\frac{1}{2}} \left|D_{ew}(\psi) - D_{hw}(\psi)\right| \qquad (8)$$

$$\text{with } \frac{1}{\bar{\varepsilon}} = \frac{1}{\varepsilon_\infty} - \frac{1}{\varepsilon_0}$$

where $D_w(\psi)$ are the Fourier transforms of $D(\psi, r')$ defined above (6).

Several important remarks should be made on $\bar{H}$ (5) ;

1) The exciton wave function is obtained by a further variational calculation by putting $\partial\bar{H}/\partial\psi = 0$. Dykman and Pekar (8) have shown that a self-trapped exciton with following type of wave function is energetically favourable if m_h/m_e or m_e/m_h is greater than 8 for large values of $(\varepsilon_\infty^{-1} - \varepsilon_0^{-1})$:

$$\psi(r_e, r_h) = \psi_1(r)\,\psi_2(R) \quad \text{where } r = r_e - r_h$$

$$R = (m_e r_e + m_h r_h)/M$$

with

$$\psi_1(r) \propto e^{-\alpha r^2} \qquad (9)$$

$$\psi_2(R) \propto e^{-\beta R^2} \qquad \beta \neq 0$$

2) Because of the motion of the two particles with regard to the centre of mass which are different in general, D_e and D_h are different, so that $D_e - D_h \neq 0$. One can easily show that for any arbitrary type of function , if $m_e = m_h$ then $D_e = D_h$.

3) For free exciton with $\psi_2(R) \propto e^{iK\cdot R}$ one can show that $D_e - D_h = 0$. This means that for free exciton the normal coordinates are not modified, while for a self-trapped exciton the displacement of the normal coordinates is non-zero : $q_{wo} \neq 0$. This probably is an oversimplified conclusion due to the form of (4) which is a crude approximation.

We will see in the next chapter that the multi-phonon structure is intimately linked with q_{wo} which is non-zero for self-trapped excitons.

CHAPTER 3 - TRANSITION PROBABILITIES

We calculate the transition probability between the vibrational states of the fundamental and excited states of the crystal in which we take into account the localized perturbation of the lattice due to the self-trapped exciton :

$$W = \frac{2\pi}{\hbar} |\langle i|\varepsilon.p|f\rangle|^2 \delta(h\nu - E_f + E_i) \qquad (10)$$

Here $|i\rangle$ and $|f\rangle$ are the initial and final states of the total system :

$$|i\rangle = \Phi_{o,n}(q)|0\rangle$$
$$|f\rangle = \Phi_{ex,n'}(q')|ex\rangle \qquad (11)$$

with $|0\rangle$ and $|ex\rangle$ the Slater determinants for the fundamental and excited state of the electronic system.
n designate the set of vibrational quantum numbers.
The application of the Condon's approximation allows to separate the electronic part from the vibrational part and consider it to be independant of the vibrational state :

$$\langle 0,\Phi_{o,n}(q)|\varepsilon.p|\Phi_{ex,n'}(q'),\ ex\rangle \Longrightarrow$$
$$\langle 0|\varepsilon.p|ex\rangle \langle\Phi_{o,n}(q)|\Phi_{ex,n'}(q')\rangle \qquad (12)$$

If we neglect the local deformation of the lattice due to the self-trapped exciton q_{wo}, then the second matrix element is equal to unity and we obtain the usual expression for excitonic transition.
Before going into the detailed calculation of this matrix which is given in the original paper of Huang and Rhys, we slightly rewrite the above expression of the transition probability.
We sum W over the final vibrational states n' in such a way that the net population change of phonons is clearly indicated :

$\sum_w \Delta n_w = p$ where p is a positive integer (emission of p phonons)

$\Delta n_w = 0, \pm 1$ the change in the mode w.

$$W = \frac{2\pi}{\hbar} \sum_{n'} |\langle 0|\varepsilon.p|ex\rangle|^2 \ |\langle\Phi_{o,n}(q)|\Phi_{ex.n'}(q')\rangle|^2 .$$
$$\delta(h\nu - E_\lambda - p\hbar\omega) \qquad (13)$$

where E_λ is the exciton level of state λ of the exciton internal motion counted from the fundamental state of the crystal.

According to Huang and Rhys the second matrix element can be expressed as follows[9] :

$$\sum_{n'} |\langle \Phi_{o,n}(q) | \Phi_{ex,n'}(q') \rangle|^2 = \frac{2\pi}{\omega} \left(\frac{\bar{n}+1}{\bar{n}}\right)^{p/2} \cdot I_p(2S\sqrt{\bar{n}(\bar{n}+1)}) \exp\left[-S(2\bar{n}+1)\right]$$

Finally we obtain the following expression for the transition probability with emission of p optical phonons (thermal average is taken over the initial vibrational states) :

$$W \propto |\langle 0|\varepsilon.p|ex\rangle|^2 \exp\left[-S(2\bar{n}+1)\right]\left(\frac{n+1}{n}\right)^{p/2} \cdot I_p(2S\sqrt{\bar{n}(\bar{n}+1)}).\ \delta(h\nu - E_\lambda - p\hbar\omega) \quad --- (14)$$

where

$\bar{n} = \left[\exp(\hbar\omega/kT) - 1\right]^{-1}$

I_p is the modified Bessel function,

S is the so-called Huang-Rhys S factor defined in our case as follows :

$$S = \frac{1}{\hbar\omega}\left[\frac{1}{2}\omega^2\ \overline{(\Delta Q)^2}\right] \quad --- (15)$$

with

$$\overline{(\Delta Q)^2} = \frac{1}{4\pi\bar{\varepsilon}\omega^2} \sum_w |D_{ew}(\psi) - D_{hw}(\psi)|^2 \quad --- (15')$$

This factor can be interpreted as the mean number of phonons intervening in the relaxation of lattice around the self-trapped exciton. We will discuss these results in the next chapter for CuCl.

CHAPTER 4 - DISCUSSION

4-§1 - Multi-Phonon Structure in CuCl

We first rewrite the preceding formula of W (14) for the case of CuCl at 4°K. In this case $\bar{n} \simeq \exp(-100)$ and the factors containing $\bar{n}$ can be developped for $\bar{n} \simeq 0$.

$$W \propto |\langle 0|\varepsilon.p|ex\rangle|^2 \exp(-S)\ S^p/p\,! \quad --- (16)$$

This expression gives the Poisson distribution for different number of emitted phonons.

The S factor can be calculated from its definition (15) if one knows the exact exciton wave functions. This being a very difficult problem, we evaluate the value of S factor so as to obtain a satisfactory agreement with experimental results. Before doing this we can make the following remarks :

1) For n = 1 exciton, the electron is moving around the hole at a distance of a few interatomic distance, so that the net polarization of the lattice, measured by $D_e - D_h$ should be very small.

2) For n = 2 and 3 exciton, the electron is respectively about 40 Å and 90 Å apart from the hole according to evaluations from exciton spectra (2) . Under such condition the polarizing effect of the hole is almost entirely felt by the lattice around the self-trapped exciton. Hence $D_e - D_h$ should be considerable.
We thus esteem that the S factor should have larger value for n = 2 and n = 3 exciton than for n = 1.

The experimental results are not precise enough to permit a quantitative evaluation of S factor. However the exciton absorption spectrum of Ref.(3) shows that the first three peaks have a comparable intensity, then decreases progressively until it disappears completely for p = 7 peak. We have obtained such a distribution with S = 1,2 (table 1). For n = 3 exciton S factor should have a comparable value, but because of the very weak intensity of these peaks no precise comparaison is feasible. For n = 1 exciton only p = 0 (zero-phonon) peak is observed so we may put S ≃ 0.

Table 1

Multi-phonon line intensity for n=2 exciton

Number of emitted phonons p	0	1	2	3	4	5	6
Transition probability (relative value)	1,00	1,20	0,72	0,29	0,09	0,02	<0,01

4-§2 - Intensity of Zero Phonon Lines

Another interesting feature of our self-trapped exciton is the intensity of the zero phonon lines corresponding to n = 1, 2 and 3 levels. According to Elliott's theory (12), which gives the transition probability (13) without the second matrix element, the intensity varies with n as follows :

$$W \propto \frac{1}{n^3 a_o^3} \qquad - - - (17)$$

where $a_o = \varepsilon_o / \mu^*$ is the exciton 'Bohr' radius

The results we obtained in the previous paragraph, however, show the following interesting features for zero phonon lines :

1) Even the zero phonon lines are modified by the local deformation of lattice through the factor exp(-S). As the S factor has different values for different exciton levels, the intensity should vary in a more complicated way with n.

2) When the transition probability W(16) is summed over p for a given exciton level, we obtain formally the Elliott's expression :

$$\sum_p W(p) \propto |\langle 0|\varepsilon.p|ex\rangle|^2 \sum_p \exp(-S) S^p/p! = |\langle 0|\varepsilon.p|ex\rangle|^2$$

These two points are qualitatively confirmed by experiment. The second point was remarked by Ringeissen (13). To check the first point we reproduce the intensity of zero phonon lines as measured by Ringeissen (2) (table2). The difficulties of precise measurements are such that we cannot make a quantitative check, but the discrepancy between the experimental values and the theory of Elliott can be accounted for by the extra factor exp(-S) as is shown in the last column of table 2.

Table 2

Intensity of zero phonon lines

1) Experimental values (with thin film)
2) Experimental values (with massive samples)
3) Elliott's theory (with ε_o constant)
4) Elliott's theory (with values of ε_o adjusted so that the exciton dissociation energy fit to the

Rydberg's formula)
5) Present model(with factor exp(-S) multiplied to values of 4).

	(1)	(2)	(3)	(4)	(5)
f_1/f_2	24	41	8	16	48
f_1/f_3	85	150	27	70	210

CONCLUSION

The results we obtained can be resumed as follows:

1) The exciton-optical phonon interaction is very important because of the extremely heavy hole mass.
2) The self-trapped exciton model allows a satisfactory explanation of the observed multi-phonon structure.
3) The important discrepency between Elliott's theory and the experiment for the zero phonon exciton line intensity can be accounted for by the extra factor due to the local lattice deformation around a self-trapped exciton.

Acknowledgments

The author wishes to thank Profs. E. DANIEL and F. GAUTIER for their advice and Prof. S. NIKITINE and Dr. J.RINGEISSEN for the discussion on their measurements.

REFERENCES

(1) J. Ringeissen, Phys. Letters 20, 571 (1966)
(2) J. Ringeissen, Thesis (D. Sc.) Strasbourg (1967), Documentation Centre C.N.R.S., Paris,n°
(3) J. Ringeissen, A. Coret and S. Nikitine, Present Conference
(4) A. Coret, J. Ringeissen,Present Conference
(5) K.S. SONG, J.Phys. 28, 195 (1967)
(6) K.S. SONG, J. Phys.Chem. Solids (under press)
(7) K.S. SONG, Thesis (D.Sc.) Strasbourg (1967), Documentation Centre CNRS, Paris, n° AO 1682
(8) I.M. Dykman and S.J. Pekar, Dokl. Akad. Nauk. SSSR 83 , 825 (1952)

(9) K. Huang and A. Rhys, Proc. Roy. Soc. (London), 204 A, 406 (1950)
(10) K.S. SONG, Symposium (Strasbourg) on " Transitions électroniques dans les solides non-conducteurs" J. Phys. (under press)
(11) S.I. Pekar, Untersuchungen über die Elektronentheorie der Kristalle (Akademic Verlag, Berlin) (1954)
(12) R.J. Elliott, Phys. Rev. 108, 1384 (1957)
(13) J. Ringeissen, Private Communication.

VIBRATIONAL STRUCTURE IN THE EXCITONIC ABSORPTION SPECTRUM OF CuCl AT 4,2°K

J. Ringeissen, A. Coret, S. Nikitine

Laboratoire de Spectroscopie et d'Optique du Corps Solide, Université de Strasbourg, France

I. INTRODUCTION

The optical properties of CuCl at low temperatures have been extensively studied (1) (2) (3) (4). Nikitine and Reiss (1) have shown that at 4,2°K the spectrum of thin layers of CuCl is composed of two strong lines, one sharp and one diffuse, and of a number of weak lines. The two strong lines have been suggested to be the first lines (n = 1) of two different exciton line series separated by spin orbit splitting. The weak lines have been tentatively suggested to be high order lines of the two series.

Recently it has been possible to prepare samples composed of single crystal platelets of suitable thickness. With these samples, new lines are observed. Some of these are equidistant. So the previous interpretation has to be reconsidered.

Further on it has been possible to determine for the sharp series both m_e^* and m_h^* and to show that m_h^* is unusually high. In this paper we describe new results and we give a new interpretation of the spectrum of CuCl. This interpretation leads to a new type of excitonic spectrum, the excitons being strongly localised and the exciton lines showing vibronic satellites.

This paper is connected to the paper of Song (5) who has given the theory of the band structure of CuCl and of vibronic satellites of exciton lines when $m_h^* \gg m_e^*$ and to the paper of Coret, Ringeissen and Nikitine (6) in which an oscillatory structure of photocurrent spectrum, in agreement with the absorption measurements, is reported.

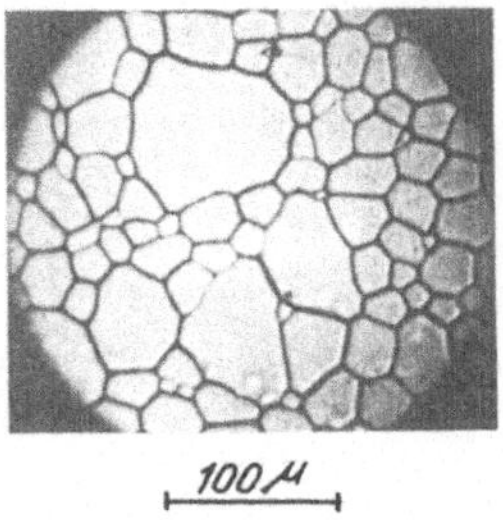

Fig.1 Thin Recrystallized Layer of CuCl
(thickness about 1,4 micron).

II. EXPERIMENTAL DATA AND DESCRIPTION OF THE EXCITON SPECTRUM

The samples used in these experiments are thin layers evaporated on fused quartz support. Subsequently these layers are recrystallized according to a method described elsewhere (7). Such a layer is represented on Fig.1. The monocrystalline domains can reach fifty to hundred microns for a thickness of about 1,4 micron f.i.

The samples are immersed directly in the liquid helium. The absorption spectrum is observed with a Baush and Lomb spectrograph which has a dispersion of 4 A° per mm. The spectrograms are recorded with a Hilger microdensitometer.

As stated above, the first two strong absorption lines λ_{1f} (f for sharp) and λ_{1d} (d for diffuse) are suggested to be first lines of two different series of lines, the sharp series and the diffuse series. The separation is due to spin orbit splitting. In Song's recent paper (8) on band structure of CuCl the valence band is split in two bands Γ_7 and Γ_8. The calculated splitting (0,082eV) is very close to the observed one (0,069eV).

The sharp series is obviously composed of the lines λ_{2f} and λ_{3f} (Fig.2). These lines can be recognised by their sharp character. The lines λ_{2s} and λ_{3s} will be discussed later in this paper. The lines of higher energy are diffuse : they are equidistant and will be also discussed later. The first of these weak diffuse lines λ_{2d} has been previously classified (1) as the n = 2 line of the diffuse series.

III. DISCUSSION OF THE SHARP EXCITONIC SERIES OF LINES

The three lines λ_{1f}, λ_{2f}, λ_{3f} of the sharp series do not satisfy a hydrogenlike series. This is not surprising as Haken's correc-

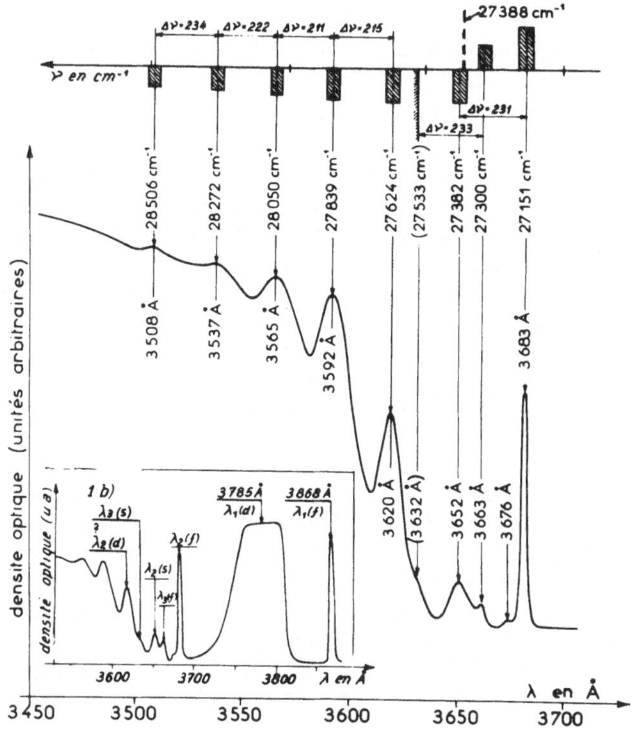

Fig.2 Absorption Spectrum of CuCl at 4,2°K. The upper diagram gives the intervals in energy of the different lines.

tion has to be applied.

a) Haken's Correction. Evaluation of m_e^* and m_h^*.

According to Haken (9) the potential energy in the Schroedinger equation is given by the following relation :

$$V = -\frac{e^2}{\varepsilon r} = -\frac{e^2}{\varepsilon_o r} + \frac{e^2}{r}\left(\frac{1}{\varepsilon_o} - \frac{1}{\varepsilon_s}\right)\left(1 - \frac{e^{-u_1 r_{12}} + e^{-u_2 r_{12}}}{2}\right) \qquad (1)$$

where $u_n = \sqrt{2m_n^* \frac{\omega}{\hbar}}$, ε_o is a dielectric constant for high frequencies, ε_s the static dielectric constant, ω the circular frequency of the optical longitudinal phonon, r_{12} the radius of the exciton and m_n^* the apparent mass of the electron or the hole.

It can be seen that for large orbits ε is close to ε_s, for small orbits the polarisation of the lattice being small ε is close to ε_o. This explains why an excitonic series is not hydrogen like.

Rydberg's constant is given by the well known relation :

$$R_e = R_H \frac{\mu}{\varepsilon_n^2 m_o} \text{ with } \nu = \nu_\infty - \frac{R_e}{n^2} \text{ and } r_n^2 = \frac{e^2}{2\varepsilon_n Rch} \quad (2)$$

ε_n being a dielectric constant characteristic from a given excitonic level, μ the effective mass. If the three lines would fit in an excitonic series, the value of ε_n must satisfy at the same time the relations (1) and (2). But we cannot determine ε_n because ν_∞ and μ are unknown. We have solved this problem by a graphical method which allows to obtain an approximate value of ν_∞ , m_e^* and m_h^*. For these calculations we have taken ε_o = 3,73 and ε_s = 7,43 (7). By emission experiments we have been able to determine the value of the longitudinal optical phonon LO = 220 cm^{-1} (8).

For the determination of ν_∞ we assume that the radius of the n = 3 orbit is great enough so that the contribution of ε_o can be neglected. We have than given to μ arbitrary values included between 0,25 m_o and 0,50 m_o. Since the same effective mass is obtained for different values of m_e^* and m_h^*, we have drawn curves ε_n = f(m_e^*) for different values of m_h^*, ε_n being alternatively calculated by the relation (2) or (1). Thus we obtain two families of curves. Their intersection gives the parameters given in the table I.

μ=0,406m_o m_e^*=0,415m_o $m_h \simeq$ 20m_o ν_∞=27388cm^{-1}

n	Γ_ν in cm^{-1}	R in cm^{-1}	ε	r in Å
1	25865	1523	5,41	7,03
2	27151	956	6,84	35,52
3	27300	809	7,43	86

Table I : Parameters of the Sharp Series

So it seems very likely that the three lines can be classified in an excitonic series corrected by Haken's theory and assuming the values of the electron and hole mass given in the table I. Also it appears that the mass of the hole has an unusually high value. On account of that, the exciton must be strongly localised. This is a first important result of the above analysis of experimental data. The determination of the f factor of an absorption line is another way of characterising an excitonic transition.

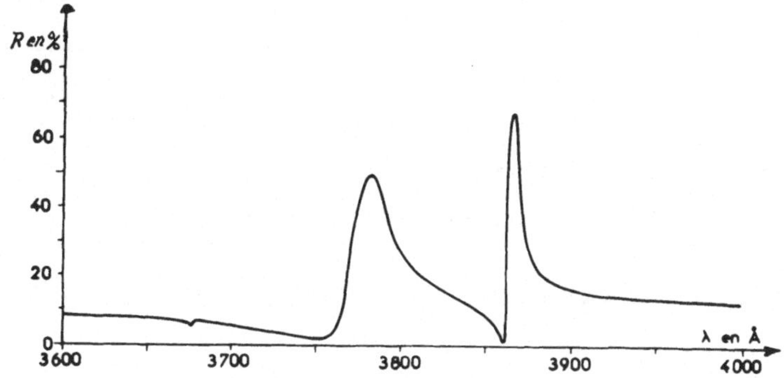

Fig. 3 Reflexion Spectrum of CuCl at 4,2°K (bulk crystal) The λ_{1f} and λ_{1d} absorption lines are accompanied by a strong anomaly of the reflexion.

b) Oscillator Strength of the Lines of the Sharp Series.

According to Elliott (11), for first class transitions, the f factors are given by

$$f_n = \frac{C}{\varepsilon_n^3 n^3}$$

The n = 1 line should be 8 times stronger than the n = 2 line, 27 times more stronger than the n = 3 line.

But we have seen that each level is characterised by an appropriate value of the dielectric constant. With the values of ε calculated above, the ratios of the f factors are given in the first column of table III. In order to determine the f values, the absorption coefficient K has been measured as a function of ν on thin films. From the Krawetz relation we can calculate f_{1f}, f_{2f} and f_{3f}. For the n = 1 level these measurements are very difficult because the very high absorption is accompanied by a strong anomaly of reflexion. This is illustrated in Fig. 3. We have taken in account the intense reflexion for the calculation of K.

The measurements are also difficult since for the n = 1 lines the layers must be very thin. In very thin layers the grains are very small. It has been shown that when the grains are too small the exciton lines become weaker and for very small grains can disappear completely (12). For CuCl the f factors from the n = 1 lines measured on such layers are smaller than those calculated from Kramers Kronig analysis on the reflexion spectrum of bulk crystal. A comparison with the theoretical reflexion given by a classical oscillator has also been made. These results are recapitulated in table II.

f_{1f}	3×10^{-3}	thin films
	$5,2 \times 10^{-3}$	bulk crystal (K.K. or classical dispersion)
f_{2f}	$0,125 \times 10^{-3}$	thin recrystallized layers
f_{3f}	$0,035 \times 10^{-3}$	thin recrystallized layers
f_{2s}	$0,10 \times 10^{-3}$	thin recrystallized layers

Table II : f Values for Sharp Lines (and Satellite).

The order of magnitude of the f factors are in good agreement with the predicted values for first class transitions (13). As we can see in table III this agreement is less good when we compare the relative intensity of the different lines.

	theoretical	n = 1 measured on thin film	n = 1 measured on bulk crystal
f_{1f}/f_{2f}	16	24	41
f_{1f}/f_{3f}	70	85	150

Table III : Ratios of the f Factors from the Sharp Series

It has to be noted that in all these measurements we have run into a peculiar anomaly, the reflexion coefficient being always considerably stronger than it should be from absorption measurements.

IV. VIBRONIC STRUCTURE OF THE SPECTRUM OF CuCl.

It has been seen in the above sections that the line λ_{1d} is interpreted as the n = 1 line of a diffuse series of exciton lines, the lines λ_{1f}, λ_{2f}, λ_{3f} as the successive lines of a sharp series of exciton lines. Other lines have however not yet been interpreted.

a) The Vibronic Satellites of the Sharp Series.

The lines λ_{2s} and λ_{3s} are separated by 230 cm^{-1} from respectively λ_{2f} and λ_{3f}. The LO phonon is 220 cm^{-1}. We have suggested that these two lines are vibronic satellites of the n = 2 and n = 3 lines of the sharp series (7). The satellites correspond to a higher transition probability to a continuum if the excited state can by emission of a LO phonon reach a final sharp quantum state, say an exciton state below the continuum. Apart the above considerations three other arguments are in favour of this interpretation.

- i) Cordona (3) has shown that when CuCl is doped with Br the first lines of both exciton series are shifted in a different way since the spin orbit separation is reduced. By doping experiments of recrystallized layers we have shown that λ_{2s} shifts in the same way as λ_{2f}. λ_{3f} and λ_{3s} are unfortunately broadened and disappear at low concentration of Br before measurable effects of a doping shift can be detected. These experiments show that λ_{2s} is connected to λ_{2f}.

- ii) Song (5) has shown that the theory predicts the above suggested process of absorption when $m_h^* \gg m_e^*$. We have shown in the preceding section that this is precisely the case for CuCl. Therefore Song's theory gives a strong support to the process of absorption we have deduced from experimental observations.

- iii) The addition of the oscillator strengths f_{2f} and f_{2s} improves considerably the agreement with Elliott's theoretical value. Further on, recent calculations of Song (unpublished, private communication) show that there is a kind of sum rule for direct transitions to an excitonic state and to satellite states. We have obtained two different values for f_{1f} for very thin evaporated films and for absorption curves obtained from Kramers Kronig treatment of the reflectivity. The values of f_{1f} are 5,2 x 10^{-3} and 3 x 10^{-3} leading to f_{2f} values (taking account of the correction of the dielectric constant ε_n) of 0,32.10^{-3} and 0,18 x 10^{-3}. The addition of the experimental values gives $f_{2f} + f_{2s} = 0{,}22 \times 10^{-3}$.

It is believed that on account of the above arguments the interpretation of λ_{2s} and λ_{3s} is very plausible.

b) Multiphonon Satellite Lines

Up from λ_{2d} five lines are diffuse and equidistant. They are about 215 cm^{-1} apart. It is obvious that this equidistant structure strongly suggests a process of absorption analogous to the one phonon process but extended to a multiphonon process. The emission of many phonons can be simultaneous or in cascade.

Such a process has been studied theoretically by Song (5). Song's calculations are in very good agreement with the observed structure.

The interpretation is however not quite simple as the final state which we suggest to name the collector level has to be specified. This determination is bound to some difficulties.

Doping experiments are not decisive in this case because the multiphonon lines disappear by small concentrations of Br. The shift of the two first lines is smaller than that of λ_{1d} and larger than that of λ_{1f} and λ_{2f}. The following possible interpretations are considered :

- i) The collector level could be the n = 2f exciton level. Then the λ_{2d} line could be a two phonon line. However this line is 240 cm^{-1} apart from λ_{2s}. It is not clear why such a difference with the LO phonon between λ_{2s} and λ_{2d} would exist if the higher lines have a mean difference of 215 cm^{-1}. Further on the sum rule f_{2f} + f_{2s} + f_{2d} + etc would give a much to high value. It is believed that this suggestion, though possible, is rather unprobable. Song's calculations have been made in this hypothesis. However we had to consider another hypothesis.

- ii) The collector level would be the n = 2d level, λ_{2d} being the n = 2 line of the diffuse series as suggested previously by Reiss and Nikitine (1). This line could be the zero phonon line. The fact that this line is 240 cm^{-1} off λ_{2s} might be ascribed to a coincidence. This hypothesis is in much better agreement with the experiment as regards Song's f sum rule.

This second interpretation seems to be the most plausible one at present. It is strongly supported by the results of photoconductivity measurements briefly discussed in next section. But we shall see that this agreement suppose the existence of a -1 phonon line (-1 for absorption) only visible in the photocurrent spectrum. Song's theory predicts the possibility of a -1 phonon line. But at low temperature such a line should be very weak.

V. PHOTOCONDUCTIVITY

Results of photoconductivity measurements are given in a separated paper (6). As they are directly connected with our absorption measurements some of the results will be mentioned here.

A rather complicated oscillatory structure is observed in the photocurrentresponse spectrum. The minima of this curve correspond well to the maxima of absorption of presumed diffuse vibrational satellite lines. It appears that these minima start with λ_{2d}. The minimum at 3645 A° could be a -1 phonon line. The n = 1f and n = 2f lines appear as minima of photocurrent. The n = 1d line gives also a series of very weak photoconductivity minima indicating that this level is also a collector level. Peculiarly the n = 1f level does not correspond to any marked vibronic satellite structure. Further on the satellites of n = 1d are not observed in absorption. They are probably too weak. Oscillatory structures in photocurrentresponse curves have been observed previously with other compounds (14) (15) (16) but the simultaneous observation of vibronic staellites in absorption and photoconductivity is to our knowledge a new result.

VI. CONCLUSION

The results obtained can be summarized as follows. The two strong bands of CuCl arise from the spin-orbit splitting of the valence band and are n = 1 lines of two series, a sharp and a diffuse one.

The sharp line is the first term of an excitonic series including two other lines of smaller intensity. These two higher order terms are accompanied at least by one vibrational satellite, one LO phonon apart from these lines. Using Haken's correction it has been possible to obtain an evaluation of m_e^* and m_h^*. It so happens that $m_h^* \gg m_e^*$. Therefore the excitons are strongly localized in CuCl.

The n = 1 diffuse line is probably accompanied by the n = 2 diffuse line. The diffuse lines of higher energy form a vibronic structure. The agreement between absorption and photoconductivity measurements is good if the level corresponding to the n = 2 diffuse line is the collector level from a vibronic structure involving at least five LO phonons. However this interpretation suggests that a -1 phonon line is observed, but only in photoconductivity though.

REFERENCES

(1) R. Reiss, S. Nikitine, C.R. Acad. Sc. 250, 2862 (1960)
(2) S. Nikitine, J. Ringeissen, J.L. Deiss, J. Phys. 23, 890 (1962)
(3) M. Cardona, Phys. Rev. 129, 69 (1963)
(4) M. Ueta, T. Goto, J. Phys. Soc. Japan, 20, 1024 (1965)
(5) K.S. Song, Conference on Localised Excitons in Solid, Irvine
(6) A. Coret, J. Ringeissen, S. Nikitine, Conference on Localised Excitons in Solid, Irvine
(7) J. Ringeissen, Thèse d'Etat, Strasbourg (1967)
(8) K.S. Song, J. Phys. 128, 195 (1967)
K.S. Song, Thèse d'Etat, Strasbourg (1967)
(9) H. Haken, Fortsch. Phys. 38, 271 (1958)
(10) S. Nikitine, J. Ringeissen, C. Sennett, 7ème Congrès Intern. sur la Phys. des Semiconducteurs, Recombinaison Radiative dans les Semiconducteurs, Ed. Dunod, Paris, 279 (1964)
(11) R.J. Elliott, Phys. Rev. 108, 1385 (1957)
(12) R. Satten, S. Nikitine, Phys. Matière Condensée, 1, 394 (1963)
(13) S. Nikitine, Progress in Semiconductors, 6, 233 (1962)
(14) D.C. Reynolds, C.W. Litton, T.C. Collins, Phys. Stat. Solidi, 9, 671 (1965)
(15) H.J. Stocker, C.R. Stannard, H. Kaplan, H. Levinstein, Phys. Rev. Letters, 12, 7, 163 (1964)
(16) M.A. Habegger, H. Y. Fan, Phys. Rev. Letters, 12, 4, 99 (1964)

PHOTOCONDUCTIVITY SPECTRUM OF CuCl, OSCILLATORY STRUCTURES

A. Coret, J. Ringeissen and S. Nikitine

Laboratoire de Spectroscopie et d'Optique du Corps Solide, Université de Strasbourg, France

I. INTRODUCTION

The correlation between absorption and photoconductivity spectra in the domain of band to band transitions has been studied in several papers (1) (2) (3). However, a new mechanism of photoconductivity has been discovered recently by several authors : equally-spaced minima and maxima are observed on the short wavelength side of the intrinsic photoresponse spectra of InSb, GaSb and CdS. However, such structures have not yet been reported in the corresponding absorption spectra. In this paper the properties of the photoconductivity spectrum of CuCl at 4°K are described and discussed. A pronounced oscillatory structure is observed. Another paper describes the absorption spectrum of this compound at the same temperature. The analogy between these two kinds of spectra suggests a new interpretation of these equally-spaced structures.

II. THE GENERAL SHAPE OF PHOTOCONDUCTIVITY SPECTRA

The main contribution to the intrinsic photoconductivity spectrum is the band to band transition. The photocarriers created in this way increase the conductivity. It has been shown that the spectral variation of the photocurrent is not simple. When the photocarriers are created in the vicinity of the surface (for absorption coefficient of the order of magnitude of $10^5 cm^{-1}$); their recombination velocity is greater than when they are created in the bulk. De Vore (4) has calculated theoretically the spectral distribution of the photocurrent as a function of two parameters : the surface recombination velocity and the diffusion length. Fig. 1 shows the

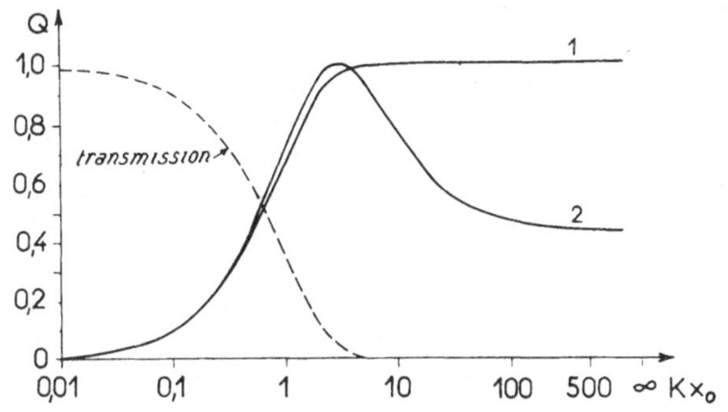

Fig.1 - Efficiency of the photoconductivity versus optical density (K is the absorption coefficient, x_0 the thickness of the sample).
Curve 1 : Low surface recombination velocity
Curve 2 : High surface recombination velocity

type of curves obtained when the recombination velocity is greater on the surface than in the bulk (curve 2); a region where the efficiency of the photocurrent decreases with increasing absorption coefficient can be observed.

The formation of excitons constitutes a second contribution to the intrinsic photoconductivity. The exciton is neutral and should not contribute to the photoconductivity. However it can recombine as dissociate, creating in this last case two photocarriers. In most cases maxima or minima of photocurrent which coincide with the excitonic absorption lines are observed. In Cu_2O, f.i., we have found that the presence of minima or maxima is well interpreted by De Vore's calculations. Minima appear when the photocurrent is a decreasing function of the absorption coefficient (5). The lifetime of the excitons being very short and their dissociation very probable, the absorption in an exciton line has a similar effect to the absorption in a continuum. However in CdS Gross and co-workers (6) have related these extrema to the existence of impurity centers on which the excitons can annihilate in a radiative or non-radiative way or dissociate.

The third contribution to the shape of the spectral photoresponse is due to the interaction of photoelectrons with optical phonons at sufficiently low temperatures. Equally-spaced minima and maxima of photocurrent appear on the short wavelength side of the spectrum. The intervals correspond to the energy of an optical longitudinal phonon. The minima happen when the photoelectrons have a high probability to fall by emission of one or several LO phonons on a sharp energy state (exciton or impurity level) or on the bottom of the conduction band. Let the energy of such a level be E_o. So if the energy of the photoelectrons is equal to $E_o + n\hbar\omega_{LO}$, where $\hbar\omega_{LO}$ is the energy of the optical longitudinal phonon and n an integer, minima are observed. The decrease of the photocurrent is explained by Reynolds and al. (7) assuming that the lifetime of photoelectrons is reduced when they can recombine with an excitonic or impurity level with emission of n phonons. Stocker and al. (8) have calculated the energy loss of photoelectrons when they fall on the bottom of the conduction band. They concluded also to a decrease of the photocurrent.

We have seen that for the band to band and excitonic contributions there is a correlation between photoconductivity and absorption spectra. However, in the case of interaction between photoelectrons and optical phonons, according to the above authors this correlation does not exist : the equally-spaced structures are not abserved by optical absorption measurements. We are going to show that for CuCl a correlation between an oscillatory structure of the absorption curve and a similar structure of the photoresponse spectrum can be established.

III. EXPERIMENTAL RESULTS : PHOTOCONDUCTIVITY SPECTRUM OF CuCl AT 4°K

1) Experimental Set Up

The photoresponse is measured by a Keithley electrometer with an input resistance higher than $10^{14}\Omega$. A voltage drop of 10 to 300 volts is applied between the electrodes. These electrodes are formed by two layers of evaporated gold, on the same side of the sample and separated by about 1mm. The sample is illuminated between the electrodes through a monochromator. The light source is a 900 watts Xenon lamp. The samples are placed in cryostat described elsewhere and are brought to helium temperatures.

2) Preparation and Quality of the Samples

Crystals prepared by a zone melting technique have been used. The illuminated surface of the sample is formed by fusion on a fused SiO_2 plane plate. The plate is separated from the sample subsequently. These samples have a very good reflecting surface with a minimum of recombination centers. If this precaution is not taken, a photoconductivity spectrum shown in Fig.2 is observed : in the excitonic line n = 1 of the sharp series we observe a minimum of

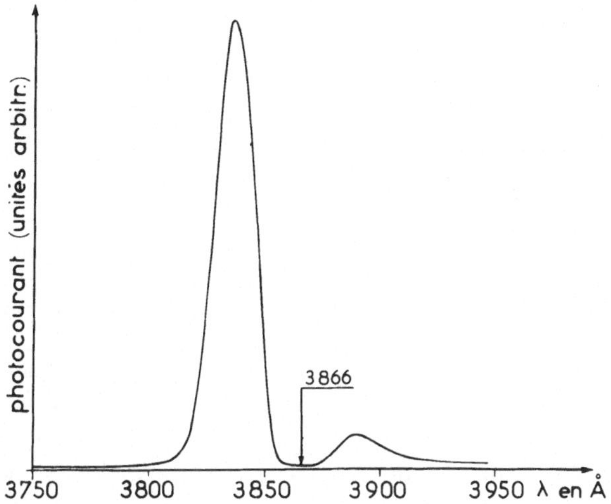

Fig. 2 - Photoconductivity Spectrum of CuCl at 20°K
(sample with a great density of surface recombination centers)

photocurrent and on the short wavelength side no photocurrent is observed. As shown in Fig.1, according to De Vore, when the photocarriers are created in the vicinity of the surface the photocurrent decreases and in our case is zero.

3) Photoresponse Curve of Samples with Low Concentration of Surface Recombination Centers.

The absorption spectrum of CuCl at 4,2°K and the band structure of CuCl have been published elsewhere (9) (10). Let us now describe the photocurrent spectrum of a crystal with good surface conditions. Starting from the low energies (Fig.3) the following accidents can be seen on the photoresponse curve :

- a maximum at 3864,5 A° preceded by a dip at 3870,5 A°. This structure corresponds to the line n = 1 of the sharp excitonic series ($\Gamma_7 \rightarrow \Gamma_6$). The minimum coincides with a maximum of reflexion.
- a large maximum at 3840 A° which we can attribute to surface conditions according to De Vore's curve (Fig.1).
- a pronounced minimum at 3787 A°. In the absorption spectrum this minimum corresponds to the line n = 1 of the excitonic diffuse series ($\Gamma_8 \rightarrow \Gamma_6$).There are three small dips on the

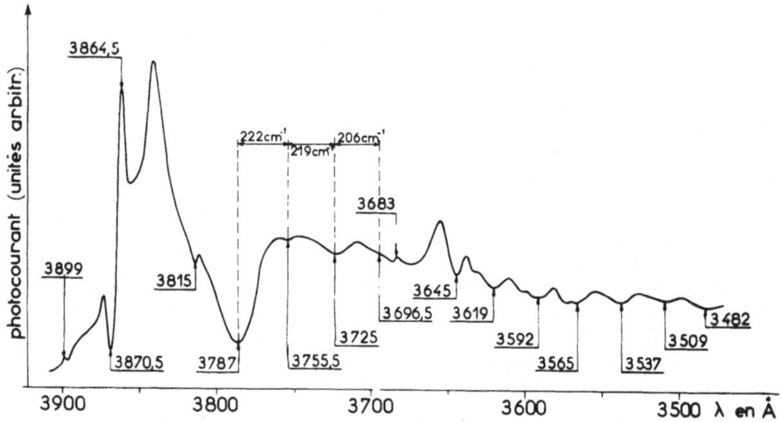

Fig.3 - Photoconductivity Spectrum of CuCl at 4°K (good surface conditions)

short wavelength side of this minimum (3755,5 A°, 3725 A° and 3696,5 A°) and one dip only on the large wavelength side. All these minima are to a good accuracy equally-spaced, the mean separation is 216 cm^{-1}. No corresponding oscillatory structure appears in the absorption spectrum (9).

- a weak maximum at 3683 A° which corresponds to the line n = 2 of the sharp series (Fig.4); an inflexion at 3663 A° is probably due to the line n = 3 of this series.

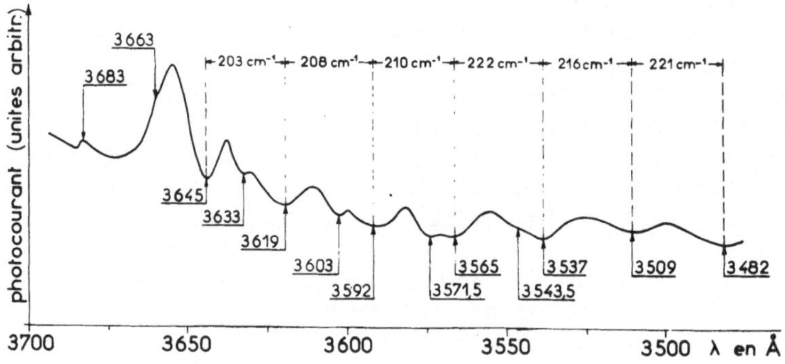

Fig.4 - Detail of the Preceding Figure

- a series of 7 equally-spaced dips, the first of them being situated at 3845 A. The mean separation is 215 cm^{-1} and the error on the determination of wavelength is $\pm$ 2 A°.
- a second series of equally-spaced dips beginning with a minimum at 3633 A°. The separation is also 215 cm^{-1}. This second series of equidistant lines is superimposed on the first. The dips are less pronounced.

The shape of this spectrum does not depend on the electric field (between 0 and 3 kV); the surface conditions only seem to have an important influence on the photoresponse. The oscillations are observed at 4,2°K but can also be observed at 20°K, they disappear at 77°K.

IV. DISCUSSION

The multiphonon satellite structure of the absorption spectrum has been discussed previously. It can be seen from the described spectrum that the photoresponse curve gives still more information and the series of equally-spaced dips can probably be explained by an analogous process.

At first, it is of importance to find the collector levels of the three multiphonon series. The collector level coincides with a zero phonon transition. It is the final state to which leads the multiphonon cascade.

For the first multiphonon series the collector level is obviously the n = 1 diffuse exciton level ($\Gamma_8^- \rightarrow \Gamma_6^+$).

For the two other series of multiphonon dips the most plausible interpretation is to assume that the collector levels are the n = 2 and n = 3 levels of a diffuse exciton series, corresponding to lines at 3620 A° and 3603 A°. The first of these two levels is observed in absorption. It has also been suggested to be the collector level of the multiphonon series in the absorption spectrum. The zero phonon line was suggested to be the n = 2 line of the diffuse series. The fact that the absorption line corresponding to the n = 3 exciton level of a diffuse series is not actually observed in absorption is not a difficulty for the above interpretation. This line is weak and the much stronger one phonon line of the multiphonon diffuse spectrum leading to n = 2 (dif.) may well hide the correspondent absorption line.

If this interpretation is to be accepted the dips at 3645 A° and 3633 A° should be considered corresponding to the - 1 phonon lines of the two series. Such a possibility is predicted by Song's theory for absorption but has not been observed. These lines are not observed in absorption.

The second possibility mentioned in the paper on satellite structure in absorption has been considered as less plausible. In this hypothesis the n = 2 and n = 3 levels of the sharp series were suggested to be the collector levels of both oscillatory structures. As regards photoconductivity this is not plausible : the higher terms of the sharp series appear very weakly on the photoresponse curve in comparison with the corresponding equally-spaced phonon structures.

The opinion can be expressed that although all the facts are not yet completely explained, the first process suggested is the most probable. The correspondence between the absorption and photoresponse curves in the multiphonon part of the spectrum is also very striking.

A final remark seems necessary. Two theories can be suggested to explain the multiphonon structure of the photoresponse curve.

a) Reynolds theory (7) or Stocker's theory (8) explain well the multiphonon structures of the response curve by assuming that for photon energies of $E_{exc} + n\hbar\omega_{LO}$ the lifetime of a free carrier created is smaller than for other values of incident photons both sizes of such a preferential value. This would reduce the photoconductivity (in Stocker's theory the momentum of a free carrier is reduced). However, this does not seem to give a good account of the absorption maxima.

b) Song has given a theory of the multiphonon absorption. When the mass $m_h^* \gg m_e^*$ the coupling with LO phonons may become very strong, it is shown that the absorption in a continuum is strongly enhanced for photon energies equal to $E_{exc} + n\hbar\omega_{LO} = \hbar\omega_{photon}$. This theory is in good agreement with absorption measurements for CuCl.

Although the a) processes are related to photoconductivity experiments, it seems that in the case of CuCl, it is not necessary to explain the multiphonon structures by a transport mechanism : maxima of absorption coincide with the dips of photocurrent. In this region of wavelength, the photocarriers are absorbed in the vincinity of the surface and to an increase of the absorption coefficient corresponds a decrease of photocurrent by a De Vore process.

V. CONCLUSIONS

a) Three multiphonon structures of the photoresponse curve have been observed with CuCl at low temperatures. The phonon concerned is the LO phonon.

b) The dips observed in this curve are related to an exciton level E_{exc} by the relation $E_{ex} + n\hbar\omega_{LO}$. It has been possible to show that the exciton levels are essentially the n = 1, n = 2 and n = 3 levels of the diffuse series. They are named collector levels.

c) A multiphonon is observed in the absorption spectrum of CuCl. This structure corresponds to the structure of the photoresponse curve. The absorption maxima correspond to minima in the photocurrent.

d) Though multiphonon structures in the photoconductivity spectrum have been observed previously, it is believed that the observation of multiphonon structures simultaniously in absorption and photoconductivity is a new effect. It is likely that its observation is bound to the very high mass of the hole in CuCl.

REFERENCES

(1) M.A. Hakegyar and H.Y. Fan, Phys. Rev. Letters, 12, 99 (1964)
(2) D.N. Nasledov, Yu.G. Popov and Yu. S. Smetanikova, Soviet Phys. Solid State 6, 2989 (1965)
(3) Y.S. Park and D.W. Langer, Phys. Rev. Letters, 13, 392 (1964)
(4) H.B. De Vore, Phys. Rev. 102, 86 (1956)
(5) A. Coret, Ann. de Phys., 1, 673 (1966)
(6) E.F. Gross, I.Kh. Akopian, F.I. Kreingold, B.V. Novikov, R.A. Titov, R.I. Shekmanetiev, Proc. of the Intern. Conf. on Semiconductors, Paris, 957 (1964)
(7) I.C. Reynolds, C.W. Litton and T.C. Collins, Phys. Stat. Sol. 9, 671 (1965)
(8) H.I. Stocker, H. Levinstein and C.R. Stannord Phys. Rev. 150, 613 (1966)
H.I. Stocker, H. Kaplan, Phys. Rev., 150, 619 (1966)
(9) J. Ringeissen, S. Nikitine, J. de Phys. (to be published)
(10) K.S. Song, J. de Phys. (to be published)

METAMORPHISM OF VAN HOVE SINGULARITIES AND RESONANCE EFFECTS IN THE ABSORPTION SPECTRA

Makoto Okazaki

Department of Applied Physics, Faculty of Engineering, University of Tokyo, Tokyo

Masaharu Inoue

Department of Physics, Tokyo Metropolitan University, Setagaya-ku, Tokyo

Yutaka Toyozawa and Eiichi Hanamura

The Institute for Solid State Physics, University of Tokyo, Roppongi, Minato-ku, Tokyo

and

Teturo Inui

Faculty of Science and Engineering, Chuo University, Kasuga-1, Bunkyo-ku, Tokyo

1. INTRODUCTION AND PRELIMINARIES

Much information has been accumulated on the electronic structures of various types of solids through the detailed investigation of the optical absorption spectra.[1] It is now of advantage that one can get spectra in a wide range of energy with high resolutions.

In narrow gap semiconductors with large dielectric constants and large energy widths of conduction and valence bands, the absorption spectrum is dominated by the band-to-band continuum with Van Hove singularities.[1),2)] In molecular crystals in which the intermolecular interaction is weak and the band width is small, sharp excitonic peaks dominate.[3] Between the two above-mentioned limiting cases, we have intermediate crystals such as alkali halides[4] and rare gas crystals[5]: the exciton peaks are as strong as band-to-band continuum. Generally speaking, the fundamental absorption spectra have two aspects: the band character represented by Van Hove singularities and the local character represented by exciton and resonance peaks. In view of this situation it seems useful to study the complementary interplay of the local and band characters in the absorption spectra.

Similar situation is seen in the impurity-induced infrared absorption of lattice vibration.[6),7] The continuous spectrum of acoustic modes of host crystal is induced by the presence of impurities. In addition, there appear more or less sharp peaks which are ascribed to the local and quasi-local (embedded in the continuum) modes.

In this paper we report on calculated line shapes of the fundamental absorption and the impurity-induced infrared absorption of lattice vibrations for a simple model.[8] We mainly focus our attention to clarifying the complementary interplay of the local and band characters. The interplay appears in different ways for different values of the defect parameters (electron-hole Coulomb potential in the electronic case, the mass and the force constant of the impurity in the vibrational case). By varying the parameters, we can survey various types of solids.

Consider a monatomic simple cubic lattice with nearest neighbor interaction, and electron-hole relative motion in the matrix representation with the Wannier functions as basis. We divide the whole lattice points of the relative coordinate space into internal (i) and external (e) points, the former consisting of origin and several of its nearby points located symmetrically around it.

We decompose the hamiltonian H of electron-hole relative motion, which consists of kinetic energy K associated with a pair band and the electron-hole interaction V, into three parts $H^{(i)}$, $H^{(e)}$ and H': $H=H^{(i)}+H^{(e)}+H'$; $H^{(i)}$ has matrix elements associated with internal points (n,n') only, $H^{(e)}$ with external points only. H' connects the internal and external points. The shift-broadening matrix $\mathcal{G}$ is defined by

$$\mathcal{G}_{\alpha\alpha'}=\langle\alpha|H'\frac{1}{\hbar\omega-i\varepsilon-H_0}H'|\alpha'\rangle=\Delta_{\alpha\alpha'}+i\Gamma_{\alpha\alpha'}\qquad(\varepsilon\to+0)\tag{1}$$

where $|\alpha\rangle$'s -- local solutions -- are the eigenfunctions of $H^{(i)}$ which have amplitudes only on the internal points, and $\hbar\omega$ is the photon energy. By diagonalizing the energy matrix for the internal states $H^{(i)}+\mathcal{G}$ (including the interaction with external states) with the transformation T, we obtain the renormalized energy $\widetilde{E}_\lambda(\hbar\omega)+i\widetilde{\Gamma}_\lambda(\hbar\omega)$. Then we get the expression for the line shape:

$$I(\hbar\omega)=\sum_\lambda \widetilde{F}_\lambda(\hbar\omega)\frac{\widetilde{\Gamma}_\lambda(\hbar\omega)+\widetilde{A}_\lambda(\hbar\omega)[\hbar\omega-\widetilde{E}_\lambda(\hbar\omega)]}{[\hbar\omega-\widetilde{E}_\lambda(\hbar\omega)]^2+[\widetilde{\Gamma}_\lambda(\hbar\omega)]^2} \tag{2}$$

where $\widetilde{F}_\lambda$(effective intensity) and $\widetilde{A}_\lambda$(asymmetry parameter) are defined by

$$\left.\begin{array}{c}\widetilde{F}_\lambda\\ \widetilde{A}_\lambda\widetilde{F}_\lambda\end{array}\right\}=\left\{\begin{array}{c}\mathrm{Re}\\ \mathrm{Im}\end{array}\right.\sum_{\alpha\alpha'}\langle\mu|\alpha\rangle\, T_{\alpha\lambda}(T^{-1})_{\lambda\alpha'}\,\langle\alpha'|\mu\rangle\ . \tag{3}$$

The procedure is quite similar for the vibrational problems if appropriate replacements of the physical quantities are made, as will be shown in Sec. 3.

2. ELECTRONIC CASE

We assume that there is one simple pair band with nearest neighbor interaction J. The interband excitation energy is given by

$$\mathcal{E}(k)=\mathcal{E}_0-2|J|(\cos k_xa+\cos k_ya+\cos k_za). \tag{4}$$

$\mathcal{E}_0$ is the center of the pair band. We introduce dimensionless energy $x=(\hbar\omega-\mathcal{E}_0)/2|J|$. Electron-hole interaction potential $v_n=V_n/2|J|$ and the transition dipole moment μ_n are taken into account for the origin (n=0) and its nearest neighbors (n=1). These points are chosen as internal points. The valence and conduction bands are assumed to consist of the atomic p-states and s-state respectively. Then the optically allowed excited states are such that the envelope function of the relative motion has s-like or d_γ-like symmetry.

For the calculation of the s-component spectrum the shift-broadening matrix is two dimensional since there are two s-like states originating from the internal region. In the expression for the line shape we have three parameters v_0, v_1 and $p\equiv(2\mu_\sigma+4\mu_\pi)/6\mu_0$, μ_σ and μ_π are transition dipole moments corresponding to the electron transfer parallel and perpendicular to the dipole moments vectors, respectively. Calculation is made for the values of $v_0=0$, -0.4, -1.0, -3.0, -5.0, $v_1/v_0=1/2$, 1/6 and p=0.0, 0.3. Some of the results are shown in Fig.1. The points x=-3.0, -1.0, 1.0 and 3.0 correspond to the Van Hove singularities of the pair band state density, of the type M_0, M_1, M_2 and M_3, respectively. p=0 means that the transition occurs within an atom while in the case of $p\neq0$

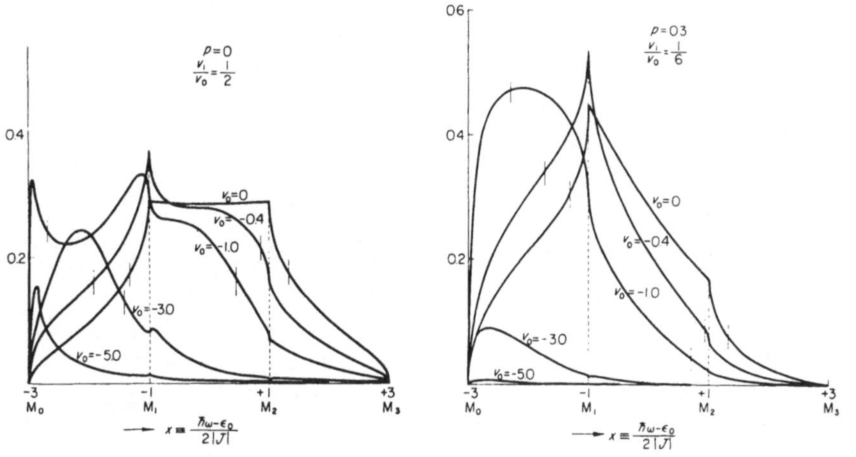

(a) (b)

Fig.1.Calculated intensity of the s-component of the fundamental absorption for the monatomic simple cubic lattice with nearest neighbor interaction in the cases of (a) p=0.0, $v_1/v_0=1/2$, (b) p=0.3, $v_1/v_0=1/6$. The quasi-localized levels are shown by the vertical lines on each absorption curve.

the transition probability depends on the wave vector k. A vertical line on each curve represents a root of the relation $\tilde{E}_\lambda(x)=x$: the real part of renormalized energy is equal to the energy considered. We call the root a quasi-localized level, although it may not necessarily appear as a prominent peak.

Here we discuss the calculated absorption spectra, taking notice of coexistence of the local and band aspects. Consider the result of the case of p=0, $v_1/v_0=1/2$. For $v_0=0$, the spectrum represents an s-component of the density of states. As one deepens v_0, the singularity at M_1 mutates to upward cusp, reversed-S type, downward cusp and so on as shown in Fig.2. We call the mutation of Van Hove singularities metamorphism. Also the singularity at M_2 mutates in the same direction as the one at M_1.

Let us get a deeper insight into the origin of metamorphism. The unrenormalized broadening function $\Gamma_{\alpha\alpha}$, that is the imaginary part of $\mathcal{G}_{\alpha\alpha}$ has singularities of the same form as those of the state density. The unrenormalized shift function $\Delta_{\alpha\alpha}$, which is related to $\Gamma_{\alpha\alpha}$ by Kramers-Kronig relation, has singularity on the opposite side of each critical energy. As the result of transformation T, the shift and broadening parts are mixed with each other, and therefore the renormalized quantities may have singularities on the both sides of the critical energies. The same is true for the line shape function, which is given by Eq.(2) in terms of the

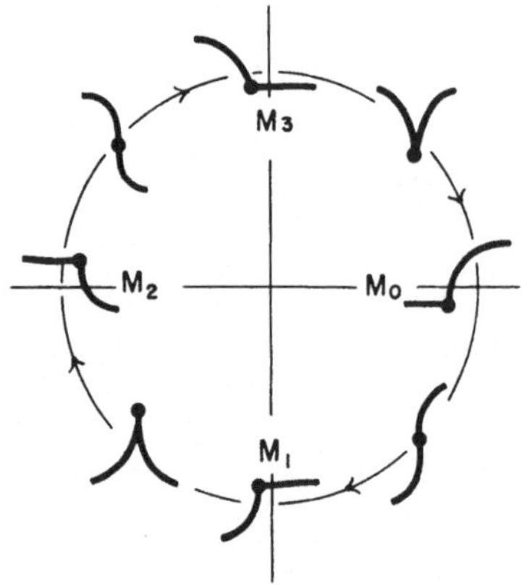

Fig.2. Metamorphism of Van Hove singularities. As one applies and increases an attractive potential between electron and hole, the shape of the singularity corresponding to any of the four types (M_0,M_1,M_2 and M_3) of the critical points changes clockwise, starting from its original shape.

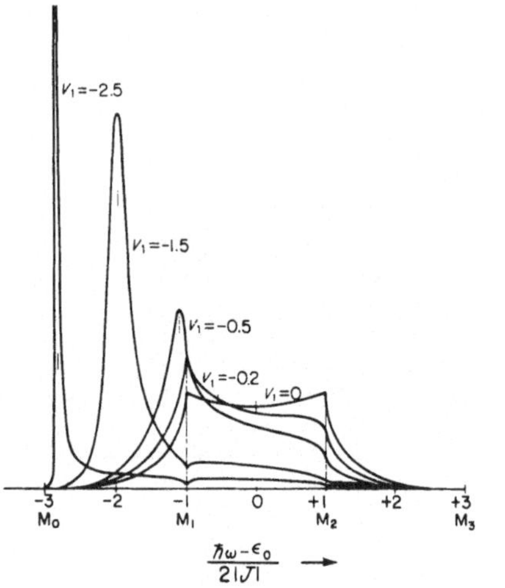

Fig.3. Calculated intensity of the d-component of the fundamental absorption for the monatomic simple cubic lattice with nearest neighbor interaction. The quasi-localized levels are shown by the vertical lines on each absorption curve.

renormalized quantities.

Metamorphism of Van Hove singularities can be considered as change of the band aspect in the spectrum due to the coexisting local characters. On the other hand, the local aspect shows itself in various resonant structures depending upon the strength of the interplay with the band characters. Resonance peak (v_0=-3.0,(a)), sharp asymmetric peak (v_0=-1.0,(a)), unusual bell-shape (v_0=-1.0, (b)) and the tail of a true bound state (v_0=-5.0,(a) and v_0=-3.0, (b)) are seen in the calculated spectra. Which of these structures appear is determined by, roughly speaking, the density of states at the quasi-localized level. For example, the condition of the appearance of the asymmetric peak near the band edge is critical as mentioned below. The absorption intensity is approximately given by inverse of $\tilde{\Gamma}_\lambda$, when the energy coincides with the quasi-localized state. Near the continuum edge, where $\tilde{\Gamma}_\lambda$ changes rapidly with energy, this situation is realized to an appreciable extent of the energy range. Thus, in Fig.1 where the quasi-localized level is located at x=-2.7 for v_0=-1.0 we get an asymmetric peak nearly proportional to $1/\sqrt{x}$.

Remarkable feature appears in the case of p=0.3 for v_0=0, that is almost two-sided cusp is seen even if there is no Coulomb potential. This is understood from the fact that the lower energy part has preponderance in overall spectrum since the transition dipole moment is larger for smaller k-values. In view of this possibility it is not necessary to have the accidental proximity of the energies of types M_1 and M_2 singularities which is proposed in the interpretation of ultraviolet spectra of Si and Ge.[2] In GaAs[9] and CdTe,[10] the similar shapes are seen in the reflectivity spectra; they are assigned to band-to-band transitions by Greenaway and Cardona. The unusual bell-shape peak may remind us of one of the structures in the absorption spectra of solid Xe[5] and KI.[4]

In the case of dγ-spectrum, more sharp resonance line appears below M_1, since the dγ-component of the state density rapidly decreases and tends to zero as 3/2 powers of the energy measured from M_0. The results are shown in Fig.3. The resonance peak below M_1 is to be associated with the local structure rather than with the particular interband edge M_1.

3. VIBRATIONAL CASE

Consider a lattice of the host atom with mass M and scalar force constant between nearest neighbors λ. The angular frequency ω_k, which is triply degenerate, is given by

$$\omega_k^2=\frac{2\lambda}{M}(3-\cos k_x a-\cos k_y a-\cos k_z a) \quad . \tag{5}$$

It is convenient to introduce the quantity $x=M\omega^2/2-3$.

When an atom of the host crystal is replaced by a foreign ion, impurity-induced infrared absorption of lattice vibrations will be observed. Consider an impurity at origin with mass M' which interacts with the adjacent host atoms through the scalar force constant λ'. The force constant between any host atoms is assumed to be unchanged by the substitution. If we assume that the induced effective charge e appears only on the impurity atom, we have only to consider those modes which have s-like envelope function. Absorption line shapes can be obtained in the same manner as in the electronic case if $\hbar\omega$, H and μ are replaced by ω^2, dynamical matrix Ω and $e/\sqrt{M'}$. Calculation was made for various values of λ'/λ and M/M'. Correspondence relations between the defect parameters in the electronic and vibrational cases are given by

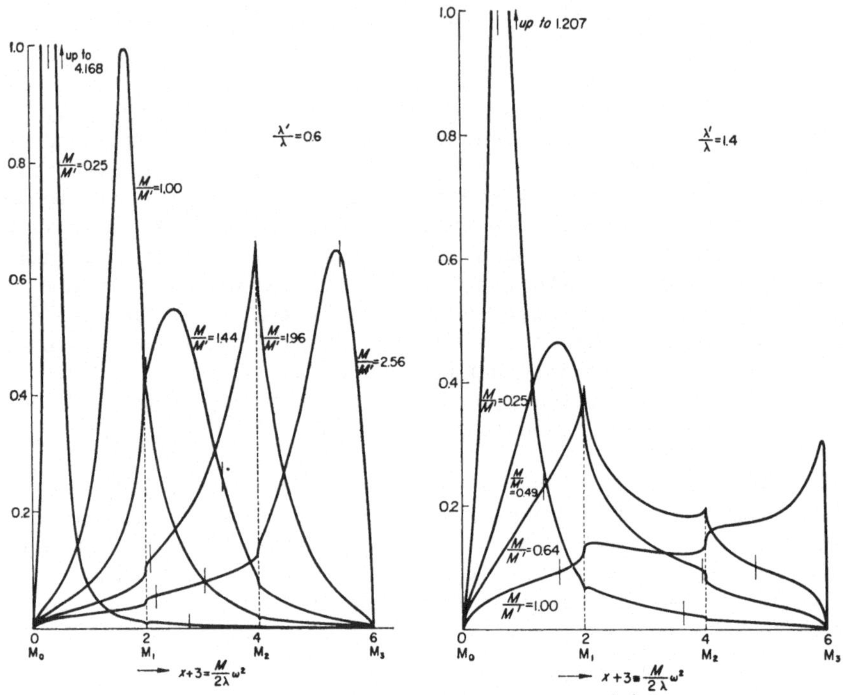

(a) (b)

Fig.4. Calculated intensity of the impurity-induced infrared absorption of lattice vibrations for the monatomic simple cubic lattice with nearest neighbor interaction in the cases of (a) $\lambda'/\lambda=0.6$ and (b) $\lambda'/\lambda=1.4$. The quasi-localized levels are shown by the vertical lines on each absorption curve.

$$v_0=3\left(\frac{\lambda' M}{\lambda M'}-1\right) \quad , \quad v_1=\frac{1}{2}\left(\frac{\lambda'}{\lambda}-1\right) \quad . \tag{6}$$

As examples, the calculated absorption spectra in the cases of $\lambda'/\lambda=0.6$ and 1.4 are shown in Fig.4. Metamorphism of Van Hove singularities as the defect parameter changes is seen at both M_1 and M_2. With increase in M' or decrease in λ' the metamorphism takes place in the direction indicated in Fig.2, as is easily seen from Eqs.(6). As regards resonance structures, there is a trend that sharp resonance can be observed for the case of smaller values of λ'/λ and M/M'. Asymmetric peak near the higher frequency edge is seen for $\lambda'/\lambda=1.4$ and M/M'=1.0. Generally speaking, it is easier to observe the quasi-localized states in the spectra of the lattice vibrations than in the electronic case.

Up to the present we have not much experimental data of the

quasi-local mode of the impurity-induced lattice vibration. In potassium bromide and iodide activated with Ag,[11] Li[12] or NO_2^-,[13] infrared absorption by the quasi-local modes are observed. Furthermore, even less observations are available, in which particular attention was paid to the modification of the singularities by the impurity. To our knowledge, we might say that an experimental evidence of metamorphism is seen in the absorption spectrum of B and P doped Si by Angress, Goodwin and Smith.[14] The reversed-S type singularity is seen at 0.0528eV, which corresponds to M_1 type singularity of the longitudinal optical phonon at point L. The observed metamorphism is considered to be consistent with our calculation for the simplified model with reasonable values of impurity mass and force constant.

REFERENCES

1) J. C. Phillips, in Solid State Physics, edited by F. Seitz and D. Turnbull (Academic Press Inc., New York, 1966), Vol.18.
2) D. Brust, Phys. Rev. 134, A1337 (1964); E. O. Kane, Phys. Rev. 146, 558 (1966).
3) D. S. McClure, in Solid State Physics, edited by F. Seitz and D. Turnbull (Academic Press Inc., New York, 1959), Vol.8; H. C. Wolf, in Solid State Physics, edited by F. Seitz and D. Turnbull (Academic Press Inc., New York, 1959), Vol.9.
4) J. E. Eby, K. J. Teegarden and D. B. Dutton, Phys. Rev. 116, 1099 (1959); K. Teegarden and G. Baldini, Phys. Rev. 155, 896 (1967); D. M. Roessler and W. C. Walker, J. Opt. Soc. Am. 57, 677 (1967).
5) G. Baldini, Phys. Rev. 128, 1562 (1962).
6) M. Balkanski, Proc. Int. Conf. Phys. Semicond. (Dunod, Paris, 1964) p.1021; M. Lax and E. Burstein, Phys. Rev. 97 , 39 (1955).
7) A. J. Sievers, A. A. Maradudin and S. S. Jaswal, Phys. Rev. 138, A272 (1965).
8) M. Okazaki, M. Inoue, Y. Toyozawa, T. Inui and E. Hanamura, J. Phys. Soc. Japan, 22, 1349 (1967).
9) D. L. Greenaway, Phys. Rev. Letters 9, 97 (1962).
10) M. Cardona and D. L.Greenaway, Phys. Rev. 131, 98 (1963).
11) A. J. Sievers, Phys. Rev. Letters 13, 310 (1964).
12) A. J. Sievers and S. Takeno, Phys. Rev.140 , A1030 (1965).
13) A.J.Sievers and C. D. Lytle, Phys. Letters 14, 271 (1965); K. F. Renk, Phys. Letters 14, 281 (1965).
14) J. F. Angress, A. R. Goodwin and S. D. Smith, Proc. Roy. Soc. A287, 64 (1965).

INFRARED ABSORPTION AND VIBRONIC STRUCTURE DUE TO LOCALIZED POLARONS IN THE SILVER HALIDES*†

Richard C. Brandt‡ and Frederick C. Brown

Department of Physics and Materials Research Laboratory

University of Illinois, Urbana, Illinois

The theoretical and experimental investigations of a new type of localized excitation in an ionic crystal, specifically a polaron bound to a positive charge center, are the subject of this paper. Although the exact nature of the positive charge center is currently unknown, there are indications that it is an intrinsic defect, for example, a self trapped hole. If this is the case, the excitation is an exciton, and our results challenge some commonly held ideas concerning the basic nature of the exciton. More specifically we have found that the Born-Oppenheimer approximation is not adequate for a description of such transitions in the silver halides, and that polaron effects must be explicitly included.

The experiments were carried out on high purity silver halide crystals. The silver halides are basically very similar to the alkali halides. They differ in that their indirect band gap is close to the visible, their static dielectric constants are higher than those of the alkali halides, and their measured polaron masses, .33 m_e for AgBr and .51 m_e for AgCl[1] are in general smaller than the polaron masses in the alkali halides.

In the experiments on the silver halides we looked for infrared absorption over a range from 33 to 2000 cm^{-1} due to transitions between excited states of the crystal. The phrase between excited states should be emphasized because normally the electronic excitations of ionic crystals are studied by examining transitions between the ground state of the crystal and its excited states. The range of the search for excited state absorption was restricted to the infrared for several reasons. First, previous investigations on very pure crystals had shown little or no color center like absorption in the visible. Second, transport experiments had revealed the existence of shallow electron traps which thermally ionize a few degrees above the temperature of liquid helium. Third, when the Bohr radius of an electron in silver bromide is calculated one obtains a radius of 7.5 Å using the optical dielectric constant (ϵ_∞=4.62) and a radius of 17 Å using the static dielectric constant (ϵ_s=10.6). These large orbits, especially the one corresponding to the static constant, give the lattice enough room to both screen the positive charge and to follow the electron's instantaneous position. In this case the electron would be a polaron, and the lattice screening would make it reasonable to employ the static dielectric constant for calculating the energy levels. The Bohr 1s-2p transition in AgBr, using the polaron mass of .33 is then 242 cm^{-1}, well into the infrared.

To observe the absorption due to transitions between excited electronic states of the crystal, an infrared spectrophotometer was used according to the following procedure. First, the optical density of a crystal that had been cooled in the dark to liquid helium temperature was measured. The crystal's optical density was then remeasured, while it was simultaneously being illuminated with light whose energy exceeds the indirect band gap. The difference between this second optical density and the first optical density is the absorption due to transitions between excited states of the crystal.

Figure 1 shows the induced optical density, that is the absorption due to transitions between excited states, over a wide range of frequency and optical density in silver bromide. To show some of the weaker structure, the results were plotted on log log paper. On the low frequency side there is a narrow zero phonon line at 168 cm^{-1} and some electronic transitions of somewhat higher energy. At higher energies there are a number of phonon side bands, identified by the numbers 1 through 6. Figure 2 shows the absorption around the zero phonon lines in silver bromide. At the lowest energy there is the strong narrow line which we will show can be identified with a 1s to 2p transition. This is followed by a weak and not very well resolved line which could correspond to 1s to 3p transitions and a relatively broad 1s to continuum region. The first prominent phonon side band is believed to represent mainly the emission of longitudinal acoustic phonons near the edge of the zone boundary. The second one phonon side band, which is considerably larger, is due to the emission of longitudinal optical phonons. Induced absorption and a zero phonon line at 272 cm^{-1} was observed in AgCl although this absorption was smaller and harder to work with than AgBr.

Three aspects of the absorption were investigated theoretically. First, the position of the zero phonon lines; second, the

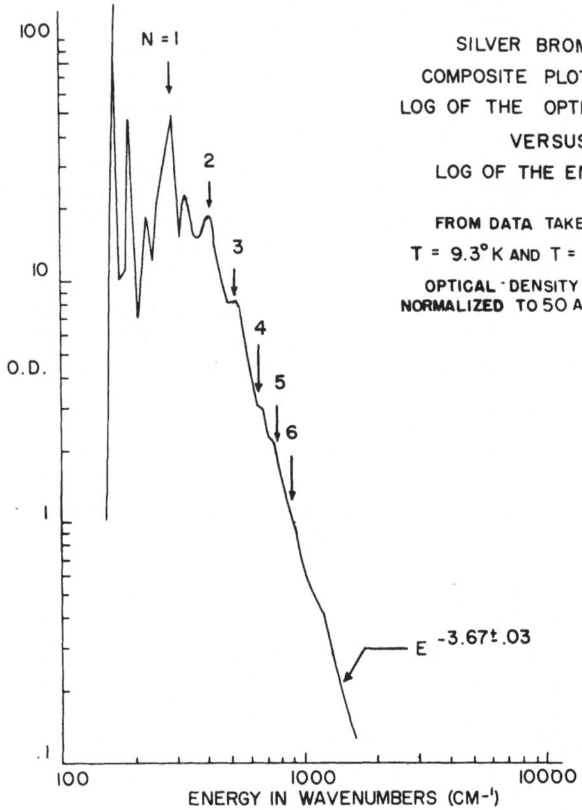

Fig. 1. Induced absorption in AgBr from 160 to 1200 cm^{-1}.

position, shape, and amplitude of the phonon side band absorption; and third, the kinetics of the excitations, in particular their decay processes and their dependence on the intensity of the exciting light. In relation to the first of these aspects the non-translationally invariant theory of the polaron due to V. Buimistrov and S. Pekar(2) was used to predict the position of the induced absorption lines. These workers employed a variational version of the small oscillation theory of the polaron. They also developed a translationally invariant variational theory suitable for application to the free polaron(3). In order to determine the accuracy of this approach the self energy of the polaron was numerically evaluated using the translationally invariant theory for a range of coupling constants. All integrations were carried out by means of a high speed digital computer rather than by the approximations used in the original work. The polaron self energy derived in this manner was found to be better, i.e. lower, than from any other polaron theory for coupling constants less than six. This result is significant since the silver halides have coupling

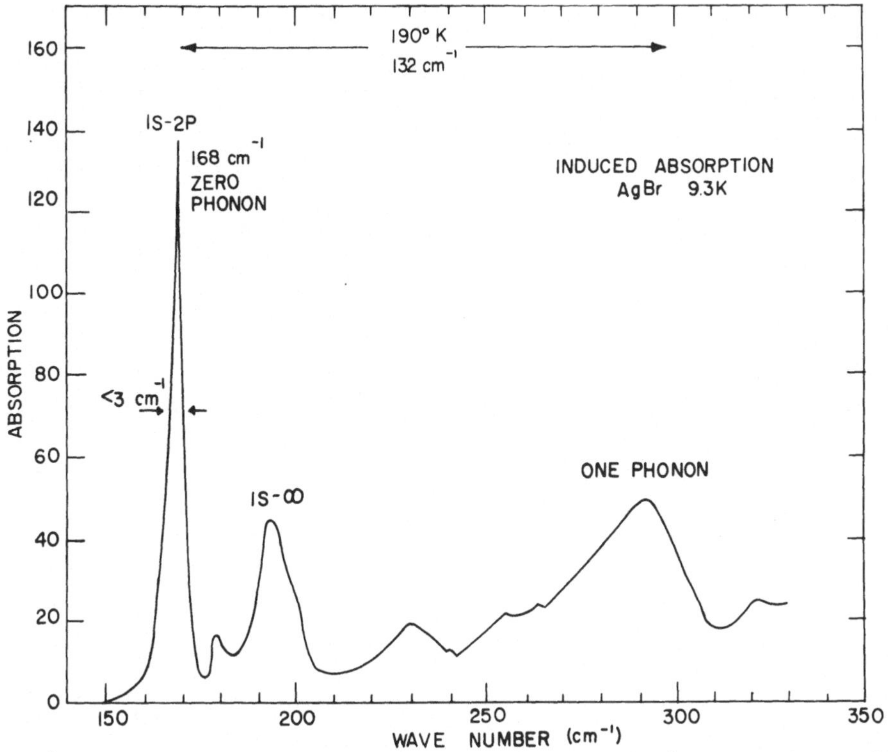

Fig.2. Induced absorption in AgBr around the 1s-2p zero phonon line.

constants about 2 and the alkali halides have constants in the range 3 to 5.[1]

The bound theory results for a hydrogenic variational wave function for silver bromide are shown in Figure 3. The transition energies were calculated as a function of reduced mass for three different kinds of positive charge distributions. The solid lines show the results for a positive point charge, the dashed lines the results for a diffuse positive charge, and the dotted line for what is referred to as the symmetrized V_k center. The last charge distribution consists of an inner sphere of negative charge surrounded by a shell of positive charge (there is twice as much positive charge as negative charge, so that the overall charge of the center is positive). In all of these models, the static dielectric constant was used and polaron effects were explicitly included. If the optical dielectric constant and the fixed lattice or bare mass had been used without including polaron effects, the predicted energies would have been at least five times larger than shown in Figure 4. We now adopt a criterion that the best potential is that potential for which a single reduced mass fits both the observed transitions and the difference between these transitions i.e. the

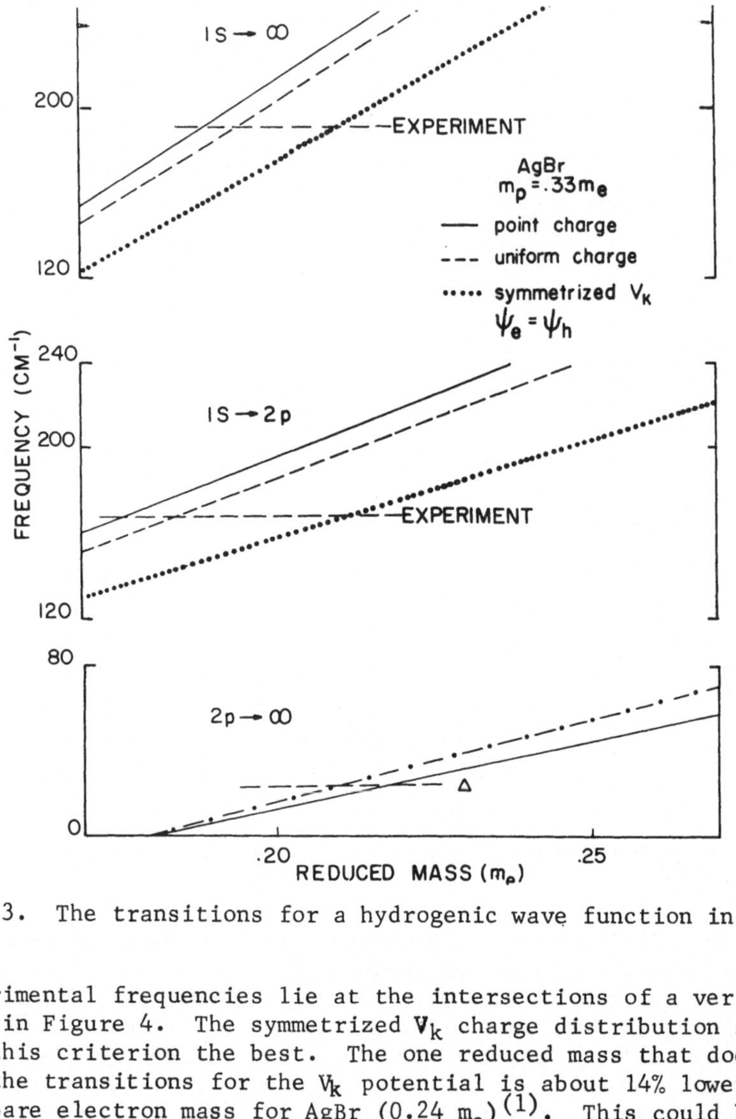

Fig. 3. The transitions for a hydrogenic wave function in AgBr.

experimental frequencies lie at the intersections of a vertical line in Figure 4. The symmetrized V_k charge distribution seems to fit this criterion the best. The one reduced mass that does fit all the transitions for the V_k potential is about 14% lower than the bare electron mass for AgBr (0.24 m_e)[(1)]. This could be taken to imply that the positive charge distribution has a finite rather than infinite mass compared to the electron, which would require it to be a hole. On the other hand, it might be related to the approximate nature of the positive potential. A similar comparison between experiment and theory can be made for AgCl.

Some information as to the nature of the coupling between the lattice and the excitation was gained by examining in detail the phonon side bands shown in Figure 4. The solid line shows the experimental results where the frequency is measured relative to the strongest pure electronic transition, the 1s to 2p. Let us

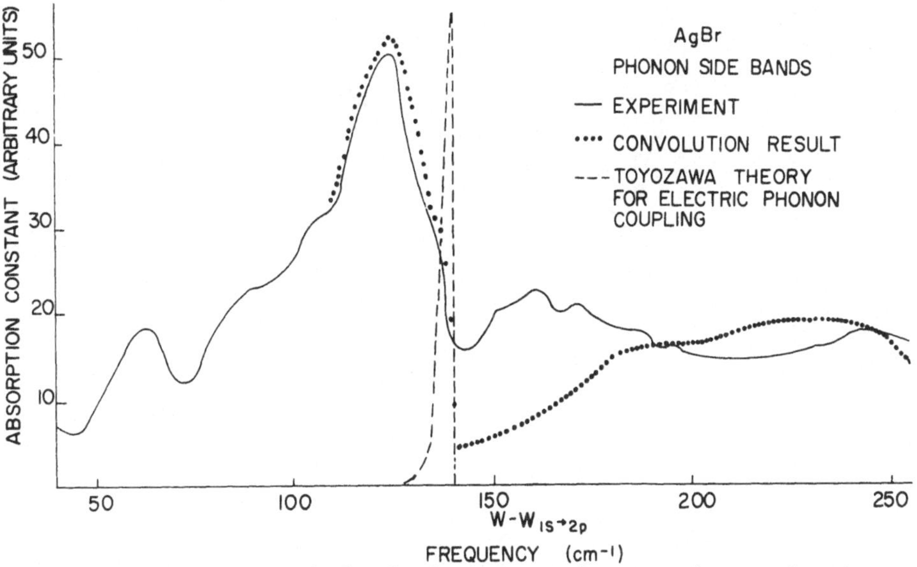

Fig. 4. The phonon side bands in AgBr: theory and experiment.

consider what kind of side bands we would expect if the coupling to the lattice were through the electron phonon coupling only. The electron states are quite diffuse, for instance for silver bromide using the Buimistrov-Pekar theory the 1s radius is 39 Å. Such wave functions will project only onto the central part of wavevector space and hence will only interact with phonons of small wavevector or equivalently long wavelength. Toyozawa's theory(4) for the coupling of a bound polaron to the lattice phonons was used to obtain a description of this phenomena. The shape of the one phonon side band obtained from this theory and appropriate to our bound polarons are shown by the dashed line. The experimentally observed lines are, on the other hand, quite broad. There is evidence here that the coupling to the lattice may be through changes in the electron density in the vicinity of the well localized positive charge center, although the actual mechanism is not yet understood.

Figure 5 shows the experimental results for the time decay of the induced optical density after the exciting light has been turned off. Here the absorption depends on the log of the time. In the figure this is shown by having the time on the vertical log scale and the induced absorption on the horizontal linear scale. Normally, decay plots are plotted in the opposite way, that is, the absorption normally depends on the time exponentially instead of vice versa. This is always true for monomolecular decay, since for monomolecular decay, the decay rate is proportional to the number left, giving a differential equation with only an exponential solution. Here the decay is certainly not a monomolecular process.

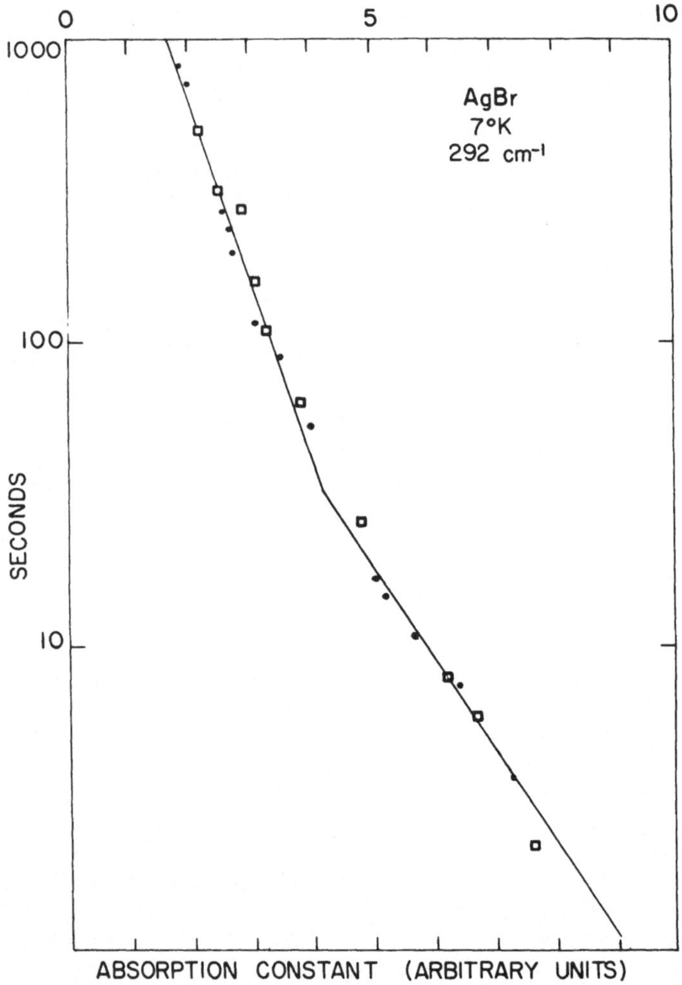

Fig. 5. The time decay of the induced absorption in AgBr at 292 cm^{-1} at 7°K.

If we assume we are observing triplet exciton decay, we must consider the effects of large hydrogen like orbits. We exclude singlet decay because it is allowed and is known to have a half life of about 40 microseconds for silver bromide at low temperatures.[5] For triplet states (also for separated donor-acceptor states) the decay could occur through the overlap of spread-out hydrogen-like wavefunctions. Where the radius of the orbits is a_0 and the separation of the exciton centers is R, this overlap is proportional to e^{-R/a_0} or a power of this function. Thomas, Hopfield, and Augustyniak[6] studied a similar case involving large orbit states in GaP, where the overlap integral had the same exponential dependence. They found experimentally, that the time decay of the emission depended on the inverse of the time, that is, the time derivative of the log of time. Thus our results are consistent with theirs.

In conclusion we make a few remarks on the nature of the center formed by band to band light in these crystals. Although the details of the observed side band spectra are not entirely understood it does appear that we are dealing with an electron polaron bound to a positive charge center. One is forced to this conclusion on the basis of the very small binding energies and large orbital radii involved. At least two models are consistent with the observed generation and decay phenomena. These are a triplet exciton model and a model consisting of separated donor and acceptor pairs. We favor the first model for two reasons: first, the application of Smakula's equation to the induced absorption shows that there are far more absorption centers and hence more positive charge centers in the crystal than can be accounted for by the impurity content revealed either through mass spectroscopy or by the analysis of low temperature transport experiments. Secondly, there is also some agreement between the energy levels seen here in the infrared and certain detailed structure seen in the indirect absorption edge in the ultraviolet.(7) This agreement strongly indicates that polaron effects should be included in the exciton problem and in fact these are apparent in the absorption spectrum. Exciton binding energies in the silver halides are obtained by the use of the static rather than optical dielectric constants and the polaron masses are more appropriate than the band masses. A similar trend has been noted in the case of the alkali halides.(1)

REFERENCES

*Supported in part by ARPA under Contract No. SD-131 and the U. S. Army Research Office (Durham) Contract No. ARP-217.
†This paper is based on the Univ. of Illinois Ph.D. thesis of RCB. A more detailed discussion will appear elsewhere.
‡Present address: Lincoln Laboratory, Lexington, Massachusetts.
1. J. Hodby, J. Borders, F. C. Brown, and S. Foner, Phys. Rev. Letters, to be published.
2. V. M. Buimistrov and S. I. Pekar, Zh ETF 32, 1193 (1957); [Soviet Phys.-JETP 5, 970 (1957)].
3. V. M. Buimistrov and S. I. Pekar, Zh ETF 33, 1271 (1957); [Soviet Phys.-JETP 6, 977 (1958)].
4. Y. Toyozawa, "Multiphonon Structures in the Absorption Spectra of Localized Electrons in Solids," Technical Report No. 238, Institute of Solid State Physics, Tokyo 1967.
5. G. C. Smith, Phys. Rev. 140, A221 (1965).
6. D. G. Thomas, J. J. Hopfield, and W. M. Augustyniak, Phys. Rev. 140A, 202 (1965).
7. F. C. Brown, T. Masumi and H. H. Tippins, J. Phys. Chem. Solids 22, 101 (1961); see also B. L. Joesten and F. C. Brown, Phys. Rev. 148, 919 (1966).

EXCITATION TRANSFER IN RADIATION-DAMAGED ORGANIC SOLIDS

Herbert B. Rosenstock and James H. Schulman

Naval Research Laboratory, Washington, D. C.

I. Introduction

The purpose of this paper is to point out that radiation damage experiments can be a useful tool towards determining the mechanism of energy transfer in organic solids. The basic approach of the study is to relate the luminescent yield of an organic solid to the previously applied dose of "damaging" radiation.

The substance for which we report quantitative results is solid anthracene; each lattice site is occupied by an anthracene molecule, which consists of three benzene rings. However, other organic solids have similar properties and have also been used in experiments as described here. The basic property of interest is the ability of such solids to luminesce upon irradiation by ultraviolet radiation. The luminescent yield is decreased if, prior to the luminescence experiment, the crystal is damaged by high energy radiation (which might be gamma rays or Mev range electron radiation).[1] Let I be the luminescent yield-that is, the number of luminescent photons emitted per incident ultra-violet photon, and let D be the damaging dose of high energy radiation to which the crystal has been subjected prior to the luminescence experiment; it is then found[2] that

$$I(D)/I(0) = [1 + D_{1/2}^{-1} D]^{-1} \qquad (1)$$

Here $D_{1/2}$ is a constant, to wit the dose needed to reduce the luminescent yield by a factor of 2. The most important fact about this experiment, aside from the perhaps unexpected form of the observed relation (1), is the very small value of $D_{1/2}$; that is, in radiation damage experiments of a different nature, using indicators other than luminescence as a tool for determining the amount of damage produced[3], a much larger dose of high energy radiation is needed to produce substantial damage. It is this fact that primarily needs to be explained in terms of energy transfer mechanisms.

II. Models for Energy Transfer

We now consider four physical models in order to explain the experimental results (1). We shall find that only one of these four models is satisfactory. This does not prove this model to correspond to reality, but does strongly suggest its validity since the other three seem to exhaust possible models that are both simple and intrinsically reasonable.

The statement, at the end of the last section, that $D_{1/2}$ is surprisingly small, actually already implies a very simple and obvious model for the radiation damage and luminescence, viz., one involving no energy transfer at all. Most simply, luminescence can be thought of as a property determined strictly by one individual molecule. The incoming ultraviolet photon is absorbed by one molecule in the crystal, it stays there, and is reemitted as luminescence, after a Stokes shift perhaps, some time later. This is what happens when the molecule which does the absorbing is intact; if on the other hand that particular molecule is a previously damaged one, then luminescence does not take place and the photon's energy is instead dissipated in the form of heat. We have already stated why this simple, completely localized model is inadequate: a much larger decrease in the luminescent yield is observed due to a small damaging dose then would be expected from other radiation damage experiments. Furthermore, the form of equation (1) cannot be reproduced. We must therefore find a mechanism which, first of all, will transmit the effect of radiation damage some distance away to molecules that are not directly damaged.

A simple way of doing this is to assume that all molecules within a radius R of a damaged molecule are somehow "poisoned", so that all the molecules within that radius R, and not only the single molecule that is actually damaged, are unable to luminesce. This leads to an interesting problem involving "probabilities of coverage". When the damaging dose is very small, the probability that two spheres of radius R centered about arbitrary points in the crystal will overlap is negligibly small, and the decrease in luminescent yield will be proportional to the damaging dose; but when the damaging dose becomes larger, overlaps of such spheres will become more probable and the yield-to-damaging-dose relationship will deviate from a linear one, and as triple and quadruple overlaps become more probable, all overlaps will finally have to be taken into account. The "probability of coverage" problem can be solved in the mean (that is, when the points at which the damaged molecules are located are suitably averaged over, as is physically realistic)[4], and the result found is that the luminescent yield should be given by

$$I(D)/I(0) = e^{-cD} \tag{2}$$

The decrease of luminescence as a function of damaging dose can now be made as large as one wants, by properly adjusting the constant c which is related to the volume of the "poisoning" sphere whose actual size is unknown; but the functional relationship (2) is in clear-cut disagreement with the experimental results (1).

A third frequently used model treats the excitation energy as a traveling exciton, or wave packet, propagating along a straight line in the crystal. One would then expect to observe luminescence only if the decay time for the luminescence is smaller than the time needed for the travelling wave packet to hit the first damaged site; if a damaged site is arrived at before luminescence has taken place, then the exciton will be absorbed at the damage site and luminescence will not be seen. This model again enables us to get an arbitrarily great amount of quenching of luminescence by even a small damaging dose, depending on the velocity of the motion of the excitonic package, but fails to reproduce the correct form of the yield to dose relationship: one would clearly expect a dependence not on the first power of the damaging dose but upon $D^{1/3}$, as the linear density of damaged molecules will be proportional to the cube root of the number of damaged molecules, or the cube root of the damaging dose.

We therefore consider the following model for excitation transfer. The ultraviolet photon is absorbed at one arbitrary lattice site, it stays there for a while and then proceeds to one of the nearest neighboring lattice sites, stays there and after a while goes on to another nearest neighbor, etc., thus performing a random walk. There are two ways of ending such a random walk: luminescent emission, and subsequent observation by the experimenter, or absorption and quenching by a damaged site; which of these two possibilities will actually occur in one particular case will depend on whether a damaged site is stepped upon before or after luminescence takes place. We are thus led to a random walk problem, and compelled to determine the probability of emission prior to the stepping onto a damaged site, for an arbitrary concentration of damaged sites. The analytical work has now been performed by Rudemo[5], and the result is

$$I(q)/I(0) = [1 + (1 - F)q\alpha^{-1}]^{-1} \tag{3}$$

Here α is the probability of luminescence per step, q is the fractional number of damaged sites in the lattice, and F is a quantity which depends on the lattice and the unrestricted random walk problem that we are considering: it is the probability that a particle performing such a random walk (in absence of both emission and absorption) return to its origin at least once. This is a known number, approximately 0.34 in case of the simple cubic lattice[6]. Equation (3) is seen to be in complete agreement, in form, with the experimental result (1), and indeed some of the physical quantities can be numerically estimated by comparing the two relations.

III. Results

Numerically, $D_{1/2}$ was found[2] to be .067 ev/molecule, and other experiments, using gas evolution rather than luminescence as a criterion for the number of molecules damaged[3] yield an estimate of 10^{-3} molecules damaged per ev for q/D. Equating the denominators of (1) and (3) then gives

$$1/\alpha = 2.25 \times 10^4 \;;$$

this is the expected number N of steps in a walk (in an undamaged crystal), for $N\alpha$ must be 1. The range of a random walk is proportional the square root of the number of steps; just prior to luminescence, the exciton can therefore be expected to be about 150 lattice distances, or, 900 Angstroms from the starting point. Since the luminescence has a time constant[2] of $3\text{x}10^{-8}$ seconds, the time spent on one site is about $1.3\text{x}10^{-12}$ seconds. In that time, the exciton averages 1 jump or 6A; the instantaneous (random) velocity is thus $4.6\text{x}10^4$. Alternatively, one can compute a mean velocity of 900A in $3\text{x}10^{-8}$ sec, or 300 cm/sec.

In summary, we have found that the results of radiation damage experiments in organic solids are inconsistent with both strict localization of the excitation on one molecule, and with linear excitonic propagation in form of a wave packet. A random walk model for excitonic motion, on the other hand, is consistent with experimental data and leads to localization of the excitation within about 1000 Angstroms of its origin. Since electromagnetic interaction among lattice sites is known to produce excited electrons, or excitons, that travel linearly with a definite momentum, our contrary result implies that this interaction is weak, and that the hopping motion is caused by another mechanism, presumably thermal activation. The fact that the time spent on one site, 10^{-12} seconds, is of the order of the period of one lattice vibration is also consistent with this view.

Discussions with Dr. Michael H. Reilly are gratefully acknowledged.

REFERENCES

1. We must clearly distinguish between the ultraviolet radiation that produces the luminescence, and the previously applied high-energy radiation that damages the crystal and thereby reduces the luminescent yield.

2. Schulman, Etzel and Allard, J. Appl. Physics 28, 792(1957).

3. M. Burton, ONR Symposium Report ACR-2 (Office of Naval Research, Washington, D. C., 1954), page 1.

4. H. B. Rosenstock and J. H. Schulman, J. Chem. Physics 30, 116(1959).

5. M. Rudemo, SIAM J. Appl. Math. 14, 1293(1966).

6. W. Feller, "Introduction to Probability Theory and its Applications", Vol. I, 2nd edition (Wiley, New York, 1957) p. 327.

PART E. LOCALIZED MAGNONS AND INTERACTIONS OF SPINS WITH LOCALIZED EXCITATIONS

LOCALIZED MAGNONS (THEORY)

Daniel Hone*

Physics Department, University of California

Santa Barbara, California

I. INTRODUCTION

The Heisenberg Hamiltonian for impure ferro - or antiferromagnets is of the class originally discussed by Lifshitz(1) in his work on the properties of the energy spectra of disordered systems. Within the simple (non-interacting) spin wave approximation, applicable in the low temperature limit, the problem is closely analogous to that of phonons in impure lattices, to which the general theory was first applied. One expects (and finds) for appropriate impurity parameters true localized modes at energies outside the pure crystal spin wave band and virtual modes (or "resonances"), with finite lifetimes, at energies within the band. However, even in this simple picture, there is no analog to the most easily treated disordered lattice - with a single isotopic impurity. The spin magnitude enters dynamically in the exchange interaction and the "force constant" associated with a magnetic impurity must therefore be changed along with its "mass." In general, the substitutional impurity may differ from the atom it replaces in the exchange constants coupling it to its neighbors as well as in spin magnitude. We shall review here some of the work which has been done in the recent active study of the effects of such impurities in Heisenberg ferro- and antiferromagnets, covering situations in which the substitutional impurity is exchange coupled to its neighbors both with the same and with the opposite sign as would be the atom which it replaces.

In Sec. II we discuss the model and approximations used to treat these systems. The necessary Green's function formalism is set up in Sec. III and results for the spectra are given in Sec. IV. The final section deals with the predicted consequences of the localized modes on inelastic neutron scattering, magnetization, and specific heat experiments.

II. MODEL

We consider a system of spins localized at fixed lattice sites and coupled pairwise by an isotropic exchange interaction. A low concentration, c, of impurity ions is distributed randomly throughout the crystal. In certain cases we wish to include the effects of a uniaxial anisotropy field (taken, for simplicity and with reasonable generality to be proportional to the magnetization at each site) and of an external magnetic field, H_o, along the anisotropy axis. In general, then, we will deal with the hamiltonian

$$H = -\mu_\beta H_o \sum g_i S_i^z - \sum K_i \langle S_i^z \rangle S_i^z - \sum J_{i\ell} \vec{S}_i \cdot \vec{S}_\ell \qquad (2.1)$$

where the subscripts label lattice sites, the spectroscopic splitting factor g_i is equal to g for a host atom and g' for an impurity, and the anisotropy constant K_i takes on the values K and K' in these two possible cases. This hamiltonian is, of course, strictly applicable only to magnetic insulators. However, there is evidence[2] that, at least outside the very low temperature region, it may be reasonable to treat a metallic ferromagnet as a collection of localized moments, with the conduction electrons merely providing a mechanism for an effective exchange interaction. To the extent that this model is suitable the consequences of the hamiltonian (2.1) may usefully be compared with experimental results in metals.

For reasons of algebraic simplicity calculations have been carried out (e.g., Refs. 4-10) with non-vanishing exchange integrals $J_{i\ell}$ between neighboring spins only. With the restriction to dilute impurity concentrations this requires specification of only two isotropic exchange parameters – J, coupling two host spins, and J', coupling an impurity to each of its (host) near neighbors. In addition, simple lattices are chosen for explicit calculations. Various pure crystal spin wave propagators, or Green's functions, are required for discussion of the impure systems, and these have been numerically evaluated only for a simple cubic lattice for ferromagnets[4] and a body-centered orthorhombic lattice for antiferromagnets.[3] Although the former model is unrealistic, one would anticipate that it would provide a sensible description of the nature of impurity effects in ferromagnets, if not reproducing all the finer experimentally observable details. In fact those comparisons which have been made of predicted and observed impurity magnetization as a function of temperature have been remarkably successful. On the other hand there is a class of antiferromagnets which is well described by a body-centered orthorhombic model. The magnetic structure consists of two sub-lattices of opposite net spin alignment - one at the centers and the other at the corners of the crystal lattice, so that the nearest neighbors of an "up" spin lie on the "down" sub-lattice and vice versa. Such a model should, for example, be quite accurate in applications to impure $Mn\,F_2$ (although for more detailed agreement with experiment a second exchange parameter should be included for the pure crystal).

Finally, the nature of the spin operator commutation relations requires two approximations which are not necessary for discussion of phonons in an impure (harmonic) lattice. The lattice vibrations are independent excitations, whose thermodynamic properties are those of a gas of non-interacting bosons. At sufficiently low temperatures that no more than a single spin deviation from the ground state can be anticipated at each lattice site an harmonic oscillator representation of the spin operators is useful,(11) and the single spin excitation spectrum - in particular, the energies of localized modes - can be determined (exactly only in the case of the ferromagnet: $J, J' > 0$). But the interesting thermodynamic consequences of the localized modes must occur at temperatures comparable to their excitation energies, where large deviations are found, at least at impurity sites, and the spin operator matrix elements are highly anharmonic. To proceed one defines independent magnons at each temperature - most simply by using the random phase approximation (RPA) - but their energies are renormalized by the other excitations present. The thermodynamic properties should be primarily determined by the real part of the spin wave self-energy shift (rather than by the imaginary part, or lifetime), and the RPA predictions of this are in good agreement with neutron inelastic scattering experiments.(12) However, because of the renormalization, no free boson hamiltonian correctly predicts the thermodynamics of these excitations. A second consequence of the anharmonicity, which persists even in the RPA, is that the behavior of the impurity depends on the states of all other spins - which are, in principle modified by the impurity even though arbitrarily far from it. This infinite extent of the effective perturbation would strictly preclude the use of Lifshitz's method, based on a local perturbation, but the effects clearly diminish rapidly with distance from the impurity; a final approximation then consists of neglecting these modifications beyond the immediate neighborhood of the impurity.

III. GREEN'S FUNCTIONS

In order to investigate the propagation of spin excitations in the impure crystals we consider the retarded Green's functions

$$G_{i\ell}(t) = -i\theta(t)\,\langle [S_i^+(t),\, S_\ell^-(0)]\rangle\,(2|J|z\,\langle S^z\rangle) \qquad (3.1)$$

where $\theta(t)$ is the unit step function, z is the number of nearest neighbors of any spin, and $S^{\pm} = S^x \pm i S^y$ are Heisenberg representation spin operators, whose time evolution is governed by the hamiltonian (2.1). Angular brackets denote the thermal expectation value of the enclosed operators, and $\langle S^z \rangle$ is $\langle S_\ell{}^z \rangle$ for the pure host (with ℓ on the up sublattice in the antiferromagnet). Within the RPA, where all operators $S_\ell{}^z$ have been replaced by their average values in the final expression, the time Fourier transform of $G_{i\ell}(t)$ obeys the equation of motion

$$\omega\, G_{i\ell}(\omega) = \frac{1}{\pi} \langle S_\ell{}^z \rangle\, \delta_{i\ell} + [\, g_i h + \alpha_i \langle S_i{}^z \rangle / \langle S^z \rangle]\, G_{i\ell}(\omega) \tag{3.2}$$
$$- \frac{1}{z} \Sigma\, (J_{ig} \langle S_i{}^z \rangle / |J| \langle S^z \rangle) G_{g\ell}(\omega) + \frac{1}{z} \Sigma\, (J_{ig} \langle S_g{}^z \rangle / |J| \langle S^z \rangle) G_{i\ell}(\omega)$$

where we have taken the energy ω in units of $2|J|z\langle S^z \rangle$ and have defined

$$h = \mu_\beta H_o / 2Jz\langle S^z \rangle ; \quad \alpha_i = K_i / 2Jz \quad (= \alpha, \alpha' \text{ for } K, K') \tag{3.3}$$

We can invert this matrix equation for the $G_{i\ell}$, making use of the locality of the perturbation in the usual way, by isolating to the right hand side all terms arising explicitly from the presence of impurities. We restrict ourselves henceforth to the nearest neighbor exchange approximation discussed in Sec. II.

A. Ferromagnetic Host

We consider first the ferromagnet (J>0), for which

$$[\omega - gh - \alpha - 1] G_{i\ell} + \frac{1}{z} \sum_\Delta G_{\ell+\Delta,\ell} = \frac{1}{\pi} \langle S_\ell{}^z \rangle \delta_{i\ell} + \sum_m R_{im} G_{m\ell} \tag{3.4}$$

where Δ is summed over nearest neighbor vectors, the matrix R_{im} vanishes in the absence of impurities, and the corresponding pure crystal solution of Eq. (3.4) is

$$G^o_{i\ell}(\omega) \equiv \frac{1}{\pi} \langle S^z \rangle L_{i\ell}(\omega) = \frac{1}{\pi} \langle S^z \rangle \frac{1}{N} \sum_q \frac{\exp i\vec{q}\cdot(\vec{r}_i - \vec{r}_\ell)}{\omega - \omega_q + i0^+} \tag{3.5}$$

where the spin wave energies are

$$\omega_q = 1 + \alpha + gh - \gamma_q ; \quad \gamma_q = \frac{1}{z} \Sigma\, e^{i\vec{q}\cdot\vec{\Delta}} \tag{3.6}$$

Then Eq. (3.4) can be rewritten in the form

$$G_{il} = G^o_{i\ell}\langle S_\ell{}^z\rangle/\langle S^z\rangle + \sum_{m,n} L_{im}R_{mn}G_{n\ell} \equiv G^o_{i\ell}\langle S_\ell{}^z\rangle/\langle S^z\rangle + \Sigma\, M_{in}G_{n\ell} \tag{3.7}$$

where

$$M_{in} = L_{in}\,[(g_n - g)h + (\alpha_n\langle S_n{}^z\rangle/\langle S^z\rangle - \alpha) + \tag{3.8}$$
$$+ \frac{1}{z}\sum_\Delta (J_{n,n+\Delta}\langle S^z_{n+\Delta}\rangle/J\langle S^z\rangle - 1)] - \frac{1}{z}\sum_\Delta L_{i,n+\Delta}(J_{n,n+\Delta}\langle S^z_n\rangle/J\langle S^z\rangle - 1).$$

The Dyson type equation (3.7) is readily interpreted. A spin excitation travels from i to ℓ either freely (by G^o) or by propagating freely to an intermediate site n, scattering there and then proceeding (with possible additional scattering) from n to ℓ. It is clear from Eq. (3.8) that M_{in} is non-zero when n is an impurity site ($g_n = g'$, $\alpha_n = \alpha'$, $J_{n,n+\Delta} = J'$), a near neighbor to an impurity ($J_{n,n+\Delta} = J'$), or whenever $\langle S_n{}^z\rangle$ or $\langle S^z_{n+\Delta}\rangle$ is different from $\langle S^z\rangle$. It is the last set of terms which rigorously can produce scattering at an arbitrary distance from the impurity, and it is here that one must make the cutoff approximation of Sec. II (neglect of the difference between $\langle S_n{}^z\rangle$ and $\langle S^z\rangle$ for n sufficiently far from an impurity), in order to take advantage of the simplifications introduced by locality of the perturbation. One expects that the local magnetization will, in fact, approach that of the pure host very rapidly as one moves away from the impurity. Even a spin in the first shell, after all, has (z-1) host neighbors and must feel an effective molecular field much like that appropriate to a spin in the pure system. Numerical calculations,(6,7,9,13) determining $\langle S_n{}^z\rangle - \langle S^z\rangle$ after neglecting it in the original equations (3.7), imply the essential validity of this conjecture.

In the dilute impurity concentration limit we can neglect coherent scattering from different impurities. (The perturbation, however, is not restricted to be weak, so that we must take account of multiple scattering from each one.) Thus we expect (and find)(14) that the self-energy accumulates incoherently as a spin excitation reaches successive scattering centers; it is proportional to the impurity concentration, c. It is therefore useful to begin with a consideration of the problem of a single impurity in the system, located at a site to be labeled 0. We define the local magnetization deviation

$$\delta_i \equiv \langle S_i{}^z\rangle/\langle S^z\rangle - 1 \tag{3.9}$$

and in lowest approximations we neglect all except δ_o. Then the scattering matrix R_{mn} is of dimension (z+1) x (z+1). In particular, for the simple cubic lattice

$$R = \frac{1}{6}\begin{pmatrix} 6(\nu+\epsilon) & -\rho & -\rho & \cdots & -\rho \\ -\epsilon & \rho & 0 & \cdots & 0 \\ -\epsilon & 0 & \rho & \cdots & 0 \\ \cdots & \cdots & \cdots & \cdots & \cdots \\ -\epsilon & 0 & 0 & \cdots & \rho \end{pmatrix} \tag{3.10}$$

where

$$\nu = (g'-g)h+[\alpha'(1+\delta_o)-\alpha]; \quad \epsilon = \frac{J'}{J}-1; \quad \rho = \frac{J'}{J}(1+\delta_o)-1 \tag{3.11}$$

in agreement with the conventional symbols(4,5) used at zero temperature. The rows and columns label the sites of the impurity and its nearest neighbors at zero temperature. (We note here that R is not the interaction matrix which results(4,5) if the boson representation is used ab initio, but it does give the same excitation energies in the zero temperature limit.)

Returning to Eq. (3.7) we point out that for fixed ℓ the right hand side of the equation involves $G_{n\ell}$ only for n in the "cluster" of the impurity and its 6 nearest neighbors. Therefore the solution can be carried out in the 7 x 7 space of this cluster. We have

$$G_{i\ell} = \sum_m \left(\frac{1}{1-M}\right)_{im} G^o_{m\ell} \frac{\langle S_\ell{}^z\rangle}{\langle S^z\rangle} \tag{3.12}$$

where the matrix inversion is greatly simplified(4) by the use of group theory:

$$\left(\frac{1}{1-M}\right)_{im} = \frac{(-1)^{i+m}\,\text{Minor}\,[(1-M)_{im}]}{D} \tag{3.13}$$

$$D = \det\,|\,1-M\,| = D_s\, D_p^3\, D_d^2 \tag{3.14}$$

The representation of the cubic group based on the 7 lattice positions in the cluster reduces to two 1-dimensional, one 3-dimensional and one 2-dimensional representation, to which the subscripts s, p, and d refer, respectively. The dimensionalities (or degeneracies of the corresponding eigenfunctions) are indicated by the exponents in (3.14). We find

$$D_s(\omega) = 1+\epsilon-\rho\nu+\rho\tilde{\omega}\,[\rho\tilde{\omega}^2 + (\epsilon-\rho-\rho\nu)\tilde{\omega} + (\rho+1)\nu]I_{00}(\omega) \tag{3.15a}$$

where $\quad \tilde{\omega} = \omega - gh - \alpha,$

$$D_p(\omega) = 1 - \frac{\rho}{6}\,[L_{00}(\omega) - L_{12}(\omega)] \tag{3.15b}$$

$$D_d(\omega) = 1 - \frac{\rho}{6}\,[L_{00}(\omega) + L_{12}(\omega) - 2L_{13}(\omega)] \tag{3.15c}$$

where sites 1 and 2 are on opposite sides of the impurity and 3 is in a direction orthogonal to the line joining 0, 1, and 2. These expressions will be investigated further in the following Section, with particular attention to their determination of the location of local modes. Similar expressions can be developed when the range of the perturbation is allowed to extend further, but the number of sites in the relevant "cluster" clearly increases very rapidly and there is a corresponding increase in the complexity of the expressions. Calculations have been performed,(6,7) however, which treat approximately the magnetization deviation in the first neighbor shell.

B. Antiferromagnetic Host

We turn now to the two sub-lattice antiferromagnetic host ($J<0$). Here one must consider simultaneously a pair of equations - one appropriate to each sub-lattice. Thus, taking the site m on the up sub-lattice and n on the down sub-lattice, we have from Eq. (3.2):

$$(\omega-gh-\alpha-1)G_{m\ell} - \frac{\sigma}{z}\sum G_{m+\Delta,\ell} = \frac{1}{\pi}\langle S_\ell^z\rangle\delta_{m\ell} + \sum P_{ms}G_{s\ell}$$

$$[(\omega-gh+\sigma(\alpha+1)]G_{n\ell} + \frac{1}{z}\sum G_{n+\Delta,\ell} = \sum Q_{ns}G_{s\ell} + \frac{1}{\pi}\langle S_\ell{}^z\rangle\delta_{n\ell} \tag{3.16}$$

where σ is the ratio of down to up sub-lattice alignment in the pure host. As before, terms arising explicitly from the presence of impurities have been isolated to the right hand side, so that the matrices P_{ms} and Q_{ns} vanish for the pure host crystal. Here the pure crystal solutions to Eq. (3.13) are

$$G^o_{m\ell}(\omega) \equiv \frac{1}{\pi}\langle S^z\rangle[(\omega-gh+\sigma(\alpha+1)]\,L'_{m\ell}(\omega) = \frac{1}{\pi}\langle S^z\rangle$$

$$\frac{[\omega-gh+\sigma(\alpha+1)]}{N}\sum_q \frac{\exp i\,\vec{q}\cdot(\vec{r}_m-\vec{r}_\ell)}{(\omega-\omega_q^+ + i0^+)(\omega-\omega_q^- + i0^+)} \tag{3.16}$$

where the antiferromagnet spin wave energies are

$$\omega_q^{\pm} = gh-\frac{1}{2}(\alpha+1)(\sigma-1)\pm\frac{1}{2}[(\sigma+1)^2(\alpha+1)^2-4\sigma\gamma_q^2]^{\frac{1}{2}} \tag{3.18}$$

For the down sub-lattice

$$G^o_{n\ell}(\omega) = \frac{1}{\pi}\langle S^z\rangle\,\frac{1}{z}\sum_\Delta L'_{n+\Delta,\ell}(\omega) \tag{3.19}$$

The external field introduces relative motion of the positive and negative frequency spin wave bands and is therefore expected to be a useful tool for exploring the behavior of localized modes. However, it introduces considerable complication into the formalism, and since finite field calculations have not yet been carried out we shall henceforth take h = 0.

Then Eq. (3.13) is readily inverted in terms of G^o (or L'):

$$G_{m\ell} = \pm \frac{\langle S_\ell^z \rangle}{\langle S^z \rangle} G^o_{m\ell} + \sum_{s,m'} L'_{mm'} [\frac{1}{z} \sum_\Delta Q_{m'+\Delta,s} + (\omega + \alpha + 1) P_{m's}] G_{s\ell} \quad (3.20a)$$

$$G_{n\ell} = \sum L'_{nn'} [(\omega - 1 - \alpha) Q_{n's} - \frac{1}{z} \sum_\Delta P_{n'+\Delta,s}] G_{s\ell} \pm G^o_{n\ell} \frac{\langle S_\ell^z \rangle}{\langle S^z \rangle} \quad (3.20b)$$

where the upper sign refers to ℓ on the up sub-lattice and the lower to ℓ on the down sub-lattice.

Again we consider the single impurity problem, neglecting all δ_i except for δ_0. Then the only non-vanishing matrix elements of the scattering matrices P and Q are

$$P_{00} = \nu + \epsilon; \quad P_{0\Delta} = \rho/z; \quad Q_{\Delta 0} = -\epsilon/z; \quad Q_{\Delta\Delta} = -\rho/z \quad (3.21)$$

where ν, ϵ, and ρ are defined again by Eq. (3.11). We treat the body centered cubic lattice (the orthorhombic lattice becomes effectively cubic if the primitive cubic lattice vectors are taken as the units of length along the coordinate axes). The values of s on the right hand side of Eqs. (3.17) then extend over the cluster of the impurity and its 8 nearest neighbors and the solution requires inversion of a 9 x 9 matrix. This can again be simplified by taking advantage of the point symmetry about the impurity. The form of the solution is, of course, identical to that of Eq. (3.12) with the 7 x 7 matrix M replaced by a 9 x 9 matrix M', and

$$D'(\omega) = \det |1-M'| = D'_s(\omega) D'^{3}_p(\omega) D'^{3}_d(\omega) D'_f(\omega) \quad (3.22)$$

is the denominator of $G(\omega)$; its behavior will determine the localized mode energies for the impure antiferromagnet. As before, the subscripts s,p,d, and f refer to the symmetry under the point group at the impurity of the functions which bring (1-M') into block diagonal form. The factors of Eq. (3.22) are given explicitly by

$$D'_s(\omega) = 1 + [\epsilon + \rho(\omega-\nu-\alpha)] L'_{10}(\omega) - (\epsilon+\nu)(\omega+\alpha+1) L'_{00}(\omega) \quad (3.23a)$$

$$D'_p(\omega) = 1 + \frac{\rho}{8} (\omega-\alpha-1) [L'_{00}(\omega) + L'_{25}(\omega) - L'_{12}(\omega) - L'_{14}(\omega)] \quad (3.23b)$$

$$D'_d(\omega) = 1 + \frac{\rho}{8} (\omega-\alpha-1) [L'_{00}(\omega) - L'_{25}(\omega) - L'_{12}(\omega) + L'_{14}(\omega)] \quad (3.23c)$$

$$D'_f(\omega) = 1 + \frac{\rho}{8}(\omega-\alpha-1)[L'_{00}(\omega)-3L'_{25}(\omega)+3L'_{12}(\omega)-L'_{14}(\omega)] \quad (3.23d)$$

with the lattice sites labeled as $2\vec{r}_1=-2\vec{r}_5=(1,1,1)$; $2\vec{r}_2=-2\vec{r}_6=(1,-1,-1)$; $2\vec{r}_3=-2\vec{r}_7=(-1,1,-1)$; $2\vec{r}_4=-2\vec{r}_8=(-1,-1,1)$. Again calculations have been performed[15] which allow δ_Δ, as well as δ_0 to take on non-zero values.

IV. LOCALIZED MODES AND RESONANCES

The time Fourier transform of the Green's function (3.1) can be written[16] in the spectral representation

$$G_{i\ell}(\omega) = \int \frac{d\omega'}{2\pi} \frac{(1-e^{-\beta\omega'})\, A_{i\ell}(\omega')}{\omega-\omega' + i0^+} \quad (4.1)$$

where β is the inverse temperature. The spectral weight function $A_{i\ell}(\omega)$ is given in terms of the exact eigenstates of the system $|\alpha\rangle$ of energy ω_α:

$$A_{i\ell}(\omega) = -\frac{2\text{Im}\, G_{i\ell}(\omega)}{1-e^{-\beta\omega}} = \frac{1}{Z}\sum_{\alpha,\gamma} e^{-\beta\omega_\alpha}\langle\alpha|S_i^+|\gamma\rangle\langle\gamma|S_\ell^-|\alpha\rangle\delta(\omega-\omega_\gamma-\omega_\alpha) \quad (4.2)$$

and Z the partition function. Thus the singularities of $G_{i\ell}(\omega)$ occur at the exact excitation energies of the system. For the diagonal terms, $i = \ell$, $A_{ii}(\omega)/2\langle S_i^z\rangle$ measures the probability that a single spin flip excitation of energy ω will flip the particular spin at the site i (we note here that $(1-e^{-\beta\omega})A_{ii}(\omega)$ is normalized[6] to $2\langle S_i^z\rangle$). Because all sites are equivalent (on a given sub-lattice in the antiferromagnet) in the pure crystal this probability is 1/N for every i for each excitation. Then an excitation localized near the site i in an impure crystal is manifested by an enhancement of $A_{ii}(\omega)/\langle S_i^z\rangle$ relative to its value in the pure host. (We deal always with the limit $N \to \infty$ before $\text{Im}\,\omega \to 0$, so that $A(\omega)$ is a smooth function rather than the sum of δ-functions of Eq. (4.2), within the host spin wave band.[6] For a single (or finite number) of impurities in this system modes outside the band still appear as δ-functions in $A(\omega)$.) The excitations are accompanied by the vanishing of the real parts of the denominators of $G(\omega)$, $D(\omega)$ (Eqs.3.13-15) and $D'(\omega)$ (Eqs.3.22,24,25) although the converse is not necessarily true; enhancement of $A(\omega)$ need not occur at these points[4,15]). Because of the orthogonality of excitations of different point group symmetry, we in fact need only look for the zeros of each of the factors $D_\mu(\omega)$ or $D'_\mu(\omega)$. The pure crystal Green's functions, $L_{ij}(\omega)$ and $L'_{ij}(\omega)$, required for evaluation of these determinants have been tabulated[3,6,17] both inside and outside the host spin wave bands.

A. Ferromagnets

The system with J and J' both positive is the simplest to discuss. The ground state is totally aligned and the spin wave band covers the range $\alpha+gh\leq\omega \leq 2+\alpha+gh$. The lower limit is determined by the energy of a decrease in net magnetization, $\Sigma\langle S_i{}^z\rangle$, by unity in the external and anisotropy fields. Thus, for sufficiently negative ν (i.e., weak net field at the impurity) one would expect the possibility of s-wave localized modes below the band; this effect is discussed briefly in Ref. 9a. (Only the s-wave mode is involved, because only it has finite amplitude at the impurity. Correspondingly, we note the absence of ν from Eqs. (3.15b and c).) If host and impurity feel the same fields (ν=0) then all energies are simply shifted uniformly by these fields, as is manifest in Eqs. (3.15). Results have been obtained[4,4a,6,7,18] for $h=\alpha=\alpha'=0$. Localized modes of s, p, and d symmetries can all appear above the band for sufficiently strongly coupled impurities. The energy of each type ($\omega_s, \omega_p, \omega_d \geq 2$) is, as expected, an increasing function of both ϵ and S'/S - i.e., of the relative impurity-host exchange coupling strength. The energies ω_p and ω_d are much smaller than ω_s for corresponding values of ϵ and S'/S. The temperature dependence of ω_s has been explored,[6,18] and it is found to increase monotonically with T, in the energy units of 12 $J\langle S^z\rangle$ we have been using (in which the spin wave band width is independent of T). This is because the relatively more strongly coupled impurity spin is harder to turn over than a host spin so that $\langle S_o{}^z\rangle/\langle S^z\rangle$, and therefore the effective relative impurity - host exchange strength, increases with T. It was found[18] that it is even possible for a localized mode to "pop out" of the top of the band at finite temperatures although none is present at T = 0°K.

The zeros of the real parts of Eqs. (3.15) also suggest resonant, or virtual modes within the band. Those of p and d symmetry are restricted[19] to the upper half of the band, but for weakly coupled impurities there can be s-wave resonances at low energies. Detailed numerical calculations[6] have been made of the spectral weight function $A_{oo}(\beta,\omega)$ (which involves only $D_s(\omega)$), showing peaks which sharpen as they move to lower energies with decreasing J'/J. The position of the resonance is nearly temperature independent (in units of $12J\langle S^z\rangle$); it is also a weak function of S'/S. The resonance occurs very nearly at the molecular field energy, J'/J, when this is small.

The introduction of an antiferromagnetically coupled impurity (J'<0) brings in a different sort of s-wave localized mode.[19] The ground state of the system (known exactly only for $S'=\frac{1}{2}$) corresponds to essentially anti-parallel alignment of impurity and host, so S_o^- tends to destroy rather than create an excitation; the corresponding frequency is negative. Semi-classically the impurity feels an exchange field in the -z direction and precesses in the opposite sense ($\omega<0$) from the host spins. The mode exists for any value of $J'<0$ and any S'/S. In fact, since $D_s(-\infty)=1$ and $D_s(0) = J'/J-\rho\nu$ it has a zero for some $\omega<0$ whenever $J'/J-\rho\nu<0$. In addition s,p, and d modes can split off the top of the band for sufficiently strong coupling. The minimum values of $|J'|/J$ and S'/S for these to appear are obtained by setting $D_\mu(\bar{\omega}=2)=0$; the resulting boundary curves for $T=0^\circ K$, $\nu=0$ are given in Fig. 1 of Ref. 9a.

B. Antiferromagnets

From Eq. (3.18) in the absence of an external field, ($h=0$; $\sigma=1$) we have for the pure antiferromagnet $\omega_q^\pm = \pm\,[(1+\alpha)^2-\gamma_q^2]^{\frac{1}{2}}$, giving symmetrical branches of the spin wave band between $\pm(2\alpha+\alpha^2)^{\frac{1}{2}}$ and $\pm(1+\alpha)$. In general localized modes can appear below the continuum ($\omega< -(1+\alpha)$), above ($\omega> 1+\alpha$) or in the gap ($-(2\alpha+\alpha^2)^{\frac{1}{2}} < \omega < (2\alpha+\alpha^2)^{\frac{1}{2}}$). For $J'<0$ and $|J'|$ sufficiently large an s-type mode (the "s_o mode") appears above the band. If $|J'|$ is large enough to produce p, d, and/or f localized modes, these appear below the band, since they have amplitude on the nearest neighbors, Δ, on the down sub-lattice (if the impurity is chosen to be on the up sub-lattice), and the corresponding frequencies are negative, as discussed above. Furthermore, the second s-type mode (the "s_1 mode"), in which the spins precess in the sense opposite to that in the s_o mode, also lies below the band. The sign of J' has little effect on the p, d, and f modes since the impurity spin does not participate in the motion (alternatively, $\rho=J'\langle S_o^z\rangle/J\langle S^z\rangle-1$ changes little on reversing the sign of J', and this is the only parameter entering the local mode criteria (3.23 b, c, d)). However, for $J'<0$ the impurity aligns parallel to its neighbors, forming an effectively ferromagnetic cluster, and only a single s-wave mode (with the energy of the s_1 mode for $J'>0$) remains.

For strong enough anisotropy, α, and small $|J'|/J$ an s_o resonance can emerge from the bottom of the $\omega>0$ band into the gap. The criterion for this to occur is that J'/J be less than the value which makes $D_s'[(2\alpha+\alpha^2)^{1/2}] = 0$. Plots of the positions and widths of localized s_o modes and resonances as functions of J'/J and S'/S and for values of α of 0.015 (appropriate to MnF_2) and 0.110 can be found in the work of Lovesey[10] (this is a low temperature - unrenormalized spin wave - calculation). Examples of the local spectral weight function $A_{oo}(\omega)$ will be found in Ref. 15.

V. PREDICTED CONSEQUENCES OF LOCAL MODES

Localized magnons can be observed spectroscopically - with lines identified as magnon sidebands of optical transitions.[20] Ishii et al[19a] have proposed a resonance experiment which takes advantage of the dependence on external field of the energy of the

s_o mode for an antiferromagnetic impurity in a ferromagnet. Perhaps the most direct means of observation is through inelastic neutron scattering. The magnetization on an impurity is strongly affected by low energy localized states and is experimentally accessible through nmr, Mössbauer, and neutron scattering experiments. Anomalies are expected in the magnetic specific heat from local modes. We consider briefly here the theory of the last three effects.

A. Impurity Magnetization

A convenient relation[21] between the impurity magnetization $\langle S_o^z \rangle$ and the spectral weight function is through the "quasiboson energy," ω_o, defined by

$$2\langle S^z \rangle/(e^{\beta\omega_o}-1) = \int_{-\infty}^{\infty} d\omega\, A_{oo}(\omega)\, e^{-\beta\omega} = -\int d\omega\, \mathrm{Im}\, G_{oo}(\omega)/(e^{\beta\omega}-1) \quad (5.1)$$

Then $$\langle S_o^z \rangle = S' B_{S'}(\beta\omega_o) \quad (5.2)$$

where B_S is the Brillouin function appropriate to spin magnitude S. Thus ω_o is the energy spacing of the levels of the isolated impurity in an effective molecular field which would produce the correct magnetization; conventional molecular field theory (field $\propto$ $\langle S^z \rangle$) corresponds to ω_o independent of temperature. These relations have been used[6,15] to obtain $\langle S_o^z \rangle$ as a function of temperature for a variety of values of impurity parameters (although only the two cases J, J' > 0 and J, J' < 0 have been treated in this way). As expected, the magnetization of weakly coupled impurities drops off faster than that of the host; reduction in J'/J is much more effective in this regard than reduction of S'/S. As one moves to lower energy and sharper s-wave resonances (weaker coupling) an increasing fraction of the total spectral weight is found in the resonance. Recalling that ω_s is almost temperature independent we see from Eq. (5.1) and the definitions (4.1) and (4.2) that for temperatures $T \gtrsim \omega_s$ the quasiboson energy will be nearly constant (and $\approx \omega_s$). Thus the usual form of molecular field theory will describe the impurity magnetization far more accurately than that of the host. For $T < \omega_s$, of course, the low energy, long wavelength spin waves will dominate the thermodynamics and these have nearly equal amplitudes at all sites (clearly this is true in the low energy limit of the uniform mode). Then molecular field theory fails; the impurity magnetization deviation approaches a limiting $T^{3/2}$ behavior, as does that of the host. However, in the antiferromagnet with a localized s mode in the gap, this mode dominates even the low T behavior. As T increases $\langle S_o^z \rangle$ decreases faster than $\langle S^z \rangle$, leading to an even weaker effective host-impurity exchange strength, so that the gap mode energy decreases. The fall-off of $\langle S_o^z \rangle$ with T is extraordinarily sharp in this case.[11]

Whenever either J or J' is negative there is incomplete spin alignment even at T = 0. From Eqs. (5.1-2) it is clear that this is determined by the negative ω spectral weight, where the Bose factor in (5.1) is always (-1). In particular, for $J>0$, $J'<0$ the s_0 mode completely determines this zero point deviation. As the impurity-host coupling increases, the local mode frequencies move further from the band. Therefore the modes become more localized; their spectral weight in the cluster increases. For negative frequency s-type modes this means increased impurity zero point deviation. Physically, the strongly coupled antiferromagnet impurity ($|J'|>|J|$) finds it energetically more favorable than does a host spin to develop transverse components of spin to reduce its exchange energy. We note that since it is "stiffer" than the host and remains more aligned at higher temperatures, the $\langle S_0{}^z\rangle$ and $\langle S^z\rangle$ curves cross. These features are observed[11] in numerical calculations.

B. Inelastic Neutron Scattering

The question of local mode observation by inelastic neutron scattering will be discussed in detail elsewhere at this conference; we limit ourselves here to a short summary of the theory as applied to spin systems. In the first Born approximation the inelastic cross section for energy transfer $\omega=E'-E$ and momentum transfer $\vec{q}=\vec{k}'-\vec{k}$ may be written[22]

$$\frac{d^2\sigma}{d\Omega dE'} = \frac{1}{2}(r_o\eta)^2 \frac{k'}{k}(1+\hat{q}_z)^2 \sum F_n(q)F_m(q)e^{i\vec{q}\cdot(\vec{m}-\vec{n})}$$

$$\times (1/2\pi)\int dte^{i\omega t}\langle S_n^+(0)S_m^-(t) + S_n^-(0)S_m^+(t)\rangle \quad (5.3)$$

where $\eta = 1.91$ is the nuclear magnetic moment in nuclear magnetons, r_o is the classical radius of the electron, and $F_n(q)$ is the neutron form factor of the atom at n. The Fourier transform of spin correlation functions can readily be expressed[16] in terms of $G_{nm}(\omega)$:

$$\frac{1}{2}\int dt\; e^{i\omega t}\langle S_n{}^+(0)S_m^-(t)+S_n^-(0)S_m^+(t)\rangle$$

$$= n(\omega)\;\mathrm{Im}\;G_{nm}(\omega)+[n(\omega)+1]\;\mathrm{Im}\;G_{mn}(\omega) \quad (5.4)$$

where the first term corresponds to magnon destruction and the second to magnon creation ($n(\omega) = 1/(e^{\beta\omega}-1)$). From this form it is already clear that peaks in the spectral weight functions at resonances or local modes may produce corresponding peaks in the cross section. To obtain correctly the coherent part of the cross-section one must consider a finite concentration of impurities; the sum over m and n is then simplified by averaging over impurity configurations. This program has been carred out by Izyumov[23] for impure ferromagnets in the low temperature unrenormalized spin wave (or boson)approximation. He finds resonances in the widths of the coherent scattering peaks near virtual modes in the band. The incoherent cross-section is, not surprisingly,simply related to the single impurity expression. In both ferromagnets(28) and antiferromagnets(10) sharp peaks in this function are predicted at local mode frequencies and near low-lying s-wave resonances. Interpretation of s-wave effects may be complicated by ignorance of the impurity form factor. For a non-magnetic impurity in a ferromagnet this problem disappears; Lovesey and Marshall(24) have explored neutron scattering in this system in some detail, using a molecular field approximation.

C. Specific Heat

If temperature-dependent energy renormalization could be ignored and magnons treated as independent boson excitations, then a calculation of their density of states $N(\omega)$ would be sufficient for determining the free energy and thence, e.g., the specific heat, C_V. This procedure is, as we have mentioned, acceptable at low enough temperature. In this approximation we have (for either sign of J)

$$N(\omega) = \frac{1}{N\pi} \operatorname{Im} \sum_i G_{ii}(\omega)/\langle S_i^z\rangle = N_o(\omega) + \frac{1}{N\pi} \operatorname{Im} \frac{dD(\omega)}{d\omega}/D(\omega) \qquad (5.5)$$

($D(\omega)$ is used here for either (3.12) or (3.22)). Thus the impurity-induced changes, $N(\omega)-N_o(\omega)$, peak near resonances, so that when T is sufficiently high to show appreciable effects from these changes the impurity is already strongly excited, and the boson approximation is invalid. A straightforward approach(25) to calculation of C_V consists of differentiating with respect to temperature the expectation value of the hamiltonian (2.1), which in the RPA involves only $\langle S_i^z\rangle$ and the Green's functions (3.1). Allowing fully for multiple scattering from each impurity, but ignoring all coherent scattering effects, one can find the self-energy (and therefore excitation energies) correctly to first order in the concentration c. Results have been obtained(25) at this time only for the simple cubic isotropic ferromagnet with J', as well as J, positive. In agreement with earlier results(14) the low energy single spin wave excitations are found to obey the dispersion relation $\omega \approx ak^2$, but with "a" in general shifted(by O(c)) from the pure host value. Thus C_V is predicted to be proportional to $T^{3/2}$, like the pure crystal, at temperatures well below any resonances. The boson density of states picture suggests Shottky-like anomalies in C_V near temperatures comparable to the energies of sharp (low-lying) resonances. The numerical calculations(25) indeed indicate anomalies in C_V at approximately the anticipated temperatures, broadening as J'/J is increased (and the resonance itself is broadened). The effects should be even more dramatic for systems in which - as discussed in Sec. IV - localized modes occur in the anisotropy-produced energy gap below the spin wave continuum.

REFERENCES

*On leave from the Physics Department, University of Pennsylvania, Philadelphia, Pa.

(1) I. M. Lifshitz, Soviet Phys. Uspekhi 7, 549 (1965).
(2) T. Moriya, Progr. Theoret. Phys. (Kyoto) 33, 159 (1965).
(3) L. R. Walker and D. Hone, to be published.
(4) T. Wolfram and J. Callaway, Phys. Rev. 130, 2207 (1963).
(4a) S. Takeno, Progr. Theoret. Phys. (Kyoto) 30, 731 (1963).
(5) Yu. A. Izyumov and M. V. Medvedev, Soviet Phys. JETP 21, 381 (1965).
(6) D. Hone, H. Callen, and L. R. Walker, Phys. Rev. 144, 283 (1966).
(7) T. Wolfram and W. Hall, Phys. Rev. 143, 284 (1966)
(8) T. Tonegawa and J. Kanamori, Phys. Letters 21, 130 (1966).
(9a) H. Ishii, J. Kanamori, and T. Nakamura, Progr. Theoret. Phys. (Kyoto) 33, 795 (1965).
(9b) Y-L Wang and H. Callen, Phys. Rev., to be published.
(10) S. W. Lovesey, Proc. Phys. Soc., (London) to be published.
(11) See, e.g., D. C. Mattis, "The Theory of Magnetism," (Harper and Row, New York, 1965), Chap. 6.
(12) K. C. Turberfield, A. Okazaki, and R. W. H. Stevenson, Proc. Phys. Soc. (London) 85, 743 (1965).
(13) L. R. Walker, private communication.
(14) Yu. Izyumov, Proc. Phys. Soc. (London) 87, 505 (1966).
(15) L. R. Walker, D. Hone, and H. Callen, to be published.
(16) See D. N. Zubarev, Soviet Phys. Uspekhi 3, 320 (1960).
(17) I. Mannari and C. Kawabata, Research Note 15, Physics Dept., Faculty of Sciences, Okayama Univ. (1964).
(18) D. Hone and H. Callen, J. Appl. Phys. 37, 1440 (1966).
(19) Yu. Izyumov and M. V. Medvedev, Soviet Phys. JETP 22, 1289 (1966).
(20) R. L. Greene, D. D. Sell, W. M. Yen, A. L. Schawlow, and R. M. White, Phys. Rev. Letters 15, 656 (1965).
L. F. Johnson, R. E. Dietz, and H. J. Guggenheim, Phys. Rev. Letters 17, 13 (1966).
(21) H. B. Callen, Phys. Rev. 130, 890 (1963).
H. Callen and S. Shtrikman, Solid State Commun. 3, 5 (1965).
(22) L. Van Hove, Phys. Rev. 95, 1374 (1954).
(23) Yu. Izyumov, Proc. Phys. Soc. 87, 521 (1966).
(24) S. Lovesey and W. Marshall, Proc. Phys. Soc. 89, 613 (1966).
(25) D. Hone and Karl Vogelsang, Proceedings of 1967 International Congress on Magnetism, to be published.

COUPLING BETWEEN IMPURITY SPINS IN A MAGNETIC HOST

Robert M. White and C. Michael Hogan

Department of Physics, Stanford University

Stanford, California

I. INTRODUCTION

The localized modes associated with crystal impurities have been studied by many investigators.[1] Localized excitons play an important role in the fluorescence of optically active materials, and localized phonons are a useful probe for studying lattice dynamics. We shall treat the localized modes associated with impurities in a magnetic host. At zero temperature an exact formulation[2] using Green's functions as well as spin wave techniques[3,4,5] have been used. The thermodynamics of such impurity modes have been studied elsewhere.[6]

In previous calculations interactions between impurities were neglected. However, it is known that nuclear spins in a magnetic host are coupled by the virtual exchange of spin waves.[7,8] Our purpose is to see whether such coupling exists between localized magnons.

To resolve this question we consider two impurities in a linear chain. It is found that the interaction between impurities lifts the degeneracy that is present for two isolated impurities. The strength of the impurity interaction is measured by the mode splitting energy due to the impurities interacting. We find that when the local modes lie outside the spin wave continuum their interaction depends upon the overlap of their localized mode wave functions. Thus, if the modes have energies far above the spin wave band and are highly localized, the interaction is quite small. In like manner, if the modes are near the spin wave band, their interaction is appreciable. These results indicate the importance of considering bound states in connection with indirect coupling via virtual magnon exchange.

II. EIGENVALUES FOR ONE IMPURITY

In the absence of an impurity the isotropic exchange Hamiltonian for a linear chain of spins is

$$\mathcal{H}_o = -J \sum_n \sum_{\delta=\pm 1} \underset{\sim}{S}_n \cdot \underset{\sim}{S}_{n+\delta} \quad . \tag{1}$$

Let us consider the eigenstates of (1) when one spin deviation is present. The matrix elements of $\mathcal{H}_o$ are calculated in a basis of orthonormal spin deviation states in which $|n\rangle$ represents the state in which one spin deviation is located at the nth site. In such a basis the matrix elements of $\mathcal{H}_o$ are

$$\langle n|\mathcal{H}_o|m\rangle = [-2JNS^2 + 4JS]\,\Delta(n,m) - 2JS \sum_{\delta=\pm 1} \Delta(n,m+\delta). \tag{2}$$

The Schroedinger equation for the eigenstates $|k\rangle$ and eigenfunctions $E_o(k)$ is

$$[-2JNS^2 + 4JS - E_o(k)]\,\langle n|k\rangle - 2JS \sum_{\delta=\pm 1} \langle n+\delta|k\rangle = 0 \quad . \tag{3}$$

Assuming periodic boundary conditions the solution is

$$\langle n|k\rangle = \frac{1}{\sqrt{N}}\, e^{ikna} \quad , \tag{4}$$

where a is the lattice spacing and k is equal to an integer times $2\pi/Na$. It is customary to choose the N values from $-(\pi/Na)(N-2)$ to π/a. The dispersion relation is the familiar spin wave spectrum,

$$E_o(k) = -2JNS^2 + 4JS\,(1-\cos ka) \equiv E_o\left(\frac{\pi}{2a}\right) - 4JS \cos ka \tag{5}$$

Now replace the spin at ℓ by an impurity spin S' which has exchange coupling J' with its host neighbors. The Hamiltonian becomes $\mathcal{H} = \mathcal{H}_o + \mathcal{H}_1$, where

$$\mathcal{H}_1 = 2(J\underset{\sim}{S}_\ell - J'\underset{\sim}{S}'_\ell) \cdot \sum_{\delta=\pm 1} \underset{\sim}{S}_{\ell+\delta} \tag{6}$$

The matrix elements of $\mathcal{H}_1$ in the basis of single spin deviation states are

$$\langle n|\mathcal{H}_1|m\rangle = -4JS^2\rho\Delta(n,m) + 4JS\epsilon\Delta(\ell,m)\Delta(n,m) + 2JS\rho \sum_\delta \Delta(\ell+\delta,m)\Delta(n,m) - 2JS\gamma \sum_\delta [\Delta(\ell+\delta,n)\Delta(\ell,m) + \Delta(\ell,n)\Delta(\ell+\delta,m)] \quad , \tag{7}$$

where

$$\rho = \frac{J'S' - JS}{JS} \quad , \quad \epsilon = \frac{J' - J}{J} \quad ,$$

and

$$\gamma = \frac{J'}{J}\sqrt{\frac{S'}{S}} - 1 \quad . \tag{8}$$

Let us redefine $\mathcal{H}_o$ and $\mathcal{H}_1$ by incorporating the first term in $\langle n|\mathcal{H}_1|m\rangle$ above into $\langle n|\mathcal{H}_o|m\rangle$. Then $\langle n|\mathcal{H}_1|m\rangle$ is an N by N matrix all the elements of which are zero except for a 3 by 3 submatrix centered at ℓ. Since all the sites are equivalent we may choose $\ell = 2$. Then $\langle n|\mathcal{H}_1|m\rangle$ has the form

$$\langle n|\mathcal{H}_1|m\rangle = 2JS \begin{bmatrix} \rho & -\gamma & 0 & \\ -\gamma & 2\epsilon & -\gamma & 0 \\ 0 & -\gamma & \rho & \\ & 0 & & 0 \end{bmatrix} \quad . \tag{9}$$

The Schroedinger equation is the matrix equation

$$(\mathcal{H}_o + \mathcal{H}_1)|\Psi\rangle = E|\Psi\rangle \quad . \tag{10}$$

Or,

$$[I - (E - \mathcal{H}_o)^{-1}\mathcal{H}_1]|\Psi\rangle = 0 \quad . \tag{11}$$

where I is the N by N unit matrix. We shall find it convenient to introduce the Green's function matrix

$$G = 2JS(E - \mathcal{H}_o)^{-1} \quad . \tag{12}$$

The eigenvalue equation then becomes

$$[I - (2JS)^{-1}G\mathcal{H}_1]|\Psi\rangle = 0 \tag{13}$$

Because of the form of $\mathcal{H}_1$ the matrix product $G\mathcal{H}_1$ only has elements in its first three columns. In particular,

$$I - (2JS)^{-1}G\mathcal{H}_1 = \begin{bmatrix} 1-\rho G_{11}+\gamma G_{12} & \gamma G_{11}-2\epsilon G_{12}+\gamma G_{13} & \gamma G_{12}-\rho G_{13} & \\ -\rho G_{21}+\gamma G_{22} & 1+\gamma G_{21}-2\epsilon G_{22}+\gamma G_{23} & \gamma G_{22}-\rho G_{23} & 0 \\ -\rho G_{31}+\gamma G_{32} & \gamma G_{31}-2\epsilon G_{32}+\gamma G_{33} & 1+\gamma G_{32}-\rho G_{33} & \\ & \cdot & & 1 \\ & \cdot & & \quad 1 \\ & \cdot & & \qquad 1 \\ -\rho G_{n1}+\gamma G_{n2} & \gamma G_{n1}-2\epsilon G_{n2}+\gamma G_{n3} & \gamma G_{n2}-\rho G_{n3} & \qquad \cdot \\ & \cdot & & \qquad \cdot \\ & \cdot & & \qquad \cdot \end{bmatrix} \tag{14}$$

This means that the first three components of $|\Psi\rangle$, corresponding to the impurity and its nearest neighbors, may be solved for independent of the others. These other components are then easily found from the remaining equations. Thus, our problem reduces to the 3 X 3 matrix equation

$$[I-(2JS)^{-1}G\mathcal{H}_1]|\Psi\rangle = 0 \ . \tag{15}$$

We notice that G_{nm} depends only upon $|n-m|$ so that we find it convenient to indicate an element of G with a single subscript

$$G_{|n-m|} \equiv G_{nm} \ . \tag{16}$$

G_n may be evaluated by expanding the spin deviation states in terms of eigenfunctions of $\mathcal{H}_o$ and converting the sum to an integral.

$$G_n = \frac{2JS}{N}\frac{Na}{2\pi}\int_{-\pi/a}^{\pi/a} \frac{e^{ikna}\,dk}{E-E_o\left(\frac{\pi}{2a}\right)+4JS\,\cos(ka)+i4JS\epsilon} \ . \tag{17}$$

Introducing

$$\mathcal{E} = \frac{E-E_o\left(\frac{\pi}{2a}\right)}{4JS} \ , \tag{18}$$

this integral may be evaluated by the method of residues and gives

$$G_n(\mathcal{E}) = \frac{(\sqrt{\mathcal{E}^2-1}\ -\mathcal{E})^n}{2\sqrt{\mathcal{E}^2\ -1}} \ . \tag{19}$$

Notice that for $\mathcal{E} > 1$ the Green's function is real, while it becomes complex when $\mathcal{E} < 1$.

We now perform a unitary transformation which block diagonalizes the upper 3 X 3 portion of (14). It is necessary to use symmetry arguments to find such a transformation. (13) and (14) then lead to

$$\begin{vmatrix} 1-2\epsilon G_o+2\gamma G_1 & \sqrt{2}(\gamma G_o-\rho G_1) & 0 \\ \sqrt{2}(\gamma G_o-2\epsilon G_1+\gamma G_2) & 1+2\gamma G_1-\rho(G_o+G_2) & 0 \\ 0 & 0 & 1-\rho(G_o-G_2) \end{vmatrix} = D_s D_p = 0 \tag{20}$$

We refer to the solution of the 1 X 1 portion of (20) as a p-mode because of its antisymmetric nature that will be displayed. We shall specialize our attention to the p-mode, whose eigenvalue equation is obtained with the aid of (19),

$$1 - \frac{1}{2\sqrt{\mathcal{E}^2\ -1}}\,[1-(\sqrt{\mathcal{E}^2\ -1}\ -\mathcal{E})^2] = 0 \quad , \tag{21}$$

which leads to

$$\mathcal{E} = \frac{\rho^2+1}{2\rho} \quad . \tag{22}$$

We can examine the p-mode eigenstate knowing the elements of G as function of ρ using (19) and (22). From the first three components of (14)

$$\langle 2|\Psi\rangle = 0 \tag{23}$$

and

$$\langle 3|\Psi\rangle = - \langle 1|\Psi\rangle \quad .$$

Also, the nth equation gives for $n > 3$

$$\langle n|\Psi\rangle = \rho(G_{n-1}-G_{n-3})\langle 1|\Psi\rangle = \rho^3(-\frac{1}{\rho})^n \langle 1|\Psi\rangle \quad . \tag{24}$$

Finally $\langle 1|\Psi\rangle$ is obtained by the condition that the total spin deviation be one, i.e., that the function be normalized. This condition gives

$$2|\langle 1|\Psi\rangle|^2+2\sum_{n=4}^{N/2} |\langle n|\Psi\rangle|^2 = 2\left[1+\rho^6 \sum_{n=4}^{N/2} (-\frac{1}{\rho})^{2n}\right]|\langle 1|\Psi\rangle|^2 = 1 \tag{25}$$

$$\langle n|\Psi\rangle = \sqrt{\frac{\rho^2-1}{2}}\,(-\frac{1}{\rho})^{n-2} \qquad n > 3 \tag{26}$$

The origin of the term "p-like" mode is now evident. Since $|\langle n|\Psi\rangle|^2$ is the probability of finding the spin deviation at site n we see that as the mode moves away from the spin wave band it becomes more localized.

It is worthwhile to calculate η_p, the change in density of states due to the p-mode.[1]

$$\eta_p = -\frac{1}{\pi}\,\mathrm{Im}\,\frac{D_p'}{D_p} = \frac{\mathrm{ImD}_p\mathrm{ReD}_p'-\mathrm{ReD}_p\mathrm{ImD}_p'}{\pi[(\mathrm{ReD}_p)^2+(\mathrm{ImD}_p)^2]} \tag{27}$$

Taking into account (20)

$$\eta_p = \frac{\rho(\mathcal{E}-\rho)}{\pi\sqrt{1-\mathcal{E}^2}\,(1-2\rho\mathcal{E}+\rho^2)} \tag{28}$$

Notice that

$$\int_{-1}^{+1} \eta_p(\mathcal{E})\,d\mathcal{E} = \begin{cases} 0 & \rho < 1 \\ -1 & \rho > 1 \end{cases} \tag{29}$$

The minus one corresponds to the fact that the localized mode appearing outside the band does so at the expense of one mode inside the band. For $\rho < 1$ the perturbation is not strong enough to produce either a localized mode or a resonant state, but merely causes a redistribution of the states.

III. EIGENVALUES FOR TWO IMPURITIES

Let us now consider, in addition to our first impurity at site 2, a second identical impurity at the site $n + 2 \geq 5$. Thus, n-1 is the number of intervening host spins. The perturbing Hamiltonian then becomes

$$\mathcal{H}_1 = 2JS \begin{bmatrix} \rho & -\gamma & 0 & & & \\ -\gamma & 2\epsilon & -\gamma & & 0 & \\ 0 & -\gamma & \rho & & & \\ & & & \rho & -\gamma & 0 \\ & 0 & & -\gamma & 2\epsilon & -\gamma \\ & & & 0 & -\gamma & \rho \end{bmatrix} \tag{30}$$

and

$$G = \begin{bmatrix} G_o & G_1 & G_2 & G_n & G_{n+1} & G_{n+2} \\ G_1 & G_o & G_1 & G_{n-1} & G_n & G_{n+1} \\ G_2 & G_1 & G_o & G_{n-2} & G_{n-1} & G_n \\ G_n & G_{n-1} & G_{n-2} & G_o & G_1 & G_2 \\ G_{n+1} & G_n & G_{n-1} & G_1 & G_o & G_1 \\ G_{n+2} & G_{n+1} & G_n & G_2 & G_1 & G_o \end{bmatrix} \tag{31}$$

Just as in the case of one impurity we can block-diagonalize the eigenvalue matrix by making use of symmetry operations. The p-like modes are determined by

$$D_p^{(+)} D_p^{(-)} = 0 \tag{32}$$

where

$$D_p^{(\pm)} = 1-\rho\, G_o - G_2 \pm [G_n + \tfrac{1}{2}(G_{n+2} + G_{n-2})] \tag{33}$$

For $\mathcal{E} > 1$ this gives

$$\rho = \frac{2\sqrt{\mathcal{E}^2-1}}{1-(\sqrt{\mathcal{E}^2-1} - \mathcal{E})^2 \pm 2\,{}^2(\sqrt{\mathcal{E}^2-1} - \mathcal{E})^n} \tag{34}$$

For a given value of ρ there are two solutions. These are shown in Fig. 1 for n = 3. The splitting of these two solutions, Δ is shown in Fig. 2 as a function of n.

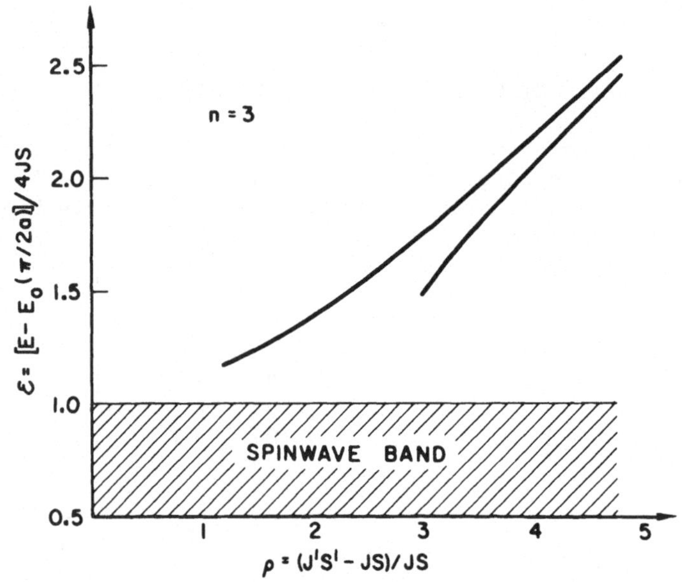

Fig. 1. Splitting of p-like mode with two impurities.

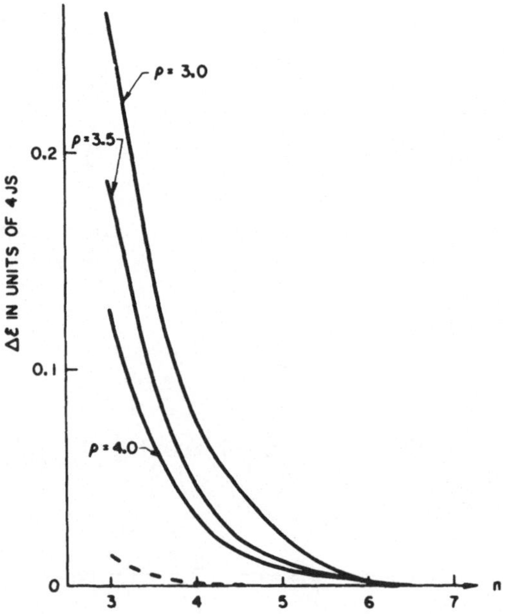

Fig. 2. Mode splitting as a function of impurity separation.

IV. PERTURBATION THEORY

It is interesting to compare our results with those obtained from a spin wave scattering approach. Let us consider impurities from the single-ion point of view. Thus, the longitudinal part of the exchange interaction between an impurity and the host produces a molecular field, while the transverse part enables the impurity to emit or absorb spin waves.

For our one dimensional chain of N spins the perturbation Hamiltonian is given by the transverse exchange.

$$\mathcal{H}_1 = -2\sqrt{\frac{2S}{N}}\, J' S_2'^{\dagger} \sum_k \gamma_k e^{-ik2a} a_k^{\dagger} + c.c. - 2\sqrt{\frac{2S}{N}}\, J' S_{n+2}'^{\dagger} \sum_k \gamma_k e^{-ik(n+2)a} a_k^{\dagger} + c.c. \qquad (35)$$

We now apply second order perturbation theory to obtain an effective interaction between the impurity spins. Identifying the exchange constant with the mode splitting energy, we arrive at

$$\Delta = \frac{1}{4S} \frac{(1+\epsilon)^2(3+\epsilon)[2+\epsilon - \sqrt{(2+\epsilon)^2-1}]^n}{\sqrt{(2+\epsilon)^2-1}} \qquad (36)$$

This result is plotted as the dashed curve in Fig. 2 as a function of n for $S' = S = \frac{1}{2}$ and $\rho = 3$.

We see that the result obtained from a spin wave scattering calculation differs appreciably from the exact solution. We feel that our exact determination of the impurity coupling illustrates that bound states, which scattering theory neglects, are very important.

REFERENCES

1. Y. Izyumov, Adv. Phys. 14, 569 (1965).
2. T. Wolfram and J. Callaway, Phys. Rev. 130, 2207 (1963).
3. H. Ishii, J. Kanamori and T. Nakamura, Prog. Theor. Phys. 33, 795 (1965).
4. Y. Wang and H. Callen (to be published).
5. T. Tonegawa and J. Kanamori, Phys. Letters 21, 130 (1966).
6. D. Hone, H. Callen, and L. Walker, Phys. Rev. 144, 283 (1966).
7. H. Suhl, J. Phys. Rad. 20, 333 (1959).
8. T. Nakamura, Prog. Theor. Phys. 20, 542 (1958).

SPIN-CLUSTER RESONANCE IN THE ISING SPIN SYSTEM

M. DATE

Department of Physics, Faculty of Science, Osaka University, Toyonaka, Osaka, Japan

A new type excitation of localized magnons in the Ising-like spin system is reported. It was found that in a strongly anisotropic ferro- or antiferromagnetic crystal such as $CoCl_2 2H_2O$, non-uniform magnetic resonances with the selection rule $\Delta m = \pm 1$, where m is the number of spins in the short range order spin cluster directed oppositely to the majority spins, can be excited. Examples of such a localized spin wave excitation observed in $FeCl_2$ and $CoCl_2 2H_2O$ is discussed using a model of the Ising spin system.

I. INTRODUCTION

Motions of spins in usual ferro- and antiferromagnetic crystals are well described by spin waves which are quantized as magnons. One of the general characters of the spin wave is that motions of spins are not localized and keep coherence over a considerable region of the crystal. In a strongly anisotropic crystals, i.e. in the Ising-like spin system, however, a different kind of excitation can exist which may be called " the spin-cluster excitation ". This excitation is described as follows: let us consider for example an Ising spin linear chain coupled by a ferromagnetic exchange interaction J_o with an external magnetic field parallel to the spin easy axis z. There is no phase transition in such a system, but almost all spins point along H_o for a sufficiently strong field at low temperatures. However, some spins point toward -z so as to minimize the free energy. Such a oppositely pointed spin-cluster is of the short range and appears randomly in crystal having a character of localized excitation different from usual spin waves. It should be pointed out that the spin-cluster state can not be an eigen state if the system consists of the Heisenberg spins because terms such as $J_oS_i^+S_j^-$ in the Heisenberg spin Hamiltonian act so as to mix the localized spin-cluster state with other states. In the Ising spin system, however, the state of the localized spin-cluster is the exact eigen state because there are no transverse component of spins.

Recently we have succeeded in observing a new type magnetic resonance of the spin-clusters with the selection rule $\Delta m = \pm 1$. This m is the number of spins in the short range order spin-cluster appearing in ferromagnetic linear chains of $CoCl_2 2H_2O$ near two metamagnetic critical fields at liquid helium temperatures. A preliminary report has already been published.(1) Afterwards, we could observed the spin-cluster resonance in a ferromagnetic Ising spin layer of $FeCl_2$ as an example of the resonance in the two dimensional spin system. These experimental results and discussions will be given in the following sections.

II. SPIN-CLUSTER RESONANCE IN $CoCl_2 2H_2O$

The crystal structure of $CoCl_2 2H_2O$ is schematically shown in Fig. 1. The unit cell contains two Co ions which form a base centered structure. Two oxygens(H_2O) and four Cl^- ions are octahedrally coordinated to each Co ion so as to form a $CoCl_4 2H_2O$ group which is connected with adjacent octahedrons along the c-axis by a polymeric $-CoCl_2-$ chain. Accordingly, the crystal can be looked

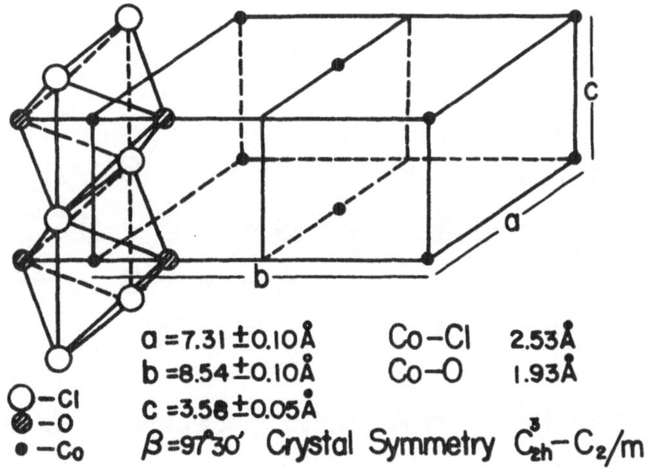

Fig. 1 Crystal structure of $CoCl_2 2H_2O$

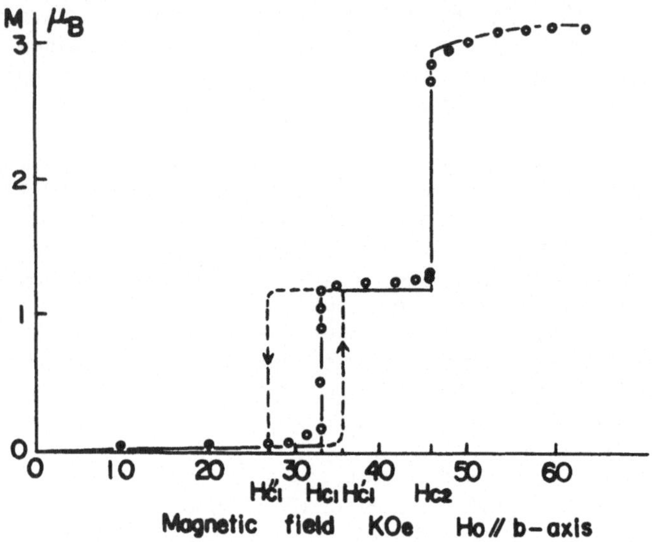

Fig. 2. Magnetization vs. external magnetic field H_o at 4.2°k. The experimental data given by open circles were obtained by Haseda et al. Dotted lines show the hysteresis effect observed in a pulse magnetic field by which the lower critical field H_{cl} shifts to H_{cl}' and H_{cl}'' under increasing and decreasing magnetic fields, respectively.

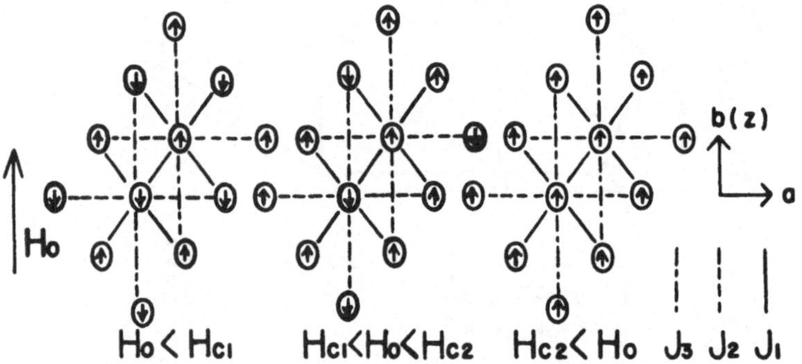

Fig. 3. Spin structures corresponding to three phases in the ab-plane. Three exchange interactions are shown by J_1,J_2 and J_3.

at as a bundle of the chains. These chains are linked together by hydrogen bonds. This compound become antiferromagnetic below 17.5 °K with an antiferromagnetic arrangement of ferromagnetic chains having the spin direction along the b-axis. When an external magnetic field is applied along the spin easy axis, the magnetization increases stepweise as is shown in Fig. 2. This interesting phenomenon was discovered by Haseda[(2)] and Narath[(3)] independently and has been analysed by several researchers. The result can be understood as follows: There is a strong ferromagnetic exchange interaction J_o ($J_o/K = 9.3°K$) in the chain which couples with adjacent chains by weak antiferromagnetic exchange constants J_1 and J_2 and very weak ferromagnetic J_3 where J_1, J_2 and J_3 are defined as are shown in Fig. 3. Moreover, the anisotropy energy favouring the b-axis is so strong that the spin system can be looked at as an Ising spin antiferromagnet. In such a case, the two step metamagnetic transition as is shown in Figs. 2 and 3 becomes possible. The Ising-like exchange constants J_1, J_2, and J_3 have been determined as -4.4, -1.0 and +0.3°K, respectively.[(4)] Two critical fields H_{c1} and H_{c2} have been determined to be 32 and 46 kOe, respectively.

As has been reported preliminary, a new type magnetic resonance was observed near two critical fields at liquid helium temperatures.[(1)] The experiment was done with microwaves of 24, 35, 50 and 70 GHz regions in a pulsed magnetic field up to 50 kOe. Considering the frequency-field diagram of the resonance, absorption intensity, and temperature dependence, we concluded that these resonances come from a localized excitation as may be called the spin-cluster resonance. For simplicity, let us consider a spin-cluster in an Ising spin linear chain as is shown in Fig. 4. Under an external field H_o and an rf-field perpendicular to it, one can expect the magnetic transition of $\Delta m = \pm 1$ where m is the total magnetic quantum number of the short range order spin cluster. It should be emphasized that the resonance frequency for such a spin-cluster transition does not depend on J_o as there is no change in exchange energy when a terminal spin of a cluster reverses direction. Moreover, this transition does not depend on the anisotropy energy if the spin system consists of S = 1/2. Accordingly the resonance condition is

$$\omega/\gamma = H_o. \tag{1}$$

Next, the absorption intensity should be considered. The absorption is proportional to the population differences between clusters of size m and m + 1, summed over all clusters. This sum is just equal to the number of m = 1 clusters, i.e., " clusters " of a single spin. Hereafter, the number of m = 1 spin clusters is called the effective spin number, written as n_e. The physical nature of the effective spin number is also expressed in the following way: For all $m \neq 1$, the transition probabilities for $\Delta m = +1$ and $\Delta m = -1$ are equal so that there is no net absorption. But for single spin clusters, while the transition $m = 1 \rightarrow m = 2$ can accompany absorp-

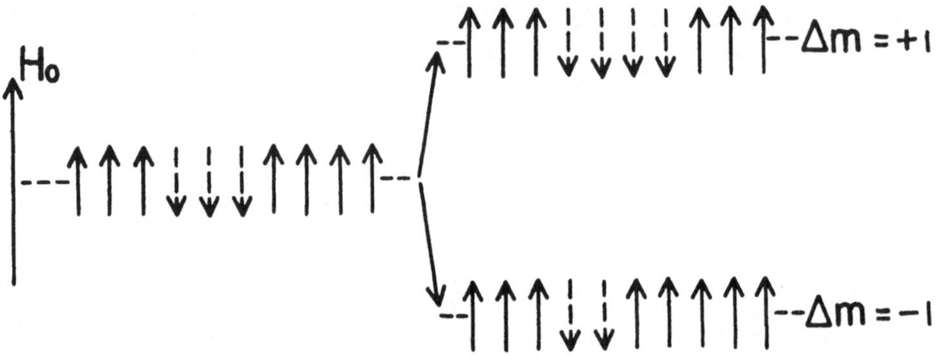

Fig. 4. Model of the spin-cluster resonance in an Ising-like ferromagnetic linear chain. Dotted arrows show the short range order spin-cluster.

tion of a photon, the transition $\Delta m = -1$ is forbidden as the annihilation of the cluster liberates exchange energy $2J_o$ so that the resonance condition shifts far from the microwave region.

The effective spin number n_e can be obtained by statistical calculation(1) and the result is

$$n_e = n_c(n_c - 1)/(n_s - 1)$$
$$\simeq n_c^2/n_s, \qquad \text{for } n_c, n_s \gg 1, \tag{2}$$

where n_c and n_s show numbers of total spin-clusters and total reversed spins, respectively and are calculated from the partition function as follows:

$$n_s/N = \{1 - \sinh(g\mu_B H_o S/kT)/R\}/2, \tag{3}$$

$$n_c/N = \exp(-2J_o/kT)/2\{\cosh(g\mu_B H_o S/kT) + R\}R, \tag{4}$$

$$R = \{\sinh^2(g\mu_B H_o S/kT) + \exp(-2J_o/kT)\}^{1/2}, \tag{5}$$

where N is the total spin number. The calculated n_e/N for $CoCl_2$-$2H_2O$ is given in Fig. 5 as functions of magnetic field and temperature.

Now we discuss the resonances observed in $CoCl_2 2H_2O$. it is assumed that for the spin-cluster resonance, effective field acting on each cluster are the sum of H_o and exchange fields H_1, H_2, and H_3 which come from the interchain exchange energy J_1, J_2 and J_3, respectively. These effective fields are given by

$$H_i = 2|J_i|S/g\mu_B, \qquad i = 1,2,3. \tag{6}$$

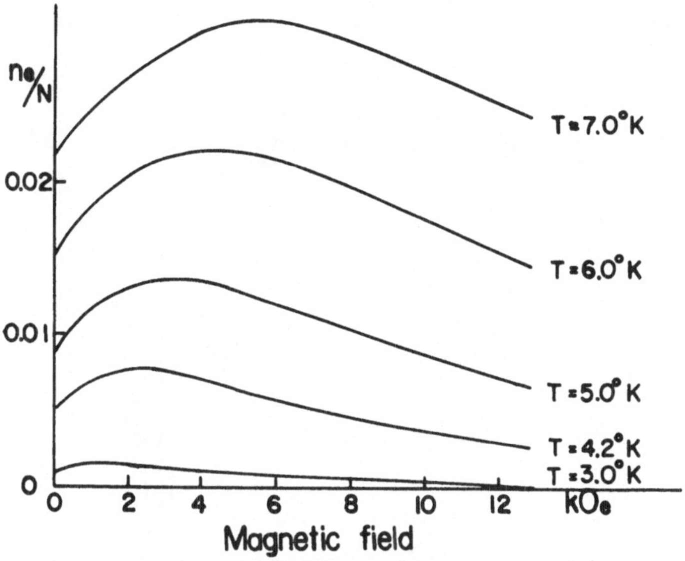

Fig. 5. n_e/N vs. magnetic field (H_o z-axis).

Using these effective fields we have five resonance conditions for the spin-cluster resonance in three phases as follows:

(1) $H_o < H_{c1}$,	(I)	$\omega/\gamma = 4H_1 - 2H_2 + 2H_3 + H_o$,	(7)
	(II)	$\omega/\gamma = 4H_1 - 2H_2 + 2H_3 - H_o$,	(8)
(2) $H_{c1} < H_o < H_{c2}$,	(III)	$\omega/\gamma = 4H_1 + 2H_2 + 2H_3 - H_o$,	(9)
	(IV)	$\omega/\gamma = 2H_3 + H_o$,	(10)
(3) $H_{c2} < H_o$,	(V)	$\omega/\gamma = 2H_3 - 4H_1 - 2H_2 + H_o$.	(11)

Of these resonance conditions, branches (II),(III) and (V) are observable. The experimental results are shown by open circles in Fig. 6. It should be noticed that the resonance points shift by n_e and the line width effect. Taking the intrinsic line width to be 4 kOe, branches (II),(III), and (V) are modified as (II'),(III') and (V') respectively as are shown in Fig. 6. Coincidence between theory and experiment is satisfactory. The temperature dependence of the absorption intensity is well explained by the exponential decrease in n_e at low temperatures. The angular dependence, relative absorption intensity are also explained by our model.

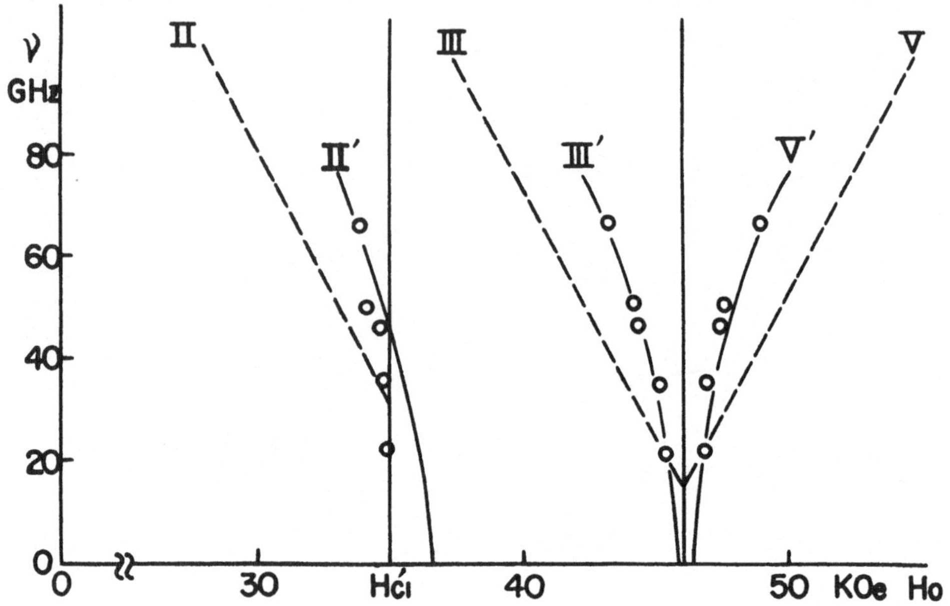

Fig. 6. Frequency-field diagram of branches (II),(III), and (V) (dotted lines) and their modified curves by n_e. Experimental points are shown by open circles.

III. SPIN-CLUSTER RESONANCE IN $FeCl_2$

The crystal structure of $FeCl_2$ is isomorphous with $CdCl_2$ having the hexagonal layer structure in which each layers of ferrous ions are separated by two layers of chlorine ions. This compound become antiferromagnetic below 23.5°K and shows a metamagnetic change at 11.6 kOe when an external magnetic field is applied along the spin easy axis(hexagonal c-axis) at low temperatures. There is a strong ferromagnetic exchange interaction among the intralayer ferrous ions, while the interlayer interaction is weak and antiferromagnetic. A strong uniaxial anisotropy favoring the c-axis is revealed by many experiment and the spin system can be looked at as the two dimensional Ising spin antiferromagnet. In such a ferromagnetic layer one can expect to observe the spin-cluster resonance slightly different from that of the linear chain. The most simple example of the resonance in the hexagonal layer is illustrated in Fig. 7. It shows a ferrous ion layer in which the short range order spin cluster (black circles) appears thermally. Now let us consider the transition from left to right in the Figure. It must be noticed that this transition does not depends on the exchange and anisotropy energies. Thus we have again the spin-cluster resonance in the microwave region if such a cluster is thermally exist.

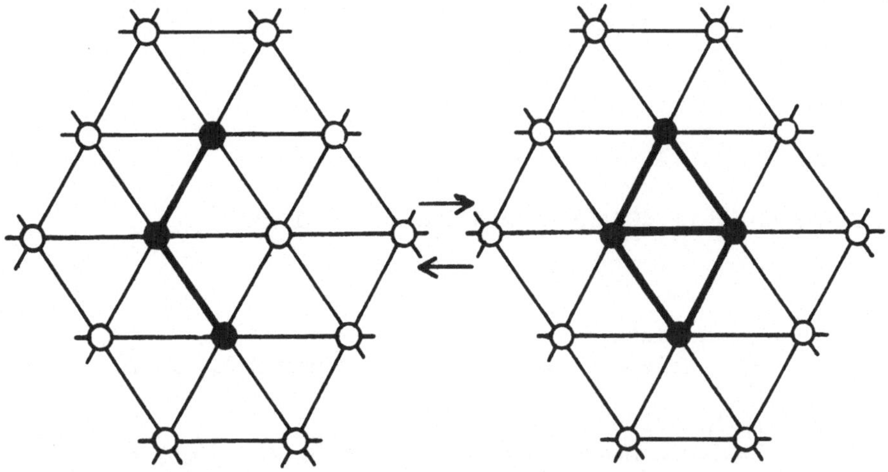

Fig. 7. A simple example of spin-cluster resonance in a hexagonal layer. External magnetic field is applied perpendicular to the plane.

We observed such a resonance in $FeCl_2$ above 11°K. From the frequency-field diagram, the effective g-value was obtained to be 4.0 which is in good agreement with the antiferromagnetic resonance data.[5] The absorption intensity was maximum at 20°K but it did not become zero near T_N. The resonance could be observed above T_N up to $T = 2T_N \sim 3T_N$. This interesting fact may be explained that the spin-cluster excitation is of the short range order so that it may survives even above T_N. We are now doing the similar experiment in $Dy_3Al_5O_{12}$ for observing the spin-cluster resonance in the three dinensional spin system.

References

(1) M. Date and M. Motokawa: Phys. Rev. Letters 16, 1111 (1966).
(2) T. Haseda, H. Kobayashi, T. Watanabe and E. Kanda: Proc. 8-th Int. Conf. on Low Temp. Phys. Butterworths, London (1963).
H. Kobayashi and T. Haseda: J. Phys. Soc. Japan 19, 765 (1964).
(3) A. Narath and D.W. Alderman: Bull. Am. Phys. Soc. 9,732 (1964).
A. Narath: Phys. Rev. 136,A766 (1964), ibid. 140,A552 (1965).
A. Narath: Phys. Letters 13, 12 (1964).
(4) T. Oguchi: J. Phys. Soc. Japan 20, 2236 (1965).
(5) I.S. Jacobs, S. Roberts and P.E. Lawrence: J. Appl. Phys. 36, 1197 (1965).

OPTICAL STUDIES OF LOCALIZED MAGNONS AND EXCITONS

R. E. Dietz
Bell Telephone Laboratories, Murray Hill, N. J.

A. Misetich
National Magnet Laboratory*, MIT, Cambridge, Mass.

I. INTRODUCTION

In this paper we discuss recent optical experiments on localized magnons and other Frenkel excitons arising from the 3d manifolds of transition ion antiferromagnetic insulators. This review is not intended to be exhaustive, but, rather, illustrative of the various types of localized excitations peculiar to magnetic crystals.

In the first section we discuss magnons and excitons bound or perturbed by various types of substitutional impurity ions. Because of the simplicity of magnons (relative to other excitons and phonons) properties of localized magnon modes such as frequency and line shape can be understood in great detail.

The second section of this paper summarizes recent experiments on intrinsic excitons and their magnon sidebands in MnF_2. It is shown that these excitons have no spectrally measurable dispersion and may be localized, and that the sideband shapes are determined by an interaction between the exciton and magnon. Thus these intrinsic pair excitations are similar in several respects to magnon modes associated with an impurity.

II. IMPURITY-BOUND EXCITONS AND MAGNONS

Excitations bound to impurities in a magnetic crystal form two general classes, distinguished by whether or not the impurity ion has a net electron spin. If the impurity does have unpaired electrons, then these electrons may exchange with those of the host ions, and consequently excitations of the host ions may be transferred to the impurity ions where under some circumstances, they may be highly localized. Another class of impurity magnons and excitons has been observed which arise from excitations bound to impurity ions which do not possess a net spin, and do not have any low-lying crystal field states. Impurity states resulting from these ions can be most easily described as perturbed host states. Magnons bound to such impurity ions appear to form impurity states lying in the intrinsic magnon continuum: i.e., "resonance states". Specific examples of both of these classes of impurity states are discussed in the following sections.

A. Magnons Bound to Impurities Having Spin

This is the case in which the molecular field theory is usually applied. The energy levels of the system are described in terms of crystal field (C.F.) levels of the impurity atom perturbed by exchange interactions with near neighbor host magnetic ions:

$$\mathcal{H}_i = \mathcal{H}_{C.F.\ i} + 2\sum_j J_{ij}\underset{\sim}{S}_i\underset{\sim}{S}_j \qquad (1)$$

Here i numbers ions on one sublattice, j ions on the other sublattice. A number of such systems have been or are being investigated, among which are Ni^{2+} ion impurities in MnF_2,[1][2] $KMnF_3$[1][2][3] and $RbMnF_3$.[1][2] The Ni-doped crystals offer the simplest and most complete interpretation, The reason being that the ground C.F. state of Ni^{2+} in octahedral coordination is $^3A_2$, an orbital singlet. Similarly, the C.F. ground state of Mn^{2+} is $^6A_1$, also an orbital singlet. Since the lowest excited C.F. state of Ni^{2+}, $^3T_2$ lies some 11,000 cm^{-1} below the lowest excited C.F. state $^4T_1$ of Mn^{2+}, it can be accurately described as an excited state of the nickel ion. Fluorescence has been observed from this state by optically pumping the Mn^{2+} $^4T_1(^4G)$ levels.[1][2] This excitation is transported with near unit quantum efficiency to nickel ions at concentrations as low as 10 ppm! The nickel ion acts as a deep trap for the excitation, with the lowest level of $^3T_2$ near 7,000 cm^{-1} being metastable and radiating to the perturbed $^3A_2$ levels. Polarized emission spectra of Ni in MnF_2[1] are shown in Fig. 1. Nearly all of the emission is magnetic dipole, including the sidebands to the strong zero-phonon, zero magnon transition to the ground state.

In the molecular field approximation the last term of Eq. (1) is written as $g\beta H_{eff}\, S_{zi} = 2\sum_j J_{ij}S_{zi}S_{zj}$ where z is the direction of sublattice magnetization. This effective field is expected to split the spin triplet into the magnetic substates $|1\rangle$ $|0\rangle$,

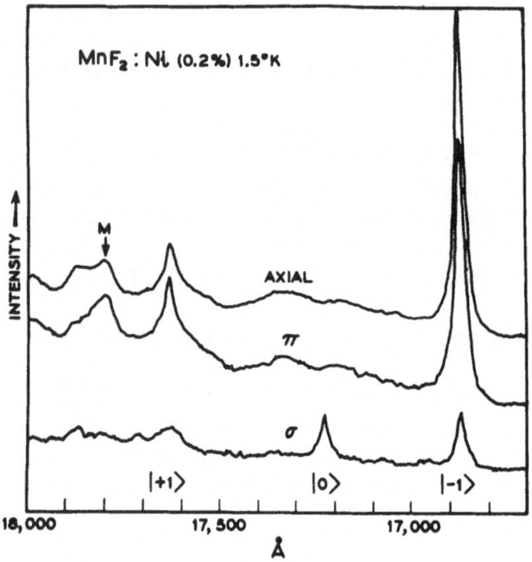

Figure 1. Polarized emission from the $^3T_2$ state of Ni^{2+} impurities in MnF_2 showing local magnon modes. The line marked "M" may be an intrinsic-type magnon sideband of the transition denoted by $|+1\rangle$.

$|-1\rangle$ which have been labelled according to the corresponding transitions on the diagram. The choice of $|-1\rangle$ as the ground state is arbitrary and refers to a particular sublattice. The molecular field experienced by a Ni^{2+} ion on the other sublattice would be of opposite sign, therefore sending $|1\rangle$ lower.

Analogous emission has also been observed from Ni^{2+} in $RbMnF_3$ and $KMnF_3$(1)(2) and the frequencies of the low lying energy levels are given in Fig 2. The exchange splitting of the levels

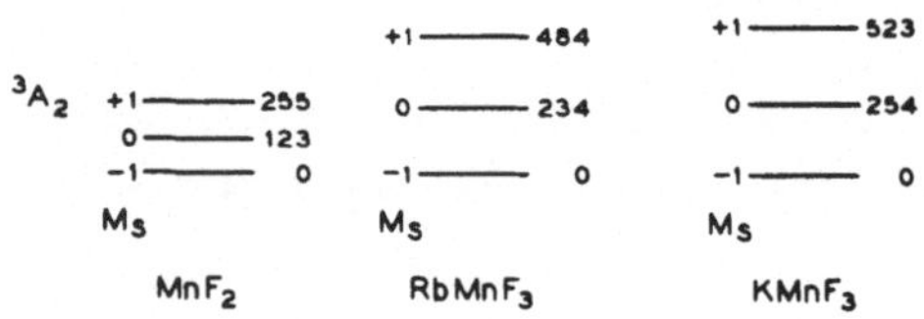

Figure 2. **Frequencies** of magnons localized on nickel ions in manganese fluorides.

is larger in the perovskite crystals since the nickel ions make 180° bonds to the ions on the other sublattice through the intervening fluoride ions, the optimum geometry for superexchange.

While the levels are split nearly equally, as one would expect from a molecular field-type interaction, there is a slight asymmetry in the splitting, with the interval between the two lowest levels being smaller than the higher pair of levels. This asymmetry has been attributed (2) to the terms in the exchange hamiltonian (1) which were neglected in the molecular field approximation. These terms represent the coupling of spins on neighboring atoms along directions normal to the sublattice magnetization:

$$\mathcal{H}' = 2\sum_{i\neq j} J_{ij}(S_{x_i}S_{x_j}+S_{y_i}S_{y_j})$$

which may be transformed to the raising and lowering operators:

$$\mathcal{H}' = -\sum_{i\neq j} J_{ij}(S_{+_i}S_{-_j}+S_{-_j}S_{+_j})$$

The effect of this term is to mix into the molecular field eigenstates $|M_{si}, M_{sj}\rangle$ other (excited) states having the same sum of $M_{si}+M_{sj}$. In other words, the effect is to deviate the spin of the nickel ion from the direction of sublattice magnetization and to spread this deviation through the crystal to neighboring M_n ions. The correction to the molecular field energy levels of the nickel impurity was computed by second order perturbation with the transverse exchange hamiltonian. The contribution $\delta'E$ arising from terms which cause spin deviations on the nickel ion and its nearest antiferromagnetically coupled M_n neighbors is computed in Fig. 3, and values of this correction are listed for the three above-mentioned crystals in Table 1. There will also be a correction $\delta''E$ from pair excitations of two M_n ions, one of which is a nearest neighbor of the nickel, but, as indicated in the Table, these contributions are small. The zero-point spin deviations of

Table I. Comparison of experimental and theoretical values of the asymmetry in the splitting of the $^3A_2$ state of nickel ions in antiferromagnetic compounds. (2)

	$\delta'E(cm^{-1})$	$\delta E=\delta'E+\delta''E$	δE (Experimental)
MnF_2	12.5	12.3	13.3 ± 0.1
$RbMnF_3$	12.7	11.3	10.9 ± 0.1
$KMnF_3$	13.7	12.2	11.9 ± 0.1

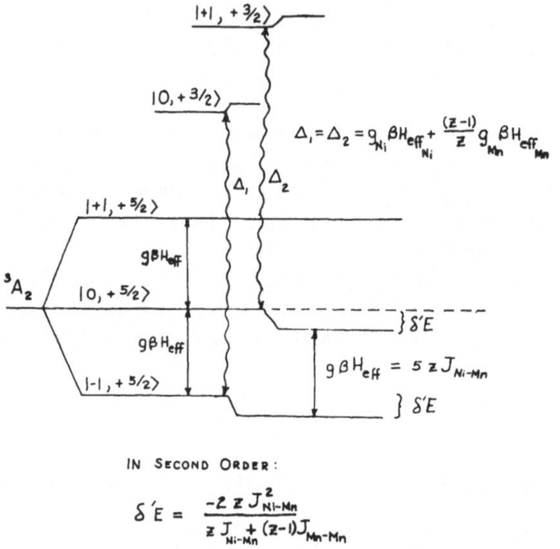

Figure 3. Computation of the second-order correction from the transverse exchange interaction to the molecular field levels of Ni^{2+}.

the ions were also calculated: For the nickel ion in $KMnF_3$ and $RbMnF_3$ the deviation is 4.1% For Ni in MnF_2 it is 2.7%. The difference between the zero point deviation on the adjacent Mn ions, and that for the pure Mn compound was small, indicating that the Ni^{2+} impurity state is a highly localized mode. These experiments are the first experimental confirmations of the existence of the zero point spin deviation in an antiferromagnet.

A number of calculations, (4), (5), and (6), have now been reported for localized or impurity spin waves. In this terminology, the excited nickel level $|0\rangle$ would correspond to a singly excited localized spin wave of S-type symmetry. The excited state labeled $|1\rangle$, on the other hand, corresponds to a doubly excited or localized 2-magnon transition, and has not yet been treated theoretically other than in our second order calculation. This transition is quite different from the intrinsic two magnon transitions observed in the far IR (7) (8) and in Raman scattering (9);both spin deviations are localized on the Ni ion in the present case, while they are created on opposite sublattices in the other experiments.

B. Excitons Bound to Impurities Having Spin

If the C. F. levels of an impurity ion in an antiferromagnetic crystal lie quite far from the energies of the excited levels of the host ion, then they may mix very weakly with the excited host states, and consequently excitation may be highly localized on the impurity ion. As seen in Fig. 4, the $^3T_2$ and $^3T_1$ states of Ni^{2+} impurities in MnF_2 lie far from any Mn^{2+} state, and their intensities and gross band shapes are very similar to those of Ni impurities in MgF_2 or ZnF_2 (10). The ^1E state, on the other hand, is much more intense than for a non-magnetic host, while the $^1T_2$ and $^3T_{1b}$ bands have lost their identity and are merged with the Mn^4T_1 and $^4T_2$ bands. Very little is known about such impurity excitons, although Ferguson, Guggenheim and Tanabe (3) have discussed their absorption strengths.

Another type of impurity exciton has been observed in cases where intrinsic exciton levels are well separated from the C.F. levels of the impurity. In such cases the impurity potential may split off localized states from the intrinsic exciton bands, and the excitation will be concentrated on the nearby host magnetic ions rather than on the impurity ion. Absorption and emission spectra from such states have been reported for transition metal impurities in $KNiF_3$, (11), but no detailed analysis of these

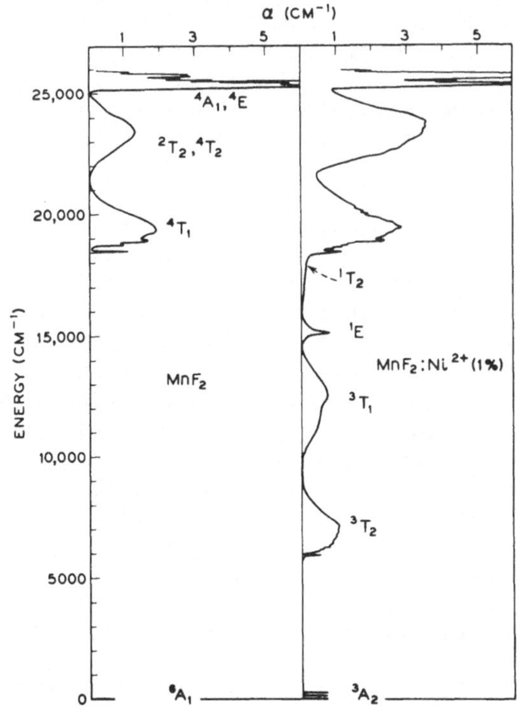

Figure 4. Comparison of the optical absorption spectrum of MnF_2 doped with Ni, and the pure compound. (1)

levels has been made. The remarkable temperature dependence of the emission intensity from these states is due to the fact that they become unstable against thermal scattering of the excitation to the intrinsic exciton band when kT is a small fraction of the energy separation. It will be seen that these states are more related to the impurity excitations bound to spinless ion impurities discussed in the next section, and, in fact, constitute an intermediate case since the excitation will be found partly on the impurity ion.

C. Excitons and Magnons Perturbed by Impurities Without Spin

The closed shell ions Ca^{2+}, Zn^{2+} and Mg^{2+} have been observed to produce impurity states in MnF_2 which can be discussed in terms of weakly perturbed intrinsic states.(12)(13) Since the lowest excited configurations of these ions lie many electron volts above the ground state, the d-shell excitation cannot reside on the impurity ion itself, but is confined to the neighboring Mn^{2+} ions. The exact nature of the impurity potential has not been determined, but it is undoubtedly a combination of the differences between the crystal and exchange fields of the impurity and host ions. Since the impurity potential on crystallographically inequivalent ions neighboring the impurity are expected to differ by amounts which are large compared to the intrinsic exciton dispersion, we expect the impurity excitons bound to each impurity atom to comprise several localized states, each of which corresponds to the excitation of a particular type of crystallographic neighbor to the impurity atom.

In Fig. 5 is shown σ-polarized light emission from MnF_2. The weak, sharper lines are impurity excitons mainly associated with Zn^{2+} impurities and derived from the intrinsic exciton E1. The more intense lines are magnon sidebands associated with a particular exciton. When the crystal at low temperatures is stressed along the (110) direction, the exciton lines split into two lines of equal intensity corresponding to the two sublattices. Thus we conclude that in general each impurity atom will have two metastable impurity states, one consisting of excited Mn^{2+} neighbors which are on the same sublattice as the impurity, and one on the opposite. The fluorescence lines in the figure are labelled according to whether the excitation is on second or on third neighbors to the Zinc impurity. Since only one of the sublattice impurity states is observed for Ca^{2+} impurities (excitons localized on first Mn neighbors of the Ca^{2+}), the other sublattice states (as excitons localized on second Mn neighbors to Ca^{2+}) are perturbed too near or above the intrinsic exciton band to be metastable. These excitons(13) are thermally scattered into the intrinsic exciton band in a manner similar to $KNiF_3$.(11) The magnon side bands correspond to the annihilation of the localized excitons with simultaneous creation of impurity magnons. Properties of these excitations are discussed in the following paper.(12)

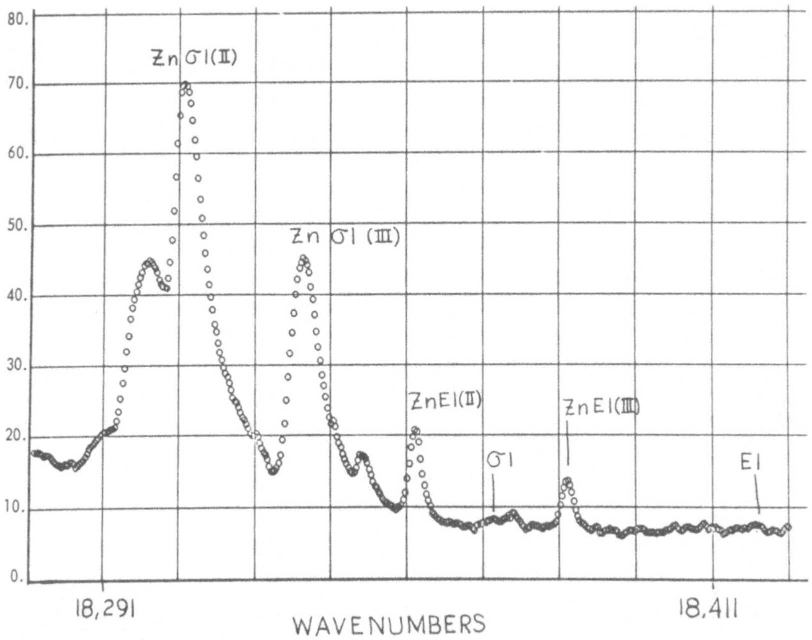

Figure 5. Emission from excitons bound to Zn impurities in MnF_2 at 1.5°K, σ polarization. The lines not labelled derive from the decay of excitons bound to Mg impurities.

III. ARE THE INTRINSIC EXCITONS IN MnF_2 LOCALIZED?

Recently the optically excited excitons of MnF_2 have been the subject of an intense investigation by several groups, following the identification of an electric dipole line as a magnon sideband of a magnetic dipole transition. (14) Stress experiments (15) have shown that this magnetic dipole absorption line, denoted E1 , and another one lying 17 cm^{-1} to higher energy, denoted E2, correspond to the creation of zone center excitons of the $^4T_1$ (^4G) state of the Mn^{2+} ion involving changes in orbital and spin quantum numbers $|M_L, M_S\rangle$ from the ground state $|0, 5/2\rangle$ to combinations of the bases $|1, 3/2\rangle$ and $|-1, 3/2\rangle$. The particular combination of bases for each exciton is determined by the energy separation of the two excitons, which can be altered by uniaxial stress. In Fig. 6 we show the effect of uniaxial (100) stress on the excitons and their magnons in both absorption and emission. Vertical lines have been drawn through the peaks of the magnon sidebands of E1 in absorption and emission at zero stress to aid in distinguishing the effect of stress on the shape and position of the sidebands. It is clear that the shapes and positions of the sidebands of E1 in absorption and emission are different, and there are three possible reasons for this difference. First, as shown

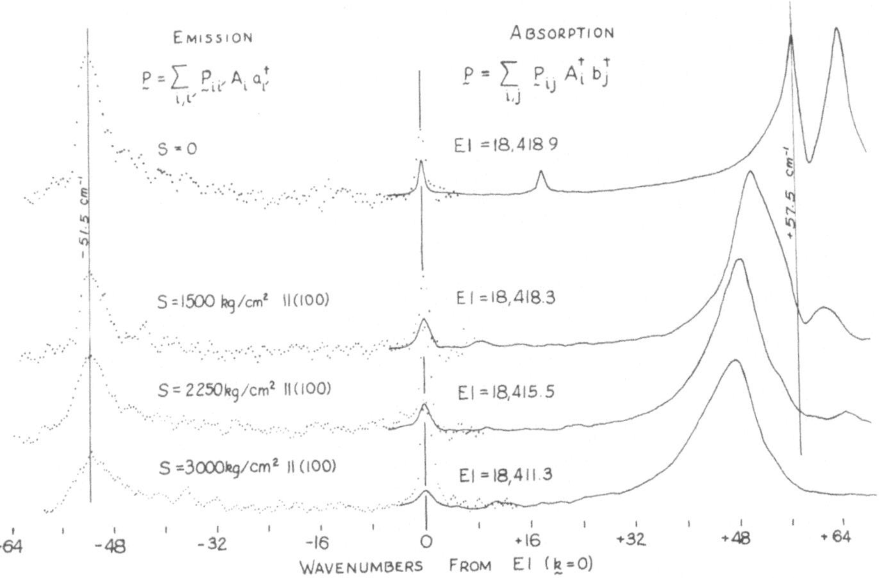

Figure 6. The effect of uniaxial (100) stress on the σ-polarized absorption and emission of the MnF_2 excitons E1 and E2 and their magnon sidebands at 1.5°K. (19) The ordinate for the absorption is proportional to the absorption coefficient, and for the emission, to the number of photons.

at the top of the Fig., the optical dipole moment operators are different; in absorption the moment is proportional to the simultaneous creation of an excitation on an ion on one sublattice and a spin deviation on a second neighbor which is on the other sublattice, while in emission the excitation and spin deviation couple on the same sublattice. These processes are well understood (16)(17)(18)(19), and it is possible to compute the shape of the intrinsic magnon sideband in emission in good agreement with experiment (19), as is shown in Fig. 7, with no adjustable parameters, and assuming that E1 has no dispersion. The reported uncertainty in the exchange parameter J_2 as determined from inelastic neutron scattering (20) is nearly large enough to account for the discrepancy between the predicted and observed zero magnon positions. This is not so for the absorption sidebands of E1 and E2, and it has been suggested (15)(16) that the absorption sideband shapes may be explained if exciton E1 has about 2.5 cm^{-1} positive dispersion, while E2 has about 7 cm^{-1} negative dispersion. This suggestion would seem to be bourne out by the effect of (100) stress on the absorption sideband shapes, since, as E1 mixes with E2 (and passes through an anticrossing(15)), the shape of the sideband of E1 converts into a shape like that of the sideband of E2 at zero stress. However this suggestion is contradicted by the lack of a similar effect of (100) stress on the emission sideband since it should also shift and change its shape as E1 acquires dispersion. It is thought that the slight broadening of the emission sideband at the higher stresses is the result of inhomogeneous strain and lower signal-to-noise. These results offer conclusive evidence that neither excitons E1 nor E2 have any spectrally measurable dispersion.

What then determines the shapes of magnon sidebands in absorption? We believe that the answer lies in the third possibility - namely, the fact that the exciton and magnon will interact strongly with each other <u>if they coexist in time</u>. Thus, the exciton and magnon may interact strongly in absorption at low temperatures since they are created simultaneously, while in emission at low temperatures they may interact only weakly, since the exciton is annihilated as the magnon is created. For transitions involving "hot" bands where a thermally populated magnon is annihilated, the interaction will be strong only for emission.

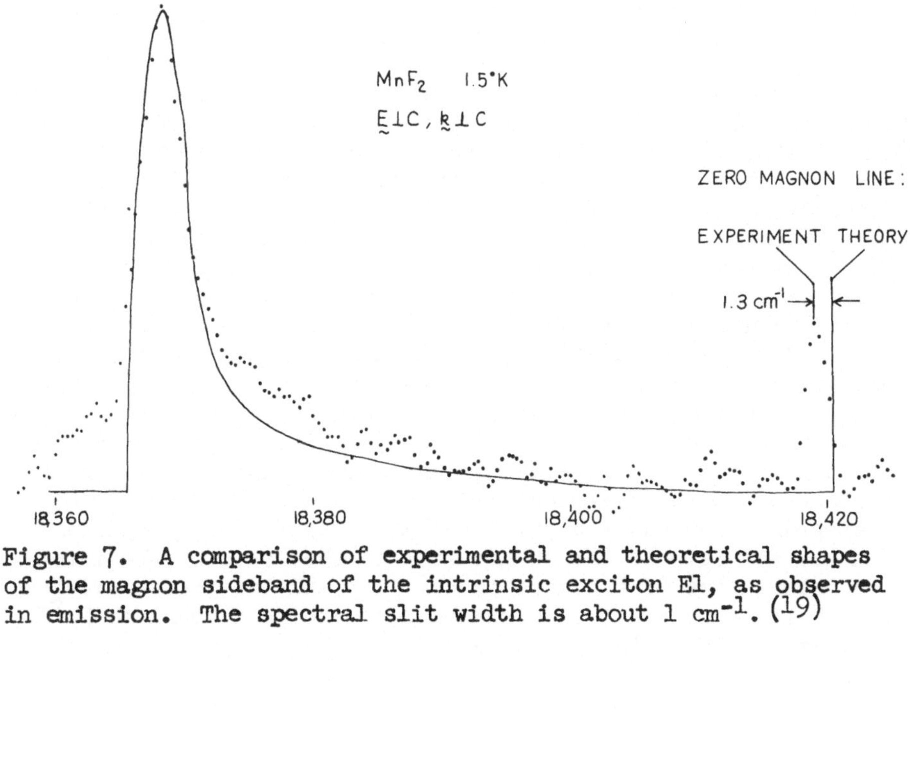

Figure 7. A comparison of experimental and theoretical shapes of the magnon sideband of the intrinsic exciton E1, as observed in emission. The spectral slit width is about 1 cm^{-1}.[19]

To compute the effect of exciton-magnon interaction on the sideband shape, we have taken the interaction hamiltonian in second order perturbation

$$\mathcal{H} = \sum_{i,j} C_{ij} (A_i^\dagger b_j^\dagger + A_i b_j)$$

which yields on Fourier transformation a wavevector dependent correction to the combined exciton-magnon energy. The effect of this wavevector dependent correction is to change the relative separation of the two exciton-magnon states (i.e., the sidebands of E1 and E2), causing them to mix: the former effect resulting in a change of sideband shape, and the latter in a change of absorption strength. Preliminary calculations of the absorption sideband shapes using only two adjustable coupling constants (C_{ij}) indicate that a single set of constants will permit the calculation of the σ polarized sideband shapes and relative intensities for any value of stress in good agreement with experiment.[21]

It should be noted that the exciton-magnon interaction will affect magnon sidebands of excitons bound to impurities under the same conditions as for the intrinsic transitions. Thus this interaction need not be considered in analysing the shapes of the low-temperature emission sidebands of excitons bound to impurities, as discussed in the following paper.[12]

Finally we may inquire why the exciton states have no measurable dispersion. Loudon[18] has given formulas for calculating the dispersion of the excitons, and if one uses his formulas, and assumes that the off-diagonal matrix element responsible for exciton dispersion is about equal to J_1, the diagonal intrasublattice exchange integral for the ground state, one would predict an exciton dispersion of about 2 cm^{-1}. However, the off-diagonal exchange will be reduced from this figure if one takes into consideration the nuclear displacements during the transition. Since the excited state has a different orbital configuration from the ground state, it is expected that these states will have different equilibrium nuclear configurations. In order to estimate how this affects the off-diagonal exchange, we approximate the true wavefunction by products of electronic and nuclear functions. Thus each electronic matrix element will be reduced by a vibronic overlap integral. Now the vibronic overlap integral for an optical matrix element is just the square root of the ratio of the integrated intensities of the zero phonon line and the phonon sideband. In Fig. 8 is shown the phonon sideband of excitons bound to Mg^{2+} impurities. Because these states are only weakly perturbed by

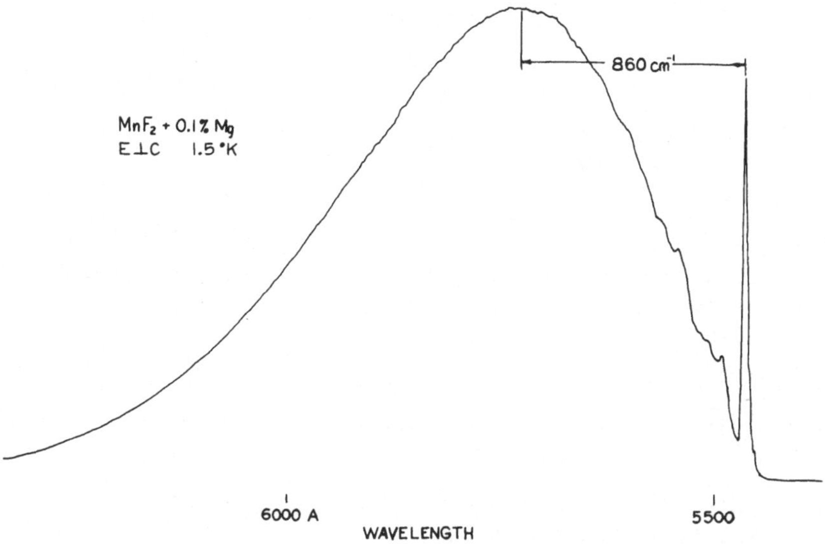

Figure 8. The electric dipole, σ-polarized, phonon sideband associated with the decay of excitons bound to Mg ions impurities in MnF_2. The strong, sharp line is the magnon sideband which constitutes the principal zero-phonon line.

the impurity potential, we expect that the vibronic overlap determined from this spectrum (0.1) is similar to that for intrinsic El excitons. This figure should be squared to produce 10^{-2}as a measure of the vibronic overlap for the off-diagonal exchange integral, in which the ground and excited state functions each appear twice. This factor reduces the predicted exciton dispersion to the order of 0.02 cm^{-1}, which is much less than the accuracy of our experiments. Such excitons would be localized at low temperatures by the random strain-modulation in a real crystal.

Not all excitons in magnetic crystals will have such small dispersion. For example, neutron scattering measurements (22) in CoF_2 indicate the low lying excitons (which have the same orbital configuration) have a few cm^{-1} of dispersion. Likewise other excited states of MnF_2, as the $^4A_1$ and ^4E, may show a measurable dispersion. However, proper allowance must be made for the exciton-magnon interaction when interpreting such data.

Acknowledgements: We are grateful for helpful discussions with M. D. Sturge, and with nearly all of those authors listed in the references.

REFERENCES

* Supported by the U.S. Air Force Office of Scientific Research

1. L. F. Johnson, R. E. Dietz, and H. J. Guggenheim, Phys. Rev. Letters 17, 13 (1966) and unpublished work.
2. A. Misetich and R. E. Dietz, Phys. Rev. Letters 17, 392 (1966).
3. J. Ferguson, H. J. Guggenheim, and Y. Tanabe, Phys. Rev. Letters 14, 737 (1965).
4. T. Tonegawa and J. Kanamori, Phys. Letters 21, 130 (1966).
5. S. W. Lovesey (to be published).
6. D. Hone and L. R. Walker (to be published).
7. S. J. Allen, Jr., R. Loudon, and P. L. Richards, Phys. Rev. Letters 16, 463 (1966).
8. J. W. Halley and I. Silvera, Phys. Rev. Letters 15, 654 (1965).
9. P. A. Fleury, R. Loudon, and S. P. S. Porto, Phys. Rev. Letters (1967).
10. J. Ferguson, J. J. Guggenheim, H. Kamimura, and Y. Tanabe, J. Chem. Phys. 42, 775 (1965).
11. R. E. Dietz, L. F. Johnson, and H. J. Guggenheim in Physics of Quantum Electronics edited by P. L. Kelley, B. Lax, and P. E. Tannenwald (McGraw-Hill Book Company, Inc., New York, 1966).
12. A. Misetich, R. E. Dietz, and H. J. Guggenheim (following paper).
13. R. L. Green, D. D. Sell, and G. F. Imbush (to be published).
14. R. L. Green, D. D. Sell, W. M. Yen, A. L. Schawlow, and R. M. White, Phys. Rev. Letters 15, 656 (1965).
15. R. E. Dietz, A. Misetich, and H. J. Guggenheim, Phys. Rev. Letters 16, 841 (1966).
16. D. D. Sell, R. L. Green, and R. M. White (to be published in Phys. Rev.)
17. Y. Tanabe and K. Gondaira, J. Phys. Soc. Japan 22, 573 (1967).
18. R. Loudon (to be published).
19. R. E. Dietz, A. Misetich, J. J. Guggenheim, and A. E. Meixner (to be published).
20. A. Okazaki, K. C. Turberfield, and R. W. H. Stevenson, Phys. Letters 8, 9 (1964).
21. A. Misetich, R. E. Dietz, A. E. Meixner, and H. J. Guggenheim (to be published).
22. R. A. Cowley, P. Martel, and R. W. H. Stevenson, Phys Rev. Letters 18, 162 (1967).

SPIN WAVE SIDE BANDS OF LOCALIZED EXCITONS

A. Misetich
National Magnet Laboratory,* MIT, Cambridge, Mass.

and R. E. Dietz and H. J. Guggenheim
Bell Telephone Laboratories, Murray Hill, New Jersey

Most of the visible emission in MnF_2 occurs at energies smaller than the corresponding transition in absorption. This emission has been shown to vary from crystal to crystal and is believed to be derived from the decay of excitons trapped on Mn^{++} ions perturbed by various impurities, such as Ca, Zn, Mg[1,2]. Since these localized impurity levels lie a few tens of wave numbers below the intrinsic levels, they form stable traps at low temperatures. This emission consists of a broad phonon side band and a number of zero phonon lines: narrow lines, identified as zero phonon zero magnon decay of localized excitons, and then spin wave side bands. The spin wave side band is a result of the creation of a spin wave simultaneously with the anihilation of a localized exciton. Pressure experiments as well as dopings with impurities were used to assign the side bands. Figure 5 in ref. 3 shows a typical spectra of a crystal doped mainly with Mg and some Zn. Table I summerizes the position of exciton lines as well as spin wave side bands with different dopings.

In this paper we report a theoretical study of the shapes of the spin wave side bands associated with the localized excitons. As a result we are able to identify the different localized excitons observed in emission and characterize the nature of the coupling between spin wave modes and localized excitons.

Essentially three different shapes for the spin wave side bands are observed: (a) having two peaks, one at 55 and another at 51 cm^{-1}, the one at 55 cm^{-1} being more intense; (b) having two peaks at 52 and at 46 cm^{-1}, the one at 46 cm^{-1} being more intense; (c) having one peak at 52 cm^{-1}. Note that the energies are measured from the corresponding zero magnon exciton transitions.

Table I. Emission lines in MnF_2 doped with impurities.

Doping	Exciton lines	Side Band peaks
Intrinsic emission	18,419.0 cm^{-1}	18,367.0
Ca^{++}	18,120.0	18,064.4 and 18,068.7
Mg^{++}?	18,371.8	18,319.8
	18,341.9	18,296.4 and 18,289.9
Zn^{++}	18,382.7	18,330.7
	18,352.7	18,307.2 and 18,300.7

Transition Probability for Spin Wave Side Bands

In the emission process we create a spin wave of momentum $\vec{k}$ at the same time that we anihilate an exciton with the same momentum. This spin wave is mainly localized on the same sublattice as that of the exciton that is anihilated. By analysis of the intrinsic emission spectra, that is, emission from non-localized excitons, we concluded that an excitation on one ion couples to spin deviations on first neighbors of it.[4] Stress experiments of the emission from localized excitation confirmed that the same mechanism is valid for this last case. Thus, if the excitation is localized on site i, we write the electric dipole moment for the emission side band as:

$$P = \sum_{i'} P_{ii'} A_i a_{i'}^{+} \tag{1}$$

where A_i is the anihilation operation of an excitation on ion i and $a_{i'}^{+}$ is the creation of a spin deviation on ion i'. Both i and i' are on the same sublattice, i' being a first neighbor of i. (There are two first neighbors, located along the z axis). If we write a_i^{+} as a sum over spin wave operators,

$$P = \frac{1}{\sqrt{N}} \sum_{k} \sum_{i'} P_{ii'} \; e^{-i\vec{k}\cdot\vec{r}_{i'}} \; u_{\vec{k}} A_i \, \alpha_{\vec{k}}^{+} \tag{2}$$

where α_k^{+} is the creation of a spin wave of momentum $\vec{k}$ mainly localized on sublattice i. Equation (1) also leads to a term proportional to β_k, anihilation of spin wave mainly on sublattice j, but this term will be important only at higher temperatures when spin waves are thermally excited.

We shall consider the different localized excitons that can be formed near an impurity and discuss which are the corresponding spin wave side band shapes we should observe in each case.

If the excitation is delocalized on n ions, so that the wave functions are linear combinations of excitations on each of the n ions:

$$\Psi = \frac{1}{\sqrt{n}} (C_1 \Psi_1 + C_2 \Psi_2 + \ldots\ldots + C_n \Psi_n)$$

we get for the electric dipole operator that creates a spin wave of momentum $\vec{k}$:

$$P_{\vec{k}} = \frac{1}{\sqrt{nN}} \sum_i C_i \sum_{i'} P_{ii'} \, e^{-i\vec{k}\cdot\vec{r}_{i'}} \, \mu_{\vec{k}} A_i \alpha^+_{\vec{k}} \tag{3}$$

where for each ion i we have to sum over the two first neighbors of it (sum on i') except when the place i' is occupied by a non magnetic impurity.

(a) Excitation localized on Mn^{++} first neighbors to the impurity. There are two Mn^{++} ions first neighbors to the impurity, located at [001] and [00⁻1], assuming the impurity is at site [000]. The exciton wave functions will be linear combinations of the excitations of these two ions:

$$\Psi_g = \frac{1}{\sqrt{2}} (\Psi_{001} + \Psi_{00^-1})$$

$$\Psi_u = \frac{1}{\sqrt{2}} (\Psi_{001} - \Psi_{00^-1}) \tag{4}$$

When the excitation is on site [001] it can couple to a spin deviation on the ion [002] (which is a Mn^{++}) but not to spin deviation on ion [000], since it is now occupied by the non magnetic impurity, like Ca^{++}. The same for excitation on site [00⁻1], since now it can only couple to spin deviations on site [00⁻2]. Therefore, the presence of the impurity on site 000 will make this case quite different than the emission from unperturbed Mn^{++} ions, where an excitation on one ion can couple to two Mn^{++} ions first neighbors of it. Making use of Eqs. (3) and (4) we get

$$P_{\vec{k},\,{}^g_u} = \frac{1}{\sqrt{2N}} \left(P_{001,002}\, e^{-i2k_z c} A_{001} \pm P_{00^-1,00^-2}\, e^{i2k_z c} A_{00^-1} \right) \times$$

$$\mu_{\vec{k}} \alpha^+_{\vec{k}}\, , \tag{5}$$

the upper sign for Ψ_g and the lower one for Ψ_u. Let us consider the case of light polarized along the ξ axis (110). In that case by symmetry

$$P^{\xi}_{001,002} = -P^{\xi}_{00^-1,00^-2}$$

$$P^{\xi}_{\vec{k},g} = -i \sqrt{\frac{2}{N}}\, P^{\xi}_{001,002} \sin(2 k_z c)\, \mu_{\vec{k}} A_g \alpha^+_{\vec{k}}$$

$$P^{\xi}_{\vec{k},u} = \sqrt{\frac{2}{N}}\; P^{\xi}_{001,002} \cos(2k_z c)\, \mu_{\vec{k}} A_u\, \alpha^{+}_{\vec{k}}$$

If we assume that Ψ_g and Ψ_u have the same energy,[5] the intensity for the creation of spin waves of momentum $\vec{k}$ will be

$$I_{\vec{k}} = \left| P^{\xi}_{\vec{k},g} \right|^2 + \left| P^{\xi}_{\vec{k},u} \right|^2 = \frac{2}{N} \left| P^{\xi}_{001,002} \right|^2 \mu_{\vec{k}}^2$$

Except for the factor $\mu^2_{\vec{k}}$ the intensity becomes independent of the k vector and therefore we expect to obtain a side band proportional to the spin wave density of states, since $\mu^2_{\vec{k}} = 1$ at the Brillouin zone boundary, where the density of states is maximum. Figure 1 shows the theoretical side band we predict in this case. This is the shape observed when we dope MnF_2 with Ca^{++}, as shown in the same figure.

The same result is obtained even if the exciton is localized only on one of the two first neighbors to the impurity, 001 or $00\bar{1}$, instead of forming an orbital involving both of them.[6]

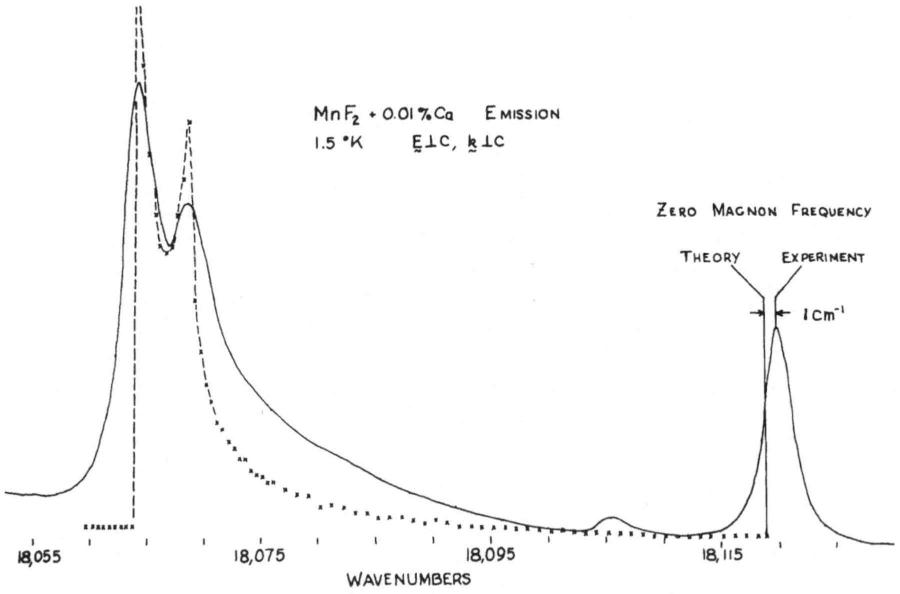

Fig. 1. Spin wave side band observed in σ emission of MnF_2 doped with Ca^{++}.

(b) Excitation localized on Mn^{++} second neighbors to the impurity. There are eight Mn^{++} second neighbors to an impurity. They are equivalent to each other in the unperturbed crystal. However, the presence of the impurity breaks the equivalency, resulting into two kinds of second neighbors (kind I formed by Mn^{++} located at $(\frac{1}{2},\frac{1}{2},\frac{1}{2})$, $(-\frac{1}{2},-\frac{1}{2},\frac{1}{2})$, $(-\frac{1}{2},-\frac{1}{2},-\frac{1}{2})$ and $(\frac{1}{2},\frac{1}{2},-\frac{1}{2})$; kind II formed by the other four Mn^{++} ions). In any of those cases we shall have four exciton wave functions, formed as linear combinations of these four ions, and we should apply Eq. (3) for each of them. However it is not so simple, as we illustrate in the following example: when the excitation is on a Mn^{++} located at $(\frac{1}{2},\frac{1}{2},\frac{1}{2})$ it couples to spin deviations on sites first neighbors to the excitation, $(3/2,\frac{1}{2},\frac{1}{2})$ and $(-\frac{1}{2},\frac{1}{2},\frac{1}{2})$. Site $(3/2,\frac{1}{2},\frac{1}{2})$ is far enough from the impurity so that we can apply the spin wave decomposition illustrated in Eq. (2). However, site $(-\frac{1}{2},\frac{1}{2},\frac{1}{2})$ is a second neighbor to the impurity and therefore it will be strongly affected by it; for instance its effective magnetic field is reduced by approx. 7. cm^{-1} and in the spin wave decomposition we have to put more weight than before for spin waves around 45 cm^{-1}. This may explain the two peaks (52 and 45 cm^{-1}) observed in one side band of MnF_2 doped with Zn or Mg (see Table I). A quantitative answer requires the solution of spin waves in an antiferromagnet with non-magnetic impurities.

(c) Excitation localized on Mn^{++} third neighbors to the impurity. We have four third neighbors located at (1, 0, 0), (0, 1, 0), ($\bar{1}$, 0, 0), (0, $\bar{1}$, 0). As before, we can form the four exciton wave functions as linear combinations of excitations at those four sites and couple each excitation to two possible spin deviations; for instance excitation at (1, 0, 0) will couple with spin deviations at (1, 0, 1) and (1, 0, $\bar{1}$). These two positions are far enough from the impurity and will not be affected by it. We get

$$I^{\xi}_{\vec{k}} = \left(P^{\xi}_{\vec{k}}\right)^2 = 4 \times \left(P^{\xi}_{001,002}\right)^2 \sin^2 (k_z c)\, u^2_{\vec{k}}$$

The same result is obtained if we assume that the excitation is localized on just one neighbor instead of forming a combination of the four third neighbors. Figure 2 shows the theoretical shape, and by comparison, one of the side bands observed in Zn.

It is interesting to point out that the shape predicted in this case is the same as the one predicted for the side band of the intrinsic emission[4], that is emission from unperturbed Mn^{++} ions. Fig 3 shows a unit cell of MnF_2, rutile structure, and the position of the different kinds of neighbors.

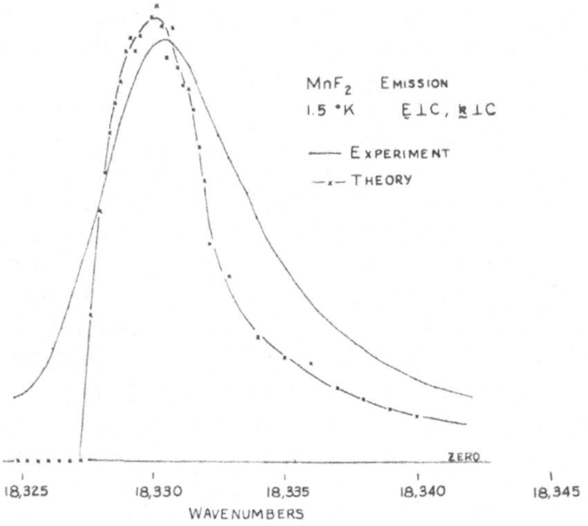

Fig. 2. Spin wave side band observed in σ emission of MnF_2 doped with Zn^{++}.

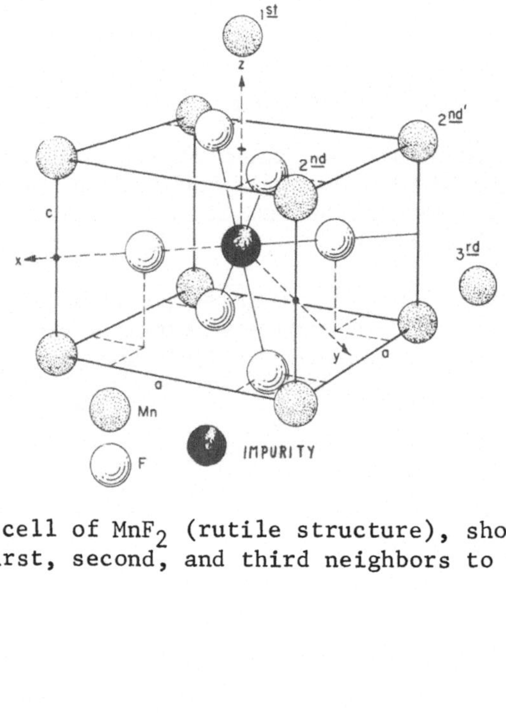

Fig. 3. Unit cell of MnF_2 (rutile structure), showing the position of first, second, and third neighbors to the impurity.

REFERENCES

*Supported by U. S. Air Force Office of Scientific Research.

[1] R. L. Greene, D. D. Sell and R. M. White, Proc. Conf. on Optical Properties of Ions in Crystals (to be published).

[2] R. E. Dietz, H. J. Guggenheim and A. Misetich, Proc. Conf. on Optical Properties of Ions in Crystals (to be published).

[3] R. E. Dietz and A. Misetich, accompanying paper.

[4] R. E. Dietz, A. Misetich, H. J. Guggenheim and A. E. Meixner (to be published).

[5] The difference in energy between ψ_g and ψ_u will be of the order of the dispersion of the intrinsic exciton EI from which they were formed. Experiments on the intrinsic exciton side band emission suggest such dispersion is less than 1 cm^{-1}.

6 D. D. Sell (Private communication) has reached similar conclusions as to the form of the selection rule for Ca in MnF_2 using symmetry considerations.

STUDIES OF LOCALIZED MODES BY SPIN-LATTICE RELAXATION MEASUREMENTS*

J.G. Castle, Jr.†

Westinghouse R & D Center

Pittsburgh, Pennsylvania

Observations of spin-lattice relaxation(1) have established the presence of vibrations localized at the site of each of several types of electron spin centers in dielectric solids. Even with the advent of Raman scattering of laser light and far infrared absorption techniques, there are some centers whose localized modes are being identified by spin relaxation. It is the purpose of this paper to discuss briefly several cases, including high frequency local modes, very low frequency resonance modes and one intermediate case which will be called gap modes. We will derive the frequency of the dominant localized vibrations in each case and note the evidence of strong anharmonicity. We omit the many interesting effects of Jahn Teller modes on spin relaxation.(2)

RAMAN SCATTERING OF PHONONS IN PERFECT LATTICE MODES

To deduce the presence of spin relaxation due to localized vibrations, we must calculate(1) what to expect from the 'perfect' harmonic lattice. Recall that phonon-induced relaxation events in a weakly coupled spin system are independent and therefore their relaxation frequencies are additive. Occasionally one spin-lattice coupling mechanism between a spin and its immediate neighbors will be a priori the most significant one, but usually several types of coupling may be effective and therefore need to be added together. We recommend in the summations(1) that a reasonable approximation to the phonon spectrum of the undisturbed lattice be used whenever the spectrum is known. So for the cases in which relaxation is described by a single relaxation time, T_1, we expect that

$$1/T_1 = A\,T + B\,X_{TA}^{-7}\,J_6(X_{TA}) + B'\,X_{TA}^{-9}\,J_8(X_{TA})$$
$$+ C\,X_{LA}^{-7}\,J_6(X_{LA}) + + C'\,X_{LA}^{-9}\,J_8(X_{LA})$$
$$+ D\,X_{TA}^{-5}\,J_4(X_{TA}) + D'\,X_{LA}^{-5}\,J_4(X_{LA})$$
$$+ E\,e^{X_{TA}}/\,(e^{X_{TA}}-1)^2 + \sum_i F_{Oi}\,e^{X_{Oi}}/\,(e^{X_{Oi}}-1)^2. \qquad (1)$$

where the summation is to be taken over the effective optical modes, $J_m(X) = \int_0^X \chi^n e^{\chi}(e^{\chi}-1)^{-2}\,d\chi$, and $X_j = hf_j/kT$. Eq.(1) is obtained for the harmonic lattice by assuming the relative displacements of the spin and its neighbors are very small in every mode, that the modes are sufficiently anharmonic to keep the phonons in a thermal distribution, that the Zeeman energy is much less than kT, and that the angular integrations over the acoustic modes that are effective in relaxation gives single Debye spectra for TA and LA modes plus an Einstein peak at X_{TA}. The first term (direct processes) involves only phonons with the energy of an electronic splitting of the spin center. The remaining terms describe inelastic (Raman) scattering of phonons by the spin.

A few cases have been compared with theory in the form of eq.(1). Many rare earth ions, having Kramer's doublets lowest, show the expected T^9 dependence but the data, to our knowledge, do not test the complete function $X_{TA}^{-9}\;J_8(X_{TA})$. For the F center in KCl, T_1 has been observed(3) to vary with T from 1000 seconds near 2^oK to 100 microseconds near 150^oK. Considering the nominal ±10% uncertainty, these data can be fitted(3) by either of two Raman terms:

$X^{-7}\,J_6(X)$ with $XT = 120^oK$ which could arise from magnetic hyperfine coupling or $X^{-9}J_8(X)$ with $XT = 170^oK$ which could arise from crystal field modulation. Even with the range of 1,000,000 in the data the two functions fit within the uncertainty in the data, possibly a stricter application of the recently measured phonon values would permit discrimination.

For the $3d^5$ Mn^{2+} ion substituted in ZnS we might expect to have the slightly larger polarizability of the Mn^{2+} ion compensate for the slightly smaller mass and allow the spin-phonon coupling to be described by Eq.(1). We have recently found[4] by monitoring recovery of equilibrium magnetization after inversion that T_1 can be reasonably well fitted by a term for direct processes plus a single Raman term for the perfect ZnS lattice. The data are given in Fig. 1, where 175°K is the top of the TA modes of ZnS.

RAMAN SCATTERING BY LOCAL MODES

Localized vibrations to the extent they are harmonic will depress the values of some of the coefficients

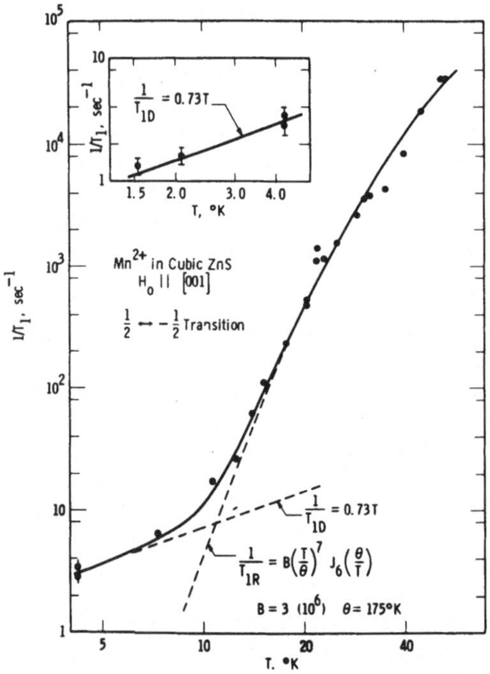

Fig. 1. Spin-Lattice Relaxation observed for Mn^{2+} in ZnS at an applied magnetic field of 3kOe. The inset shows the low temperature region on the same scales. The solid curve is the best fit for the direct process term plus a single function due to summations over the acoustical phonons of ZnS.

in Eq.(1) and add a term of the form $e^{X_L}/(e^{X_L}-1)^2$ where $X_L = hf_L/kT$. To the degree that the relative displacement of the spin and its surroundings is large within a localized mode, a given excitation of that mode will relax the spin more rapidly.

The Li^+ ion in ZnS is expected to have a local mode above the optical mode frequencies for ZnS from infra red absorption measurements on other II-Vl compounds.[5] We have observed an ESR line of the Mn^{2+} ion in ZnS containing Li to broaden very rapidly with increasing T. Fig. 2 shows the reciprocal lifetime, R, calculated from these data. The 440 cm^{-1} energy is clearly above the band modes for ZnS, and may be due to local modes involving Li^+ ions. When proper account of the relaxation shown in Fig. 1 is taken for the temperature range of Fig. 2, there may be significant contribution to the line width from the T^7 J_6 term and the local mode frequency will be greater than 440 cm^{-1}.

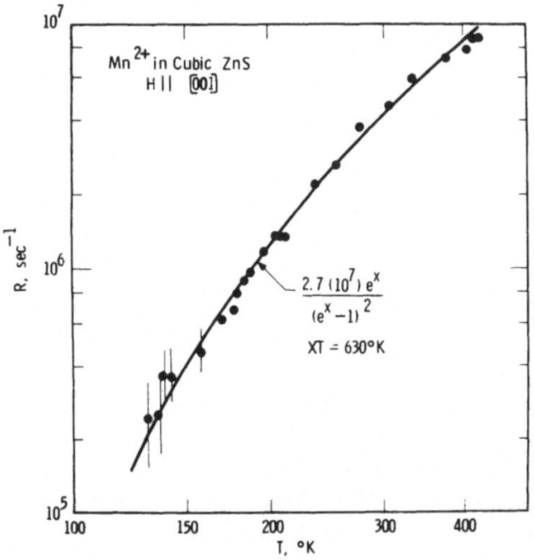

Fig. 2 Relaxation of Mn^{2+} in ZnS containing Li. R is the reciprocal of the lifetime calculated from the observed broadening of the -½ to ½ transition at 3kOe. The solid curve is the best fit to the data of a single Einstein oscillator function (shown) without subtracting any contribution to R from lattice modes of ZnS. The 440cm^{-1} is tentatively assigned to Li^+ vibrations.

Atomic hydrogen is put in the tightest container known when it is in CaF_2 surrounded by eight F^- ions. We have reported(6) measurements of T_1 over a wide range of T and interpreted the data in terms of band modes, using the specific heat Debye limit of 474°K, and local modes at 850±50°K. Fig. 3 gives the same data and an improved interpretation in terms of the observed phonon spectrum of CaF_2. The frequency of the local modes may be as high as 1000°K but 850±50°K is preferred as the best overall fit to the observations.

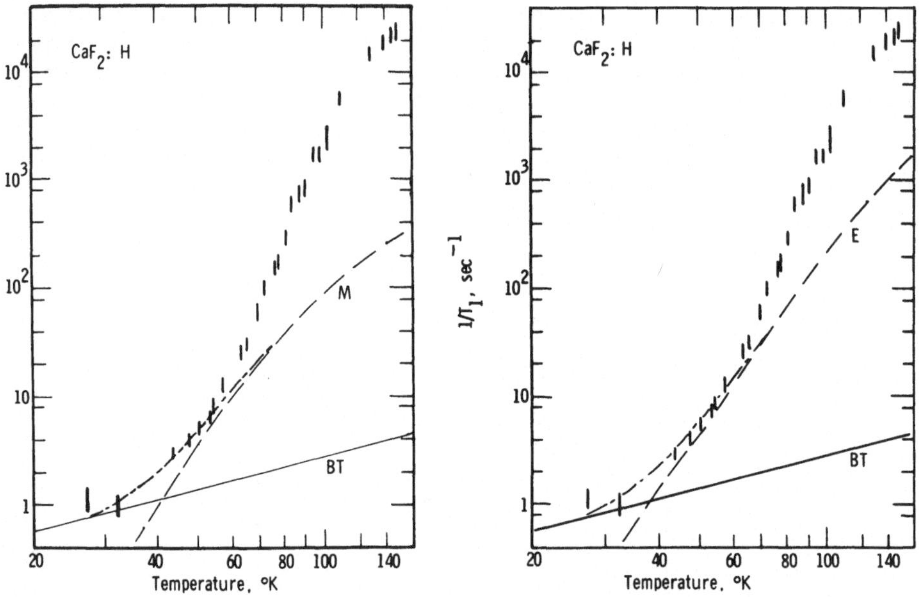

Fig. 3 Spin-Lattice Relaxation observed for Atomic Hydrogen in CaF_2 at 8.8Gc. The solid curve is the relaxation due to direct processes with $B = 0.028\ Sec^{-1}$, as determined to ±10% by measurements below 20°K. In a, the dashed curve $M = 1.5(10^4)\ X_{TA}^{-7}\ J_6(X_{TA})$ with $X_{TA}T = 400°K$ is the most relaxation that can be attributed to Raman relaxation by TA phonons of CaF_2 with a term of this form. In b, the dashed curve $E = 2(10^3)X_{TA}^{-9}J_8$

$$(X_{TA}) \quad + \quad 1(10^4)\ X_{LA}^{-9}\ J_8(X_{LA}) \quad + \quad 5(10^4)\ e^{X_{LO}}/(e^{X_{LO}}-1)^2$$

with the CaF_2 lattice values $X_{TA}\ T = 250°K$, $X_{LA} = 400°K$, and $X_{LO}\ T = 580°K$. The double dashed curve is the sum of the other two curves in each case. The excess relaxation (above the double dashed curve) is assigned to the local mode of the hydrogen with $hf_L/k = 850°k$ from a or 950°K from b; the 850±50°K value is preferred.

RAMAN SCATTERING BY HYDROGEN ATOMS IN SiO_2

Atomic hydrogen goes into SiO_2 in a site which is apparently at or near a vacancy. Our preliminary measurements of T_1 for H^o in crystalline quartz for T up to 110 °K show (7) direct processes plus a term exponential in T with an activation temperature of 170 ± 15 °K. This activation energy is presumably due to localized vibrations involving the hydrogen atom.

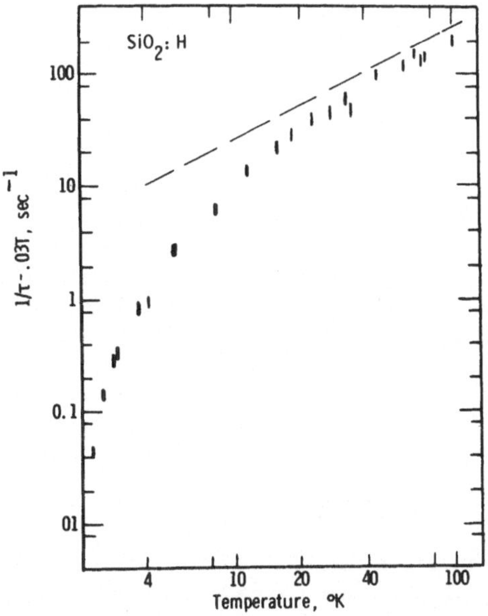

Fig. 4. Spin-Lattice Relaxation of Atomic Hydrogen in fused silica at 9 Gc due to Raman scattering of thermal phonons. The direct relaxation of 0.04 T has been subtracted from the observed data. The best fit of a single relaxation function to the data is that of a deep, square well with the first energy gap of 13.5 ± 1.5 °K; an acceptable fit is a tunnelling oscillator withan energy splitting of 16 °K. The dashed line (= 20 T) is shown to indicate that the asymptote of the relaxation data is T^1 and not the T^2 dependence required of harmonic modes.

In order to check on their anharmonicity, we would like to measure T_1 at higher T but unfortunately the H^o disappears. On the other hand H^o centers infused silica have a proton electron hyperfine coupling which indicates a somewhat larger volume for the H^o than in crystalline quartz. Our reported(8) measurements of T_1 show that very low frequency ($\sim 10cm^{-1}$) resonance modes dominate the relaxation up to T = 100^oK. These data (Fig. 4) show by approaching T^1 and not T^2 that the motion is very anharmonic near the H^o site. If one considers the mode to be a one dimensional square well in which the hydrogen atom moves, then the well is about 7A° long. In any case the relaxation apparently occurs by scattering between band mode phonons and motion in the resonance mode such that its amplitude is independent of temperature.

RAMAN SCATTERING BY GAP MODES

One paramagnetic center known to have modes with energy between the normal optical bands and the transverse acoustic bands of the host crystal is Cr^{3+} in MgO. We have recently extended our measurements(9) of relaxation to higher T and preliminary indications are that the best fit to all the Raman relaxation is a single Einstein function with the energy of 540^oK. This value agrees with a prominent peak observed in the phonon-induced sidebands of the R line but falls below the optical bands identified by neutron scattering.(10)

SUMMARY

We conclude that spin-lattice relaxation is observed to be dominated by the localized vibrations present at spin centers in a variety of cases. The frequencies of these localized vibrations can be determined from the observed dependence of relaxation on temperature. However, the value calculated for the high frequency local modes depends on how much of the relaxation is attributed to the acoustical and optical phonon scattering among the modes of the 'perfect' lattice; a consistent procedure is suggested. The role of 'gap' modes can be supported by observations of phonon-induced sidebands to the optical emission lines of the same spin center whose relaxation is observed. Gap modes appear to dominate spin-phonon scattering of Cr^{+3} in MgO over the complete range of Raman relaxation from 40°K to 400°K. The frequencies of very low 'resonance' modes can be determined with sufficient accuracy to show the presence of anharmonicity equivalent to motion in a square well or to tunnelling modes for the defect.

ACKNOWLEDGEMENTS

Collaboration with M. Abraham, D.W. Feldman, G.R. Wagner, P.G. Klemens, J. Murphy, R.W. Warren and R.A. Weeks is gratefully acknowledged.

*This work was supported in part by the USAF Aerospace Research Laboratory, WPAFB, Ohio.

† Permanent address: Dept. of Electrical Engineering, University of Pittsburgh, Pittsburgh, Pa. 15213.

REFERENCES

1. For a review of spin-lattice relaxation, an excellent collection of pertinent papers is contained in the convenient paperback volume by A.A. Manenkov and R. Orbach, "Spin-Lattice Relaxation in Ionic Solids", Harper & Row, New York City (1966).

2. c.f. V.T. Hochli & T.L. Estle Phys Rev. Lett 18,128 (1967).

3. D.W. Feldman, R.W. Warren, and J.G. Castle Jr., Phys. Rev. 135, A470 (1964).

4. G.R. Wagner, J. Murphy, and J.G. Castle Conference on II-V1 Semi-conducting Compounds, Brown University, September 6-8, 1967, proceedings to be published.

5. c.f. M. Balkanski, R.Beserman, & L.K. Vodopianov, this conference, paper C2.

6. D.W. Feldman, J.G. Castle,Jr., and J. Murphy,Phys. Rev 138, A1208(1965). The CaF_2 phonon spectrum has been reported by D. Cribier, B. Farnoux, and B. Jacrot in "Inelastic Scattering of Neutrons in Solids & Liquids" II p225, IAEA Vienna

7. J.G. Castle,Jr, D.W. Feldman, and G.R. Wagner Bull. Am.Phys. Soc 11, 907(1966).

8. D.W. Feldman, J.G. Castle,Jr., and G.R. Wagner, Phys. Rev. 145,237 (1966).

9. R.L. Hartman, A.C. Daniel, J.S. Bennett and J.G. Castle, Jr., Bull Am Phys Soc 11,313 (1966).

10. G. Peckham, Proc. Phys. Soc.(London) (1966).

STUDY OF THE RESONANT PHONON SCATTERING IN KCl:Li USING THE SPIN-PHONON INTERACTION*

D. Walton

Solid State Division, Oak Ridge National Laboratory

Oak Ridge, Tennessee

INTRODUCTION

The recent interest in defect modes has been fueled in large measure by their effect on the low temperature thermal conductivity of the host crystal. Beginning with the work of Walker and Pohl,(1) the change in thermal conductivity as a function of temperature has been used as a spectrometer to study those modes whose energies lie in the acoustic range.

A variation of this technique will be described here which, in effect, improves the resolution of the phonon spectrometer. The technique depends on the presence of magnetic ions in the crystal, and the subsequent removal from the spectrum of heat carriers, of a band of phonons close to the Larmor frequency for the spin system. Since this frequency is a function of an applied magnetic field it is possible to obtain, by subtraction, the distribution function for the carriers.(2) The distribution function, in turn, reflects the scattering of the phonons by non-magnetic defects in the crystal. The results of applying this technique to the study of the tunneling states of the Li^+ ion in KCl will be described.

The Spin-Phonon Interaction

The "gap" in the phonon spectrum can be readily understood by asking for the normal modes of the coupled spin-lattice system. The

*Research sponsored by the U. S. Atomic Energy Commission under contract with the Union Carbide Corporation.

theory of this effect was first given by Jacobsen and Stevens,[3] and later extended by Sears,[4] Tucker,[5] and Parkinson.[6] Following Sears the Hamiltonian for the system may be written as the sum of three terms

$$\mathcal{H} = H_L + H_S + H_{S-L}$$

where H_L is the contribution of the lattice. In terms of phonon creation and annihilation operators

$$H_L = \sum_{qp} (a^+_{qp} a_{qp} + 1/2)\, \hbar\omega_{qp} \quad .$$

The sum is over all wave vectors, q, and polarizations, p.

The contribution of the spin system alone is contained in H_S. We will assume that the temperature is sufficiently low that only the two lowest levels of the spin system are populated, and that they are split by the Zeeman energy, $g\beta H$. With this approximation it is possible to define two operators b^+ and b^- which have the following properties.

$$b|-\rangle = |+\rangle \qquad\qquad b^+|-\rangle = 0$$

$$b|+\rangle = 0 \qquad\qquad b^+|+\rangle = |-\rangle$$

where $|-\rangle$ and $|+\rangle$ denotes respectively the ground and excited states of the spin system.

The spin system has become one of effective spin 1/2. It is important to distinguish between this and a system of real spin 1/2, as discussed by Sears.[4] In this approximation

$$H_S = \sum_{i=1}^{N_s} (b^+_i b_i - 1/2)\, \hbar\omega_o$$

where the sum is over all magnetic ions, N_s, and $\hbar\omega_o = g\beta H$.

The spin-spin interaction is ignored specifically in the interaction part, H_{S-L}, which is limited to the spin phonon interaction. This part is written

$$H_{S-L} = N_s^{-1/2} \sum_{i=1}^{N_s} \sum_{qp} (b_i + b^+_i)[A_{qp} a_{qp} \exp.(i\vec{q}\cdot\vec{R}_i) + \text{H.C.}]$$

where R_i is the vector to atom i, and A_{qp} is given by, and again we follow Sear's treatment,

$$A_{qp} = \varepsilon \left(\frac{\hbar^2 \omega_o v q}{12} \right)^{1/2}$$

v is the velocity of sound, and ε is a dimensionless parameter which measures the strength of the spin-lattice coupling. This expression for A_{qp} is the limit which is approached as q→o, and therefore is valid only for small q.

The resultant Hamiltonian can be diagonalized in the random phase approximation. If there is one magnetic ion per unit cell in the crystal, the results of such a diagonalization yield the following dispersion relations[4]

$$(\omega^2 - \omega_o^2)(\omega^2 - \omega_q^2) = 2\varepsilon^2 < S_z > \omega_o^2 \omega_q^2 \qquad (1)$$

where $< S_z >$ represents a thermal average.

For a dilute crystal an average over configurations can be performed to restore translational invariance to the system. The procedure followed in Reference 4 for diagonalization can then be used with the result that to first order the thermal average of S_z is multiplied by the average number of magnetic ions per unit cell.[7] The correction term will be of the order of the square of the concentration and can be neglected for a dilute system.

The dispersion relations are shown schematically in Fig. 1. It can be seen that excitations in the spin-phonon system suffer anomalous dispersion in the neighborhood of the Larmor frequency of the spin system, ω_o. This is nothing more than the acoustical analogue of the phenomenon of anomalous dispersion of light traveling through an absorptive medium.

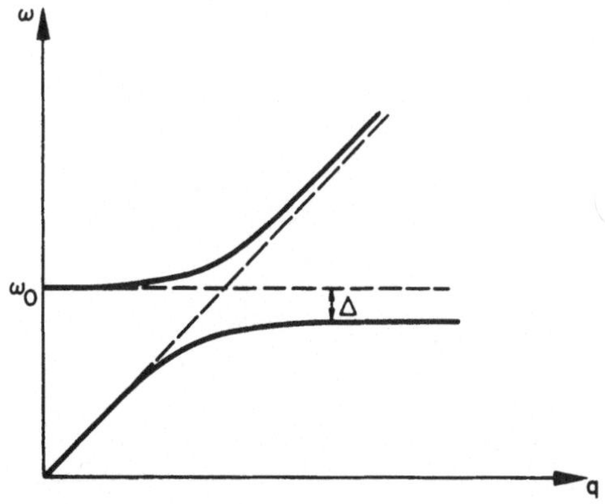

Fig. 1. Dispersion relation for effective spin 1/2.

In addition to this, excitations in the spin system of large q have their frequency reduced by an amount Δ where

$$\Delta = \omega_o[1 - (1 - 2\varepsilon^2 < S_z >)^{1/2}] \quad . \tag{2}$$

Therefore, excitations where energy lies between $\omega_o - \Delta$ and ω_o are no longer present. Because of the absence of these carriers the thermal conductivity will be reduced.

The Thermal Conductivity

The thermal conductivity may be written[8]

$$K = \frac{1}{2\pi^2} \int_o^{q_m} F(q)\, dq \tag{3}$$

where q_m is the maximum value of the acoustic wave-vector, and

$$F(q) = \frac{e^{\hbar\omega_q/kT}}{(e^{\hbar\omega_q/kT} - 1)^2} \left(\frac{\hbar\omega_q}{kT}\right)^2 \frac{v_q^2\, q^2}{\Sigma 1/\tau_q} \quad . \tag{4}$$

In this expression $\Sigma 1/\tau_q$ is a sum of reciprocal relaxation times, and the sum is over all the scattering processes in the crystal.

A schematic plot of F(q) is shown in Fig. 2 for a crystal without magnetic ions and in Fig. 3 for a crystal with magnetic ions. Because of the anomalous dispersion referred to above, the conductivity of the crystal which contains the magnetic ions will be reduced.

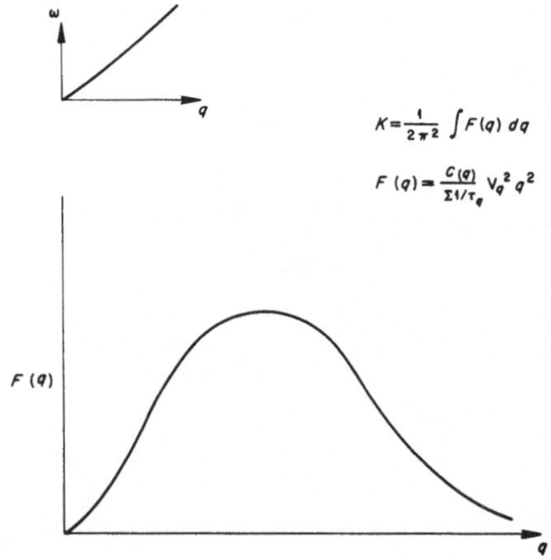

Fig. 2. The distribution of carriers as a function of wave-vector for a crystal in the absence of a spin-phonon interaction.

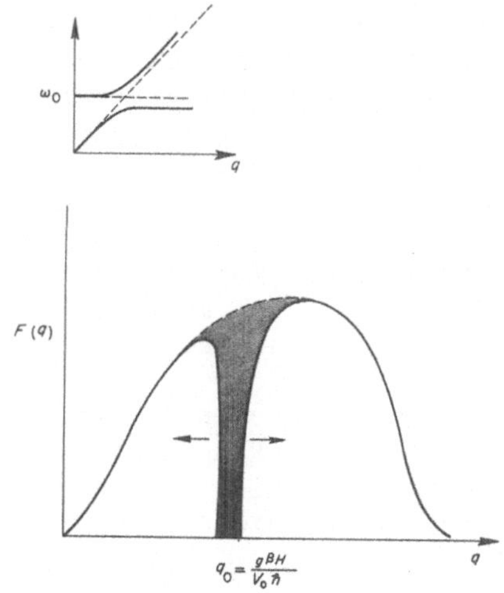

Fig. 3. The distribution of carriers for a crystal with a spin-phonon interaction present.

The hatched area in Fig. 3 shows schematically the result of suppression of the carriers near the Larmor frequency by the magnetic ions. This area, and hence the change in thermal conductivity, is proportional to F(q). Thus, the change in thermal conductivity as a function of applied magnetic field is a measure of F(q). Unfortunately, the width of the hatched region in general is also a function of field, and the relationship is not direct.

Reference to Eq. (2) reveals that this width is also a function of $< S_z >$. Since a change in the concentration of magnetic ions can change this average value, it can be seen that the resolution of the "spectrometer" is adjustable.

The Technique

Equation (4) contains the term $\Sigma 1/\tau_q$ which is proportional to the sum of the independent scattering cross-sections for phonons in the crystal. The presence of defects which scatter phonons in the crystal modifies F(q). Therefore, the scattering cross-section as a function of frequency for a non-magnetic defect can be obtained by introducing a magnetic defect in addition to the non-magnetic impurity, and by observing the change in thermal conductivity as a function of magnetic field. The magnetic impurity acts as a probe with which the scattering by the non-magnetic impurity is determined.

Application of this technique in general requires a knowledge of the change in the width of the hatched area in Fig. 3 with magnetic field, i.e. the change in spectrometer resolution. However, if the scattering by the defect being studied does not change F(q) too drastically, this requirement may be circumvented by essentially calibrating the "spectrometer."[9] The change in conductivity with field is obtained first for a crystal containing only the magnetic ion. Then the experiment is repeated with a crystal which contains, in addition, the non-magnetic ion. The cross-section of the non-magnetic ion can then be obtained by subtraction.

These difficulties are not present if resonant phonon scattering by a non-magnetic defect is taking place and the frequencies of the resonances are desired. This is a particularly simple application of the spin-phonon technique which is being used to study the Li^+ ion in KCl.[10]

The Li^+ ion in KCl is expected to occupy a position displaced from the center of the unit cell[11,12] in a [111] direction. If the potential is isotropic, the ground state is split by tunneling into four equally spaced levels.[12] If phonons are scattered resonantly by transitions from the ground state to these levels, F(q) for Li^+ doped KCl would be expected to look something like Fig. 4. If a magnetic impurity is now added to the Li doped KCl and a magnetic field is applied, the anomalous dispersion due to the spin-phonon interaction should remove the hatched band shown. Coincidence of the Larmor frequency with a phonon resonance would reveal itself by recovery of the conductivity towards its zero field (or very high field) value.

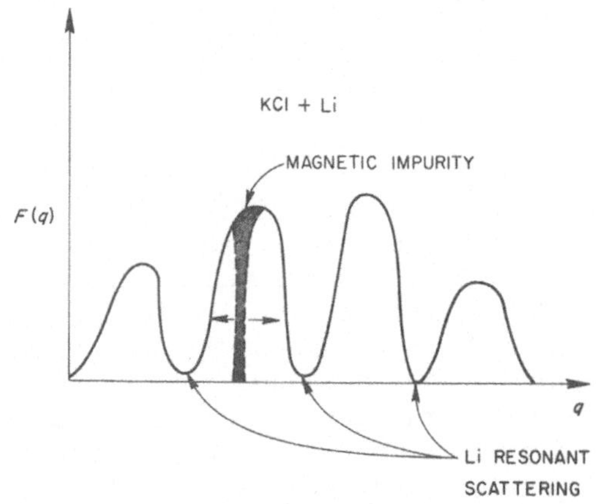

Fig. 4. Expected carrier distribution for Li doped KCl. The hatched area represents the additional effect of a magnetic impurity.

The Experiment

We have found that irradiated KCl exhibits a magnetic field dependent decrease in thermal conductivity. Since the effect scales roughly with the concentration of R-centers, we have concluded that the R-center is the magnetic defect responsible for the effect. Thus the R-center can be used as a convenient "probe."

High purity KCl obtained from the pure materials program at ORNL was used to grow Li doped single crystals. These were then irradiated in a Co^{60} gamma source at liquid nitrogen temperature. After irradiation they were bleached in room light to maximize the R-center concentration. Subsequently, the crystals were mounted in a ^{3}He cryostat and the field dependent thermal conductivity was measured.

The results obtained for a crystal containing .09 μg/g Li are shown in Fig. 5. They reveal the three expected recoveries of the thermal conductivity towards its zero field value. The fields at which the minima in Δk occur are very nearly multiples of each other. However there appears to be a slightly greater separation between the highest and next highest minima than between the others. Thus it may be concluded that the level spacings appear to be almost, but not quite, identical. The potential is very nearly isotropic, but not exactly.

The level spacing itself, assuming a g value of 2 appropriate to the R-center, is 0.7 ± 0.1 cm^{-1} for the first two and 0.8 cm^{-1} $\pm$ 0.1 cm^{-1} for the highest transition. This is in agreement with results obtained by Lakatos and Sack.(13)

Finally, a comment about the scattering strength of the various transitions is in order: the lowest energy transition appears to be weaker than the others. However, a sample taken from the same ingot

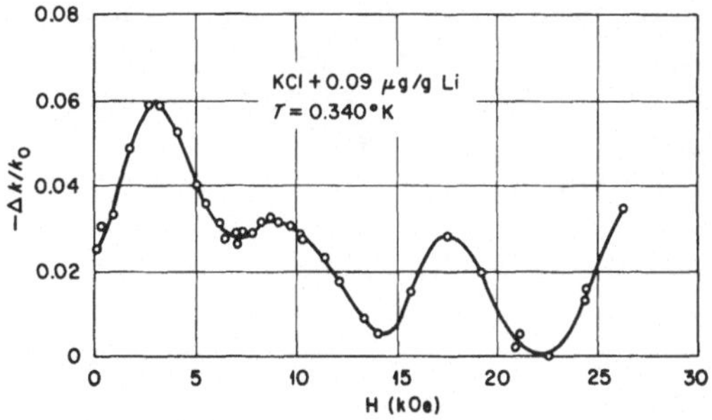

Fig. 5. Experimental results for KCl:Li plotted on the fractional change in thermal conductivity as a function of magnetic field.

has revealed a much stronger transition at 0.7 cm^{-1} and a weaker effect at 1.4 cm^{-1}. Since the specimens were taken from adjacent slices of the ingot, it does not appear to be an impurity or concentration effect. A possible explanation is that the crystal was strained, since the 1.4 cm^{-1} transition corresponds to tunneling along the [110] face diagonal of the unit cell. This explanation was supported by observation of a larger effect at 1.4 cm^{-1} upon annealing the specimen in question.

CONCLUSION

The spin-phonon technique can resolve defect modes and in principle can yield detailed information about them. This is so only in principle because quantitative measurements with this technique cannot be made unless the variation in "spectrometer resolution" with field is understood. Although there is a relatively complete theoretical treatment which provides all the ingredients necessary for this understanding,(6) at the present time experimental confirmation is lacking. However, this should be easy to rectify and the improved resolution of the spin-phonon technique can then be fully exploited.

REFERENCES

1. C. T. Walker and R. O. Pohl, Phys. Rev. 131, 1433 (1963).
2. R. Berman, J. C. F. Brock and D. J. Huntley, Phys. Letters 3, 310 (1963).
3. E. H. Jacobsen and D. W. H. Stevens, Phys. Rev. 129, 2036 (1963).
4. V. F. Sears, Proc. Phys. Soc. 84, 951 (1964).
5. J. W. Tucker, Proc. Phys. Soc. 85, 559 (1965).
6. J. B. Parkinson, Thesis, Oxford University (unpublished).
7. R. J. Elliott, private communication.
8. P. A. Carruthers, Rev. Mod. Phys. 33, 92 (1961).
9. D. Walton, Phys. Rev. 151, No. 2, 627 (1966).
10. D. Walton, Phys. Rev. Letters 19, 305 (1967).
11. G. J. Dienes, R. D. Hatcher, R. Smoluchowski and W. Wilson, Phys. Rev. Letters 16, 25 (1966).
12. M. Gomez, S. P. Bowen and J. A. Krumhansl, Phys. Rev. 153, 1009 (1967).
13. A. Lakatos and H. S. Sack, Solid State Comm. 4, 315 (1966).

PART F. EXCITATIONS LOCALIZED AT SURFACES

SURFACE OPTICAL MODES OF VIBRATION IN CUBIC CRYSTALS

R. F. Wallis,* D. L. Mills, and A. A. Maradudin

University of California, Irvine

Theoretical treatments of surface optical modes of vibration have been given by a number of workers,(1-4) but to date explicit results have not yet been reported for wave vectors (parallel to the surface) significantly different from zero using three-dimensional models which are rotationally invariant. In the present paper such results are obtained for a model of a NaCl-type lattice with nearest and next-nearest neighbor central forces. Surface optical modes are also investigated for a continuum model for which assumptions of short-range interactions are unnecessary.

We first take up the lattice-dynamical treatment of surface optical modes associated with a (001) free surface of an NaCl-type crystal. Only nearest and next-nearest neighbor central interactions characterized by force constants a and b, respectively, are assumed. No variation of the force constants near the surface is considered.

Let the x, y, z-components of displacement of an atom at lattice site $(\ell r_o, mr_o, nr_o)$ be $u(\ell mn)$, $v(\ell mn)$, $w(\ell mn)$, respectively, where r_o is the lattice constant, ℓ and m take on all integral values, while n is restricted to non-negative integral values. The equation for the x-component of force acting on an atom not in the surface layer may be written as

$$
\begin{aligned}
-\omega^2 M_i u(\ell mn) &= a\{-2u(\ell mn) + \Sigma_\lambda\, u(\ell+\lambda,m,n,)\} \\
&+b\{-8u(\ell mn) + \Sigma_{\lambda,\mu}\, [u(\ell+\lambda,m+\mu,n) + u(\ell+\lambda,m,n+\mu) \\
&\qquad + \lambda\mu\, v(\ell+\lambda,m+\mu,n) + \lambda\mu\, w(\ell+\lambda,m,n+\mu)]\} \qquad (1)
\end{aligned}
$$

*Present address: Naval Research Laboratory, Washington, D. C.

where ω is the frequency, M_i is the mass of the i-th type of atom, i = 1 or 2 for $\ell+m+n$ even or odd, respectively, and λ,μ take on the values ± 1. Corresponding equations for the y- and z-components of force are obtained by cyclic permutations of u, v, w and of the increments in ℓ, m, n.

For atoms in the surface layer, n=0, the equations of motion must be augmented by the following boundary conditions:

$$-2u(\ell m0) + \Sigma_\lambda \, [u(\ell+\lambda,m,-1) - \lambda w(\ell+\lambda,m,-1)] = 0 \tag{2a}$$

$$-2v(\ell m0) + \Sigma_\lambda \, [v(\ell,m+\lambda,-1) - \lambda w(\ell,m+\lambda,-1)] = 0 \tag{2b}$$

$$a[w(\ell,m,-1) - w(\ell m0)] - 4bw(\ell m0)$$
$$+b\Sigma_\lambda [-\lambda u(\ell+\lambda,m,-1) - \lambda v(\ell,m+\lambda,-1) + w(\ell+\lambda,m,-1)$$
$$+ w(\ell,m+\lambda,-1)] = 0 \quad . \tag{2c}$$

We seek solutions to the equations of motion of the form

$$(u,v,w) = (U,V,W) \exp [-qn + i(\ell\varphi_1 + m\varphi_2)] \tag{3}$$

where U,V,W are amplitudes, φ_1 and φ_2 are wave-vector components parallel to the surface and q is the attenuation constant. Substitution of Eq. (3) into the equations of motion (1) leads to a non-trivial solution only if a 6X6 determinantal equation is satisfied:

$$|D(\omega^2, \cosh q)| = 0 \quad . \tag{4}$$

For given ω there are in general six values of cosh q which satisfy Eq. (4). Denoting these values by cosh q_j, j = 1,2, ... 6, we seek a solution to the boundary conditions (2) of the form

$$(u,v,iw) = \sum_{j=1}^{6} (\sigma_{1j},\sigma_{2j},\sigma_{3j}) \, K_j \exp [-q_j n + i(\ell\varphi_1 + m\varphi_2)] \tag{5}$$

where the σ_{ij} are appropriate cofactors of $D(\omega^2$, cosh $q_j)$ and the K_j are new amplitudes. Equation (5) provides a non-trivial solution of Eqs. (2) only if a second 6X6 determinantal equation is satisfied:

$$|T_{jk}| = 0 \quad . \tag{6}$$

The simultaneous solution of Eqs. (4) and (6) gives the attenuation constants q_j and the frequency ω for the surface mode. For general φ_1 and φ_2, high-speed computer solutions are being calculated. For $\varphi_1 = \pi$ and $\varphi_2 = 0$, an analytic solution has been found. The surface optical mode frequency is a solution of the quadratic equation

$$\mu[(1-\rho)X-\mu](\mu-2\rho X)+\mu-\rho X=0 \tag{7}$$

where $\mu = 1+4(b/a)$, $\rho = M_1/(M_1+M_2)$ and $X = (M_1+M_2)\omega^2/2a$. The other solution is a Rayleigh (acoustical) surface wave. The attenuation constant is specified by

$$e^{-2q} = (\mu-2\rho X)/\mu \tag{8}$$

The atomic displacement amplitudes are shown diagramatically in Fig. 1

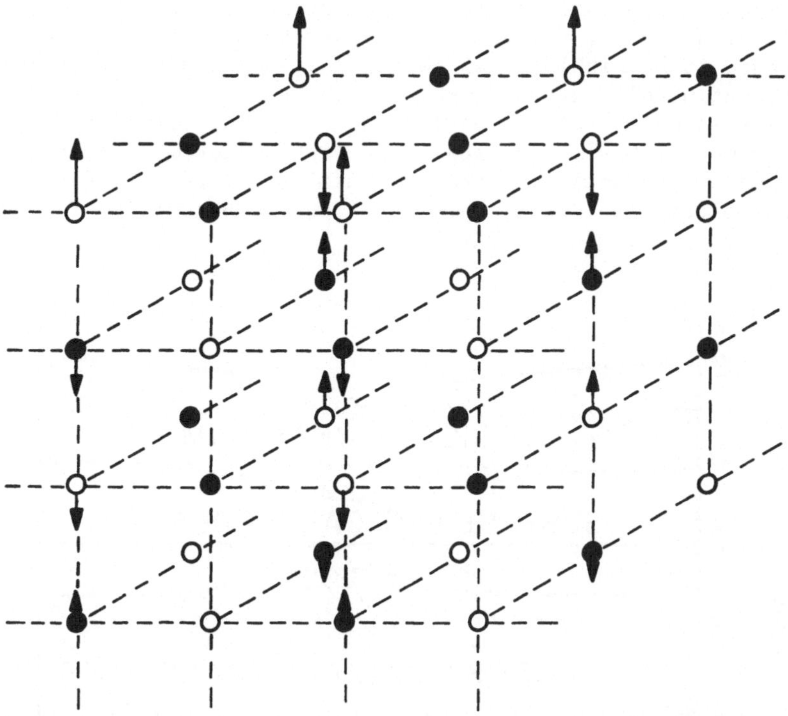

Fig. 1. Displacement field for the surface optical mode with $\varphi_1 = \pi$, $\varphi_2 = 0$. The open circles denote the lighter atoms and the solid circles denote the heavier atoms.

An alternative approach may be followed by considering a finite plate and imposing boundary conditions similar to Eqs. (2) on the N-th layer. Letting

$$M_i^{\frac{1}{2}}(u,v,w) = [F_i(n),G_i(n),H_i(n)]\exp[i(\ell\varphi_1+m\varphi_2)] \tag{9}$$

and substituting into the equations of motion combined with the boundary conditions, we obtain a non-trivial solution only if a 6N×6N determantal equation is satisfied. Using a high-speed computer, this equation has been solved for the normal mode eigenfrequencies and eigenvectors for values of N up to 18.

Some results of the computer calculations for N = 18 are presented in Fig. 2, where the upper and lower bounds of the "bulk" vibrational modes and the surface optical phonon dispersion curve are given along the line $\varphi_2 = 0$, $0 \leq \varphi_1 \leq \pi$. The values (b/a) = 0.1 and $(M_1/M_2) = 2$ were assumed. Both surfaces of the slab were assumed free, so for a given value of φ_1 there are two surface modes whose fractional frequency separation is the order of 10^{-5}. For all values of φ_1 the upper and lower bounds of the bulk modes were found to be within a few percent of the corresponding bounds for the infinitely extended medium. The surface optical mode frequency is only weakly dependent on φ_1, with $(\omega_s(\pi) - \omega_s(0))/\omega_s(0)$ of the order of 0.02. The analytic solution for $\varphi_1 = \pi$ previously mentioned agrees well with the numerical results. The vibrational spectrum along the line $\varphi_1 = \varphi_2$ was found to be qualitatively similar to the results of Fig. 2.

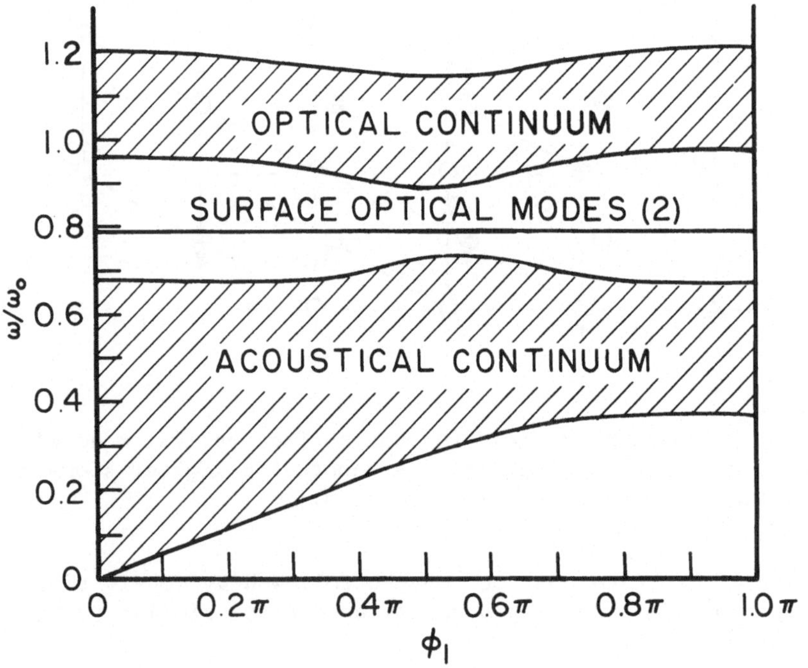

Fig. 2. Excitation spectrum for $\varphi_2 = 0$, $0 \leq \varphi_1 \leq \pi$ for the 18 layer slab with two free surfaces. ω_0 is the Raman frequency of the infinite medium.

In a crystal constructed from uncharged atoms, the surface layer may acquire an electric dipole moment during vibration, since the local crystalline fields lack inversion symmetry. We have calculated the infrared absorption spectrum of the finite slab assuming a spacially uniform radiation field with electric vector $\underset{\sim}{E}$ at an angle θ to the surface. The results for $\theta = 0°$ and $\theta = 45°$ are shown in Fig. 3. The displacement field for the $\varphi_1 = \varphi_2 = 0$ surface optical mode is normal to the surface; hence, there is no contribution to the absorption spectrum from the sufrace mode when $\theta = 0$. In Fig. 3b, it may be seen that the integrated absorption by the surface mode is comparable to the bulk contribution for $\theta = 45°$

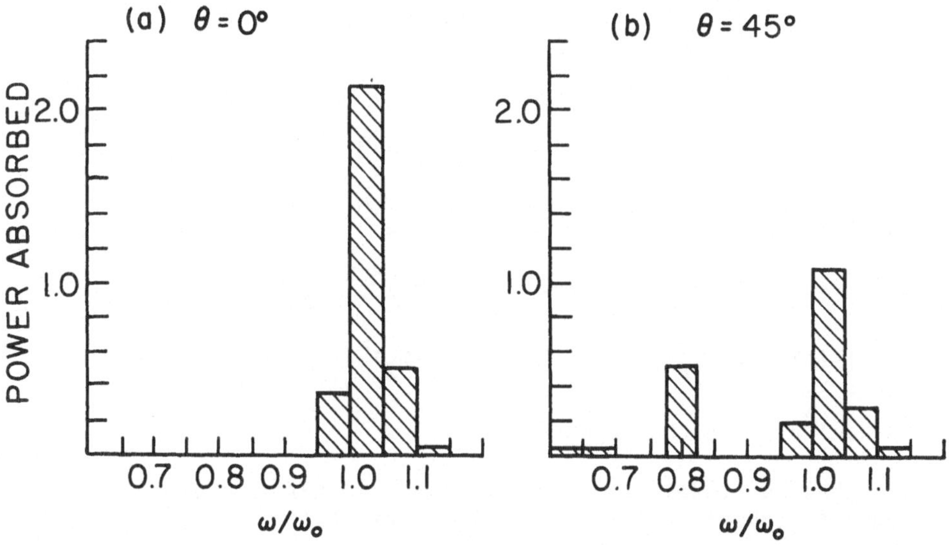

Fig. 3. Infrared absorption spectrum of the slab. The $\underset{\sim}{E}$ field is assumed polarized in a (100) plane. The angle θ is the angle between the $\underset{\sim}{E}$ field and the film surface. The units on the vertical scale are arbitrary.

Considering next the continuum theory, we write the equations of motion for an arbitrary crystal containing two atoms per primitive unit cell in the form

$$M_\kappa \omega^2 u_\alpha(\ell\kappa) = \sum_{\ell'\kappa'\beta} \Phi_{\alpha\beta}(\ell\kappa;\ell'\kappa')\, u_\beta(\ell'\kappa') \tag{10}$$

where $u_\alpha(\ell\kappa)$ is the α Cartesian component of displacement of the κ-th atom in the ℓ-th unit cell from its equilibrium position, M_κ is the mass of the κ-th kind of atom and the $\{\Phi_{\alpha\beta}(\ell\kappa;\ell'\kappa')\}$ are the atomic force constants. By introducing the mass-weighted displacements $v_\alpha(\ell\kappa) = M_\kappa^{\frac{1}{2}} u_\alpha(\ell\kappa)$, we can make a unitary transformation to two new independent variables for each unit cell according to

$$\begin{pmatrix} \eta_\alpha(\ell) \\ \xi_\alpha(\ell) \end{pmatrix} = (1/M_c^{\frac{1}{2}}) \begin{pmatrix} M_1^{\frac{1}{2}} & M_2^{\frac{1}{2}} \\ M_2^{\frac{1}{2}} & -M_1^{\frac{1}{2}} \end{pmatrix} \begin{pmatrix} v_\alpha(\ell 1) \\ v_\alpha(\ell 2) \end{pmatrix} , \tag{11}$$

where $M_c = M_1 + M_2$, $\eta(\ell)$ is the displacement of the center of mass of the ℓ-th unit cell, and $\xi(\ell)$ is the relative displacement of the two atoms in the ℓ-th unit cell. Substituting Eq. (11) into Eq. (10) we obtain as the equations for $\eta_\alpha(\ell)$ and $\xi_\alpha(\ell)$

$$\omega^2 \eta_\alpha(\ell) = \sum_{\ell'\beta} C_{\alpha\beta}^{(11)}(\ell\ell')\eta_\beta(\ell') + \sum_{\ell'\beta} C_{\alpha\beta}^{(12)}(\ell\ell')\, \xi_\beta(\ell') \tag{12a}$$

$$\omega^2 \xi_\alpha(\ell) = \sum_{\ell'\beta} C_{\alpha\beta}^{(21)}(\ell\ell')\eta_\beta(\ell') + \sum_{\ell'\beta} C_{\alpha\beta}^{(22)}(\ell\ell')\, \xi_\beta(\ell') \tag{12b}$$

where
$$C_{\alpha\beta}^{(11)}(\ell\ell') = (1/M_c) \sum_{\kappa,\kappa'} \Phi_{\alpha\beta}(\ell\kappa;\ell'\kappa') \tag{13a}$$

$$C_{\alpha\beta}^{(12)}(\ell\ell') = (1/M_c) \sum_{\kappa,\kappa'} (-1)^{\kappa'-1}(M_1/M_2)^{\kappa'-\frac{3}{2}}\, \Phi_{\alpha\beta}(\ell\kappa;\ell'\kappa') \tag{13b}$$

$$C_{\alpha\beta}^{(21)}(\ell\ell') = (1/M_c) \sum_{\kappa,\kappa'} (-1)^{\kappa-1}(M_1/M_2)^{\kappa-\frac{3}{2}}\, \Phi_{\alpha\beta}(\ell\kappa;\ell'\kappa') \tag{13c}$$

$$C_{\alpha\beta}^{(22)}(\ell\ell') = (1/M_c) \sum_{\kappa,\kappa'} (-1)^{\kappa-\kappa'}(M_1/M_2)^{\kappa+\kappa'-3}\, \Phi_{\alpha\beta}(\ell\kappa;\ell'\kappa'). \tag{13d}$$

and κ,κ' take on the values 1 and 2.

When Eq. (12a) is solved for $\eta_\alpha(\ell)$ in terms of $\xi_\alpha(\ell)$ and the result is substituted into Eq. (12b), we obtain

$$\omega^2 \xi_\alpha(\ell) + \sum_{\ell'\beta} A_{\alpha\beta}(\ell\ell')\, \xi_\beta(\ell') \tag{14}$$

where
$$A_{\alpha\beta}(\ell\ell') = C_{\alpha\beta}^{(22)}(\ell\ell') + \sum_{\ell''\gamma} \sum_{\ell'''\delta} C_{\alpha\beta}^{(21)}(\ell\ell'') \times$$

$$\times \left[\omega^2 \underset{\sim}{I} - \underset{\sim}{C}^{(11)}\right]^{-1}_{\ell''\gamma;\ell'''\delta} C_{\alpha\beta}^{(12)}(\ell'''\ell'). \tag{15}$$

For long wavelength vibrations the relative displacement amplitude $\xi_\alpha(\ell)$ varies slowly from one lattice site $\underset{\sim}{x}(\ell)$ to the next. The discrete variable $\underset{\sim}{x}(\ell)$ may then be regarded as a continuous variable.

We assume that the crystal is semi-infinite and occupies the half-space $x > 0$. This assumption can be incorporated into Eq. (14) by using the Heaviside unit step function $\theta(x)$:

$$\omega^2 \xi_\alpha(\underset{\sim}{x}(\ell)) = \theta(x_3(\ell)) \sum_{\ell'\beta} A_{\alpha\beta}(\ell\ell')\theta(x_3(\ell'))\xi_\beta(\underset{\sim}{x}(\ell')) \quad . \tag{16}$$

This procedure is correct only for two-body interactions between atoms and must be modified when many-body forces are taken into account. The atomic force constants entering into the matrix element $A_{\alpha\beta}(\ell\ell')$ are assumed to be those appropriate to an infinitely extended crystal.

Expanding $\xi_\beta(x(\ell'))$ and $\theta(x_3(\ell'))$ in Taylor series about $\underset{\sim}{x}(\ell)$ and $x_3(\ell)$, respectively, we can rewrite Eq. (16) after some manipulation in the form

$$\begin{aligned}\omega^2 \xi_\alpha(\underset{\sim}{x}) &= \sum_\beta S_{\alpha\beta}\xi_\beta(\underset{\sim}{x}) + \tfrac{1}{2}\delta(x_3)\sum_{\beta\mu} S_{\alpha\beta\mu 3}\frac{\partial \xi_\beta(\underset{\sim}{x})}{\partial x_\mu} \\ &\quad + \tfrac{1}{2}\sum_{\beta\mu\nu} S_{\alpha\beta\mu\nu}\frac{\partial^2 \xi_\beta(\underset{\sim}{x})}{\partial x_\mu \partial x_\nu}, \quad x_3 > 0\end{aligned} \tag{17}$$

where

$$S_{\alpha\beta} = S_{\beta\alpha} = -[(M_1+M_2)/M_1M_2] \sum_{\ell'} \Phi_{\alpha\beta}(\ell 1;\ell' 2) \tag{18a}$$

$$\begin{aligned}S_{\alpha\beta\mu\nu} &= (1/M_c)\{(M_2/M_1)\, W_{\alpha\beta\mu\nu}(11) - W_{\alpha\beta\mu\nu}(12) \\ &\quad -W_{\alpha\beta\mu\nu}(21) + (M_1/M_2)\, W_{\alpha\beta\mu\nu}(22)\}\end{aligned} \tag{18b}$$

$$W_{\alpha\beta\mu\nu}(\kappa\kappa') = \sum_{\ell'} \Phi_{\alpha\beta}(\ell\kappa;\ell'\kappa')\, x_\mu(\ell\kappa;\ell'\kappa')\, x_\gamma(\ell\kappa;\ell'\kappa') \tag{18c}$$

$$x_\mu(\ell\kappa;\ell'\kappa') = x_\mu(\ell\kappa) - x_\mu(\ell'\kappa') \quad . \tag{18d}$$

Only those terms have been retained in Eq. (17) which lead to a macroscopic wave equation. The simple result given by Eq. (17) obtains only for crystals every atom of which is at a center of inversion symmetry.

The solution of Eq. (17) is equivalent to solving the system of equations

$$\omega^2 \xi_\alpha(\underset{\sim}{x}) = \sum_\beta S_{\alpha\beta}\xi_\beta(\underset{\sim}{x}) + \tfrac{1}{2}\sum_{\beta\mu\nu} S_{\alpha\beta\mu\nu}\frac{\partial^2 \xi_\beta(\underset{\sim}{x})}{\partial x_\mu \partial x_\nu} \tag{19a}$$

for $x_3 \geq 0$, subject to the boundary conditions

$$\sum_{\beta\mu} S_{\alpha\beta\mu 3}\frac{\partial \xi_\beta(\underset{\sim}{x})}{\partial x_\mu} = 0 \text{ at } x_3 = 0 \quad . \tag{19b}$$

Using Stoneley's method[5] we have solved Eqs. (19) for optical surface waves propagating in the [100] direction along the (001) plane of a NaCl-type lattice with nearest and next-nearest neighbor central forces. For this model,

$$S_{\alpha\beta} = \delta_{\alpha\beta}\, \omega_0^2 = 2a[(M_1+M_2)/M_1M_2] \tag{20}$$

while the non-zero components of $S_{\alpha\beta\mu\nu}$ are

$$S_{\alpha\alpha\alpha\alpha} = (4r_0^2/M_c)\, [a-2(\tau b + (c/\tau))],\ \alpha = 1,2,3 \tag{21a}$$

$$S_{\alpha\alpha\beta\beta} = S_{\beta\beta\alpha\alpha} = S_{\alpha\beta\alpha\beta} = S_{\beta\alpha\beta\alpha} = S_{\alpha\beta\beta\alpha} = S_{\beta\alpha\alpha\beta} =$$
$$- (4r_0^2/M_c)(\tau b+(c/\tau)),\ \alpha,\beta = 1,2,3;\ \alpha \neq \beta \tag{21b}$$

where a is the nearest neighbor force constant, b and c are the next-nearest neighbor force constants between atoms of types 1 and 2, respectively, r_0 is the nearest neighbor separation, and $\tau = M_2/M_1$. The elastic constants are given by

$$r_0 C_{11} = a+2(b+c) \tag{22a}$$

$$r_0 C_{12} = r_0 C_{44} = b+c \quad . \tag{22b}$$

Assuming b = c, we can write the dispersion relation for the surface optical mode in the [100] direction as

$$(\omega_s/\omega_0)^2 = 1 + \gamma(b/a)\ [(\tau+\tau^{-1})/(\tau+\tau^{-1}+2)]\ \varphi_1^2 \tag{23}$$

The value of γ was found by solving Eqs. (19) using values of a,b, M ,M for which the infinitely extended crystal is stable. For example, it was found that $\gamma = 0.5455$ for $(a/b) = 5.3799$ and $\tau = 0.64755$. The dispersion curve for surface optical modes in this case lies below the dispersion curve for the doubly degenerate transverse optical modes,

$$(\omega_T/\omega_0)^2 = 1 + (b/a)[(\tau+\tau^{-1})/(\tau+\tau^{-1}+2)]\ \varphi_1^2\ , \tag{24a}$$

but lies above the dispersion curve for the longitudinal optical modes

$$(\omega_L/\omega_0)^2 = 1 - [1-2(b/a)(\tau+\tau^{-1})]\ \varphi_1^2/(\tau+\tau^{-1}+2) \quad . \tag{24b}$$

REFERENCES

1. I. M. Lifshitz and L. N. Rosenzweig, J. Exp. Theor. Phys. 18, 1012 (1948); I. M. Lifshitz and S. I. Pekar, Uspekhi Fiz. Nauk 56, 531 (1955).
2. R. F. Wallis, Phys. Rev. 105, 540 (1957); 116, 302 (1959).
3. R. Fuchs and K. L. Kliewer, Phys. Rev. 140, A2076 (1965).
4. R. Englman and R. Ruppin, Phys. Rev. Letters 16, 898 (1966).
5. R. Stoneley, Proc. Roy. Soc. (London) A232 447 (1955).

OPTICAL PHONONS IN IONIC CRYSTALS IN THE PRESENCE OF SURFACES OR DEFECTS

Robert Englman and Raphael Ruppin

Israel Atomic Energy Commission

Soreq Nuclear Research Centre, Yavne, Israel

INTRODUCTION

We have investigated the properties of long wave optical phonons in ionic crystals in the presence of surfaces or defects. Because of the existence of long range Coulomb forces the localized phonon modes in these crystals differ in many respects from those appearing in crystals in which the atoms interact through short range forces only. Whereas in the latter case the displacement amplitudes decay exponentially or even faster with increasing distance from the surface or the impurity site, it is found that in the presence of long range forces the localized optical modes usually decay much more slowly.

DIATOMIC IONIC CRYSTALS

An infinite diatomic ionic crystal (or one with periodic boundary conditions) has only two optical phonon frequencies near $\underline{k} = 0$; the longitudinal frequency ω_L and the transverse frequency ω_T. If only short range forces were present these two frequencies would degenerate into a single frequency ω_o. The existence of the long range Coulomb forces causes the splitting, which can be written in the following form: $\omega_T^2 = \omega_o^2 - 4\pi/3; \quad \omega_L^2 = \omega_o^2 + 8\pi/3$

where ω^2 is measured in units of $e^2/\mu V$; μ is the reduced mass of the two ions and V is the volume of the unit cell.

We define $\underline{f}(\ell)$ to be the relative displacement of the positive and the negative ions in the ℓ-th unit cell. Since we limit our discussion to phonons whose wavelength is long compared with the lattice constant, we can replace $\underline{f}(\ell)$ by a continuous function $\underline{f}(\underline{r})$. For finite diatomic ionic crystals the frequencies and the vibration amplitudes of the atoms in the long wave optical modes are then obtained from the integral equation[1]

$$(\omega^2 - \omega_o^2 + 4\pi/3)\ \underline{f}(\underline{r}) = -\ \mathrm{grad\ div} \int \frac{\underline{f}(\underline{r}')}{|\underline{r} - \underline{r}'|}\ d^3r',$$

where the integration is over the volume of the crystal. No further boundary conditions, at the surface of the crystal, need be imposed on the solutions of this equation since any solution automatically satisfies the following natural boundary conditions: the tangential components of the electric field and the normal component of the displacement field are continuous at the surface of the crystal. The integral equation was obtained by assuming the Coulomb interaction to be instantaneous, i.e. neglecting retardation effects.

The equation has three types of solution:

a) $\mathrm{div}\ \underline{f} = 0 \quad , \quad \omega^2 = \omega_o^2 - 4\pi/3$

These are transverse modes and their frequency is equal to ω_T, the transverse optical frequency of the infinite crystal.

b) $\mathrm{curl}\ \underline{f} = 0 \quad , \quad \omega^2 = \omega_o^2 + 8\pi/3$

These are longitudinal modes and their frequency is equal to ω_L, the longitudinal optical frequency of the infinite crystal.

c) $\mathrm{div}\ \underline{f} = 0$ and $\mathrm{curl}\ \underline{f} = 0$

These modes are neither transverse nor longitudinal and their frequencies are found to lie in the range between ω_T and ω_L. They will be called surface modes since their number is proportional to the surface area of the specimen and also since the vibration amplitude in these modes decreases with increasing distance from the surface of the crystal. This decrease in amplitude is however not very fast.

For a spherical crystal, for example, the surface mode amplitudes are given by

$$\underline{f}^S_{\ell m} = A\ \mathrm{grad}\ [r^\ell Y_{\ell m}(\theta,\phi)], \quad \ell = 1,2,3, \ldots$$

where r,θ,ϕ are spherical coordinates . The surface mode of lowest order, with $\ell=1$, describes a vibration having a constant amplitude over the whole crystal. This is the mode which was first discussed by Fröhlich[2]. In all the higher modes ($\ell = 2,3,\ldots$) the amplitude of vibration decreases as $r^{\ell-1}$, with increasing distance from the surface.

The frequencies of the surface modes crowd, when $\ell \to \infty$, to ω_S, an intermediate value between ω_T and ω_L, given explicitly by

$$\omega_S^2 = \omega_o^2 + 2\pi/3$$

In terms of the dielectric constant this frequency is just

$$\varepsilon(\omega_S) = -1$$

The surface modes appearing in finite crystals of other shapes have properties similar to those found in the case of the spherical crystal. The vibration amplitudes decrease rather slowly with increasing distance from the surface and the frequencies form a series converging to the value ω_S.

COMPLEX CUBIC STRUCTURES

The preceeding discussion was restricted to finite diatomic ionic crystals. We now extend the treatment to include the case of finite ionic crystals with more than two atoms per unit cell, but still having cubic symmetry. This will also enable us to go over from the rigid-ion model, which we have used so far, to the more realistic shell-model, since this involves just an increase in the number of "ions" (cores and shells) per unit cell.

It is found that for finite ionic crystals having a unit cell of cubic symmetry the equations of motion

$$M_\kappa \omega^2 u_\alpha \begin{pmatrix} \ell \\ \kappa \end{pmatrix} = \sum_{\ell' \kappa' \beta} \Phi_{\alpha\beta} \begin{pmatrix} \ell\ell' \\ \kappa\kappa' \end{pmatrix} u_\beta \begin{pmatrix} \ell' \\ \kappa' \end{pmatrix}$$

have, in the long wavelength limit, solutions which are separable in the form

$$u_\alpha \begin{pmatrix} \ell \\ \kappa \end{pmatrix} = f_\alpha^\lambda (\ell)\, Q_\alpha (\kappa)$$

where $\underline{f}^\lambda(\underline{r})$ is a solution of the integral equation

$$(\lambda + 4\pi/3)\, \underline{f}^\lambda(\underline{r}) = - \text{grad div} \int \frac{\underline{f}^\lambda(\underline{r}')}{|\underline{r} - \underline{r}'|} d^3r'$$

with an eigenvalue λ. $f(\underline{r})$ describes the spatial variation of the vibration amplitude over the crystal in exactly the same way as for a diatomic crystal, whereas $\underline{Q}(\kappa)$ gives the displacements of the different atoms in the unit cell. The latter are obtained from the following equations

$$M_\kappa \omega_\lambda^2 Q_\alpha (\kappa) = \sum_{\kappa'} F_{\alpha\alpha} (\kappa\kappa')\, Q_\alpha (\kappa')$$

F is the long wave ($\underline{k}=0$) limit of the usual dynamical matrix, modifie in the following way: the long range Coulomb term appearing in this matrix (which is equal to $-4\pi/3$ for transverse modes and to $8\pi/3$ for longitudinal modes) has to be replaced by λ, the eigenvalue of the integral equation.

To derive the equations it was assumed that only central forces act between the atoms. We have calculated the frequencies of the surface modes and the displacements of the atoms in these modes for finite specimens of $SrTiO_3$. The lattice dynamics of the infinite crystal has been thoroughly studied by Cowley[(3)], and we use his forc constants for the rigid-ion model.

Fig. 1 shows all the long wave optical frequencies for the case of a spherical crystal. The longer lines denote the frequencies of bulk modes (transverse and longitudinal). These have the same values as in the infinite crystal. The shorter lines denote the surface mode frequencies. The small letters on the right classify the modes according to their displacement pattern. In modes of type a the Sr moves in a direction opposite to that of all the other ions. In type b modes two oxygen ions move in opposite senses (in a flapping motion) while the third oxygen and the cations do not move. These are non-polar modes; their longitudinal and transverse frequencies are degenerate. In type c modes all the oxygen ions move in one direction and the Sr and the Ti ions in the opposite direction. In type d modes two oxygen ions move in the same direction as the Sr and Ti ions while the third oxygen ion moves in the opposite direction.

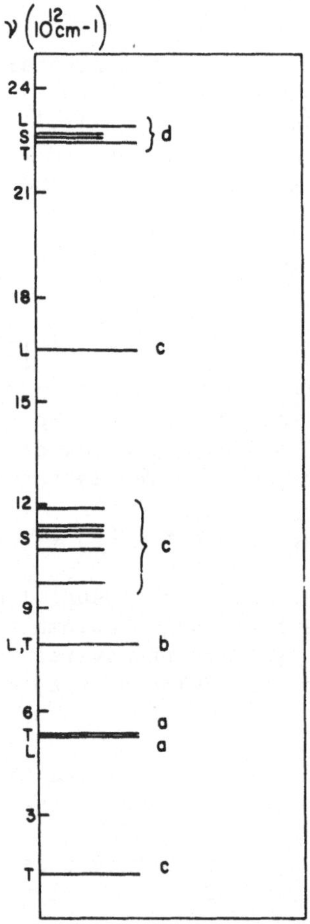

FIG. 1. The long wave optical frequencies of a spherical $SrTiO_3$ crystal (at 90° K). L,T and S denote longitudinal, transverse, and surface modes respectively.

INCLUSION OF RETARDATION

If the retardation of the Coulomb force is to be taken into account, one has to solve the equations of motion of the ions together with Maxwell's equations(4). It is then found that the transverse optical lattice vibratins interact with the light waves to form polaritons.

The polariton modes in finite diatomic ionic crystals are obtained from the equations

$$[\Delta + \varepsilon(\omega)\,\frac{\omega^2}{c^2}\,]\;\underline{f}(\underline{r}) = 0; \qquad \text{div}\;\underline{f}(\underline{r}) = 0 \quad ,$$

where

$$\varepsilon(\omega) = \varepsilon_\infty + \frac{\varepsilon_0 - \varepsilon_\infty}{1 - \omega^2/\omega_T^2}$$

inside the crystal and $\varepsilon(\omega)=1$ outside the crystal. Δ is the Laplace operator. The boundary conditions, which we impose on the solutions at the surface of the crystal, are those dictated by Maxwell's equations (continuity of the tangential components of the electric and the magnetic fields, which are derivable from $\underline{f}(\underline{r})$). All the modes given by the above equation are transverse.

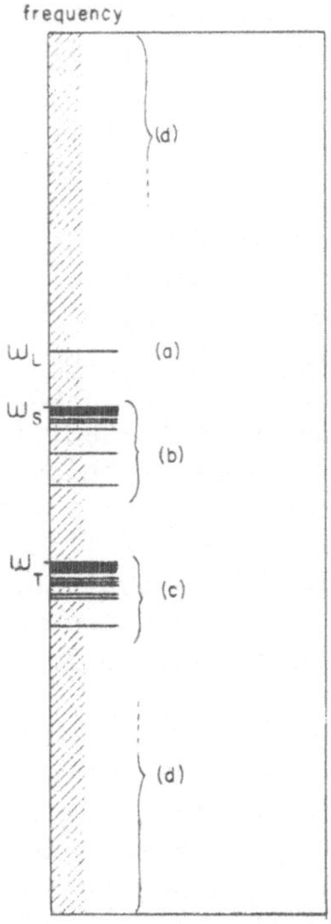

FIG. 2. The normal modes of a spherical diatomic crystal.
(a) Longitudinal phonon bulk modes; (b) Polariton surface modes;
(c) Polariton bulk modes; (d) Vacuum modes.

The longitudinal lattice vibrations do not interact with the electromagnetic waves and their frequency ω_L is the same as without retardation.

We now discuss in some detail the results for the case of a spherical crystal. As our quantization system we have taken a big hollow sphere, whose radius is much larger than that of the spherical specimen. We assume the boundary of the big sphere to be perfectly reflecting. The exact choice of this boundary condition is immaterial since the properties of the polariton modes at the crystal should not depend on the conditions at very remote distances.

Four types of modes are found to exist; they are shown in Fig.2:
(a) Longitudinal modes. These are pure lattice vibrations having no photon component.
(b) Polariton surface modes. Their properties are similar to those of the surface modes which were found to exist in the absence of retardation, only that now they have a photon component. They again converge to a frequency ω_S, given by $\varepsilon(\omega_S) = -1$.
(c) Polariton bulk modes. These converge to ω_T from below.
(d) Vacuum modes. These are essentially photon modes, similar in their properties to the electromagnetic waves that would exist in the big sphere if the small crystal was absent. They form a quasi-continuum.

We have also calculated the absorption cross section of the spherical crystal as a function of frequency (Fig. 3). The calculations were made for a sphere of NaCl whose radius R is given by $R=c/\omega_T$ (which means $R\sim 10\mu$). To obtain absorption a small damping term was added to the dielectric constant $\varepsilon(\omega)$. This damping constant was taken to be independent of the frequency, so that the structure that is found in the absorption spectrum is not due to anharmonic effects. The peaks in the absorption cross section are due to the polariton modes. The peaks below ω_T arise from the bulk polaritons ((c) in Fig.2). The peaks in the region between ω_T and ω_L arise from the surface polaritons ((b) in Fig.2). Increasing the radius of the crystal would move the positions of the peaks and would also cause the absorption by the surface polaritons to decrease and the absorption by the bulk polaritons to increase.

It is of interest to compare the absorption properties of a small spherical crystal with those of a thin film[5,6]. A thin film absorbs (at non-normal incidence) in a region below ω_T and also in a region above ω_L. If the film is thin enough one obtains two sharp absorption peaks at ω_T and ω_L. No absorption peaks occur in the region between ω_T and ω_L. On the other hand, for very small spherical crystals one obtains an absorption peak at ω_T and another (stronger) one at a frequency lying at about the middle of the range between ω_T and ω_L.

Results of infrared absorption measurements by Tsuboi, Terada and Shimanouchi[7] seem to be consistent with our results. These workers found, for fine powdered samples of UO_2, strong, broad absorption in the region between ω_T and ω_L, which should be attributed to the polariton surface modes. Axe and Pettit[8] have indeed suggested that this absorption is a manifestation of size and shape effects. They, however, interpreted it in terms of the Fröhlich frequency[2] alone, which is the frequency of the surface mode of longest wavelength. From inspection of the absorption spectrum it seems that other surface polaritons also contribute to the absorption (see our Fig. 3). The absorption curve has also a slight bump in the region below ω_T, which can be attributed to the polariton bulk modes.

EXTRAPOLATION TO DEFECT MODES

We shall next consider a small spherical dielectric of one material (denoted by I) embedded in an infinite dielectric of another material (II). Both the short range and the long range

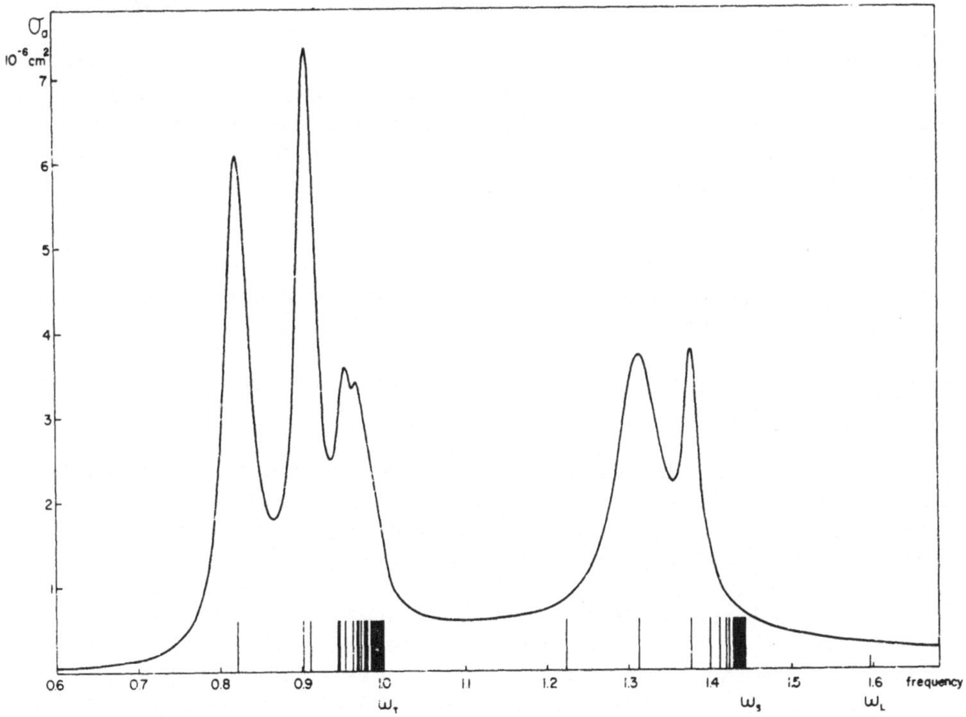

FIG. 3. Absorption cross section of a spherical specimen of NaCl of radius c/ω_T ($\sim 10\mu$). The frequency is measured in units of ω_T. The vertical lines denote the polarition frequencies.

(Coulomb) force constants in the two materials may be different. We can easily derive the surface modes in this macroscopic situation. We can also go on to extrapolate our results to study first the localized modes around an extended microscopic defect (such as R, M, etc. - centers) in a polar solid and then those around a point defect. This extrapolation takes us beyond the validity of the continuum approximation used in this work, which holds only for wavelengths of oscillations longer than interatomic distances. Nevertheless, an indication of the behavior of defect modes can be obtained by this approach.

We shall compare the localized modes formed by long range forces with those formed by the more commonly studied short range forces. We assume that two parameters (ω_o^I and K^I) satisfactorily characterize the mechanical and the dielectric properties of the defect, in analogy to the parameters ω_o^{II} and K^{II} ($= e^2/\mu V$) of the medium.

The following features are found for the localized modes:

1) The position of the localized mode level with respect to the optical branch (LI, LII, TI or TII) from which it grows out is proportional to the differences between the values of the parameters, ω_o or K, in the two media (for small differences). Accordingly, in this model, there is no critical value for the difference, below which the discrete localized levels do not appear. (If the short wavelength part of the dispersion curve overlaps the localized levels these may get damped).

2) In general, the localized vibrations will penetrate on either side of the boundary to a length of the order of the dimension of the inner sphere.

3) The angular variation of the vibration amplitudes follows a spherical harmonic.

4) For small differences between the values of the parameters in the two media there will be narrow gaps between ω_L^I and ω_L^{II} and again between ω_T^I and ω_T^{II}. The localized levels will fall in these gaps, converging toward the middle of the gap.

5) For large differences in ω_o in the two materials ($|\omega_o^I - \omega_o^{II}| \gg K^I \sim K^{II}$) there will be independent localized vibrations in the inner sphere, with levels in the lower half of its forbidden band, and independent vibrations in the outer medium (and no vibration in the inner sphere), whose levels are in the upper half of the $\omega_L^{II} - \omega_T^{II}$ gap. These last modes probably resemble localized polar vibrations about small microscopic defects.

6) As already implicit in the previous points, there will be long range effects due to the inner sphere also when this differs from the outer part only through short range force constants.

REFERENCES

(1) R. Englman and R. Ruppin, Phys. Rev. Letters, 16, 898 (1966).
(2) H. Fröhlich, Theory of Dielectrics (Oxford University Press, London, 1958) ch. IV.
(3) R.A. Cowley, Phys. Rev. 134, A981 (1963).
(4) M. Born and K. Huang, Dynamical Theory of Crystal Lattices (Oxford University Press, London, 1956) ch. II.
(5) R. Fuchs, K.L. Kliewer, and W.J. Pardee, Phys. Rev. 150, 589 (1966).
(6) D.W. Berreman, Phys. Rev. 130, 2193 (1963).
(7) M. Tsuboi, M. Terada, and T. Shimanouchi, J. Chem. Phys. 36, 1301 (1962).
(8) J.D. Axe and G.D. Pettit, Phys. Rev. 151, 676 (1966).

LOCALIZED LATTICE RESONANCE NEAR SURFACE PITS OR BUMPS

Dwight W. Berreman

Bell Telephone Laboratories, Incorporated

Murray Hill, New Jersey

Small regions in the neighborhood of pits or bumps on the surface of an ionic crystal will resonate under the influence of infrared radiation at frequencies intermediate between the frequencies of long wavelength transverse optic and associated longitudinal optic modes. Absorption due to similar intermediate frequency resonance was first intentionally observed in transmission through thin grating shaped films by Hass.[1] The effect is due to distortion of electric displacement fields caused by the presence of curved boundaries. Maxima in resonant polarization occur at points on the curved surfaces of the pits or bumps. We have computed the electric polarization associated with hemispherical pits and bumps on an otherwise plane surface. From this we have deduced the approximate effect of a random array of such bumps or pits on the reststrahl reflectance of an otherwise flat surface on an isotropic medium. With normally incident radiation, either bumps or pits will produce a dip in the top of an ordinary reststrahl reflectance band, together with a rather sharp peak of enhanced reflectance just below the frequency of the longitudinal optic mode. Anomalous band shape similar to this is observed in nearly all ionic crystals. Such structure in reststrahl bands is ordinarily attributed to second order of multi-phonon resonance effects. Although such explanations may usually be correct, it is quite possible that some of the data reported show additional structure due to unsuspected surface roughness effects.

The method that was used to compute change of reflectance is described in detail elsewhere.[2] The method was to solve, approximately, the boundary value problem of a quasi-static electric field around a single sphere, half submerged in the flat surface of a semi-infinite medium, using a truncated series of spherical harmonics. Here we shall present the results of that analysis in simplified form for the special case of radiation incident normal to the surface.

In a quasi-static approximation, the total electric field including the effect of polarization induced in the imperfection, may be written as the negative gradient of an electrical potential V. Let θ be the polar angle measured from the normal to the flat dielectric surface, φ be the azimuthal angle, and r be the distance from the center of an imperfection. The potential above the sphere, caused by a uniform electric field applied parallel to the surface in the $\varphi = 0$ direction, may be expanded in a series of spherical harmonics:

$$V_{above} \approx V_o \cos \varphi (r \sin \theta + \sum_{j=1}^{M} B_{1j} P_j^1(\cos \theta)/r^{j+1}). \quad (1)$$

A similar series expansion is made for the field in the region below the sphere, using constants B_{2j} in place of B_{1j}. A series with positive powers of r is made for the region inside the sphere:

$$V_{inside} \approx V_o \cos \varphi \sum_{j=1}^{M} B_{3j} r^j P_j^1(\cos \theta). \quad (2)$$

The series for each region is truncated at M so that a finite set of simultaneous equations is obtained using electrostatic boundary conditions. If the sphere is taken to have unit refractive index, the imperfection is equivalent to a hemispherical pit. If it has the refractive index of the semi-infinite medium below, it is equivalent to a hemispherical dome. At distances large compared to the radius of the sphere but small compared to the wavelength of the radiation, the field is approximately uniform and parallel to the flat surface. The sphere results in induced multipole moments, B_{1j}, as seen in the vacuum above the sphere, that depend on the dielectric constant or refractive index of the sphere and of the substrate. Only the dipole moment, B_{11}, is involved in first order perturbation of reflectance by a surface with spherical imperfections, if the imperfections are sufficiently far apart to ignore the effect of one imperfection on the field measured at another.

Under the conditions of the approximation made here, the effect on reflectance caused by a random array of spheres can be expressed as

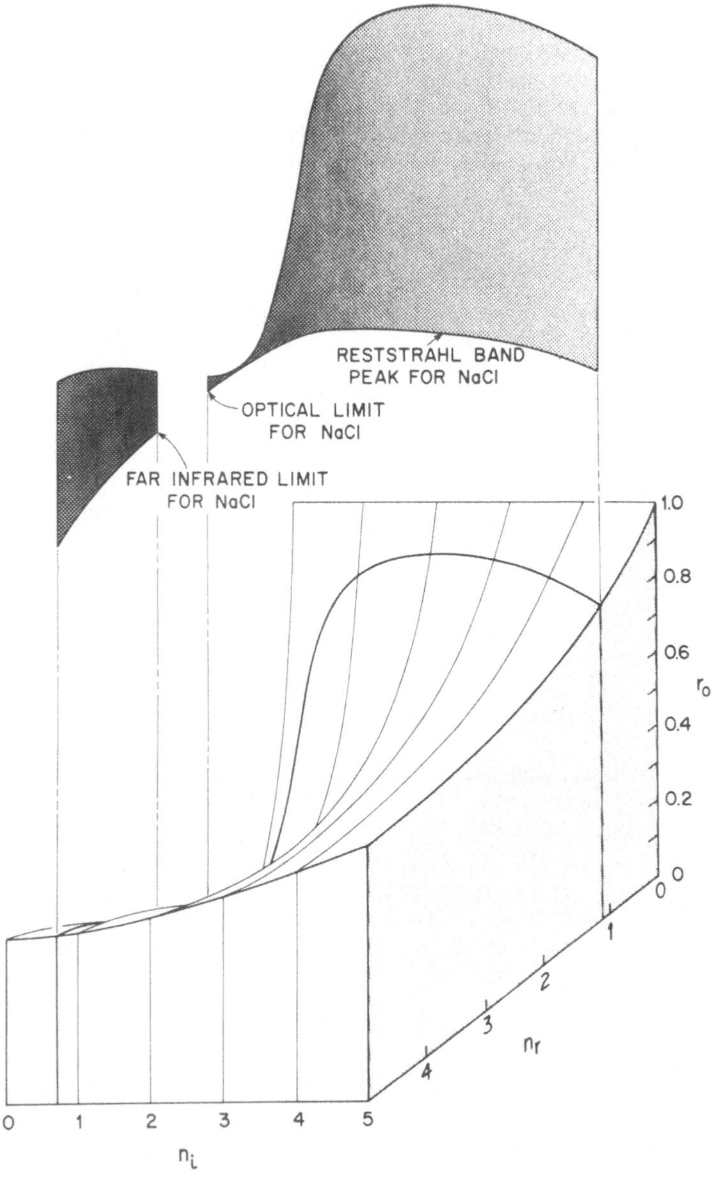

Fig. 1 - Reflectance, r_0, of a flat surface for radiation at normal incidence, as a function of complex refractive index. The raised cylindrical curves show the values of r_0 on part of the loop of values of refractive index through which a model of NaCl passes in going through its reststrahl band.

$$r \approx r_o + r_1(Na^3/\lambda) \qquad (3)$$

where r_o is the reflectance of a flat surface, a is the radius of each sphere, N is the number of spheres per unit area and λ is the wavelength of the incident radiation. Incoherent scattering is zero to this order of approximation. Note that (Na^3/λ) is a measure of the ratio of a sort of "mean thickness of surface roughness" to wavelength. The result for normally incident radiation can be expressed as

$$r_o = \left| \frac{n-1}{n+1} \right|^2 \qquad (4)$$

$$r_1 = 16\pi^2 r_o \, \mathrm{Im}[B_{11}/(n-1)]. \qquad (5)$$

Im [ ] means the imaginary part of the quantity in brackets. The refractive index of the semi-infinite medium is n. For bumps or pits the dipole coefficient B_{11} has large absolute values when the refractive index, n, is mostly imaginary and has values between about 0.5i and 2i. The details of the resonance spectrum depend on whether the spherical irregularity is a bump or a pit but the overall effect is similar whether there are bumps, pits, or both in arbitrary ratio. The effect is to make a dip in the top of a reststrahl band and a subsidiary peak at the high frequency end of the band.

The reflectance of a flat surface, r_o, and the correction coefficient, r_1, for pits, are plotted as functions of the real and imaginary components of refractive index, n_r and n_i, in figures 1 and 2, respectively. As one sweeps through a reststrahl band or other optical resonance of a medium, the refractive index passes over a loop in the complex refractive index plane. The curves generated by such a sweep, projected onto the r_o and r_1 surfaces, are shown in the two figures. The curves are derived from a simplified model for the reststrahl band of NaCl. (If we had used the complex dielectric constant $\varepsilon = n^2$, instead of n as the base plane in these plots, the loop would have been an arc of a perfect circle. This is true because a single, linearly damped oscillator model was chosen to approximate the reststrahl band. However, the contours of r_o and r_1 show up better on the refractive index plots.) The contours of the plot of r_1 for hemispherical domes is very similar to that shown for pits in figure 2, except that on the reststrahl band contour the peak is higher, the dip is shallower, and both are somewhat displaced.

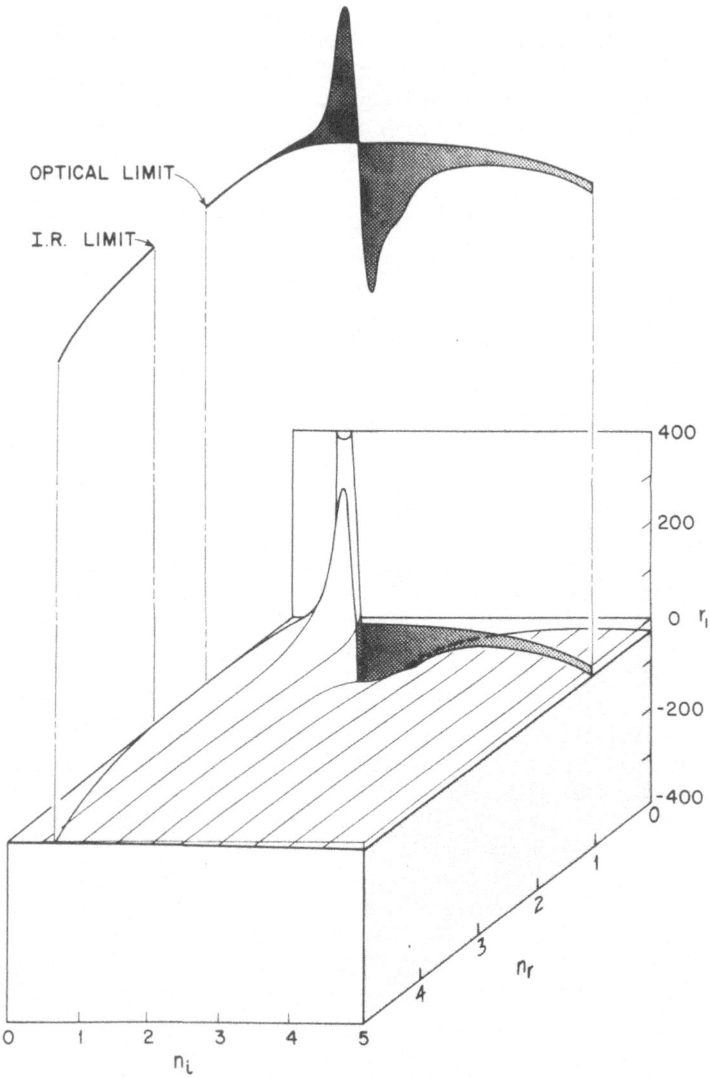

Fig. 2 - Correction term, r_1, for reflectance of a surface with small hemispherical pits, as a function of refractive index. Parts of the loop for the Reststrahl band of the simplified model of NaCl are shown on the raised cylindrical segments.

We have found it very difficult to make the surface of a crystal of any alkalai halide smooth enough that a reasonable measure of the volume of small scale surface roughness per unit area is less than one hundredth of the wavelength of reststrahl radiation. Crystals may look quite smooth and shiny even when they are considerably rougher. Since r_1 goes to values of the order of 100 or more in the region of the reststrahl peak, the effect of surface roughness must seldom be negligible unless extraordinary precautions are taken to smooth the surfaces of samples. More often, the roughness is probably so great that first order approximations such as were made in this computation are not accurate. However, the approximation certainly indicates that the magnitude of the effect of such roughness is large.

1. M. Hass, Phys. Rev. Letters 13, 429 (1964).
2. D. W. Berreman, Phys. Rev. October, 1967 (to be published).

SURFACE SPIN WAVES

D. L. Mills

Department of Physics
University of California
Irvine, California

I. Introduction

In the study of lattice dynamics, it is well known that associated with free surfaces there exist normal modes characterized by a displacement field that varies in a wave like manner in the directions parallel to the surface, and which decay to zero exponentially fast as one moves toward the interior of the crystal.[(1),(2)] It has recently been shown[(3)] that in a Heisenberg ferromagnet at temperatures low compared to the Curie temperature, surface spin wave modes exist with properties similar in many respects to the lattice dynamical counterparts. Within the spin wave approximation, the surface modes are eigenstates of the Heisenberg Hamiltonian. Thus, the modes contribute to the specific heat of the system. If one approaches the crystal surface from within, the mean spin deviation will increase because of the presence of the thermally excited surface waves.

It is the purpose of this paper to present a brief summary of some recent studies of the properties of surface spin waves and their influence on a number of properties of magnetic crystals.

II. Properties of Surface Spin Waves in the Ferromagnet

(a) The Semi-Infinite Heisenberg Ferromagnet

Consider an array of spins interacting via an exchange interaction of the Heisenberg form. Suppose the array to be semi-infinite in extent, with the surface in the x-y plane. We shall ignore the possibility of surface reconstruction, as well as any changes in the exchange constants in the vicinity of the surface. The exchange interactions will be supposed to be of short range, in the sense that a spin in a layer characterized by a particular value of ℓ_z interacts only with spins in the layers ℓ_z-1, ℓ_z and ℓ_z+1.

Let the frequency of a bulk spin wave of wave vector $\underset{\sim}{k} = (k_x,k_y,k_z)$ be denoted by $\Omega = \Omega_B(k_x k_y k_z)$. Then if the equations of motion of the semi-infinite material[3,4] are examined for solutions localized near the surface,

$$S_{\ell}^{(+)} = \exp(i\underset{\sim}{k}_{\parallel}\cdot\underset{\sim}{\ell}_{\parallel})\exp(-q\ell_z)\exp(i\Omega t),$$

one finds the surface spin wave frequency $\Omega_s=\Omega_B(k_x,k_y,iq)$. The quantity q is determined from

$$\exp(-qa) = \alpha(k_x,k_y)/[\alpha(o)-\omega_s] ,$$

where $\alpha(\underset{\sim}{k}) = 2\,S \sum_{\underset{\sim}{\delta}(\delta_z<o)} J(\underset{\sim}{\delta})\exp(i\underset{\sim}{k}_{\parallel}\cdot\underset{\sim}{\delta})$.

We consider some special cases. Since the frequency $\omega_s<<(\alpha(o)-\alpha(\underset{\sim}{k}))$ in many cases, suppose first $\omega_s=0$. Then since it is necessary that q>0, solutions localized near the surface will exist only if in creating the surface, bonds non-normal to the surface are cut. As a special case, consider a simple cubic array of spins, with nearest neighbor interactions J_1 and next nearest neighbor coupling J_2. Then(with $\omega_s=0$), one easily shows the frequency of a surface save with $k_y=0$ is

$$\Omega_s(k_x) = \Omega_B(k_x oo)-32SJ_2\sigma\sin^4(k_x a/2),$$

where $\sigma = J_2/(J_1+4J_2)$. For $k_x a<<1$, the surface spinwave frequency differs from the frequency of a bulk spinwave only by terms of order $(k_x a)^4$. Also note $q=\sigma a\, k_x^2$ for $(k_x a)$ small.

For ω_s small but finite, no surface solutions exist $\underset{\sim}{k}_{\parallel}$ is so small that $S\sum_{\delta<0} J(\underset{\sim}{\delta})(\underset{\sim}{k}_{\parallel}\cdot\underset{\sim}{\delta})^2<\omega_s$. As discussed in reference (3), as $\underset{\sim}{k}_{\parallel z}$ is decreased from above until the two side of the last inequality are equal, the surface spinwave frequency approaches the bulk value $\Omega_B(k_x,k_y 0)$.

(b) The Effect of Dipolar Interactions

In the preceding discussion, the spins were assumed to interact only via short range, istropic exchange interactions. In bulk materials, it is known that exchange interactions make the dominant contribution to the energy of short wave length spinwaves in most materials. When the magnon wave becomes sufficiently long, magnetic dipole interactions influence the dispersion relation in an important way, while the exchange contribution becomes negligible.[7] The purpose of this section is to examine the nature of the surface waves in the dipolar dominated regime.

Consider a semi-infinite array of spins interacting via magnetic dipole interactions. The sample is placed in an external magnetic field H; both the magnetization and H are parallel to the surface. The surface is assumed located in the x-z plane, with the magnetization parallel to the z axis. We begin with the equations of motion for the operators $S_x(\ell)$, $S_y(\ell)$. The equations are linearized in the spirit of the spinwave approximation, then the sums over lattice sites are replaced by integrations treating the spin deviations as continuous variables. After some partial integrations, one finds

$$\dot{S}_x(\tilde{r})=\omega_B S_y(\tilde{r})-\frac{\omega_M}{4\pi}\int' d\tilde{r}'\Big[\Big\{\Gamma_{xx}(\tilde{r}\tilde{r}')+\Gamma_{zz}(\tilde{r}\tilde{r}')\Big\}S_y(\tilde{r}')+ + \Gamma_{yx}(\tilde{r}\tilde{r}')S_x(r')$$

and

$$\dot{S}_y(\tilde{r})=-\omega_H S_x(\tilde{r})-\frac{\omega_M}{4\pi}\int' d\tilde{r}'\Big[\Gamma_{xx}(\tilde{r}\tilde{r}')S_x(\tilde{r}')+\Gamma_{yx}(\tilde{r}\tilde{r}')S_y(\tilde{r}')\Big]$$

In these equations, $\omega_H=g\mu_B H, \omega_M=4\pi g\mu_B M_s, \omega_B=\omega_H+\omega_M$, and $\Gamma_{ij}(rr')=\frac{\partial}{\partial x'_i}\frac{1}{|\tilde{r}-\tilde{r}'|}\frac{\partial}{\partial x'_j}$. The integration on $\tilde{r}'$ is taken over the sample, excluding a small sphere centered at $\tilde{r}$. If the solutions are assumed to vary in a wave like manner parallel to the surface, one may derive a set of coupled integral equations for the spatial dependence of S_x and S_y in the y direction:[9]

$$\frac{\partial}{\partial t}S_x(y)=+\omega_B S_y(y)-\frac{\omega_M}{2}\int_0^\infty dy'\Big[k_\parallel\gamma_s S_y(y')+ik_x\gamma_a S_x(y')\Big] \quad \text{(a)}$$
$$\text{(1)}$$
$$\frac{\partial}{\partial t}S_y(y)=-\omega_H S_x(y)-\frac{\omega_M}{2}\frac{kx}{k_\parallel}\int_0^\infty dy'[k_x\gamma_s S_x(y')-ik_\parallel\gamma_a S_y(y')], \quad \text{(b)}$$

where $k_\parallel=\sqrt{k_x^2+k_z^2}$ $\gamma_s=\exp(-k_\parallel|y-y'|)$, and $\gamma_a=\text{sgn}(y-y')\exp(-k_\parallel|y-y'|)$.

For the infinite crystal, these equations admit solutions that vary as $\exp(ik_y y)$, provided the frequency Ω is chosen so $\Omega^2=\omega_H(\omega_H+\omega_M[1-\cos^2\theta])$, where $\cos\theta = (k_z/k)$. This well known result describes the effect of demagnetizing fields on the long wave length spin waves. As θ varies from 0 to $\pi/2$, the frequency Ω varies from

ω_H to $[\omega_H\omega_B]^{\frac{1}{2}}$.

The equations may be examined for solutions localized near the surface. Suppose S_x and S_y vary with y as exp(-qy). If this form is inserted into Eqs.(1), one obtains a solution with frequency

$$\Omega_s^2=\omega_H(\omega_H+\omega_M\ [1-k_z^2/(k_\parallel^2-q^2)]), \tag{2}$$

provided q is chosen so that

$$q = k_x[(k_\parallel\Omega_s-k_x\omega_H)/(k_x\Omega_s-k_\parallel\omega_H)].$$

Upon eliminating q^2 from Eq.(2), one obtains the simple result(10)

$$\Omega_s = \frac{1}{2}\frac{\omega_H}{\cos\beta} + \frac{1}{2}\,\omega_B \cos\beta \ ; \ \cos\beta = (k_x/k_\parallel). \tag{3}$$

Since the quantity Ω_s^2 must be positive, one finds surface waves only when

$$\cos\beta > (\omega_H/\omega_B)^{\frac{1}{2}} .$$

As cos β varies from 1 to $(\omega_H/\omega_B)^{\frac{1}{2}}$, notice that the surface wave frequencies vary from $(\omega_H\omega_B)^{\frac{1}{2}}$ to $\frac{1}{2}(\omega_H+\omega_B)$. Thus in the long wave length regime, where the exchange interactions have little influence, the surface spin waves lie above the bulk spin waves in frequency. This result should be contrasted with the result of the previous section, where it was found that (neglecting the dipolar interactions), the surface spin wave frequency associated with a given wave vector $k_\parallel$ is less than that of any bulk wave characterized by the wave vector component $k_\parallel$ parallel to the surface. Notice also that in the dipolar-dominated region, the surface wave penetrates into the crystal by an amount directly proportional to the wave length, while in the exchange-dominated regime the attenuation constant q varies for $k_\parallel^2$ for $|\underset{\sim}{k}_\parallel|a<<1$.

It is clearly desirable to formulate an analytical scheme that will allow one to pass smoothly from the "dipole-dominated" region to the "exchange dominated" region. This problem is currently under investigation; it appears likely that it will be necessary to employ a high speed computer to properly include the effect of the modifications that arise in the equations of motion of the spins in the surface layer from the short range interactions.

However, it is possible to find the curvature of the surface spin wave dispersion relation near $k_\parallel$=0 by a perturbation theoretic technique. One may introduce the effect of exchange into the equations of motion by replacing ω_B in Eq.(1a) by $\omega_B+D(k_\parallel^2-\frac{\partial^2}{\partial y^2})$, and by making a similar replacement in Eq.(1b). The frequency may be expressed in terms of the exchange constant D and the attenuation length q. One may find the change in q to first order in D, to obtain an expansion of Ω_s to first order in D. We find

$\Omega_s=\Omega_s^{(o)}-D_s k_\parallel^2$, where $\Omega_s^{(o)}$ is the result of Eq.(3)

and

$$D_s=\frac{D}{2}\left\{\frac{\omega_M^2 \sin^2\beta\cos\beta}{\Omega_s^{(o)2}-\omega_H^2}\right\}\left\{\frac{\omega_B\cos^2\beta+\omega_H}{\omega_B\cos^2\beta-\omega_H}\right\}^2 .$$

The expression for D_s diverges as $\cos\beta\to(\omega_H/\omega_B)^{\frac{1}{2}}$. The perturbation theoretic derivation is invalid near this value of cos β.

This treatment indicates that the curvature of the surface spin wave branch is negative near $k_\parallel=0$ in the "dipolar dominated" region. (Recall that the present discussion is valid only when $Dk_\parallel^2<<\Omega_s^{(o)}$.) The negative curvature has its origin in the fact that in the "dipolar dominated" region, $q>k_\parallel$, so $D(k_\parallel^2-q^2)<0$.

The negative curvature near $k_\parallel=0$ suggests that an improved treatment will show that the surface branch crosses below the $k_y=0$ bulk curve to join the result of the previous section in the "exchange dominated" region. As mentioned above, this point is currently under investigation.

III. Influence of the Surface Modes on the Properties of the System

As mentioned above, thermal excitation of the surface spin waves will cause the mean spin deviation to increase as one approaches the surface of the crystal from within. Also, these modes will contribute to the specific heat. As we shall see in the discussion below, it is necessary to realize that the presence of the surface alters the frequency distribution of the bulk modes. The modification in the frequency distribution of the bulk modes leads to contributions to the thermodynamic quantities proportional to the surface area of the crystal, and of the same order of magnitude as the surface wave contribution.

We have calculated the contribution to the specific heat proportional to the surface area, along with the dependence of the mean spin deviation on distance from the crystal surface[4]. The calculation is valid at low temperatures, where the Holstein-Primakoff transformation may be employed. Since in many materials, the wave length of thermal magnons is sufficiently short that the effect of small surface pinning fields and dipolar interactions may be ignored, the spins were assumed to interact only via short-range Heisenberg exchange interactions.

At low temperatures, we have found the total specific heat of the crystal is

$$C(T) = \frac{V}{a^3} \frac{15\zeta(5/2)}{32\pi^{3/2}} k_B \left(\frac{k_B T}{\hbar D}\right)^{\frac{3}{2}} + \frac{S}{a^2} \frac{\zeta(2)}{4\pi} k_B \frac{kT}{\hbar D}$$

where $\hbar D = 2S(J_1 + 4J_2)$. In this equation, the volume of the crystal is V, and its surface area S. The second term is equal in magnitude to one-half the contribution from the surface waves alone, since there occurs a partial cancellation between the surface wave contribution and that from the change in frequency distribution of the bulk modes. A similar cancellation evidently occurs in the computation of the surface contribution to the specific heat from lattice vibrations(11). The very elegant mathematical analysis of Maradudin and Wallis yields only the total surface specific heat, without isolating the contribution from the two sources discussed above. However, upon computing the contribution from the Rayleigh waves alone, one sees that their total specific heat is less than this portion. In the ferromagnet, the origin of the partial cancellation may be seen to occur because (for a slab geometry), for a given value of $\underset{\sim}{k}_{\parallel}$, there is one continuum mode with $k_z << |\underset{\sim}{k}_{\parallel}|$ there are two surface modes(12).

The change in the mean spin deviation associated with a lattice plane ℓ planes from the surface has been shown to be

$$\delta\Delta_{\ell} = \frac{1}{4\pi n\alpha} \left[1 + 2 \exp(-n\sqrt{m\pi/\alpha}) \cos (n\sqrt{m\pi/\alpha})\right]$$

where $n = 2\ell - 1$, and $\alpha = 4S(J_1 + 4J_2)/k_B T$. Again, one must combine the contribution from the perturbed bulk modes with the surface waves, since both contributions are comparable.

IV. Surface Magnons in a Simple Anti-ferromagnet

Anti-ferromagnets appear to be especially attractive systems for the study of surface effects, since the excitation energy of a $\underset{\sim}{k}=0$ bulk spin wave is generally finite and often the order of a few tens of degrees Kelvin. If the exchange field H_E is large compared to the anisotropy field H_A, the $\underset{\sim}{k}=0$ AF magnon frequency is $(2H_EH_A)^{\frac{1}{2}}$. For MnF_2, this excitation energy is the order of $12^\circ k$.

We have investigated[5] the occurrence of surface magnons in anti-ferromagnets of the MnF_2 structure. For a free (100) surface, a surface magnon has been found with a limiting frequency of $(H_EH_A)^{\frac{1}{2}}$ for $H_A << H_E$.[13] At the zone boundary, the surface wave frequency rises to $(H_E/2)$.

The existence of this mode in the gap below the main AF resonance mode can have an important influence on the thermodynamic properties of the system when $k_BT << (2H_EH_A)^{\frac{1}{2}}$, since the occupation number associated with the surface mode will be larger than that associated with the k=0 bulk mode by a factor $\exp((\sqrt{2}-1)(H_EH_A)^{\frac{1}{2}}/k_BT)$.

We have calculated the surface contribution to the specific heat at low temperatures, when $(k_BT) << (H_EH_A)^{\frac{1}{2}}$. The result is

$$C_V(T) = k_B \frac{16 \cdot 2^{\frac{1}{4}}}{\pi^{3/2}} \frac{V}{a^3} \left(\frac{\omega_A}{\omega_E}\right)^{5/4} \left(\frac{\omega_A}{k_BT}\right)^{\frac{1}{2}} \exp(-\sqrt{2\omega_E\omega_A}/k_BT) +$$

$$+ k_B \frac{4}{\pi} \frac{S}{a^2} \left(\frac{\omega_A}{\omega_E}\right)\left(\frac{\omega_A}{k_BT}\right)^{\frac{1}{2}} \exp(-\sqrt{\omega_E\omega_A}/k_BT).$$

If one employs the parameters relevant to MnF_2, then for a film N layers thick the surface and bulk contributions become equal when the temperature T satisfies

$$N = \frac{18.5}{T^{\frac{1}{2}}} \exp(3.8/T).$$

For $T=1^\circ$, the surface and bulk contributions become equal when the film thickness is $\approx$ 850 layers. This estimate indicates that the surface spin wave contribution to the specific heat may be readily observable if measurements on thin films are carried out at low temperatures.

Surface contributions to the specific heat have been examined when $\sqrt{2\omega_E\omega_A} << k_BT << k_BT_N$. In addition, the spatial variation of the mean spin deviation near surface and the contribution of the layers near the surface to the parallel susceptibility are being studied. The results of these investigations will be reported in a forthcoming publication.[5]

References

1) Lord Rayleigh, Proc. London Math. Soc. 17,4(1887).
2) D. C. Gazis, R. Herman, and R. F. Wallis, Phys. Rev. 119,533(1960).
3) R. F. Wallis, A. A. Maradudin, I. P. Ipatova and A. A. Klochikhin, Solid State Comm. 5,89(1967).
4) D. L. Mills and A. A. Maradudin, J. Phys. Chem.Solids (to be published).
5) D. L. Mills and W. Saslow (to be published).
6) C. Kittel, Phys. Rev. 110,1295(1958).
7) T. Holstein and H. Primakoff, Phys. Rev. 58,1098(1940).
8) In this geometry, the effect of demagnetizing fields is most pronounced.
9) Fuchs and Kliewer have employed a similar procedure to study surface optical phonons in ionic crystals. See Phys. Rev. 140, 2076(1965). In order to derive the integral equations from the microscopic Hamiltonian, these authors found it necessary to introduce an unphysical cut off parameter in the spatial integrations. (See the Appendix). This parameter was determined by demanding the resulting equations be consistent with a set of macroscopic equations. The derivation sketched above avoids this difficulty.
10) This same result has been obtained previously by Damon and Eshbach by a rather different method. See J. Eshbach and R. Damon, Phys. Rev. 118,1208(1960).
11) A. A. Maradudin and R. F. Wallis, Phys. Rev. 148, 962 (1966).
12) Loosely speaking, for a given value of $\underset{\sim}{k}_{\parallel}$, one has two surface waves, one localized near each surface. See the discussion in ref. (4).
13) I am indebted to R. F. Wallis for pointing out an algebraic error in an early study of this problem.

PART G. THERMAL PROPERTIES OF DEFECT MODES

LOCALIZED EXCITATIONS IN THERMAL CONDUCTIVITY*

R. O. Pohl

Laboratory of Atomic and Solid State Physics

Cornell University, Ithaca, New York

In this paper we wish to demonstrate on a few examples the use of lattice thermal conductivity measurements for the study of low frequency impurity modes. We shall see that this method is complementary to the far infrared absorption measurements, because the thermal conductivity is sensitive to impurity modes which, although infrared active, are too strongly damped to be detected through infrared absorption. Furthermore, we shall see that there is evidence that phonons can also be coupled to even parity impurity modes, and hence thermal conductivity can also detect these infrared inactive modes. Finally, we wish to gain some insight into the nature of impurity modes by comparing their phonon scattering with that caused by spatially well localized motional states, associated for instance with the rotational degrees of freedom of molecular impurities.

For experimental reasons, our knowledge so far is confined to doped alkali halide crystals. Clearly, an extension of the work to other electrical insulators would be very desirable.

I. RESONANCES CAUSED BY QUASI-ISOTOPIC DEFECTS

Fig. 1 shows the influence of small concentrations of several alkali or halogen ions substituted into the KCl host lattice.[1,2] In addition to the overall decrease of the thermal conductivity of the doped crystals, a very marked indentation in the curves is observed at certain temperatures which are the lower, the larger the mass of the defect ion. It was suggested by Wagner[3] that these "dips" were caused by an interaction between the plane wave phonons and impurity modes occurring inside the acoustic continuum of the

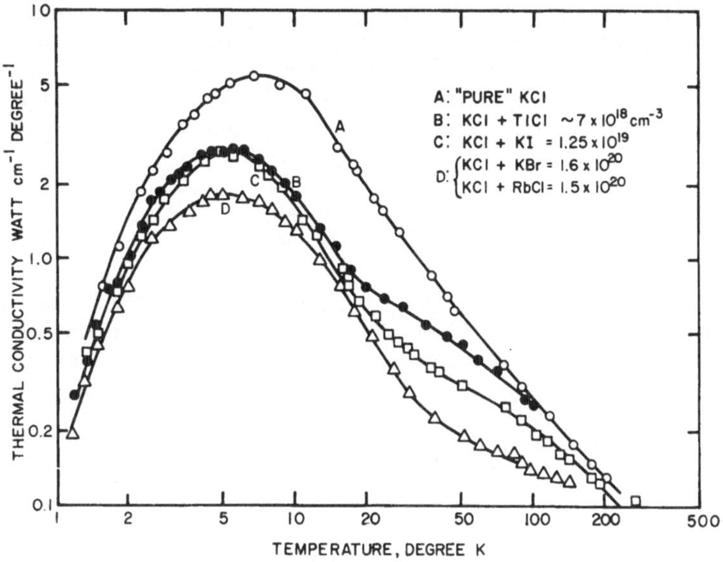

Figure 1: Comparison of thermal conductivity of KCl containing Tl^+, I^-, Rb^+, and Br^-. Concentrations $n_{Tl} \sim 7 \times 10^{18}$ cm^{-3}; $n_I = 1.25 \times 10^{19}$ cm^{-3}; $n_{Rb} = 1.5 \times 10^{20}$; $n_{Br} = 1.6 \times 10^{20}$ cm^{-3}. The data for KCl:Rb^+ and KCl:Br^- were identical over the entire temperature range. This graph shows the "dips" in the thermal conductivity curves typical for resonance scattering. Note that the temperature of which the dips occur decreases with increasing mass of the impurity. Ref. 2.

host lattice. Guided by the observation that these dips always occurred on the high temperature side of the conductivity curve, where anharmonic three-phonon processes are predominant in the pure crystal, he assumed that the scattering by the impurity modes also was through anharmonic processes, i.e. that two plane wave phonons combined to excite the impurity mode. In the temperature region where this process was most likely to occur, the conductivity was strongly decreased. Although there is good evidence that this process is indeed important, it is difficult to test experimentally, because this theory contains as yet too many unknown parameters. An alternative explanation was suggested by Takeno.[4] He demonstrated that in a lattice containing heavy isotopic impurities elastic phonon resonance scattering was possible in the harmonic approximation. This scattering is exactly analogous to the classical scattering of waves (light, sound) by particles capable of resonating. In the case of the isotopic defect, the resonating center was the heavy atom, oscillating with respect to its surroundings and emitting spherical waves into the crystal. This scattering mechanism has since been discussed by a number of authors.[5] The phonon relaxation rate τ^{-1} in the Debye approximation quoted in the form given by McCombie and Slater[6] is given by:

$$\tau^{-1} = \frac{N_s}{N} \frac{\beta^2\omega^3 Q}{(1 - \beta\omega^2 P)^2 + \beta^2\omega^4 Q^2} \tag{1}$$

Here N_s/N is the number density of impurity atoms divided by that of the host crystal, $\beta = (M'-M)/M$, with M' and M the mass of impurity atom and host atom respectively. In the Debye approximation:

$$Q = \frac{3\pi}{2\omega_D^3}\,\omega \qquad (\omega_D \text{ is the Debye frequency}), \tag{2}$$

and

$$P = \frac{3}{2\omega_D^3}\left(2\omega_D + \omega \ln \frac{|\omega_D - \omega|}{\omega_D + \omega}\right) . \tag{3}$$

For frequencies considerably smaller than ω_0 $(=(\beta P)^{-1/2})$, Eq. (1) is identical to the well known[7] isotopic Rayleigh scattering rate. The dip in the thermal conductivity arises because the phonons which are the dominant carriers of heat in that temperature range have frequencies near ω_0. At high frequencies the phonon scattering rate diminishes rapidly. Eq. (1) contains no adjustable parameters, and McCombie and Slater were able to produce a good fit to the experimental data obtained on KCl:I by simply considering the mass difference between the I^- and the Cl^- and by taking the host lattice to be monatomic. In fact, the fit in the Rayleigh region is even better[2] than was shown by McCombie and Slater, presumably because of a poor fit to the pure crystal conductivity used in their calculation. Equally good fits, as far as the position of the dips as well as the strength of the phonon scattering in the low temperature region are concerned, have since been obtained for the other impurities shown in Fig. 1.[2] From this the conclusion was drawn that at least as far as thermal conductivity is concerned, Tl^+, I^-, Br^- and Rb^+ in KCl can be considered to be essentially isotopic defects, i.e. the scattering results from the difference in mass alone. The resonance frequencies ω_0 are 45, 57, 86 and 86 cm^{-1} respectively, essentially in agreement with the value of the resonance frequency as computed from Eq. (1). Note that the calculated resonance dips are not as pronounced as those observed. We believe, however, that this disparity can be removed only by a theory considering anharmonicities and not by including elastic scattering off other impurity modes, as they are

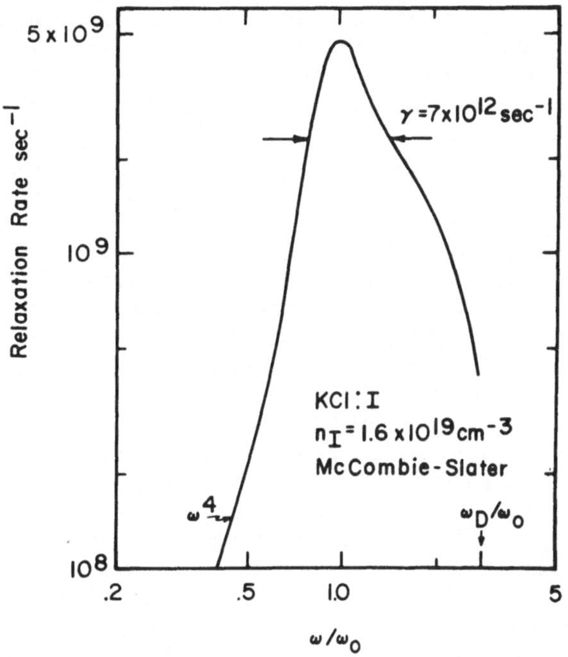

Figure 2: Single mode phonon relaxation rate calculated for elastic scattering by an isotopic defect of the mass of the I^- ion in KCl, which was taken to be a monatomic crystal. $n_I = 1.6 \times 10^{19}$ cm^{-3}. Ref. 6.

expected to occur if the defect has not only a different mass, but is also held by different forces. Since the mass difference alone seems sufficient to describe the resonances in these cases, these modes are believed to be predominantly odd, i.e. of symmetry T_{1u}, which should also be infrared active. In none of these four crystals, however, was an absorption band found in the frequency range between 10 cm^{-1} and 100 cm^{-1}.[8] How can this be reconciled with the result obtained from thermal conductivity?

Fig. 2 shows the phonon relaxation rate τ^{-1} as computed from Eqs. (1), (2), and (3) for KCl:I (M_{Cl} = 35.5, M'_I = 126.9 a.m.u.), with the resonance frequency ω_0 at 1.08×10^{13} rad sec^{-1} (57 cm^{-1}). The half width $\Gamma = \omega_0^2\pi/(2\omega_D) = .7 \times 10^{13}$ rad sec^{-1}, or $\omega_0/\Gamma = 1.55$. Its detection in the IR, therefore, may not have been possible because of its large breadth.[8] Thermal conductivity, on the other hand, is more sensitive to moderate scattering occurring over a wide frequency range than it is to strong scattering occurring over a narrow frequency range. Note that we have been comparing phonon relaxation rates and photon absorption constants. Both quantities, however, have the same frequency dependence and half width, at least in the framework of the simple model used here.

II. RESONANCES CAUSED BY FOREIGN ATOMS, GENERAL CASE

If our conclusion, that broad resonances are seen more easily in thermal conductivity than in the IR, is correct, than conversely we might expect that in crystals in which narrow resonances are seen through infrared absorption, phonon resonant scattering should be absent or at least much weaker. This, however, is not true. A number of narrow resonances have been detected optically, e.g. KBr:Li[9] ($\omega_{0,IR}$ = 16.3 cm^{-1}, $\omega_{0,IR}/\Gamma$ = 20) and KCl:Ag[10] ($\omega_{0,IR}$ = 38.8 cm^{-1}, $\omega_{0,IR}/\Gamma$ = 7). These crystals do show strong phonon resonance scattering, similar to that found with "isotopic" defects.[2,11] It was found, though, that the resonance frequencies $\omega_{0,th}$ determined from thermal conductivity were consistently larger than the corresponding $\omega_{0,IR}$. Moreover, an isotopic shift of 10% found in the IR in Li^6 and Li^7 doped KBr was absent in thermal conductivity.[11] A similar absence of an isotope effect was also recently reported by Radosevich and Walker[12] in KI containing substitutional H^- and D^-. Hence, it appears that the resonance scattering in these cases is not caused by T_{1u} impurity modes. The T_{1u} modes probably do not show up in thermal conductivity, as indicated already by the smaller half-width found in the IR.

For an explanation of the origin of these phonon resonances, we turn to the optical absorption studies. There it was found, from peak position and also from the half-width of the resonances, that the force constants between the impurity and its nearest neighbors were weakened drastically in most cases. If a defect is characterized by a difference in mass as well as in force constant, one has to expect more impurity modes to be created than in the "isotopic" case. The new modes will have even parity and be of the symmetry A_{1g} and E_g. Hence, we believe that in crystals with narrow T_{1u} mode resonances the phonon scattering observed is caused by such even parity impurity modes. Fig. 3 shows a comparison of the infrared active resonance frequency $\omega_{0,IR}$ and the resonance frequency $\omega_{0,th}$. $\omega_{0,th}$ is about 60% larger than $\omega_{0,IR}$. It is reasonable that an even mode should have a higher frequency than an odd one[13] although a simple relation should not be expected to exist, since the T_{1u} mode is determined by mass and force constant changes, whereas the even modes should be independent of the mass change. The question whether the E_g mode or the fully symmetric breathing mode A_{1g} should be the more important scatterer has been studied by Caldwell and Klein.[14] In Fig. 4 is shown the thermal

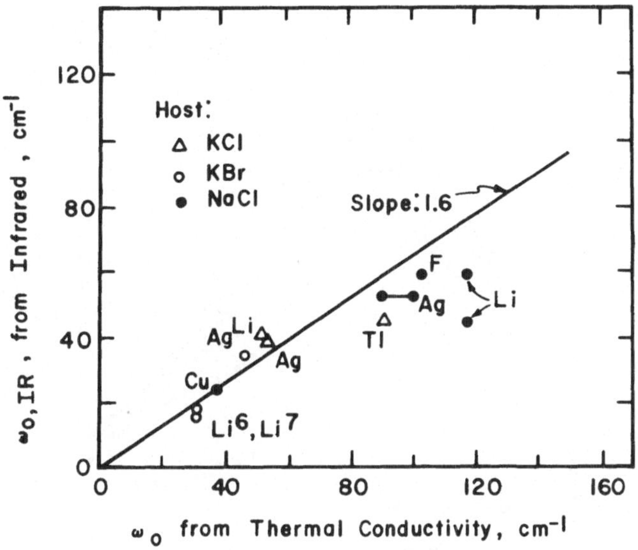

Figure 3: Comparison of the T_{1u} impurity mode frequency with the impurity mode frequency derived from thermal conductivity. The latter seems to be consistently larger than the former by about 60%. Also plotted along the horizontal axis is the E_g mode frequency in KCl:Tl as determined by Martienssen and co-workers, ref. 17. Along the vertical axis we have plotted the KCl:Tl resonance frequency determined from thermal conductivity which we believe to be of symmetry T_{1u}. Infrared data taken from ref. 14, 15, 16, and A. J. Sievers, private communication.

conductivity of NaCl:Ag.[14] From the position of the dip we deter- $\omega_{0,th}$ = 90 cm^{-1}. The phonon relaxation rate τ^{-1} calculated[14] for resonance scattering by resonances produced by changes of mass and of force constant are shown in Fig. 5. The force constant changes were chosen in such a way as to make the τ_{max}^{-1} for the T_{1u} mode scattering occur at the same frequency as the infrared absorption peak[15] ($\omega_{0,IR}$ = 52.5 cm^{-1}). A considerable contribution to the phonon scattering is caused by E_g mode scattering (around 105 cm^{-1}), A_{1g} mode scattering is much weaker. Although the calculated relaxation rates certainly must be taken with a grain of salt, since they were obtained in the harmonic approximation and contain adjustable parameters as the nearest neighbor relaxation, it affords a simple test of our model according to which only broad resonances ($\omega_0/\Gamma \sim 1$) show up in thermal conductivity. The halfwidth of $\tau_{T_{1u}}^{-1}$ is $\Gamma = 1.35 \times 10^{12}$ rad sec^{-1}, or $\omega_0/\Gamma = 7.4$ (the value derived from optical absorption is 5[15,16]), and indeed, the resonance dip does not occur at the temperature corresponding to 52.7 cm^{-1} (22°K), but at a higher temperature, 38°K, corresponding to 90 cm^{-1} which

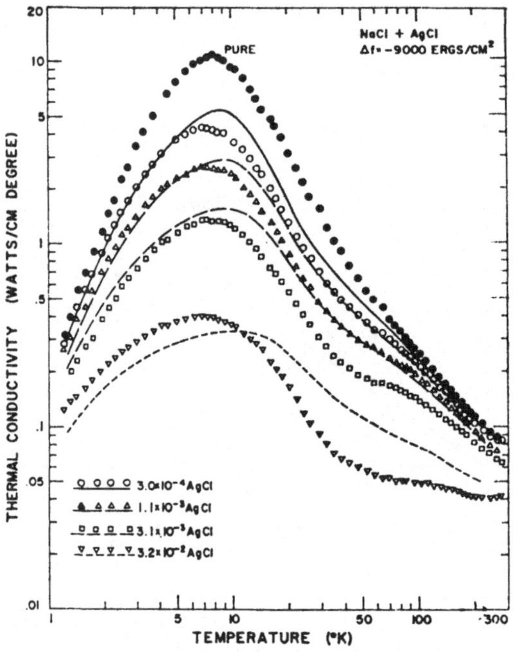

Figure 4: Thermal conductivity of NaCl:Ag. Solid curves, computed with relaxation rate as shown in Fig. 5. Ref. 14.

is close to the maximum of the E_g mode scattering rate. Further evidence that the resonance frequency $\omega_{0,th}$ is determined predominantly by the E_g impurity mode is shown in Fig. 3: We have claimed that the thermal conductivity sees the odd mode T_{1u} in KCl:Tl, so its frequency, 45 cm^{-1}, is plotted on the vertical axis. The E_g mode frequency (90 cm^{-1}) was determined through measurements of the temperature dependence of the electronic transition in KCl:Tl by Fussgaenger.[17] This value is plotted on the horizontal axis in Fig. 3. The KCl:Tl point agrees well with the general dependence found between the two types of resonances ascribed to T_{1u} and E_g modes.

The conclusion we reach then is that thermal conductivity is a useful tool for the detection of the infrared inactive, even parity modes. Obviously, one must be cautious in ascribing the resonance to one impurity mode alone, as demonstrated in Fig. 5. On the other hand, the sharpness of the dip in the thermal conductivity curves seems to imply that one impurity mode is indeed the dominant scatterer and we believe to have evidence that this is an E_g mode. It is unfortunate, though, that a detection of the same modes as seen in thermal conductivity by optical spectroscopy, e.g. Raman scattering, should be difficult because of their strong damping.

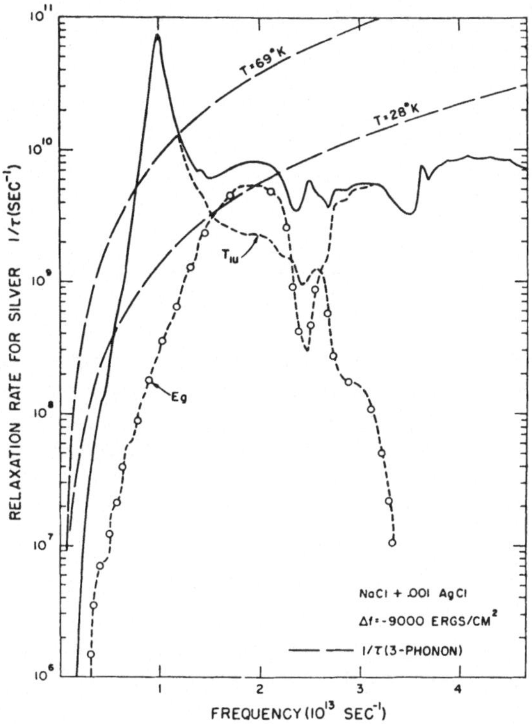

Figure 5: Theoretical relaxation rate used to calculate the conductivity of NaCl:Ag. Force constant change adjusted so that the T_{1u} mode frequency coincides with the one determined from IR (ref. 15). Ref. 14.

III. THE PHONON SCATTERING PROCESS

So far in our discussion, we have used thermal conductivity mostly for the detection of impurity modes. In this section, we want to consider the phonon scattering process. We want to compare the phonon scattering by various impurity modes.

In Section I we had given the relaxation rate for elastic resonant scattering by an in-band resonance caused by a heavy impurity mass (isotopic impurity). Inserting Eqs. (2) and (3) into Eq. (1) and with $\omega_0 = (\beta P)^{-1/2}$, we obtain:

$$\tau^{-1} = N_S \pi 4 \, v^3 \, \frac{\pi^2}{4\omega_D^{\,2}} \, \frac{\omega^4}{(\omega_0^{\,2} - \omega^2) + \dfrac{\pi^2}{4\omega_D^{\,2}} \cdot \omega^6} \qquad (4)$$

The halfwidth of τ^{-1} (width at half the maximum relaxation rate) is given in good approximation by

$$\Gamma = \frac{\pi^2}{2\omega_D} \omega_0^2 = \alpha\omega_0^2 , \tag{5}$$

hence $\alpha = \frac{\pi}{2}\frac{1}{\omega_D}$. Upon inserting α in (4), we obtain:

$$\tau^{-1} = N_S \, 4\pi v^3 \, \alpha^2 \frac{\omega^4}{(\omega_0^2 - \omega^2)^2 + \alpha^2\omega^6} \tag{6}$$

It is interesting to note that an identical expression is obtained if one calculates the attenuation of sound waves by elastic dipoles with dipole moment p_0 and rotational inertia I, performing angular motions $\theta(t)$ in a medium of stiffness c_s and sound velocity v. In this case α is determined to:

$$\alpha = \frac{2p_o^2}{3(4\pi c_s^{-1})v^3 I} \tag{7}$$

Finally, elastic scattering by harmonically bound electrons is also described by an expression of the form of Eq. (6), with

$$\alpha = \frac{2e^2}{3(4\pi\varepsilon_0)c^3 m} \tag{8}$$

with v changed to the velocity of light c. We conclude that Eq. (6) is quite generally true in the case of classical scattering.

It has been found experimentally in a large number of cases, however, that resonances occurring at temperatures below the temperature of the thermal conductivity maximum of the pure crystal could not be described with the relaxation rate as given by Eq. (6) regardless of the modes causing the scattering. Two examples are shown in Fig. 6 and 7 for resonant scattering by tunneling states. In Fig. 8 we see an example of our efforts to describe the experimental data with an elastic scattering rate, Eq. (6). At temperatures below the dip, the calculated conductivity recovers more quickly towards that of the pure crystal than the measured one does. For these machine fits we used the Debye model of thermal conductivity, whose validity has been checked repeatedly[19] and which should be particularly well suited to describe the thermal conductivity at very low temperatures, where the phonon density of states is very well approximated by the Debye density and where the scattering in the pure crystal is entirely by the crystal surfaces (boundary scattering). Using the Debye expression

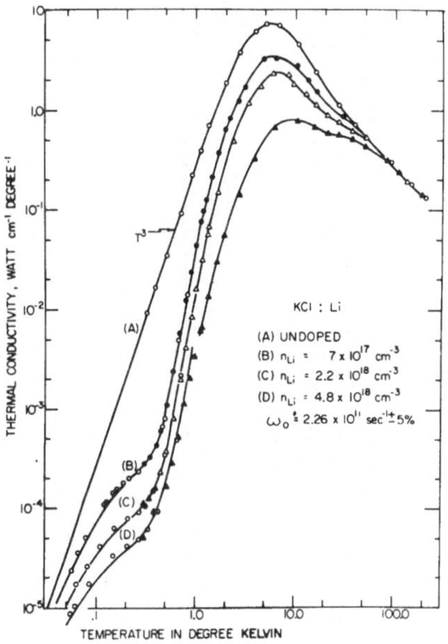

Figure 6: Thermal conductivity of KCl:Li, ref. 11. The solid curve through the pure KCl is a machine fit. In the doped crystals the machine fits (solid curves) extend only up to 1°K.

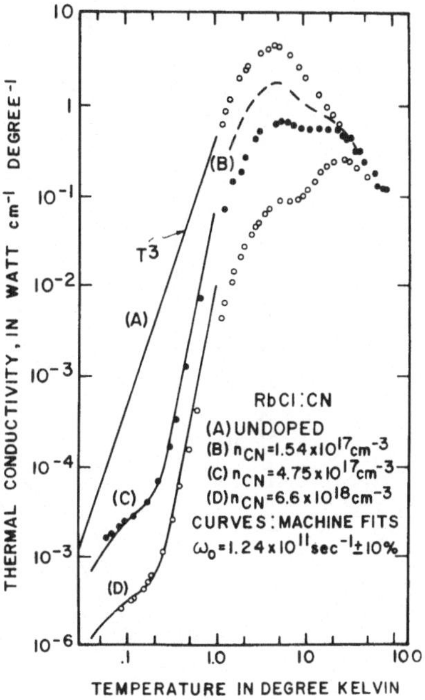

Figure 7: Thermal conductivity of RbCl:CN. Solid curves are machine calculations. Ref. 18.

$$K(T) = \frac{1}{3} \int_0^{\omega_D} \frac{dc}{d\omega} v^2 \tau_{tot} \, d\omega \qquad (9)$$

and

$$\tau_{tot}^{-1} = \sum_i \tau_i^{-1} , \qquad (10)$$

with τ_{def}^{-1} of the form of Eq. (6), one of the best fits that could be accomplished is shown in Fig. 8, curve (B), see the caption for details.

On the other hand, consistently good fits have been obtained using a different, phenomenological relaxation rate of the form

$$\tau^{-1} = n A \frac{\omega^2}{(\omega_0^2 - \omega^2)^2} \qquad (11)$$

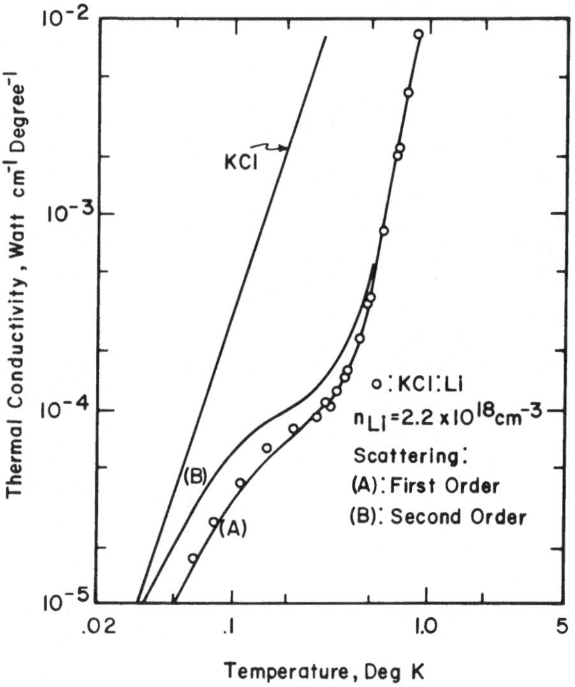

Figure 8: Best fits to KCl:Li, $n_{Li} = 2.2 \times 10^{18}$ cm^{-3} (Fig. 6) using (A) an inelastic (first order) scattering process, with a relaxation rate of the form of Eq. (11) and using (B) an elastic (second order) scattering process, with a relaxation rate of the form of Eq. (6). Both factors, the one in front of the numerator in front of the ω^4 term, and the α^2 in front of the ω^6 term in the denominator, where adjusted separately to improve the quality of the machine fit to the data. Curve (B) was computed using $\alpha = 4.05 \times 10^{-14}$ sec in the numerator, and $\alpha^* = 1.2 \times 10^{-11}$ sec in the denominator.

whose basic difference is that it varies as ω^2 at low frequencies. The solid curves in Figs. 6 and 7 were computed with such a term, which was found to scale with the defect concentration over several orders of magnitude in concentration. The resonance frequencies ω_0 were found to agree with the frequencies of the tunneling states of the molecular or the monatomic defects or with the frequencies of the states of free rotation in the case of molecular defects. Hence, it is believed that these states are strongly coupled to the phonons. Direct experimental evidence for this picture to be correct was provided by measurements by Chau et al.[20] who showed that the resonance frequency ω_0 found in KCl:OH decreased by about 30% when the OH^- was replaced by OD^-. An increase of 40% in the resonance frequency associated with the tunneling motion of Li^+ in

KCl when Li^6 was introduced instead of Li^7 has recently been found in thermal conductivity by Peressini.[21] This shift agrees with the one determined through specific heat measurements by Harrison *et al.*[22]

In order to obtain a phonon relaxation rate of the form found in the experiments from the model of the classical harmonic oscillator, a different type of damping has to be introduced. This damping has to be viscous, i.e. the damping force must be assumed to go proportional to $\dot{x}$ or $\dot{\theta}$ rather than proportional to $\dddot{x}$ or $\dddot{\theta}$ as was the case in the derivation of Eq. (6) where the damping was entirely through radiation. If we carry out the calculation for a damping force containing both $\dot{x}$ and $\dddot{x}$ damping, we obtain the relaxation rate

$$\tau^{-1} = n\, 4\pi v^3(\alpha\omega^2 + \delta)\alpha \frac{\omega^2}{(\omega_0^2 - \omega^2)^2 + (\alpha\omega^2 + \delta)^2\omega^2} \qquad (12)$$

where δ is the reciprocal time during which the oscillator loses all but 37% of its energy E through the viscous ($\propto \dot{x}$) damping:

$$E = E_0\, e^{-t\delta}\ . \qquad (13)$$

For the low temperature resonances, then, it appears that $\delta >> \alpha\omega^2$, hence:

$$\tau_{def}^{-1} = n\, 4\pi v^3(\varepsilon\delta) \frac{\omega^2}{(\omega_0^2 - \omega^2)^2 + \delta^2\omega^2} \qquad (14)$$

where δ is the halfwidth. In order to distinguish between the coupling in this case and the coupling in the case of isotopic scattering, we have re-labeled the constant α to ε. Since the term $\delta^2\omega^2$ in the denominator affects only a narrow band of phonons with frequencies near ω_0, it is not surprising that it remained unnoticed in thermal conductivity experiments, where the heat is carried by a broad phonon spectrum. Hence Eq. (14) is indeed identical to the phenomenological Eq. (11), and from the quantity A determined by a best fit to the experimental curves the product $(\varepsilon\delta)$ can be determined. The results are plotted in Fig. 9 versus $(\omega_0/\omega_D)^2$. $(\varepsilon\delta)$ is seen to increase practically as $(\omega_0/\omega_D)^2$, regardless of the impurity ion and the host. This is a very remarkable observation, because one would have expected a stronger influence of both defect and host on the phonon coupling. This situation becomes even more mysterious if we plot the product of α and Γ, the halfwidth of the resonance scattering relaxation rate, as computed for isotopic impurity resonances (Eq. (5)) on the same graph. This straight line $\alpha\Gamma \propto \omega_0^2$ agrees very closely with the

Table I

Determination of the quantity ($\varepsilon\delta$), see text, from the empirical scattering strength A. The lifetime δ^{-1} of the excited state is tentative, see text. The values δ are essentially identical to the transition probabilities γ given in the paper by Pohl and Lombardo, see Bulletin of this Conference.

System	Type of mode	ω_0 (10^{11}rad sec^{-1})	Debye ω_D (10^{13}rad sec^{-1})	Vel. of sound (10^5cm sec^{-1})	A (cm^3/sec^3)	($\varepsilon\delta$)	δ (sec^{-1})
RbCl:CN	tunnel	1.28	2.13	1.85	1.88×10^{12}	2.36×10^{-5}	6.2×10^{8}
KCl:Li	tunnel	2.26	3.0	2.45	1.86×10^{12}	$1\ \times10^{-5}$	7.2×10^{8}
KCl:CN	tunnel	3.0	3.0	2.45	1.04×10^{13}	5.65×10^{-5}	2.3×10^{9}
KBr:CN	tunnel	3.0	2.29	1.94	1.61×10^{13}	1.75×10^{-4}	$4\ \times10^{9}$
KI:CN	tunnel	3.0	1.71	1.55	$5.5\ \times10^{12}$	1.18×10^{-4}	3.4×10^{9}
KCl:CN	rot.	35.4	3.0	2.45	9.38×10^{15}	$5.1\ \times10^{-2}$	$8\ \times10^{11}$
KBr:CN	rot.	35.4	2.29	1.94	1.25×10^{16}	1.36×10^{-1}	1.5×10^{12}
KCl:NO_2	rot.	38.0	3.0	2.45	4.77×10^{15}	2.58×10^{-2}	6.1×10^{11}
RbCl:CN	rot. [a]	66.0	2.13	1.85	$3.9\ \times10^{16}$	4.92×10^{-1}	$5\ \times10^{12}$
KBr:Li	Imp. [b]	68.0	2.29	1.94	$\sim3.9\ \times10^{15}$	$\sim4.26\times10^{-2}$	$\sim1.5\times10^{12}$

a: It is believed that this mode is a combination of a rotation of the CN^- ion and a motion of the surrounding ions of the breathing type. Ref. 18.

b: This impurity mode is believed to be of symmetry E_g, see ref. 11. A is correct only to within a factor of 5.

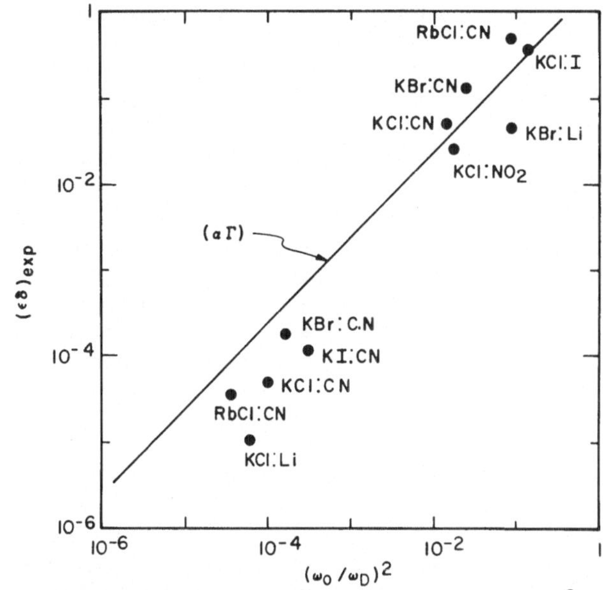

Figure 9: This figure shows a plot of $(\varepsilon\delta) = A/4\pi v^3$, see Eq. 14, with A as obtained from experiment through a best fit using a relaxation rate of the form of Eq. (11), plotted vs. $(\omega_0/\omega_D)^2$. It can be seen that $(\varepsilon\delta) \propto (\omega_0/\omega_D)^2$. Plotted also is the quantity $(\alpha\Gamma)$ as calculated from Eq. (6). $(\varepsilon\delta)$ and $(\alpha\Gamma)$ are almost equal. The experimental A and ω_0 were taken from Ref. 11, 18, 23, and 24.

empirical values $(\varepsilon\delta)$, although the damping in both processes is quite different (radiative versus viscous in the classical picture).

We are thus led to the conclusions: (1) that the dominant role in the scattering is played by the resonance frequency, and not so much by the type of mode (tunneling vs. rotations, vs. inband resonant states), and (2) that although the frequency dependence is different for different scattering states, the product of the coupling times the halfwidth is not.

Let us go one step further: If $(\varepsilon\delta)$ is practically equal to $(\alpha\Gamma)$ over such a wide frequency range and for that many different impurity modes, we may tentatively assume that $\alpha = \varepsilon$ and $\delta = \Gamma = \alpha\omega_0^2$. With this speculation the values for δ, the reciprocal lifetimes of the oscillators was determined, see Table I. From this, it appears that the lifetime increases strongly with decreasing resonance frequency. Furthermore, the speculation $\delta = \Gamma = \alpha\omega_0^2$ implies that the viscous damping is equal to the radiative damping at the frequency ω_0. We can translate this into the quantum picture: Because of its lifetime, phonons of a wide frequency spectrum can be absorbed by the oscillators, but the oscillators get rid of their energy through radiation at the frequency ω_0.

It appears premature to carry the speculations much farther as long as the lifetimes δ^{-1} have not been measured by an independent method. Nevertheless, we would like to emphasize the question which suggests itself when one tries to view the phonon impurity mode interaction as a phenomenon determined almost exclusively by the resonance frequency: The low frequency modes scatter phonons through an absorption-like process, the high frequency resonances, like KCl:I, scatter them through resonance fluorescence, the high temperature resonances in RbCl:CN and KBr:Li seem to scatter by a mixture of both. What is the cause for this transition? We do not know.

IV. SUMMARY AND CONCLUSIONS

Phonon resonance scattering observed in the thermal conductivity of alkali halide crystals containing foreign alkali or halogen ions in solid solution can be explained through excitation of resonant states. In some cases, like in KCl:I, the strength of the scattering and also the resonance frequency ω_0 itself can be explained by considering the impurity ion merely as a mass defect (isotopic impurity). The breadth of this resonance is of the order of ω_0 even in the harmonic approximation, hence it is possible that these modes could escape detection through infrared absorption measurements.

In cases where an infrared active impurity induced resonance has beeen observed in addition to the one observed in thermal conductivity, the resonance frequency ω_0 obtained from thermal conductivity measurements is larger than the infrared impurity mode of symmetry T_{1u} by about 60%. We have presented evidence that these phonon active resonances are even ones, probably of symmetry E_g.

Finally, we have considered resonant phonon scattering observed at temperatures of which negligible phonon scattering in the bulk of the pure crystal is observed (boundary regime). The resonant scattering in this region can be ascribed to excitation of states of the impurity ion which are highly localized in the vicinity of the defect. The frequency dependence of the phonon relaxation rate in these cases differs from that observed for scattering by high frequency resonances (e.g. KCl:I), and it appears that the relaxation occurs through absorption (inelastic scattering) rather than through elastic resonance scattering. A comparison of the experimentally determined quantities describing the scattering strength of the low frequency impurity modes (e.g. $(\varepsilon\delta)$) with the equivalent quantities found to describe the elastic resonance scattering by high frequency resonances indicates that the two scattering mechanisms are actually quite similar. This leaves the question unanswered, why the frequency dependence of the phonon relaxation rates appear to be different. It points, however, towards a great similarity between the impurity modes which are spatially localized (e.g. tunneling states) and the high frequency resonances (e.g. KCl:I resonance of symmetry T_{1u}), the main difference between these two types of modes remaining, at least as far as phonon scattering is concerned, the high frequency of the latter types of modes and hence their greater radiative damping.

V. ACKNOWLEDGMENTS

I have benefited greatly from many discussions with my colleagues, in particular with D. Channin, J. P. Harrison, J. A. Krumhansl, G. Lombardo, V. Narayanamurti and A. J. Sievers, all of whom I would like to thank at this place. I also want to thank H. Haken from the Institut for Theoretical Physics of the University of Stuttgart, Germany, for the hospitality extended to me while working on this paper. I am also grateful to H. Haken and M. Wagner for several stimulating discussions.

References

* Research supported in part by the U. S. Atomic Energy Commission under contract AT(30-1)-2391, Technical Report No. NYO-2391-61, and by the Advanced Research Projects Agency through the use of the central facilities of the Cornell Materials Science Center.

1. C. T. Walker and R. O. Pohl, Phys. Rev. 131, 1433 (1963).
2. F. C. Baumann and R. O. Pohl, Phys. Rev. (to be published).
3. M. C. Wagner, Phys. Rev. 131, 1443 (1963).
4. S. Takeno, Prog. Theor. Physics (Japan) 29, 191 (1963).
5. For recent reviews we refer to the articles by A. A. Maradudin in Solid State Physics, Vols. 18 and 19, edited by F. Seitz and D. Turnbull (Academic Press, Inc., New York, 1966), and by M. V. Klein, in Physics of Color Centers, edited by W. B. Fowler (Academic Press, Inc., New York, 1967).
6. C. W. McCombie and J. Slater, Proc. Phys. Soc. (London) 84, 499 (1964).
7. P. G. Klemens, Proc. Phys. Soc. (London) A68, 113 (1955).
8. A. J. Sievers, private communication.
9. A. J. Sievers and S. Takeno, Phys. Rev. 140, A1030 (1965).
10. A. J. Sievers, Phys. Rev. Letters 13, 310 (1964).
11. F. C. Baumann, J. P. Harrison, W. D. Seward, and R. O. Pohl, Phys. Rev. 159, 691 (1967).

12. L. G. Radosevich and C. T. Walker, Phys. Rev. 156, 1030 (1967).
13. J. A. Krumhansl, Proceedings of the 1963 International Conference on Lattice Dynamics, Copenhagen, Pergamon Press, 1964, p. 523.
14. R. F. Caldwell and M. V. Klein, Phys. Rev. 158, 851 (1967).
15. R. Weber, Phys. Letters 12, 311 (1964).
16. A. J. Sievers, NATO Advanced Study Institute, Cortina d'Ampezzo, Italy, 1966, p. VI. 1 (unpublished).
17. K. Fussgaenger, private communication.
18. J. P. Harrison, P. P. Peressini, and R. O. Pohl, submitted to Phys. Rev.
19. J. Callaway, Phys. Rev. 113, 1046 (1959); R. O. Pohl, NATO Advanced Study Institute, Cortina d'Ampezzo, Italy, 1966 p. VIII, 1; and P. D. Thacher, Phys. Rev. 156, 975 (1967).
20. C. K. Chau, M. V. Klein, and B. Wedding, Phys. Rev. Letters 17, 521 (1966).
21. P. P. Peressini, private communication.
22. J. P. Harrison, P. P. Peressini, and R. O. Pohl, Proceedings of this conference.
23. W. D. Seward and V. Narayanamurti, Phys. Rev. 148, 463 (1966).
24. V. Narayanamurti, W. D. Seward, and R. O. Pohl, Phys. Rev. 148, 481 (1966).

THE ROLE OF LOCALIZED VIBRATIONAL MODES IN HEAT CONDUCTION*

D. N. Payton, III,[†] Marvin Rich, William M. Visscher

University of California, Los Alamos Scientific

Laboratory, Los Alamos, New Mexico

I. INTRODUCTION

In this paper we shall report the results of some numerical experiments which we have performed on certain simple lattice models. Our aim has been to study the effects of anharmonicity and of disorder on thermal resistance, separately and in combination. Our principal result is that an increase in anharmonicity usually produces a decrease in thermal resistance.

According to existing theories of thermal conductivity,[1] which stem from the ideas of Debye[2] and Peierls,[3] thermal resistance is caused by scattering of phonons. Impurities and anharmonicities provide two independent scattering mechanisms. One would therefore expect, and indeed existing theory[1] predicts, that their contributions to the thermal resistance should be more or less additive.

Our results are not consistent with this prediction, except possibly in the limit of very small impurity concentration, where the impurities and the anharmonicity may be considered to be perturbations on the ordered harmonic lattice. In this limit the idea of a phonon gas with weak impurity and phonon-phonon interactions is valid, because the lifetimes of the phonons are long compared to their periods.

A Boltzmann equation for the phonon distribution function, with collision terms arising from anharmonicity and from impurities, leads to expressions for the thermal resistance which increase with increasing anharmonicity. We believe that the opposite behavior of our numerical results indicate that the phonon gas is a useful concept only for very low impurity concentrations.

Our approach, which is purely classical, uses a specific model for the lattice in interaction with thermal reservoirs at the two ends. The model and the method by which we numerically solve the equations of motion for the system until a steady state is reached are described in Sec. II.

II. DYNAMICAL MODEL FOR THE CRYSTAL AND RESERVOIRS

In one dimension our lattice model is a chain of N atoms with nearest-neighbor interactions. Rigid boundary conditions are assumed (i.e., the zeroth and (N+1)st atoms have infinite mass), and the first and Nth atoms interact with thermal reservoirs. N is usually 100; however, we have taken it as large as 2000.

In two dimensions (2D) we use an array of N × M (usually 10 × 50) atoms with nearest-neighbor central and non-central forces. Rigid boundary conditions are used in the long dimension; the atoms adjacent to the fixed boundary are in contact with thermal reservoirs. Periodic boundary conditions are used in the short dimension. Our 2D lattice may be considered to be a tubular lattice connecting the hot and cold reservoirs.

We shall numerically solve the classical equations of motion for this system with nearest-neighbor potentials of the form

$$V(x_i-x_{i-1}) = -\epsilon_o + \tfrac{1}{2}\gamma(x_i-x_{i-1})^2 - \tfrac{1}{3}\mu(x_i-x_{i-1})^3 + \tfrac{1}{4}\nu(x_i-x_{i-1})^4 \quad (1)$$

in the 1D case, and reasonable values, corresponding to parameters appropriate to rare-gas solids, for γ, μ, and ν. In the 2D case central and non-central potentials, again of the form of Eq. (1), are assumed. The arguments then are either differences in displacements in the x-direction or in the y-direction; motions in the two perpendicular directions are independent.

The end atoms interact with the external systems through a random force

$$F_i(t) = \sum_{s,\alpha} \delta_{i\alpha} P^s_\alpha \, \delta(t-t_s). \quad (2)$$

$P^s_\alpha = 0$ unless α is the label of an end atom; it is the momentum transferred to the end atom by an instantaneous binary elastic collision at time t_s with an atom of the perfect-gas thermal reservoir at the end of the chain or array.

Equation (2) gives the force exerted on the end atoms by the collisions they suffer with the reservoir particles. Simple kinematics gives

$$p_\alpha^s = \frac{2M_\alpha m_\alpha}{M_\alpha + m_\alpha}\,(w - \dot{x}_\alpha) \tag{3}$$

for the momentum transferred to the αth atom which has mass m_α and velocity $\dot{x}_\alpha$ just before colliding when struck by a reservoir atom of mass M_α and initial velocity w.

The method for solving the equations of motion is as follows:
(1) A random set of initial positions and velocities is chosen for the lattice atoms.

(2) The equations of motion are used to advance the time t by an increment δt. An implicit integration procedure was used for the harmonic lattices, and a fourth-order Runge-Kutta procedure was used for the anharmonic lattices. The increment δt is preassigned and should be small compared to the minimum vibration period of the lattice.

(3) With some probability $\lambda\delta t$, each of the end atoms suffers a collision with an atom in its respective reservoir and impulsively acquires a momentum increment given by Eq. (3). The collision probability λ is a preassigned constant, independent of the velocities w and $\dot{x}_\alpha$. The initial velocity of the reservoir atom w is chosen according to a Maxwellian distribution characterized by temperature T_α:

$$n(w) = \sqrt{\frac{2kT_\alpha}{M_\alpha \pi}}\;\exp\left(-\frac{M_\alpha w^2}{2kT_\alpha}\right). \tag{4}$$

(4) Steps (2) and (3) are repeated until the system has achieved a steady state. For a random initialization (1) this takes typically 10^3 to 10^4 time steps. Probabilistic decisions required in steps (1) and (3) were made with the aid of a random number generator.

III. LOCAL TEMPERATURE, HEAT FLOW, AND THERMAL CONDUCTIVITY

Energy flow was calculated at the ends of the linear chain and across the end rows in the square lattice. We define local temperature to be the sum of the kinetic and potential temperatures, i.e.

$$k_B T_i = \frac{1}{2} m_i \overline{v_i^2} + \frac{1}{2} \sum_{j \neq i} \overline{V(x_i - x_j)} \tag{5}$$

where the bars denote time averages.

Thermal conductivity is defined to be the quotient of heat flow and temperature gradient. The calculation of the heat flow is straightforward. The most convenient criterion for the attainment of the steady-state is that the heat flow into one end of the lattice becomes equal to that out of the other end. Figure 1 shows an example. The slope of the upper curve is the instantaneous energy flow into the lattice at the hot end; that of the lower curve is the flow into the cold end.

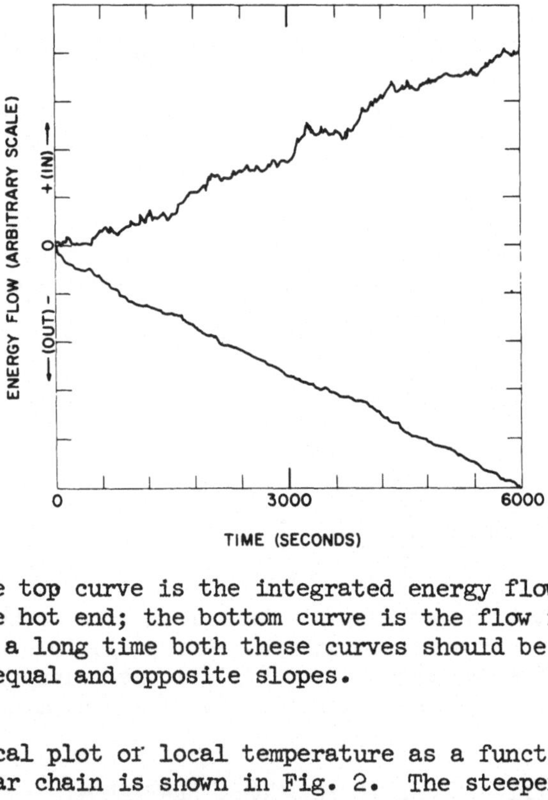

Fig. 1. The top curve is the integrated energy flow into a linear chain at the hot end; the bottom curve is the flow in at the cold end. After a long time both these curves should be linear in the time, with equal and opposite slopes.

A typical plot of local temperature as a function of position in the linear chain is shown in Fig. 2. The steeper straight line is the imposed temperature gradient, defined by the reservoir temperatures. The other straight line is a visual fit to the local temperature. Its slope gives the internal temperature gradient.

The irregularities are caused by the particular impurity configuration of the disordered chain. If a configuration average were taken the curve would be smooth, but computer time consumption would then be prohibitive.

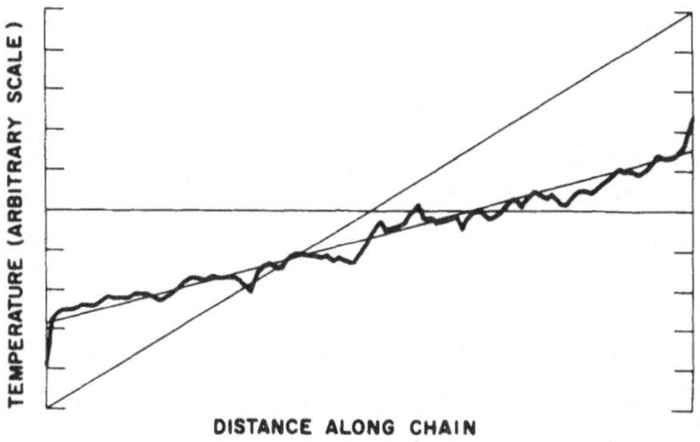

Fig. 2. Local temperature in a 100-atom anharmonic 50-50 disordered chain of masses 1 and 3. The time-average here is over 3000 "seconds."

IV. PARAMETERS AND UNITS

The variable parameters used in these numerical experiments are as follows:

(1) The masses of the atoms in the arrays. Our lattices are always disordered binary mixtures with no alteration of the forces by the impurities. The two masses and their concentrations are preassigned. A random number determines whether the atom at a given site has one mass or the other.

(2) The force parameters defined in Eq. (1). These are dimensional and the following arbitrary choice of units was used:

$$\begin{aligned} \text{Length} &= 10^{-9}\ \text{cm} \\ \text{Mass} &= 10^{-22}\ \text{g} \\ \text{Energy} &= 10^{\circ}\text{K}. \end{aligned} \tag{6}$$

A physically reasonable choice of Lennard-Jones potential parameters appropriate to noble gas solids

$$\begin{aligned} \epsilon_o &= 175 \times 10^{-16}\ \text{erg} \\ r_o &= 3 \times 10^{-8}\ \text{cm}, \end{aligned} \tag{7}$$

then gives

$$\begin{aligned} \gamma &= 1400\ \text{erg/cm}^2 \\ \mu &= 0.35\ \gamma \\ \nu &= 0.069\ \gamma. \end{aligned} \tag{8}$$

Our time unit is then one "second" = $2/\omega_L$ = 2.673×10^{-13} sec., which is π^{-1} times the minimum period of a harmonic chain composed of unit masses. The harmonic force constant is unity in the system of units given by Eq. (6); therefore $\omega_L = (4\gamma/m)^{1/2} = 2$ for a monatomic harmonic chain of unit masses. The anharmonic lattices use the force constants given by Eq. (8), except when otherwise stated.

(3) The temperatures of the end reservoirs. In order to attain a steady state in a reasonable time, and to be able to measure a thermal gradient above background noise in a graph like that of Fig. (2), it has been necessary to use large temperature gradients.

(4) The probability per unit time, λ, of collision of a reservoir particle with an end particle. This collision probability is proportional to the density of the reservoir gas. Variation of λ does not seem to affect the calculated thermal conductivity.

(5) The time step δt used in the integration scheme. The time step was made small enough that the integration conserves energy to better than 0.1% over 10^5 time steps.

(6) The masses of the reservoir atoms. These have been set equal to the lighter of the lattice masses.

V. RESULTS AND DISCUSSION

Figure 3 shows the calculated heat conductivities for the linear chain. Monatomic harmonic lattices (concentration of light atoms = 0 or 1) have infinite κ. The heat conductivity decreases rapidly as random mass impurities are added. This effect is enhanced with increasing mass difference.

The addition of anharmonicity to the monatomic chains, as expected, decreases their thermal conductivities to finite values. However, all of the disordered chains studied were affected in the opposite way by anharmonicity, i.e., the addition of the anharmonic scattering mechanism decreased the thermal resistance.

Conducive to the understanding of these results is a knowledge of the nature of the normal modes of disordered harmonic lattices.[4] Two of us[5,6] have recently published the results of a calculation of the normal modes of isotopically disordered binary lattices in one, two, and three dimensions, certain features of which are particularly relevant here. For example, in a disordered lattice composed of a mixture of heavy and light atoms, the upper part of the frequency spectrum has a very peaky structure and the normal modes to which these frequencies correspond are all localized if the ratio of heavy to light mass is large enough. The lower part of the frequency spectrum (quasicontinuum) is smooth and is associated with extended normal modes.

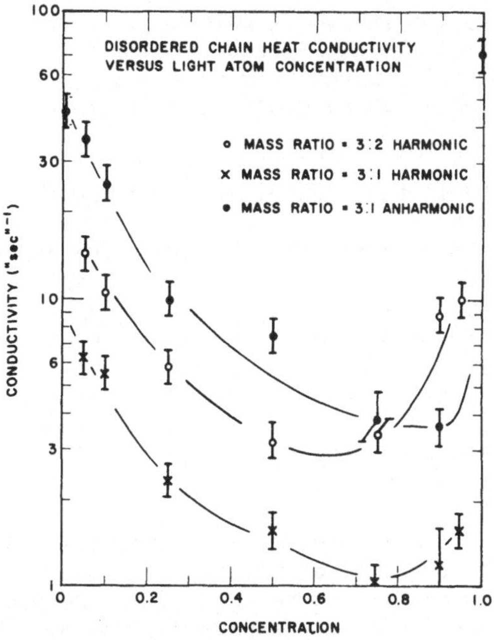

Fig. 3. Heat conductivity coefficient $\varkappa$ for the linear chain. The unit of the ordinate is inverse time, because temperature gradient is in units of energy per lattice point. The thermal conductivity in (cal/cm^{2}sec$^\circ$K) of a 3D lattice composed of a square array of parallel linear chains can be obtained by dividing our values of the thermal conductivity by 770a where a is the lattice spacing in Angstroms. The error bars are estimates of the accuracy of our reading of the temperature gradient.

Our view of the physics underlying the results shown in Fig. 3 is as follows:

The thermal conductivity of the monatomic lattice is infinite because the thermal gradient vanishes. The thermal gradient vanishes because all of the phonons propagate in the lattice without attenuation. The addition of impurities changes this situation qualitatively; now translational symmetry is lost and some of the modes become localized. The localized modes do not contribute appreciably to the heat current from one end of the crystal to the other because to do so a mode must have a sizable amplitude at both ends. Only the non-localized modes in the quasicontinuum below the monatomic heavy lattice maximum frequency satisfy this requirement, so only they can transport energy in the disordered harmonic lattice. On the other hand, the localized modes can make a significent, perhaps dominant, contribution to the thermal gradient. The numerator, i.e., energy flow, in the heat conductivity is then mostly composed of contributions from quasicontinuum modes; the denominator, i.e., the thermal gradient, is mostly attributable to localized modes from the high-frequency end of the spectrum. The dependence of thermal conductivity on isotopic composition in our numerical experiments can be qualitatively understood on the basis of these ideas.

Anharmonicity in the isotopically disordered system enhances thermal conductivity. This result becomes understandable when one adds to the above discussion the fact that the anharmonic coupling allows the localized modes to decay into continuum modes. The localized modes are no longer frozen out of energy transport, and their energy may now flow away in lower frequency phonon-like modes. Therefore the energy flow, which is the numerator in the heat conductivity, is increased. The denominator, the internal thermal gradient, is decreased by this same mechanism. The heat conductivity should therefore be enhanced by the anharmonic relaxation of the localized modes, as observed.

Since the localized modes must be excited by collision of reservoir atoms with the ends of the crystal, one might think that if the localized modes are to be held solely responsible for the apparent thermal gradient in the harmonic chain, a long chain would be unable to support a thermal gradient at its center. To test this, the thermal conductivity and temperature gradient were calculated in a disordered chain 2000 atoms long - twenty times the usual length. We found no significant difference in the thermal conductivity between this and the corresponding 100 atom case.

For a finite lattice, any localized mode will have some small but finite amplitude at the ends and so will ultimately be excited by the heat baths. The length of the chain (or multi-dimensional lattice) will only affect the time required to reach steady state.

For the 50-50 disordered chain with mass ratio 3:2, we have studied the effects of varying the anharmonicity parameters μ and ν from those given in Eq. (8). As these parameters are decreased, so is the deviation of the thermal conductivity from the harmonic value. If they are increased, the deviation is increased but not so markedly. The behavior of the energy flow, heat conductivity, and temperature gradient for varying strengths of the anharmonic forces are shown in Fig. 4 for a 50-50 disordered chain with masses in the ratio of 3:2.

Corresponding results for the square lattice, not shown here, exhibit the same qualitative features as the linear chain, and the same interpretations apply.

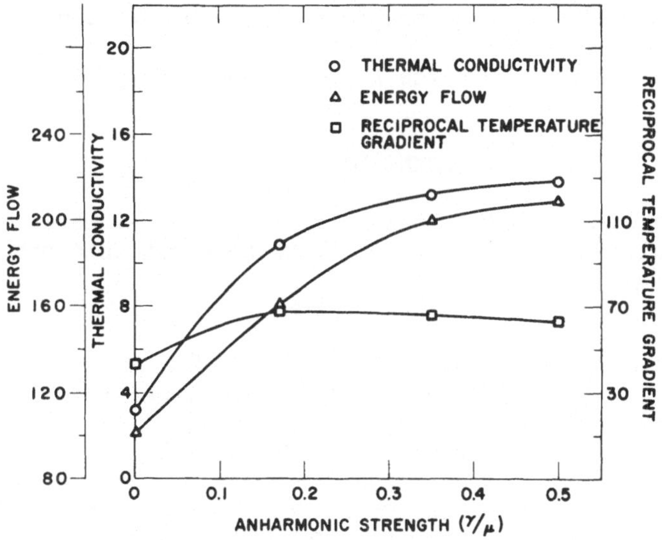

Fig. 4. Energy flow, thermal conductivity, and reciprocal temperature gradient as a function of the strength of the anharmonic terms in Eq. (1). The ratio of μ/γ is the abscissa and μ/ν is held constant.

The rapid decrease of the heat conductivity at the right-hand side of Fig. 3 can be understood qualitatively in terms of the normal modes for a lattice with a few heavy impurities. All of the high frequency modes with wavelength shorter than the mean heavy atom spacing tend to become localized and are effectively blocked by the heavy impurities from contributing to heat conduction (see e.g., Fig. 2 of Ref. (4)). This is in contrast to the case of a few light impurities in a heavy host lattice where the number of localized modes equals the number of impurities and the majority of high frequency modes can contribute to the heat current. This difference appears to be responsible for the marked asymmetry about the middle of Fig. 3.

Predictions about the behavior of 3D lattice thermal conductivities can be made with confidence by an extrapolation of our 1D and 2D results. The square lattice contains many of the features of simple cubic lattices which are missing in the linear chain, e.g., low frequency resonance modes associated with heavy impurities and multiple paths between lattice points. The frequency spectra and the normal modes are qualitatively similar in the square and simple cubic lattices. It is therefore reasonable to expect that many features of the thermal conductivity which have been found in the square lattice will appear also in the simple cubic lattice.

* Work performed under the auspices of the U. S. Atomic Energy Commission.
† Present address: Air Force Weapons Laboratory, Kirtland AFB, New Mexico.

ACKNOWLEDGEMENTS

The authors wish to thank J. R. Pasta, A. A. Maradudin, and G. P. Devault for stimulating discussions about this problem.

REFERENCES

1. J. Callaway, Phys. Rev. 113, 1046 (1959).
2. P. Debye, Vorträge über die kinetische Theorie der Materie und der Elektrizität (Teubner, Berlin, 1914).
3. R. E. Peierls, Ann. Physik 3, 1055 (1929).
4. The normal modes of the disordered chain were elucidated by P. Dean and M. D. Bacon, Proc. Phys. Soc. 81, 642 (1963).
5. D. Payton and W. M. Visscher, Phys. Rev. 154, 802 (1967).
6. D. Payton and W. M. Visscher, Phys. Rev. 156, 1032 (1967).

COMPARISON OF PHONON-ACTIVE AND INFRARED-ACTIVE LATTICE RESONANCES FOR U CENTERS IN ALKALI HALIDES*

Lee G. Radosevich and Charles T. Walker

Physics Department, Northwestern University

Evanston, Illinois

I. INTRODUCTION

We have previously published low temperature thermal conductivity data for KI doped with H^- and D^- ions.(1) In addition we have also discussed(1,2) but not published, our data for H^- and D^- ions in KCl, KBr, and RbCl. Of these systems only the KI data exhibited a resonant phonon interaction. The data for KI with H^- and D^- ions exhibited a resonant dip on the high temperature side of the conductivity maximum at a temperature of about 18°K. Moreover, the position or shape of the dip did not depend on the dopant used, and the resonance is therefore mass independent. In this paper we will confine our attention chiefly to KI. A detailed account of all our thermal conductivity data and analysis by means of Green's functions will be published separately in a more lengthy paper.

We were able to fit the data for doped KI by using Krumhansl's vacancy-type scattering model.(3) In this model both nearest neighbor central and non-central force constants are allowed to change. By the use of a continuum approximation to the perfect crystal Green's functions Krumhansl calculated the scattering cross-sections for the A_{1g} "breathing" and F_{1g} "shear" configurations of the full cubic group. For these modes the cross-section depends only on the impurity induced force constant change and not on the impurity mass. When this model was applied to our data we deduced a resonant frequency of 31 cm^{-1} for the phonon resonance in KI with U centers. However, it was necessary to multiply the Krumhansl cross-section by a temperature dependent exponential function in order to fit the data.

Experimental data on the infrared spectra for U centers in alkali halides[4] show a main absorption peak accompanied by weaker sidebands. For KI the splitting between the main peak and the first high frequency sideband is 29 cm^{-1}, which is tantalizingly close to the 31 cm^{-1} value needed for the phonon-active resonance. The question is--what relation, if any, is there between the phonon resonant phenomena and the photon resonant phenomena?

Sideband calculations have been made by Timusk and Klein[5] (TK model), Gethins et al.[6] (GTW model), Bilz et al.[7] and Nguyen.[8] Except for Bilz et al., all these calculations involve lattice resonances.

II. CALCULATION OF GREEN'S FUNCTIONS

In an attempt to correlate our thermal conductivity data with theory and hopefully with the available infrared data and theories we have used the model of Nguyen et al.[9] in order to calculate the perturbed bound states of KCl, KBr, and KI containing U centers. For this calculation both nearest neighbor central and non-central force constants are allowed to change in a cubic crystal. The model as such is more general than either the TK or GTW models since possible shear mode resonances are now included.

For the A_{1g}, E_g, F_{1g}, F_{2g}, and F_{2u} configurations basis vectors, Ψ^s, can be chosen which will diagonalize the product matrix, $\underline{g\delta l}$, giving the eigenvalue, λ_s, viz.,

$$\underline{g\delta l}\ \Psi^s = \lambda_s(\omega^2)\ \Psi^s \qquad (1)$$

Here, λ_s is a function of ω, the phonon frequency, and resonances appear in each configuration when

$$\lambda_s(\omega^2) = 1. \qquad (2)$$

g is the perfect crystal partitioned Green's function matrix, and $\delta\underline{l}$ is the defect matrix whose elements describe the changes in mass and force constants induced by the substitutional defect.

For the above mentioned configurations the eigenvalue λ_s depends only on the force constant changes and not on the impurity mass. For the infrared active F_{1u} mode it is necessary to solve a 3x3 determinant to find the eigenvalues, λ_s. The eigenvalues for the F_{1u} mode depend on both the force constant changes and mass change. Resonances for the F_{1u} mode are found by calculating the real part of the 3x3 determinant, $|D|$, as a function of frequency for λ=1. A resonance occurs when Real $|D|$ is zero provided Imag $|D|$ is small.

The eigenvalue equation, Eq. (2) depends on β, the central force constant change for the A_{1g} and E_g configurations, and on γ, the non-central force constant change for the F_{1g}, F_{2g}, and F_{2u}

configurations. $|D|$ depends on both of these along with the mass change. The explicit expressions for all the eigenvalue equations are given in the paper by Nguyen et al.[9] and will not be repeated here. β and γ are defined as

$$\beta = \frac{1}{\omega_L^2} (M_o M_n)^{-1/2} \Delta\Phi_{xx}(00,1n) \tag{3a}$$

$$\gamma = \frac{1}{\omega_L^2} (M_o M_n)^{-1/2} \Delta\Phi_{yy}(00,1n), \tag{3b}$$

where the $\Delta\Phi_{\alpha\beta}(lK;l'K')$ are the force constant changes induced by the presence of the impurity. The cell index, l, takes on the value 0 for the impurity and 1,2,3 and $\bar{1},\bar{2},\bar{3}$ for the nearest neighbors located on the positive and negative branches of the coordinate axes respectively. The atom species index K is given the value 0 if it refers to the host atom at the origin and n if it refers to any of the nearest neighbors. M_n is the mass of the atoms which are nearest neighbors to the impurity, and M_o is the mass of the atom of the perfect host crystal which has been replaced by the impurity. ω_L is the maximum lattice frequency of the host crystal.

With these definitions for β and γ the Green's function matrix elements have the following form:

$$g_{\alpha\beta}(lK,l'L';x) = \frac{1}{N} \sum_{\bar{k},j} \frac{W_\alpha(K|\bar{k}j)\, W_\beta(K'|\bar{k}j)}{x^2 - x_j^2(k)}\, e^{2\pi i\bar{k}\cdot[\bar{x}(lK)-\bar{x}(l'K')]} \tag{4}$$

with

$$x = \frac{\omega}{\omega_L}, \qquad x_j(\bar{k}) = \frac{\omega_j(\bar{k})}{\omega_L}$$

The $\omega_j(\bar{k})$ and $W_\alpha(K|kj)$ are the perfect crystal eigenfrequencies and eigenvectors respectively. They depend on the wave vector $\bar{k}$ and polarization j. $\bar{x}(lK)$ specifies the position of the Kth atom located in the lth cell. There are N unit cells, and the sum is taken over the 1st Brillouin zone of the crystal.

To evaluate the Green's function matrix elements, $g_{\alpha\beta}$, we have used the 0^oK deformation dipole eigenfrequencies and eigenvectors of Karo and Hardy for KCl and the shell model eigenfrequencies and eigenvectors for KBr and KI. These matrix elements were summed using the Kellermann sampling[10] of 1000 k vectors in the first Brillouin zone. The maximum lattice frequencies needed to obtain the reduced frequencies, $x_j(\bar{k})$, were the frequencies for the k=0 longitudinal optic mode. The procedure for the actual numerical evaluation of these matrix elements has been described by Sievers et al.[11] and will not be repeated here.

With the Green's function matrix elements known, the ratios λ/β for A_{1g} and E_g, and λ/γ for F_{1g}, F_{2g}, and F_{2u} were evaluated as a function of the reduced frequency. To determine whether or not the condition $\lambda_s(x^2)=1$ is satisfied and to compute $|D|$ for F_{1u} it is still necessary to know the force constant changes.

III. CALCULATION OF THE FORCE CONSTANT CHANGES

Values for the force constant changes were obtained by substitution of the experimental infrared local mode frequencies for H^- and D^- into the 3x3 determinant for the F_{1u} mode. If one neglects the polarizability of the impurity ions and also assumes that the force constant changes are the same for both hydrogen and deuterium impurities, two sets of force constant changes, (β,γ), will be obtained. Substitution of the local mode frequencies for the KCl and KI systems gave real values for these sets, but this same procedure applied to the KBr systems gave complex values for (β,γ). Such a behavior was found by Nguyen et al.[9] and MacDonald[12] in trying to determine the force constant changes for KCl-KH. The force constant sets (β,γ) for KI are $\beta=.1583$, $\gamma=-.0250$, and $\beta=.0396$, $\gamma=.0343$.

It is now necessary to choose between the two sets of (β,γ). As will be seen shortly resonance modes occur for large values of β or γ. It is these values of β or γ which are of interest to us here, since we are looking for low frequency resonances which might be observable in thermal conductivity data.

With the force constant changes known, it is now possible to apply the condition, $\lambda_s(x^2)=1$, for each configuration. The ratios, λ/β for A_{1g} and E_g are given in Fig. 1 for KI with a negative impurity. These quantities are plotted as a function of reduced frequency. Complete sets of graphs for KBr, KCl, and KI will be published elsewhere[13] as will all the sets of (β,γ). The horizontal line near the top of Fig. 1 is the inverse of β, where $\beta=.0396$. The second horizontal line near $\lambda/\beta \approx 6$ is the inverse of β for $\beta=.1583$. Their intersections with the curves $\lambda_{A_{1g}}/\beta$ and λ_{E_g}/β determine the resonance frequencies for these configurations. The computed values of γ^{-1} were found to be too large to intersect with the comparable sets of $\lambda_{F_{1g}}/\gamma$, $\lambda_{F_{2g}}/\gamma$, or $\lambda_{F_{2u}}/\gamma$. Hence no resonances are predicted for these modes.

For the F_{1u} mode the eigenvalue condition, $\lambda=1$, is applied, and the determinant is evaluated as a function of frequency. A resonance is found whenever the real part of $|D|$ passes through zero and $|D|^2$ is small. This latter quantity is given in Fig. 2 for $\beta=.1583$, $\gamma=-.0250$.

IV. PERTURBED BOUND STATES AND RELAXATION RATES

As is evident from Figs. 1 and 2 and the discussion of the last section resonant modes for KI-KH appear in the A_{1g}, E_g, and F_{1u} configurations, but not for the others. For the A_{1g} configuration the lowest frequency resonance mode occurred at 93.5 cm^{-1} for $\beta=.1583$. No resonances occur in the A_{1g} mode for $\beta=.0396$.

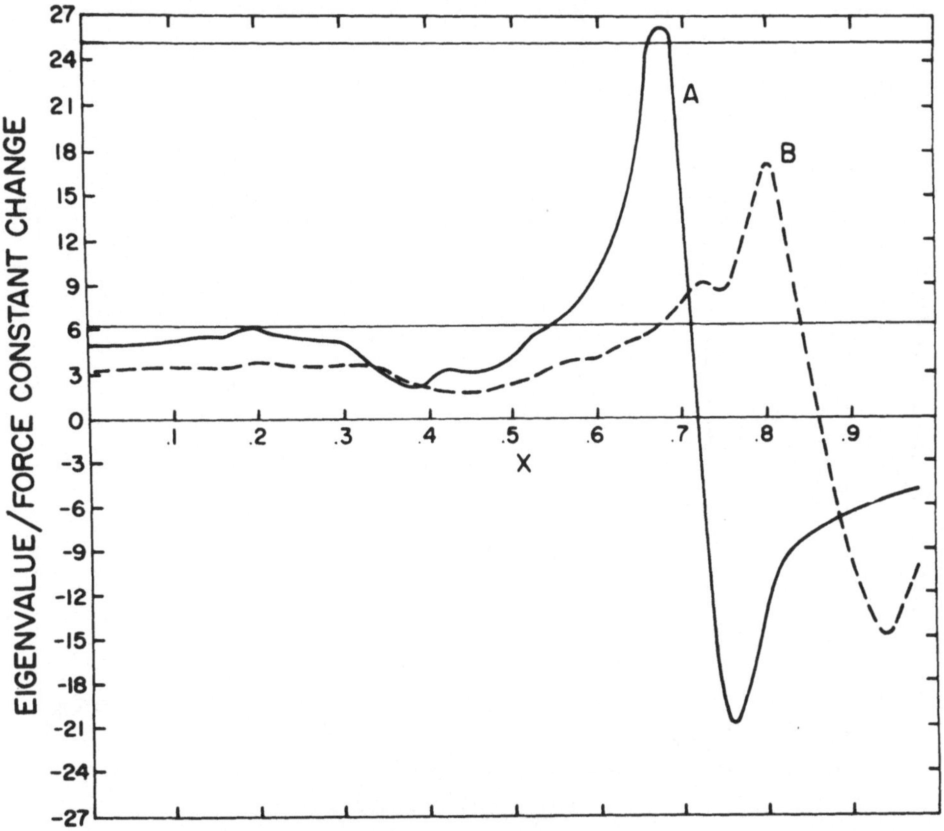

Figure 1. Ratio of eigenvalue to force constant change vs. reduced frequency. Curves A and B are E_g and A_{1g} configurations.

For the E_g configuration low frequency resonance modes appear at 76.0 cm^{-1} for β=.1583 and 93.5 cm^{-1} for β=.0396. Moreover a near resonance appears in the E_g mode at 28 cm^{-1} for β=.1583. A 2% uncertainty in the calculation of λ or β would make this a real resonance. Such a distinction between real and incipient resonances is not significant to the interpretation of our results, however, as will be seen shortly. For the F_{1u} mode resonances are predicted at 15.5 cm^{-1}, 36.5 cm^{-1}, and 63 cm^{-1}.

The 28 cm^{-1} E_g resonance is in close agreement with the 31 cm^{-1} resonance observed in thermal conductivity studies and the 29 cm^{-1} infrared sideband maximum, and leads us to suggest a possible correlation between phonon-active and infrared-active lattice resonances for KI doped with H^- or D^-.

In order to evaluate the influence of these predicted resonances on the thermal conductivity it is necessary to calculate the scattering matrix (T matrix), and then the relaxation rate.

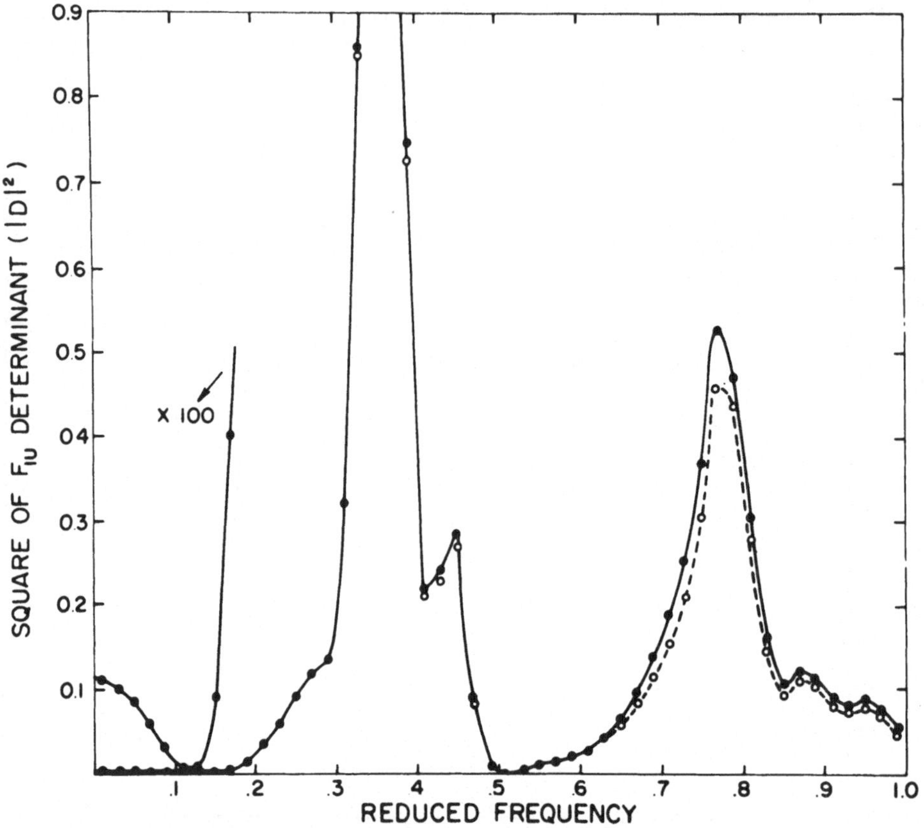

Figure 2. Square of determinant for F_{1u} configuration vs. reduced frequency. Low frequency part also shown expanded a factor of 100. Closed circles, H^-; open circles, D^-.

This extension of the theory has been discussed by Klein.(14) We have calculated the relaxation rates for the A_{1g}, E_g, F_{1g}, F_{2g}, and F_{2u} configurations in KI and find enhancements in the relaxation rate for several of the predicted resonances. The strongest enhancement is displayed by the 28 cm^{-1} near-resonance for the E_g mode. At this low frequency the enhancement is about an order of magnitude greater than peaks for any other resonances. The relaxation rate for the E_g mode in KI is shown in Fig. 3. Similar enhancements are found at about 48 cm^{-1} for KBr and 64 cm^{-1} for KCl; these frequencies are very close to the sideband splittings for KBr and KCl. When scaled to the actual H^- impurity concentrations the E_g relaxation rate for KI is at least a factor of 3 greater than the pure crystal relaxation rate. For KBr and KCl the enhanced E_g rate is less than the pure crystal relaxation rate. This comparison of relaxation rates is in agreement with the observation of a resonance dip in the thermal conductivity of KI and the lack of dips in the data for KBr and KCl. It is

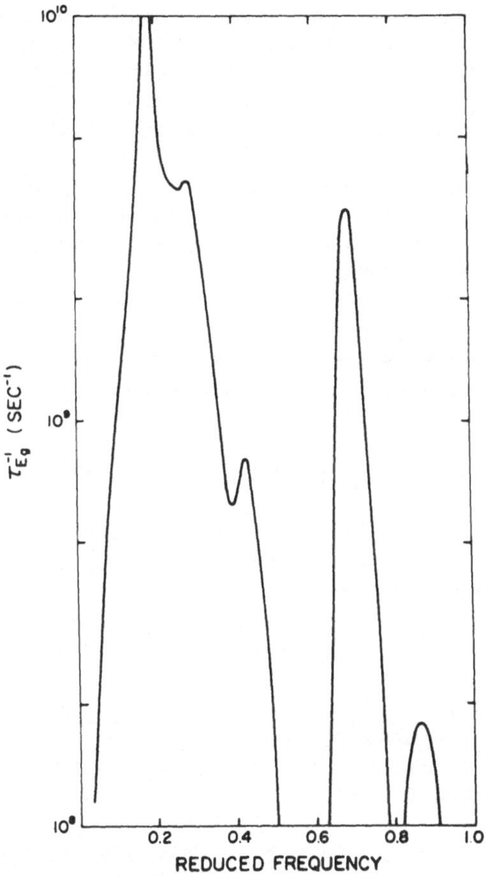

Figure 3. Calculated relaxation rate vs. reduced frequency for E_g configuration of H^- and D^- in KI.

to be emphasized that in the theoretical treatment this agreement was attained without the use of any adjustable parameters.

Thus, the results of this work suggest a correlation of the thermal conductivity data and the first infrared sidebands as both being due to incipient resonances in the E_g configuration. The theory also predicts a 93.5 cm^{-1} A_{1g} in-gap resonance observed experimentally by Fritz[15] and a 63 cm^{-1} far infrared resonance close to the 62 cm^{-1} triplet observed by Sievers.[16]

Since the calculations of Fieschi et al.[17] and Bilz et al[7] indicate that consideration of the H^- polarizability is important to theoretical calculations of the local mode frequencies it is of interest to note how this might affect the interpretation of our data. Bilz et al. suggest that the nearest neighbor force constant change for KCl and KBr is almost 0; such a result is not incompatible with the observed thermal conductivity data for KBr and KCl. If the force constant change, β of KI, is reduced greatly, however, the resonance at 28 cm^{-1} will shift to higher frequencies where the competing phonon-phonon processes will be stronger. Other means would then have to be found to explain our data.

The actual values of the H^- and D^- shell charges and polarizabilities in alkali halides are still not well known so that calculations of the nearest neighbor force constant changes up to the present time are somewhat in doubt.

We would like to thank Dr. A. M. Karo for sending us the KCl eigenfrequencies and eigenfunctions and Dr. R. A. Cowley for send- us the KBr and KI eigenfrequencies and eigenfunctions.

REFERENCES

*Supported by the U. S. Army Research Office (Durham) and the Advanced Research Projects Agency through the Northwestern University Materials Research Center.

1. L. G. Radosevich and C. T. Walker, Phys. Rev. 156, 1030 (1967).
2. L. G. Radosevich and C. T. Walker, Bull. Am. Phys. Soc. 12, 279 (1967).
3. J. A. Krumhansl, in Proceedings of the International Conference on Lattice Dynamics, Copenhagen, 1963, edited by R. F. Wallis (Pergamon Press, Oxford, 1965), p. 523.
4. B. Fritz, Lattice Dynamics, J. Phys. Chem. Sol. Suppl. 1, 485 (1965).
5. T. Timusk and M. V. Klein, Phys. Rev. 141, 664 (1966).
6. T. Gethins, T. Timusk, and E. J. Woll, Phys. Rev. 157, 744 (1967).
7. H. Bilz, D. Strauch, and B. Fritz, J. Phys. Radium 27, Suppl. C2-3 (1966).
8. Nguyen Xuan Xinh, Sol. State Comm. 4, 9 (1966).
9. Nguyen Xuan Xinh, A. A. Maradudin, and R. A. Coldwell-Horsfall, J. Phys. Radium 26, 717 (1965).
10. E. W. Kellermann, Trans. Roy. Soc. (London) 238, 513 (1940).
11. A. J. Sievers, A. A. Maradudin, and S. S. Jaswal, Phys. Rev. 138, A272 (1965).
12. R. A. MacDonald, Phys. Rev. 150, 597 (1966).
13. L. G. Radosevich and C. T. Walker, (to be published).
14. M. V. Klein, Phys. Rev. 141, 716 (1966).
15. B. Fritz, Private communication.
16. A. J. Sievers, Low Temperature Physics, Daunt, Edwards, Milford, and Yaqub, Eds. (Plenum Press, New York, 1965), LT9, Part B, p. 1170.
17. R. Fieschi, G. F. Nardelli, and N. Terzi, Phys. Rev. 138, A203, (1965).

ENHANCEMENT OF THE LATTICE HEAT CAPACITY DUE TO LOW FREQUENCY RESONANCE MODES IN DILUTE Al-Ag ALLOYS*

H. V. Culbert and R. P. Huebener

Argonne National Laboratory, Argonne, Illinois

The lattice properties of a solid are changed appreciably due to the presence of impurities which differ from the host lattice in mass or in the nearest neighbor force constant.(1) The introduction of isolated heavy impurities into a relatively light host lattice causes a change in the phonon spectrum, which is characterized by the existence of low-frequency resonance modes localized at the impurity site. If we neglect the change in the force constant, the resonance frequency associated with a heavy impurity is given by

$$\omega_o = \frac{\omega_D}{\left(3 \cdot \frac{M' - M}{M}\right)^{\frac{1}{2}}} , \qquad (1)$$

using a Debye model. Here, ω_D is the Debye frequency, M' and M are the mass of the impurity and of the atoms of the host lattice, respectively. The presence of low-frequency resonance modes enhances the low temperature heat capacity of the lattice. This enhancement has been calculated recently by Kagan and Iosilevskii(2) and by Lehman and De Wames.(3) The theory of Kagan and Iosilevskii leads to the following expression for the enhancement ΔC_L in the lattice heat capacity caused by

*Based on work performed under the auspices of the U. S. Atomic Energy Commission.

low-frequency resonance modes:

$$\Delta C_L = \frac{1.5}{2}\,\eta\,k_B \cdot \left(\frac{\Theta}{T}\right)^2 \int_0^1 dx \; \frac{x^{5/2} \cdot \left(\frac{3x_o}{x} - 1\right)}{(x-x_o)^2 \quad \frac{2.25\pi^2}{9} x^3} \cdot \frac{\exp\left(\frac{\Theta}{T}\sqrt{x}\right)}{\{\exp\left(\frac{\Theta}{T}\sqrt{x}\right)-1\}^2} \qquad (2)$$

Here, η, k_B, and Θ are the impurity concentration, Boltzman's constant, and the Debye temperature, respectively, $x = \omega^2/\omega_D^2$ and $x_o = \omega_o^2/\omega_D^2$ (with ω_o given by Eq. (1)). Again, in Eq. (2) the Debye model is assumed. The effect predicted by Kagan and Iosilevskii and by Lehman and De Wames has been observed recently by Panova and Samoilov[4] in Mg-Pb alloy, by Cape *et al.*[5] in Mg-Pb and Mg-Cd alloys, and by Chernoplekov *et al.*[6] in Ti-U alloy.

We have measured the heat capacity of polycrystalline samples of pure aluminum (99.9999%) and of the alloys Al + 0.95 at.% Ag and Al + 0.50 at.% Ag (made from 99.9999% pure Al and 99.999% pure Ag) between 1.3 and 25 K. Whereas for the first alloy the temperature dependence of the heat capacity has been measured rather carefully, for the second alloy, data have been taken only at a few selected temperatures. The specimens had been annealed in a helium atmosphere for seven days at 600° C.

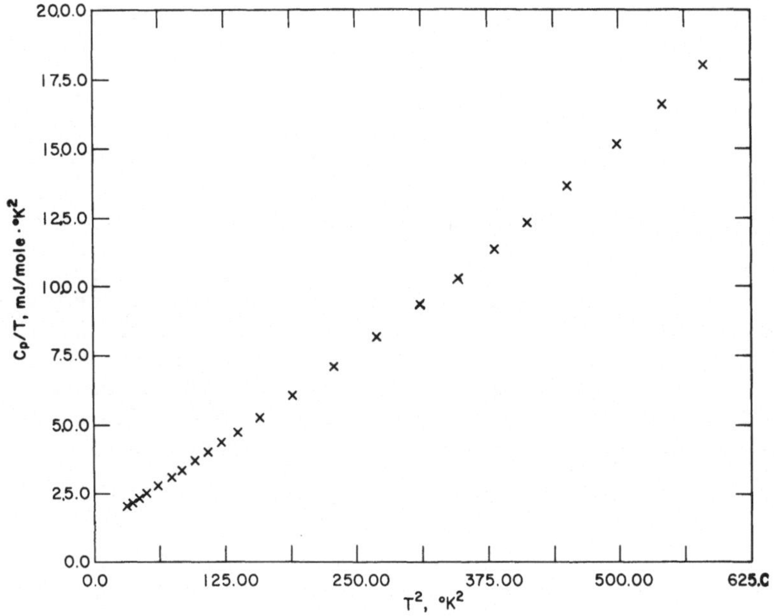

Fig. 1. Data for Al + 0.95 at.% Ag sample.

A description of the cryostat and of the experimental technique may be found elsewhere.(7) The specific heat data were corrected for curvature. The corrected data were then fitted by least squares to a polynomial expansion in odd powers of temperature up to T^7. Below 4°K the data were fitted by the function $\gamma T + \alpha T^3$ to extract the electronic specific heat γT and the Debye temperature of pure aluminum which is contained in the coefficient α of the pure aluminum specimen. The Debye temperature for pure Al obtained in this way was $\Theta = 435.5$°K. γT was subtracted from the polynomial expansion which fitted the original data. The resulting polynomial for pure aluminum was then subtracted from the polynomial for the alloy to obtain the difference ΔC_L in the lattice heat capacity of the alloy and of pure aluminum ($\Delta C_L = C_L(\text{alloy}) - C_L(\text{Al})$).

In Fig. 1 we show as an example of our original data the total specific heat C_p of the alloy Al + 0.95 at.% Ag in a plot of C_p/T versus T^2. Figure 2 shows the experimental ΔC_L values for both alloys plotted versus temperature. The points given for the alloy Al + 0.95 at.% Ag are the ΔC_L values obtained for integer temperatures in the way described above. The diameter of these points is approximately one standard deviation of the

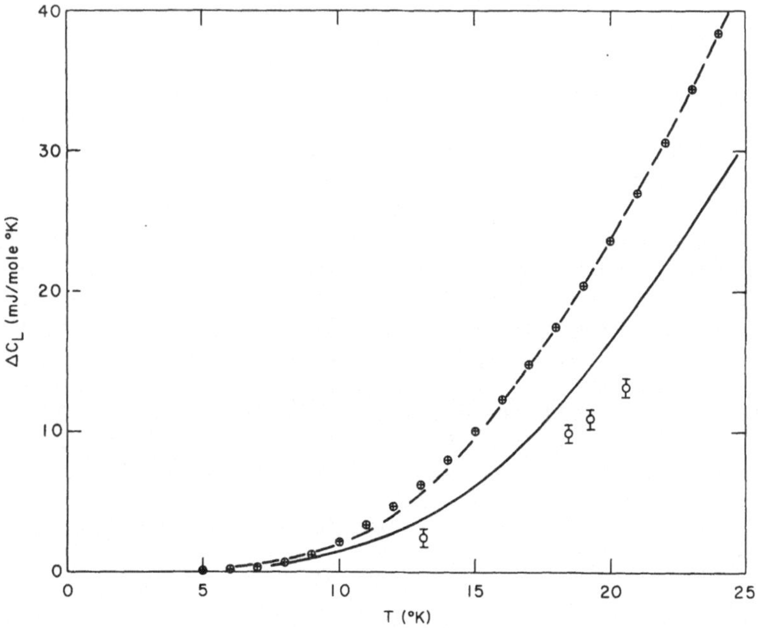

Fig. 2. ⊕ -ΔC_L data for Al + 0.95 at.% Ag. O-ΔC_L data for Al 0.5 at.% Ag. The solid curve was calculated for Al + 0.95 at.% Ag using equations (1) and (2). The dashed curve was calculated using equation (2) with x_o adjusted for a fit of the Al + 0.95 at.% data.

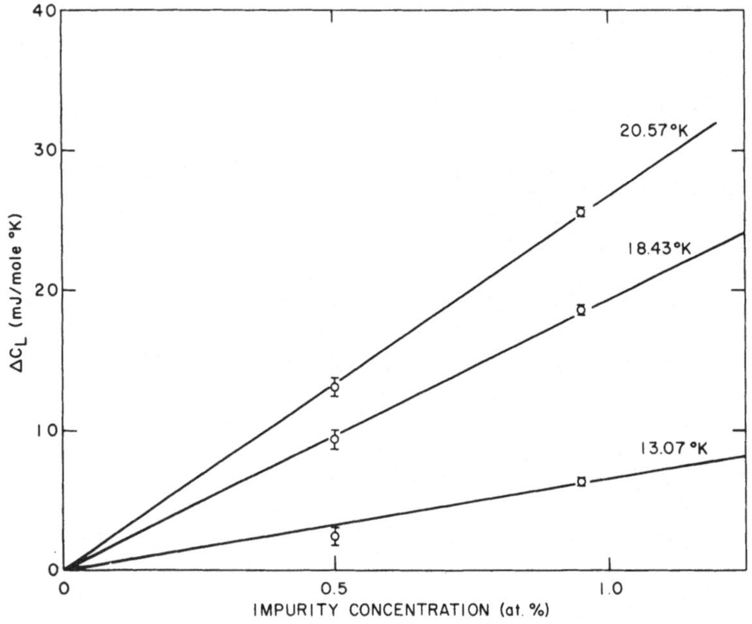

Fig. 3. ΔC_L versus impurity concentration at three temperatures.

original specific heat data from the fitted polynomial. For the alloy Al + 0.50 at.% Ag the data are given for the actual temperatures at which they were obtained. The error for these points is indicated by the vertical bars. The solid line and the dashed line are calculated from Eq. (2) for the case Al + 0.95 at.% Ag using for aluminum Θ = 435.5 K, given above. For the solid line the resonance frequency ω_0 was taken from Eq. 1 yielding x_0 = 0.111. The dashed line was obtained after varying ω_0 in order to fit the experimental data. The dashed line which fits the data reasonably well, was obtained using the value x_0 = 0.0833. The lattice specific heat in the alloys is clearly enhanced due to the silver admixture. The enhancement ΔC_L is in excellent agreement with the theory of Kagan and Iosilevskii, provided ω_0 is treated as an adjustable parameter. The shift of the resonance frequency to a value slightly smaller than that given by Eq. (1) may be caused by a softening of the interatomic forces around the impurities.

In Fig. 3 the variation of ΔC_L with impurity concentration is shown for three temperatures. For the range of impurity concentration investigated, the relation between ΔC_L and the impurity concentration apparently is linear. It is a pleasure to acknowledge the help of Z. Sungaila during the measurements.

REFERENCES

[1] A. A. Maradudin, Solid State Physics (Edited by F. Seitz and D. Turnbull, Academic Press, New York, 1966) Vol. 18, p. 273.

[2] Yu. M. Kagan and Ya. A. Iosilevskii, Zh. Eksperim. i. Teor. Fiz. 45, 819 (1963); Soviet Phys.--JETP (English translation) 18, 562 (1964).

[3] G. W. Lehman and R. E. De Wames, Phys. Rev. 131, 1008 (1963).

[4] G. Kh. Panova and B. N. Samoilov, Zh. Eksperim. i. Teor. Fiz. 49, 456 (1965); Soviet Phys.--JETP (English translation) 22, 320 (1966).

[5] J. A. Cape, G. W. Lehman, W. V. Johnston, and R. E. De Wames, Phys. Rev. Letters 16, 892 (1966).

[6] N. A. Chernoplekov, G. Kh. Panova, M. G. Zemlyanov, B. N. Samoilov, and V. I. Kutaitsev, Phys. Stat. Sol. 20, 767 (1967).

[7] O. V. Lounasmaa, Phys. Rev. 143, 399 (1966).

STUDY OF AN IMPURITY MODE THROUGH SPECIFIC HEAT MEASUREMENTS BETWEEN .05 AND 2.5°K [+ *]

J. P. Harrison, P. P. Peressini and R. O. Pohl

Physics Department

Cornell University, Ithaca, New York

A low frequency impurity mode in KCl:Li first observed through phonon resonance scattering in thermal conductivity measurements, has since been explained through tunneling states associated with a quasi-rotational motion of the small Li^+ ion between its 8 equilibrium positions located in the ⟨111⟩ crystallographic directions. A recent review of the extensive work which led to this model is given in ref. 1. We have investigated these tunneling states through specific heat measurements in a demagnetization cryostat.[(2)]

Figure 1 shows the influence that a small concentration of LiCl has on the specific heat of KCl. The specific heat anomaly obtained by subtracting the pure KCl specific heat from that measured in the doped samples, see Fig. 2, can be described very accurately for this small concentration (Ca. 10 p.p.m. Li) with a Schottky specific heat anomaly of a system of four equally spaced levels with the degeneracies 1, 3, 3, 1 as predicted by Krumhansl and co-workers.[(4)] No evidence for additional levels being above these four levels was found. For crystals containing the (naturally abundant) Li^7, the level spacing ΔE_7 was found to be .82 cm^{-1} ± 5% (ω'_7 = 1.55 x 10^{11} rad sec^{-1}), in agreement with microwave and dielectric work by Sack and co-workers.[(5)] If the Li^7 is replaced by Li^6, the shape of the anomaly remains unaltered, but the anomaly is shifted up in temperature by (40 ± 5)%.[(6)] Hence the level spacing now is $\Delta E_6 = 1.4 \times \Delta E_7 = 1.15\ cm^{-1}$ (ω'_6 = 2.17 x 10^{11} rad sec^{-1}). An isotope effect of such a magnitude is almost impossible to explain with a model of the Li^+ performing simple harmonic motion in a single potential well centered in the K^+ vacancy. Although a very large reduction in the binding forces to the nearest neighbors needed to explain the low frequency of

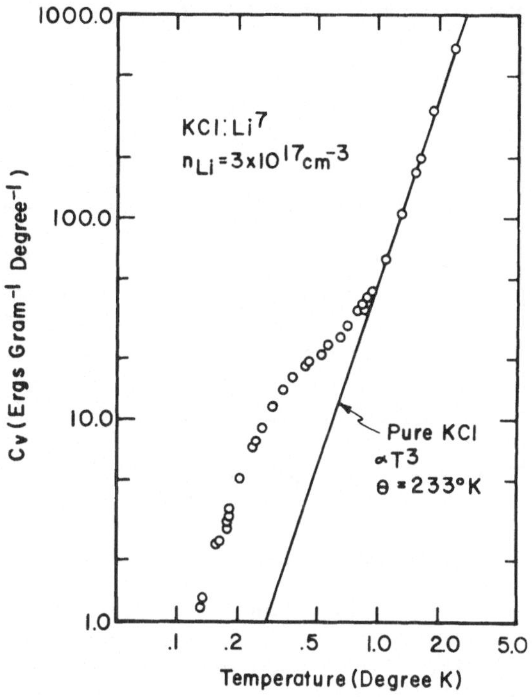

Figure 1: Specific heat of pure KCl and of KCl containing 3 x 10^{17} cm^{-3} Li^+ ions in solid solution. For the pure KCl sample, the data points were omitted for clarity. Sample mass was about 15 gram. The specific heat of the pure KCl is C_v = 42 erg $gram^{-1}$ deg^{-1} T^3 deg^{-3}, from which a Debye temperature θ = 233°K was determined, in agreement with earlier work, see ref. 2. The samples were grown by seedpulling under a protective atmosphere of chlorine, see ref. 3. We want to thank Mr. D. A. Bower from the Crystal Growing Facility of the Cornell Materials Science Center for preparing these samples. The Li^+ concentrations given in this paper were determined through a flame photometric method in the Analytic Facility of the Materials Science Center by Dr. R. Skogerboe, whom we would like to thank for his constant help and advice.

oscillation appears to be possible, and has indeed been observed for Li^+ in KBr(7), the isotope effect for such a case should be close to 1.08 $= \omega'_6/\omega'_7 = (m_{Li^7}/m_{Li^6})^{1/2}$, (as was indeed observed in KBr:Li by Sievers and co-workers).(7) Equally unlikely is the picture of the Li^+ behaving like a particle-in-a-box, because this would result in an isotope effect of 1.17 = m_{Li^7}/m_{Li^6}, still much

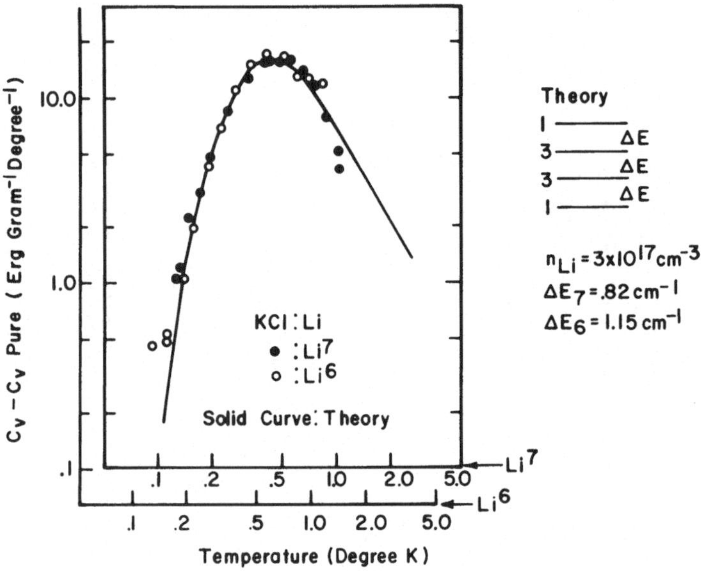

Figure 2: Excess specific heat caused by the tunneling states of LiCl dissolved in KCl and the influence of the isotopic mass of the lithium ion on it. Closed circles: Li^7 (actually normal isotopic mixture, 92.6% Li^7, 7.4% Li^6); open circles: isotopically pure Li^6. The concentration in both samples was 3×10^{17} cm^{-3}. Note that the anomaly for Li^6 occurs at 40% higher temperatures than that for Li^7, indicating a 40% larger tunnel splitting for Li^6. In order to demonstrate this clearly, the temperature scale used for the Li^6 anomaly (lower scale) was shifted to the left by 40% relative to that used for the Li^7 anomaly until the two anomalies coincided exactly. Doubly logarithmic plot. The solid curve was computed for a system of four equally spaced levels of spacing E with degeneracies 1,3,3, and 1, according to the model by Bowen <u>et</u>. <u>al</u>. From the best fit to the experimental data, one finds $E_7 = .82$ cm^{-1}, $E_6 = 1.15$ $cm^{-1} = 1.4 \times E_7$. Energy measured in wavenumbers.

smaller than the observed one. The tunneling model, on the other hand, gives an isotope effect of the right magnitude, as can be seen with a calculation using a simple one-dimensional double-well harmonic potential.[8] In the notation of ref. 8 the tunnel splitting ω' is

$$\omega' = 2\omega\sqrt{\frac{2V_o}{\omega\pi}}\ \exp\left(-\frac{2V_o}{\omega}\right), \qquad (1)$$

where ω is the harmonic oscillator frequency and V_o the potential barrier, all measured in wavenumbers. If we ascribe the band observed by Sievers[9] in KCl:Li^7 at 40 cm^{-1} to ω, V_o can be determined from (1), using the known ω'_7. We find $V_o^o = {}_o96\ cm^{-1}$, and from this a separation of the two wells of 2 x .76 Å.) In a three-dimensional calculation for eight wells, Bowen et. al.[4] found 2 x .87 Å, and Quigley and Das[10] and Dienes et. al.[11] found a separation of 2 x .66 Å between adjacent wells. The latter authors also found a potential barrier between two nearest ⟨111⟩ potential wells of 56 cm^{-1}. An isotopic increase of 8% of the harmonic oscillator frequency ω for Li^6 in the one-dimensional model then results in a tunnel splitting $\omega'_6 = 1.5\omega'_7$, i.e. a 50% increase. In view of the crudeness of the model, this is considered very satisfactory agreement with the increase found in our work. The same model[8] results in a tunnel splitting of 9 cm^{-1} for the first excited state of the harmonic oscillator. The half-width of the 40 cm^{-1} band measured by Sievers is 16 cm^{-1}.[12] From the specific heat anomaly the total entropy of the tunnel system, which should be $n_{Li}k \ln 8$ (n_{Li} = Lithium concentration) can be determined. n_{Li} determined in this way is 30% smaller than the n_{Li} determined chemically.

As the Li concentration is increased the anomaly broadens; the anomaly becomes similar to that found by Huiskamp and co-workers.[13] The exponential temperature dependence below the peak gives way to a more slowly varying temperature dependence. The steepness decreases with increasing Li^+ concentration. For $n_{Li} = 3 \times 10^{18}\ cm^{-3}$ the specific heat below the maximum varies roughly as T^3. This phenomenon is ascribed to a broadening of the tunneling states caused by concentration dependent inhomogeneous strain in the samples. Although the levels have a relative half-width of 0.75 for $n_{Li} = 3 \times 10^{18}\ cm^{-3}$, very little change in the level spacing is observed.

The effect of applying an electric field to the sample is to split the levels (Stark effect) and hence to change the specific heat. The change in specific heat caused by electric fields applied in the ⟨100⟩ and ⟨111⟩ crystallographic directions is in good agreement with a dipole moment of 2.5 Debye observed in dielectric[14] and electrocaloric work.[15] We associate this dipole moment with the displacement of the Li^+ ion from the center of the cavity as calculated above. For a displacement of .76 Å, the effective charge of the Li^+ ion then appears to be .7 x e, where e is the electronic charge. Again, this result is changed very little if instead of the one-dimensional caculation a three-dimensional one is used. From dielectric work on pure alkali halide crystals, it is known that the effective ionic charge is typically .8 x e.

A more detailed report on the concentration dependence and on the influence of electric fields on the specific heat anomaly is presently in preparation.

In summary then, our data provide unambiguous proof for the model of Lombardo and Pohl,[15] according to which the low frequency impurity modes observed in KCl:Li are indeed associated with a tunneling motion of the lithium ion. The shape of the specific heat anomaly associated with these impurity modes is of the Schottky type.

References

* Research supported in part by the U. S. Atomic Energy Commission under contract AT(30-1)-2391, Technical Report No. NYO-2391-59, and by the Advanced Research Projects Agency through the use of the central facilities of the Cornell Materials Science Center.

1. F. C. Baumann, J. P. Harrison, R. O. Pohl, and W. D. Seward, Phys. Rev. 159, 691 (1967).
2. J. P. Harrison, Rev. Scien. Instr., to be published. This paper gives details of the experimental techniques used in measuring specific heat and thermal conductivity at low temperatures.
3. J. M. Peech, D. A. Bower, and R. O. Pohl, J. Appl. Physics 38, 2166 (1967).
4. S. P. Bowen, M. Gomez, J. A. Krumhansl, and J. A. D. Matthew, Phys. Rev. Letters 16, 1105 (1966); M. Gomez, S. P. Bowen, J. A. Krumhansl, Phys. Rev. 153, 1009 (1967).
5. A. Lakatos and H. S. Sack, Solid State Comm. 4, 315 (1966), and H. Bogardus and H. S. Sack, Bull. Am. Phys. Soc. 11, 229 (1966).
6. In the Bulletin of this Conference we had given this value as 30%. The value given in the present paper resulted from a more careful comparison of the data. An isotope effect of 40% has also recently been observed in thermal conductivity by one of us (P.P.P.).
7. A. J. Sievers and S. Takeno, Phys. Rev. 140, A1030 (1965), and B. P. Clayman, R. D. Kirby, and A. J. Sievers, to be published.
8. See, for instance, E. Merzbacher, Quantum Mechanics, John Wiley, New York, 1961, Chapter 5, Section 6.
9. A. J. Sievers, in Lectures on Elementary Excitations and Their Interactions in Solids, NATO Advanced Study Institute, Cortina d'Ampezzo, Italy, G. F. Nardelli, ed., Instituto Documentazione Associaz. Meccanica Italiana, Milan 1966, pg. VI, 1.
10. R. J. Quigley and T. P. Das, Solid State Comm. 5, 487 (1967).
11. G. J. Dienes, R. D. Hatcher, R. Smoluchowski, and W. D. Wilson, Bull. Am. Phys. Soc. 12, 351 (1967).

12. More recent work with higher resolution indicates a somewhat smaller halfwidth. R. D. Kirby and A. J. Sievers, private communication.
13. R. F. Wielinga, A. R. Miedema, and W. J. Huiskamp, Physica 32, 1568 (1966).
14. H. S. Sack and M. C. Moriarty, Solid State Comm. 3, 93 (1965).
15. G. Lombardo and R. O. Pohl, Phys. Rev. Letters 15, 291 (1965).

PART H. DEFECT MODES ASSOCIATED WITH COLOR CENTERS

THE VIBRATIONAL PROPERTIES OF U CENTERS: EXPERIMENTAL ASPECTS

B. Fritz*

II. Physikalisches Institut, Universität Stuttgart, Germany

I. INTRODUCTION

U centers are negative hydrogen ions replacing proper lattice anions in an alkali halide crystal. These centers give rise to an electronic impurity absorption as well as to vibrational defect spectra. The electronic absorption consists mainly of a broad band(1) in the ultraviolet region ("U band", in KBr at 228 mμ). Vibrational absorption as investigated to the present time occurs on both the short-wavelength side and the long-wavelength side of the reststrahlen peak of the crystal. The short-wavelength absorption(2) has been attributed to the local mode created by the relatively light impurity mass. To a good approximation, this local mode displacement is restricted to the motion of the H^- ion inside a cubic (O_h) potential well. The long wavelength absorption(3) is due to the distortion of the plane wave modes of the perfect crystal near the defect site, which makes some of these modes infrared active.

By far the largest amount of information of the vibrational properties of U centers has been derived from investigations of the local mode absorption. Our review will comprise these as well as the far-infrared and phonon scattering experiments. The importance of Raman scattering experiments will be discussed briefly.

---

*Presently visiting with the Physics Department at the University of Utah, Salt Lake City, Utah.

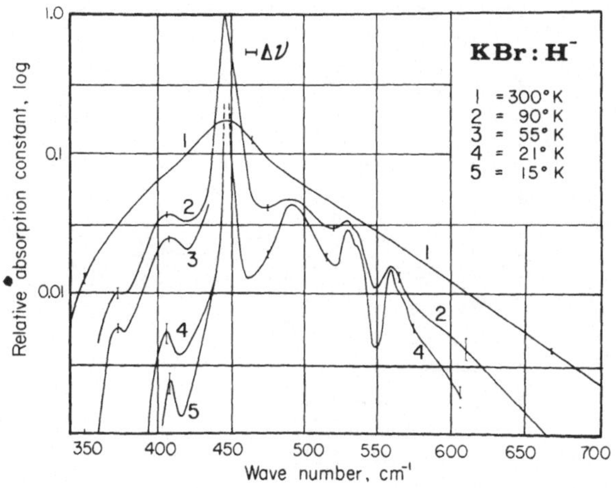

Fig. 1: Local mode absorption region in $KBr:H^-$; the absorption constant is given on a logarithmic scale. $\Delta\nu$=slit width. (ref.[6]).

It should be mentioned here that U centers in the strict sense are by no means the only species of hydrogen centers in cubic ionic crystals. Especially closely related are the substitutional H^- centers in the alkali earth flori des[4] (of T_d symmetry), and the interstitial H^- centers ("U_1 centers") in alkali halides, which may be created by low-temperature irradiation of U centers[5]. Some important results about the former ones will be contained in this article.

II. LOCAL MODE FREQUENCY AND OSCILLATOR STRENGTH

Fig. 1. shows a typical example of the U center local mode absorption (KBr)[6]. At room temperature there is but one broad and relatively structureless band. This sharpens up considerably on cooling the crystal. At 90°K we have a prominent peak in the center, the main band, and a broad region of sidebands at a level of absorption which is three to one orders of magnitude lower than the main band peak. Some insight into the origin of this spectrum is immediately derived from a reference experiment using "heavy" (D^-) U centers. The spectra obtained with $KCl:H^-$ and $KCl:D^-$ are given on Fig. 2. One finds the main band frequencies of the two isotopes in a ratio $\omega_L(H^-)/\omega_L(D^-) = 1.40$. Quite differently, the separation of the one sideband peak by about 60 cm^{-1} from the main band is not reduced on D^- substitution, but stays the same. The conclusion is that the main band is due to excitation of the highly localized vibration expected from a very light impurity. The sideband absorption is due to a coupling of the local mode to other phonons and will be discussed in some detail later on.

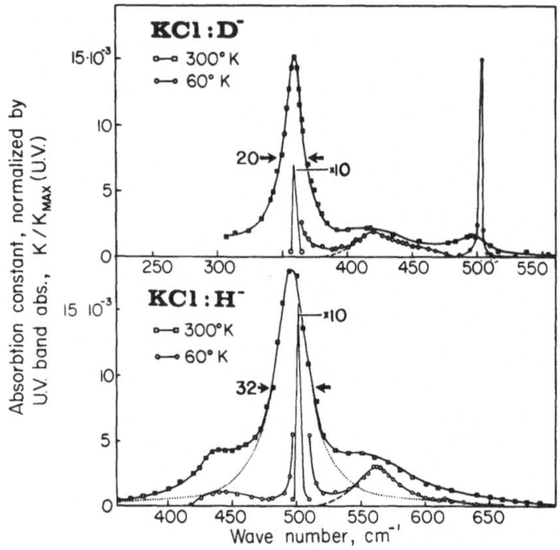

Fig. 2: Local mode absorption in KCl:H$^-$ and KCl:D$^-$. The absorption constant is normalized by the UV absorption (giving the U center concentration). The peak near 500 cm^{-1} in the KCl:D$^-$ spectra is due to the presence of H$^-$ centers. The dotted curve represents a Lorentzian fit of the main peak at room temperature (B. Fritz, unpublished).

Table 1 gives typical examples for the positions of the main band peaks, ν_R, at low temperatures. Schäfer who made the first measurements on these bands(2) stated that on varying, say, the anion through one set of compounds, the frequencies follow an Ivey relation $\nu_R \cdot d^x$ = const. where d is the lattice constant of the host material. In the case of the sodium family in Table 1, this relation is well fulfilled with an exponent x = 2.1.

Theoretical calculations of the local mode frequency to be expected from a purely isotopic defect (considering the mass

Table 1

Substance:	ν_R(cm^{-1}):	T:	$\nu_R(H^-)/\nu_R(D^-)$:	Reference:
NaF	858.9	70°K	1.397	(7)
NaCl	563	90°K	1.38	(2)(6)
NaBr	498	90°K	1.38	(2)(6)
NaI	426.8	10°K	1.34	(8)
KCl	502	90°K	1.40	(2)(6)(9)(10)
RbCl	476	90°K	1.40	(2)(6)
CsCl	357	100°K	——	(11)
CaF_2	965.6	20°K	1.393	(4)

change only, but leaving force constants between the ions unchanged) arrive at values which are 40% to 60% too high[12][13]. The experimental values of Table 1 thus indicate that a considerable lowering of the force constants for the local mode displacement takes place. The interpretation of this phenomenon in terms of microscopic models is still in progress[12][14].

There is, however, fair agreement between experiment and theory[13] as far as the isotope ratio of frequencies $\nu_R(H)/\nu_R(D)$ is concerned, since most model calculations lead to a figure for this ratio which is close to 1.40 and thus slightly less than $\sqrt{2}$. The case of NaI, where a quite unusual value of 1.34 was recently observed[8], points to one serious difficulty encountered in interpreting the ν_R values in general. Fairly large anharmonic shifts of the harmonic local mode frequency may be present in all observed values. In the case of NaI these may accidentally contribute to the isotope frequency ratio and thus become manifest.

From the foregoing, we would expect the U center to absorb infrared radiation with a cross section appropriate to a vibrating particle of hydrogen mass m_H and (negative) effective charge of the order of 1. The integrated absorption would be given by

$$(1) \quad \frac{1}{N}\int K(\omega)d\omega = \frac{2\pi^2}{n}\frac{e^2}{m_H}\cdot\left[\frac{n^2+2}{3}\right]^2\cdot f_{IR},$$

where we define an effective oscillator strength f_{IR} by setting the charge e equal to unity and assume the local field correction (in brackets) to be valid. The concentration of the U centers, N, may be obtained from a measurement of the UV band, whose oscillator strength has been calibrated. Fig. 2 gives the infrared absorption constant already reduced by the height of the UV band measured at room temperature. It turns out that the total band area in the D^- case is smaller than in the H^- case by a factor of $\frac{1}{2}$, as it should be. The area stays the same on going from room temperature to 90°K, (in KBr, NaBr, RbCl) or 60°K(KCl), within limits of about ± 15%[6][9]. Futhermore, using a value $f_{uv} = 0.8$ for the oscillator strength[15] of the UV band, we obtain $f_{IR} = 0.6 \pm 20\%$. At least this is in support of a strongly ionic character of the defect and may be explained with a dynamical effective charge $e^* \simeq 0.8$.

The lowering of force constants for the local mode and the effective charge smaller than unity are related to some degree. This becomes clear from shell model treatments of the H^- ion, which take into account the deformation of the charge distribution which follows a displacement of the ion from the equilibrium position[12][14][15].

Fig. 3 gives the relative variation of the main band and the

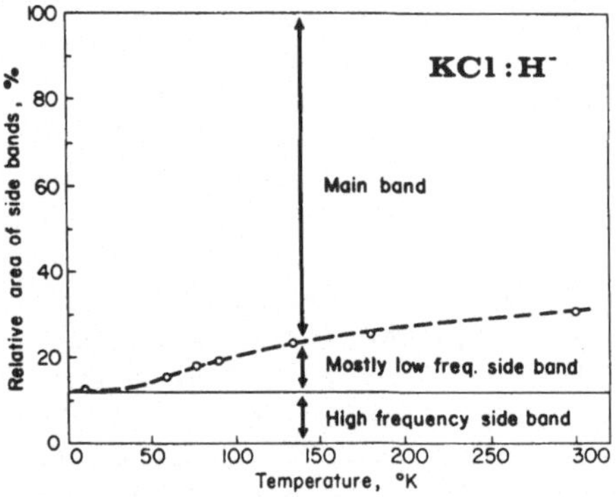

Figure 3.

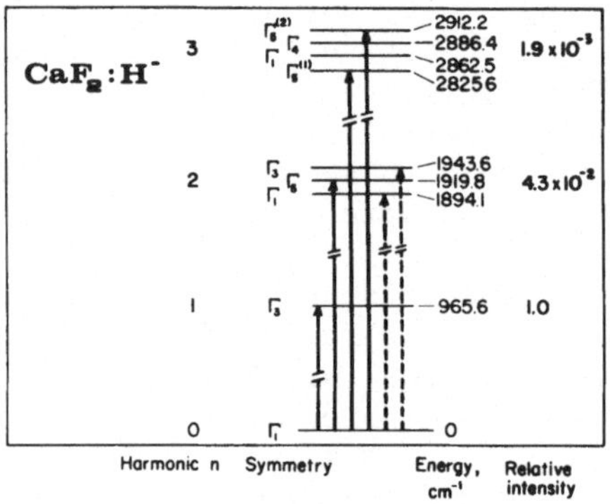

Figure 4.

Fig. 3: Relative intensity of side band and main band area versus temperature (B. Fritz, unpublished).
Fig. 4: Energy levels n=0 to n = 3 of the H^- local mode oscillator in CaF_2. Transitions observed with T_d symmetry are indicated by full arrows. The dotted arrows indicate transitions which are induced by shear strain; the positions of levels are as calculated from Eq. 2 (ref.[4]).

side band areas if we make the assumption of exactly temperature independent oscillator strengths. The points refer to measurements of the side band area, whose temperature dependence may be followed up to room temperature in KCl, if one extrapolates the main band by a Lorentzian line of the appropriate half-width (Fig. 2). Our results are in qualitative agreement with those obtained by Elliott et. al.[4] for CaF_2, SrF_2 and BaF_2. A variation of the main band strength in $KCl:H^-$ and $KCl:D^-$ has been reported[11], which amounted to a fourfold increase in area on going from 300°K to 90°K, which we have not been able to verify. The existing disagreement in the literature regarding the contents of Fig. 3 is reflected in varying statements about the possibility to fit the temperature dependence of the main band to an expression similar to a Debye-Waller factor[16][17]. Fig. 3 cannot be plotted in the desired manner.

III. ANHARMONIC EFFECTS IN THE LOCAL MODE ABSORPTION

1. Higher Harmonic Transitions

The potential energy of the H^- ion inside a rigid potential well of tetrahedral symmetry may be written

$$(2)\quad \Phi(x,y,z) = \frac{k}{2}(x^2+y^2+z^2) + bxyz + c_1(x^4+y^4+z^4) + c_2(x^2y^2+y^2z^2+z^2x^2) \ldots$$

With cubic (O_h) symmetry, b =0. Since the third order term makes transitions between the ground state (n=0) and the second excited state (n=2) possible, a second harmonic absorption may be observed with H^- centers in CaF_2, but not with alkali halide U centers. In both cases, the quartic terms allow for two absorption lines in which n=3 levels become excited.

Second and third harmonic transitions were observed(4) in $CaF_2:H^-$ and similar crystals. The frequencies measured at 20°K are given in Fig. 4. (Full arrows indicate the observed transitions). They may be used to calculate the values of the parameters b, c_1, c_2, and thus the whole level diagram of Fig. 4. Given on the same figure are the observed relative intensities of the higher harmonics. These values, however, are larger than one would deduce from the values of the anharmonic parameters fitted to the observed frequencies. It was not possible so far to observe third harmonic lines with U centers in alkali halides. From several experiments with KBr and KI one may conclude that there the third harmonic absorption is below 10^{-4} times the main peak strength(6).

2. Main Line Width and Shift.

Anharmonic terms in the potential energy provide for interaction processes between the local mode and other modes of the crystal. In general, further terms which contain combinations of H^- displacements and the displacements of ions on nearby lattice sites have to be added to Eq.(2) in order to account for all the observed interactions. (See(4) for a complete discussion). It is well known that such processes are observable as strongly temperature dependent effects, since they depend on the thermal occupation numbers of the phonon levels involved.

Fig. 5 presents the temperature dependence of the width of the main band in KCl as an example. A double logarithmic scale is used to bring out the typical T^2-dependence of the half-width which is found with most of these centers in a certain temperature region(4)(6). Again, the variation observed on substituting D^- instead of H^- helps to understand the underlying processes. In the region immediately below 300°K, the D^- points are shifted downwards by roughly a factor of 2, maintaining the T^2-like behavior. Near 50°K there is a crossover of the two curves, and the low temperature half-width assumed by the D^- band is about 5 times larger than that of the H^- band. The explanation is as follows:

Since the D^- local mode phonon corresponds to 360 cm^{-1}, its lifetime at T =0 is governed by a decay into two ordinary lattice phonons whose spectrum reaches up to 200 cm^{-1}. The temperature

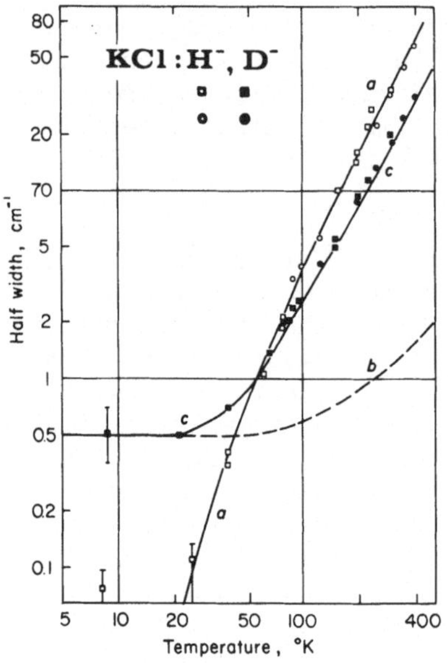

Figure 5.

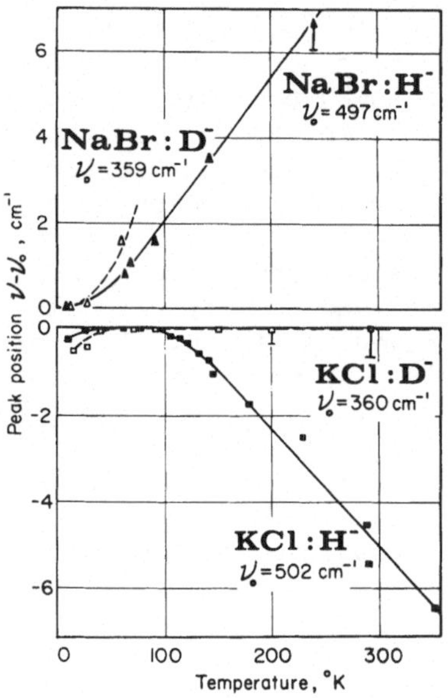

Figure 6.

Fig. 5: Half-width of H^- and D^- main bands in KCl (double log scale). Measurements after ref.[6] (squares) and ref.[9]. (circles). curve a = theoretical fit (phonon scattering process); curve b = temperature dependence of two-phonon decay; curve c = sum of contributions from phonon scattering and two phonon decay (ref.[15]). Lowest H^- band width indicated is based on estimate from measurement with 0.3 cm^{-1} slit width (U. Gross, unpublished).
Fig. 6: Temperature dependent shift of main band peaks.

dependence for this third-order process is given by $2\bar{n} + 1$, where $\bar{n}$ is the thermal occupation number of the typical phonons which are created in the decay. The H^- local mode quantum cannot decay into two phonons but must decay into at least three. This is a fourth-order process of temperature dependence $3\bar{n}^2 + 3\bar{n} + 1$. The absence of a two-phonon decay contribution explains the longer lifetime and thus the smaller half-width of the H^- band. In the T^2- region, another fourth-order process takes over which has been proposed by Elliott[4], and may be described as an elastic scattering of a phonon leading to a change in the local mode state. Since this process depends on the square of the local mode amplitude, it brings about the factor of 2 in the halfwidth of the two isotope lines which is observed.

It may be seen from Fig. 5 that an attempt to fit the D^- data (curve c) using the scattering and the two phonon decay contributions only, is fairly successful. The two phonon decay curve (b) represents the variation of $2\overline{n} + 1$. The scattering curve is taken from the experimental H^- data curve (a) by taking into account the appropriate factor $\frac{1}{2}$. The theoretical discussion of the details of curve a is given in[15].

The temperature-dependent shift of local mode bands has been observed by several authors[4][6][7]. It is usually found to be linear at temperatures not too far below room temperature. In the alkali halides the situation is not simple, because an increase of frequency may result as well as a decrease, when the temperature is lowered. In KCl even a reversal of sign takes place (Fig. 6). It has been proposed[15] that two effects may contribute to the shift actually present; firstly, the direct anharmonic contributions as discussed in connection with the more simple CaF_2 results[4], and secondly, the influence of the thermal lattice expansion which changes the potential parameters in Eq.(2). It may be concluded from experiments using mixed crystals[18] and uniaxial stress[19], that the thermal lattice expansion should always result in a decrease of the local mode frequency.

3. Sidebands.

Examples of the sideband structure were given already in Fig. 1 and Fig. 2. The existence of sidebands may be interpreted (although not exclusively) as arising from anharmonic local mode-lattice mode coupling. The experiments establish two main facts:

1. The difference between sideband and main band frequencies is not affected by D^- substitution (Fig. 2).
2. The sidebands are almost symmetrically displaced from the main band to higher and lower frequencies. The low-frequency part freezes out completely (Fig. 1).

From there we arrive at an interpretation of these bands in terms of two-phonon-absorption. A local mode quantum is (virtually) excited by one absorbed photon. This state decays into a final state in which one local mode quantum is created and one lattice mode quantum is either created (summation band) or destroyed (difference band). In this picture any peak in the density of phonons at ν_i, provided that there is strong enough coupling for these phonons, would lead to sideband peaks at

$$\nu \pm = \nu_R \pm \nu_i \quad (3)$$

The displacements of several sideband peaks as measured at 90° are given on Table 2. It turns out that the low frequency sidebands are found at slightly smaller separation than the high frequency side bands.

On Fig. 7 we have compared the high frequency sideband

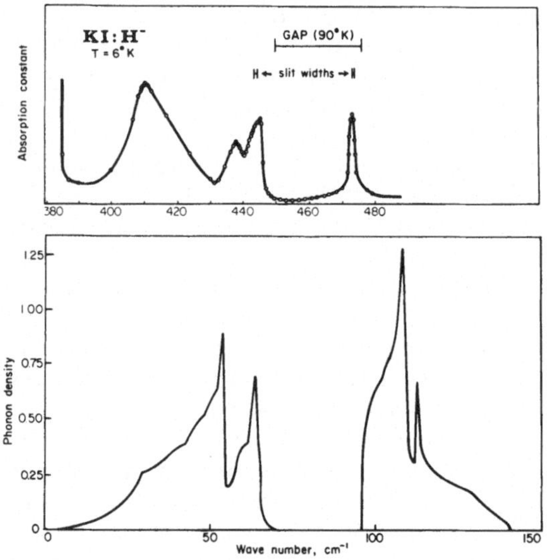

Fig. 7: High-frequency sideband in KI:H⁻ (upper half, ref.[19]) and one phonon density of KI (lower half, ref.[20]).

structure of KI:H⁻ with the phonon density of the pure material as obtained from neutron scattering data[20]. In this case as in all other cases, the relation between the two curves is not too obvious. The general decrease of absorption area in the sideband with increasing frequency is most remarkable. In the region of the phonon gap there appears a sharp sideband peak which is strong evidence for the existence of an in-gap-localized mode due to the presence of the U center.

The theoretical discussions[21] of the sidebands lead to the following qualitative results:

1) Only even modes of the perturbed lattice, namely those which have cubic(A_{1g}), tetragonal(E_g), or trigonal(T_{2g}) symmetry appear in the sideband spectrum.

2) The degree of perturbation of these modes by the force constant relaxation around the H⁻ion (See section on LOCAL MODE FREQUENCY) is of great influence on the shape of the sideband.

Table 2.

Substance	ν_R	$(\nu^+ - \nu_R)$	$(\nu_R - \nu^-)$
KCl	502 cm^{-1}	63 cm^{-1}	58 cm^{-1}
KBr	446	46,84	40,75±2
RbCl	476	44	41
RbBr	425	36	31

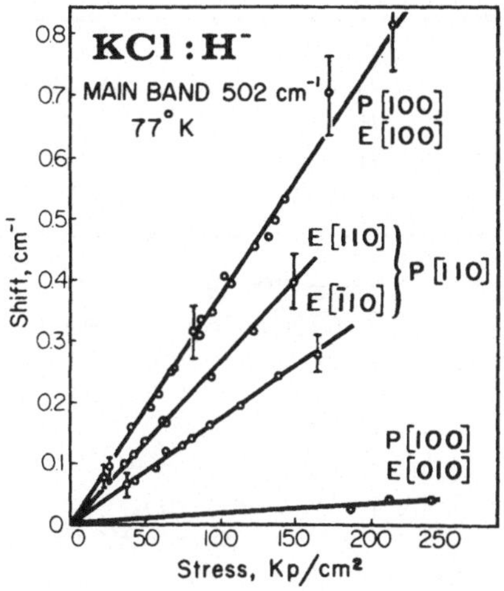

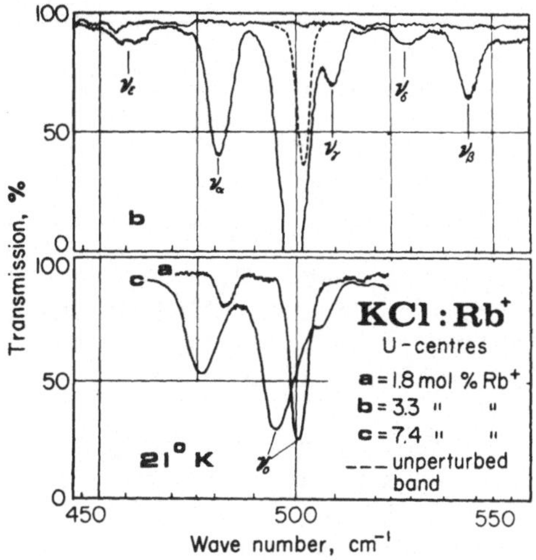

Figure 8 Figure 9

Fig.8: Splitting of main band in KCl:H$^-$ under uniaxial stress (ref.(13)).

Fig.9: Splitting of U center local mode frequency in KCl:Rb$^+$. New bands are denoted by $\nu_\alpha \ldots \nu_\epsilon$.

The creation of an in-gap-mode in KI is a direct demonstration of such an effect.

3) There are two principally different coupling mechanisms, anharmonic coupling through third order terms in the potential, and second order dipole moment coupling. The rapid decrease of the experimentally observed sideband spectra with frequency is evidence in favor of anharmonic coupling being the stronger contribution. It is interesting to note that the results of Table 2 also find an explanation based on the anharmonic coupling mechanism which says that the local mode frequency ω_L which is to be taken in analyzing the position of a sideband is not equal to the main band peak position ν_R.

IV. PERTURBATION BY EXTERNAL FIELDS

1. Uniaxial Stress.

The lowering of the site symmetry produced by uniaxial stress lifts the 3 fold degeneracy of the n= 1 level of the local mode. With light propagating along one of the symmetry axes perpendicular to the stress axis, usually two different bands may be observed which are polarized parallel and perpendicular to the direc-

Table 3.

$KCl(O_h)$	$CaF_2(T_d)$
$-B(A_{1g}) = 2.6 \cdot 10^4$ erg/cm^2	$-B(A_1) = 5.2 \cdot 10^4$ erg/cm^2
$-B(E_g) = 0.8$	$-B(E) = 0.6$
$-B(T_{2g}) = 0.35$	$-B(T_2) = 6.3$

tion of stress. This may be seen on Fig. 8 for KCl:H$^-$ with stress applied parallel to the [100] or [110] directions. The shift of the bands appears to be linear with stress within limits of experimental accuracy.

The theoretical analysis of this experiment shows (22)(23) that in any case the strain field contains only a cubic(A_{1g}), tetragonal and orthorhombic (E_g), and shear(T_{2g}) component. Given the magnitude of the strain of each type, one may from the measurements evaluate three coupling parameters B which are to a good approximation linear combinations of third order potential derivatives of the type Φ_{xxx_i}, where x refers to the coordinate of the U center, and x_i to a different lattice ion. Thus, the stress experiment provides the most direct access to some anharmonic potential parameters for the defect. The way in which the results are related to the sideband absroption is briefly discussed in another paper in this volume(19).

The expressions for the local mode shift under stress for four modes of observation are given in ref.(19). Table 3 contains the numerical values of the parameters which best fit the measurements in KCl(19) and in CaF_2(23)*. Note that the dimension of the parameters B comes about by defining the displacement of ions under stress in terms of a dimensionless strain.

In the present evaluation of the experiments, the strain has been calculated from the compliance constants of the unperturbed crystal. So far there seems to be no simple model which could explain the exact values of the splitting parameters. Their rough magnitude compares favorably with estimates of the third-order anharmonic parameters derived from the low-temperature line-width of the D$^-$line or from the integrated sideband intensity(4)(6).

Measurements of the splitting of a sideband under stress may lead to important information about the "sideband phonon". The first results of this kind are reported in ref.(19), where the

* We have recently learned about stress splitting measurements by H. Dötsch on NaF:H$^-$ (31).

splitting of the sharp sideband in KI:H^- (Fig. 7) has been investigated. The conclusion is that the mode excited in this sideband is an A_{1g} breathing mode.

The H^- centers of T_d symmetry in CaF_2 allow for observation of another effect of externally applied stress. This is the admixture of T_2 states to any of the infrared inactive levels belonging to n=2 in Fig. 4, which takes place under shear strain. Hayes and Macdonald[23] were thus able to observe stress induced absorption leading to the two forbidden n=2 levels (broken arrows in Fig. 4). In the alkali halides a similar activation of second harmonic levels would only be possible under the influence of a perturbation which is of odd parity itself, e.g. an external electric field.

2. Electric Fields.

The only results available so far are on CaF_2:H^-. Here a splitting of the degenerate first excited levels may be expected which is linear in the field strength E because of the lack of inversion symmetry. It was found that fields of the order of 10^5 V/cm when applied in [111] produced a splitting of at most 18 percent of the halfwidth of the local mode line[23].

V. U CENTERS IN MIXED CRYSTALS

The observations on the U center local mode in alkali halide mixed crystals are related to the results obtained with external fields. First of all, new lines appear besides the normal main band which indicate that a certain fraction of the U centers has a foreign cation or anion as a first or second nearest neighbor respectively, with an environment of lowered symmetry resulting. The splitting patterns obtained with various mixed crystals, usually containing a few percent doping, have been measured and interpreted in ref.[18][24]. An example is given in Fig. 9. The new lines α and β are most likely due to U centers which are perturbed by Rb^+ ion on a nearest neighbor site, while the γ line supposedly comes from a perturbation in the third-neighbor (cation) shell. The lines labeled by δ and ε have intensities which are proportional to the square of the doping concentration and thus represent U centers having a pair of foreign cations at adjacent sites. Apart from the appearence of split local mode lines, the doping produces a shift of the entire spectrum to lower frequencies. It is suggested that this is an effect of the increase of the lattice parameter. From ref.[19], Table 1 one deduces the following relation between the band shift and the fractional volume change 3 da/a_o:

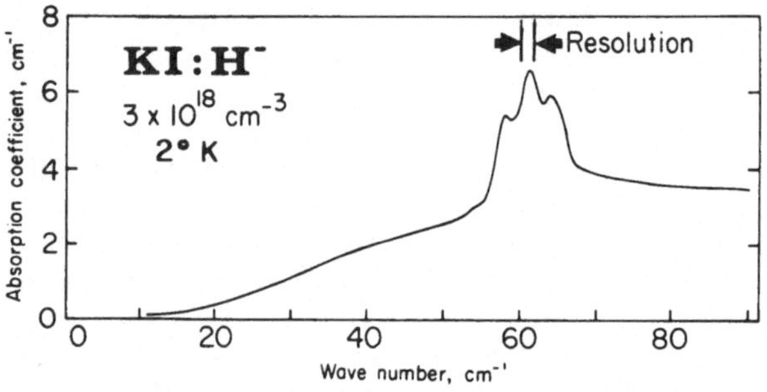

Figure 10: Far-infrared absorption of KI:H⁻ (ref.[3]).

(4) $$\Delta\omega = 3\,\frac{da}{a_o}\,\frac{B(A_{1g})}{2m_H\omega_R}\ .$$

Extrapolating the lattice parameter change from x-ray diffraction results on mixed crystals, the band shift in KCl:Rb^+ may be evaluated to give $B(A_{1g}) = 1.65\cdot10^4$ erg/cm^2. The agreement with the uniaxial stress result (Table 3) is fair regarding the fact that the lattice parameter change in the mixed crystal experiment is about an order of magnitude higher than in the stress experiment.

FAR - INFRARED ABSORPTION

Sievers[3] has measured the additional absorption which is produced on the long wavelength side of the reststrahlen peak of KI by introducing a high U center concentration ($3\cdot10^{18}$/cm^3) into the sample. This absorption has a total area which is about one-tenth of that in the local mode region[25]. Of the spectra given in Fig. 10, only the triplet structure centered around 62 cm^{-1} may be attributed to the presence of U centers as such as was established by U center - F center conversion experiments[3]. There is no resemblance between the sideband spectra in Fig.7 and the band-mode absorption in Fig. 10. This is to be expected. The sidebands arise from the A_{1g}, E_g and T_{2g} symmetry phonons. The defect-induced far-infrared absorption has to come from T_{1u} phonons of the perturbed lattice (considering only one-phonon processes), which may directly interact with the light. In this sense the sideband results and the far-infrared results are complementary to each other.

PHONON SCATTERING

Heat conductivity measurements at low temperatures on crystals containing U centers have been made by Walker and Radosevich[26][28]. In KI:H⁻ and KI:D⁻ the heat conductivity curve exhibits a dip in the region above the maximum (at 18°K), which seems to depend on the concentration of U centers present, other than the depression of the curve below the maximum, for which it is obviously unrelated to the presence of U centers as such. Measurements of KBr and KCl did not result in a curve of similar shape. Such dips are an indication of some type of resonant phonon scattering. They occur in many different alkali halides containing monatomic substitutional impurities[27]. In the present case it is suggestive to establish a connection between the observed dip and an even mode with resonant amplitude near the defect, because the dip does not depend in a measurable way on the isotope mass of the U center. Since the sideband spectrum displays all the even modes, one might expect that e.g. the 29 cm^{-1} peak and the conductivity dip have a common origin. This suggestion, however, is unsatisfactory for many reasons as discussed in a recent paper[28].

FINAL REMARKS

U center investigations provide a means of studying certain effects of the lattice dynamics of perturbed lattices: the existence of local modes due to mass and force constant changes, and a number of interesting interaction processes between local modes and band modes. These observations and their analysis result in a picture of the U center which is not a simple one. The H^- ion is not a purely isotopic defect; furthermore, it seems not to behave similar to a substitutional halogen ion characterized by a fairly constant ionic radius and a repulsive potential close to a Born-Mayer type[19][23]. These complications make a quanitative theory of all observable features a difficult task which has not yet been solved. The insight into the lattice dynamical processes may still be improved by doing more accurate experiments. In some places this attempt may meet serious difficulties, e.g. that of preparing otherwise pure samples with unusually high concentrations of U centers, which would be desirable for far-infrared and heat conductivity experiments.

At this point one should mention another type of experiment which has not yet been done successfully in connection with U centers: Raman Scattering. The expectation is that one would see a defect-induced one-phonon scattering superimposed on the two-phonon background of the pure crystal. The contributions to the one-phonon scattering would be from A_{1g}, E_g, and T_{2g} phonons only[29]. Thus the Raman active phonons correspond exactly to the ones showing up in the local mode absorption sidebands. Since the two modes of observation of the same phonons differ by the coupling mechanism (anharmonic local mode - band mode coupling in one case, electron - band mode in the other), a comparison of the two types of spectra would be highly interesting. A recent

Raman Scattering experiment by the author, using He-Ne-laser excitation and KCl and KI crystals with relative U center concentrations of $4 \times 10^{18}/cm^3$ has shown that the scattering cross section of U centers under these conditions is so small that no reliable information may be separated from the two-phonon lattice scattering background. A great improvement of the situation is expected from excitation under resonance-like conditions[30].

Addendum: The effects of finite concentrations leading to interactions between localized oscillators have not been considered here. Recently, H.Dötsch[31] has performed some measurements on NaF:H^- which indicate that such effects become sizeable with concentrations as high as 9% molar fraction. A splitting of the local mode under these conditions has been observed.

REFERENCES

(1) Color Centers in Solids, by J.H. Schulman and W.D. Compton Pergamon Press, New York, 1962.
(2) G. Schäfer, J. Phys.Chem. Solids 12, 233 (1960).
(3) A.J. Sievers, Lecture Notes of the NATO Summer School on Elementary Excitations and Their Interactions, Cortina d'Ampezzo 1966, Report #562, Materials Science Center, Cornell University, Ithaca, N.Y.
(4) R.J. Elliott, W.Hayes, G.D.Jones, H.F. Macdonald and C.T.Sennett; Proc. Roy. Soc. A 289, 1 (1965).
(5) B.Fritz, J. Phys. Chem. Solids 23, 375 (1962).
(6) B.Fritz, U. Gross and D. Bäuerle, phys. stat.sol. 11 231(1965).
(7) H.Dötsch, W. Gebhardt and Ch.L. Martins, Solid State Comm. 3, 297 (1965).
(8) D. Bäuerle and B. Fritz, phys. stat. sol. (To be published).
(9) D.N. Mirlin and I.I. Reshina, Soviet. Phys. Solid State (transl.) 6, 2454 (1965) and references given there.
(10) A.Mitsuishi and H. Yoshinaga, Theoret. Physics, Suppl. 23, 241 (1963).
(11) W.C.Price and G.R. Wilkinson, Final Technical Report No.2 (Dec. 1960) US Army Contract No. DA-91-591-EUC 1308 01-4201-60 (R. and D. 260).
(12) R.Fieschi, G.F. Nardelli, and N.Terzi, Phys. Rev. 138, A 203 (1965).
(13) S.S. Jaswal and D.J. Montgomery, Phys. Rev. 135, A 1257(1964).
(14) J.B. Page and D. Strauch, phys. stat. sol.(To be published).
(15) H.Bilz, D.Strauch and B.Fritz, Journal de Physique, 27 (Supplement) C2-3(1966).
(16) S.Takeno and A.J.Sievers, Phys.Rev.Letters 15, 1020 (1965).
(17) S.S. Mitra and R.S. Singh, Phys. Rev. Letters 16, 694 (1966).
(18) W.Barth and B.Fritz, phys. stat. sol. 19, 515 (1967).
(19) B.Fritz, J.Gerlach and U.Gross, Contrib. paper (this volume).

(20) G.Dolling, R.A.Cowley, C.Schittenhelm and I.M. Thorson Phys. Rev. 147, 577 (1966).
(21) H.Bilz, Review article ,(this volume).
(22) W.Gebhardt and K.Maier, Phys. Stat. Sol. 8, 303 (1965).
(23) W.Hayes and H.F.Macdonald,Proc. Roy. Soc. A 297, 503 (1967).
(24) D.N. Mirlin and N.N. Reshina, Soviet Phys. Solid State (Transl.) 8; 116 + (1966).
(25) According to our estimate based on local mode absorption measurements on $KI:H^-$ of the same H^- concentration.
(26) L.G. Radosevich and C.T. Walker, Phys.Rev. 156, 1030 (1967).
(27) F.C. Baumann, and R.O. Pohl, Report #671, Materials Science Center, Cornell University, Ithaca, N.Y.
(28) L.G. Radosevich and C.T. Walker , Contrib. paper, (this vol.).
(29) X.X. Nguyen, A.A. Maradudin and R.A. Coldwell-Horsfall, Journal de Physique 26 717 (1965).
(30) J.M. Worlock and S.S. Porto, Phys. Rev. Letters, 15, 697 (1965); B. Fritz , contrib. paper,(this volume).
(31) H. Dötsch, Thesis, July 1967, Frankfurt.

Resonance Raman Scattering from F_A(Li)-centers in KCl‡

B. Fritz*

Dept. of Physics, University of Utah

I. INTRODUCTION: Raman scattering due to very small concentrations of point defects in crystals may become observable by using light for the excitation whose energy is close to some resonance energy of the defect system. This has been clearly demonstrated by Worlock and Porto[1] in their investigation of F-center Raman scattering in NaCl and KCl. We apply this method in a study of F_A(Li) centers in KCl. We expect to observe Raman spectra of those lattice vibrations which interact with the excited states of the defect[2]. The presence of the defect in the crystal destroys the translation symmetry of the lattice and thereby makes one-phonon scattering allowed for certain types of perturbed lattice modes throughout the Brillouin zone. In the unperturbed NaCl structure one-phonon scattering processes are forbidden by symmetry and only a relatively weak two-phonon scattering is present.

The F_A(Li) center may be expected to show some interesting effects due to the presence of the Li^+ ion: these may be either small-mass-effects (U center-like high frequency local modes) or small-radius-effects (low frequency resonance modes, tunneling states in a multiple-well potential) or a combination of such effects. The structure of an F_A center is described on Fig. 1. It is that of an F center which is perturbed by one smaller foreign cation (in our case a Li^+ ion) on a nearest neighbor site. The symmetry is C_{4v}. This structure has been established by optical[3] and ENDOR[4] studies. All known F_A centers exhibit the dichroic

‡This work has been supported by the Committee for Institutional Funds, University of Utah and by AFOSR Contract #1141-66 (F. Lüty).
*On leave from II. Physicalisches Institut, University of Stuttgart Germany.

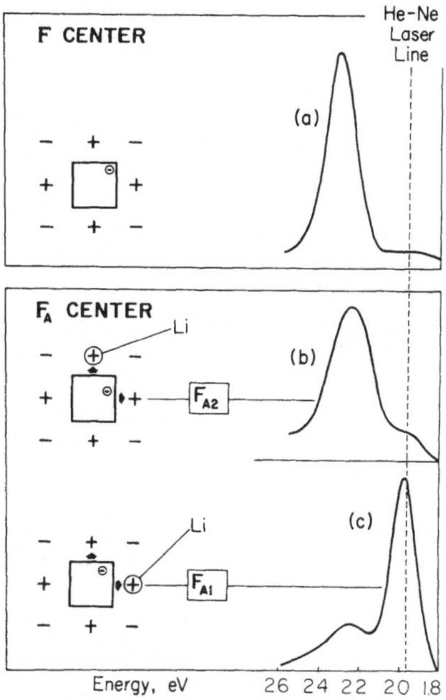

Fig. 1. Structure of F and F_A(Li)-centers in an alkali halide. curve a) F absorption band; curve b) F_{A2}-absorption, measured with light polarized normal to 4-fold axis (90°K); curve c) F_{A1}-absorption measured with light polarized along 4-fold axis.

absorption indicated in the figure (twofold degenerate transition F_{A2}, non-degenerate transition F_{A1}).

A very important property of F_A centers is their reorientation which may take place upon excitation in the F_{A1} or F_{A2} band. In the case of F_A(Li) centers in KCl one observes reorientation of centers independent of temperature at least above 2°K[5]. This phenomenon may be used to populate one particular orientation for all centers, in which they do not absorb the pumping light. An example is given by curve c of Fig. 1, which may be measured in [001] direction after shining unpolarized F_{A1} light into the crystal in the same direction. In a Raman scattering experiment using a He-Ne laser (6328 Å) the laser beam itself serves to maintain one particular state of alignment. For unpolarized laser light propagating in the [001] direction, the fourfold axis of the aligned center system would be parallel to [001], and the beam would pass through the sample under minimum absorption loss. Note that the electronic dipole matrix elements involved in the Raman scattering process are of the F_{A2} type in this case.

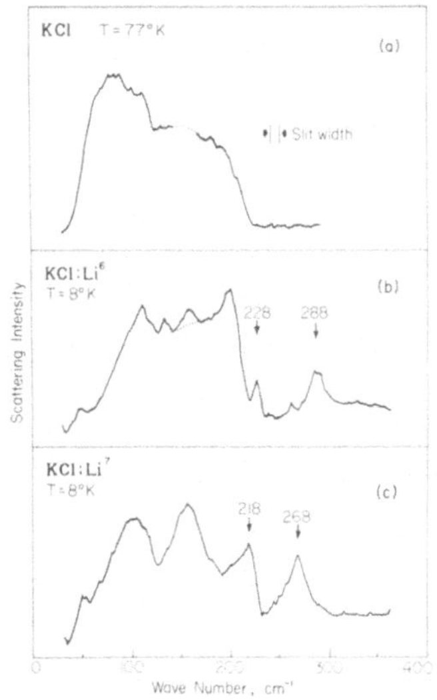

Fig. 2

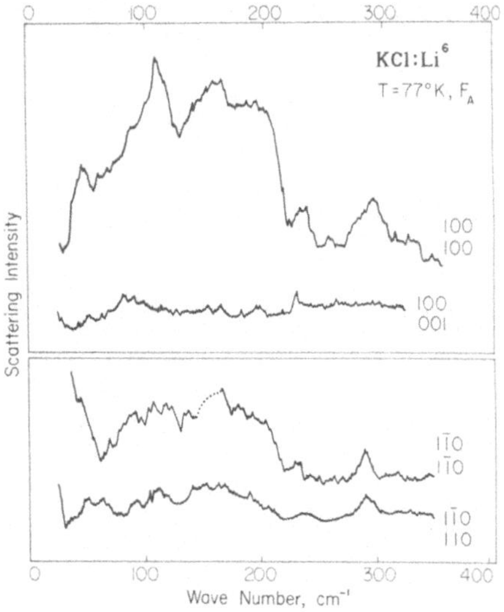

Fig. 3

Fig. 2. Stokes component of F and F_A(Li) center scattering. Unpolarized spectra for [001] direction of laser beam. Zero of wave number scale = position of He-Ne laser line. From curve a, stray light has been subtracted in the region around 155 cm^{-1} (dotted part of curve). The true Raman scattering intensity is uncertain in the same region of curve b.

Fig. 3: $F_A(Li^6)$-scattering under defined conditions of polarization.

II. <u>EXPERIMENTAL RESULTS</u>: The following results were obtained using a He-Ne laser (Spectra Physics 125; 50 mW) and a Spex double monochromator (1400) with photoelectric detection and standard phase-sensitive a.c. amplification techniques. The laser beam passed through the crystal mounted in a helium cryostat in upward direction and was brought to a focus inside the sample. The crystals were grown from KCl with $5 \cdot 10^{-3}$ molar fraction of Li^6Cl or Li^7Cl added to the melt. They were additively colored with $2 \cdot 10^{16}/cm^3$ F centers which were quenched down from 500°C and converted to F_A centers under F light at about 250°K.

Raman scattering spectra (Stokes component only) of F centers and F_A(Li) centers in KCl are shown in Fig. 2. The resolution is 4 Å or 10 cm^{-1}. The F center scattering (77°K, curve a) consists of two broad bands in the region between 50 and 220 cm^{-1}, in agreement with the first observation in ref.(1). At 77°K, the two-phonon scattering from the KCl lattice peaks at about 310 cm^{-1} and is too weak to be seen on these curves. Curve b was obtained with $F_A(Li^6)$ centers. Besides considerable changes of the spectrum in the F-center scattering region (the continuum) there are two new peaks above the continuum cut-off frequency at 228 cm^{-1} and 288 cm^{-1} ($\pm$ 2 cm^{-1}, 8°K).

The isotope shift of these two lines may be seen in curve c of Fig. 2. On substituting Li^7 instead of Li^6, the high-frequency line is shifted to 268 cm^{-1}. The other line seems to merge with the upper edge of the continuous spectrum, where now a peak at 218 cm^{-1} appears.

The temperature dependence of the spectra was investigated using a spectral slit width of 8 cm^{-1}. On going from He temperature to 77°K the separate lines broaden. For the Li^7 line at 268 cm^{-1} we measured a halfwidth of 16 cm^{-1} at 8°K and 21 cm^{-1} at 77°K; the Li^6 line behaves similarly. In the continuum region below about 100 cm^{-1} one observes the expected increase in intensity on going to 77°K, and in the same region on the high energy side of the laser line there appears an anti-Stokes scattering spectrum. In the latter the 50 cm^{-1} peak visible in both the Li^6 and Li^7 spectra of Fig. 2 has been identified.

The polarization behavior of the scattering has been measured for two different orientations of the aligned center system with respect to the directions of the exciting and the observed scattered light. We denote by [001] that symmetry axis of the crystal which is parallel to the fourfold axis of the aligned F_A(Li) centers. The first mode of observation is that in which the laser beam propagates along [001] and is polarized in [100] direction. The scattered light is observed in [010] direction. We may assume that there are not many centers oriented along the [010] direction, since a small portion of depolarized light from the beam is sufficient to rapidly deplete this orientation. The second case is that in which the beam passes through the sample in [110] direction, the electric vector being along [$1\bar{1}0$]. This results--even in the presence of some depolarized light--in an accumulation of all centers in the [001] direction, which is in this case also the direction of observation. The curves on Fig. 3 are labeled by the directions of the exciting field vector (upper symbol) and the polarization of the outgoing light (lower symbol). We mention the result that both the 228 and 288 cm^{-1} lines of the $F_A(Li^6)$ center are present in the [100]-[100]-curve. The 288 cm^{-1} line is not seen in the [100]-[001] spectrum, but in both [110]-type curves. This behavior of the 288 cm^{-1} line has been clearly established to hold also for the 268 cm^{-1} line in $F_A(Li^7)$.

We have to mention two points which are not directly related to the proper F_A(Li) scattering features. Firstly, every additively colored crystal with F or F_A centers gives rise to some background intensity in a wide region on the long wavelength side of the laser line (6328 Å), beginning at about 20 Å separation. It has been established with the aid of appropriate spike and edge filters that this is some fluorescence or inelastically scattered light. Secondly, the spectrometer produces some ghost intensity at 6390 Å which adds to the Raman spectra depending on the intensity of elastic diffuse scattering inside the crystal or at surfaces (see remark to Fig. 2).

III. DISCUSSION:

1) Analysis of the C_{4V}-model. We begin by discussing the nature of the separate lines at 228 and 288 cm^{-1} (Li^6). The observed isotope effect relates them closely to modes which are localized or have a pronounced resonance behavior at the Li site. In fact, the isotope frequency ratio is 288/268 = 1.075 for the high frequency line, which is close to the square root of the mass ratio (1.080). If we assume the Li^+ to be in a tetragonal potential well centered on the symmetry axis, we have only to discuss transitions between vibrational levels. In this picture the isotope frequency ratio and the line intensities are strongly in favor of the assumption of a fundamental excitation, i.e. a one-phonon scattering process. If we were trying to assign the lines to higher harmonic transitions, then we would have to look for corresponding sharp fundamental lines of higher intensity and precisely the same isotope shift in the continuum region, which are not observed.

The knowledge of the excited vibrational state is sufficient to predict the polarization properties (see Appendix). In C_{4V} symmetry there is one non-degenerate Li local mode fundamental of type A_1 and one twofold degenerate of type E. One then would try to identify the two "local mode peaks" with these two types of vibration. If we analyze the [100]-excitation experiment (upper half of Fig. 3), in which the centers are assumed to be aligned along the direction of the laser beam, we find that only the A_1 local mode can appear in the [100]-polarized local mode. The observation of the two peaks of 228 cm^{-1} and 288 cm^{-1} is thus unexplained in terms of the foregoing assumptions. If we analyze the [110]-excitation experiment, in which the fourfold ([001])-axis points along the direction of observation, we find that an A_1 mode would appear only in the [110]-polarized spectrum, and an E-mode in no case. The fact that the 288 cm^{-1} line appears in both cases is again in contradiction to the assumptions.

Extending this analysis to the higher harmonic levels, one finds that the non-degenerate vibration contains only the representation A_1 for any level, while the degeneracy increases with every higher level of the twofold degenerate mode. In fact the second harmonic for this mode contains $A_1 + B_1 + B_2$ which could lead to an explanation of the main polarization effects discussed above. But since the Li^+ displacements involved are within the plane perpendicular to the fourfold axis, it is unlikely that they lead to strong enough coupling to the F_{A2} electron states to be observable in a higher-order effect.

2) Influence of F_{A1}-Scattering. The discussion given so far is based on the assumption that the details of the spectra are entirely determined by the majority of the center population which is known to be oriented in some defined manner. Since it is known from the absorption spectra of aligned center systems that there is always some residual absorption left in the F_{A1} band region (Fig. 1), we have to discuss the possible scattering contributions from centers whose fourfold axis F_{A1} direction is parallel to the electric field vector. Unknown are the following important properties:

a) The magnitude of the F_{A1} scattering cross section relative to the F_{A2} scattering cross section, and the changes in the electron-phonon coupling which take place on going from F_{A2} to F_{A1} excitation

b) Possible contributions of the electron-phonon interaction potential to the measured "local mode" frequencies, and their dependence on F_{A1} or F_{A2}-excitation.

The complications arising from a) can be taken into account to some extent when an analysis is made of the polarization behavior. It turns out that the inclusion of F_{A1} scattering of arbitrary spectral distribution and intensity but involving the same eigenfrequencies as F_{A2} scattering does not explain the presence of two local mode fundamentals in the [100]-excited spectra polarized in the [100] direction. The more serious point is b). Nothing is known about the size of such effects, but it is felt that they should be largest in a case of strong coupling between highly localized electronic and vibrational states. The question then is, whether the "local mode" peaks at 228 cm^{-1} and 288 cm^{-1} could result from the two different excited states involved in F_{A2} and F_{A1} scattering from the same A_1-type local mode.

To clarify the situation, an experiment was done in which the laser beam propagated in [111] direction, the $[\bar{1}\bar{1}2]$-oriented field exciting all the centers in their F_{A1}-transition regardless of orientation. Consequently the laser light was absorbed within a few mm of crystal thickness, but the Raman scattering from the illuminated volume could be observed. The intensity of the scattering was at least four times smaller than in a case of [001]-illumination (F_{A1} region transparent); furthermore none of the local mode lines of $F_A(Li^6)$ was enhanced relative to the total spectrum. We conclude from these observations that F_{A1}-scattering plays a minor role, if any, in the spectra of Fig. 2 and 3, and that it is very unlikely that the presence of two local mode lines id due to mechanism b).

3) <u>Off-axis model</u>. Since no satisfying explanation of the "local mode" lines in terms of fundamental transitions in a simple single-well potential of C_{4V} symmetry can be reached, we propose that the Li^+ ion actually occupies off-axis positions. The fact that an overall C_{4V} symmetry has been established for the F_A center by ENDOR experiments(4) leads to the following restrictions: The off-axis position of the Li^+ must be at least fourfold degenerate about the center's axis; either quantum mechanical tunneling or thermally activated jump processes among the potential "pockets" take place at so high a rate that the nuclear spin transitions observed in ENDOR see the field average over all pocket states. This sets a lower limit for the jump rate at about 10^8 sec^{-1}.

The Raman scattering selection rules are then determined by one of the following conditions:

a) The instantaneous symmetry given by one occupied pocket state, if the motion of the Li^+ may be described classically. This symmetry is monoclinic.

b) The symmetry of the tunneling eigenstates, if tunneling predominates over classical diffusion.

In the latter case, the new states may be built up as linear combinations of the vibrational states in one pocket of monoclinic symmetry. A non-generate vibrational level (e.g. the ground state) then becomes a fourfold level with symmetries A_1, B_1 and E. If we make the fairly reasonable assumption that the tunneling splitting between these levels is of the order of 0.1 - 1 cm^{-1}, then the splitting would not be resolved in our observations of Raman local mode lines, and any such line would show the polarization behavior of all the representations contained in the level manifold. The classical description a) leads to a similarly effective relaxation of polarization selection rules, for reasons of the low symmetry.

The off-axis model for the F_A(Li) center has the one additional advantage, that it may in principle lead to higher local mode frequencies than an axis-centered model, as a result of the separating potential barriers incorporated in the vacancy. Thus even the assumption that the "local mode" peaks are due to fundamental transitions seems better justified in an off-axis situation.

The question is still open as to how the structure of the F_A(Li) center is related to the structure of the simple Li^+-center in KCl, which is very likely to have eightfold degenerate pocket states(6). An attempt to obtain first-order Raman scattering from KCl:Li^+ samples (containing relative concentrations of 10^{-3} Li^+) using the same techniques as with the F (Li) center has failed; in particular, it has not been possible to reproduce the observation of a 200 cm^{-1}-line made by Stekhanov et al.[7], who used mercury light excitation.

ACKNOWLEDGEMENTS: The author wants to express his thanks for the hospitality extended by the Physics Department of the University of Utah. He wishes especially to thank Professor F. Lüty for his help with equipment and for many clarifying discussions.

REFERENCES

(1) J.M. Worlock and S.S. Porto, Phys. Rev. Letters 15, 697 (1965).
(2) C. Henry, Phys. Rev. 152, 699 (1966).
(3) F. Lüty, Z. Physik 165, 17 (1961), and other references quoted in our ref.[5].
(4) R.L. Mieher, Phys. Rev. Letters 8, 362 (1962).
(5) B. Fritz, F. Lüty and G. Rausch, Phys.Stat.Sol. 11, 635 (1965).
(6) R.O. Pohl, Review article in this volume.
(7) A.I. Stekhanov and M.B. Eliashberg, Fiz. Tver. Tela 5, 2985, (1963).
(8) R. Loudon, The Raman Effect in Crystals, Adv. Phys. 13, 423 (1964).

APPENDIX: The scattering intensity $S = A\cdot[\sum_{\rho,\sigma} \vec{e}_{i,\rho}\, P_{\rho,\sigma}\, \vec{e}_{sc,\sigma}]^2$ where A is a constant, $P_{\rho\sigma}$ are the elements of the Raman scattering tensor, and $\vec{e}_i$ and $\vec{e}_{sc}$ are unit vectors pointing in the directions of the exciting field vector and the field vector of the outgoing light respectively. The scattering tensors for the five Raman active modes in C_{4v} symmetry may be taken from ref.[8]. A purely diagonal one is that for A_1 symmetry:

$$\begin{pmatrix} a & & \\ & a & \\ & & b \end{pmatrix}$$

The following table gives the contribution from all five representations which are to be observed under the conditions which lead to the results of Fig. 3.

$\vec{e}_i$	$\vec{e}_{sc}$	$S(A_1)$	S(E)(2-fold)	$S(B_1)$	$S(B_2)$
[100]	[100]	a^2	0	c^2	0
	[001]	0	e^2	0	0
$\frac{1}{2}\sqrt{2}[1\bar{1}0]$	$\frac{1}{2}\sqrt{2}[1\bar{1}0]$	a^2	0	0	d^2
	$\frac{1}{2}\sqrt{2}[110]$	0	0	c^2	0

STRESS SPLITTING OF INFRARED ABSORPTION BANDS DUE TO U CENTERS

B. Fritz*, J. Gerlach and U. Gross

II. Physikalisches Institut der

Universität Stuttgart, Germany

Introduction. We are reporting on measurements of the splitting under uniaxial stress of the U center local mode absorption bands in KCl and KI. These infrared bands consist of two different components: Firstly, the main band which corresponds to direct excitation of the threefold degenerate first vibrational level of the local mode; and secondly, sidebands to the main band, which arise from transitions to a final state in which a local mode quantum is excited and another phonon is either created or destroyed. This combined absorption may be caused by the anharmonic coupling or nonlinear dipole moment coupling of the two vibrations involved.[(1)(2)] Main band splittings under stress were measured in KCl and KI. They give the magnitude of the separate contributions of cubic, tetragonal, and shear strain to the shift of the local mode frequency. On the basis of a most simple model, this information may be used to obtain the contributions to the sidebands of phonons of the same symmetries. The behavior under stress of a sharp sideband peak at 93.7 cm^{-1} separation from the main band in KI was also observed, and the symmetry of the 93.7 cm^{-1} phonon directly deduced from the splitting pattern.

Experiments. The KCl and KI crystals used in our experiments were grown from the melt under inert gas atmosphere. It was necessary to dope the material with divalent cations to increase the proportional limit. Thus KCl contained about 0.1% Sr doping, and KI, 0.1% Ca doping, added to the melt. Under these conditions KCl single crystals show proportional behavior up to 300 kp/cm^2 at 77°K, and KI crystals up to 150 kp/cm^2 at 6°K. (1kp = 1 kilogram-weight). The U center

*Presently visiting with the Physics Dept., University of Utah, Salt Lake City, Utah.

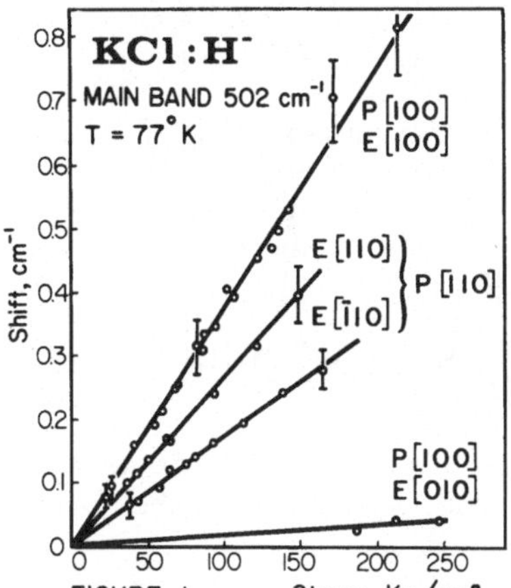

FIGURE 1

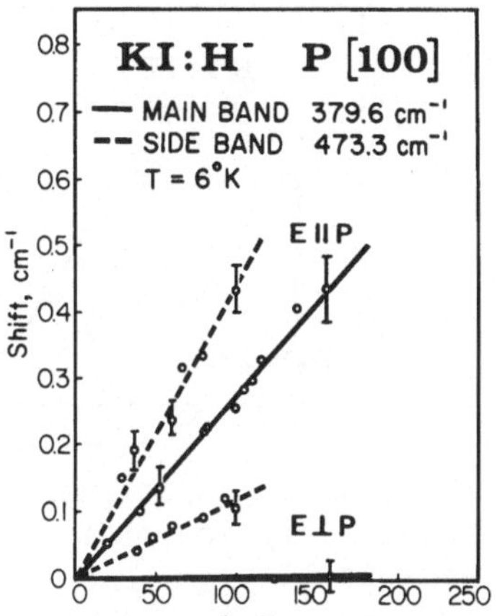

FIGURE 3

Splitting of U center local mode bands under uniaxial stress

concentrations were between 10^{16} and $10^{18}/cm^3$. Optical studies ensure that the U center spectra in these crystals are not affected by local strain effects which are known to occur if a large fraction of the U centers are preferentially formed in the vicinity of other defects.(3) Samples of ⟨100⟩ and ⟨110⟩ orientations were cut to size (13 x 6 x 2 mm) and polished if necessary. The two faces subject to pressure were polished and padded with indium. The static pressure produced at a sample inside the cryostat was calibrated using the known elasto-optic effect of alkali halide crystals. The infrared spectra were measured by means of a foreprism-grating spectrometer. The incorporation of a polarizer leads to a serious cut-down of transmitted energy for certain polarizer settings, and results in a useful spectral width of 1.3 cm^{-1} at 380 cm^{-1} frequency.

The application of uniaxial stress produced a shift of the U center absorption band, which depended on the direction of polarization of the measuring light with respect to the stress axis. To the accuracy of our measurements, we have not noticed genuine changes of the band shape and area under pressure. We therefore have determined the shift of the first moment of the band from the change in absorption at half-height found upon applying or removing a certain stress. The main band shifts measured in KCl:H^- at 77°K are given in Fig. 1. Upon applying pressure along [100] a splitting of the band is observed which is linear with pressure. The two components are polarized along [100] and [010]. Furthermore, the average absorption frequency (transitions along [010] and [001] are degenerate) increases with pressure. This is a direct consequence of the overall volume compression of the lattice under stress. Application of pressure along [110] results in a splitting pattern different from that with [100] pressure. This type of

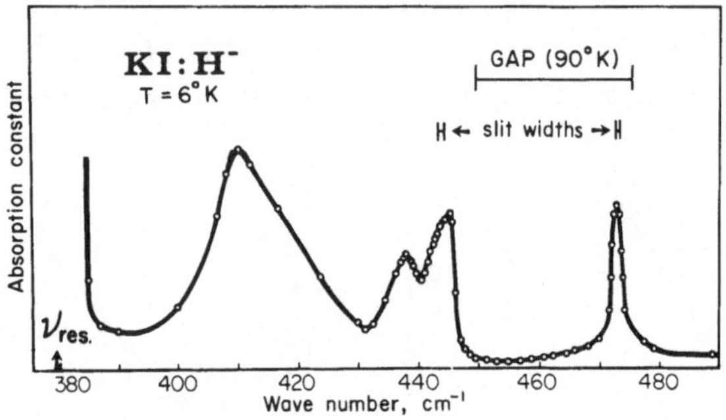

Fig. 2. The sideband spectrum on the high energy side of the local mode absorption in KI:H⁻. The position of the main band peak is at ν_{res}.

stress lifts the threefold degeneracy of the local mode completely. The third component which is polarized along [100] has not been measured, however.

The results obtained for the main band in KI:H⁻ are presented together with those for the sideband at 93.7 cm^{-1}. The whole absorption structure as measured at 6°K is given on Fig. 2. At liquid nitrogen temperature the sideband structure is covered up by the main band which has then broadened heavily. On the same figure we have indicated the position of the phonon gap(4) in KI. The fact that the peak at 93.7 cm^{-1} is quite narrow and lies within the gap region suggests that it may be due to an in-gap localized mode produced by a force constant change at the defect. One would expect this local mode to be of a well-defined symmetry type rather than a composite of those symmetry types which in general contribute to the sidebands. (See our discussion). We expect that a stress experiment will help to clarify this situation. The results in Fig. 3 show that the main band shifts measured in [100] direction are qualitatively similar in KI and KCl. The main band shifts were measured with samples whose optical density would not allow for simultaneous observation of the sideband shift. The results obtained for the sideband may be described by saying that the difference of band peak frequencies, for observation parallel and perpendicular to the stress axis [100] is found to be the same as with the main band; but that both components undergo a large additional shift towards higher frequencies. Under observation with polarized light the sideband did not show any change in area or shape upon application of pressure. We have determined the main band shift as accurately as possible in the same samples where the sideband measurements were made to eliminate any influence of sample preparation on the relative position of the two sets of curves. We have also made measurements with stress applied in the [110] direction which confirm the basic results of the [100] pattern in Fig. 3.

These measurements lead to a complete picture of the main band splitting as analyzed below.

Theoretical remarks. Both the main features of the infrared U band and its behavior under uniaxial stress may be explained using a linear triatomic molecule as a model of the defect. With the central atom (the H^- ion) at a center of inversion symmetry, the odd vibrational mode of this molecule represents the local mode which is restricted to a displacement of the H^- ion itself, denoted by x_o. The symmetric displacement x_1 of two identical heavier outer atoms represents any even mode in which the nearest neighbors of the U center may participate. The total potential energy may be written

$$(1) \qquad \Phi = \frac{k}{2}(x_o^2+x_1^2) + b(x_o^2 x_1 + x_1^3) + c(x_o^2 x_1^2 + x_o^4 + x_1^4) + \ldots$$

Odd terms in x_o drop out because of the inversion symmetry. The only infrared active oscillator is the local mode oscillator (x_o) absorbing at ω_L, the position of the main band. The first anharmonic term couples it to the even mode in a way which gives rise to the sidebands of the main band. The biquadratic term gives rise to broadening of the main line. These two and the other anharmonic terms contribute also frequency shift effects. All these have been extensively treated with respect to U centers in ref (5)(6).

If we subject the molecule to an external stress resulting in a shift of the coordinate x_1, the potential (1) will change to

$$(2) \qquad \Phi' = \left(\frac{k}{2} + b\Delta x_1\right)x_o^2 + \left(\frac{k}{2} + 3b\Delta x_1\right)x_1^2 + (b + 2\Delta x_1 \cdot c)x_o^2 x_1 + \ldots$$

We have dropped a term linear in Δx_1 which describes the elastic energy. The other linear terms in Δx_1 have the following effects. Firstly, changes in the harmonic force constants which are determined by the third order term, b, and secondly, a change in the third order coupling term which determines the coupling to any symmetric strain. This second effect affects in two ways the shift of the absorption bands, since it not only produces nonlinear shifts, through the usual redefinition of force constants, but also affects the truly anharmonic self-energy shift of the local mode band[6]. However, we may estimate the magnitude of the parameters b and c from experimental values of the sideband area[1] and the main band width[1] at low temperatures, giving $b = 3\times10^{12}$ erg/cm^3 and $c = 5\times10^{20}$ erg/cm^4 for KCl:H^-. This shows that the relative change in b responsible for the last mentioned effects is of the order of $10\Delta x_1/r_o$, where r_o is the nearest neighbor distance. It therefore seems justified to drop these terms as unimportant.

We may now analyze the linear shift of the main band in terms of third-order parameters of the type b. Furthermore, we may assume that the side band shift is merely the sum of the main band shift and the shift of the second phonon whose frequency ω_i fulfills the relation

$$\omega_{SB} = \tilde{\omega}_L \pm \omega_i$$

Although strictly speaking the local mode frequency $\tilde{\omega}_L$ contains the frequency dependent anharmonic self energy contributions mentioned, we again drop the stress effects on that part, using the argument given above. Note that the absence of a change in sideband intensity upon application of stress directly shows the small size of the effect in question here.

Derivations of the strain Hamiltonian for the U center local mode have been given in ref (7)(8). In three dimensions the expression that stands for $bx_o^2 \cdot \Delta x_1$ may be written

$$\Delta\Phi = \sum_{\rho} e(\rho) \cdot \hat{B}(\rho) \tag{3}$$

The strain field is given in terms of strain components e (ρ) transforming like one of the representations of the cubic group A_{1g}, $2E_g$, and $3T_{2g}$. The form of the B's, first derivatives of the potential with respect to the strain components, is given in ref. (7). The connection between the strain components and an arbitrary uniaxial stress P is established by Hooke's law by means of the three cubic compliance constants, s_{11}, s_{12}, and s_{44}.

Here the question arises as to whether or not the displacements of atoms close to the defect may be calculated from the compliance constants of the bulk material. Leaving this question open for the moment, we can derive expressions for the local mode shift for a given pressure using the unperturbed compliance constants and the perturbation (3). Table 1 gives these expressions for the symmetry directions which were used in our experiments.

The anharmonic parameters $B(\rho)$ are combinations of third derivatives of the potential Φ of the U center. There is only one such parameter for each representation:

$$B(A_{1g}) = \frac{1}{6} \sum_i (\Phi_{xxx_i})x_i + 2(\Phi_{xxy_i})y_i \tag{4}$$

$$B(E_g) = \frac{1}{12} \sum_i (\Phi_{xxx_i})x_i - (\Phi_{xxy_i})y_i \; ; \qquad B(T_{2g}) = 2\sum_i (\Phi_{xyx_i})y_i$$

x, y, and z are the displacements of the U center along the cubic axis, and x_i, y_i, and z_i denote the position of any ion in the cubic structure, the origin being at the site of the U center. It is useful to write down Eq.(4) for the case of nearest neighbor interaction only. They read

$$B(A_{1g}) = \frac{1}{3} r_o(\alpha_1' + 2\gamma_1'), \quad B(E_g) = \frac{1}{6} r_o(\alpha_1' - \gamma_1'), \text{ and } B(T_{2g}) = 4 r_o \beta_1' \tag{4a}$$

with r_o = distance of nearest neighbor (i = 1)

Table I

P:	Polarization:	$-(\Delta\omega/P)\cdot 2m\omega_L$:
[100]	[100]	$2B(A_{1g})(s_{11}+2s_{12})+8B(E_g)(s_{11}-s_{12})$
[100]	[010]	$2B(A_{1g})(s_{11}+2s_{12})-4B(E_g)(s_{11}-s_{12})$
[110]	[110]	$2B(A_{1g})(s_{11}+2s_{12})+2B(E_g)(s_{11}-s_{12})+\frac{1}{2}B(T_{2g})\cdot s_{44}$
[1$\bar{1}$0]	[110]	$2B(A_{1g})(s_{11}+2s_{12})+2B(E_g)(s_{11}-s_{12})-\frac{1}{2}B(T_{2g})\cdot s_{44}$

and $\alpha_1' = \Phi_{xxx_1}$, $\beta_1' = \Phi_{xyx_1}$, and $\gamma_1' = \Phi_{xxy_1}$.

In the nearest neighbor interaction model we can directly compare the stress splitting parameters $B(\rho)$ and the sideband coupling parameters. It is evident that only phonon modes of any one of the representations A_{1g}, E_g or T_{2g} can contribute to the sideband absorption,(9) analogous to the situation found with static stress. However, there is no simple relation between the B's and the phonon coupling parameters because the strain field produced by a phonon may have a spatial behavior completely different from a uniform strain field. If we discuss the sideband coupling in a nearest-neighbor coupling model also, this complication of course vanishes. The formula for anharmonic coupling on such a model may be written according to Page and Dick(9):

$$(5)\quad \epsilon''\omega = \text{const.}\,\omega_i^{-3}\left[\tfrac{3}{2}(\alpha_1'+2\gamma_1')^2\rho(A_{1g})\ell^2(A_{1g}) + (\alpha_1'-\gamma_1')^2\rho(E_g)\ell^2(E_g) + 4(\beta_1')^2\rho(T_{2g})\ell^2(T_{2g})\right]$$

The phonon density of states $\rho(\omega)$ and the normalized amplitude $\ell^2(\omega)$ for each symmetry type is to be calculated from the equations of motion of the perturbed lattice. The result of the stress experiment then gives the exact weighting factors which determine the contribution of A_{1g}, E_g, and T_{2g} phonons separately.

The behavior of the sidebands under stress is more complicated than that of the main band. The splitting of the first excited level 1 of the "sideband phonon" ω_i under the influence of the perturbation Eq.(3) is determined by the matrix elements

$$(6)\qquad \Delta E_\rho = \langle 1|e(\rho)\cdot\hat{B}(\rho)|1\rangle$$

The nonvanishing matrix elements are obtained from a reduction of the direct product

$$(7)\qquad \Gamma_i \times \Gamma_\rho \times \Gamma_i$$

where Γ_i is the representation of the sideband phonon which is either A_{1g} or E_g or T_{2g}. The A_{1g} phonons will be affected by the cubic part of the strain $e(A_{1g})$ only; the only nonvanishing shift contribution coming from $A_{1g} \times A_{1g} \times A_{1g}$. Both E_g and T_{2g} phonons

Table II

Cubic type	Effect of [100] stress	New levels (D_{4h})	Coupling to local mode in [100]	[010]
A_{1g}	Shift	A_{1g}	Yes	Yes
E_g	Shift and Splitting	A_{1g}	Yes	Yes
		E_g	No	Yes
T_{2g}	Shift and Splitting	B_{2y}	No	Yes
		E_g	Yes	Yes

will experience a splitting of their degenerate states by E_g and T_{2g} type strains. We give the splitting pattern for the case of [100] stress (A_{1g}+E_g strain) in Table II. This table also shows the vanishing of the anharmonic coupling between the local displacement and one of the sideband phonon levels for certain directions of polarization. Here it is assumed that the polarization of a sideband component is given by the polarization of the local mode state, which is initially created by the absorbed photon. The magnitude of the shift and splitting of these levels is generally different for the different normal modes of the perturbed lattice and remains to be calculated from a fairly realistic model potential.

Discussion of Results. From the measurements of the main band splitting, we may obtain the values of the three strain coupling parameters given in Table III. We have attempted a discussion of this result using a Born-Mayer potential to describe the repulsive forces between nearest neighbors:

$$\Phi(0,1) = A\cdot e^{-r(0,1)/\rho} \tag{8}$$

Equations (4a) and the corresponding equation for the harmonic force constant due to nearest neighbor interaction

$$m_H\omega_L^2 = 2\left(\frac{d^2\Phi(0,1)}{dr^2} + \frac{2}{r_o}\cdot\frac{d\Phi(0,1)}{dr}\right) \tag{9}$$

should give a value of the parameter ρ/r_o based on experimental observation. We must state that only the KCl values, together with the value of ω_L provide solutions for ρ/r_o (0.22 and 0.27), while the KI values don't. Another inconsistency arises from the fact that Eq.(8) predicts that $\alpha_1 < 0$ but $\gamma_1 > 0$. This gives (Eq.4a)

$$|B(A_{1g})| \leq |2B(E_g)|$$

Table III

	KCl	KI
$-B(A_{1g})$	2.5 ± 10%	0.70 ± 17% x 10^4 erg/cm^2
$-B(E_g)$	0.80 ± 10%	0.29 ± 20%
$-B(T_{2g})$	0.35 ± 10%	0.51 ± 20%

which is certainly not observed by the KCl data. The conclusion is that the most simple description based on Eq.(8) fails completely. However, it seems to be in order here to recall the assumption of unperturbed elastic constants of the crystal near the U center; which leads to the values of Table III. If, as was pointed out by Gebhardt(7), it is the cubic displacement of the nearest neighbors of the defect, in particular, which is underestimated by using the cubic compressibility $(s_{11} + 2s_{12})$ the values of $B(A_{1g})$ come out too large. At this moment it is not clear to which degree the values of Table III reflect the influence of second neighbor coupling. As long as one assumes nearest-neighbor coupling to be predominant, we can derive the weighting factors for the three phonon types appearing in the sideband formula Eq.(5). In KI, for example, the expressions $\rho(\omega)\cdot\ell^2(\omega)$ for the symmetry types A_{1g}, E_g, and T_{2g} appear to be weighted by factors which are in ratios 104:47:1. Our results thus seem to favor the assumption made in several sideband calculations(10)(11)(12) that the coupling to T_{2g} modes is of very little importance.

The results on the 94 cm^{-1} sideband, namely the absence of a splitting of this band into more than two components, and a relative shift against the main band which is practically independent of polarization or stress axis seem to show clearly that the 94cm^{-1} phonon itself is of symmetry A_{1g}. This agrees with a calculation by Gethins et al.(11) which predicts a local mode of this type to appear in the gap as a consequence of the relaxed force constants pertinent to nearest neighbors of the U center.

We wish to thank Prof. H. Pick for his interest and support of this work. It is our pleasure to acknowledge valuable discussions with Profs. W. Bron, B.G. Dick, R.O. Pohl, and M. Wagner. This work has received financial aid from the Deutsche Forschungsgemeinschaft.

REFERENCES

(1) B. Fritz, Review article in this volume.
(2) H. Bilz, Review article in this volume.
(3) W. Barth and B. Fritz, Phys. Stat. Sol. 19, 515 (1967).
(4) G. Dolling, Cowley, Schittenhelm and Thorson, Phys. Rev. 147, 577 (1966).
(5) R.J. Elliott, Hayes, Jones, MacDonald, Sennett, Proc. Roy. Soc. A289, 1 (1965).
(6) H. Bilz, D. Strauch and B. Fritz, Jour. de Physique, 27, (Suppliment) C2-3 (1966).
(7) W. Gebhardt and K. Maier, Phys. Stat. Sol. 8, 303 (1965).
(8) W. Hayes and H. F. MacDonald, Proc. Roy. Soc. A297, 503 (1967).
(9) T. B. Page and B. G. Dick, Phys. Rev. (To be published).
(10) T. Timusk and M. V. Klein, Phys. Rev. 141, 664 (1366).
(11) T. Gethins, T. Timusk and E.J.Woll, Phys. Rev. 157, 744 (1967)
(12) X. X. Nguyen, Phys. Rev. (To be published).

LOCALISED VIBRATION AND ELECTRONIC SPECTRA OF RARE-EARTH HYDRIDE ION PAIRS IN CALCIUM FLUORIDE

G.D. Jones,[+] S. Yatsiv,[*] S. Peled[*] and Z. Rosenwachs[*]

+University of Canterbury, Christchurch, New Zealand

*Hebrew University of Jerusalem, Israel

Negative hydride ions can readily be introduced as substitutional impurities into the alkaline earth fluorides. The crystals then possess an infra-red absorption spectrum due to localised vibrations of the hydride ion. In the case of calcium fluoride there is a strong triply degenerate line at 965 cm^{-1} with associated second and third harmonics.[(1)]

Calcium fluoride crystals have a body centred cubic structure; each calcium ion is surrounded by eight fluorine ions at the corners of a cube and every second cube of fluorines is empty. When these crystals are doped with trivalent rare-earths the rare-earth ions substitute for the calcium ions and have a site symmetry which depends on the exact way charge compensation is achieved.[(2)] In the case of crystals containing both rare-earth and hydride ions the hydride ions could serve as charge compensators with the formation of well defined rare-earth - hydride ion pairs. In the present work we report results for the electronic, vibronic and vibrational spectra of such ion pairs.

Crystals of calcium fluoride containing less than 0.1% concentration of rare-earth were hydrogenated by the method of Hall and Schumacher[(3)] and then rapidly quenched to room temperature. In this way, the formation of complex clusters of hydride and rare-earth ions is minimised and the spectra show almost only simple hydride - rare-earth ion pairs. A preliminary account of the spectral behaviour of the crystals for different thermal treatments and different rare-earth concentrations has already been presented.[(4)] This paper is restricted to the relatively simple spectra obtained when the crystals are treated as above.

The crystals possess, after hydrogenation, an infra-red spectrum of four lines. One of these is the 965 cm^{-1} line which also appears in hydrogenated pure calcium fluoride. The other three lines occur only in the rare-earth doped crystals. One of these is present in appreciable intensity only in crystals containing rare-earth ions at the middle and end of the rare-earth series and is barely detectable for the remaining ions. It has a frequency independent of the particular rare-earth ion present. At room temperature it has a frequency of 1298 cm^{-1} and a width of 50 cm^{-1}. On cooling to 20°K its frequency shifts to 1312 cm^{-1} and its width becomes 15 cm^{-1}. The corresponding deuterium line has a frequency of 938 cm^{-1} at room temperature giving a hydrogen to deuterium frequency ratio of 1.39. The closeness of this to the square root of two shows that one has a mode strongly localised around the light ion and, to a good approximation, the neighbouring ions may be considered static. This line is assigned to the localised vibration mode of a hydride ion in an interstitial cubic site remote from any rare-earth ion. Because the mode is very localised, the separation of the hydride ion from the rare-earth ion need not be more than a few lattice spacings for negligible effect on the hydride ion's frequency. The large residual width of the line at low temperature is a consequence of large random variations in this separation.

The remaining two lines of the infra-red spectra are observed in all crystals. They have frequencies which depend on the particular rare-earth present. The absorption spectra are shown in Figure 1 and the frequencies and line-widths listed in Table 1. The frequency separation of the lines is greatest for lanthanum and decreases along the rare-earth series; the lines have an approximate intensity ratio of 2:1. The mean weighted frequency is almost constant over the rare-earth series apart from small shifts at cerium, neodymium and terbium. Three weak second harmonic lines are observed in thick crystals; their frequencies for the gadolinium crystals are 2029.2, 2208.4 and 2189 cm^{-1}.

From the large variation of the frequencies of the lines with rare-earth ion and from the occurrence of vibronic transitions described below, it is evident that these lines arise from hydride ions adjacent to the rare-earth ions. There are two possible sites for the hydride ion: one is for the hydride ion to replace a fluorine ion in the cube surrounding the rare-earth ion; the other is for the hydride ion to occupy an interstitial site in an adjacent empty cell and

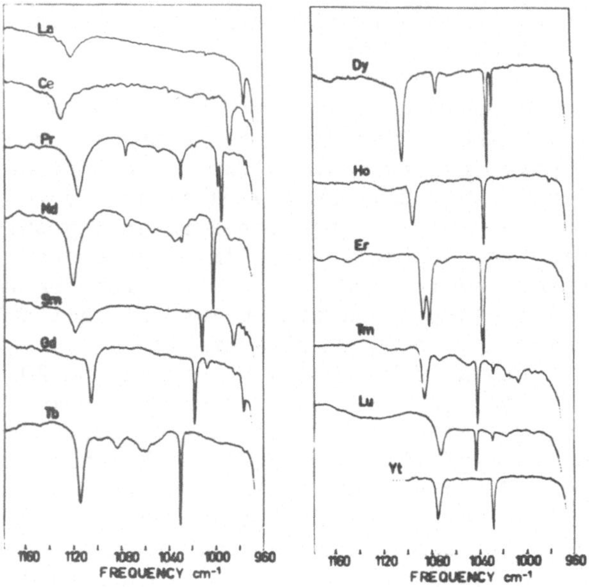

Figure 1. Absorption spectra of hydrogenated crystals of calcium fluoride containing rare earths recorded at 20°K.

Table 1. Frequencies and line-widths of infra-red absorption lines observed at 20°K in hydrogenated crystals of $CaF_2:Re^{3+}$

Rare-earth	Low frequency line.		High frequency line.	
La	976.9	1.0	1121.7	8.9
Ce	988.9	2.4	1130.1	6.6
Pr	994.7	1.2	1115.0	8.8
Nd	1001.5	1.4	1119.4	6.8
Sm	1011.6	1.3	1117.3	7.4
Gd	1017.0	1.0	1104.2	3.5
Tb	1029.3	1.0	1112.8	4.1
Dy	1033.1	1.0	1103.3	3.0
Ho	1035.9	1.1	1094.6	3.6
Er	1036.2	0.8	1086.6	4.3
Er*	1037.6	1.0	1081.2	3.5
Tm	1043.5	0.8	1087.3	4.1
Lu	1042.5	0.8	1072.0	6.3
Yt	1028.9	1.1	1074.8	3.5

*Erbium is anomalous in having two sets of lines. See Figure 1.

charge compensate the rare-earth ion. The respective site symmetries are C_{3v} and C_{4v}. The lines observed here are ascribed to the C_{4v} site on the basis of the following:-

(1) The lines occur at relatively large intensity. For a rare-earth concentration of 0.05% and a hydride ion concentration of the order of 0.01% the lines have an intensity 1/10 of the 965 cm^{-1} line. This is much larger than would be expected from a random distribution of rare-earth and hydride ions in the lattice.

(2) Three second harmonic lines are observed and three are predicted for a site of C_{4v} site symmetry.

(3) The vibronic transitions discussed below are consistent with the C_{4v} site symmetry.

(4) Electron spin resonance measurements on the cerium doped crystals containing hydrogen reveal a new electron spin resonance spectrum of tetragonal symmetry which can be correlated with the absorption lines in the infra-red.

In the C_{4v} site the triply degenerate vibration of a hydride ion is split into a doubly degenerate and a non-degenerate line. The splitting is a measure of the perturbation from cubic symmetry produced by the nearby rare-earth ion, and is linearly dependent on this ion's radius. The centre of gravity, 1070 cm^{-1}, of the lines is different from the frequency, 1312 cm^{-1}, of an isolated hydride ion in an interstitial cubic site. Such a lowering in the mean frequency can be interpreted as evidence of transfer of electronic charge from the hydride to the rare-earth ion with a consequent reduction in the forces binding the hydride ion to its neighbours. The extent of such charge transfer determines the degree of charge compensation achieved. The line widths of the two lines differ: the lower frequency non-degenerate one has a very similar line-width behaviour to the 965 cm^{-1} line; the upper frequency doubly degenerate one has a larger low temperature residual line-width which varies along the rare-earth ion series. It is greatest near the ends where the rare-earth ions are of different ionic size compared to the substituted calcium ion.

The gadolinium doped crystals were also investigated by examining the fluorescence emission due to the transition $^6P_{7/2} \rightarrow {}^8S_{7/2}$ of the gadolinium ion. The fluorescence was excited by radiation from a high pressure mercury lamp and a rotating drum chopper [5] used to time resolve the fluorescence spectra from the exciting radiation. The hydrogenated crystals show extra electronic lines not present in the parent crystals. Similar lines appear in the deuterated crystals at slightly different frequencies. There is thus an isotope shift for these lines. Coupled to these lines at low intensity are vibronic transitions with frequency intervals closely matching the localised mode frequencies measured in the infra-red. Table 2 gives the electronic, vibronic and infra-red frequencies measured for the one electronic line visible at liquid nitrogen temperature. On warming the crystal to higher temperature other electronic lines appear and at 195°K a total of four lines are present. Their frequencies are listed in Table 3.

To all of these four lines is associated a pair of vibronic lines, each of which corresponds to the 1013 cm^{-1} and 1097 cm^{-1} line pair observed in the infra-red absorption spectrum at the same temperature.

It is evident from frequency differences that the extra electronic lines are parents for the local mode vibronics and so must arise from transitions of the gadolinium ion situated in a site perturbed by a nearby hydride ion. The shift of the extra electronic lines from those of the parent crystals is a consequence of the changed environment of the gadolinium ion. There are two ways in which the hydride ion can affect the energy levels of the gadolinium ion: through an altered crystal field and through a change in the covalent bonding of the gadolinium ion to its neighbours. For comparison the four frequencies for a gadolinium ion in a tetragonal site charge compensated by an interstitial fluoride ion are listed in Table 3. Relative to these lines, the centre of gravity of the four lines of the hydrogen perturbed gadolinium ion is displaced 60 cm^{-1} to lower energy. This shift reflects the change in covalent bonding for the gadolinium ion and arises from the different ionic radius and electro-negativity of a hydride ion compared to a fluoride ion. The shift is small because only one of the neighbouring ions around the gadolinium is altered.

There is a crystal field change accompanying the introduction of the hydride ion and the structure of the multiplet of the four lines is changed. The extra electronic lines also display an isotope shift when hydrogen is replaced by deuterium. This arises through the coupling of the localised mode vibrations to the electronic states of the gadolinium ion; the widely different amplitude for the hydrogen and deuterium ion results in a different perturbation of the parent electronic state. Such isotope effects have previously been observed in hydrogenated and deuterated praseodymium and neodymium trifluorides.[7] The observed isotope shift is the difference in isotope shifts for the ground $^8S_{7/2}$ and the excited $^6P_{7/2}$ states; its magnitude is close to those obtained for praseodymium and neodymium fluorides.

Preliminary temperature dependence studies on the fluorescence spectra show that the electronic lines decrease very rapidly in intensity at 220°K and are completely absent at room temperature. The temperature of maximum decrease for the deuterated crystals appears to be approximately 10° higher than for the hydrogenated crystals, while the transition region extends over a range of 60°. These spectra are being studied further to determine the kinetics of the motion of the hydride ion relative to the gadolinium ion and the influence of this on the environment of the gadolinium ion.

If the gadolinium doped crystals containing hydrogen are irradiated at liquid nitrogen temperature with ultra-violet

Table 2. Frequencies of infra-red absorption and ultra-violet fluorescence lines measured at 77°K in hydrogenated and deuterated crystals of $CaF_2:Gd^{3+}$.

Infra-red absorption frequencies.	Parent electronic line frequency.	Vibronic line frequencies.	Vibronic interval.
$CaF_2:Gd^{3+}:H^-$.			
1016.8 cm^{-1} 1103.7 cm^{-1}	31927 cm^{-1}	30911 cm^{-1} 30823 cm^{-1}	1017 cm^{-1} 1104 cm^{-1}
$CaF_2:Gd^{3+}:D^-$.			
735.5 cm^{-1} 799.2 cm^{-1}	31925 cm^{-1}	31187 cm^{-1} 31126 cm^{-1}	735 cm^{-1} 798 cm^{-1}

Table 3.

Extra four electronic lines observed in the hydrogenated crystals at 195°K.	Four lines of the F^- interstitial tetragonal site in the parent gadolinium crystal at 195°K.
31933 cm^{-1}	32037 cm^{-1}
32032 cm^{-1}	32109 cm^{-1}
32099 cm^{-1}	32147 cm^{-1}
32149 cm^{-1}	32158 cm^{-1}

radiation of shorter wavelength than 2500 A, the 31927 cm^{-1} electronic line of the Gd^{3+} - H^- ion pair decreases in intensity and an electronic line at 31981 cm^{-1} appears. To this new line is coupled a weak vibronic line at 31213 cm^{-1} which corresponds to a vibrational interval of 768 cm^{-1}. The deuterated crystals show an analogous behaviour with the formation of an electronic line at 31982 cm^{-1}, and an associated vibronic line at 31416 cm^{-1} giving a vibrational interval of 566 cm^{-1}. The vibration frequency ratio is 1.36. The localised modes thus extend a little and involve some of the neighbouring ions. The low frequencies indicate considerably reduced force constants of interaction of the light atoms with their neighbours.

The infra-red spectra of the irradiated crystals were carefully examined at low temperatures in the 768 cm^{-1} region for evidence of direct infra-red absorption, but none was found. At the spectrometer sensitivity used, this negative result sets an upper limit of 1/100 for the oscillator strength of these modes as compared with the hydride localised modes at 1017 and 1104 cm^{-1}. To explain these results, it is postulated that the ultra-violet radiation converts the hydride ion to neutral hydrogen atoms, with the formation of a Gd^{3+} - H^o ion pair. Such an entity would have little infra-red absorption yet vibrations of the hydrogen can still modulate the crystal field acting on the gadolinium to give a vibronic spectrum. Because the irradiation is at low temperatures, it is assumed that no transfer of atoms to different sites occurs and the ion pair consists of a gadolinium ion and a hydrogen atom in adjacent interstitial sites. The site symmetry of either ion has not yet been established definitely by experiment. The vibronic spectrum shows only one line associated with each of the three electronic lines. These facts suggest that both ions are in cubic symmetry; however, this is inconsistent with the physical environment of the two ions and it is possible that other lines are present that could be detected under more favourable conditions of observation.

With the present data, one has a hydrogen atom vibration at 768 cm^{-1}. This value may be compared to the value of 590.5 ± 41.7 cm^{-1} determined on the basis of spin relaxation measurements for neutral hydrogen atoms in cubic interstitial sites in calcium fluoride.(6) Both values show that a hydrogen atom is less tightly bound in an ionic lattice than a hydride ion. Although the localised vibration involves a neutral atom, one would expect to see some infra-red absorption through the Lax-Burstein mechanism.(8) Our negative result shows such induced absorption to be small. Direct observation of the localised vibrations predicted by Feldmann et. al.(6) is thus likely to be difficult.

Temperature dependence studies on the 31981 cm^{-1} electronic line show that this line largely vanishes above approximately 150°K and the 31927 cm^{-1} line regains its former intensity. The hydrogen atom pair is thus unstable and readily reconverts to the original ion pair.

Preliminary fluorescence lifetime measurements have been made on the various fluorescence transitions. At 77°K the tetragonal F^- interstitial site of the parent crystal has a fluorescence lifetime of 13 milliseconds, the C_{4v} hydrogen charge compensated site has 4 millisecs, and the deuterium site 3 milliseconds. The hydrogen and deuterium atom pairs have fluorescence lifetimes of 7 milliseconds. The trigonal site obtained in oxygenated crystals shows a lifetime of 2 milliseconds. These results seem consistent with the lifetimes of the upper electronic states being determined by multiphonon relaxation mechanisms involving the localised phonons.

We wish to express our thanks to Mr. M. Presland for making lifetime measurements on the fluorescence lines and to Mr. Avia for making electron spin resonance measurements on the cerium doped crystals. One of us (GDJ) wishes to thank the U.S. Office of Naval Research and the Erskine Bequest of the University of Canterbury for making his attendance at this Conference possible.

REFERENCES

(1) R.J. Elliott, W. Hayes, G.D. Jones, H. Macdonald and C.T. Sennett, Proc. Roy. Soc. A289, 1 (1965).
(2) M.J. Weber and R.W. Bierig, Phys. Rev. 134, 1492 (1964).
(3) J.L. Hall and R.T. Schumacher, Phys. Rev. 127, 1892 (1962).
(4) S. Yatsiv, S. Peled, Z. Rosenwachs and G.D. Jones, Johns Hopkins University Conference, September 1966.
(5) S. Peled, U. El-Hanany and S. Yatsiv, Rev. Sci. Instrum. 37, 1649 (1966).
(6) D.W. Feldmann, J.G. Castle Jr. and J. Murphy, Phys. Rev. 138, A1208 (1965).
(7) G.D. Jones and R.A. Satten, Phys. Rev. 147, 566 (1966).
(8) M. Lax and E. Burstein, Phys. Rev. 97, 39 (1955).

LOCAL MODE ABSORPTION FROM HYDROGEN YTTERBIUM PAIRS IN CALCIUM FLUORIDE

R.C. Newman and D.N. Chambers

J.J. Thomson Physical Laboratory,

University of Reading, Berks., U.K.

INTRODUCTION

The calcium fluoride structure consists of a cubic lattice of fluorine ions in which every second body-centre position is occupied by a divalent calcium ion. Infra-red absorption due to the localized modes of vibration of hydrogen and deuterium ions diffused into pure calcium fluoride has been studied in detail by Elliott et al[1]. In hydrogen diffused crystals, a single, triply degenerate fundamental absorption line was observed at 965 cm^{-1} [77°K] together with one second harmonic and two third harmonics. It was concluded that the hydrogen was present as H^- ions on substitutional fluorine sites, which have T_d symmetry.

In this paper we shall discuss the satellite structure which appeared around the fundamental band at 965 cm^{-1} in hydrogen treated calcium fluoride crystals which had been doped with ytterbium ions. Rare-earth ions can be incorporated into the lattice on substitutional calcium sites and, in the case of ytterbium, these may be present in either the divalent or trivalent charge state. In the latter case some form of charge compensation is necessary and this may be effected in a variety of ways as shown by the detailed study of Kirton and McLaughlan[2]. If hydrogen ions occupy sites close to an ytterbium ion, their local site symmetry is reduced and satellite lines should be produced near the fundamental peak in an infra-red absorption spectrum. The system calcium fluoride-ytterbium was chosen for investigation since the electronic and ESR spectra of trivalent ytterbium in calcium fluoride are now well understood[2]. Moreover, McLaughlan and Newman[3] have shown that a new ESR spectrum which appears in these crystals after hydrogen diffusion is due to hydrogen-ytterbium pairs.

EXPERIMENTAL DETAILS

Both pulled crystals, grown by Barr and Stroud Ltd., and Stockbarger crystals, obtained from the Mervyn Company, have been examined. Ytterbium was introduced by adding 0.05% to 0.2% by weight of the oxide or fluoride to the melt. Hydrogen or deuterium ions were introduced into the crystals by the method first described by Hall and Schumacher(4), the samples being placed on a bed of aluminium metal and heated in an atmosphere of the gas. Most of the samples were heated for 19 hours at a temperature of about 900°C, after which they were polished for optical examination.

Infra-red measurements in the range 2-16μm (5000-625 cm^{-1}) were made on a Grubb-Parsons Spectromaster grating instrument which has a resolution of 1.0 cm^{-1} in the region of 10μm where most of the detailed measurements were made. All the spectra were recorded with the samples fixed to the cold finger of a standard liquid nitrogen cell.

RESULTS

The spectra obtained from hydrogen treated Stockbarger and pulled crystals are shown schematically in Figures 1 and 2. In pulled crystals, four new satellites were seen which could be ascribed to hydrogen ions associated with ytterbium. The energies of these satellite lines, all of which had half-widths of about $1\frac{1}{2}$ cm^{-1}, are given in Table 1. In samples which had been diffused with deuterium an analogous spectrum was produced at lower energies, the positions of the corresponding satellites being given in Table 2. Further structure seen near the fundamental hydrogen band in pulled crystals (see Fig. 2) was also present in undoped pulled crystals and could not therefore be attributed to ytterbium ions.

TABLE 1. Energies of hydrogen satellites.

LINE	P_{1H}	P_{2H}	Q_{1H}	Q_{2H}
ENERGY (cm^{-1})	1048.4	1030.4	996.2	943.8

TABLE 2. Energies of deuterium satellites.

LINE	P_{1D}	P_{2D}	Q_{1D}	Q_{2D}
ENERGY (cm^{-1})	753	727	717	680

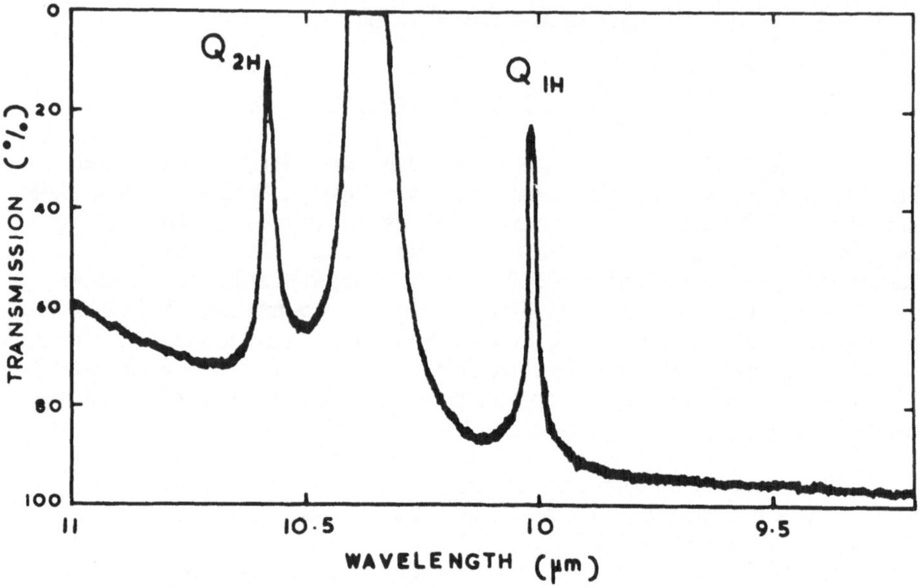

FIGURE 1. Local Mode Absorption: Yb-doped Stockbarger Crystals.

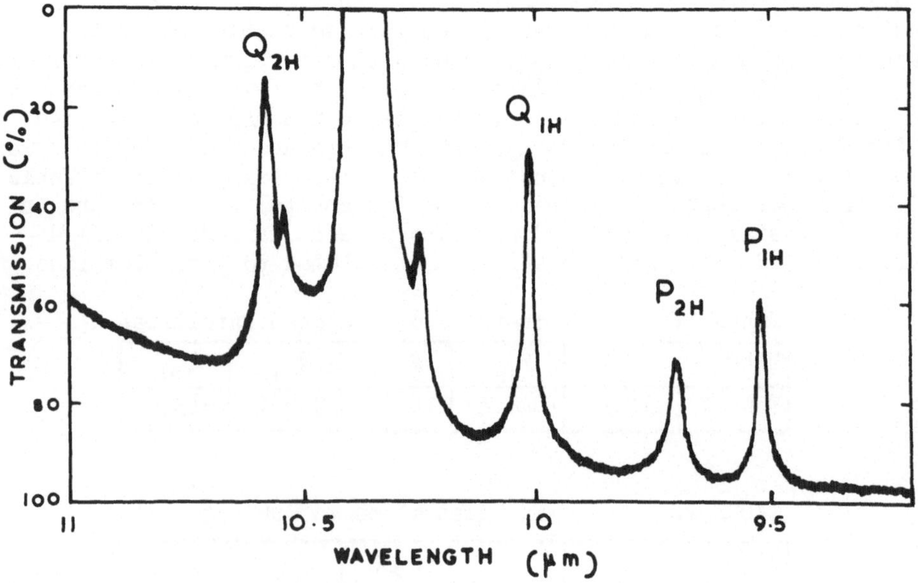

FIGURE 2. Local Mode Absorption: Yb-doped Pulled Crystals.

The fact that this structure did not appear in Stockbarger crystals, which are known to contain a very low oxygen content compared with pulled crystals, suggests that the extra lines may arise from the vibrations of hydrogen ions located near divalent oxygen ions.

In Stockbarger crystals (Fig. 1) only two satellites, labelled Q_{1H} and Q_{2H}, were seen. The ratio of integrated absorption (the peak absorption in cm^{-1} multiplied by the half-width in cm^{-1} for Q_{2H} to Q_{1H} was found to be constant and equal to 2.15 ± 0.05. The separations of the lines from the unperturbed fundamental band were 21 cm^{-1} and 31 cm^{-1} respectively. Their strength varied from crystal to crystal in direct proportion to the ytterbium concentration showing that the centre involved only one ytterbium ion. Four second harmonic satellite lines have been found which correlate with Q_{1H} and Q_{2H}, indicating that the centre has trigonal symmetry. In these samples very little absorption was found in the lines attributable to the electronic transitions of trivalent ytterbium in the region of 1μm showing that most of the ytterbium was present in the divalent state. It is therefore proposed that the lines Q_{1H} and Q_{2H} correspond to the normal modes of vibration of a hydrogen ion occupying a nearest neighbour site to a divalent ytterbium ion. Such a centre would have C_{3V} symmetry about a $\langle 111 \rangle$ crystal axis. The relative intensities of the two lines indicate that Q_{1H} corresponds to vibrations along the axis and Q_{2H} to the doubly degenerate transverse vibrations.

In pulled crystals, two extra satellites designated P_{1H} and P_{2H} appeared, for which the integrated intensity ratio was $P_{1H}/P_{2H} = 1.37 \pm 0.05$. These lines were always more intense in oxide-doped crystals than in those doped with ytterbium fluoride. They could also be made to appear in Stockbarger (oxygen-free) material by heating the crystals in dry air for several hours at a temperature of about 750°C before hydrogen diffusion. These observations suggest that an oxygen ion is one of the constituents of the defect centre which gives rise to the P lines. Four second harmonic satellites were found for the centre, indicating that the hydrogen ion was in a site of trigonal symmetry.

Low and Ranon[5] have shown that heating ytterbium doped calcium fluoride crystals in vacuum converts all the trivalent ytterbium ions to the divalent state. Consequently a sample of hydrogen treated pulled crystal was heated at 500°C for two hours, in an atmosphere of hydrogen (but in the absence of aluminium metal) to suppress the outdiffusion of hydrogen ions. After this treatment, it was found that P_{1H} and P_{2H} had disappeared, leaving only Q_{1H} and Q_{2H}. Examination of the crystal in the region 9,000 Å to 10,000 Å showed that the electronic spectrum due to trivalent ytterbium had disappeared. This observation lends further support to the proposed interpretation of the Q lines.

After hydrogen diffusion of ytterbium doped calcium fluoride, a new electronic absorption line, labelled μ, due to trivalent ytterbium is produced at 10,179 cm^{-1} . By diffusing crystals with a mixture of hydrogen and deuterium, it has now been established that this transition arises from an ytterbium ion paired with a hydrogen ion; the absorption around 10,179 cm^{-1} appeared as a partially resolved doublet, which could be synthesized from the line at 10,179 cm^{-1} together with a new line, due to ytterbium-deuterium pairs, shifted by about 0.3 cm^{-1} to higher energy. The absorption in the line μ has been measured at 77°K in several different hydrogen diffused crystals and compared with the total absorption in the local mode satellites P_{1H} and P_{2H}. A good correlation was obtained showing that the two types of absorption were attributable to the same defect centre. The strength of the absorption in P_{1H} and P_{2H} has also been correlated with the new ESR spectrum from trivalent ytterbium, labelled T_{3H}, which has been observed in hydrogen diffused samples but which is absent in the same samples prior to this treatment . This centre T_{3H}, which has been correlated previously with the electronic line μ(2), has trigonal symmetry and the presence of a hydrogen ion can be inferred by the splitting of the ESR absorption line with the magnetic field along a trigonal ⟨111⟩ axis. It was verified that this splitting was due to a superhyperfine interaction with the proton spin by replacing hydrogen with deuterium, since the deuteron has a nuclear spin I = 1 and a much smaller magnetic moment than the proton. Fig. 3 shows the ESR spectrum obtained from a crystal which had been diffused with a mixture of deuterium and hydrogen, and it is seen that the line T_{3D} is not resolved

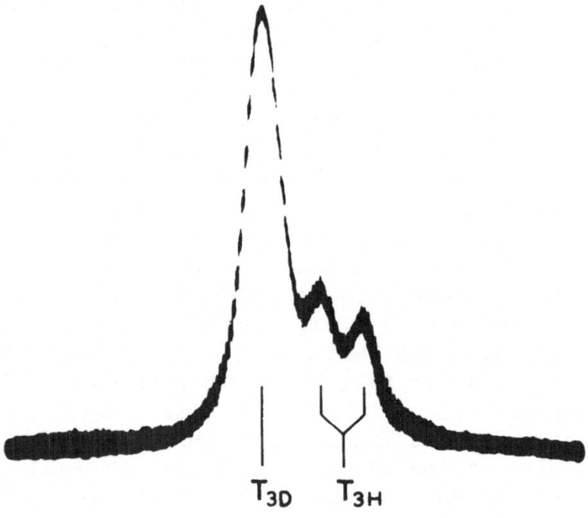

FIGURE 3. Appearance of T_{3D} and T_{3H} ESR absorption lines when the magnetic field is applied along a ⟨111⟩ axis.

into a triplet. However, the splitting of the components of the triplet is estimated to be only about 1.5 gauss as determined from the observed splitting of 10 gauss due to the proton magnetic moment. Resolution of the triplet components is not therefore expected since the individual components in T_{3H} had widths of 7 gauss. It will also be noted from Fig. 3 that the value of $g_{\parallel}$ for T_{3D} (1.512 ± 0.002) is slightly different from the value for T_{3H} (1.516 ± 0.002 at the centre of the split line). The value of $g_{\perp}$ (4.147 ± 0.005) is the same for both spectra.

CONCLUSION AND DISCUSSION

The g values for T_{3H} are similar to those from another trigonal spectrum T_1 with $g_{\parallel} = 1.323$ and $g_{\perp} = 4.389$ thought to be due to a trivalent ytterbium ion compensated by a divalent oxygen ion in a nearest neighbour fluorine site(5). Furthermore the only paramagnetic centres which remained in crystals after hydrogen diffusions at temperatures above about 950°C were those giving rise to spectra T_1 and T_{3H}. It is therefore proposed that these two centres are related in the manner shown in Fig. 4.

Both centres have over-all charge compensation. The site symmetry of the ytterbium ion is C_{3v} which agrees with the trigonal symmetry of the ESR spectra. Replacing the hydrogen ion by a deuterium ion would lead to displacements in the positions of the

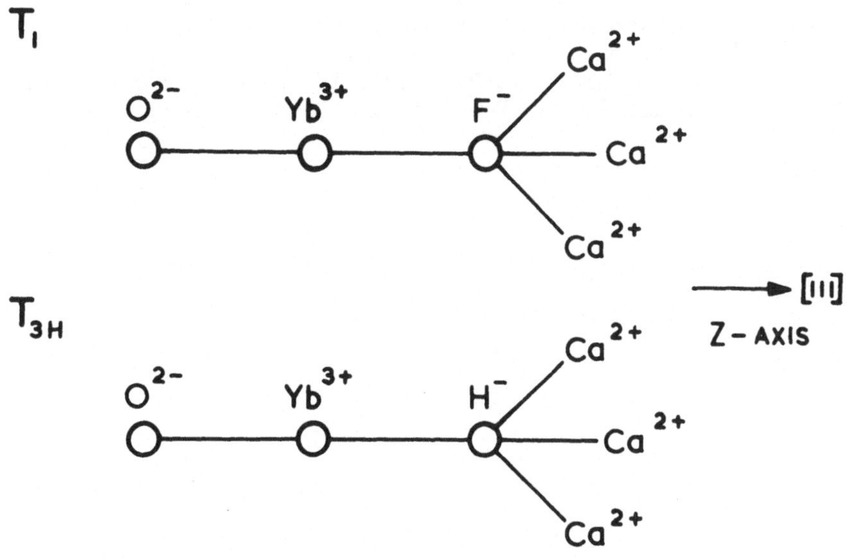

FIGURE 4. Proposed models for T_1 and T_{3H} centres.

ions surrounding the ytterbium ion, due to differences in the vibrational zero point energy between hydrogen and deuterium. The corresponding change in the ytterbium wave functions and energy levels is reflected in the modified ESR g-values and the shift in the electronic transition energy. The hydrogen ion also has C_{3V} symmetry and hence two fundamental local mode satellites and four second harmonic satellites are to be expected, in agreement with the observations. The shift of both these fundamental vibrational satellites to energies greater than that of isolated hydrogen ions is indicative of increased force constants, presumably due to the presence of the trivalent ion. That the integrated absorption ratio of these two satellites is not 2:1 means that the vibrational modes must have different effective dipole moments. It is not clear which of the infra-red satellites is to be associated with which vibrational mode.

ACKNOWLEDGEMENTS

Thanks are due to the Ministry of Defence for financial support for this work.

REFERENCES

1. R.J. Elliott, W. Hayes, G.D. Jones, H.F. Macdonald and C.T. Sennett, Proc. Roy. Soc. 289, 1 (1965).
2. J. Kirton and S.D. McLaughlan, Phys. Rev. 155 279 (1967).
3. S.D. McLaughlan and R.C. Newman, Phys. Letters 19, 552 (1965).
4. J.L. Hall and R.T. Schumacher, Phys. Rev. 127 1892 (1962).
5. W. Low and U. Ranon, in Paramagnetic Resonance, ed. W. Low (Academic Press, New York and London, 1963).

ANHARMONIC BROADENING OF RESONANT MODES

K.H.Timmesfeld and H.Bilz

Institut für Theoretische Physik der Universität

Frankfurt/Main, Germany

Introduction

Far infrared measurements on NaCl : Cu^+ performed by Weber and Nette[1] show, at low temperatures, a sharp absorption line at 23.7 cm. The width of this line broadens strongly with increasing temperature (fig.1). Recent measurements by Sievers[2] show, that the low-temperature absorption measured by Weber consists of two lines corresponding to the two copper isotopes, having a single width of ca. 0.3 - 0.4 cm^{-1} and a distance of ca. o.3 cm^{-1}. In order to reduce the halfwidths measured by Weber to the corresponding ones for only one copper isotope it is therefore necessqry to substract an amount of ca. 0.3 cm^{-1}. To explain the sharp absorption line at low temperatures and the strong temperature dependence of its width, the Cu^+ion is assumed to be coupled to its nearest neighbours with a weak spring and a cubic anharmonic coupling-constant. The cubic anharmonicity has been treated by solving a Dyson-equation with a mass-operator of second order in the cubic anharmonicity. The unperturbed crystal is described by a Debye-approximation in such a way that the phonon density in the low-frequency acoustic region is the same as that given by Schroeder's[3] breathing shell model.

1. Absorption Theory

The additional absorption of the perturbed crystal is mainly caused by the dipole moment $\underline{M}$ of the elongation $\underline{u}(0)$ of the Cu^+ ion:

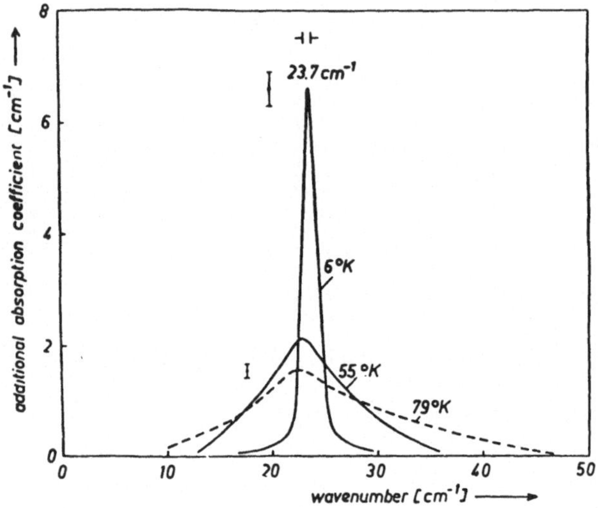

Fig. 1. Temperature dependence of the resonant mode in $NaCl:Cu^{+}$ (ref. 1).

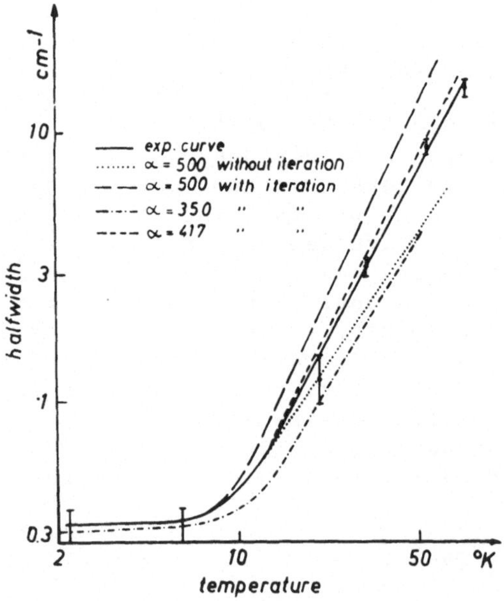

Fig. 2. Plot of the halfwidth of the resonance band versus temperature.

(1.1) $$\underline{M}(t) = e\,\underline{u}(0,t)$$

Using the absorption formalism of Kubo[4] and Zubarev[5] the dielectric susceptibility is given by

(1.2) $$\chi_{\alpha\beta}(\Omega) = e^2 G_{\beta\alpha}(0,0,-\Omega)$$

where G is the Fourier transform of the retarded Green's function

(1.3) $$G_{\alpha\beta}\left(\begin{smallmatrix} l & l' \\ k & k' \end{smallmatrix}, t\right) = \frac{2\pi}{i\hbar}\vartheta(-t)\langle\langle\left[u_\alpha\left(\begin{smallmatrix} l \\ k \end{smallmatrix}, t\right), u_\beta\left(\begin{smallmatrix} l' \\ k' \end{smallmatrix}, 0\right)\right]\rangle\rangle$$

2. Description of the harmonic crystal

The Green's function of the unperturbed harmonic crystal has already been derived by many authors, e.g. Elliott[6]. Its form is

(2.1) $$G^H_{\alpha\beta}\left(\begin{smallmatrix} l & l' \\ k' & k \end{smallmatrix}, \Omega\right) = \frac{1}{N(M_k M_{k'})^{\frac{1}{2}}} \sum_{q j} \frac{e^*_\alpha\left(k\middle|\begin{smallmatrix} q \\ j \end{smallmatrix}\right) e_\beta\left(k'\middle|\begin{smallmatrix} q \\ j \end{smallmatrix}\right)}{\omega^2\left(\begin{smallmatrix} q \\ j \end{smallmatrix}\right) - (\Omega + i\varepsilon)^2} e^{2\pi i q\left[x\left(\begin{smallmatrix} l' \\ k' \end{smallmatrix}\right) - x\left(\begin{smallmatrix} l \\ k \end{smallmatrix}\right)\right]}$$

where the $\omega\left(\begin{smallmatrix} q \\ j \end{smallmatrix}\right)$'s are the eigenfrequencies and the $e_\alpha\left(k\middle|\begin{smallmatrix} q \\ j \end{smallmatrix}\right)$'s the eigenvectors of the crystal normal modes.

Let the substitution of one sodium ion by the copper ion give rise to an additional potential matrix C containing the effect of the altered mass and spring constants; then - following Elliott[6] - the Green's function Γ of the perturbed harmonic crystal is given by the matrix

(2.2) $$\Gamma = (I + G^H C)^{-1} \cdot G^H$$

Because C is a 3 x 3-matrix, $G^H C$ has only three nonvanishing columns. Hence it is possible to obtain Γ by inverting only a 3 x 3-matrix, namely the "central" matrix of $I + G^H C$. In the special case of a pure mass-defect one arrives at the well known formulae[7]

(2.3) $$\Gamma_{\alpha\beta}\left(\begin{smallmatrix} l & l' \\ k & k' \end{smallmatrix}, \Omega\right) = G_{\alpha\beta}\left(\begin{smallmatrix} l & l' \\ k & k' \end{smallmatrix}, \Omega\right) + \frac{\sum_\gamma \Delta M \cdot \Omega^2 G^H_{\alpha\gamma}\left(\begin{smallmatrix} l \\ k \end{smallmatrix} 0, \Omega\right) G^H_{\gamma\beta}\left(0 \begin{smallmatrix} l' \\ k' \end{smallmatrix}, \Omega\right)}{1 - \Delta M \Omega^2 G^H_{\alpha\alpha}(00, \Omega)}$$

In particular, for $\left(\begin{smallmatrix} l \\ k \end{smallmatrix}\right) = 0$ we have

(2.4) $$\Gamma_{\alpha\beta}\left(0 \begin{smallmatrix} l \\ k \end{smallmatrix}, \Omega\right) = \frac{G^H_{\alpha\beta}\left(0 \begin{smallmatrix} l \\ k \end{smallmatrix}, \Omega\right)}{1 - \Delta M \cdot \Omega^2 G^H_{\alpha\alpha}(00, \Omega)}$$

3. The Effect of Anharmonicity

Developing the crystal Hamiltonian into a power-series in terms of the ionic displacements one obtains

(3.1) $$H = H_0 + H_3 + H_4 + \ldots$$

where H_0 contains the harmonic contributions, H_3 the cubic, H_4 the quartic anharmonicity, and so on. We want to consider now the effect of cubic anharmonicity for the special case where only the coupling coefficients of the substituted copper-ion with its nearest neighbours are non-zero:

$$H_3 = \frac{3\alpha}{3!}\Big[u_x^2(0)\{u_x(100)-u_x(\bar{1}00)\}+u_y^2(0)\{u_y(010)-u_y(0\bar{1}0)\}+ + u_z^2(0)\{u_z(001)-u_z(00\bar{1})\}\Big] \tag{3.2}$$

Here α means the constant coupling coefficient

$$\alpha = \frac{\partial^3\Phi}{\partial u_x(0)^2\,\partial u_x(100)} = \frac{\partial^3\Phi}{\partial u_y(0)^2\,\partial u_y(010)} = -\frac{\partial^3\Phi}{\partial u_x(0)^2\,\partial u_x(\bar{1}00)}, \text{ etc.} \tag{3.3}$$

The effect of anharmonicity has already been treated in normal mode space[8)]. One gets for the Green's function of the anharmonic crystal a Dyson-equation:

$$G(ss',\Omega) = \delta_{ss'}\Gamma(s,\Omega) + \Gamma(s,\Omega)\sum_{s''}\Pi(s,s'',\Omega)G(s'',s',\Omega) \tag{3.4}$$

where the mass-operator Π in the lowest order of cubic anharmonicity has the form

(3.5) 

meaning

$$\Pi(s,s',\Omega) = \Pi^{(1)}(s,s',\Omega) + \Pi^{(2)}(s,s',\Omega) \tag{3.6}$$

Here

$$\Pi^{(1)}(s,s',\Omega) = -\frac{18}{\hbar^2}\sum_{s_1 s_2} V(s,s_1,s')V(s_1,s_2,s_2)[2n(s_2)+1]\Gamma(s_1,0,\Omega) \tag{3.7}$$

and

$$\Pi^{(2)}(s,s',\Omega) = -\frac{18}{\hbar^2}\sum_{s_1 s_2} V(s,s_1,s_2)V(s_1,s_2,s')\times \times\{[n(s_1)+n(s_2)+1]\Gamma(s_1+s_2,\Omega)+[n(s_2)-n(s_1)]\Gamma(s_1-s_2,\Omega)\} \tag{3.8}$$

The $V(s,s',s'')$ are the cubic anharmonic coupling coefficients in normal mode space and $n(s)$ is the Bose occupation number of the phonon having the quantum-number s.
Transforming the Dyson-equation and the mass-operator from the normal-mode-space into the ordinary space one arrives at the formulae:

$$G_{\alpha\beta}\binom{l\ l'}{k\ k'},\Omega) = \Gamma_{\alpha\beta}\binom{l\ l'}{k\ k'},\Omega) + \sum_{l_1 k_1 \gamma}\sum_{l_2 k_2 \delta}\Gamma_{\alpha\gamma}\binom{l\ l_1}{k\ k_1},\Omega)\times \times\Pi_{\gamma\delta}\binom{l_1\ l_2}{k_1\ k_2},\Omega)G_{\delta\beta}\binom{l_2\ l'}{k_2\ k'},\Omega), \tag{3.9}$$

$$\Pi^{(1)}_{\alpha\beta}\begin{pmatrix} l & l' \\ k & k' \end{pmatrix},\Omega\Big) = -\frac{18\hbar}{\pi} \sum_{\substack{l_1 l_2 l_3 l_4 \\ k_1 k_2 k_3 k_4 \\ \gamma\delta\varepsilon\zeta}} \Phi_{\alpha\gamma\beta}\begin{pmatrix} l & l_1 & l' \\ k & k_1 & k' \end{pmatrix} \cdot \Gamma\begin{pmatrix} l_1 & l_2 \\ k_1 & k_2 \end{pmatrix}, 0\Big) \times \tag{3.10}$$

$$\times \Phi_{\delta\varepsilon\zeta}\begin{pmatrix} l_2 & l_3 & l_4 \\ k_2 & k_3 & k_4 \end{pmatrix} \int_0^\infty du \left[2n(u)+1\right] \mathrm{Im}\,\Gamma_{\varepsilon\zeta}\begin{pmatrix} l_3 & l_4 \\ k_3 & k_4 \end{pmatrix}, u\Big)$$

and

$$\Pi^{(2)}_{\alpha\beta}\begin{pmatrix} l & l' \\ k & k' \end{pmatrix},\Omega\Big) = -\frac{18\hbar}{\pi^2} \sum_{\substack{l_1 l_2 l_3 l_4 \\ k_1 k_2 k_3 k_4 \\ \gamma\delta\varepsilon\zeta}} \Phi_{\alpha\gamma\delta}\begin{pmatrix} l & l_1 & l_2 \\ k & k_1 & k_2 \end{pmatrix} \int_0^\infty du \int_0^\infty dv \times \tag{3.11}$$

$$\times \mathrm{Im}\,\Gamma_{\gamma\varepsilon}\begin{pmatrix} l_1 & l_3 \\ k_1 & k_3 \end{pmatrix}, u\Big) \Big\{ \left[n(u)+n(v)+1\right] \frac{2(u+v)}{(\Omega+i\varepsilon)^2-(u+v)^2} +$$

$$+ \left[n(v)-n(u)\right] \frac{2(u-v)}{(\Omega+i\varepsilon)^2-(u-v)^2} \Big\} \mathrm{Im}\,\Gamma_{\delta\zeta}\begin{pmatrix} l_2 & l_4 \\ k_2 & k_4 \end{pmatrix}, v\Big) \Phi_{\varepsilon\zeta\beta}\begin{pmatrix} l_3 & l_4 & l' \\ k_3 & k_4 & k' \end{pmatrix}.$$

The first part in the brackets of (3.11) describes the sum-processes and the second one the difference-processes. Because of equation (3.2), $\Pi_{xx}\begin{pmatrix} l & l' \\ k & k' \end{pmatrix}$ contains only a 3 x 3-matrix of nonvanishing components as do Π_{yy} and Π_{zz}. Hence equation (3.9) can be solved easily in the same manner as equation (2.2).

In the internal lines of the mass-operator Π (3.5), we have inserted the harmonic Greens function Γ. Of course it is much better to use there the anharmonic Greens function G. Doing this one gets an integral-equation for G which can be solved in an iterative way.

4. Numerical Calculation

For the description of the unperturbed harmonic crystal the following assumptions are made:

I. Only the acoustic branches of the crystal are included, because the measured resonance lies in the very far infrared region.

II. The acoustic branches are described by using a Debye approximation with one Debye-frequency which is chosen so that the phonon-density coincides with that calculated from U. Schröder's[3] breathing shell model for low frequencies. This of course means that the Brillouin-zone of the diatomic NaCl-lattice is replaced by a sphere with the same volume.

III. A long-wave-approximation is made for the normal mode eigenvectors.

With these approximations various calculations have been done for the harmonic Greens function Γ (2.2) and the anharmonic Greens function G (3.9.). For a small value of the altered spring-constant the results show in the harmonic case a sharp resonance line at low frequencies. Taking a value of 3% of the unperturbed spring-constant one gets a line at 23.5 cm^{-1} with a half-width of 0.28 cm^{-1}, which is in good agreement with the measured values for very low temperatures.[2)]

For the anharmonic crystal, until now only the part $\Pi^{(2)}$ (3.11) of the mass-operator hass been taken into account, because this part is mainly responsible for the broadening of the resonant line. The results for various values of the anharmonic coupling coefficient and for calculations both with and without an iteration of equation (3.9) are shown in fig.2. One sees that an iteration of equation (3.9)leads to a better description of the experimental facts, as is to be expected. The best agreement with the experimental data is obtained with a value for the anharmonic coupling coefficient α of 4.6 10^{11} [erg/cm^3] . This corresponds to a value of $\bar{\alpha} = \alpha \cdot r_0/f^* = 417$, where r_0 is the lattice constant of the NaCl lattice; f^* stands for the harmonic spring constant of the substitution to its nearest neighbours and has a value of 3% (see above) of the value of a typical spring constant f of the unperturbed lattice given by $2f = m\omega_0^2$ with the Debye-frequency ω_0 and the reduced mass of the lattice m.

The neglect of $\Pi^{(1)}$(2.10) leads to a much too large shift of the calculated absorption line to lower frequencies with increasing temperature. We suppose that this discrepancy can be removed by taking into account the effect of the part $\Pi^{(1)}$(3.10) of the mass-operator.

References

1) R.Weber and P.Nette, Phys.Letters 20, 493 (1966)
2) A.J.Sievers, private communication
3) U.Schröder, Sol.State Comm. 4, 347 (1966)
4) R.Kubo, J.Phys.Soc.Japan 12, 570 (1957)
5) D.N.Zubarev, Sov.Phys.Uspekhi 3, 320 (1960)
6) R.J.Elliott in "Phonons in perfect lattices and in lattices with point imperfections" edited by R.W.H. Stevenson (Oliver and Boyd, Edinburgh and London, 1966), page 377
7) R.Brout and W.A.Visscher, Phys.Rev.Letters 9, 54 (1962)
8) A.A.Abrikosov, L.P.Gorkov and I.E.Dyzaloshinski, Methods of Quantum Field Theory in Statistical Physics (Prentice Hall Inc., London, 1963)

Acknowledgements: Financial support by the Deutsche Forschungsgemeinschaft is gratefully acknowledged.

USE OF REALISTIC SHELL MODEL PHONONS IN INTERPRETING DEFECT SPECTRA*

T. Timusk†, E. J. Woll, Jr., and T. Gethins

McMaster University, Hamilton, Ontario, Canada

The understanding of the infrared (IR) absorption spectrum of alkali halide crystals containing substitutional H^- ion defects has been progressively improved, using a combined experimental-theoretical approach. The absorption spectrum of these crystals consists of a direct absorption in the far IR including contributions from the defects as well as the reststrahlen absorption, and a direct absorption in the vicinity of 400 cm^{-1} due to excitation of the high-frequency localized vibrational mode of the H^- ion.(1) This local mode peak is accompanied by an anharmonic sideband which has been extensively investigated.(2,3,4,5,6) A line due to double excitation of the local mode is forbidden by symmetry, but an anharmonic sideband of this line should be observable, including an absorption peak at three times the local mode frequency.

Sievers(7) has measured far IR absorption for crystals of KI containing a variety of impurity ions, including H^-. In the present work we present further experimental as well as theoretical investigations of the far IR absorption for KBr:KH and KI:KH. In addition, we make some theoretical comments about the as yet unobserved sideband of the (forbidden) two phonon line.

*Research supported in part by the National Research Council and by the Defence Research Board of Canada.
†Alfred P. Sloan Research Fellow.

EXPERIMENT

Experiments were carried out on single crystals of KBr and KI obtained from the Harshaw Chemical Company. H^- ions were introduced by heating the crystals in potassium vapor at 680°C, followed by a prolonged (20 hour) heating at 450°C in 100 atms. of hydrogen. The crystals were polished, and measured in a metal cryostat with polyethylene windows. The temperature was measured by a differential thermocouple attached directly to the crystals.

A commercial double beam spectrometer, the Perkin Elmer 301, was used in all the measurements. It was found that, at a scanning rate of 10 cm^{-1}/hour, a resolution of 1 cm^{-1} could be attained with this instrument in the 100 cm^{-1} - 40 cm^{-1} region when used with the cryostat.

The absorption spectrum of KBr:KH was measured at 7 ± 2°K. The result, for the region between 60 and 100 cm^{-1}, is shown in Figure 1a. Above this region the crystal becomes completely opaque, due to the onset of the strong reststrahlen absorption. Below 60 cm^{-1} down to 30 cm^{-1} there is essentially no measurable absorption. The absorption above 96 cm^{-1} is probably due to difference processes in the host crystal, and the line at 95 cm^{-1} has been identified as a Cl^- impurity absorption by Levine.[8] Both of these features are present in undoped crystals. The broad absorption peak between 80 cm^{-1} and 92 cm^{-1} is an effect of the H^- impurity.

A similar spectrum is shown for KI in Figure 3a. Here again the spectrum above 75 cm^{-1} is due to the host crystal, and the weak peak at 77 cm^{-1} is probably due to Cl^- impurity.[7] The strong triplet between 55 cm^{-1} and 68 cm^{-1} is attributed to H^- impurities. The peaks do not seem to be limited in width by experimental resolution. The resolution of these measurements is approximately 1 cm^{-1}, and the accuracy of the wave number scale is ± 0.5 cm^{-1}.

THEORY

In interpreting these defect spectra, it is necessary to know the lattice dynamics of the host crystal, and to have a realistic model for the changed force constants and masses associated with the defect. In the present work, the lattice dynamics are taken from neutron-derived shell models for KBr and KI[9,10]. For the change of force constants, we use the model recently introduced by the authors[5] to describe the anharmonic sidebands in KBr and KI. This model has two parameters: the change in nearest neighbor force constant Δf, and the change of nearest neighbor - fourth neighbor force constant Δg. The first of these has been determined fitting the local mode frequency, while the second has been adjusted to give reasonable agreement for the E_g and A_{1g} resonances which occur in the sideband spectrum of the local mode. The present calculations can thus be carried through without the introduction of any further parameters.

Calculation of the far IR absorption for crystals containing defects has been discussed by several authors.(11,12,13,14) In particular, calculations of the present type have been performed by Benedek and Nardelli(15) and by Patnaik and Mahanty(16) for the case of Li and Ag impurities in KBr. In a recent review by Klein(17), an expression for the absorption constant is given, which is essentially

$$\alpha(\omega) = \frac{(n^2(\infty)+2)^2}{9n(\omega)} \frac{4\pi}{c} \frac{e^2}{\mu} \frac{N}{V} I_m\{\bar{G}_{Rx,Rx}\}. \tag{1}$$

Here $n(\omega)$ is the index of refraction of the crystal at frequency ω, N is the number of unit cells in the crystal of volume V, e is the electronic charge, c the speed of light, and μ the reduced mass of the alkali and halide ion pair. The quantity $\bar{G}_{Rx,Rx}$ is a diagonal element of the Green's function matrix $\bar{G}(\omega^2)$ for the crystal containing impurities. This matrix is to be taken in the representation of eigenmodes of the perfect lattice; the element required corresponds to the reststrahlen frequency; that is, a mode of wave vector 0 in the transverse optical branch, polarized in the x direction.

In evaluating $\bar{G}$ we use the T matrix, defined by

$$\begin{aligned} T &= \Gamma(1 + G\Gamma)^{-1}; \\ \bar{G} &= G - GTG, \end{aligned} \tag{2}$$

where Γ is the matrix of force constant changes and G is the Green's function matrix of the perfect lattice. In our case, Eq. (2) yields

$$\bar{G}_{Rx,Rx} = (\omega_R^2-\omega^2)^{-1} - (\omega_R^2-\omega^2)^{-2} T_{Rx,Rx} \tag{3}$$

where ω_R is the reststrahlen frequency, and the required matrix element of T is

$$T_{Rx,Rx} = \frac{1}{N} \sum_{\substack{\underset{\sim}{L}\ \underset{\sim}{L}' \\ \pi\ \pi' \\ \alpha\ \alpha'}} \epsilon_\alpha^\pi(Rx)\epsilon_{\alpha'}^{\pi'}(Rx) T_{\underset{\sim}{L}\pi\alpha,\underset{\sim}{L}'\pi'\alpha'} \tag{4}$$

where $\underset{\sim}{L}$ is the position of the ion, π the sign of the ionic charge, and α a cartesian coordinate. The polarization vectors $\epsilon_\alpha^\pi(Rx)$ are given by

$$\epsilon_{\alpha}^{\kappa}(Rx) = \kappa\sqrt{\frac{\mu}{M_{\kappa}}}\,\delta_{\alpha,x}, \tag{5}$$

with M_{κ} the mass of the ion of charge κe.

In our model the matrices Γ and T have non-zero elements only in a 15 x 15 subspace. Moreover, if cubic symmetry is exploited, the matrix to be inverted for evaluation of $T_{Rx,Rx}$ can be reduced to 3 x 3: namely, one of the three matrices corresponding to T_{1u} symmetry.

The required 6 elements of G were evaluated using the model VI shell model phonons of references 9 and 10. In summing over the Brillouin zone, the technique due to Gilat and Raubenheimer(18) was used, taking an effective 32,000 points in the zone. (Eigenvectors were evaluated once for each point.)

The calculated spectra are shown in Figure 1b and Figure 3b for KBr and KI, respectively. The large reststrahlen peak begins at the right side of each figure. The remaining contribution comes from the second term of Eq. (3). This contribution comes from a resonance (of T_{1u} symmetry) which arises where the determinant $\det|(1+G\Gamma)_{T_{1u}}|$ becomes small, since this quantity occurs as a denominator in the matrix inversion required to determine $T_{Rx,Rx}$.

It should be noted that, because the symmetry of the present problem is odd rather than even, and because the coupling to the radiation field arises from the force constant changes - extending to fourth neighbors - rather than from anharmonic connection to nearest neighbors, certain Van Hove singularities (VHS) are observable in these curves which fail to appear distinctly in the sideband spectrum of the local mode. The appearance of VHS has been noted by Benedek and Nardelli(15).

We have also used the model of force constant changes to calculate the shape of the sideband of the (forbidden) line corresponding to double excitation of the local mode. The detailed calculation will be described elsewhere; for the present, it is sufficient to point out that elements of T_{1u} symmetry are also required, and, consequently, certain VHS appear.

RESULTS AND DISCUSSION

The absorption below 92 cm^{-1} in KBr, shown in Figure 1a, is not present in the undoped crystal and follows the strength of the local mode line at 440 cm^{-1}; it is therefore taken to be due to the H^- ion. This absorption is to be compared with the theoretically calculated absorption curve shown in Figure 16. The most

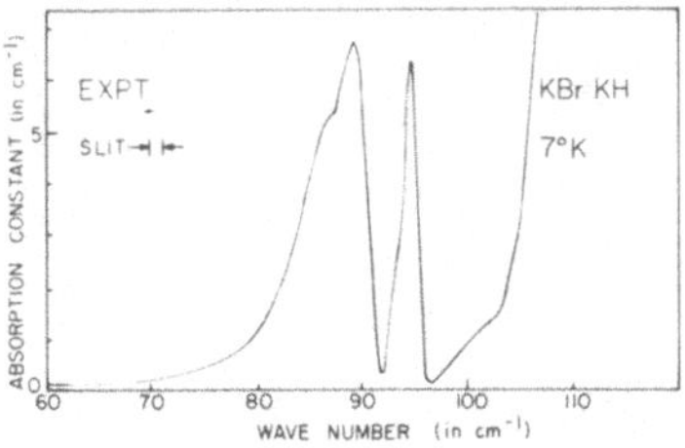

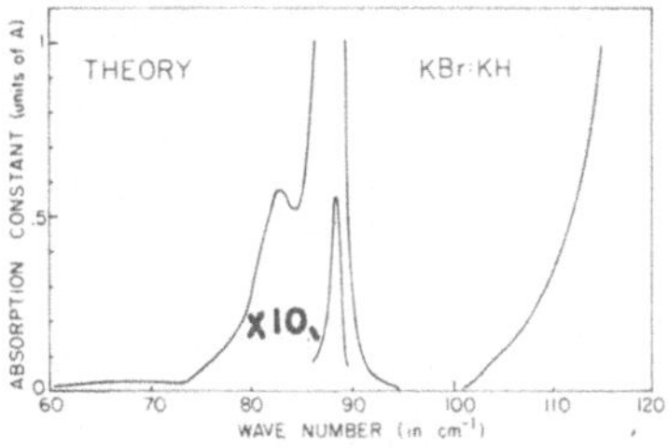

Figure 1a. Experimental absorption spectrum of KBr:KH. The structure below 93 cm^{-1} is attributed to H^- impurities.
Figure 1b. Theoretical absorption spectrum of KBr:KH. (The unit A is $[4\pi e^2\, q(n^2(\infty)+2)^2/9n(\omega)\mu c] \times 10^{-12}$ sec, where q is the relative density of H^- ions.)

prominent feature of this curve is the resonance predicted at 89 cm^{-1}; this agrees well with the strong peak observed experimentally at 89 cm^{-1}. This agreement is better than one should expect, since the force constant model reproduces the resonances in the sideband of the local mode to at best a few wavenumbers.(5)

The shoulder at 86 cm^{-1} in the experimental curve is about 3 cm^{-1} above the corresponding peak in the theory. The theoretical peak arises from a very pronounced VHS at an off-symmetry point in the LA branch of the shell model dispersion curve. The position of this peak in our theory curve is independent of the details of our change-of-force constant model; depending only on the lattice dynamics of the host crystal. We are led to conclude that the VHS actually occurs at 86 cm^{-1} in the host crystal. Such a disagreement with the predictions of the shell model is not unreasonable, since the experimental temperature is 7°K while the neutron data were taken at 90°K,(9) and, moreover, since the peak

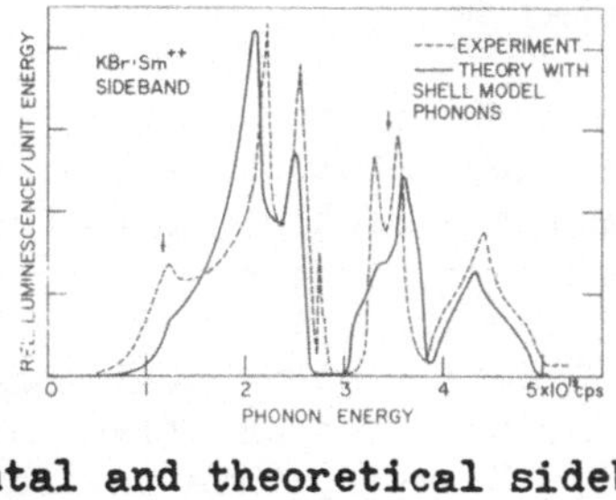

Figure 2. Experimental and theoretical sideband spectra of the Sm^{++} florescense in KBr. Note that the high experimental peak near 2.6 x 10^{12}/sec (86 cm^{-1}) disagrees by a few wavenumbers with the theoretical prediction.

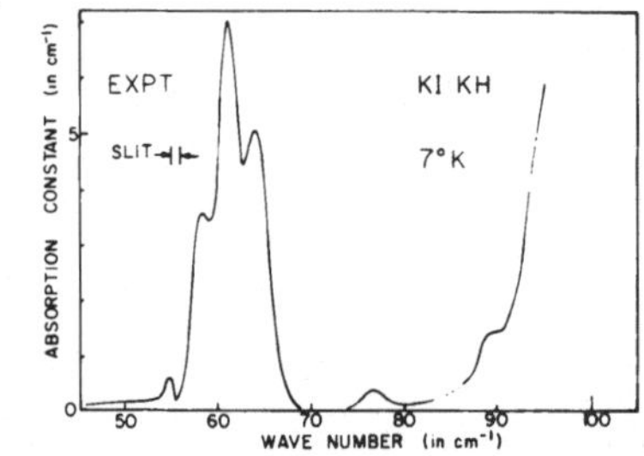

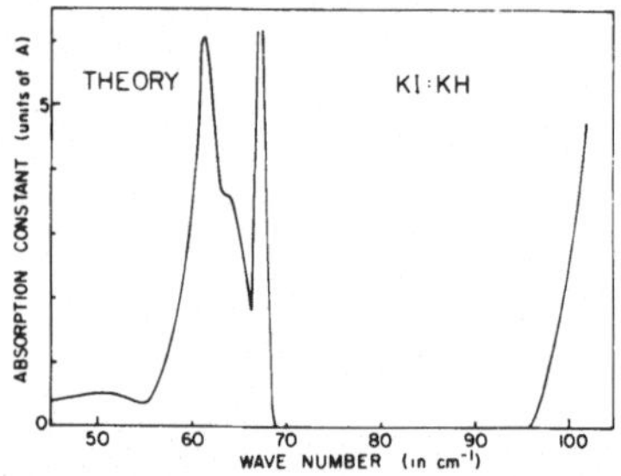

Figure 3a. Experimental absorption spectrum for KI:KH. The structure below 70 cm^{-1} is attributed to H^- impurities.
Figure 3b. Theoretical absorption spectrum for KI:KH. (See caption of Figure 1b for unit A.) Our model has evidently predicted the resonance too high by 6 - 10 cm^{-1}.

in question is from an off-symmetry direction, and therefore is not observed directly by neutrons. As supporting evidence, we present in Figure 2 a spectrum of the sideband of the luminescense of Sm^{++} in KBr. Details of this work are reported elsewhere(19), but, for the present, it is interesting to note that the peak at 83 cm^{-1} in the theory, based on shell model phonons, lies a few wave numbers below the experimental peak, in agreement with the discrepancy found here. It is also interesting to note that only the two high peaks - both associated with off symmetry directions - are in disagreement between the Sm^{++} experiment and theory. This suggests that the discrepancy is more likely to be due to errors of the shell model in predicting off-symmetry properties than to thermal shifts.

Figure 3 displays the experimental and theoretical absorption curves for KI. The experiment agrees with measurements of Sievers(7). Again, the absorption below 80 cm^{-1} is taken to be due to the H^- ion. Comparison of the two figures clearly shows that the position of the resonance, predicted at 67.5 cm^{-1}, is in disagreement with experiment, though the 6-10 cm^{-1} discrepancy is not out of line with the discrepancy for the Eg mode in the side-band spectrum of the local mode.

The peak of the theoretical curve at 64 cm^{-1} agrees well with the 64 cm^{-1} peak in the experiment. The theoretical peak arises from VHS in a high-symmetry direction, which has therefore been measured directly by neutrons. The remaining peaks at 61 cm^{-1} and 58 cm^{-1} are more difficult to identify. It would seem at first natural to assign the 61 cm^{-1} experimental peak to the theoretical peak at 61 cm^{-1}, which arises from an off-symmetry VHS analogous to that observed in KBr. However, this predicted peak could well be in error by 3 cm^{-1}, and might as well be identified with the experimental peak at 58 cm^{-1}. The remaining peak in the experimental curve, whichever it proves to be, can be assigned to the resonance.

To check this point further, we have also measured the far IR spectrum of a KI crystal containing deuterium. There was a slight shift to lower frequency of the 61 cm^{-1} peak, while the other two maxima remained unshifted. Such a shift can be expected because of different anharmonic contributions to the resonance frequency; this would seem to strongly favor the assignment of the 61 cm^{-1} peak as the resonance.

It is interesting to refer to the experiments of Sievers for KI with impurities(6). In a large number of absorption spectra, particularly for F^- and Cl^-, (whose structure falls in the same frequency region as the H^- absorption) peaks at 58 and 64 cm^{-1} appear. We suggest, tentatively, that the recurrance of these peaks implies observation of VHS of the host crystal. (The further implication is that the peak observed by us at 61 cm^{-1} is indeed the resonance.)

CONCLUSION

We have shown that measurements of the far infra-red absorption in KBr (and, less satisfactorily, in KI) can be understood in terms of a simple model of the force constant changes, provided realistic shell model phonons are used. The predicted spectrum, a narrow region of absorption at the top of the acoustical branch, is in good agreement with experiment. Investigation of very fine details within this region reveals discrepancies of the order of a few per cent in the positions of Van Hove singularities derived from the shell model, and of the order of 6 - 10% in the position of the resonance peaks derived from our force constant model.

REFERENCES

1. G. Schaefer, J. Phys. Chem. Solids 2, 233 (1960).

2. B. Fritz, U. Gross, D. Bäuerle, Phys. Stat. Sol. 11, 231 (1965).

3. T. Timusk, M. V. Klein, Phys. Rev. 141, No2, 664 (1966).

4. Nguyen Yuan Xinh, Sol. State. Comm. 4, 9 (1966).

5. T. Gethins, T. Timusk, and E. J. Woll, Jr. Phys. Rev. 157, 744 (1967).

6. H. Bilz, D. Strauch, B. Fritz, J. Phys. Radium 27, Supp. C2, 3 (1966).

7. A. J. Sievers, in Low Temperature Physics, J. G. Daunt et al., eds. (Pleneum Press, Inc., New York, 1965), Vol. LT9, Part B.

8. M. Levine (private communication to M. V. Klein).

9. R. A. Cowley, W. Cochran, B. N. Brockhouse and A. D. B. Woods, Phys. Rev. 131, 1030 (1963).

10. G. Dolling, R. A. Cowley, C. Schittenhelm, I. M. Thorson, Phys. Rev. 147, 577 (1966).

11. I. M. Lifschitz, J. Phys. U.S.S.R. 7, 211, 249 (1943); 8, 89 (1944).

12. P. G. Dawber, R. J. Elliott, Proc. Roy. Soc. (London) A273, 222 (1963); Proc. Phys. Soc. (London) 81, 453 (1963).

13. A. A. Maradudin in Astrophysics and the Many Body Problem, (Benjamin, New York 1963); Rev. Mod. Phys. 36, 417 (1964); Solid State Physics, F. Seitz and D. Turnbull. (Academic Press, New York, 1966), Vols. 18 and 19.

14. K. K. Rebane, N. N. Kristofel, G. D. Trifonov, V. V. Khyzhnyakov, Izv. Akad. Nauk. Estonian SSR 13, 87.

15. G. Benedek and B. F. Nardelli, Phys. Rev. 155, 1004 (1967).

16. K. Patnaik, J. Mahanty, Phys. Rev. 155, 987 (1967).

17. M. V. Klein, Physics of Color Centers, W. B. Fowler ed. (to be published).

18. G. Gilat, L. J. Raubenheimer, Phys. Rev. 144, 390 (1966).

19. T. Timusk, M. Buchanan, Phys. Rev. (to be published).

QUANTUM MECHANICAL CALCULATIONS OF THE U-CENTER FORCE CONSTANTS*

R. F. Wood

Solid State Division, Oak Ridge National Laboratory

Oak Ridge, Tennessee

INTRODUCTION

It is now well known that a hydride ion can replace a halide ion in alkali-halide crystals. The resulting defect, the U center, gives rise to absorption lines or bands in both the ultraviolet and infrared regions of the spectrum. Absorption in the former region is associated with electronic transitions between localized levels, while that in the latter region is associated with a localized vibrational mode whose frequency lies well above those of the transverse optical branch. Properties associated with electronic transitions have already been extensively investigated,(1) whereas the vibrational properties associated with the local mode have only recently received much attention.(2)

During the past two years, we(3) have carried out extensive calculations on the electronic structure of the U center in the potassium(4) and sodium halides. The major effort in this work has been directed at the calculation of the force constants and local mode frequencies from "first principles." The U center is well suited for this type of calculation, primarily because of the relatively simple structure of the H^- ion and the large difference between its mass and the masses of the ions of the host lattice. The latter condition means that the "static well" approximation in which the host ions are held fixed while the H^- ion vibrates should be a good one. Also, the highly ionic nature of alkali-alkali crystals together with the fact that the first nearest neighbors

*Research sponsored by the U. S. Atomic Energy Commission under contract with Union Carbide Corporation.

(1nn) of the H^- ion are positive alkali ions should make it a good approximation to use free ion Hartree-Fock functions to represent them in the host lattice. We formerly thought that the second nearest neighbor (2nn) ions should be relatively unimportant in determining the local mode frequency, but more recent and complete results indicate that this may not be entirely true.

STATIC WELL APPROXIMATION

The equation of motion for the j-th component of the displacement of the α-th ion in the ν-th unit cell of a crystal is given by

$$M_\alpha \ddot{u}_{j\alpha}(\nu) = - \sum_{k\beta\mu} A_{j\alpha,k\beta}(\nu,\mu)\; u_{k\beta}(\mu). \tag{1}$$

The force constants, A, are given by

$$A_{j\alpha,k\beta}(\nu,\mu) \equiv [\partial^2 V/\partial u_{j\alpha}(\nu)\; \partial u_{k\beta}(\mu)]_o, \tag{2}$$

in which V is the potential energy of the system and the derivatives are evaluated at the equilibrium positions.

In the static well approximation, the local mode frequency is given by the first term on the right of the following equation:

$$\omega^2 = (A_{j1,j1}(0,0) + \Delta A_{j1,j1}(0,0))/M_o'$$

$$+ \varepsilon \sum_{k\beta\mu}{}' A^2_{j1,k\beta}(0,\mu)/M_\mu(\varepsilon A_{j1,j1}(0,0) + \Delta A_{j1,j1}(0,0)). \tag{3}$$

Here, $\varepsilon = (M_o - M_o')/M_o$ is the relative mass change of the defect (mass M_o') and $\Delta A_{j1,j1}(0,0)$ is the change in force constant. The summation (prime indicates the diagonal term is omitted) gives a correction of the order of $1/M_\mu$ to the static well approximation. This should be about one or two percent in the potassium and sodium halides.

The force constant of interest here is thus given by

$$k_j \equiv A_{j1,j1}(0,0) + \Delta A_{j1,j1}(0,0) = \left[\frac{\partial^2 V}{\partial u^2_{j1}(0)}\right]_0 \tag{4}$$

where now V is the potential energy of the system with the defect present. Although k's are the actual quantities to be calculated, the results will be expressed in terms of ω ($\omega = (k/M_o')^{1/2}$) for convenient comparison with experimental results.

QUANTUM MECHANICAL CALCULATIONS

Hamiltonian

We write the electronic part of the Hamiltonian for the crystal as

$$H = H_U + H_{cr} + H_{int} \tag{5}$$

with

$$H_U = \sum_{i=1}^{2} \{-\tfrac{1}{2}\nabla_i^2 - \underline{r}_i^{-1}\} + r_{12}^{-1} \tag{6}$$

and

$$H_{int} = \sum_{i=1}^{2} \{-\sum_{\nu} Z_\nu |\underline{r}_i - \underline{R}_\nu|^{-1} + \sum_{j=3}^{M} |\underline{r}_i - \underline{r}_j|^{-1}\}. \tag{7}$$

In these equations, Z_ν is the nuclear charge, $\underline{R}_\nu$ the position of the $\underline{\nu}$-th ion and $\underline{r}_i$ is the position vector of the i-th electron. The coordinate origin is taken at the position of the proton of the H^- ion. H_{cr} is the Hamiltonian of the rest of the crystal, including nuclear interactions, but since the corresponding energy is treated classically, we need not write out H_{cr} explicitly.

Wave Function

The electronic part of the wave function for a crystal containing a single U center can be written in what should be an adequate approximation as

$$\Psi(\underline{1},\ldots,\underline{M}) = \{M!/2!(M-2)!\}^{1/2} A\psi_U(\underline{1},\underline{2})\psi_c(\underline{3},\ldots,\underline{M}). \tag{8}$$

$\psi_U(\underline{1},\underline{2})$ is an antisymmetrized two-electron group function for the H- ion and ψ_c is an antisymmetrized (M-2)-electron group function for the rest of the crystal. A is then an appropriate antisymmetrizer for the entire function. It is also assumed that "strong orthogonality" holds, i.e.,

$$\int\psi_U(\underline{1},\underline{2})\psi_c(\underline{3},\ldots\underline{k-1},1,\underline{k+1},\ldots\underline{M})d\tau_1 = 0 \tag{9}$$

and that ψ_U and ψ_c are separately normalized to unity.

In the calculations, ψ_U is approximated by

$$\psi_U(\underline{1},\underline{2}) = N_U[\psi_a(1)\psi_b(2) + \psi_a(2)\psi_b(1)]\theta(s_1,s_2) \tag{10}$$

in which ψ_a and ψ_b are one-electron spatial orbitals and N_U is a normalizing factor. $\theta(s_1,s_2)$ is a normalized singlet spin function. When it is necessary to have an explicit form for ψ_c (as in the calculation of the expectation value of H_{int}), a single determinant composed of doubly occupied free ion orbitals is assumed. ψ_a can be written as

$$\psi_a(\underline{r}) = N_a[\phi_a(\underline{r}) + \sum_{\nu,j} c_{\nu,j}\phi_{\nu,j}(\underline{r})], \tag{11}$$

with

$$\phi_a = (\beta_a^3/\pi)^{1/2} \exp(-\beta_a r), \qquad c_{\nu,j} = -(\phi_{\nu,j}|\phi_a) \tag{12}$$

and N_a another normalizing factor. An exactly similar form is taken for ψ_b. $\phi_{\nu,j}$ is the j-th free ion HF orbital on the ν-th ion. The parameters β_a and β_b are to be determined by a variation procedure.

Energy Expressions

The expectation value of H is given by

$$\langle\Psi|H|\Psi\rangle = \langle\psi_U|H_U|\psi_U\rangle + \langle\psi_c|H_{cr}|\psi_c\rangle +$$

$$+ (M!/2!(M-2)!)\langle\psi_U\psi_c|H_{int}A|\psi_U\psi_c\rangle \; . \tag{13}$$

As mentioned above, the second term on the right is treated by classical ionic crystal theory in a fairly straightforward manner and it will not be discussed further here. After considerable algebra, the other two terms give

$$\langle\Psi|H_U + H_{int}|\Psi\rangle = 2N_U^2 \{(a|h|a) + (b|h|b) +$$

$$+ 2s_{ab}(a|h|b) + (ab|ab) + (ab|ba)\} \tag{14}$$

in which

$$h(1) = -\tfrac{1}{2}\nabla_1^2 - r_1^{-1} - \sum_\nu Z_\nu|\underline{r}_1 - \underline{R}_\nu|^{-1} +$$

$$+ \int |\underline{r}_1 - \underline{r}_3|^{-1}(\rho_c(3;3) - 2^{-1}\rho_c(1;3)P_{13})d\tau_3 \tag{15}$$

is an effective one-electron Hamiltonian including exchange. The symbol (ab|ab) in Eq. (14) is short for the integral

$$\int\psi_a^*(1)\psi_b^*(2)r_{12}^{-1}\ \psi_a(1)\psi_b(2)d\tau_1 d\tau_2$$

with a corresponding definition for (ab|ba). The Fock-Dirac density matrix is given by

$$\rho_c(1;3) = 2 \sum_{\nu,j} \phi_j^*(\underline{r}_1 - \underline{R}_\nu)\ \phi_j(\underline{r}_3 - \underline{R}_\nu). \tag{16}$$

Substituting ψ_a from Eq. (11) into the first term of Eq. (14) gives

$$(a|h|a) = N_a^2\ \{(\phi_a|h|\phi_a) + 2 \sum_{\nu,j} c_{\nu,j}(\phi_a|h|\phi_{\nu,j})$$

$$+ \sum_{\nu,j} c_{\nu,j}^2(\phi_{\nu,j}|h|\phi_{\nu,j})\}. \tag{17}$$

For the calculations discussed here the sum over ν runs over the six 1nn ions only and j runs over all of the free ion HF orbitals on each ion. The orbitals were taken from the work of Bagus.(5) The orbitals on different ions are assumed not to overlap. From the first term in Eq. (17), one obtains the kinetic, point ion, Coulomb and exchange energy for the trial function ϕ_a. The second and third terms give the effects of orthogonalization to the ion core functions and are the primary source of the repulsive interaction between the H^- ion and the neighboring ions of the host lattice.

Corresponding to Eq. (13), we now have

$$E(\beta_a,\beta_b,\underline{R}) = E_U(\beta_a,\beta_b,\underline{R}) + E_{int}(\beta_a,\beta_b,\underline{R}) + E_{cr}(\underline{R}) \tag{18}$$

for the electronic energy including the Coulomb interaction between nuclei. $\underline{R}$ is meant to represent the positions of the 1nn ions and of the H^- ion. First, $E(\beta_a,\beta_b,\underline{R})$ is minimized with respect to β_a, β_b, for various positions of the six 1nn ions in a breathing mode displacement with the H^- ion fixed at the defect site. This establishes new equilibrium positions for the 1nn ions when the defect is present. With the ions in their new positions, the energy is recalculated for a number of small displacements of the H^- ion in a [100] direction. According to the Born-Oppenheimer theorem this maps out an effective potential energy for the vibration of the H^- ion. The force constants calculated in this way are generally higher than the measured values and this tends to confirm the intuitive feeling that the polarization of the H^- ion may be an important effect.

Polarization of H^- Ion

The polarization of the hydrogen ion has been included in these calculations by writing

$$\psi_{pol}(\underline{1},\underline{2}) = N_{pol}\{\psi_U(\underline{1},\underline{2}) + \eta\psi_U'(\underline{1},\underline{2})\} \tag{19}$$

with $\psi_U(\underline{1},\underline{2})$ given by Eq. (10) and

$$\psi_U'(\underline{1},\underline{2}) = N_U'[\psi_a'(1)\psi_b'(2) + \psi_a'(2)\psi_b'(1)]\theta(s_1,s_2). \tag{20}$$

Here, ψ_b' is a p-function of the form of Eq. 11 but with

$$\phi_b' = N_b' \; r \exp(-\beta_b' r) \cos\gamma \tag{21}$$

and ψ_a' has the same form as ψ_a. γ is the angle between the principal symmetry axis of the p-function and the vector $\underline{r}$. The inclusion of polarization effects via Eq. (19) introduces three additional parameters into the calculation, i.e., β_a', β_b', and η. In practice, we have found that β_a and β_a' can be kept equal and that it is unnecessary to vary β_a and β_b from their values when polarization is not included. This leaves only β_b' and η to be determined by the new variation calculation.

Polarizability

We have calculated a mechanical polarizability defined, in analogy to the electrical polarizability, by

$$\underline{\mu} = \alpha_m \underline{F}. \tag{22}$$

$\underline{F}$ is a mechanical force per unit charge whose magnitude is taken to be

$$F = k\delta R_c/100. \tag{23}$$

δ is the displacement of the H^- ion given as a percentage of the new nearest neighbor distance, R_c. The dipole moment can be calculated using the wave function in Eq. (19) with the result that

$$\alpha_m = 2\eta(k\delta R_c)^{-1} N_{pol}^2 \langle\psi_U(1,2)|\underline{r}_1 + \underline{r}_2|\psi_U'(1,2)\rangle \times 100. \quad (24)$$

The diagonal terms vanish because of symmetry.

Second Nearest Neighbor Effects

As we shall see, the results given below seem to indicate that 2nn effects are important. An attempt was made to include 2nn interactions by using a Born-Mayer potential of the usual exponential form with parameters for the perfect crystal taken from Fumi and Tosi.[6] Polarization of the H^- ion was included by reducing the 2nn interaction in the same ratio that the 1nn interaction was reduced by polarization effects.

RESULTS AND DISCUSSION

Figures 1 and 2 have been taken from reference 4. Figure 1 shows the potential energy curves for the displacement, δ_c, of the

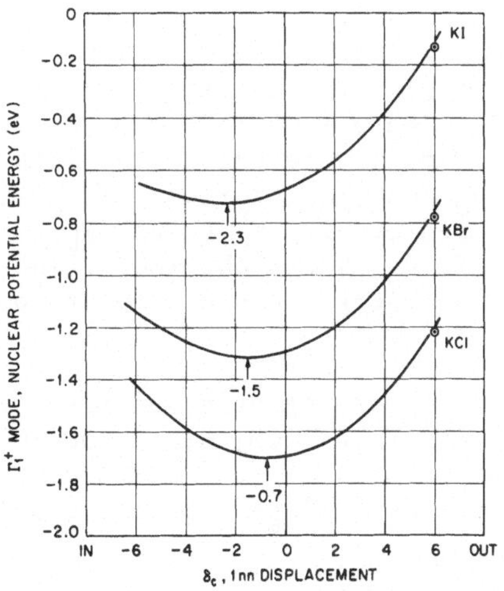

Fig. 1. The nuclear potential energy curve for the six nearest neighbor ions in the Γ_1^+ mode. δ_c is given as a percent of the nearest neighbor distance in the perfect crystal.

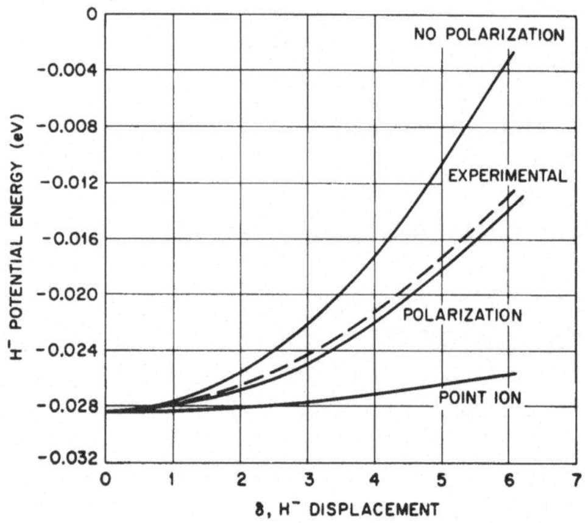

Fig. 2. Potential energy curves for the displacement of the H^- ion in KCl in various approximations. The curves are adjusted to have the same values at zero displacement. Here δ is given as a percent of the nearest neighbor distance in the crystal distorted by the presence of the U center.

1nn ions in a Γ_1^+ mode with the hydrogen ion fixed at its equilibrium position. δ_c is given as a percentage of the 1nn distance in the perfect crystal. The arrows indicate the minima. All points along the curves in Fig. 1 were calculated with values of β_a and β_b appropriate for δ_c near equilibrium. The validity of this approximation was checked by varying β_a and β_b at $\delta_c = 6$ until the actual minimum in $E(\beta_a, \beta_b, \underline{R})$ was obtained. The encircled points in Fig. 1 indicate the energy change brought about by reminimization.

Figure 2 shows various potential energy curves for the displacement of the H^- ion in a [100] direction in KCl. The dashed line gives the "experimental" curve obtained by using the "experimental" force constant ($k = M_o' \omega_{exp}^2$) in the equation

$$\Delta E = \frac{1}{2} k[(1 + \delta_c/100)R_o(\Delta\delta)]^2.$$

$(1 + \delta_c/100)R_o$ is the 1nn distance when the U center is present. The point ion curve results from a calculation in which the electronic structure of all the ions of the host crystal is neglected when calculating $E_U + E_{int}$.

Various values of the local mode frequency and the polarizability for seven crystals are shown in Table I. ω_{exp} is the experimental frequency, ω_{np} the frequency without polarization of the H^- ion, ω_p that with polarization, and ω_p^{2nn} the frequency when 2nn effects are included in the manner briefly described above.

The results given in Fig. 2 show clearly that the point ion model is inadequate, as was to be expected. They also indicate the importance of including the polarization of the H^- ion. The results in Table I, on the other hand, indicate that 2nn effects are important especially in those crystals with large ratios of anion to cation radii. They also suggest, however, that our treatment of 2nn effects is too simple. The discrepancies between ω_p^{2nn} and ω_{exp} in KI, NaBr, and NaI do not seem very consistent.

The results on the polarizability show that it changes considerably from crystal to crystal. Generally speaking, it increases with increasing nearest neighbor distance, although the size of the positive ion also appears to play a role. The suggestion is that the Madelung potential plays the most important role in determining the compactness and rigidity of the H^- ion in a crystal.

An attempt is now being made to improve the calculation of 2nn effects in the fluorides and chlorides. Unfortunately, free ion HF orbitals of the Bagus type do not, to my knowledge, exist yet for the bromine and iodine negative ions.

	ω_{exp}	ω_{np}	ω_p	ω_p^{2nn}	α_m (in Å^3)
NaF	16.19	19.35	15.14	15.71	2.12
NaCl	10.61	11.31	9.32	10.74	3.82
NaBr	9.39	9.71	8.04	9.90	4.38
NaI	8.12	7.73	6.39	8.77	5.39
KCl	9.29	11.87	9.05	9.37	6.95
KBr	8.27	10.27	7.99	8.44	8.00
KI	7.31	8.37	6.65	7.37	10.04

TABLE I. Results for the local mode frequency in various approximations given in the text and for the mechanical polarizability. All frequencies are given in units of 10^{13} radians/sec.

REFERENCES

1. See J. H. Schulman and W. Dale Compton, Color Centers in Solids (Pergamon Press, Inc., New York, 1962).

2. See, for example, G. Schaefer, Phys. Chem. Solids 12, 233 (1960); B. Lengler and W. Ludwig, Z. Physik 171, 272 (1963); M. V. Klein, Phys. Rev. 131, 1502 (1963); R. Fieschi, G. F. Nardelli, and N. Terzi, Phys. Rev. 138, A203 (1965); H. Bilz, D. Strauch, and B. Fritz, Journal de Physique 27 (Supplement), C2-3 (1966).

3. The work discussed here was done in collaboration with U. Öpik and R. L. Gilbert.

4. R. F. Wood and R. L. Gilbert, Phys. Rev. (in press).

5. P. S. Bagus, Phys. Rev. 139, A619 (1965).

6. M. P. Tosi and F. G. Fumi, J. Phys. Chem. Solids 25, 45 (1964).

THE JAHN TELLER EFFECT FOR THE U-CENTER LOCAL MODE

Max Wagner

1. Institut für Theoretische Physik

University of Stuttgart, Germany

1. Introduction

The U-center local mode has a much higher frequency than those of the lattice oscillators, whence we may regard the coupled system as a Jahn-Teller system. In previous calculations [1,2] of the sidebands in the infrared U-center absorption the coupling has been restricted to the A_{1g}- and E_g-type modes of the lattice. For this model no Jahn-Teller effect can appear, because the local mode one-quantum states do no mix dynamically. However, symmetry arguments do also allow a coupling to T_{2g}-type lattice modes. Such a coupling is essential for the Jahn-Teller effect to be present. In a forthcoming paper [3] it is shown that this additional coupling is indeed of the same order as the coupling to the A_{1g}- and E_g- type modes.

Apart from the major effect of a T_{2g}-contribution to the sidebands the Jahn-Teller coupling will give rise to alterations in the structural form of the high-energy sideband, which is due to the dynamical mixing of degenerate terms of the local oscillator (dynamical Jahn-Teller effect). This is, because the final state for the upper sideband absorption is a 2-quantum state (1 local + 1 lattice quantum), which splits into the irreducible states of the products $T_{1u} \times (A_{1g}, E_g, T_{2g})$ yielding the Jahn-Teller states T_{1u} ; T'_{1u}, T_{2u} ; A_{2u}, E_u, T''_{1u}, T_{2u}. The transition is to the 3 T_{1u}-states only, but since they lie in the same energy region, they are strongly mixed by the Jahn-Teller coupling. This means that the A_{1g}, E_g and T_{2g}

contributions to the upper sideband are no longer superposed linearly. Instead, they are combined to one single new structural form, which is different from the linear superposition. To account for this one would be forced to employ degenerate perturbation theory for the quasi- continuous sequence of lattice states, which is extremely cumbersome. To avoid this procedure, a modified method of moments is presented, which allows the exact evaluation of all moments of the upper sideband, and reveals the strength of the dynamical Jahn-Teller effect. No Jahn-Teller states are involved in the low-energy sideband absorption, whence the structural form remains a linear superposition of the contributions from A_{1g}, E_g and T_{2g}.

2. Anharmonic Interaction and Absorption

The absorption in the one-phonon sideband can be predominantly attributed to a coupling, which is quadratic in the normal coordinates x_i of the local oscillator and linear in the other normal coordinates $Q(\Gamma rj)$. (Γ=irreducible basis, r = index of the multiplicity of Γ, j = index of degeneracy). We may write this coupling in the form

$$W = \sum_{\varkappa} \hbar\omega(\varkappa)p(x;\varkappa)A(\varkappa), \qquad \varkappa \equiv \Gamma rj \tag{1}$$

where $A(\varkappa)$ is the dimensionless normal coordinate, $A(\varkappa) = (2\omega(\varkappa)/\hbar)^{1/2} . Q(\varkappa) = a(\varkappa) + a(\varkappa)^+$, i.e. the sum of a phonon creation- and annihilation operator. Group theory imposes a strong restriction onto the form of the quadratic function $p(x;\varkappa)$ of x_i, which follows from the invariance of W under the operations of the group O_h and yields

$$p(x;\Gamma rj) = C_r(\Gamma)\, w_{\Gamma j}(x) \tag{2}$$

where

$$w_{A_{1g}}(x) = (x^2+y^2+z^2)/\sqrt{3} \;;$$

$$w_{E_{g,1}}(x) = (2z^2-(x^2+y^2))/\sqrt{6} \;;\; w_{E_{g,2}}(x) = (x^2-y^2)/\sqrt{2}$$

$$w_{T_{2g,1}}(x) = yz/\sqrt{2};\; w_{T_{2g,2}}(x) = zx/\sqrt{2} \;;\; w_{T_{2g,3}}(x) = xy/\sqrt{2} \tag{3}$$

Hence only modes of the types A_{1g}, E_g and T_{2g} can couple to the local mode. In view of the fact that the light mass of the H^--ion has a big amplitude for the local mode, we may assume the essential anharmonicity to be located between the central ion and the nearest neighbour ions [1,2]. Then the number of not yet defined parameters in (2) is only 3,

$$C_r(\Gamma) = C(\Gamma) \langle \sigma(\Gamma j) | \eta(\Gamma r j) \rangle \quad , \quad \Gamma = A_{1g}, E_g, T_{2g} \qquad (4)$$

where $\sigma(\Gamma j)$ are the orthonormal symmetry vectors belonging to the nearest neighbours and $\eta(\Gamma rj)$ are the eigenvectors of the lattice modes. The parameters $C(\Gamma)$ are easily found for any analytic coupling law [3]. E.g. for $W \sim |\underset{\sim}{x} - \underset{\sim}{X}|^{-n}$, ($\underset{\sim}{X}$=position of the lattice ions), we have [3]

$$C(A_{1g}) = -S_n(n-1)/2, \; C(E_g) = -S_n(n+2)/2, C(T_{2g}) = 2S_n \qquad (5)$$

where S_n is a common factor. For n=1 (Coulomb law) we have $C(A_{1g}):C(E_g):C(T_{2g})=0 : -3 : +4$, and for a Born-Mayer potential with n=10 one would have $C(A_{1g}):C(E_g):C(T_{2g}) = -9:-12:+4$, which reveals the fact that in both extremal cases $C(T_{2g})$ cannot be neglected.

Without considering the complications due to the Jahn-Teller effect, the functional form of the infrared absorption of the U-center for T=0 can now shown to be [3]

$$\tau(\omega) = \frac{1}{N}\left[\delta(\omega) + \sum_i \langle p_{xi}(\varkappa) | \delta(\omega - \omega(\varkappa)) | p_{ix}(\varkappa) \rangle + \cdots\right] , \qquad (6)$$

where $\langle p_{xi} | \Phi | p_{ix} \rangle$ abbreviates $\sum_\varkappa p_{xi}(\varkappa) \, \Phi(\varkappa) p_{ix}(\varkappa)$ and $\tau(\omega)$ is normalized to one, i.e. $N = \exp(-\sum_i \langle p_{xi} | p_{ix} \rangle)$. The origin of energy is taken to be the position of the main line. Further on, $p_{xi}(\varkappa)$ is the difference of matrix elements

$$p_{xi}(\varkappa) = \langle i | p(x;\varkappa) | x \rangle - \langle 0 | p(x;\varkappa) | 0 \rangle \delta_{xi} \qquad (7)$$

where $|i\rangle$ are the one-quantum states of the local oscillator (i=x,y,z), and $|0\rangle$ is its groundstate. The functional form of the first sideband, as given by (6) and including the T_{2g}-contribution, has been calculated recently [3] and is shown in Fig. 1 for KBr. In this calculation the model for the dynamical disturbance of the lattice has been adopted from Gethins et al [2] with a suitable choice of the spring constants, whereas the free parameters of (4) have been chosen to yield $C(A_{1g}):C(E_g):C(T_{2g})=-1:-1:+1$. The dotted line is the experimental curve, modified according to the investigation of Bilz et al. [4] .

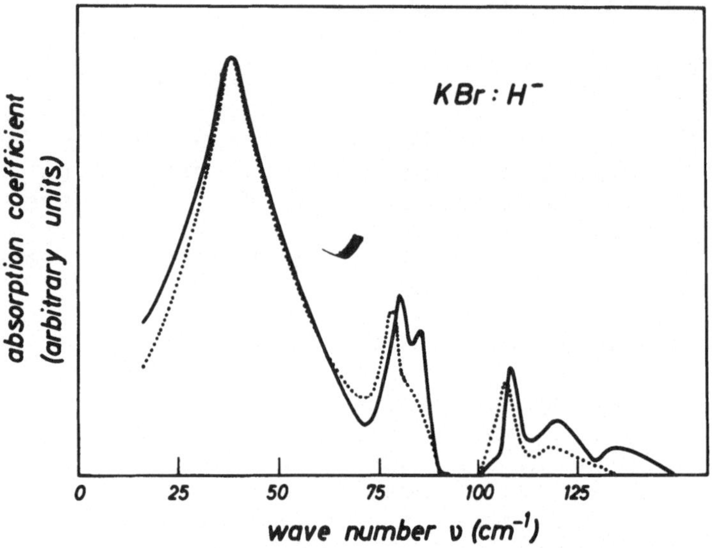

Fig.1. Sideband absorption of the U-center local mode in KBr.———— theoretical absorption curve for all parameters C(Γ) chosen to be equal and a suitable choice of the dynamical disturbance. ········ modified experimental absorption curve [4] .

3. Method of Moments for Jahn-Teller States.

For a general set of final states F and one single initial state I (i.e. for T=0) the total moments of the absorption may be written in the form

$$M_n = \Sigma_F \langle I|P_x(H-E_I-\varepsilon_o)^n|F\rangle\langle F|P_x|I\rangle \qquad (8)$$

where P_x is the dipole operator of the light field, which is considered to be polarized in the x-direction. In our case the set of final states is the sequence belonging to the first excited state of the local oscillator. Being Jahn-Teller states they may be written in the form

$$|F\rangle = \Sigma_i |i\rangle|\chi^{(i)}_{\gamma j}(Q)\rangle \qquad (9)$$

where γ designates the Jahn-Teller state and j its degeneracy. In close analogy to the Jahn-Teller effect for electronic states [5] it is assumed in (9) that the eigenvalues of the local oscillator are so far apart from each other that states of different energies are not mixed by the coupling (1). Then $\chi^{(i)}_{\gamma j}(Q)$ are the eigenfunctions of the equation

$$\Sigma_{i'} H_{ii'}(Q)\chi^{(i')}_{\gamma j}(Q) = E_\gamma \chi^{(i)}_{\gamma j}(Q) \qquad (10)$$

where integration over the coordinates x_i of the local oscillator has been performed,

$$H_{ii'}(Q) = \langle i|H(x,Q)|i'\rangle \qquad (11)$$

The functions $\chi^{(i)}_{\gamma j}(Q)$ may be chosen to satisfy orthonormality and closure properties of the form

$$\Sigma_i \langle \chi^{(i)}_{\gamma j}|\chi^{(i)}_{\gamma' j'}\rangle = \delta_{\gamma\gamma'}\delta_{jj'} \qquad (12a)$$

$$\Sigma_{\gamma j} \chi^{(i)}_{\gamma j}(Q)\chi^{(i')}_{\gamma j}(Q')^* = \delta_{ii'}\delta(Q-Q') \qquad (12b)$$

The latter may be employed to simplify the expression (8) for the moments,

$$M_n = M_0 \sum_{i_1 \dots i_{(n-1)}} \langle 0|[H_{xi_1} - (E_I+\varepsilon_0)\delta_{xi_1}]\cdots[H_{i_{(n-1)}x} - (E_I+\varepsilon_0)\delta_{i_{(n-1)}x}]|0\rangle \qquad (13)$$

where ε_0 is the position of the main line and $M_0 = |\langle x|P_x|o\rangle|^2$. The state $|o\rangle$ in (13) designates the groundstate of the lattice operator $H_{00}(Q)$, where $H_{00}(Q)$ is defined analogous to (11) for the groundstate of the local oscillator, such that the total initial state is $|I\rangle = |o\rangle|o\rangle$. The matrix element in (13) is easily evaluated, whence all moments can be calculated exactly. However, to choose the origin of energy at the position of the main line, as indicated in (13), we have to know also ε_0 explicitly, $\varepsilon_0 = E_{10} - E_{00}$, where E_{10} is the energy of the state with one quantum of the local mode and no lattice quanta. E_{10} can be calculated in a straightforward way by non-degenerate perturbation theory. This is firstly, because E_{10} is a singular state with the density of lattice states turning to zero in the immediate neighbourhood of E_{10}, and secondly, because the wave-functions of E_{10} are already properly symmetrized in zeroth approximation, being a basis to the representation T_{1u} . The result for ε_o is, up to sixth order approximation

$$\varepsilon_o = \hbar\omega_L - \Delta\varepsilon_o - \Delta\Delta\,\varepsilon_o + O(\Delta\Delta\Delta\,\varepsilon_o) \tag{14}$$

where

$$\Delta\varepsilon_o = \sum_i < p_{xi}(x)|\hbar\omega(x)|p_{ix}(x) > \tag{15}$$

and $\Delta\Delta\varepsilon_o$ is the Jahn-Teller term

$$\Delta\Delta\varepsilon_o = \sum_i \langle\langle |p_{xx}(x_1) - p_{ii}(x_1)|^2 |p_{xi}(x_2)|^2 \hbar\omega(x_2)\frac{\omega(x_2)}{\omega(x_1)+\omega(x_2)}\rangle_1\rangle_2$$
$$+\sum_{ii'i''} < p_{xi}\, p_{ii'}|\hbar\omega|p_{ii''}\, p_{i''x}>_2 - \sum_i < p_{xi}|p_{ix}>_2 \sum_{i'} < p_{xi'}|\hbar\omega|p_{i'x}>_2 \tag{16}$$

Here the index 1 stands for the A_{1g}- and E_g-vibrations and the index 2 for the T_{2g}-vibrations. $\langle\ldots\rangle_1$ and $\langle\ldots\rangle_2$ indicate the respective summations. In the same way, employing the disturbed eigenfunctions of E_{1o}, we are able to write down the zeroth moment of the main line, which is the total transition probability to E_{1o},

$$\mu_o^{(o)} = \frac{M_o^{(o)}}{M_o} = \exp\left[-\sum_i < p_{xi}|p_{ix}> - \Delta\Delta\right] + O(\Delta\Delta\Delta) \tag{17}$$

where $\Delta\Delta$ is again the Jahn-Teller term,

$$\Delta\Delta = \sum_i \langle\langle |p_{xx}(x_1) - p_{ii}(x_1)|^2 |p_{xi}(x_2)|^2 \frac{\omega(x_1)(\omega(x_1)+2\omega(x_2))}{[\omega(x_1)+\omega(x_2)]^2}\rangle_1\rangle_2$$
$$-\frac{1}{2!}\left[\sum_i < p_{xi}|p_{ix}>_2\right]^2 + 3\sum_i < p_{xi}|\frac{1}{\hbar\omega}|p_{ix}>_2 \sum_{i'} < p_{xi'}|\hbar\omega|p_{i'x}>_2$$
$$-\sum_{ii'} \langle\langle |p_{xi}(x_2)|^2 |p_{ii'}(x_2')|^2 \frac{\omega(x_2')}{\omega(x_2)+\omega(x_2')}\rangle_2\rangle_{2'}$$
$$-3\sum_{ii'i''} < p_{xi}\, p_{ii'}|p_{i'i''}\, p_{i''x}>_2 - \frac{1}{4}\sum_{ii'} <|p_{xi}\, p_{ii'}|^2>_2 \tag{18}$$

Both in (16) and (18) the first term arises from a mixing of the (A_{1g}, E_g)-bands with the T_{2g}-band. Thus these bands are no longer separable. The other terms come from an autocorrelation of the T_{2g}-band and tend to change its structural form. All terms of $\Delta\Delta\,\varepsilon_o$ and $\Delta\Delta$ originate from a dynamical mixing of the local mode state $|x\rangle$ with the states $|y\rangle$ and $|z\rangle$, which means that the Jahn-Teller states are no longer a product of a single local mode state $|i\rangle$ with a state of lattice modes, but a superposition of the form (9).

Employing the knowledge of ε_0 and $\mu_0^{(0)}$ and the argument that the Jahn-Teller correction of the one-phonon sideband is of the order of the two-phonon sideband, etc., we are in the position to extract from the total moments those of the one-phonon band. The result is

$$\mu_0^{(1)} = \mu_0^{(0)}\left[\sum_i \langle p_{xi}|p_{xi}\rangle + \Delta\Delta\right] + \text{(6th order terms)}, \quad (19a)$$

$$\mu_1^{(1)} = \mu_0^{(0)}\left[\sum_i \langle p_{xi}|\hbar\omega|p_{ix}\rangle + \Delta\Delta\,\varepsilon_0\right] + \text{(6th order terms)}, \quad (19b)$$

$$\mu_2^{(1)} = \mu_0^{(0)}\left[\sum_i \langle p_{xi}|(\hbar\omega)^2|p_{xi}\rangle\right] + \text{(6th order terms)}, \quad (19c)$$

$$\mu_3^{(1)} = \mu_0^{(0)}\left[\sum_i \langle p_{xi}|(\hbar\omega)^3|p_{xi}\rangle\right] + \text{(6th order terms)}, \quad (19d)$$

$$\mu_4^{(1)} = \mu_0^{(0)}\left[\sum_i \langle p_{xi}|(\hbar\omega)^4|p_{ix}\rangle + \Delta\Delta\,\varepsilon^4\right] + \text{(6th order terms)} \quad (19e)$$

For each moment the first term is the one which would be present alone, if the bands of the 3 symmetry types were completely disentangled, i.e. the functional form of the sideband given by the second term in eq.(6). It is interesting to note that the second and third moments do not have a Jahn-Teller term of fourth order. The term $\Delta\Delta\varepsilon^4$ consists of parts, which correspond to the contributions to $\Delta\Delta\varepsilon_0$ and $\Delta\Delta$ (eqs.(16) and (18)), i.e. there are terms, which belong to the mixing of the(A_{1g}, E_g)-bands with the T_{2g}-band, and other ones mixing the T_{2g}-states themselves, and again all parts of $\Delta\Delta\varepsilon^4$ originate from the dynamical mixing of the local mode states $|x\rangle$, $|y\rangle$, $|z\rangle$. Taking account of the specific forms (2) and (3) there are several relations between the distributions $p_{ii'}(\varkappa)$, e.g. $p_{xy}(T_{2g}r3)=p_{yz}(T_{2g}r1)=p_{zx}(T_{2g}r2)$, etc. These are used to simplify $\Delta\Delta\varepsilon^4$ to the form

$$\Delta\Delta\varepsilon^4 = 6\left[\langle p_{xx}|(\hbar\omega)^2|p_{xx}\rangle_{(A_{1g})} - \langle p_{xx}|(\hbar\omega)^2|p_{xx}\rangle_{(E_g)}\right]\langle p_{xy}|(\hbar\omega)^2|p_{yx}\rangle_{(T_{2g})} - 2\left[\langle p_{xy}|(\hbar\omega)^2|p_{yx}\rangle_{(T_{2g})}\right]^2 \quad (20)$$

4. Summary and Discussion

From a recent investigation [3] we have concluded that a coupling of the local oscillator to T_{2g}-type lattice oscillators is of the same order of magnitude as that to the (A_{1g}, E_g)-type modes. This new coupling gives rise to a dynamical Jahn-Teller effect, consisting in a dynamical mixing of local mode states, by means of which the high energy sidebands of different symmetry species are correlated. This correlation produces structural changes, the magnitude of which has been investigated by means of a new method of moments for Jahn-Teller states. It is found that the relative change of the moments (with the exception of the 2nd and 3rd ones) is of the same order as the relation of the integrated intensities of the sideband to the main line, which is of the order of 10% [6]. The disagreement of numerical sideband calculations [1,2,3] with experiment is still too great to reveal the Jahn-Teller effect, but the accuracy may be expected to be improved sufficiently in the near future. The knowledge of moments may then be used to establish approximate modifications of the functional form (6) for the sideband. In conclusion it should be emphasized that our method is not restricted to an anharmonic coupling of the form (1). A fourth order coupling of the type x^2Q^2 may also be assumed to make a considerable contribution to the Jahn-Teller effect and play its role in the future investigations.

References:

[1] Th. Timusk a.M.V. Klein, Phys.Rev. 141,664 (1966)
[2] T.Gethins, Th.Timusk a. E.J. Woll, Phys.Rev. 157, 744 (1967)
[3] D.Kühner a. M.Wagner, Z.Phys., in print
[4] H.Bilz, R.Zeyher a.R.R.Wilmer, phys.stat.sol. 20 K167 (1967)
[5] H.C.Longuet-Higgins, Adv.in Spectroscopy 2,429 (1961)
[6] B.Fritz, U. Gross a. D.Baeuerle, phys.stat.sol. 11, 231 (1965)

SHELL MODEL TREATMENT OF THE VIBRATIONS OF SUBSTITUTIONAL IMPURITIES IN CRYSTALS. I. U-CENTER LOCALIZED MODES

John B. Page, Jr., and Dieter Strauch

Institut für Theoretische Physik der Universität

Frankfurt/Main

INTRODUCTION

In this paper we will derive a shell model theory for the harmonic vibrations of crystals containing substitutional defects and will discuss some of the results of applying the theory to the problem of the localized modes of U-centers in alkali halide crystals. The method is an extension of Lifshitz's formal treatment [1] of the vibrations of perturbed lattices, the main difference being that our treatment deals directly with the altered shell model parameters associated with the presence of defects rather than with altered formal force constants. This results in, among other features, the ability to treat changed long-range Coulomb forces arising from defects having different core and shell charges than those of the ions they replace. Furthermore, the method is equally applicable to perturbed band modes and localized modes, the application to the study of the perturbed band modes responsible for the sidebands in the infrared spectrum of U-centers being given in the next paper. The effects of "breathing," that is, of adiabatic spherical deformations of the ions are included in the theory as we have used it, without increasing the number of independent parameters. Schröder [2] has emphasized the importance of such effects for describing the vibrations of unperturbed alkali halides. Owing to space limitations, however, we will derive here only the results for the case of a system described by the ordinary shell model [3], the inclusion of breathing involving a straightforward generalization. The results appropriate to the breathing shell model are given in a forthcoming publication [4].

THEORY

The harmonic Hamiltonian for a system described by the ordinary shell model may be written

$$(1) \quad H_H = \frac{1}{2}\left[\tilde{\underline{u}}_c \underline{\underline{M}}_c \dot{\underline{u}}_c + (\tilde{\underline{u}}_c \; \tilde{\underline{u}}_s) \begin{pmatrix} \underline{\underline{\Phi}}_{cc} & \underline{\underline{\Phi}}_{cs} \\ \tilde{\underline{\underline{\Phi}}}_{cs} & \underline{\underline{\Phi}}_{ss} \end{pmatrix} \begin{pmatrix} \underline{u}_c \\ \underline{u}_s \end{pmatrix} \right] ,$$

where $\tilde{\underline{u}}_c$ and $\tilde{\underline{u}}_s$ are the transposes of the configuration space vectors giving, respectively, the displacements of the cores and shells from their equilibrium positions, the dots denote time differentiation, $\underline{\underline{M}}_c$ is a diagonal matrix of the core masses, and the $\underline{\underline{\Phi}}$'s contain the harmonic force constants. The shells are massless in accordance with the Born-Oppenheimer approximation. This Hamiltonian is easily transformed to the diagonal form $H_H = \frac{1}{2}\sum_f (\dot{d}_f^2 + \omega_f^2 d_f^2)$ by the linear transformations $\underline{u}_c = \sum_f \underline{\chi}_c(f) d_f$ and $\underline{u}_s = \sum_f \underline{\chi}_s(f) d_f$ with the ω_f^2 's and $\underline{\chi}(f)$'s being the eigenvalues and normalized eigenvectors of the eigenvalue equation

$$(2) \quad \begin{pmatrix} \underline{\underline{\Phi}}^R_{cc} + \underline{\underline{X}}\,\underline{\underline{C}}\,\underline{\underline{X}} - \omega_f^2 \underline{\underline{M}}_c & \underline{\underline{\Phi}}^R_{cs} + \underline{\underline{X}}\,\underline{\underline{C}}\,\underline{\underline{Y}} \\ \tilde{\underline{\underline{\Phi}}}^R_{cs} + \underline{\underline{Y}}\,\underline{\underline{C}}\,\underline{\underline{X}} & \underline{\underline{\Phi}}^R_{ss} + \underline{\underline{Y}}\,\underline{\underline{C}}\,\underline{\underline{Y}} \end{pmatrix} \begin{pmatrix} \underline{\chi}_c(f) \\ \underline{\chi}_s(f) \end{pmatrix} = 0 .$$

Here we have split $\underline{\underline{\Phi}}_{cc}$, $\underline{\underline{\Phi}}_{cs}$, and $\underline{\underline{\Phi}}_{ss}$ into their short-range repulsive and long-range Coulomb parts; thus $\underline{\underline{\Phi}}^R_{cc} + \underline{\underline{X}}\,\underline{\underline{C}}\,\underline{\underline{X}}$ is equal to $\underline{\underline{\Phi}}_{cc}$, etc., where $\underline{\underline{X}}$ and $\underline{\underline{Y}}$ are diagonal matrices representing the core and shell charges (in units of e) and $\underline{\underline{C}}$ is the matrix representing the Coulomb interactions. The $\underline{\chi}_c(f)$'s are normalized according to $\tilde{\underline{\chi}}_c(f)\underline{\underline{M}}_c \underline{\chi}_c(f') = \delta_{ff'}$.

Now equation (2) is in the form $(\underline{\underline{\Phi}} - \omega_f^2 \underline{\underline{M}})\underline{\chi}(f) = 0$, and for a crystal containing substitutional impurities we could rewrite this equation as

$$\underline{\chi}(f) = -(\underline{\underline{\Phi}}_0 - \omega_f^2 \underline{\underline{M}}_0)^{-1}(\Delta\underline{\underline{\Phi}} - \omega_f^2 \Delta\underline{\underline{M}})\underline{\chi}(f)$$

where $\underline{\underline{\Phi}}$ and $\underline{\underline{M}}$ have been written as $\underline{\underline{\Phi}}_0 + \Delta\underline{\underline{\Phi}}$ and $\underline{\underline{M}}_0 + \Delta\underline{\underline{M}}$, $\underline{\underline{\Phi}}_0$ and $\underline{\underline{M}}_0$ being appropriate to the perfect lattice. This would be similar to the usual Lifshitz procedure [1] except that the $\underline{\underline{\Phi}}$ for equation (2) is not the formal force constant matrix, that matrix being the effective core-core force constant matrix obtained by first eliminating $\underline{\chi}_s(f)$ from the shell model equations of (2). The Lifshitz procedure, however, is a useful one only when the non-zero elements of the perturbing matrix $\underline{\underline{C}}^f \equiv \Delta\underline{\underline{\Phi}} - \omega_f^2 \Delta\underline{\underline{M}}$ are localized about the defects, and the Coulomb terms in equation (2) generally prevent this from being so, even though $\Delta\underline{\underline{X}}$ and $\Delta\underline{\underline{Y}}$

may themselves be localized. To avoid this difficulty, we can rewrite equation (2) as

$$(3) \qquad \begin{pmatrix} \underline{\underline{\Phi}}^R_{cc} - \omega_f^2 \underline{\underline{M}}_c & \underline{\underline{\Phi}}^R_{cs} & \underline{\underline{X}} \\ \underline{\underline{\tilde{\Phi}}}^R_{cs} & \underline{\underline{\Phi}}^R_{ss} & \underline{\underline{Y}} \\ \underline{\underline{\tilde{X}}} & \underline{\underline{\tilde{Y}}} & -\underline{\underline{e}}^{-1} \end{pmatrix} \begin{pmatrix} \underline{\chi}_c(f) \\ \underline{\chi}_s(f) \\ \underline{\chi}_N(f) \end{pmatrix} = 0,$$

this being easily verified by using the third equation to eliminate $\underline{\chi}_N(f)$. Now for the case when the Coulomb matrix is unperturbed by the presence of defects, we can apply the Lifshitz procedure to equation (3) and obtain a perturbing matrix that is localized to the same extent that the core and shell charge changes and the short-range force constant changes are localized. Thus we may write equation (3) as $\underline{\chi}(f) = -\underline{\underline{G}}^f \underline{\underline{C}}^f \underline{\chi}(f)$, where the perturbing matrix is

$$(4) \qquad \underline{\underline{C}}^f = \begin{pmatrix} \Delta\underline{\underline{\Phi}}^R_{cc} - \omega_f^2 \Delta\underline{\underline{M}}_c & \Delta\underline{\underline{\Phi}}^R_{cs} & \Delta\underline{\underline{X}} \\ \Delta\underline{\underline{\tilde{\Phi}}}^R_{cs} & \Delta\underline{\underline{\Phi}}^R_{ss} & \Delta\underline{\underline{Y}} \\ \Delta\underline{\underline{\tilde{X}}} & \Delta\underline{\underline{\tilde{Y}}} & \underline{\underline{0}} \end{pmatrix}$$

and the amplitude vector $\underline{\chi}(f)$ and the Green's function matrix are

$$(5) \qquad \underline{\chi}(f) = \begin{pmatrix} \underline{\chi}_c(f) \\ \underline{\chi}_s(f) \\ \underline{\chi}_N(f) \end{pmatrix}, \quad \underline{\underline{G}}^f = \begin{pmatrix} \underline{\underline{\Phi}}^R_{occ} - \omega_f^2 \underline{\underline{M}}_{oc} & \underline{\underline{\Phi}}^R_{ocs} & \underline{\underline{X}}_o \\ \underline{\underline{\tilde{\Phi}}}^R_{ocs} & \underline{\underline{\Phi}}^R_{oss} & \underline{\underline{Y}}_o \\ \underline{\underline{\tilde{X}}}_o & \underline{\underline{\tilde{Y}}}_o & -\underline{\underline{e}}^{-1} \end{pmatrix}^{-1}.$$

The procedures from this point on are the same as those of the usual Lifshitz theory; in particular, the perturbed frequencies are the solutions of the secular equation $\left| \underline{\underline{I}} + \underline{\underline{G}}^f_{II} \underline{\underline{C}}^f_{II} \right| = 0$, where $\underline{\underline{G}}^f_{II}$ and $\underline{\underline{C}}^f_{II}$ are the submatrices of $\underline{\underline{G}}^f$ and $\underline{\underline{C}}^f$ in the impurity space, defined by those particles with which non-zero elements of $\underline{\underline{C}}^f$ are associated.

It remains to evaluate the elements of $\underline{\underline{G}}^f$. Using the defining equation for $\underline{\underline{G}}^f$, namely

$$\begin{pmatrix} \underline{\underline{G}}^f_{cc} & \underline{\underline{G}}^f_{cs} & \underline{\underline{G}}^f_{cN} \\ \underline{\underline{\tilde{G}}}^f_{cs} & \underline{\underline{G}}^f_{ss} & \underline{\underline{G}}^f_{sN} \\ \underline{\underline{\tilde{G}}}^f_{cN} & \underline{\underline{\tilde{G}}}^f_{sN} & \underline{\underline{G}}^f_{NN} \end{pmatrix} \begin{pmatrix} \underline{\underline{\Phi}}^R_{occ} - \omega_f^2 \underline{\underline{M}}_{oc} & \underline{\underline{\Phi}}^R_{ocs} & \underline{\underline{X}}_o \\ \underline{\underline{\tilde{\Phi}}}^R_{ocs} & \underline{\underline{\Phi}}^R_{oss} & \underline{\underline{Y}}_o \\ \underline{\underline{X}}_o & \underline{\underline{Y}}_o & -\underline{\underline{C}}^{-1} \end{pmatrix} = \begin{pmatrix} \underline{\underline{I}} & \underline{\underline{O}} & \underline{\underline{O}} \\ \underline{\underline{O}} & \underline{\underline{I}} & \underline{\underline{O}} \\ \underline{\underline{O}} & \underline{\underline{O}} & \underline{\underline{I}} \end{pmatrix},$$

the following results are readily obtained:

$$\underline{\underline{G}}^f_{cc} = (\underline{\underline{\Phi}}_{occ} - \underline{\underline{\Phi}}_{ocs}\underline{\underline{\Phi}}^{-1}_{oss}\underline{\underline{\tilde{\Phi}}}_{ocs} - \omega_f^2\underline{\underline{M}}_{oc})^{-1}, \quad \underline{\underline{G}}^f_{cs} = -\underline{\underline{G}}^f_{cc}\underline{\underline{\Phi}}_{ocs}\underline{\underline{\Phi}}^{-1}_{oss},$$

$$\underline{\underline{G}}^f_{ss} = \underline{\underline{\Phi}}^{-1}_{oss}\underline{\underline{\tilde{\Phi}}}_{ocs}\underline{\underline{G}}^f_{cc}\underline{\underline{\Phi}}_{ocs}\underline{\underline{\Phi}}^{-1}_{oss} + \underline{\underline{\Phi}}^{-1}_{oss},$$

(6)
$$\underline{\underline{G}}^f_{cN} = (\underline{\underline{G}}^f_{cc}\underline{\underline{X}}_o + \underline{\underline{G}}^f_{cs}\underline{\underline{Y}}_o)\underline{\underline{C}}, \quad \underline{\underline{G}}^f_{sN} = (\underline{\underline{\tilde{G}}}^f_{cs}\underline{\underline{X}}_o + \underline{\underline{G}}^f_{ss}\underline{\underline{Y}}_o)\underline{\underline{C}},$$

$$\underline{\underline{G}}^f_{NN} = (\underline{\underline{\tilde{G}}}^f_{cN}\underline{\underline{X}}_o + \underline{\underline{\tilde{G}}}^f_{sN}\underline{\underline{Y}}_o - \underline{\underline{I}})\underline{\underline{C}}.$$

The first equation in (6) tells us that $\underline{\underline{G}}^f_{cc}$ is just the usual Lifshitz Green's function since $\underline{\underline{\Phi}}_{occ} - \underline{\underline{\Phi}}_{ocs}\underline{\underline{\Phi}}^{-1}_{oss}\underline{\underline{\tilde{\Phi}}}_{ocs}$ is the unperturbed formal force constant matrix for the shell model. As is well known, $\underline{\underline{G}}^f_{cc}$ may therefore be written in terms of the unperturbed eigenfrequencies and eigenvectors as

(7)
$$\underline{\underline{G}}^f_{cc} = \underline{\underline{S}}^f_{cc},$$

where $\underline{\underline{S}}^f_{PQ}$ is given by

$$\underline{\underline{S}}^f_{PQ} = \sum_{qj} \underline{\chi}_P\left(\tfrac{q}{j}\right)\underline{\chi}^{+}_Q\left(\tfrac{q}{j}\right)\left(\omega_j^2(q) - \omega_f^2\right)^{-1}.$$

For generality, we have written the unperturbed eigenvectors as complex quantities. Now the second of the shell model equations in (2), when applied to the perfect lattice, gives $\underline{\chi}_s\left(\tfrac{q}{j}\right) = -\underline{\underline{\Phi}}^{-1}_{oss}\underline{\underline{\tilde{\Phi}}}_{ocs}\underline{\chi}_c\left(\tfrac{q}{j}\right)$, while the third equation in (3) gives the unperturbed $\underline{\chi}_N$'s as $\underline{\chi}_N\left(\tfrac{q}{j}\right) = \underline{\underline{C}}\left(\underline{\underline{X}}_o\underline{\chi}_c\left(\tfrac{q}{j}\right) + \underline{\underline{Y}}_o\underline{\chi}_s\left(\tfrac{q}{j}\right)\right)$. Using (7) together with these expressions for $\underline{\chi}_s\left(\tfrac{q}{j}\right)$ and $\underline{\chi}_N\left(\tfrac{q}{j}\right)$ in equations (6) we obtain the results

$$\underline{\underline{G}}^f_{CS} = \underline{\underline{S}}^f_{CS}\,, \quad \underline{\underline{G}}^f_{CN} = \underline{\underline{S}}^f_{CN}\,, \quad \underline{\underline{G}}^f_{SS} = \underline{\underline{S}}^f_{SS} + \underline{\underline{\Phi}}^{-1}_{0SS}\,,$$

$$(8) \qquad \underline{\underline{G}}^f_{SN} = \underline{\underline{S}}^f_{SN} + \underline{\underline{\Phi}}^{-1}_{0SS}\,\underline{\underline{Y}}_0\,\underline{\underline{C}}\,,$$

$$\underline{\underline{G}}^f_{NN} = \underline{\underline{S}}^f_{NN} + \underline{\underline{C}}\,\underline{\underline{Y}}_0\,\underline{\underline{\Phi}}^{-1}_{0SS}\,\underline{\underline{Y}}_0\,\underline{\underline{C}} - \underline{\underline{C}}\,.$$

These expressions are in a practical form for numerical work, their evaluation being straightforward when the unperturbed eigenfrequencies, eigenvectors, and dynamical matrices are at hand. The inclusion of breathing merely involves adding an extra degree of freedom to the Hamiltonian (1) and proceeding as above. The results in that case are given in [4].

APPLICATION TO THE LOCALIZED MODES OF U-CENTERS

In reference 4 we have discussed at some length the results of applying the breathing shell model version of the foregoing theory to the problem of the infrared-active localized modes of U-centers, which consist of H^- ions located at anion vacancies in alkali halide crystals, and we will now discuss a few additional results for this problem. We will not consider here the even-parity localized gap mode observed [5] in the sideband spectrum of U-centers in KI, since, as will be seen in the following paper, an extended impurity space must be used for these modes. This extended impurity space will not affect the high-frequency Γ'^{15} modes discussed here since they are so highly localized.

To describe a U-center, we have used the four perturbed shell model parameters shown in figure 1(a). The defect's ionic charge has been assumed to be equal to that of the replaced anion so that the changes in the core and shell charges at the defect site may be written $-\delta Y$ and δY, respectively; δq is the change in the isotropic core-shell spring at the impurity site, δM is the mass change, and δk represents the change in each of the six longitudinal force constants between the defect's shell and those of its nearest neighbors.

The impurity space for this parameterization is twenty-two dimensional; however, because the H^- occupies a site of full cubic (O_h) symmetry this number can be greatly reduced, as discussed in reference 4. For the infrared-active localized vibrations, which are of Γ'^{15} [6] symmetry, it turns out that there are but five independent components of $\underline{\chi}$ in the impurity space, these being, for an x-polarized mode, $\chi_C(^0_x|\Gamma'^{15})$, $\chi_S(^0_x|\Gamma'^{15})$, $\chi_N(^0_x|\Gamma'^{15})$, $\chi_S(^1_x|\Gamma'^{15}) = \chi_S(^{-1}_{\ x}|\Gamma'^{15})$, and $\chi_B(1|\Gamma'^{15}) = -\chi_B(-1|\Gamma'^{15})$. The defect itself cannot breathe in Γ'^{15} modes.

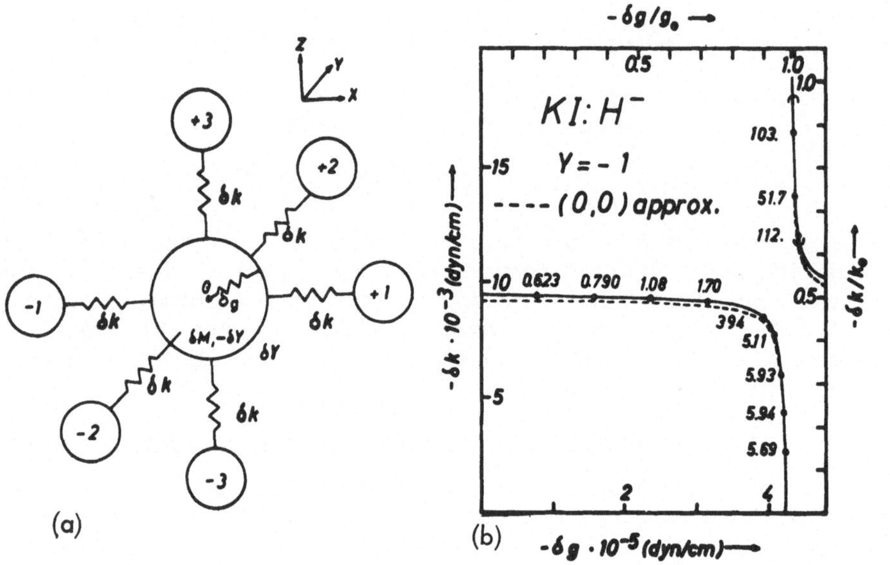

Figure 1. (a) The assumed perturbed parameters of the U-center (b) δk vs. δg for a U-center in KI. The defect's shell charge is -e, and $\omega_L = 382\ cm^{-1}$.[5] The (0,0) approximation is discussed in the text. Values of α (in $\AA^3$) are given for some points along the curve.

Using the observed local mode frequencies together with the unperturbed phonons as given by the breathing shell model[2] to evaluate the Green's function elements appearing in the five-dimensional secular equation for the Γ^{15} modes, we have obtained curves relating δk to δg for different values of δY , for U-centers in nine different alkali halides consisting of the Cl, Br, and I salts of Na, K, and Rb. The breathing shell model input data are given in reference 4. Some of our results for U-centers in NaCl, KCl, KBr, and KI are shown in figures 1(b) and 2.

In figure 1(b), we see the results for U-centers in KI, assuming that the defect's shell charge is -e, the value used by Fieschi et al.[8] We are only showing a limited portion of our curve, the omitted parts being just linear extensions of the parts given here, so that our complete curve appears as a hyperbola. Values for $\alpha = (Ye)^2/[g + 2(k + 2k_{OT})]$, the "polarizability" of the defect, are indicated at various points along the curve. Here k_{OT} is the value of the unperturbed non-central shell-shell force constants between the defect and its nearest neighbors. For the upper "branch" of the curve, the minimum positive value of α is 51.7 $\AA^3$, and α assumes positive values on just the portion enclosed by half-circles. Moreover, the upper branch contains portions where either δk or δg decrease by more than 100 percent. Similar results occurred in all of the host lattices investigated,

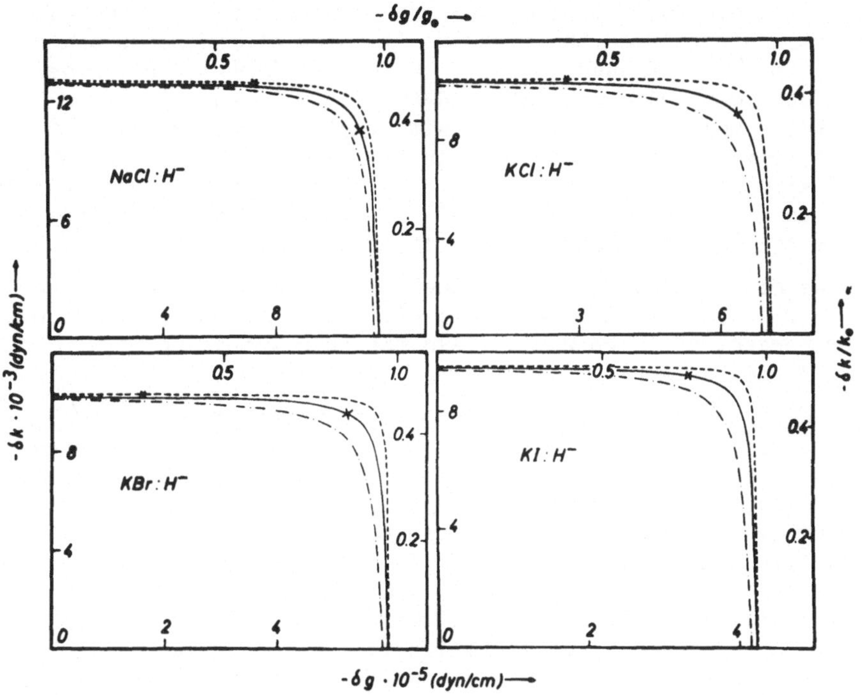

Figure 2. δk vs. δg for U-centers in NaCl, KCl, KBr, and KI, for ω_L equal to 565 cm^{-1} (5), 502 cm^{-1} (5), 444 cm^{-1} (7), and 382 cm^{-1} (5). —·—: Y = 0, ——: Y = -1, - - -: Y = -2. The crosses indicate the points where α is 1.9 Å^3.

and in view of these results, the upper branches of our curves seem unphysical. We will see this conclusion further born out by the sideband calculations of part II. Henceforth just the lower branches of our curves will be considered.

In figure 1(b) we also show the results obtained when only the defect-site elements of the perturbing matrix are retained, this approximation being expected to hold for highly localized modes. Had we included changes in k_T, δk would have been replaced by $\delta(k+2k_T)$ in the (0,0) elements of $\underline{\underline{C}}^f$. In view of this and because the (0,0) approximation holds so well, we could label the ordinate of figure 1(b) by $\delta(k+2k_T)$. Again, similar results hold in each of the host lattices considered, and we will return to these results in part II.

In figure 2, we see portions of our curves for all four host crystals, for defect shell charges of 0, -e, and -2e. In order to compare our results with those of Fieschi et al., who assumed α to be 1.9 Å^3, we have indicated the points on our curves where α takes on this value; since α is proportional to Y^2, these points vary considerably from curve to curve for a given host lattice even though the curves themselves are insensitive to δY . This latter property results from the high localization of the modes under consideration [4] . Fieschi et al., using a different shell model treatment, concluded that $\delta(k+2k_T)=0$ in NaCl and KCl for $\alpha = 1.9 \text{ Å}^3$ and Y=-1. Our results disagree with theirs and show that for these values of α and Y , both δk (or $\delta(k+2k_T)$, in view of the (0,0) approximation) and δg are highly weakened in these and, in fact, in all nine host lattices [4] . Since the sidebands are sensitive to δk , this discrepancy is important.

We notice that because of the "flat" nature of our δk versus δg curves, an a priori estimate of k that turns out to be even a few percent too high would yield the observed value of ω_L only for a highly weakened g .

Owing to the uncertainty in our knowledge of α and Y , we may attempt to learn which points on our curves actually correspond to real U-centers by studying both the U-center sideband spectrum and the one-phonon absorption due to perturbed Γ'^{15} band modes, keeping δg and δk consistent with the results given here. The former problem is interesting within our framework since the breathing of the defect in the Γ' modes allows δg as well as δk to be involved in the sidebands, while in the latter problem the Γ'^{15} band modes should be sensitive to the long-range Coulomb effects arising from δY , and these are easily treated in our scheme. In the following paper of these proceedings, we turn to the sideband problem.

In summary then, we have given a formalism within which one may treat substitutional impurities in a system described by the breathing shell model. Long-range Coulomb effects, except those arising from relaxation, which changes the Coulomb matrix $\underline{\underline{C}}$, have been included. Some results from the application of the theory to the infrared-active localized modes of U-centers have been given, and for a more complete discussion of this problem the reader is referred to reference 4.

REFERENCES

1. I. M. Lifshitz, Nuovo Cimento Suppl. 3, 716 (1956).
2. U. Schröder, Sol. State Comm. 4, 347 (1966).
V. Nüsslein and U. Schröder, phys. stat. sol. 21, 309 (1967).
3. R. A. Cowley, W. Cochran, B. N. Brockhouse, and A. D. B. Woods, Phys. Rev. 131, 1030 (1963).
4. J. B. Page, Jr. and Dieter Strauch, to be published.
5. B. Fritz, U. Gross, and D. Bäuerle, phys. stat. sol. 11, 231 (1965).
6. In the notation of: L. P. Bouckaert, R. Smoluchowski, and E. Wigner, Phys. Rev. 50, 58 (1936).
7. T. Timusk and M. V. Klein, Phys. Rev. 141, 664 (1966).
8. R. Fieschi, G. F. Nardelli, and N. Terzi, Phys. Rev. 138, A203 (1965).

SHELL MODEL TREATMENT OF THE VIBRATIONS OF SUBSTITUTIONAL IMPURITIES IN CRYSTALS. II. U-CENTER SIDEBAND SPECTRUM

Dieter Strauch and John B. Page, Jr.

Institut für Theoretische Physik der Universität

Frankfurt/Main

INTRODUCTION

In part I* a modified Lifshitz theory was presented and was applied to the localized modes of U-centers. Using the breathing shell model form of this theory[1] we will now study the perturbed even-parity band modes responsible for the sidebands in the infrared spectrum of U-centers in alkali halides.

After the results of the theory of the absorption constant are given, the defect parameterization is considered, and finally numerical results for U-centers in KCl, KBr, and KI will be presented and discussed. A more complete description of our work will be published elsewhere.

THE SUSCEPTIBILITY

As is known[2], the U-center sidebands arise from the simultaneous excitation of a local mode phonon and an even-parity band mode phonon. The well-known[3),(4),(5] expression for the complex dielectric susceptibility tensor in the frequency region of the sidebands reads

$$(1) \quad \chi_{\alpha\beta}(\omega) = \lim_{\varepsilon \to 0^+} \hbar p (2\omega_L)^{-3} \omega^{-2} \sum_{L\gamma} \sum_{P,Q} \underline{\tilde{N}}_P^{\alpha}(L_\gamma)\, \underline{\underline{S}}_{PQ}(\omega + i\varepsilon)\, \underline{N}_Q^{\beta}(L_\gamma).$$

Here ω_L is the local mode frequency, p is the concentration of defects, and $\underline{N}_Q^{\alpha}(L_\gamma)$ and $\underline{S}_{PQ}(\omega)$ are given by

---

*Hereafter equations and figures from part I (preceding paper) will be prefixed by I.

$$\underline{N}_Q^{\alpha}(L_\gamma) = \sum_{P,P',P''} \underline{M}_P^{\alpha} \underline{\chi}_P(L_\alpha) \underline{\underline{\Phi}}_{P'P''Q} \underline{\chi}_{P'}(L_\alpha) \underline{\chi}_{P''}(L_\gamma) \quad \text{and}$$

$$\underline{\underline{S}}_{PQ}(\omega) = \sum_b \underline{\chi}_P(b) \underline{\tilde{\chi}}_Q(b) (\omega_b^2 - \omega^2)^{-1} .$$

As in part I, the $\underline{\chi}(f)$'s are the perturbed eigenvectors, b denotes a band mode, and L_γ denotes a localized mode polarized along the γ-direction. The tensors $\underline{\underline{\Phi}}_{PQR}$ and $\underline{M}_P^{\alpha}$ contain third and first order Taylor series expansion coefficients of the crystal potential and of the α-component of the crystal dipole moment, respectively. These expansions, usually written in terms of core displacements (formal picture), are taken here in terms of core, shell, and breathing displacements (model picture); thus P, Q, etc. may refer to any of these kinds of displacements.

The matrices $\underline{\underline{S}}_{PQ}(\omega)$ are related to the Green's function matrices $\bar{\underline{\underline{G}}}_{PQ}(\omega)$ of the perturbed crystal by

$$\text{(2)} \qquad \operatorname{Im} \underline{\underline{S}}_{PQ}(\omega + i\varepsilon) = \operatorname{Im} \bar{\underline{\underline{G}}}_{PQ}(\omega + i\varepsilon) ,$$

and $\bar{\underline{\underline{G}}}_{PQ}(\omega)$ can be expressed in terms of the perturbation matrix $\underline{\underline{C}}$ and the Green's function matrix $\underline{\underline{G}}(\omega)$ of the unperturbed lattice by

$$\text{(3)} \qquad \bar{\underline{\underline{G}}}_{PQ}(\omega) = \left[(\underline{\underline{I}} + \underline{\underline{G}}(\omega)\underline{\underline{C}})^{-1} \underline{\underline{G}}(\omega) \right]_{PQ} .$$

The expression giving the elements of $\underline{\underline{G}}(\omega)$ in terms of unperturbed eigenvectors is found in equations (I-7) and (I-8).

DEFECT PARAMETERIZATION AND SYMMETRY CONSIDERATIONS

For the calculation of $\bar{\underline{\underline{G}}}(\omega)$ in equation (3) we use the same defect parameters as used in part I. These are shown in figure I-1(a). Following a recent paper by Gethins, Timusk and Woll[6] we also allow for a change $\delta k'$ in the longitudinal force constants between the shells of the defect's nearest and fourth nearest neighbors. This represents the effects of a relaxation around the impurity and should not affect the highly localized modes. We neglect any change of the Coulomb matrix due to relaxation.

For the even-parity modes, then, the impurity space is a twenty-five dimensional subspace, consisting of the longitudinal displacements of the shells of the defect's six nearest and six fourth nearest neighbors and the breathing displacement of these same particles and the defect. Since in the even-parity modes the defect's core and shell are undisplaced, neither the mass nor the charge changes at the defect influence these modes.

As in part I and reference 1 symmetry arguments can be used to further reduce the dimensionality of the problem. Of the even-parity modes which can couple to the local mode, only the ones longitudinal on the nearest neighbors, these being of Γ^1 and Γ^{12} symmetry (in the same notation as in part I), are perturbed, owing to our assumption of perturbed longitudinal force constants.

Symmetry arguments show that the defect may breathe only in the Γ^1 modes, so that these are the only even-parity modes which can be affected by changes in the defect's core-shell force constant. Using the forms of the Γ^1 and Γ^{12} modes in the impurity space, we arrive at a five and four dimensional problem, respectively.

Since the local mode is so highly localized, we restrict the anharmonic interaction to that between the impurity and its nearest neighbors. Furthermore, we assume the shell-shell anharmonicities to be the most important ones. Then symmetry arguments yield the following independent anharmonic coefficients:

$$A = \Phi_{sss}\begin{pmatrix} 0 & 0 & 1 \\ x & x & x \end{pmatrix},\ B = \Phi_{sss}\begin{pmatrix} 0 & 0 & 1 \\ x & y & y \end{pmatrix},\ \text{and}\ C = \Phi_{sss}\begin{pmatrix} 0 & 0 & 1 \\ y & y & x \end{pmatrix},$$

and the imaginary part of the susceptibility can be written in the form

$$\chi_{\alpha\beta}(\omega) \propto \omega^{-2}\left[(A+2C)^2 g_1(\omega) + (A-C)^2 g_{12}(\omega) + (2B)^2 g_{25'}(\omega)\right], \tag{4}$$

which exhibits the combinations of anharmonic parameters appropriate to the different symmetry types of band modes responsible for the sidebands.

Since the defect's nearest neighbors are rather unpolarizable, we can assume that to a good approximation the core and shell displacements of these ions are equal. In this case core-shell and core-core anharmonicities are easily incorporated into the present scheme by simply replacing, for instance,

$$\Phi_{sss}\begin{pmatrix} 0 & 0 & 1 \\ x & x & x \end{pmatrix} \chi_s\begin{pmatrix} 0 \\ x \end{pmatrix}\Big|L_x\Big) \chi_s\begin{pmatrix} 0 \\ x \end{pmatrix}\Big|L_x\Big) \chi_s\begin{pmatrix} 1 \\ x \end{pmatrix}\Big|b\Big)$$

by

$$\sum_{PQ} \left[\Phi_{PQs}\begin{pmatrix} 0 & 0 & 1 \\ x & x & x \end{pmatrix} + \Phi_{PQc}\begin{pmatrix} 0 & 0 & 1 \\ x & x & x \end{pmatrix}\right] \chi_P\begin{pmatrix} 0 \\ x \end{pmatrix}\Big|L_x\Big) \chi_Q\begin{pmatrix} 0 \\ x \end{pmatrix}\Big|L_x\Big) \chi_s\begin{pmatrix} 1 \\ x \end{pmatrix}\Big|b\Big),$$

in which case the form of equation (4) remains the same.

NUMERICAL RESULTS AND DISCUSSION

Before proceeding to our numerical calculations, we point out that among the approximations made in deriving equation (1) was that of omitting the frequency-dependent shift and damping of the excited phonons. Bilz, Zeyher, and Wehner[9] explicitly took account of the anharmonicity of the local mode phonon and calculated "harmonic" absorption curves from the experimental "anharmonic" ones. Hence we will be comparing our calculated sideband curves not directly with the experimental ones, but with the "harmonic" sidebands

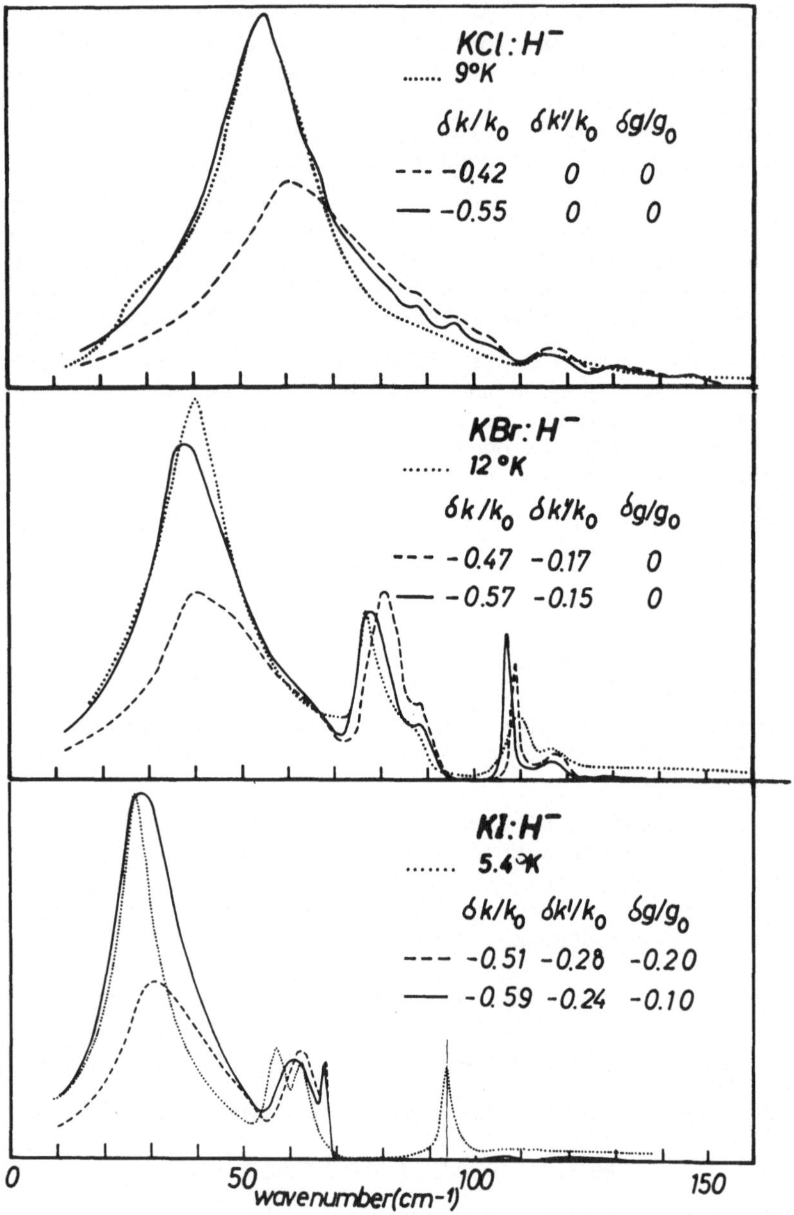

Figure 1. Comparison of our sideband calculations with experiment. The dotted curves are taken from reference 9, as explained in the text. Dashed (full) curves are calculated without (with) contributions from δk_T to the local mode frequency. The dashed and full curves have the same scale; the scale of the dotted curves is chosen so as to facilitate a convenient comparison with the full curves.

of reference 9 (dotted lines in figures 1 and 3).

We have used Schröder's breathing shell model[7] with the input data for KCl, KBr, and KI as given in reference 1 to evaluate $\underline{\underline{G}}(\omega)$ of equation (3) for the different symmetry types in these crystals. The number of points taken in the Brillouin zone was 64,000.

First we consider the results obtained by including only the anharmonic parameter $A = \Phi_{sss}(\dot{x}\ \dot{x}\ \dot{x})$. An equivalent assumption, within the formal picture, was made by Timusk and Klein[3] and Gethins, Timusk, and Woll[6] , and it results in the sidebands being determined by just the Γ^1 and Γ^{12} modes. The harmonic perturbation parameters δg , δk, and $\delta k'$ were treated as follows: the changes δg and δk were kept consistent with the observed local mode frequencies (see figures I-1(b) and I-2 for Y = -1), and for each (δg , δk) point, $\delta k'$ was determined so as to give a good overall agreement with the "experimental" sidebands (see above). Carrying out this procedure, we have found that the main band in KCl and the first two of the three principal bands in KBr and KI are due mainly to Γ^{12} modes. Fritz, Gerlach, and Gross[8] have experimentally determined that the observed third band (gap mode peak at 93.5 cm^{-1}) in the U-center sideband spectrum for KI is due to a resonant mode of Γ^1 symmetry, and we have assumed that this is also true of the third band in KBr. Because the Γ^1 modes are the only ones in which the defect can breathe, these modes are the only ones which in addition to being dependent upon δk and $\delta k'$ are also dependent upon δg . On the other hand, the observed double-peaked structure of the second bands of KBr and KI has been attributed[6] to the presence of a Γ^{12} resonance occurring in that frequency region, and we have found that this structure itself is rather sensitive to variations in δk and $\delta k'$. Thus in KBr and KI the ranges of the three parameters within which good agreement may be obtained may be fairly well determined.

Our best curves for KCl, KBr, and KI are shown by dashed lines in figures 1 and 2 and were in each case obtained for (δg , δk) points on the horizontal portion of the δk versus δg curves given in figure I-2. In KI the core-shell spring constant is weakened by 20%; in KBr g must be less weakened or even stiffened in order to improve the agreement for the third (Γ^1) sideband, while in KCl the calculations are not so sensitive to the Γ^1 modes. Accordingly, we can at present say that, within our model, $\delta g/g > -0.2$ in the three crystals, more weakening bringing in features which are not observed in the sidebands. Also, the curves for KCl are relatively insensitive to a weakening of k' up to 20%.

The curves in reference 6 were compared with the true "anharmonic" experimental ones and when our curves are so compared we obtain quite good agreement as can be seen in figure 2. Of course, we have an extra parameter δg in our model compared to that of reference 6 owing to our inclusion of breathing, but this parameter must be consistent with ω_L and δk . Although our curves are

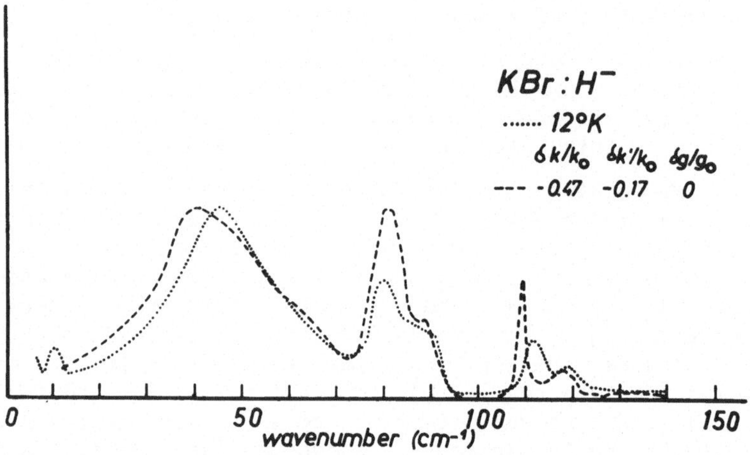

Figure 2. Comparison of theory with "anharmonic" experiment for KBr[3]. The curves are scaled so that the heights of the first band maxima are equal.

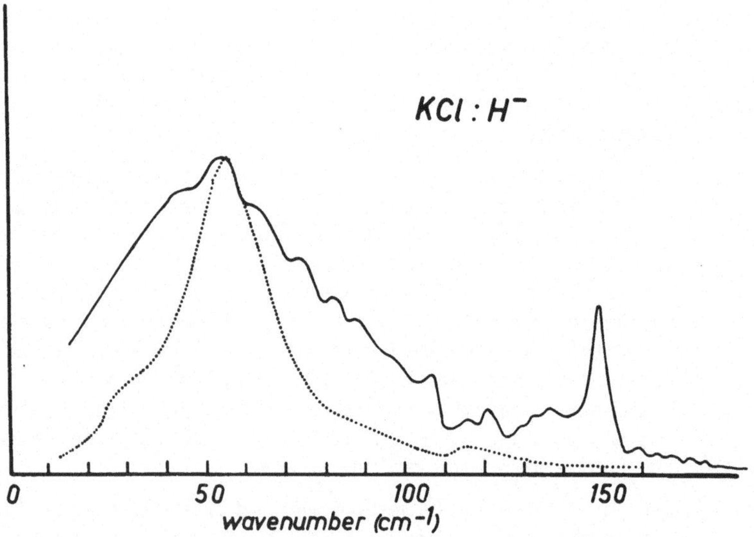

Figure 3. Absorption due to unperturbed $\Gamma^{25'}$ modes for KCl, compared with the "harmonic" absorption curve of reference 9.

improved over those of reference 6, we see that the agreement with the "experimental" curves of reference 9 is still not complete. In each of the three cases the main band has too low an intensity compared to the second band, is too broad, and has its maximum at too high a frequency, the latter discrepancy typically amounting to 5 wavenumbers.

Although the upper "branches" of our δk versus δg curves as shown in figure I-1(b) seem unphysical as discussed in part I, we have looked at the sidebands for ($\delta g, \delta k$) points on these branches. Because of the then drastic weakening of g, it turns out that the predicted Γ^1 peaks in KBr and KI occur at frequencies that are too low by about 10 wavenumbers. From this result together with the discussion given in part I, we conclude that points on the upper branches do not correspond to the actual situation of a U-center.

If we take account of all three parameters, A, B, and C, the relative sideband contributions from the Γ^1 and Γ^{12} modes are changed, and contributions from the (unperturbed) $\Gamma^{25'}$ modes are added. We note (a) that for any central potential $V(|\underline{R}|)$, B is equal to C and (b) that even small values of C/A can result in a remarkable difference of $(A+2C)^2$ compared to $(A-C)^2$ (5), these being the combinations of anharmonic coefficients governing the relative contributions from Γ^1 and Γ^{12} modes in equation (4).

The sideband contribution from the $\Gamma^{25'}$ modes alone was calculated first by Page and Dick[5], but did not give improvement in the approximation without relaxation; the possible importance of these modes for the defect model including relaxation has recently been pointed out by Kühner and Wagner[10]. Although increasing the height of the first absorption band, the inclusion of unperturbed $\Gamma^{25'}$ modes in our calculations also broadens this band and does not shift it as desired, except in KBr. It turns out that the contribution from the $\Gamma^{25'}$ modes is negligable in all three crystals, provided that $|B/A| < 0.05$. An example of the absorption due to unperturbed $\Gamma^{25'}$ modes is shown in figure 3 for KCl. In summary, the incorporation of $\Gamma^{25'}$ modes into our calculations does not seem to yield a satisfactory improvement of our results.

Among the neglected quantities which could enter into our calculations is the change δk_T of the nearest neighbor non-central shell-shell force constant, which influences the local mode frequency. In view of the accuracy of the (0,0) approximation, we could as well label the ordinate of the curves in figure I-2 by $\delta k + 2\delta k_T$ instead of δk. This point is discussed more extensively in part I. Since $\delta k/k_0 < 0$, we expect $\delta k_T/k_{0T} < 0$. Now with $\delta k + 2\delta k_T$ instead of δk describing the local mode frequency, but with a weakening of k by about 55%, 57%, and 59% for KCl, KBr, and KI, respectively, the sidebands can be described quite well without any contributions from the $\Gamma^{25'}$ modes, as shown by the full lines in figure 1; the required changes of k_T are comparable to those of k, the ratio $(\delta k_T/k_{0T}):(\delta k/k_0)$ being about 1.6, 1.2, and 1.0 for the respective crystals*. Owing to the different δk, the changes $\delta k'$ and δg are slightly altered from their earlier values.

The results given here might be affected slightly by the inclusion of other harmonic perturbations such as charge transfers, Coulomb relaxation effects, and further overlap potential changes. Since the sidebands are due mainly to the "quadrupole-like" Γ^{12} vibrations, the question naturally arises as to whether or not we should include, in addition to the s- and p-type excitations (breathing and shell displacements), some d-type excitations for the U-center.

In view of the many harmonic effects which can be treated, as far as the dynamical part of the problem is concerned, within our method, and because there is a rather large number of anharmonic coupling parameters entering the theory, the need for a detailed electronic wave function calculation becomes apparent. Nevertheless it can be seen that within our fairly simple assumptions for the perturbed harmonic parameters and the assumption of one dominant anharmonic parameter, the sidebands can be described quite well.

REFERENCES

1. J. B. Page, Jr. and D. Strauch, to be published.
2. B. Fritz, U. Gross, and D. Bäuerle, phys. stat. sol. 11, 231 (1965).
3. T. Timusk and M. V. Klein, Phys. Rev. 141, 664 (1966).
4. H. Bilz, D. Strauch, and B. Fritz, J. Phys. Radium 27, Suppl. C2, 3(1966).
5. J. B. Page, Jr. and B. G. Dick, Phys. Rev., to be published.
6. T. Gethins, T. Timusk, and E. J. Woll, Jr., Phys. Rev. 157, 744 (1967).
7. V. Nüsslein and U. Schröder, phys. stat. sol. 21, 309 (1967).
8. B. Fritz, G. Gerlach, and U. Gross, these proceedings.
9. H. Bilz, R. Zeyher, and R. K. Wehner, phys. stat. sol. 20, K167 (1967).
R. Zeyher, Diplom-Arbeit, Frankfurt/M. (1967), unpublished.
10. D. Kühner and M. Wagner, preprint.

Acknowledgements: We are gratefully indebted to the Deutsche Forschungsgemeinschaft, who provided financial support for this and the preceding work.

*The $\Gamma^{25'}$ modes are, in principle, perturbed by changes in k_T. However, results given in reference 5 indicate that these modes are relatively insensitive to δk_T, and in this case our earlier conclusions regarding the unimportance of the $\Gamma^{25'}$ modes still hold.

ANALYSIS OF R AND N_1 CENTER ABSORPTION STRUCTURES IN NaF

F. Lanzl, W. von der Osten, U. Röder, and W. Waidelich

I. Physikalisches Institut der Technischen Hochschule

Darmstadt, Germany

INTRODUCTION

The shape of the optical absorption structure caused by an electronic transition of a color center depends on the strength of the electron-phonon coupling parameter S. S can be interpreted as the most probable number of phonons involved in the transition. In the case of weak coupling ($S \ll 1$) only a sharp zero-phonon line occurs, whereas for strong coupling ($S > 8$) only a broad and smooth multiphonon band is observed. Intermediate coupling gives rise to both a zero-phonon line and a detailed multiphonon structure [1] as it can be seen in figure 2.

THEORY

The absorption coefficient $K(\omega)$ of the whole structure can be written [2-5] essentially as a Fourier transform of a function $e^{g(t)}$.

$$\frac{K(\omega)}{\text{const}\cdot\omega} = \int_{-\infty}^{\infty} dt\; e^{i(\omega-\bar{\omega})t}\, e^{g(t)} \tag{1}$$

$g(t)$ depends on the linear and quadratic coupling coefficients V_s and $V_{ss'}$, respectively, and is given as a sum

$$g(t) = - S - i\Omega t + g_1(t) + g_2(t) \tag{2}$$

The frequency $\bar{\omega}$ of the centroid of the whole structure and the shift Ω result in the energy E_o of the zero-phonon line.

$$E_o = \hbar(\bar{\omega} + \Omega) \tag{3}$$

The Fourier transform $g_1(t)$ of the effective one-phonon spectrum coupling to the electronic transition and the coupling parameter S are determined by the linear coupling coefficients V_s.

$$g_1(t) = \frac{1}{\hbar^2} \sum_s V_s^2/\omega_s^2 \cdot [(n_s+1)e^{-i\omega_s t} + n_s e^{i\omega_s t}] \tag{4}$$

$$S = \frac{1}{\hbar^2} \sum_s V_s^2/\omega_s^2 \cdot (2n_s+1) \tag{5}$$

Further

$$g_2(t) = g_2(V_{ss'}, t) + \tilde{g}_2(t) \tag{6}$$

$g_2(V_{ss'}, t)$ depends on the quadratic electron-phonon coupling [3,4] and contains a contribution to a temperature dependent halfwidth of the zero-phonon line. We introduce a term $\tilde{g}_2(t)$ so that the Fourier transform of $e^{\tilde{g}_2(t)}$ corresponds to the temperature independent shape of the zero-phonon line, resulting mainly from the broadening due to internal strain. This contribution is the dominant one at temperatures near 0°K. We therefore neglect contributions of $g_2(V_{ss'}, t)$ to the absorption structure in the vicinity of the zero-phonon line and furthermore over the whole spectrum in comparison to that of $g_1(t)$. The exponential in (1) can be expanded in powers of $g_1(t)$ resulting in the absorption coefficient $K_o(\omega)$ of the zero-phonon line [6] as the first term in the expansion

$$\frac{K_o(\omega)}{\text{const}\cdot\omega} = \int_{-\infty}^{\infty} dt\; e^{i(\omega-\bar{\omega})t}\; e^{-S - i\Omega t + \tilde{g}_2(t)} \tag{7}$$

and the absorption coefficient $K_1(\omega)$ of the effective one-phonon spectrum, which is now folded with the shape of the zero-phonon line, as the second term.

$$\frac{K_1(\omega)}{\text{const}\cdot\omega} = \int_{-\infty}^{\infty} dt\; e^{i(\omega-\bar{\omega})t}\; e^{-S - i\Omega t + \tilde{g}_2(t)} \cdot g_1(t) \tag{8}$$

The n-th term in the expansion corresponds to the n-phonon contribution.

The problem is to find the effective one-phonon spectrum coupling to a center from an experimentally measured structure as in figure 2. To solve this task we first insert the experimental $K(\omega)$ in (1) and Fourier transform both sides. Secondly we separate the shape of the line $K_0(\omega)$ from the whole spectrum as in figure 2, which procedure is not completely free from arbitrariness, and Fourier transform (7). The quotient of these results is $e^{g_1(t)}$, from which $g_1(t)$ can be calculated. With $g_1(t)$ and $e^{-S + \tilde{g}_2(t)}$ from the Fourier transform of (7) we can determine the effective one-phonon spectrum $K_1(\omega)$ by (8).

EXPERIMENTAL RESULTS

The program described above was carried out explicitly by means of a computer for the R_2 and the N_1 absorption structure in x-irradiated NaF. These structures are due to transitions of two kinds of F aggregate centers, the R and N_1 center, consisting of 3 and most probably 4 F centers, respectively (figure 1) [7,8].

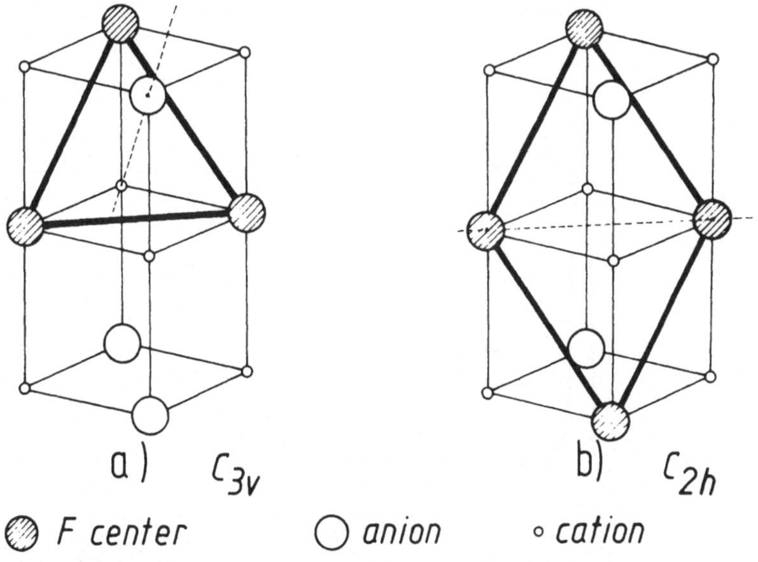

Figure 1: R center (a) and N_1 center (b) models.

Figure 2 shows the carefully measured R_2 absorption which exhibits a detailed vibrational structure. The analysis of the whole absorption by the method discussed before results in the one-, two- and three-phonon contribution seen in the figure. In a similar way the N_1 absorption is treated in figure 3. However,

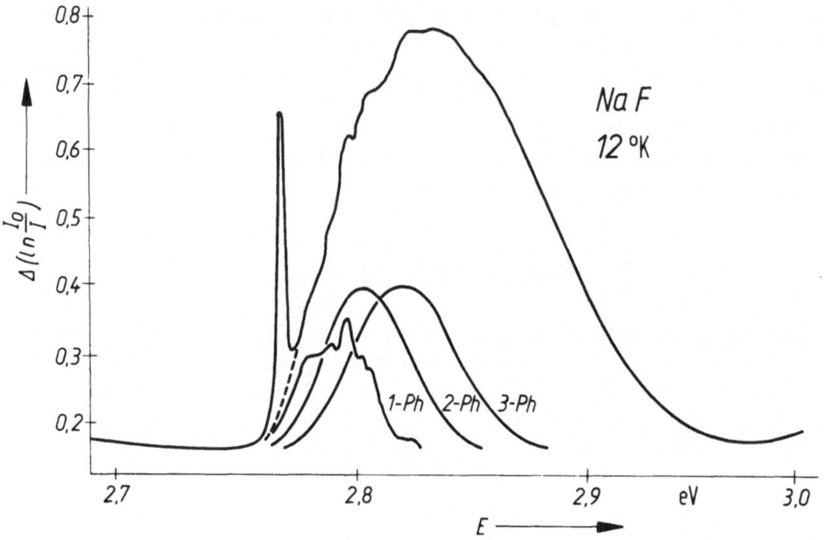

Figure 2: R_2 absorption structure.

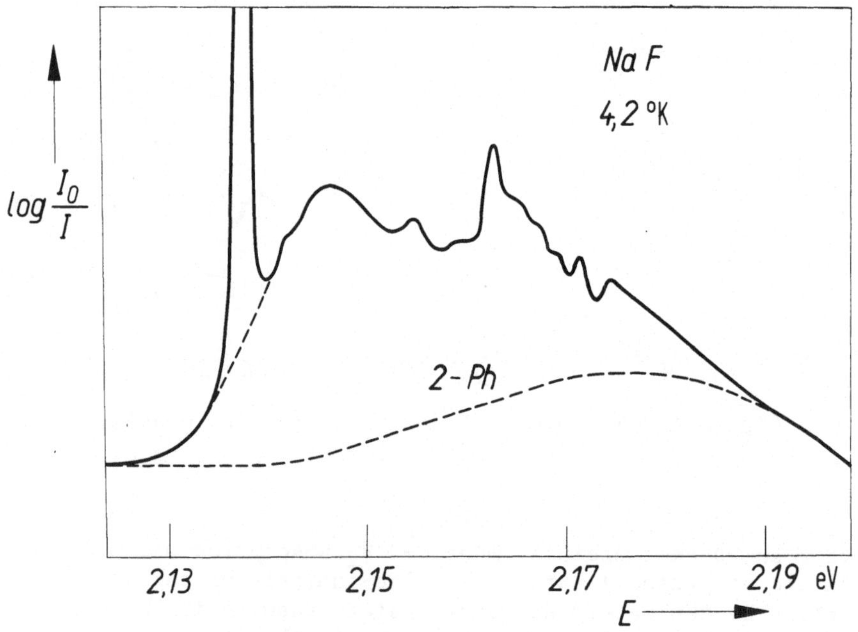

Figure 3: N_1 absorption structure.

this structure consists mainly of a one-phonon part while the higher phonon processes do not contribute very much.

From (4) and (5) one notices that $g_1(0) = S$. Using the calculated values of $g_1(t)$ we find $S = 3.8$ in the case of the R center and $S = 1.2$ for the N_1 center. It can be seen from figures 2 and 3 that the detailed structure on the bands stems entirely from the one-phonon contribution, while the higher phonon contributions are all smooth.

The effective one-phonon spectra coupling to the R and N_1 center are plotted in figure 4 on the same frequency scale. The

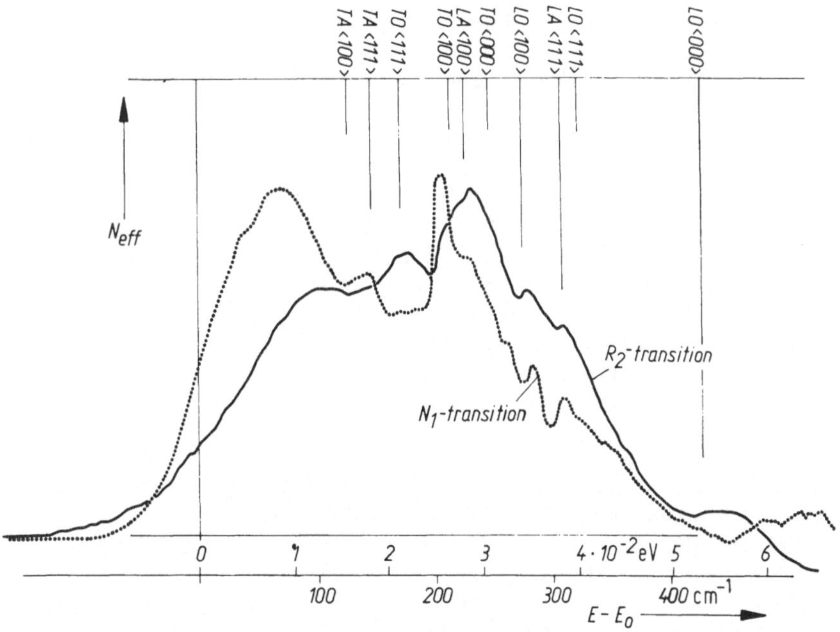

Figure 4: Effective one-phonon spectra coupling to the R and N_1 center.

maxima of both zero-phonon lines are adjusted at $E - E_o = 0$. The reason for the spectra to extend to negative values of $E - E_o$ is the folding with the shape of the zero-phonon line. For comparison all zone-boundary frequencies of the ideal lattice are indicated in figure 4 as they have been calculated by KARO and HARDY [9] for the D.D. model. The positions of the peaks in the spectra are in good agreement with most of these frequencies.

In both one-phonon spectra peaks corresponding to TO <100>, LA <100>, LO <100> and LA <111> are observed. The differences in the heights of these peaks and also in the positions of some others may be due to the different coupling of the particular phonon branches to the R and N_1 center.

In the frequency range below TA <100>, in which no other zone-boundary frequency exists, in both spectra a pseudo-localized mode is observed at about 70 cm^{-1}. In the case of the N center the localized mode appears rather pronounced while for the R center it seems to be shifted to higher frequencies because of its lower intensity. These modes are apparently caused by the R and N_1 centers themselves. Because of the similarity of these centers it seems reasonable that the modes occur at about the same frequency.

Moreover a number of other F aggregate centers in NaF have been found to exhibit transitions with phonon assisted structure in the frequency range of localized modes [10].

REFERENCES

[1] FITCHEN, D. B., R. H. SILSBEE, T. A. FULTON, and E. L. WOLF: Phys. Rev. Letters 11, 275 (1963).

[2] LAX, M.: J. Chem. Phys. 20, 1752 (1952).

[3] MARADUDIN, A. A.: Solid State Physics 18, 273 (1966).

[4] KRIVOGLAZ, M. A.: Fiz. Tverd. Tela 6, 1707 (1964); Soviet Phys.-Solid State 6, 1340 (1964).

[5] PRICE, M. H. L.: Phonons in Perfect Lattices and in Lattices with Point Imperfections, Edinburgh-London: Oliver & Boyd 1966.

[6] TRIFONOV, E. D.: Dokl. Akad. Nauk SSSR 147, 826 (1962); Soviet Phys.-Doklady 7, 1105 (1963).

[7] VAN DOORN, C. Z.: Philips Res. Repts. 12, 309 (1957).

[8] PICK, H.: Z. Physik 159, 69 (1960).

[9] KARO, A. M. and J. R. HARDY: Phys. Rev. 129, 2024 (1963).

[10] BAUMANN, G.: Z. Physik 203, 464 (1967).

PART I. MIXED CRYSTALS

OPTICAL VIBRATIONS OF IMPURITY AND HOST IONS IN MIXED CRYSTALS

A. S. Barker, Jr.

Bell Telephone Laboratories, Incorporated

Murray Hill, New Jersey

INTRODUCTION

The aim of this paper is to examine simple local mode theories applicable to cases where the concentration y of the impurity ion ranges from less than 0.001 to 1.00. That is, we trace the mode from the range of true local-mode behavior through to the range where the "impurity" ion has become a major constituent (y=1.0). The ion being replaced as y approaches 1.0 in turn becomes the impurity ion and may form a local mode. All of the crystals studied have two or more atoms in the primitive cell. In most of the crystals the impurity atom substitutes on one of the sublattices and the x-ray cell parameters remain well defined. Infrared, Raman, and neutron spectra for zone-center phonons are discussed. For zone-boundary phonons infrared combination bands, neutron and in one case phonon assisted tunneling spectra are examined and compared with the predictions of the models.

The models used to fit the experiments are derived by considering equations of motion for each significant coordinate x_i. Averaging over various possible distributions of ions is accomplished by assuming that the restoring force acting on ion i is simply $k_{ij}P_j(x_j-x_i)$. That is, we assume a bond strength k_{ij} between ions i and j but weight it with the probability P_j that ion j occurs at the end of this bond. The number of coordinates which we define as significant depends on the complexity of the data to be fit and the number of model parameters we are willing to work with. Obviously if we define $\sim 10^{23}$ coordinates we should be able to describe the vibration spectra quite well. We show below however that we can describe many experimental results with less than twenty coordinates and can often do quite well with as few as three.

GALLIUM ARSENIDE-PHOSPHIDE

$GaAs_yP_{1-y}$ has been studied optically by Raman and infrared methods.[1] For $y \simeq 1.0$ the P ion vibrates as a local mode near 352 cm^{-1} above all fundamental phonon frequencies of the GaAs host.[1,2] For $y \simeq 0$ the As impurity causes a sharp mode near 268[3] cm^{-1}. It appears likely that this is a true local mode also vibrating in a gap in the host GaP spectrum. For y near 0.5 both local modes have increased in strength and are major bands. Figure 1a shows the experimental results[4] as solid triangles for y=0.01, 0.44, and 0.99. The horizontal extent of the triangles represents the dimensionless mode strengths $4\pi\rho$. The strengths are presented on a logarithmic scale. The P and As local modes (shown at y=0.01 and .99) have also been observed at much lower concentrations (y=0.001[3] and 0.99998[2] respectively). For these much lower concentrations the local mode remains at the frequency shown in Fig. 1a but decreases in strength like y (or 1-y for the P mode).

Figure 1a shows that at y=0.44 both the high and low frequency bands have structure. The simple model we develop below has 3 coordinates and will yield the main experimental features of the modes but not the structure (i.e., splitting) observed in each of the two major bands at y=0.44.[4]

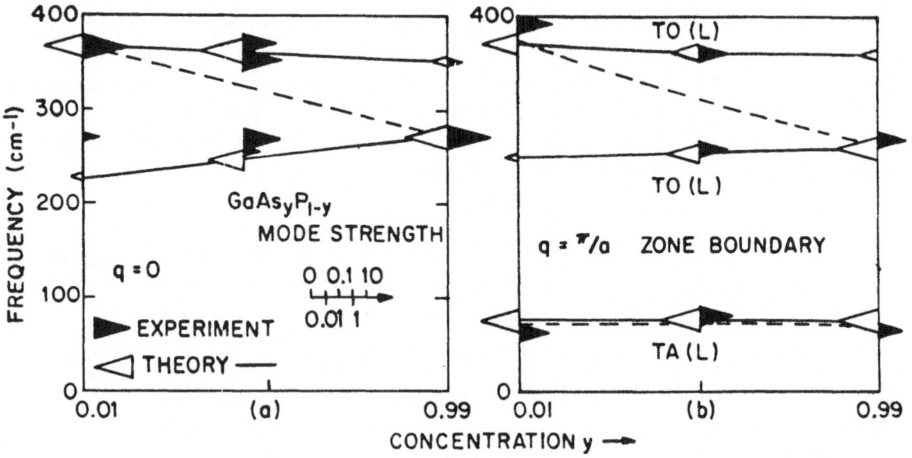

Fig. 1 (a) - Frequencies and strengths of the long wavelength transverse optic (TO) modes in $GaAs_yP_{1-y}$. The open triangles and solid lines are given by the model. The dashed line results from the virtual ion model (see text). (b) - Zone boundary modes.

Numbering the ions in the order As, P, Ga, we write for the long wavelength ($q \simeq 0$) vibrations:[5]

$$m_1\ddot{x}_1 = -((1-y)k_{12}+k_{13})x_1+(1-y)k_{12}x_2+k_{13}x_3+z_1E_{loc} \quad (1)$$

$$m_2\ddot{x}_2 = yk_{12}x_1-(yk_{12}+k_{23})x_2+k_{23}x_3+z_2E_{loc} \quad (2)$$

$$m_3\ddot{x}_3 = yk_{13}x_1+(1-y)k_{23}x_2-(yk_{13}+(1-y)k_{23})x_3+z_3E_{loc}. \quad (3)$$

For the phonon contribution to the dielectric polarization we write:

$$P = \frac{1}{v}(yz_1x_1+(1-y)z_2x_2+z_3x_3), \quad (4)$$

where m_i, z_i are the ion mass and charge and v is the volume of the primitive cell. Each x_i can be thought of as a sublattice coordinate. For $q \simeq 0$ all ions of the same type move in phase in this model. We have used probability coefficients appropriate to random ion distributions. If we suspect that there is clustering of like cations, the (1-y) factor in Eq. 1 (and similarly the y factor in Eq. 2) should be reduced since this is the factor describing the number of P ions around a given As ion.[1]

For E_{loc}, the local electric field at an ion site, we use the usual $E_{loc} = E+4\pi P/3$[7] for transverse vibrations where E is the macroscopic field. Substituting (4) into Eqs. (1-3) and solving we obtain 3 modes for each value of y. One mode is the acoustic mode (ω=o) and the other two are optic modes with **different** relative motions of the As, P and Ga ions. The modes are shown as open triangles in Fig. 1a. We have choosen the three force contants and the two independent charges to fit the main GaAs and GaP modes and the P local mode. The predicted As local mode at y=.01 comes out with a frequency 15 percent low but a strength about correct. Since the lattice contracts, the force constants should generally increase for $y \to 0$. Such an increase (using an additional Grüneisen parameter) can be easily incorporated into Eqs. (1-3) and both main modes and local modes frequencies (y=.01 and .99) can be fitted exactly.[5] We will not attempt this improvement here.

At y = .44 the model gives optic modes at 245 and 362 cm^{-1}. The eigenvectors show that the higher mode is predominantly P vibrating against Ga while the lower mode is somewhat mixed with As vibrating against P and Ga. Since we have choosen only 3 coordinates, we can get at most 3 modes and so are unable to fit the structure observed at y=.44. A serious deficiency of the model is the mode strength at y=.44. The sum of the two predicted mode strengths is 2.0. Experimentally we observe about 1.8. The mode strengths together with the transverse mode frequencies determine the frequencies of the longitudinal modes.

This 10 percent discrepancy causes the longitudinal modes to fall far from where they are observed. Thus we shall call a model successful only if it predicts both the strength and the transverse frequencies in a polar crystal. The strength deficiency can be overcome by introducing more degrees of freedom. We have used two degrees for the two types of cation (As and P) because they are obviously different by virtue of their masses. It seems evident that various Ga ions are different also depending on whether their four nearest neighbors are 4 As, 3 As1P etc. Here it is not a mass difference but a difference in the forces binding the ion to its neighbors which differentiates coordinates. This approach of recognizing several distinct Ga coordinates has been developed and successfully fits the observed mode frequencies and (also important) the mode strengths including fine structure.[1] The essential ideas used to accurately describe $GaAs_yP_{1-y}$ using 13 coordinates rather than 3 are just a generalization of Eqs. (1-4) and are contained in Ref. 1. At very low values of y or 1-y (i.e. the local mode regime) the minority ion resonance is always found to have eigenvectors of a local mode type — i.e. large amplitude for the minority ion and very small amplitudes for the surrounding ions. As the concentration increases, a second local mode at a nearby frequency usually becomes stronger also. This second mode consists of two (or sometimes 3 or 4) minority ions vibrating together with large amplitude against the surroundings. It is clearly important to set up coordinates for each of these types of possible ion motion in the model if they are to be described properly.

The equations (1-3) for our simple model can be modified to predict the zone boundary phonon behavior. With this one dimension model we cannot of course identify L point or X point phonons. Our purpose is to establish the existence of separate As like and P like bands and study their concentration dependence. Looking at Eq. 1 we note that the force $k_{13}x_3$ will be absent for the wavevector $\dot{q} = \pi/a$ since the Ga ions on each side of an As ion are 180 degrees out of phase applying a net force of zero to the intervening As. We rewrite Eq. (1) as

$$m_1\ddot{x}_1 = (-k_{13}-2k_{11}y-(1-y)k_{12})x_1-(1-y)k_{12}x_2, \tag{5}$$

where we have used the second neighbor force k_{12} as before and introduced the new second neighbor force k_{11}. In the equations for x_2 and x_3 we introduce second neighbor forces k_{22} and k_{33}. For simplicity we omit effective field corrections and set all second neighbor forces equal. Figure 1b shows the results for the zone boundary modes. Table I gives the values of the parameters used. We identify our modes as transverse modes though this is somewhat arbitrary in a one-dimensional model. The lowest mode has the Ga ion (x_3) vibrating against other second neighbor Ga ions with other ions remaining stationary. The two highest modes are again mixed though the highest is mostly P vibrating and the middle mode is mostly As vibrating. In Fig. 1b and following zone boundary figures we define the mode strength from the magnitude of the eigenvector. The experimental modes are sketched in with no attempt to normalize the strengths. Local modes have not been observed experimentally in zone boundary phonon spectra. For very low concentrations we would expect these modes to have about the same frequency as the zone center local modes. In Fig. 1b we show some zone boundary results taken from Neutron spectra[6] and infrared combination band spectra.[5] The modes which have been observed agree reasonably well with our simple model. Better agreement can be obtained by using three different values for the second neighbor force constants. Our main result is the appearance at all concentrations of an As-like band and a P-like band as is observed experimentally.

MIXED CRYSTALS WITH CLOSELY SPACE FREQUENCIES

We now examine briefly the case of a mixed crystal whose pure constituents have quite similar lattice frequencies. The mixed crystal $Co_yNi_{1-y}O$[8] is an example. Here no local mode is formed, the infrared mode goes over continuously from that of NiO to that of CoO. To study this behavior we take our Eq. (1-4) and apply them to the imaginary crystal $GaAs_yM_{1-y}$ where M is an ion like P in every way except that its mass is 60 a.m.u. instead of 31 a.m.u. Figure 2a shows the $q \simeq 0$ optic mode behavior. There is one main mode for all concentrations. The weaker mode is so close to the main mode at y=0.99 that it would be within the T.O. band and hence not vibrate as a true local mode. At y=0.5, the lower mode is still very weak and might appear as structure on the low frequency side of the single reststrahlen band of this crystal. Except for such weak structure we can describe the optical properties quite well by one mode. We find that the main mode eigenvector consists of As and M moving together against Ga. This suggests that the virtual ion (V.I.) model is appropriate here.[9] For the V.I. model we define one cation coordinate x_1, but use an average mass. For the V.I. model, Eq. 1 and 2 are replaced by

$$(ym_1+(1-y)m_2)\ddot{x}_1 = -[yk_{13}+(1-y)k_{23}]x_1+[yk_{13}+(1-y)k_{23}]x_3 + [yz_1+(1-y)z_2]E_{loc}. \qquad (6)$$

x_2 is set equal to x_1 in Eq. 3 and 4 giving only two coordinates in the problem. The solution to the V.I. equations are shown as dashed lines in Fig. 1 (a and b) and Fig. 2a. It is obvious that

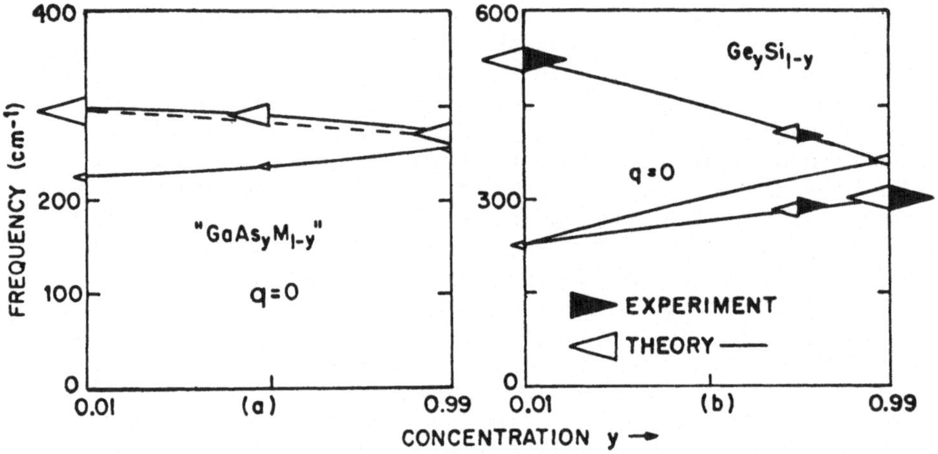

Fig. 2 (a) - Long wavelength modes for a mixture of GaAs with a hypothetical crystal with very similar modes. One mode is predominant over the entire concentration range. (b) q=o modes in Ge_ySi_{1-y}. The triangles show Raman strength on the same scale used in Fig. 1. The center mode has zero Raman strength.

only in the case of Fig. 2a is the V.I. model justified and then only as long as we do not enquire into fine structure. Solution for the $q=\pi/a$ zone boundary vibrations of $GaAs_yM_{1-y}$ yields two closely spaced optic modes and one acoustic mode at all concentrations. We should emphasize that in Fig. 2a, at y=.50, the lowest mode is weak in the dipole sense of P (Eq. 4) being small. For neutron measurements this mode might be strong. Thus, the V.I. model may be appropriate for describing optical results at $q\simeq 0$ in a mixed crystal with closely spaced modes but completely inappropriate for explaining other features of the lattice spectra. Studies of more complicated models[9] give the same result for the optical properties. Mixed crystals whose pure constituents have closely spaced frequencies have one main optically active mode which roughly obeys the simple V.I. equations.

GERMANIUM-SILICON

Figure 2b shows the q=0 optic modes of pure Ge[10] (300 cm^{-1}) and pure Si[11] (523 cm^{-1}). Since these modes are spaced rather far apart we might expect both Si-like and Ge-like optic modes in the mixed crystal. Large samples of Ge_ySi_{1-y} are difficult to prepare and only a few spectra of the q=0 optic modes are available. Figure 2b shows y=0.75 results[12] which confirm the existence of two main modes. (An additional weak mode is omitted.) Earlier phonon tunneling results[13] seem to indicate that the mixed crystal should have one mixed mode rather than two or more. We will discuss the tunneling experiments below. Our equation for Ge_ySi_{1-y} are again one dimensional; however, we must look at a two or three-dimensional model of the crystal to construct the probability coefficients for the bond strengths, otherwise spurious results are obtained. Figure 3b shows a projection of two neighboring planes of atoms onto the (111) crystallographic plane. A plane 'unit cell' is outlined and the two sublattices 1 and 2 defined. We take all first and second neighbors to be linked by simple central-force springs. Writing x_1 for Ge on sublattice 1, x_2 for Si on sublattice 1, x_3 for Ge on sublattice 2, and x_4 for Si on sublattice 2 we obtain

$$m_1\ddot{x}_1 = -(yk_{11}+(1-y)k_{12}-(1-y)2k_n)x_1+2(1-y)k_nx_2+yk_{11}x_3 + (1-y)k_{12}x_4 \quad (7)$$

$$m_2\ddot{x}_2 = 2yk_nx_1-(yk_{12}+(1-y)k_{22}-2yk_n)x_2+yk_{12}x_3 + (1-y)k_{22}x_4 \quad (8)$$

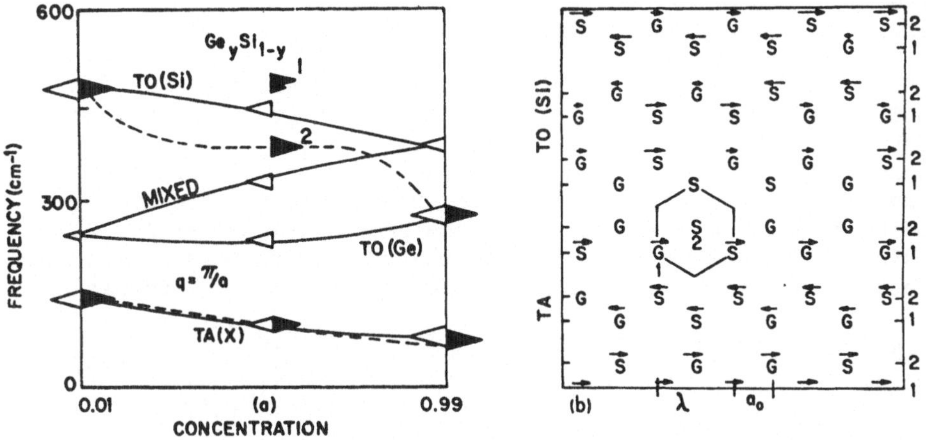

Fig. 3 (a) - Zone-boundary modes in Ge_ySi_{1-y}. The solid curves and open triangles are given by the model. Neutron results for pure crystals are shown at y=.01 and .99. Some electron-tunneling results of Logan <u>et al</u> are shown at y=.54. The dashed curve gives the concentration dependence they observe for their TA and TO modes. (b) - (111) plane of $Ge_{.5}Si_{.5}$ with random ion arrangement. Arrows show eigenvectors for the TA and upper TO modes in (a).

$$m_1\ddot{x}_3 = yk_{11}x_1+(1-y)k_{12}x_2-(yk_{11}+(1-y)k_{12}+2(1-y)k_n)x_3 + 2(1-y)k_nx_4 \quad (9)$$

$$m_2\ddot{x}_4 = yk_{12}x_1+(1-y)k_{22}x_2+2yk_nx_3-(yk_{12}+(1-y)k_{22}+2yk_n)x_4. \quad (10)$$

We use k_n for each of the 3 second neighbor bonds which appear. Each of the k's here is 1.5 times the Hooks constant of an individual spring. Figure 2b shows the solutions of these equations for the force constants given in Table I. The acoustic mode comes at $\omega=0$ as in $GaAs_yP_{1-y}$. Of the three optic modes, the center one has zero Raman strength for all y and is degenerate with the other modes at y=0 and 1. We adjust our force parameters to fit the pure Si, and pure Ge modes, and for the best fit possible to the modes at y=0.75. From the figure we note that the model predicts for y=.75 the Ge-like mode 6 cm^{-1} low and the Si-like mode 9 cm^{-1} high compared with the experimental data. Our model predicts that the local mode for Ge in the Si host should come at 226 cm^{-1}. This is a very rough prediction since we have included no lattice contraction effect on the Ge-Si bond. Since Ge appears to have no gap in its phonon bands the mode would appear as an in-band resonance. The intermediate optic mode has finite polarizability in some cells but the sign alternates depending on whether a Si or Ge is at the center of the cell. The average Raman strength for this mode is zero.

For the zone boundary vibrations we phase the atom motions as plane waves with the wavelength shown at the bottom of Fig. 3b. The force matrix contained in Eq. (7-10) now changes considerably. We must consider more second neighbor bonds since atoms which were moving in phase at q=0 are now moving against each other. Also some second neighbors are moving with and some against a given atom causing some fractional bond forces to appear. For $q = \pi/a$ Eq. 7 becomes:

$$m_1\ddot{x}_1 = -(yk_{11}+(1-y)k_{12}+\tfrac{4}{3}yk_n+2(1-y)k_n)x_1+\tfrac{2}{3}(1-y)k_nx_2 + yk_{11}x_3+(1-y)k_{12}x_4 \quad (11)$$

The Eq. 8-10 are modified in a similar way. Figure 3a shows the solutions for the zone boundary vibrations. There are four modes at each concentration. We choose our force constants to fit TO and TA data at the X point in pure Si[11] and Ge.[10] The strengths shown are taken as proportional to the modulus of the eigenvector. The three upper modes have eigenvectors with Si mostly moving (top mode), Ge and Si vibrating (center mode) and Ge mostly vibrating (lower mode). The upper and lower become the pure Si and pure Ge optic modes at y=0 and 1. Figure 3b shows two eigenvectors for $q = \pi/a$. The arrows in the upper part of the figure show the atom motions for the TO (Si-like) mode at 346 cm^{-1} for y=.50.

The lowest frequency mode in Fig. 3a becomes the TA mode of pure Si (or pure Ge) at y=0 (or 1). The unusual feature of this mode is that at y=.50 it still has the same form of eigenvector that it has for the pure crystals. Figure 3b (lower half) shows this y=.50 eigenvector. The atoms follow the displacement pattern whether they are Si or Ge. That is, this lowest mode is like a virtual ion mode in contrast to the optic mode behavior.(14) For y=.54 we show the lowest frequency mode seen by a tunneling technique. There is good agreement. The higher modes seen in tunneling lie in the range of our upper modes but their concentration dependence is quite different. The peaks selected by Logan et al.(13) show a gradual transition from Ge-like to Si-like. In Fig. 3a we show as a dashed line the mode they identify as TO in their spectra. Electron tunneling involves a band to band transition with phonon assistance.(15) Over most of the range the electron bands are silicon-like(16) which sharply selects phonons with wavevector about 85 percent of the way out to the zone boundary point. The most striking feature in the tunneling spectra is the TA mode. It alone retains a sharply peaked appearance and does not merge with other bands. On the basis of our simple model it is tempting to say that the TA mode in Ge_ySi_{1-y} does smoothly vary its frequency from Ge to Si but the optical modes do not. This latter conclusion is confirmed by the zone center optical work(12) and the zone boundary two-phonon combination band spectra.(17) We would then conclude that Ge-like, Si-like, and mixed mode tunneling peaks can appear. Logan et al. have selected four peaks (which they label TA, LA, LO, TO) which appear to follow a smooth variation with y. They do note however an additional strong peak (with structure) near 480 cm^{-1} at the Si-rich end which shifts very little with y. This peak is labeled 1 in Fig. 3a. We believe this peak should be labeled TO (Si-like). There should also be a TO(Ge-like) band, however it would lie under the LO and LA peaks they have observed near y=1. The tunneling peaks observed between y=.2 and .8 and labeled TO by Logan et al. (eg. Mode 2 in Fig. 3a) may be the mixed modes. In our simple model there is one such mixed mode. It has a fairly strong y dependence because it must be degenerate with the TO modes at y=0 and 1. In a model with more coordinates this restriction disappears and all modes can be quite flat.(1) Summarizing the tunneling situation, our simple model suggests that there is one TA mode which varies linearly with composition as is observed. There should be three or more TO modes (Ge, Si, and modes with mixed eigenvectors). The strengths of the TO(Si) should drop off as y increases, and one or several mixed modes should increase in strength. These mixed modes should drop in strength again as y approaches 1 and a TO(Ge) should become prominent. Tunneling peak 1 may correspond to TO(Si) and peak 2 (Fig. 3a) to TO(mixed). Because of the strength changes, the most prominent mode in the tunneling spectrum would be TO(Ge) or TO(Si) at the endpoints but could very well correspond to a different type of eigenvector (eg. a mixed mode) near the center of the concentration range. The same would hold for the other modes except for the TA mode. In our model the TA mode has only one type of eigenvector and is virtual ion-like in agreement with experiment.

CONCLUSIONS

We have shown that a simple model can explain the main features of the zone-center and zone-boundary phonon spectra of mixed crystals. For certain crystals which form local modes we find a smooth increase in the strength of the local mode with concentration. This increase lead to the formation of a major band at high concentrations. In all of the systems discussed there is fine structure even at y=0.5. This structure can be explained by introducing more coordinates in a natural way. This has been done for example in $Ba_ySr_{1-y}F_2$. Neutron measurements presently in progress[18] have confirmed the presence of the weak additional modes predicted by a model with 18 coordinates developed for this material.[9] Since very few extra modes are seen, a model with 18 coordinates appears useful and adequate to explain the spectra.

Acknowledgment - The author acknowledges with pleasure many helpful discussions concerning this work with D. E. McCumber and H. W. Verleur.

TABLE I

Reference	Ion	Mass (amu)	Charge (e)	Force Constants (*denotes second neighbor) k_{ij} (10^5 dyne/cm)	
Fig. 1a	As 1	74.9	-2.24	(1,3)=1.57	
	P 2	31.0	-2.14	(2,3)=1.70	
	Ga 3	69.7	2.24y+2.14(1-y)	(1,2)-.42*	
Fig. 1b	As 1		0	(1,3)=2.77	
	P 2		0	(2,3)=2.27	
	Ga 3		0	(1,1)=(2,2)=(3,3)=(1,2)=0.12*	
Fig. 2a	IDENTICAL TO FIG. 1a EXCEPT Mass (2) = 60.0				
Fig. 2b	Ge 1,3	72.6	0	(1,1)=1.93	(1,2)=1.30
	Si 2,4	28.1	0	(2,2)=2.27	
Fig. 3a	IDENTICAL TO FIG. 2b EXCEPT Force Constants			(1,1)=1.5(1,2)=2.05(2,2)=1.72 (1,1)=(2,2)=(1,2)=0.23*	

REFERENCES

1. H. W. Verleur and A. S. Barker, Jr., GaAs-P, Phys. Rev. 149, 715 (1966); CdSe-S Phys. Rev. 155, 750 (1967).
2. W. G. Spitzer, J. Phys. Chem. Solids 28, 33 (1967).
3. A. S. Barker, Jr., Phys. Rev. (to be published).
4. Only the strongest modes are shown in Fig. 1. There is evidence of considerable weak fine structure near these modes (Ref. 1).
5. Y. S. Chen, W. Shockley and G. L. pearson, Phys. Rev. 151, 648 (1966). These authors have introduced equations similar to 1-3 but neglect local field coupling and the mode strength equation (4).
6. G. Dolling and J. L. T. Waugh, Lattice Dynamics, Proceedings of the International Conference at Copenhagen 1963 (Pergamon Press, Oxford, 1965) p. 19.
7. M. Born and K. Huang, Dynamical Theory of Crystal Lattices, (Clarendon Press, Oxford, England, 1954), Ch. 2.
8. P. J. Gielisse, J. N. Plendl, L. C. Mansur, R. Marshall, and S. S. Mitra, J. Appl. Physics 36, 2446 (1965).
9. A. S. Barker, Jr. and H. W. Verleur, Solid State Comm. (to be published); also H. W. Verleur and A. S. Barker, Jr., Phys. Rev. (to be published).
10. W. Cochran, Phys. Rev. Letters 2, 495 (1959); also Proc. Roy. Soc. 253A, 260 (1959).
11. B. N. Brockhouse, Phys. Rev. Letters 2, 256 (1959).
12. D. W. Feldman, M. Ashkin, and J. H. Parker, Jr., Phys. Rev. Letters 17, 1209 (1966).
13. R. A. Logan, J. M. Rowell, and F. A. Trumbore, Phys. Rev. 136, 1751 (1964).
14. We have observed similar behavior in $K Ni_yMg_{1-y}F_3$. Here the highest and lowest infrared-active TO modes follow virtual-ion model behavior over the entire concentration range. The intermediate TO mode is split into two peaks at y=0.5 however. For $y \to 0$ or 1, one of these peaks becomes a major band and the other a local mode.
15. The form of the electron bands in mixed crystals is usually quite different from the form of the phonon bands. Physically, we picture the electrons moving rapidly and sampling the ion potentials over many unit cells. In Ge_ySi_{1-y} this prevents the formation of separate Ge-like and Si-like electron bands in the mixed crystal.
16. R. Braunstein, A. R. Moore, and F. Herman, Phys. Rev. 109, 695 (1958); also R. Braunstein, Phys. Rev. 130, 869 (1963).
17. R. Braunstein, Phys. Rev. 130, 879 (1963).
18. G. Shirane, J. Leake, Y. Yamada, and A. S. Barker, Jr. (to be published).

OPTICAL LATTICE MODES OF MIXED POLAR CRYSTALS

G. Lucovsky*
Research Laboratories, Xerox Corporation
Rochester, New York

M. Brodsky
Night Vision Laboratory
Fort Belvoir, Virginia

E. Burstein†
Laboratory for Research on the Structure of Matter and Physics Department,
University of Pennsylvania,
Philadelphia, Pennsylvania

I. Introduction

In this paper we discuss the infra-red and Raman active lattice modes of mixed crystals. We review the experimental results reported to date and discuss models which have been used to describe the variation of the observed vibrational frequencies with composition. We mention new experimental results recently reported for $Cd_{1-x}Zn_xS$ [(1)] and $Se_{1-x}Te_x$ [(2)], indicating how these studies have helped us develop a unified point of view. Finally we suggest criteria, based on whether or not localized modes and gap modes form, for predicting the behavior of the optical lattice modes of mixed crystals.

Two types of behavior are observed in mixed crystals. In one class of mixed crystal systems the phonon frequency (of each of the modes, infra-red or Raman active or both) varies continuously from the frequency characteristic of one end member to that of the other end member; the strength of the mode remaining approximately constant. We define this as "one mode" behavior. In a second class of mixed crystal systems, two phonon frequencies are observed to occur at frequencies close to those of the end members; the strength of each mode being approximately equal to the fractional formula weight of each component. We define this as "two mode" behavior.

Until recently one mode behavior has been observed only in ionic materials, such as alkali halide crystals,(3,4) $Ni_{1-x}Co_xO$(5) and $(Ca,Ba)_{1-x}Sr_xF_2$(6,7,8). An earlier report of simple one mode behavior in evaporated thin films of $GaAs_{1-x}Sb_x$(9) is now questionable in the light of recent studies which indicate two mode behavior over a part of the composition range.(10) The recently observed single mode behavior in $Cd_{1-x}Zn_xS$(1) and $Se_{1-x}Te_x$(2) is the first report of one mode behavior in mixed covalent crystals. In all of the "one mode" systems with the exceptions of $GaAs_{1-x}Sb_x$ (which has "two mode" behavior for large x) and $Se_{1-x}Te_x$, there is frequency overlap between the reststrahlen bands (which extend from the transverse optical to the longitudinal optical frequencies) of the end member components. We may also divide the "diatomic" crystals into two groups according to the nature of the mass or isotopic substitution. For one group, $Ni_{1-x}Co_xO$, $(Ca,Ba)_{1-x}Sr_xF_2$ and $Cd_{1-x}Zn_xS$, the mass substitution is being formed on the heavier element, whereas in a second group including, e.g., $Na_{1-x}K_xCl$, the masses of the three components are comparable.

In the two mode systems results have been reported only for covalent materials. Included in this class are $GaAs_{1-x}P_x$,(11,12) $CdS_{1-x}Se_x$,(13,14) $InAs_{1-x}P_x$(15) and $Ge_{1-x}Si_x$(16). For the compound material, the reststrahlen bands of the end members are well separated in frequency space and the substitution is made for the lighter element of the compound.

Several models have been advanced to describe the observed variation of frequency with composition. We are in this paper more concerned with the conditions which give rise to one or two mode behavior than with a detailed discussion of the observed frequency variation. In general, the models which have been discussed do not consider the entire frequency spectrum of the optical and acoustical branches, but focus on a single vibrational mode. In this paper we develop a model based first on low concentration "isotopic" substitution in a linear diatomic chain and then extend the result to the higher concentrations by considering mixed diatomic chains. We then discuss the conditions necessary for the existence of localized and gap modes in three dimensional crystals and based on this indicate the behavior that is to be expected in mixed crystal systems.

II. Discussion of Models Used To Describe Frequency Variation in Mixed Crystal Systems

A. Virtual Crystal

Qualitatively, the almost linear variation of frequency with concentration, i.e., one mode behavior, as observed in $Ni_{1-x}Co_xO$, $(Ca,Ba)_{1-x}Sr_xF_2$, etc. can be treated by means of a virtual crystal approximation [17,18] in which all masses and spring constants are taken as averages weighted by the mixed crystal composition. However, this model is only a descriptive one and can not predict the type of behavior.

B. Cluster Model

Verleur and Barker[11] have considered a model, based on short range clustering, to account for the two mode behavior of $GaAs_{1-x}P_x$ and $CdS_{1-x}Se_x$. Their model can also account for the observed fine structure in the reflectivity spectra. However, in considering the one mode behavior observed in $Ba_{1-x}Sr_xF_2$[8], the model requires the absence of clustering to account for the observed spectra.

C. Random Element Isodisplacement (REI) Model

The REI model[12] assumes that each of the atomic species vibrates with the same phase and amplitude. This is exactly true for a completely ordered lattice and can be extended to disordered lattices by using weighted force constants. As such the model has features similar to both features similar to both the virtual crystal approximation and the linear diatomic chain. To fit the results obtained for $GaAs_{1-x}P_x$, it was necessary to include a large force constant between the As and P sublattices. If this force constant is set equal to zero then two modes are predicted, one varying continuously in frequency between that of GaAs and GaP, and one at a frequency lower than that of either material. Simultaneous application of this model to $Cd_{1-x}Zn_xS$ and $CdS_{1-x}Se_x$ leads to obvious difficulties. To generate two mode behavior in $CdS_{1-x}Se_x$ one needs a large force constant between Se and S atoms, whereas to generate one mode behavior in $Cd_{1-x}Zn_xS$ one must set the Cd-Zn force constant equal to zero and further assume that the low frequency mode is too weak to be observable.

D. Linear Chain Models

Matossi[19] has used a linear diatomic chain model which considers only nearest neighbor force constants to discuss the one mode behavior of $Na_{1-x}K_xCl$. A second low frequency infrared active mode is also predicted for $Na_{1-x}K_xCl$, but has not been observed. Differences in the calculated frequency spectra in this model, are related to the mass differences in the isotopic substitution which generates the mixed crystal. The same mode has been used by Langer, et al.[20] to describe the two mode behavior of the longitudinal optical frequencies in $CdS_{1-x}Se_x$. Of all the models mentioned, it is suggested that the linear chain model holds the most promise for developing a criterion for determining a priori whether a system will display one or two mode behavior. As such we focus on the implications of this model.

III. Localized and Gap Modes in a Diatomic Linear Chain

The phonon dispersion curves for a linear diatomic chain which considers only nearest neighbor force constants and in which the masses of the components M and m differ substantially (M>>m) are well known. To a first approximation the frequency dispersion curve of the optical branch is determined largely by the mass of the lighter element, whereas the gap between the optical and acoustical branches depends strongly on the differences in the masses; i.e., the greater the disparity in mass, the larger the gap.

Mazur, Montroll and Potts[21] have discussed the types of impurity modes that occur due to isotopic substitution in either constituent. The substitution can be characterized by the mass defect parameter ϵ, which is the fractional change in mass normalized to the host chain atoms. For example, if we substitute for the heavier element M an atom of mass M', then

$$\epsilon = 1 - \frac{M'}{M} \qquad (1)$$

If $M' < M$, then $1 > \epsilon > o$ whereas if $M' > M, \epsilon < o$. The results of the impurity mode calculation are also shown in ref. 21. When substituting for the lighter mass m, a lighter impurity ($\epsilon > o$) gives a localized mode which rises out of the top of the optical band and a heavier mass ($\epsilon < o$) causes an optical branch gap mode to fall out of the bottom of the same optical branch. However, the situation is different when substituting for the heavier mass, M. In this case one gets two modes when substituting with a lighter mass ($\epsilon<o$), a local mode rising out of the top of the optical branch and a gap mode rising out of the top of the acoustical branch. When substituting with a heavier impurity ($\epsilon<o$) for the heavier mass, no new modes, gap or localized, are generated.

It is interesting to apply this model to $CdS_{1-x}Se_x$ and $Cd_{1-x}Zn_xS$. In the linear chain model, substituting S for Se in CdSe generates a localized mode which rises out of the top of the optical branch and substituting Se for S in CdS produces a gap mode which falls out of the bottom of the optical branch. For the $CdS_{1-x}Se_x$ system these modes are well separated from their "parent" branches. Therefore two mode behavior is possible at either end of the system. On the other hand, for $Cd_{1-x}Zn_xS$, the situation is markedly different. The addition of Cd to ZnS produces neither a localized nor a gap mode. The addition of Zn to CdS produces both an acoustical branch gap mode and an optical branch localized mode; however the frequencies of these modes lie very close to their "parent" branches. For $Cd_{1-x}Zn_xS$, therefore, two mode behavior is at most possible at only one end of the system; however the inclusion of long range forces and the dispersion of the resultant transverse optical and longitudinal optical branches prevent a distinct second mode forming close to the host lattice modes.

At this point we indicate what happens with increasing concentration by considering ordered mixed linear chains. Matossi[19] found the vibrational frequencies of the infrared active modes of a linear ordered diatomic 50% mixed chain ...xzyzxzyz... where nearest neighbor force constants, obtained from the frequencies of the end member compounds, are used. It is illustrative to consider the results obtained for two extreme cases. Consider first the mass of x to be very much heavier than the masses of y and z. For this case the two infrared frequencies are very close to those of the end member materials. In the other extreme when the masses of both y and z are very much greater than x, the model predicts one infrared active mode close to zero frequency and one at a frequency about half way between the frequencies of the end member compounds. This calculation is meant to be illustrative and is a useful guideline to follow in developing a model for one or two mode behavior that does rely heavily on mass or isotopic substitution.

It should be pointed out that the one or two mode behavior discussed above is not restricted to ordered 50% mixed linear chains. Langer, et al.[20] for example, have shown that a linear chain model applied to $CdS_{1-x}Se_x$ will yield two mode behavior over the entire composition range. Similarly, the same model applied to $Cd_{1-x}Zn_xS$[1] yields a result which is compatible with the observed one mode behavior. However for $Cd_{1-x}Zn_xS$, like the REI model, the linear chain model also predicts a second low frequency infrared active mode (which has never been observed). We now go on to discuss localized and gap modes in three dimensions and then develop the conditions which lead to one or two mode behavior in three dimensional crystals.

IV. Localized and Gap Modes in Three Dimensions

A formalism, based on isotopic substitution, and used for the calculation of gap mode and localized mode frequencies in three dimensional lattices has been developed by Dawber and Elliot,[22,23] who applied the results of their calculation to Si. Jaswal[24] extended their formalism to a diatomic lattic and calculated the frequencies of gap and localized modes for NaI. To perform the calculation it is necessary to have a complete description of the eigen-frequencies and eigen-vectors of the host lattice. This complete description is available only for a limited number of semiconducting and insulating crystals, e.g., Si[25], GaAs[26], NaI[27] and CdS[28].

Jaswal's work is most useful in discussing diatomic lattices where the masses of the elements differ substantially. The results of Jaswal's calculation for NaI, a material in which there is a gap between the optical and acoustical branches, parallel those of Mazur, Montroll and Potts for the diatomic chain. Specifically replacing the lighter mass, in this case Na, with a heavier or lighter mass yields respectively, a gap mode dropping out of the optical branch or a localized mode arising out of the optical branch. Replacing the heavier mass, in this example, I, with a heavier mass gives neither a gap nor a localized mode, whereas replacing the heavier mass with a lighter mass produces a gap mode for one range of $\epsilon (o<\epsilon<0.8)$ and a localized mode for another range $0.8<\epsilon<1$. This is to be contrasted with the linear chain model where substitution of a lighter mass for the heavier mass gives rise to localized as well as gap modes for all values of ϵ. It is not clear whether this difference is due to three dimensionality of the lattice, i.e., the inclusion of depolarization forces, or whether it is peculiar to NaI because of the eigen-values and eigen-vectors associated with its phonon spectrum. Unfortunately, there is no formalism which allows us to start with the low concentration isotopic substitution model and develop the phonon spectrum of the mixed crystal from it. Hardy[29] has however, shown how a gap mode can develop into a strong second mode in $LiH_{1-x}D_x$ mixed crystal. At this point we therefore suggest that the conditions that lead to the development of either one mode or two mode behavior are indeed derived from the criteria which allow the development of localized and gap modes out of the optical branch.

V. Conditions for One or Two Mode Behavior in Mixed Crystal Systems

Consider the mixed crystal system $A_{1-x}B_xC$, where A is heavier than B. If the substitution of A for B in BC produces a gap mode and if the substitution of B for A in AC produces a localized mode, then we suggest that two mode behavior will result in the mixed crystal system. This is usually the case if A and B are both lighter than C and is never the case if both A and B are heavier than C. A condition necessary, but not sufficient for two mode behavior is the existence of a gap in the phonon spectrum of the lighter mass compound, in our example, BC. If neither gap nor localized modes are possible, then one mode behavior results. The mechanism for the development of single mode behavior may be related to the generation of in-band resonance modes,[22] which as the concentration is increased, serve to perturb and shift the lattice bands of the crystal.

For crystals, such as $Na_{1-x}K_xCl$, where all of the masses are about equal we treat the heavier compound as a host and consider the possibility of localized mode formation by the lighter element, i.e., in this case we are concerned with the local mode of Na in KCl. The mass defect for the substitution of Na for K is 0.4. Using Dawber and Elliot's calculation for Si we predict the existence of a strong localized mode for Na in KCl; however, there is no gap in the phonon spectrum of NaCl, so that a K gap mode is not possible. Therefore we expect one mode behavior with the occurrence of a localized mode at the K rich end of the composition range. A similar situation is expected for $GaAs_{1-x}Sb_x$ and is indeed consistent with the recent observations of Stierwalt and Potter[10]. It is worthwhile to note that the absence of a gap between the acoustical and optical branches in GaAs precludes simple two mode behavior in $GaAs_{1-x}Sb_x$, but has no bearing on the two mode behavior of $GaAs_{1-x}P_x$. For the latter system a gap is required in the frequency spectrum of GaP.

It is tempting to try to tie the occurrence of one or two mode behavior to overlap of end member restrahlen bands. Here we note that the occurrence of frequency overlap usually implies the existence of one mode behavior, however, the lack of frequency overlap is not always a clear indication of two mode behavior. For example, the reststrahlen bands of GaAs and GaSb do not overlap, however the absence of a gap in GaAs precludes two mode behavior. By a similar argument we predict one mode behavior for $Ga_{1-x}In_xAs$. Another interesting situation occurs in the reststrahlen behavior of $Se_{1-x}Te_x$. For this uniaxial crystal the reststrahlen bands of the end members for the two principle directions of polarization are separated in frequency space and one mode behavior results. It is not completely obvious as to whether the observed behavior is associated with crystal structure or with the detailed behavior of Isotopic substitution as determined by the dispersion curves of the end member elements.

There are several interesting experiments which would be useful in helping to resolve completely the question of the occurrence of one or two mode behavior in mixed crystal systems. Firstly it would be of interest to observe two mode behavior in an alkali halide crystal. An examination of the list of crystals with gaps and of the observed TO and LO frequencies of end member compounds suggests that this is most difficult. In particular for the alkali halides low values of high frequency dielectric constant result in large depolarization fields and hence wide reststrahlen bands making it difficult to find ternary systems with well separated reststrahlen bands.

It would also be of interest to demonstrate two mode behavior in metal substituted covalent systems. Here we suggest $Al_{1-x}In_xSb$ or $Al_{1-x}Ga_xSb$. The reststralen bands of the end member of either of the systems are well separated, so that if AlSb has a gap (which is a likely possibility in view of the large disparity in mass between Al and Sb) we would expect two mode behavior to occur.

Acknowledgement

The authors thank Dr. R.F. Potter and Dr. D.L. Stierwalt for communicating some of their unpublished data.

References

*Present Address, School of Engineergin, Case-Western Reserve University

†Research supported in part by the U.S. Office of Naval Research.

(1) G. Lucovsky, E. Lind and E.A. Davis, presented at the "International Conference on II-Vl Semiconducting Compounds", Brown University, Providence, Rhode Island, September 6-8, 1967.

(2) G. Lucovsky and R.C. Keezer, to be published.

(3) F. Krueger, O. Reinkober A.E. Koch-Holm, Ann. Physik 85, 110 (1928).

(4) R.M. Fuller, C.M. Randall and D.J. Montgomery, Am. Phys. Soc. 9, 644 (1964).

(5) P.J. Gulisse, J.N. Pendl, L.D. Mausar, R. Marshall and S.S. Mitra, J. Appl. Phys. 36, 2447 (1965).

(6) R.K. Chang, B. Lacina and O.S. Pershan, Phys. Rev. Lett. 17,755 (1966).

(7) H.W. Verleur and A.S. Barker,Jr.,Bull.Am.Phys.Soc., 12,81 (1966).

(8) H.W. Verleur and A.S. Barker,Jr., to be published.

(9) R.F. Potter and D.L. Stierwalt, in Proceedings of the International Conference of Semiconductors, Paris, 1964 (Dunod, Paris, 1964), p.1111.

(10) D.L. Stierwalt and R.F. Potter, private communication.

(11) H.W. Verleur and A.S. Barker,Jr., Phys. Rev. 149, 715 (1966).

(12) Y.S. Chen, W. Shockley and G.L. Pearson, Phys. Rev. 151,648 (1966).

(13) M. Balkanski, R. Besserman and J.M. Besson, Solid State Commun. 4, 201(1966).

(14) H.W. Verleur and A.S. Barker, Jr., Phys. Rev. 155, 750 (1967).

(15) F.Oswald, Z. Naturforsch. 14A,374 (1959).

(16. D.W. Feldman, M. Ashkin and J.H. Parker, Jr., Phys. Rev. Lett. 17, 1209 (1966).

(17) J.S. Langer, J. Math. Phys. 2,584 (1961).

(18) Hin-Chiu Poon and A. Bienenstock, Phys. Rev. 147, 710 (1966) and Phys. Rev.142,466 (1966).

(19) F. Matossi, J. Chem. Phys., 19,161 (1951).

(20) D.W. Langer, Y.S. Park and R.N. Euwena, Phys.Rev. 152, 788 (1966).

(21) P. Mazur, E. Montroll and R.B. Potts, J. Wash. Acad. Sci. 46, (1956).

(22) D.G. Dawber and R.J. Elliott, Proc. Roy. Soc. (London) A273, 222 (1963).

(23) D.G. Dawber and R.J. Elliott, Proc. Phys. Soc. (London) 81, 453 (1963).

(24) S.S. Jaswal, Phys. Rev. 137, 302 (1965).

(25) F.A. Johnson and W. Cochran, 1962 Proc. Int. Conf. on Physics of Semiconductors, Exeter, 1962, p.498, London: Inst. of Physics and Physical Soc.

(26) G. Dolling and J.L.T. Waugh, Proc. Int. Conf. on Lattice Dynamics, J. Phys. Chem. of Solids 21, (Suppl.) 19 (1965).

(27) A.M. Karo and J.R. Hardy, Phys. Rev. 129, 2024 (1963).

(28) M.A. Nusimovici and J.L. Birman, Phys. Rev. 156, 925 (1967).

(29) J.R. Hardy, Phys. Rev. 136, A1745 (1964).

OPTICAL PHONONS IN ZnS_xSe_{1-x} MIXED CRYSTALS

O. Brafman, I. F. Chang, G. Lengyel, S. S. Mitra and E. Carnall, Jr.*

University of Rhode Island, Kingston, R.I., *Eastman Kodak Company, Inc. (A and O Division), Rochester, N.Y.

I. INTRODUCTION

The infrared transmittance and reflectance of pure crystals of ZnS and ZnSe have been reported several years ago[1-4]. The optically active phonon processes in ZnS_xSe_{1-x} crystals have been investigated here by infrared reflection, transmission and Raman spectroscopy for nine values of x between 0 and 1. We believe this is the first mixed crystal system to receive such a thorough treatment. Previous studies[5] on mixed zincblende- and wurtzite-type crystals were confined chiefly to reflection measurements and were thus limited to the elucidation of the transverse optic (TO) phonons at $\underset{\sim}{k} \sim 0$. Direct evidence on $\underset{\sim}{k} \sim 0$ longitudinal optic (LO) phonons are obtained from present Raman measurements. It became evident that the behavior of the present system is by far the most interesting case inasmuch as it neither behaves like ionic mixed crystals nor like covalent mixed crystals. In mixed systems composed of primarily ionic crystals[6], e.g. $Ni_xCo_{1-x}O$, KCl_xBr_{1-x}, $Na_xRb_{1-x}I$, $Ca_xSr_{1-x}F_2$ etc., due to the presence of long range Coulomb interactions only a single reststrahlen band occurs and the long wavelength LO and TO modes shift continuously and linearly with concentration. On the other hand, in a mixed system composed of covalent crystals like Ge:Si the triply degenerate $\underset{\sim}{k} \sim 0$ optic mode frequency of Ge shifts slightly downward with increasing Si content[7], rather than shifting upward continuously from the heavier to the lighter component's frequency.

The present investigation, which includes Raman measurements, reveals the behavior of both LO and TO phonons as functions of x. Along with the behavior of the mixed crystals CdS_xSe_{1-x} reported elsewhere[8], the present system constitutes by far the most complete and interesting system studied so far.

II. RESULTS AND DISCUSSIONS

The reflection spectra at near normal incidence of the crystals were measured by means of a Perkin-Elmer model 301 far infrared spectrophotometer. The reststrahlen spectra of a few of the mixed crystals of ZnS_xSe_{1-x} are shown in Figure 1. The infra-

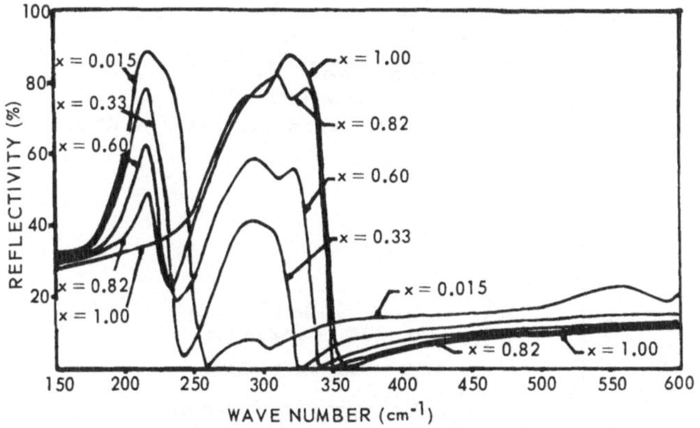

Fig. 1 - Reststrahlen spectra of ZnS_xSe_{1-x} for a few values of x.

red-active phonons occur in two main bands with frequencies near those of pure Zns and pure ZnSe, characteristic of a system in which short range forces strongly influence the lattice dynamics. Kramers-Kronig dispersion analyses[9] of the reflection spectra were performed to obtain the real (ε') and imaginary (ε'') parts of the complex dielectric constant and other optical constants. A typical result is shown in Figure 2. For all values of x only two distinct peaks with occasional minor structures were noted in the curves for extinction coefficient, conductivity or ε'' versus frequency. The peak frequency of the conductivity curves are assigned to TO (ZnS) and TO (ZnSe) depending on their positions. Verlur and Barker, on the other hand, have used a model of lattice dynamics for mixed zincblende and wurtzite type crystals, which utilizes seven distinct TO phonons at $\underset{\sim}{k} \sim 0$. A closer inspection of their values, however, reveals that for each mixed crystal studied, two out of the seven modes considered stand out in strength over the rest. We find a two-resonance damped oscillator model is adequate to describe the essential features of the reststrahlen spectra of the mixed crystals reported here. A typical example is shown in Figure 3.

The selection rules for the zincblende structure indicate

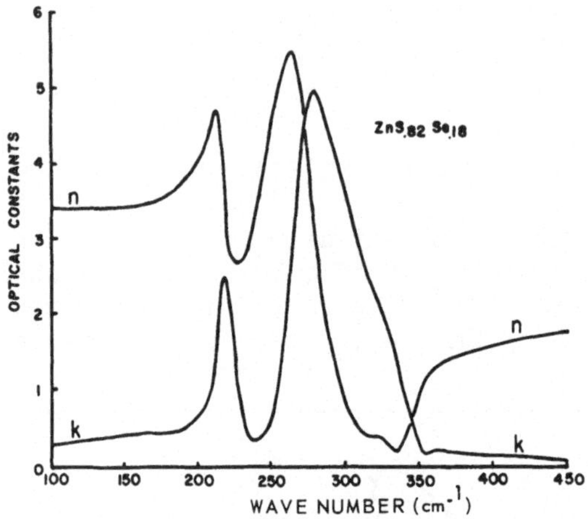

Fig. 2 - Optical constants of $ZnS_{.82}Se_{.18}$ from a Kramers-Kronig analysis.

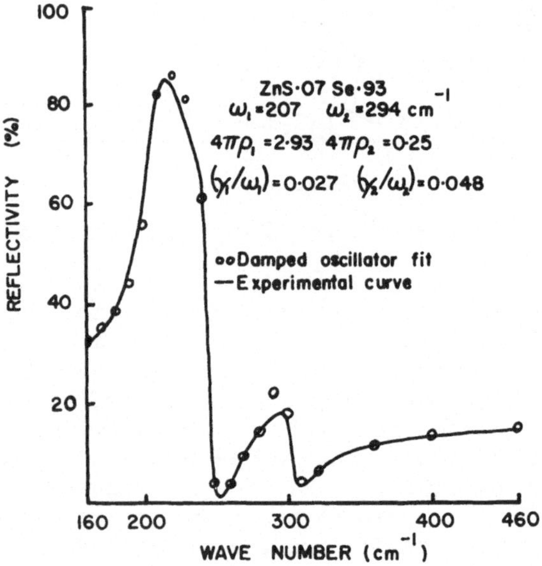

Fig. 3 - Damped oscillator fit for the reflection spectrum of $ZnS_{.07}Se_{.93}$.

that both the long wavelength LO and TO modes are Raman-active. Moreover, in the perpendicular scattering configuration the LO intensity is expected[10] to be approximately an order of magnitude higher than that of the TO. For ZnSe a strong LO and a weak TO Raman bands are observed[4]. In the case of cubic ZnS, only the LO mode is observed in the Raman spectrum. Three first order Raman bands may thus be expected for each mixed crystal. The Raman spectra were excited with a He-Ne laser (6328A) in the perpendicular scattering mode and were analyzed by means of a Spex 1400 double monochromator. A few representative spectra are shown in Figure 4. The two stronger bands are assigned to the LO modes of

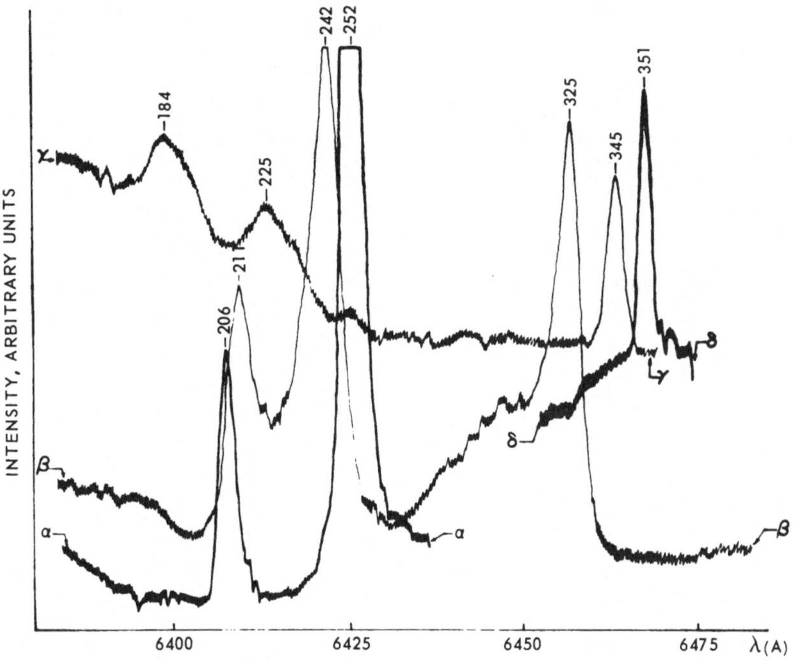

Fig. 4 - Raman spectra of ZnS_xSe_{1-x}: x = 0 (α), x = 0.33 (β), x = 0.82 (γ).

ZnS and ZnSe, while the weaker one corresponds to the TO mode of ZnSe and agrees with the values obtained from the Kramers-Kronig analyses and the damped oscillator calculations within $\pm$ 3 cm^{-1}.

The one-phonon optic frequencies are plotted as functions of the mole fraction x of ZnS in Figure 5. It may be noted that the LO frequency of ZnS monotonically decreases, while the TO frequency increases with increasing concentration of ZnSe. These two lines intersect at a point in the region of 100°/o ZnSe. This triply degenerate phonon at 297 cm^{-1} may be considered as the quasi-localized mode of S in ZnSe. The local mode frequency may be approximately evaluated by using the Dawber and Elliott theory[11]. A Debye distribution with a cut-off frequency equal to the top of the optic band (LO at $k \sim 0$) was assumed for the host lattice. A value of 305 cm^{-1} is obtained for the local mode of S in ZnSe, which is in surprisingly good agreement, in spite of the approximations involved, with the value extrapolated from the mixed crystal data. It may be noted in Figure 1 that the quasi-localized mode and its second harmonic are present in the room temperature reststrahlen spectrum of $ZnS_{0.015}Se_{0.985}$. In the Raman spectrum

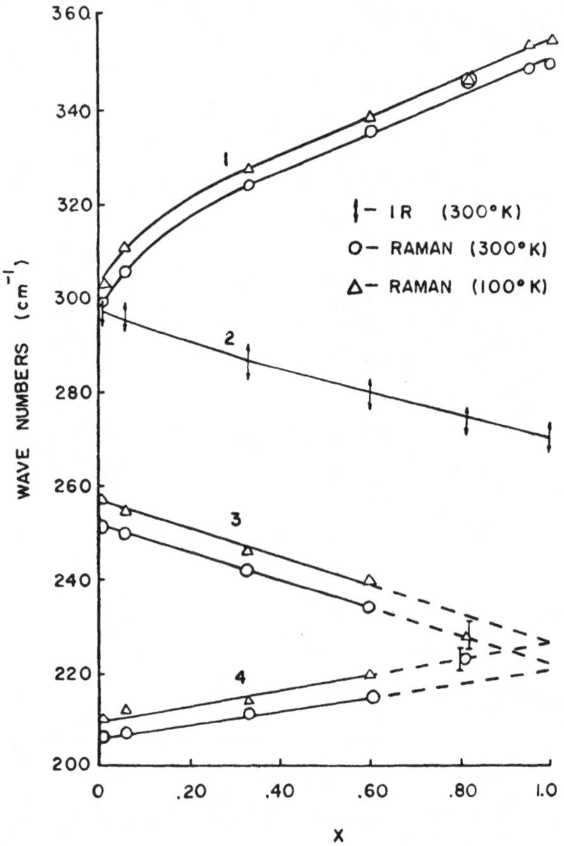

Fig. 5 - $\underset{\sim}{k}\sim 0$ Optical Phonons of ZnS_xSe_{1-x} as functions of x. 1:LO(Zns); 2:TO(ZnS); 3:LO(ZnSe); 4:TO(ZnSe).

of this crystal a band is observed at 299 cm^{-1} which sharpens considerably at 80°K. However, the local mode spectrum is perhaps best demonstrated by the absorption spectrum of $ZnS_{0.015}Se_{0.985}$. A thin sample (~ 100μ) shows a band at 295 cm^{-1} [Figure 6] which sharpens as the temperature goes down. The second harmonic at 589 cm^{-1} and its temperature dependence are also shown in Figure 6. The temperature dependence of half-width and peak intensity are not as dramatic as in the case of U-center (H^- ion in alkali halide)[12]. This is only to be expected, since while the mass ratio of H to K or Cℓ is approximately 1:35, that of S to Zn or Se is roughly only 1:2. Thus the latter impurity mode has been termed "quasi-localized". Moreover, the local mode half-width is, to some extent, concentration dependent[13]. A mole-fraction of 15 x 10^{-3} of ZnS in ZnSe may not be considered a suitable dilution for local mode condition. This is further evident from the slight difference in the local mode frequency measured by the Raman and

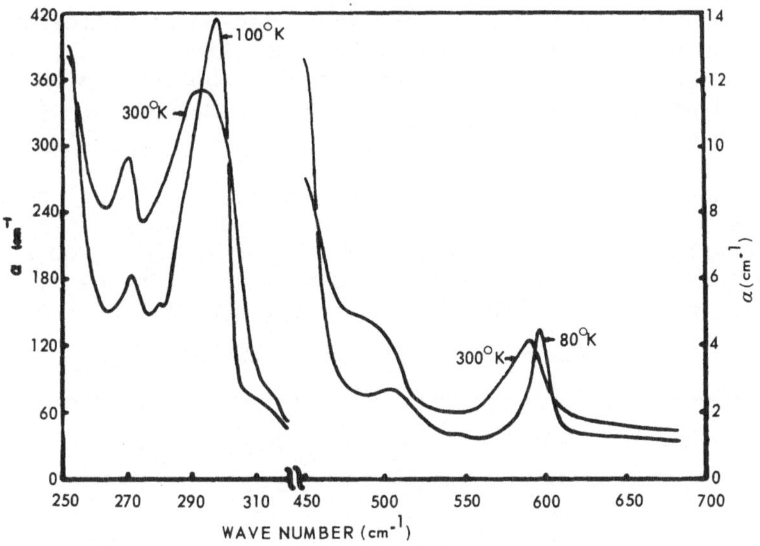

Fig. 6 - Infrared absorption due to local mode of S in ZnSe and its first overtone (left hand ordinate for the fundamental, right hand ordinate for the overtone.

infrared techniques. While the former indicates an 'LO-type' mode, the latter a 'TO-type', the complege degeneracy can occur only at the condition of infinite dilution.

The behavior of the k~0 optic modes of the heavier component may be summarized as follows. The LO of ZnSe decreases and its TO increases with increasing concentration of ZnS. The point of intersection which occurs in the region of 100°/o ZnS may be regarded as the resonant gap mode of Se in ZnS. This occurs approximately at 220 cm^{-1}. From Jaswal's theory(14) an estimated value of 225 cm^{-1} is obtained(15) for the gap mode of Se in ZnS. It is worth noting that while the variation of the TO modes with concentration is linear, that of the LO mode of the lighter component shows pronounced deviation from linearity in the region of the local mode.

The variation of the long wavelength optic phonon frequencies with concentration as presented in Figure 5 is unique in the sense that it is different from either that of the ionic(6) or the covalent(7) crystals. Yet in some respects it reflects the pertinent features of either type of solids. The lines (1) and (4) of Figure 5 are in some ways reminiscent of ionic crystals(6). Whereas line (3) is similar to that found for covalent crystals. The II-VI compounds may thus be regarded in many ways behaving as ionic crystals (e.g., the separation of the LO and TO branches at k~0), yet in some other respects they are like covalent crystals (e.g., the short range forces that give rise to characteristic frequencies for both the components in the mixed crystal, Ge:Si). The latter observation is in agreement with the conclusion recently arrived at from the study of the pressure dependence of long wavelength optic modes of ZnS[16].

Another striking feature of the present results is worth noting. The sum of one phonon optic frequencies at any concentration properly weighted for degeneracy (1 for an LO, 2 for a TO and 3 for a local or a gap mode) is a constant with a value of 1559 ± 3 cm^{-1} as indicated in Table 1. Moreover, the 'average' one-

Table I. SUM RULE FOR THE MIXED SYSTEM ZnS_xSe_{1-x}

X	LO_1	TO_1	LO_2	TO_2	$\sum_{i=1}^{6} \nu_i$	$\frac{1}{6} \Sigma \nu_i$
0.015	299	295	252	206	1553	259
0.060	306	293	250	207	1556	259
0.33	324	287	242	211	1562	260
0.60	336	280	234	215	1560	260
0.81	345	274		223	1562	260
1.00	350	271		221	1555	259
					Average	260

(Local Mode + Gap Mode)/2 = 259

$$\text{Local Mode + Gap Mode} = \left(\frac{LO+2TO}{3}\right)_{ZnS} + \left(\frac{LO+2TO}{3}\right)_{ZnSe}$$

(518) (519)

phonon frequency of the mixed system under consideration, which occurs at 259.8 ± 0.5 cm^{-1}, is to a very good approximation halfway between the quasi-local mode frequency of S in ZnSe and the resonant gap mode of Se in ZnS. Another consequence of this sum rule is that in the mixed crystal system AB_xC_{1-x} ($m_B < m_C$) the sum of the local mode of B in AC and the gap mode of C in AB can be predicted from the $\underline{k} \sim 0$ optic phonons of the pure crystals AB and AC alone:

$$\text{Local mode + gap mode} = \left(\frac{2TO+LO}{3}\right)_{AB} + \left(\frac{2TO+LO}{3}\right)_{AC} \qquad (1)$$

The left hand side of Equation (1) in the present case is 518 cm^{-1} whereas the right hand side is 519 cm^{-1}. This sum rule is not unique to ZnS_xSe_{1-x}. It is also obeyed by CdS_xSe_{1-x}.

The present study shows that the local mode and the gap mode which become operative at a very low concentration of one component in the other, follow as extensions of the general behavior of the mixed system. We venture to remark here that investigations on the lattice dynamics of mixed systems in the past may only be regarded as partial inasmuch as only a few of the optical branches were determined. For example, in the case of the ionic crystals, although it was previously noted[6] that the LO and TO modes shift continuously and linearly with concentration, no branch was noted that could predict the local or gap mode for nominal concentration of one component in the other. The study on mixed covalent crystals[7] is also limited inasmuch as the local modes and the "proper" band modes do not follow a continuous pattern. We feel some branches of the optically active phonons, although expected, were not seen, perhaps due to oversight.

ACKNOWLEDGMENT

We are very grateful to Dr. J. N. Plendl for many helpful discussions and his keen interest in the investigation. This research was supported in part by the Air Force In-House Laboratory Independent Research Fund under Contract AF19(628)-6042. An equipment grant from the ARPA (Grant No. DA-ARO-D-31-124-G754) is gratefully acknowledged.

REFERENCES

1. T. Deutsch, Proc. Internatl. Conf. on Semiconductors, Exeter, 1962 (The Inst. of Physics and Physical Society, London, 1962), p. 505.

2. M. Aven, D.T.F. Marple and B. Segall, J. Appl. Phys. 32, 2261 (1961).

3. R. Marshall and S. S. Mitra, Phys. Rev. 134, A1019(1964).

4. S. S. Mitra, J. Phys. Soc. (Japan) 21 (Supplement), 61 (1966).

5. M. Balkanski, R. Beserman and J. M. Besson, Solid State Commun. 4, 201 (1966); H. N. Verleur and A. S. Barker, Jr., Phys. Rev. 149, 715 (1966); ibid. 155, 750 (1967).

6. P. J. Gielisse, J. N. Plendl, L. C. Mansur, R. Marshall, S. S. Mitra, R. Mykalajewycz and A. Smakula, J. Appl. Phys. 36, 2446 (1965); A. Mitsuishi in "U.S.-Japan Cooperative Seminar on Far Infrared Spectroscopy", Columbus, Ohio, September 1965 (unpublished); R. M. Fuller, C. M. Randall and D. J. Montgomery, Bull. Am. Phys. Soc. 9, 644 (1964); R. K. Chang, B. Lacina and P. S. Pershan, Phys. Rev. Letters 17, 755 (1966).

7. D. W. Feldman, M. Ashkin and J. H. Parker, Jr., Phys. Rev. Letters 17, 1209 (1966).

8. J. F. Parrish, C. H. Perry, O. Brafman, I. F. Chang and S. S. Mitra, Proc. Internatl. Conf. on the Physics of II-VI Semiconductors, Providence, R. I., 1967 (in press).

9. S. S. Mitra and P. J. Gielisse, Progress in Infrared Spectroscopy, Vol. 2, Plenum Press, Inc., New York, 1964.

10. R. Loudon, Advances in Phys. 13, 423 (1964).

11. P. G. Dawber and R. J. Elliott, Proc. Roy. Soc. (London) A 273, 222 (1963); Proc. Phys. Soc. (London) 81, 459 (1963).

12. S. S. Mitra and R. S. Singh, Phys. Rev. Letters 16, 694(1966).

13. W. G. Spitzer, J. Phys. Chem. Solids 28, 33 (1967).

14. S. S. Jaswal, Phys. Rev. 137, 302 (1965).

15. The replacement of an $S^{=}$ ion with an $Se^{=}$ ion corresponds to a mass defect parameter of -1.47. Assuming that Jaswal's model for NaI is valid for ZnS, when properly scaled down, one would expect that the gap mode of Se in ZnS to be situated approximately in the center of the gap. The zone boundary acoustic phonon frequencies of ZnS are from reference 3.

16. S. S. Mitra, C. Postmus and J. R. Ferraro, Phys. Rev. Letters 18, 455 (1967).

LOCALIZATION OF EIGENMODES IN DISORDERED ONE-DIMENSIONAL SYSTEMS. I. GENERAL THEORY AND ISOTOPICALLY DISORDERED DIATOMIC CHAIN

Jun-ichi Hori and Sakae Minami*

Department of Physics, Hokkaido University, Sapporo, Japan and *Faculty of Engineering, Gumma University, Kiryu, Gumma, Japan

1. INTRODUCTION

The arguments hitherto presented on the localization of eigenmodes of the one-dimensional disordered system(1)-(3) are to a large extent intuitive and qualitative. Borland(4) presented a quantitative theory but it depends on the assumption of ergodicity of the phase distribution and on the restrictive condition of Fréchet, which is not fulfilled e. g. for an isotopically disordered diatomic chain. In this paper we present a general theoretical framework which is convenient to discuss the phenomenon of localization without being restricted by any assumption or condition, and upon this basis demonstrate numerically that the eigenmodes of the isotopically disordered chain must always be localized. At the same time it is shown that there exists a unique ensemble phase-density function in spite of the fact that the condition of Frechet is violated, and that the phase distribution has the ergodic property.

2. THEORY

Consider any one-dimensional chain which can be described by the transfer-matrix formalism.(3) Let the transfer matrix at a site of the chain be

$$\underset{\sim}{Q} = \begin{pmatrix} A & B \\ B^* & A^* \end{pmatrix} , \qquad (2.1)$$

where the elements contain an energy parameter λ . Denote the state vector and the state ratio at the site by $X\equiv(x,y)^T$ and $z\equiv x/y$, respectively, and the corresponding quantities at the next site by $X'=(x',y')^T$ and $z'=x'/y'$. Then these quantities are related by

$$X' = \underset{\sim}{Q}X \quad \text{and} \quad z' = (Az+B)/(B^*z+A^*). \tag{2.2}$$

Defining the phase δ of a state ratio z by $z=e^{i\delta}$, we obtain from (2.2) a simple relation

$$d\delta/d\delta' = |Az+B|^2 = \|X'\|^2/\|X\|^2. \tag{2.3}$$

This means that the length of the state vector, or the envelope of the solution (which satisfies the boundary condition at the end of the system from which one starts the transfer), increases or decreases at the site under consideration according as $d\delta/d\delta'>1$ or <1.

The function Az+B gives a circle-to-circle mapping on the complex plane. Since z can be considered as lying on the unit circle, Az+B lies on the circle with the centre at B and the radius $|A|$. As $|A|^2=|B|^2+1$, this circle passes through the ends C and D of the diameter of the unit circle which is orthogonal to the vector B (Fig. 1). The quantity $d\delta/d\delta'$ is smaller than unity on the arc CED while larger than unity on CFD. Therefore when z is on the arc GIH the length of the state vector decreases, while when z is on GJH, it increases. Thus the arcs GIH and GJH give the "shortening or S-" and "lengthening or L-" regions, respectively. It is important to note that the L-region is always wider than the S-region. If $\underset{\sim}{Q}$ is hyperbolic, i.e. if $|\mathrm{Re}A|>1$, its two eigenvalues $\theta_\pm$ are real ($|\theta_+|<1, |\theta_-|>1$), and the limit points $z_\pm$ of the transformation (2.2) given by $z_\pm=B(\theta_\pm-A)$ lie on the unit circle. Since at $z_\pm$ we have $d\delta/d\delta'=\theta_\pm^2$, the source point z_+ and sink point z_- must lie in the S- and L-region, respectively.

If the direction of the transfer is reversed (backward trans-

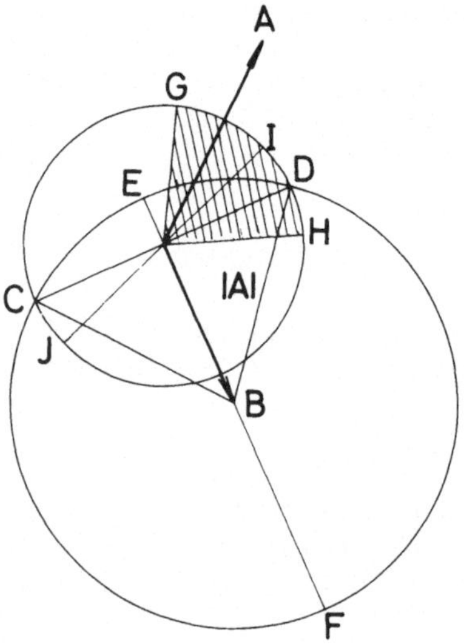

Fig. 1. Geometrical construction of S- and L-regions

fer), we have to replace A and B by A^* and $-B$, respectively. It can easily be shown that the motion of the state ratio on the unit circle becomes a mirror image of that for the forward transfer with respect to the diameter orthogonal to B (CD in Fig. 1). Since z_+ and z_- now interchange their role, they must be the mirror images of each other. The S- and L-regions for the backward transfer must also be the mirror images of those for the forward transfer.

In a regular chain described by only one matrix $\underset{\sim}{Q}$, z must move more slowly in the S-region than in the L-region, so that the phase distribution $f(\delta)$ becomes denser in the former and the average value of $\log\{d\delta/d\delta'\}$ (the sample degree of localization) vanishes. In the case of a disordered system, however, such an adjustment cannot be expected. For such a system must necessarily be described by a set of matrices $\underset{\sim}{Q}^{(i)}(i=1,2,\ldots,S;\ S>1)$. The state ratio is successively transformed by a random sequence of $\underset{\sim}{Q}^{(i)}$'s, and its motion becomes a random walk. Thus the overall distribution of the phase will either become more or less uniform, or, when some of the $\underset{\sim}{Q}^{(i)}$'s are hyperbolic, become particularly dense in their L-regions, inasmuch as z is strongly attracted by their sink points. Since it can be shown that in most cases the L-regions of $\underset{\sim}{Q}^{(i)}$'s fairly overlap one another, we can conclude that the sample degree of localization must have a positive value, and the state vector must be lengthened on the average.

If the direction of the transfer is reversed, the situation does not change, and it is again expected that the state vector becomes longer and longer on the average. Thus one is led to the conclusion that every eigenmodes of the disordered system must be localized: The two solutions, constructed from the opposite ends of the system will not match in general, i.e. their phases will not coincide at any site, except for some particular values of λ. Since these exceptional values of λ give the eigenfrequencies or eigenenergies, and the corresponding matched solutions are eigenmodes, these must be more or less localized.

3. NUMERICAL CALCULATIONS

In order to demonstrate that the expectation obtained above is correct for the case of the vibration of isotopically disordered chains, we computed numerically the phase-distribution function $f(\delta)$ together with the sample degree of localization for a chain containing two isotopes of mass ratio 2 with equal concentration. The total number of atoms was chosen to be 10^4, and $f(\delta)$ was calculated as the histograms with interval $\Delta\delta=2\pi/200$. We used the representation in which the transfer matrix $\underset{\sim}{Q}^{(0)}$ of the light atom is diagonal with the elements $A^{(0)}=\exp(2i\beta)$, $B^{(0)}=0$, and the matrix $\underset{\sim}{Q}^{(1)}$ of the heavy atom has the elements $A^{(1)}=(1+iQ\tan\beta)\times\exp(2i\beta)$, $B^{(1)}=iQ\tan\beta\exp(-2i\beta)$, where β is the wave-number para-

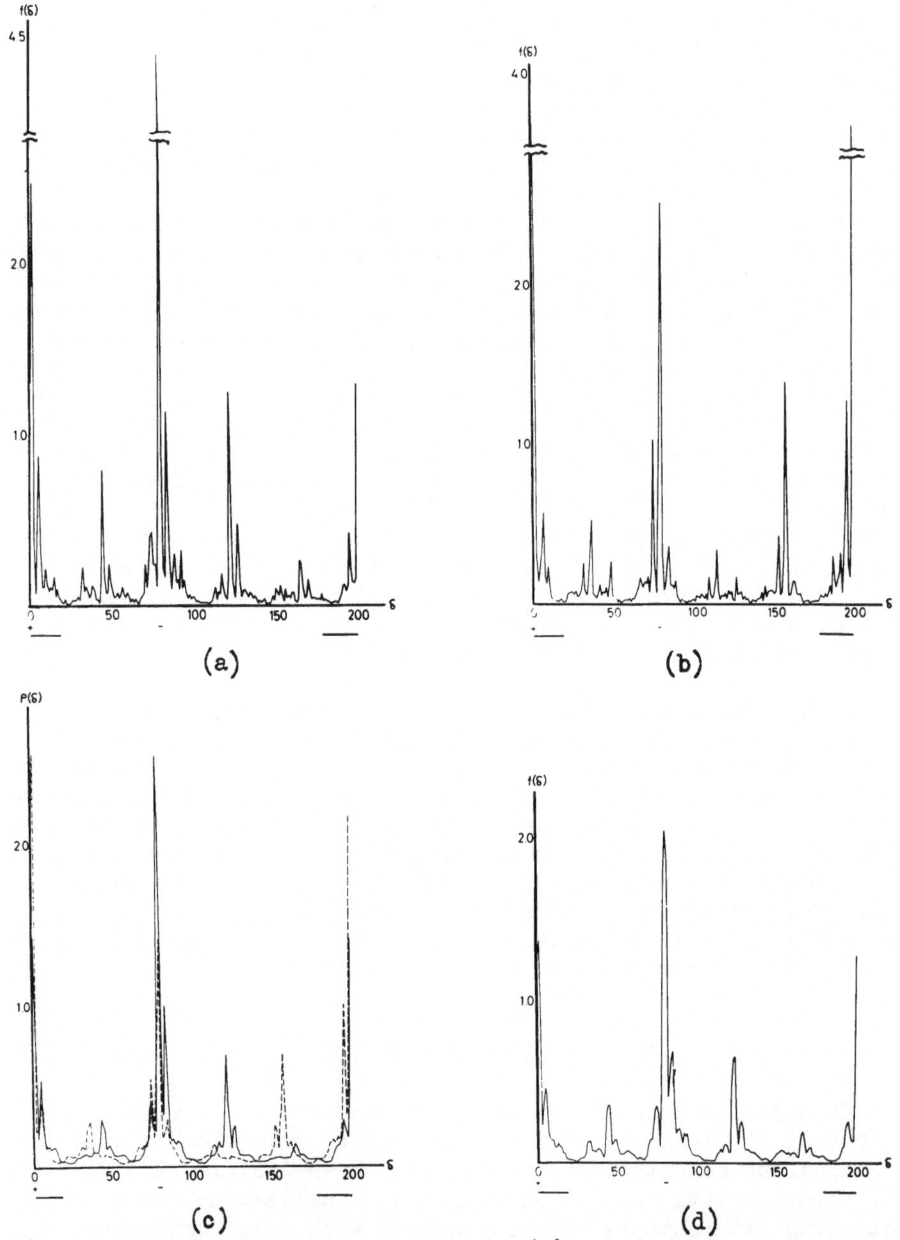

Fig. 2. Sample phase distributions f(δ) and ensemble phase densities (δ) calculated for β=0.95532. The symbols - and + and the horizontal bar indicate the positions of the sink- and source phases and the S-region, respectively, of the transfer matrix of the heavy atom, for the forward transfer. The unit of δ is $\pi/100$.

meter defined by $\omega^2=4K\sin^2\beta/m^{(0)}$, $Q\equiv(m^{(1)}/m^{(0)})-1$ (K is the force constant and $m^{(0)}$ and $m^{(1)}$ are the masses of the light and neavy atoms respectively). In this case the phase distributions for the forward and backward transfers must be mirror images of each other, since for $\underset{\sim}{Q}^{(0)}$ all the diameters of the unit circle become the axes of symmetry.

Fig. 2a shows the result for β=0.95532. The calculation was carried out for ten different initial values of δ, but the results were almost identical. This means that $f(\delta)$ is practically independent of the initial value. The sample degree of localization was found to be 0.4481±0.0001, the second term being the fluctuation owing to a small dependence on δ_{init}. The phase-values at the final 100 atoms were printed out, and it was found that all of the series of these 100 δ-values for different initial phase-values completely coincide with one another. This means that, in harmony with Borland's argument, the phase rapidly forgets its initial value during the transferring process. For $\delta_{init}=0$, $f(\delta)$ was calculated for five different sample chains, and it turned out that it does not depend upon the individual sample, except for relatively small fluctuations. We could find no marked difference between the distribution $\langle f(\delta)\rangle$ averaged over five sample chains and the individual $f(\delta)$.

Fig. 2b shows the distribution function calculated by transferring towards the backward direction. One sees at once that it indeed represents the mirror image of the curve in Fig. 2a within small fluctuations. The sample degree of localization was found to be 0.4492, which agrees well with the value found for the forward transfer.

In Figs. 2a and 2b the two most conspisuous peaks in $f(\delta)$ lie just at the sink- and source-phases $\delta_\pm$ of the transfer matrix of the heavy atom. It is a matter of course that in Fig. 2a the strongest peak lies at δ_-, since the phase is strongly attracted by δ_- in the forward transfer. That the second strongest peak lies at δ_+ is a peculiar feature for the present case, however. It has been brought about by that for the β-value chosen the matrix of the light atom incidentally transfers the phase near δ_- into the neighbourhood of δ_+. In Fig. 2b the positions of the strongest and the next strongest peak are interchanged. This is a natural consequence of the mirror-image symmetry between the forward and backward transfers.

For the same value of β we calculated also the ensemble phase-density function $\rho(\delta)$ and the ensemble degree of localization, solving the integral equation for $\rho(\delta)$ by iteration. The full curve in Fig. 2c shows the result for the uniform initial density. The value of the ensemble degree of localization came out to be 0.4892. The same calculation was carried out also for several delta-function-like initial densities, but no perceptible deviation from Fig. 2c was found. This means that there exists a unique limiting phase density, in spite of the fact that Fréchet's condition is not fulfilled in our case.

From Fig. 2c it is seen that $\rho(\delta)$ is very similar to $f(\delta)$. although there is a difference which is not trivial. The curve of $f(\delta)$ has a very jagged appearance whereas $\rho(\delta)$ is relatively smooth. Moreover, the value of the ensemble degree of localization obtained is somewhat larger than the corresponding sample degree of localization. The jaggedness can only partly be attributed to the sample fluctuation and the computational error, inasmuch as it does not disappear, as was mentioned above, even if $f(\delta)$ is averaged over samples. It almost disappears, however, if one replaces each frequency-value by the average of the nearest three frequency-values. Fig. 2d shows the curve obtained from Fig. 2a by this moving-average procedure. The similarity between Figs. 2d and 2c is remarkable, and suggests that the function $\rho(\delta)$ obtained represents a coarse-grained distribution. This is naturally expected from the discrete nature of the numerical computation. It is also expected that the value of $\rho(\delta)$ in the neighbourhood of δ_- is overestimated, since the coarse-graining at the source phase would accelerate the phase to flow into the L-region. The excess strength of the peak at δ_- in Fig. 2c and the too large value of the obtained ensemble degree of localization can be ascribed to such a situation. The broken curve in Fig. 2c shows the density $\rho(\delta)$ obtained by the backward transfer. The mirror image symmetry between the distributions for the forward and backward transfers is much more beautifully demonstrated than in the case of $f(\delta)$.

Thus we can conclude that, if it were not for the sample fluctuation in $f(\delta)$ and the coarse-graining in $\rho(\delta)$, the functions $f(\delta)$ and $\rho(\delta)$ should coincide with each other, in other words, the distribution of the phase is ergodic.

In spite of its coarse-grained nature, the function $\rho(\delta)$ serves well for investigating qualitatively how the phase distribution and the degree of localization vary with the frequency. For this purpose we calculated $\rho(\delta)$ for some other values of β. Figs. 3a-3c show the densities $\rho(\delta)$ for $\beta=3\pi/16$, $5\pi/16$ and $3\pi/8$, respectively. In each figure the full curve corresponds to the forward, and the broken one to the backward transfer. The positions of δ_- and δ_+, if any, are indicated by the symbols - and +, respectively. The horizontal bar indicates the S-region of the matrix of the heavy atom for the forward transfer. The values of the degree of localization obtained are also written in each figure (the upper corresponds to the forward, and the lower to the backward transfer). As was expected, the degree of localization is always positive. At $\beta=3\pi/16$, the distribution tends to be concentrated in the S-region, but this tendency must have been weakened compared to

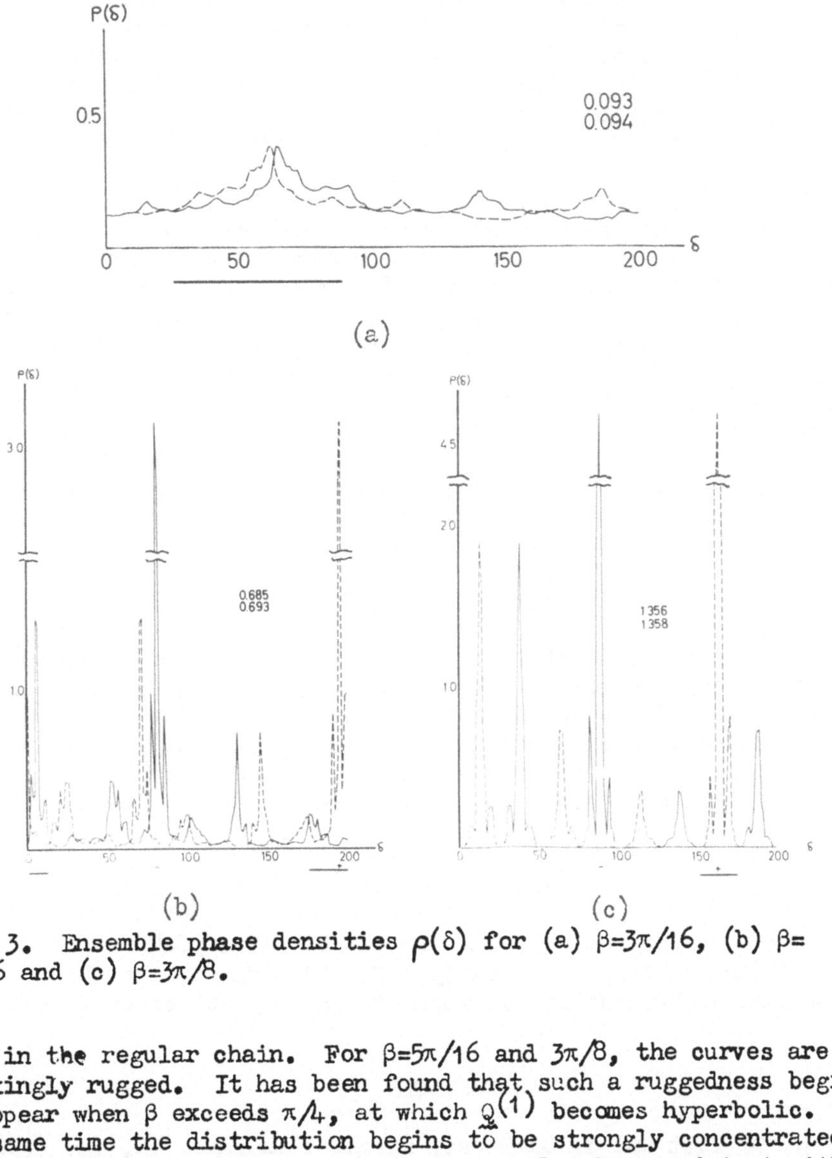

Fig. 3. Ensemble phase densities $\rho(\delta)$ for (a) $\beta=3\pi/16$, (b) $\beta=5\pi/16$ and (c) $\beta=3\pi/8$.

that in the regular chain. For $\beta=5\pi/16$ and $3\pi/8$, the curves are strikingly rugged. It has been found that such a ruggedness begins to appear when β exceeds $\pi/4$, at which $Q^{(1)}$ becomes hyperbolic. At the same time the distribution begins to be strongly concentrated into the L-region. These features are completely consistent with the observation made at the end of §2. The degree of localization rapidly increases with the frequency, as a natural consequence of the fact that the sink phase attracts the phase more and more strongly and moreover the L-region becomes rapidly wider as the frequency increases. Thus the localization of eigenmodes must become particularly conspisuous in the high-frequency region. This is in harmony with the result obtained by Dean and Bacon[5].

We carried out similar calculations also for other values of the mass ratio, and found that the ruggesness in $f(\delta)$ becomes more and more conspisuous and the degree of localization increases rapidly as the mass ratio increases. This is again a natural consequence of the fact that both the width of the L-region and the attractive power of δ_- rapidly increases with the mass ratio.

It would be worthwhile to note that in Fig. 3c the phase distributions for the forward and the backward transfers are completely non-overlapping. The same phenomenon was found also for $\beta=\pi/4$. This is natural since $\beta=\pi/4$ and $3\pi/8$ correspond to the special frequencies[3] at which the spectral density must vanish. At such frequencies the matching of the phase may not occur, and therefore the phase distributions must be non-overlapping. This means that we can conversely characterize the spectral gaps as the intervals in which the phase distributions for the forward and backward transfers do not overlap each other. It should be remarked further that at these frequencies the function $\rho(\delta)$ for the forward transfer completely vanishes in the S-region of $\underset{\sim}{Q}^{(1)}$.

4. CONCLUSIONS

(1) In the case of the isotopically disordered diatomic chain, the distribution of the phase on a chain is ergodic. (2) The condition of Fréchet is not essential. (3) The eigenmodes of the isotopically disordered diatomic chain are always localized, and the localization becomes stronger with the increase of the frequency as well as mass ratio, just as expected from our theory.

ACKNOWLEDGMENTS

The computations were carried out by HITAC 5020E of the Computer Centre of Tokyo University. The authors wish to thank Professors H. Matsuda and E. Teramoto for helpful discussions.

REFERENCES

(1) N. F. Mott and W. D. Twose, Advances in Phys. 10, 137 (1961).
(2) A. P. Roberts and R. E. B. Makinson, Proc. Phys. Soc. 79, 630 (1962).
(3) J. Hori, Prog. Theor. Phys. Suppl. 36, 3 (1966); Spectral Properties of Disordered Chains and Lattices (Pergamon Press, Oxford, 1967).
(4) R. E. Borland, Proc. Roy. Soc. 274, 529 (1963).
(5) P. Dean and M. D. Bacon, Proc. Phys. Soc. 81, 642 (1963).

VIBRATIONAL SPECTRUM OF AN ISOTOPICALLY DISORDERED CHAIN WITH ARBITRARY IMPURITY CONCENTRATION*

H. Hartmann

Department of Physics

New York University, New York, New York

I Introduction

The frequency spectrum of a one dimensional crystal with a low concentration of impurities has been studied extensively. The problem of large concentration of impurity atoms was first studied by Davis and Langer[1] who made a self consistent calculation and more recently Taylor[2] has applied Lax's multiple scattering theory to the same problem. Extensive numerical work has been done by Dean[3] and collaborators and more recently by Payton and Visscher[4]. Although the numerical work provides much information on the density of states it is still of great interest to look for an analytic solution from which dynamical effects such as lifetimes of the phonons may be studied. In this paper we report on an extension of Langer's theory for the frequency spectrum of a one dimensional crystal with a random distribution of isotopic impurities of arbitrary concentration. Langer expands the phonon self energy in a power series in the concentration of impurities and the first and second order terms have been studied. We have summed the largest contribution to the self energy from every order in the impurity concentration. The host and impurity atoms are treated equivalently to ensure that the results are symmetric in the two types of atoms present in the crystal.

II Phonon Self Energy

Langer[5] has shown that the spectral distribution function for the average chain (i.e. averaged over all mass configuration) is

*Research sponsored by the Air Force Office of Scientific Research Office of Aerospace Research, United States Air Force, under AFOSR grant number 1011-66.

given by

$$g(\omega) = \frac{2\omega}{\pi} \lim_{\substack{N\to\infty \\ \varepsilon\to 0^+}} \frac{1}{N} \operatorname{Im} \sum_k D_k (\omega^2 + i\varepsilon)$$

$$\text{where } D_k(\omega^2) = \frac{1}{\omega_k^2 - \omega^2 + G_k(\omega^2)}$$

and $G_k(\omega^2)$ is the self energy function, ω_k^2 is the frequency of vibration of an ordered crystal of atoms of mass μ where $\mu = (1-q) M + q M'$. The concentration of impurities is q, M is the mass of the heavy atom and M' is the mass of the light atom. The self energy function, $G_k(\omega^2)$, can be expanded in a power series in concentration and a set of proper graphs can be associated with each term in the series. These graphs can be interpreted as representing the scattering of a phonon of momentum k by atoms at different lattice sites. The contribution from successive scatterings at the same lattice site is easily summed giving so called reduced diagrams in which each interaction represents any number of consecutive scatterings at a given lattice point. The contribution of any diagram with n interactions at p distinct lattice sites to the self energy function is obtained from a simple set of rules, namely: 1) a factor of $-f(\omega,v)$ for each internal phonon line. For the ith phonon line v is the distance between the atoms at which the ith and $(i+1)^{th}$ scatterings occur.

2) a factor of $[q\,\rho_1^s + (1-q)\,\rho_2^s]$ for each different atom at which there are s scatterings.

3) a factor of $\omega_k^2 \exp[2\pi i k v_n/N]$ where v_n is the distance between the first and last scattering atom.

4) Sum over $v_1 \ldots v_{p-1}$ with the proper restrictions to insure that the p atoms are at different lattice points.

$$f(\omega,v) = \frac{1}{N} \sum_k \frac{\omega_k^2}{\omega_k^2 - \omega^2} \exp[2\pi i k v/N],$$

$$\rho_1 = \frac{\lambda(1-q)}{1 + f(\omega,0)(1-q)\lambda}, \quad \lambda = \frac{M}{M'} - 1$$

$$\rho_2 = \frac{-\kappa q}{1 + f(\omega,0)(-\kappa q)} \qquad \kappa = 1 - M'/M$$

Consider now a diagram with n scatterings at n-1 lattice sites. There is only one proper diagram of this type as shown in figure 1.

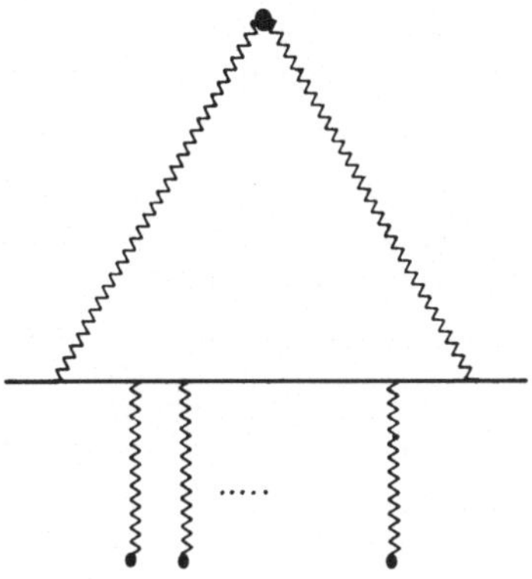

Figure 1 Diagram which corresponds to $G_k^{n,n-1}(\omega^2)$

The n-1 different lattice points may be anywhere in the crystal and a sum is performed over all possible sites for the different atoms. This same sequence of scatterings can be represented in a diagram in which the location of the atoms are designated by their position along the chain and the order of scattering is indicated by a directed line connecting the scattering sites. Figure 2 illustrates two types of atomic arrangements which correspond to 7 scatterings and 6 atoms.

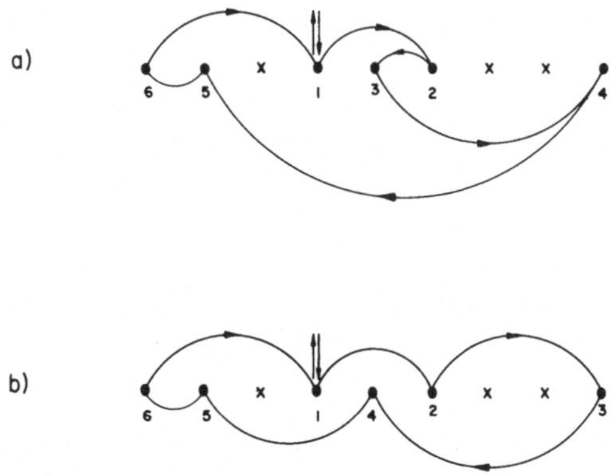

Figure 2 Example of an arrangement of scatterings which contribute to $G_k^{n,n-1}(\omega^2)$

It can be shown that the arrangement of the type illustrated in figure 2b in which there are only two lines between any two lattice sites gives a larger contribution to G than for any other atomic arrangement as for example the one illustrated in figure 2a. Thus to obtain the largest contribution to $G_k^{n,n-1}(\omega^2)$ the number of atomic arrangements of type 2b for which there are n scatterings at n-1 sites are counted and the contribution to $G_k^{n,n-1}$ from each of these arrangements is computed.

Next consider diagrams which correspond to n interactions at (n-r) atoms where there are r atoms with two interactions and (n-2r) atoms with one interaction. An example of this type of diagram is illustrated in figure 3a. The atomic arrangements for this diagram which give the largest contribution to G is illustrated in figures 4a and 4b. Again notice that these correspond

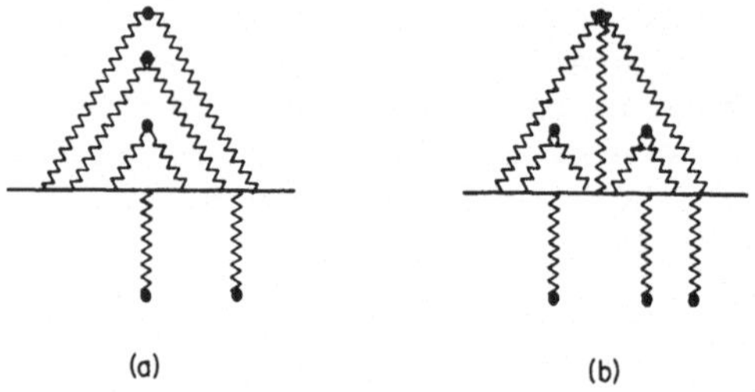

Figure 3 a) Example of a diagram which contributes to $G^{n,n-r}(\omega^2)$
b) Example of a diagram which contributes to $G^{n,n-r}(\omega^2)$

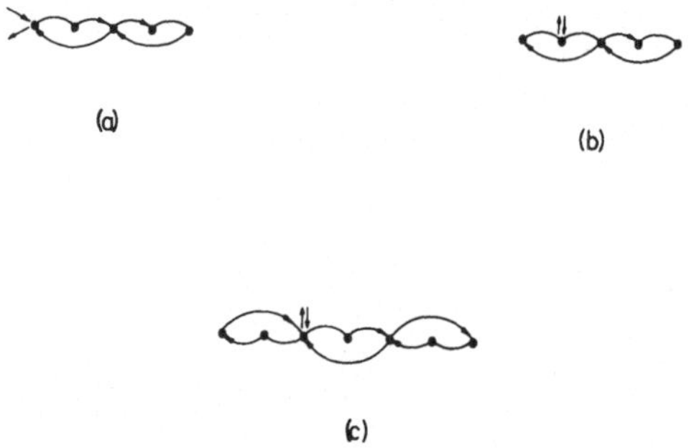

Figure 4 a,b) Arrangements of scatterings for figure 3a
c) Arrangement of scatterings for figure 3b

to arrangements in which there are no more than two lines between any pair of atoms. In figure 4 the position of the atoms which do not contribute to the scattering are not indicated so that only the relative position to the scattering atoms is given. There are of course other diagrams corresponding to n scatterings at n-r atoms but the contribution to G from these diagrams are of the same order as the ones we have already neglected.

There is one more diagram of the same order as the diagrams previously considered. It has n interactions at (n-r-2) different atoms with one atom having three interactions, r atoms having two interactions and (n-2r-3) atoms with one interaction and is illustrated in figure 3b. The atomic arrangement which gives the largest contribution to G is illustrated in figure 4c. All other diagrams with three or more scatterings at one site give contributions to G of the order already neglected. The contribution from the graphs discussed above can be summed over r (the number of sites with two interactions) and n (the total number of scatterings). Details of this calculation will be published elsewhere.

One drawback of this calculation is that we are not able to count the number of atomic arrangements for the neglected terms and thus are unable to estimate the error in the truncation of the perturbation series. In the impurity band a sum over all positions of the scattering sites for a given atomic arrangement is made. In the band modes this is not possible and only the arrangements which correspond to scatterings by a cluster of adjacent atoms are used. The only justification for this selection is that one would expect that the scattering from a cluster of atoms would contribute the most important terms.

Selecting the graphs and using the approximations discussed above an expression for the phonon self energy function is obtained in closed form, namely:

$$G'(\omega^2) = \frac{2\omega_k^2\, C^2(\omega)\, F\, q_1 q_2}{[1+C(\omega)F(2q_1-C(\omega)q_2)]^2}\left[1+C(\omega)F\, q_1+C^2(\omega)F\left(\frac{q_1q_3}{q_2}-q_2\right)\right]$$

where $$F = \frac{\alpha^2}{1-\alpha^2}$$

and $$q_i = q\,\rho_1^i + (1-q)\,\rho_2^i$$

and $$\alpha = 1-2x^2 + 2x\,(x^2-1)^{1/2} \quad ; \quad x = \omega/\omega_\mu$$

$$C(\omega) = -x/(x^2-1)^{1/2}$$

III Density of States

In figures 5,6, and 7 the density of states $\bar{g}(\omega)$ times the concentrated weighted mass is plotted against $z = \omega/\omega_{M'}$, where $\omega_{M'}$ is the max frequency of an ordered lattice of the light atoms of mass M'. The results are given for different concentrations, q, (q = 0.3,0.6,0.9) and various mass ratios, λ,(λ = 2,5,9). The arrow indicates the position of the maximum frequency of an ordered lattice of the heavy atoms. The graphs were computed for frequency which correspond to $.05 \leq x .95$ and $x \geq 1.05$ where $x = \omega/\omega_\mu$ where ω_μ = max frequency of an ordered crystal of atoms with mass μ. This restriction was necessary since there is a singularity at x = 1 (i.e. $\omega = \omega_\mu$)

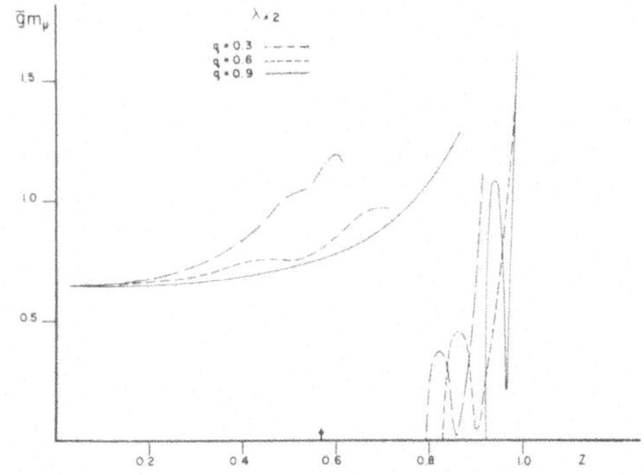

Figure 5 Density of states as a function of frequency for λ = 2

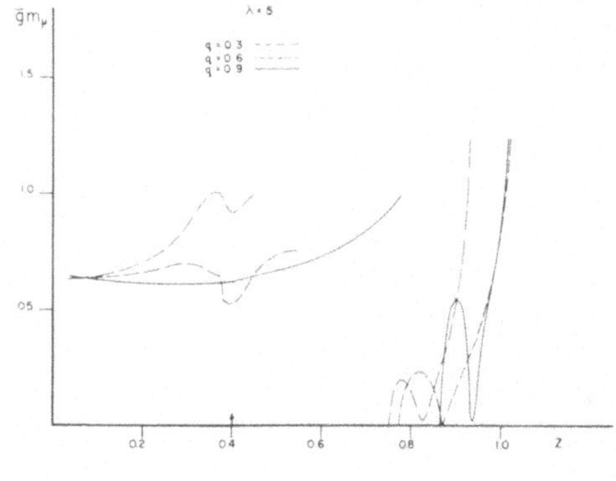

Figure 6 Density of states as a function of frequency for λ = 5

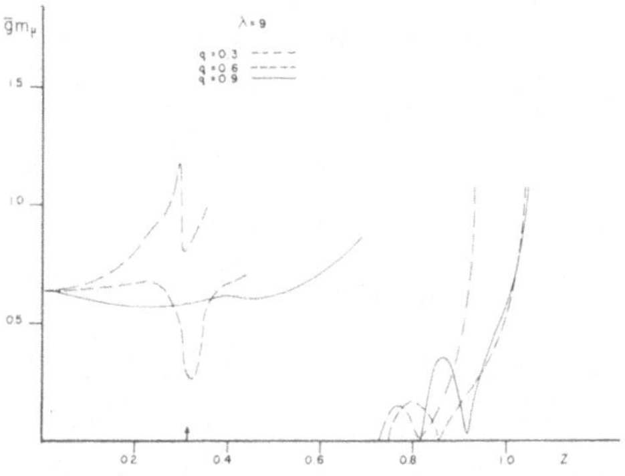

Figure 7 Density of states as a function of frequency for $\lambda = 9$

The computed density of states has the following features:-

1) as the concentration of impurities increase the width of the gap where the density of states is zero, decreases.

2) with increasing impurity concentration the number of states in the band mode decreases while the total number of states in the impurity mode increases.

3) for large q there is no break at $\omega = \omega_M$. Thus the density of states, except for high frequency, approaches a smooth curve appropriate for the pure light atom chain.

4) note as the ratio of heavy mass to the light mass increases the break at $\omega = \omega_M$, becomes more pronounced.

These qualitative features are in agreement with the machine calculations of Payton and Visscher[5] and the earlier work of Dean[4]. Their work showed that for small impurity concentration the density of states has a maximum at ω_M, smaller peaks appear at frequencies just above ω_M which are followed by a gap and highly spiked spectrum at high frequencies. As the concentration of impurities increases the maximum at ω_M decreases, the density of states at frequencies just above ω_M increases, the width of the gap decreases, and finally the highly spiked part of the spectrum becomes smooth.

The unphysical features of the computed density of states which are due to the approximations used are:

1) a singularity at $\omega=\omega_\mu$ and at $\omega=\omega_M$,

2) the existence of a gap in the spectrum even for large q.

3) nonzero values of the density of states for frequencies

slightly above $\omega = \omega_M$, for large values of q. Upon close examination of the perturbation series one sees that this last difficulty is due to the fact that only diagrams of the type illustrated in figures 1 and 3 were retained.

REFERENCES

1) R. W. Davis and J.S. Langer, Phys. Rev. 131, 163 (1963)
2) D. W. Taylor, Phys. Rev. 156, 1017 (1967)
3) P. Dean, Proc. Roy. Soc. A254, 507 (1960); ibid A260, 264 (1961)

   P. Dean and J.L. Martin, Proc. Roy. Soc. A259, 409 (1960)
   J. L. Martin, Proc.Roy. Soc. A260, 139 (1961)
4) D. N. Payton III and W.M.Visscher, Phys. Rev. 154, 802 (1967); ibid Phys. Rev. 156, 1032 (1967)
5) J. S. Langer, J. Math. Phys. 2, 584 (1961)

ON THE THERMODYNAMIC EQUILIBRIUM OF A SYSTEM OF GAS PLUS CRYSTAL WITH ISOTOPIC DEFECTS

G. Benedek
Istituto di Fisica dell' Università
Milano, Italy
R. F. Wallis and A.A. Maradudin*
University of California, Irvine, U.S.A.
I.P. Ipatova and A.A. Klochikhin
A.F. Ioffe Physico-Technical Institute
Leningrad K-21, U.S.S.R.
W.C. Overton, Jr.*
Los Alamos Scientific Laboratory
Los Alamos, New Mexico, U.S.A.

1. Introduction

The introduction of a small number of impurities into a crystal lattice significantly alters its vibrational spectrum, leading in a number of cases to the introduction of localized[1] and resonance[2] vibration modes. Such a change in the vibrational spectrum is reflected in changes in the thermodynamic properties of a perturbed crystal.[3-9] The investigation of an isotopic impurity in a crystal turns out to be particularly simple, inasmuch as in this case the dynamical and thermodynamical problems can be solved completely.

In particular, it is of some interest to investigate the thermodynamic equilibrium of a system consisting of a gas and a crystal containing isotopic defects. This problem has been investigated many times,[10-13] and it has been shown that in the classical or high temperature limit the difference between the chemical potentials of the different isotopes does not depend on the state of aggregation. Correspondingly, the vapor pressures of the different isotopes are equal, and the distribution coefficient of the isotopes is equal to unity.

Departures from these classical results were obtained by taking account of quantum effects. In the gaseous phase quantum effects are connected with the existence of vibrational and rotational degrees of freedom in a polyatomic gas. In the crystalline phase quantum corrections were evaluated by the use of thermodynamic perturbation theory, in which the expansion proceeds in powers of the parameter $\hbar\omega_L/T$ where T is the absolute temperature, and $\hbar\omega_L$ is the largest characteristic vibrational energy of the crystal.

However, there exist crystals for which the melting temperature is lower or of the order of the maximum vibrational frequency of the lattice. For example, LiH has a melting temperature $T_M=961^\circ K$[17], and $\hbar\omega_L=1380^\circ K$.[18] Thermodynamic perturbation theory in this case is inapplicable. Nevertheless, it is still possible to carry out the necessary calculations by expanding thermodynamic functions in powers of the concentration of the minority impurity specie.[5,6]

In the present paper this method is applied to the investigation of the thermodynamic equilibrium of a system consisting of a gas plus a crystal containing defects. For simplicity we investigate the particular case of a monatomic crystal which is in equilibrium with a classical monatomic gas. In this case quantum effects arise only on account of the crystal with isotopic defects. The difference between the masses of the isotopes is assumed to be arbitrary. We evaluate the free energy of the perturbed crystal, the chemical potentials, the partial vapor pressures and the dependence of the isotopic separation coefficient on the difference between the isotopic masses and the temperature.

2. The Free Energy of a Crystal Containing Impurities

The vibrational contribution to the partition function of a crystal containing impurity atoms, for a given distribution of the impurities over the lattice sites which we denote by the index λ, can be written in the form

$$Z = \exp\left\{-3N \int_0^{\omega_L(\lambda)} g_\lambda(\omega) \ln\left[2\sinh\frac{\hbar\omega}{2\theta}\right] d\omega\right\} . \quad (2.1)$$

In this expression $g_\lambda(\omega)$ is the frequency distribution function for the imperfect crystal (normalized to unity) corresponding to the impurity configuration λ, and $\omega_L(\lambda)$ is the largest normal mode frequency of the crystal.

Let us assume that the crystal consists of N atoms of which N_{II} are impurity atoms, while $N_I=N-N_{II}>>N_{II}$ are atoms of the host crystal. Averaging Eq.(2.1) over all possible configurations of the N_{II} impurities over the lattice sites of the crystal, assuming that there are no correlations between the impurity sites, we obtain for the partition function of the crystal

$$Z=\frac{N!}{N_I!N_{II}!}\exp\left\{-3N\int_0^{\omega_L} g(\omega)\ \ln\left[2\sinh\frac{\hbar\omega}{2\theta}\right]d\omega\right\} \qquad (2.2)$$

where ω_L is the largest normal mode frequency of the perturbed crystal, and $g(\omega)=\langle g_\lambda(\omega)\rangle$ is the corresponding frequency distribution function, averaged over all possible impurity configurations.

It is well known[6,9] that to obtain the frequency spectrum of the imperfect crystal correct to the first power of the concentration of the minority constituent it is sufficient to add to the frequency spectrum of the pure host crystal N_{II} times the change in the frequency spectrum of the host crystal which results from introducing a single impurity atom into it:

$$g(\omega) = g_o(\omega) + N_{II}\Delta g(\omega) + O(c_{II}^2) \qquad (2.3)$$

where $c_{II}=N_{II}/N$ is the impurity concentration. For the case of a substitutional isotopic impurity of mass M' in a cubic Bravais host crystal composed of atoms of mass M' the function $\Delta g(\omega)$ is given by

$$\Delta g(\omega) = \frac{1}{N\pi}\frac{d}{d\omega}\tan^{-1}\frac{-\pi\varepsilon\omega^2 G_o(\omega^2)}{1-\varepsilon\omega^2\tilde{G}_o(\omega^2)}, \quad o\leq\omega<\omega_L^o- \qquad (2.4a)$$

$$= \frac{1}{N}\delta(\omega-\omega_o) \qquad \omega>\omega_L^o- \ . \qquad (2.4b)$$

In these expressions ω_L^o is the largest normal mode frequency of the perfect host crystal, $G_o(\omega^2)$ is the distribution of squared frequencies of the perfect host crystal

$$G_o(\omega^2) = \frac{1}{3N}\sum_{\underset{\sim}{k}j}\delta(\omega^2-\omega_j^2(\underset{\sim}{k})) , \qquad (2.7)$$

and $\tilde{G}_o(\omega^2)$ is its Hilbert transform

$$\tilde{G}_o(\omega^2) = \frac{1}{3N}\sum_{\underset{\sim}{k}j}\frac{1}{(\omega^2-\omega_j^2(\underset{\sim}{k}))_P} , \qquad (2.8)$$

where $\omega_j(\underset{\sim}{k})$ is the frequency of the normal mode of the perfect host crystal described by the wave vector $\underset{\sim}{k}$ and branch index j. The parameter ε is given by $\varepsilon=1-(M_{II}/M_I)$. The contribution to $\Delta g(\omega)$ given by Eq.(2.4b) is present only if the isotope of type II gives rise to a localized vibration mode whose frequency ω_o is the solution of the equation

$$1 = \varepsilon\omega_o^2\ \tilde{G}_o(\omega_o^2) \qquad (2.9)$$

for $\omega_o > \omega_L^o$. Because $G_o(\omega^2)$ vanishes for $\omega=o$ and $\omega=\omega_L^o$, we see from Eq. (2.4) that the integral of $\Delta g(\omega)$ vanishes, as it must to preserve the normalization of $g(\omega)$.

Substituting Eq. (2.3) into Eq. (2.2) we obtain

$$Z=\frac{N!}{N_I!N_{II}!}\left\{e^{-\frac{\mu}{\theta}}\right\}^N\left\{e^{-\frac{\mu'}{\theta}}\right\}^{N_{II}} \tag{2.10}$$

where

$$\mu=3\theta\int_0^{\omega_L^o} g_o(\omega)\ln\left[2\sinh\frac{\hbar\omega}{2\theta}\right]d\omega \tag{2.11}$$

$$\mu' = -\frac{3\hbar}{2\pi}\int_0^{\omega_L^o}\coth\frac{\hbar\omega}{2\theta}\tan^{-1}\left\{\frac{-\pi\varepsilon\omega^2 G_o(\omega^2)}{1-\varepsilon\omega^2\tilde{G}_o(\omega^2)}\right\}d\omega$$
$$+3\theta\ln\frac{\sinh(\hbar\omega_o/2\theta)}{\sinh(\hbar\omega_L^o/2\theta)}. \tag{2.12}$$

Again, the last term on the right hand side of Eq.(2.12) is present only if the impurity isotopes give rise to localized modes. The free energy of the perturbed crystal can therefore be written to first order in the concentration c_{II} as

$$F = N\theta\left[c_I\ln c_I+c_{II}\ln c_{II}\right]+N\left[\mu+c_{II}\mu'\right]. \tag{2.13}$$

Differentiating F with respect to N_I and N_{II}, we obtain the chemical potentials of atoms of type I and type II, respectively

$$\mu_I=\theta\ln c_I+\mu \tag{2.14a}$$

$$\mu_{II}=\theta\ln c_{II}+\mu+\mu'. \tag{2.14b}$$

Consequently, the quantity μ can be interpreted as the chemical potential of the atoms comprising the pure host crystal $\mu_I^{(o)}$ while μ' gives the change in the chemical potential associated with the change in the frequency spectrum of the crystal due to the presence of impurities.

3. Partial Vapor Pressures and the Isotopic Distribution Coefficient

We must add to the values of the chemical potentials given by Eqs. (2.14) an electronic part μ_e, which we will take to be the same for the different isotopes in the same state of aggregation, but different for the crystal and for the gas.

Phase equilibrium is defined by the equality of temperatures, pressures, and chemical potentials

$$(\mu_I + \mu_e)_{cryst.} = (\mu_I + \mu'_e)_{gas} \quad (3.1)$$

$$(\mu_{II} + \mu_e)_{cryst.} = (\mu_{II} + \mu'_e)_{gas}. \quad (3.2)$$

For a monatomic classical gas (see, for example[10])

$$(\mu_i)_{gas} = \theta \ln\left[\frac{C_i P}{\theta^{5/2}}\left(\frac{2\pi\hbar^2}{M_i}\right)^{3/2}\right] \quad (3.3)$$

where i labels the different isotopes, and P is the pressure of the gas.

Substituting Eqs. (2.14) and (3.3) into Eqs. (3.1) and (3.2) we find for the partial vapor pressures

$$P_i = (C_i)_{gas} P = P_{oi}(C_i)_{cryst.}, \quad (3.4)$$

where the vapor pressures for each of the isotopes are given by

$$P_{OI} = \frac{\theta^{5/2} M_I^{3/2}}{(2\pi\hbar^2)^{3/2}} \exp\left\{\frac{(\mu_e)_{cryst.} - (\mu'_e)_{gas} + \mu}{\theta}\right\} \quad (3.5)$$

$$P_{OII} = \frac{\theta^{5/2} M_{II}^{3/2}}{(2\pi\hbar^2)^{3/2}} \exp\left\{\frac{(\mu_e)_{cryst.} - (\mu'_e)_{gas} + \mu + \mu'}{\theta}\right\}. \quad (3.6)$$

Consequently we see that the vapor pressure P_{OI} coincides with the vapor pressure of the pure host isotope I, but that P_{OII} is not equal to the vapor pressure of the pure isotope II. In what follows we will study the quantity

$$\frac{P_{OI} - P_{OII}}{P_{OI}} = 1 - (1-\varepsilon)^{3/2} e^{\frac{\mu'}{\theta}} \quad (3.7)$$

which is independent of μ_e and depends on the change in the frequency spectrum of the perturbed crystal through the quantity μ'.

Combining Eqs. (3.1) and (3.2) with Eqs. (2.14) and (3.3) we obtain for the isotopic distribution coefficient

$$\ln\alpha = \ln \frac{(C_{II}/C_I)_{cryst.}}{(C_{II}/C_I)_{gas}} = -\ln\left\{(1-\varepsilon)^{3/2} e^{\mu'/\theta}\right\} \quad (3.8)$$

It is not difficult to show that in the classical limit of high temperatures $(\mu+\mu')$ tends to the chemical potential of the pure isotope of type II, i.e., to $\mu_{II}^{(o)}$.

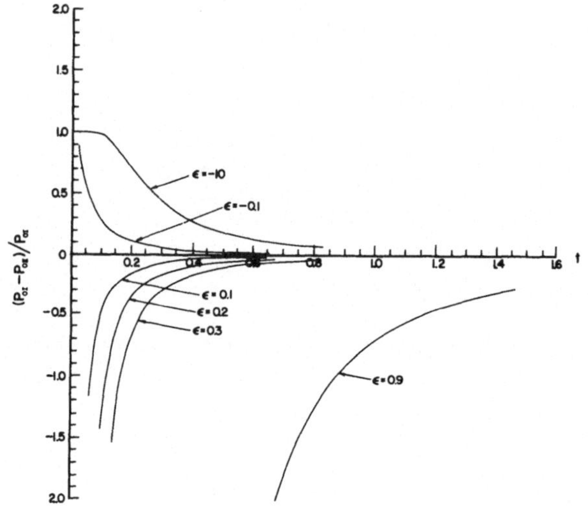

Fig.1. A plot of $(P_{OI}-P_{OII})/P_{OI}$ as a function of temperature for several values of the mass defect parameter ε. The dimensionless temperature $t=\theta/\hbar\omega_L^0$.

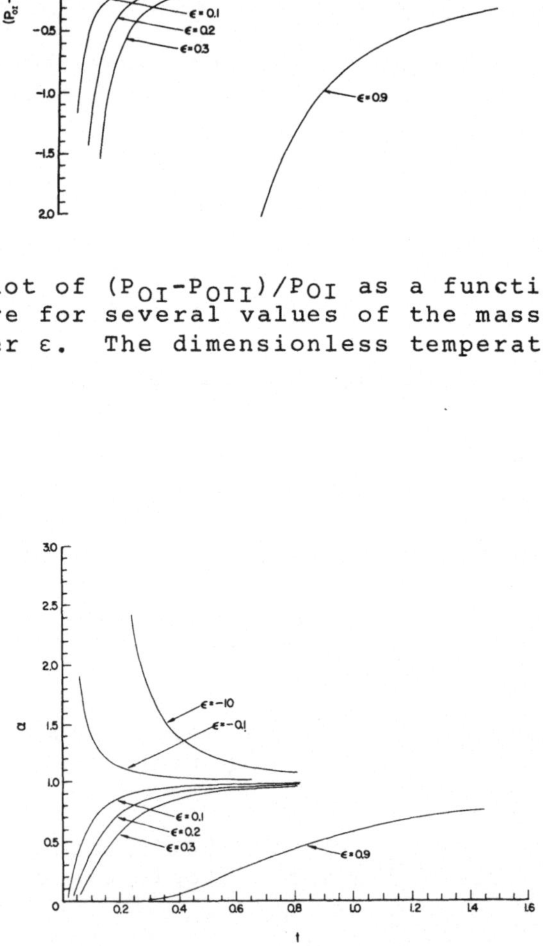

Fig.2. A plot of the isotopic distribution coefficient α as a function of temperature for several values of ε.

At the same time P_{OI} and P_{OII} tend to the vapor pressures of pure isotopes of types I and II, respectively. Inasmuch as $\mu_I^{(o)}-\mu_{II}^{(o)}=(3\theta/2)\ell n(1-\varepsilon)$ in the classical limit, then in this limit $P_{OI}=P_{OII}$ and $\alpha=1$.

We have computed $(P_{OI}-P_{OII})/P_{OI}$ and α over a wide range of temperatures for several values of ε. The calculations were based on a nearest neighbor, central force, model of a face centered cubic crystal for the host crystal. The results are shown in Figs. 1 and 2, respectively. The critical value of ε, ε_{cr}, for which a localized mode first appears is $\varepsilon_{cr}\cong 0.24$ for the present crystal model. Thus the curves for $\varepsilon=0.2$ and 0.3 correspond to impurities which just fail to give rise to localized modes and which just do give rise to localized modes, respectively. The frequency of the localized mode corresponding to $\varepsilon=0.3$ is $x_o(0.3)=1.0078$ while the value of $x_o(0.9)$ is 2.2654. The value of the lowest resonance mode frequency associated with the heaviest impurity ($\varepsilon=-10$) is $x_R(-10)\cong 0.175$, where $x = \omega/\omega_L^o$.

We can draw several qualitative conclusions from the results shown in Figs. 1 and 2. In the first place we see that irrespective of the value of ε $(P_{OI}-P_{OII})/P_{OI}$ tends to zero and α tends to unity at high temperatures, in agreement with theoretical predictions. The effective temperature at which the classical limits are attained for these functions depends significantly on the value of ε. Second, we note that in all cases when ε is positive ($M_{II}<M_I$), $P_{OI}<P_{OII}$ and $\alpha<1$. This result has its origin in the fact that for light impurity atoms the normal mode frequencies of the crystal are shifted to higher values. It signifies that the lighter isotope has the larger vapor pressure and that the relative concentration of the light isotope is greater in the gas than it is in the crystal. In other words, there is a tendency toward isotopic separation with the lighter isotope leaving the crystal. Correspondingly, in all cases when ε is negative ($M_{II}>M_I$) we have that $P_{OI}>P_{OII}$ and $\alpha>1$. This result is due to the lowering of all normal mode frequencies by the introduction of heavy impurities into a crystal, and signifies again that the light isotope has a tendency to leave the crystal. Third, we see that for all values of ε, α is a monotone function of temperature.

References

* A portion of this work was done under the auspices of the U.S. Atomic Energy Commission

1) I.M. Lifshitz, Zhur. Eksper. i Teor. Fiziki 17,1017 (1947);17,1076(1947);18,293(1948);Nuovo Cimento,Suppl.(10)3,716(1956). E.W. Montroll and R.B. Potts, Phys. Rev. 100,525(1955).

2) Yu. M. Kagan and Ya. A. Iosilevskii, Zhur. Eksper.i Teor. Fiziki 42,259(1962);44,284(1963) {English translation:Soviet Physics-JETP 15,182(1962) 17,195 (1963)}. R. Brout and W.M. Visscher, Phys. Rev. Letters 9,54(1962). S. Takeno, Progr.Theor.Phys. (Kyoto)29,191(1963).

3) I.M. Lifshitz, Usp. Mat. Nauk 7,170(1952).

4) I. Prigogine and J. Jeener,Physica 20,516(1954).

5) I.M. Lifshitz and G.I. Stepanova, Zhur. Eksper.i Teor. Fiziki 30,938(1956);31,156(1956);33,485(1957); {English translation:Soviet Physics-JETP 3,656 (1956);4,150(1957);6,379(1958)}.

6) E.W. Montroll, A.A. Maradudin, and G.H.Weiss, in The Many-Body Problem,edited by J.K. Percus,(Interscience, New York,1963) p.353.

7) A.A. Maradudin, E.W. Montroll, and G.H. Weiss, Theory of Lattice Dynamics in the Harmonic Approximation (Academic Press, Inc., New York, 1963).

8) A.A. Maradudin, in Astrophysics and the Many-Body Problem (W.A. Benjamin, Inc., New York,1963)p.107.

9) A.A. Maradudin, in Phonons and Phonon Interactions (W.A. Benjamin, Inc., New York, 1964) p.424.

10) L.D. Landau and E.M. Lifshitz, Statistical Physics (Addison-Wesley Publishing Co., Inc., Reading,1958), Chap.IX.

11) J. Bigeleisen, J. Chem. Phys. 34,1485(1961).

12) G. Boato and G. Casanova, Isotopic Cosmic Chemistry, 1964. p.16.

13) B.N. Esel'son and B.G. Lazarev, Zhur, Eksper. i Teor. Fiziki 20,742(1950).

14) C.E. Messer, E.B. Damon, P.C. Maybury, J. Mellor, and R.A. Seals, J. Phys. Chem. 62,220(1958).

15) G. Benedek, Solid State Comm. 5,101(1967).

THERMODYNAMIC EQUILIBRIUM IN THE SYSTEM LITHIUM HYDRIDE-HYDROGEN CONTAINING ISOTOPIC IMPURITIES

G. Ya. Ryskin and Yu. P. Stepanov

A.F. Ioffe Physico-Technical Institute

Leningrad K-21, U.S.S.R.

At the present time the physical properties of real crystals, i.e. crystals containing defects, are being studied very intensively. In particular, the influence of substitutional isotopic defects on the vibrational properties of crystalline lattices, light absorption, electrical conductivity, specific heat, etc., has been studied theoretically in considerable detail. However, the influence of substitutional isotopic defects on crystal properties has not yet been sufficiently studied experimentally. This is explained first of all by the fact that isotopes as a rule differ little in their masses, and correspondingly their influence on crystal properties is hard to measure.

From the experimental standpoint it is of considerable interest to study the thermodynamic equilibrium with the gaseous phase of a crystal containing isotopic defects. The quantity measured in this case, the isotopic distribution coefficient, is a characteristic only of a crystal containing impurity atoms, so that this is different from other properties of the crystal (for example specific heat, thermal conductivity, electrical conductivity), which exist both for crystals with defects as well as for crystals without defects. Recently, a theory of the thermodynamic equilibrium of a system of a gas and crystal containing defects was constructed taking into account the details of the dynamics of the perturbed crystal(1). In this work the connection was derived between the isotopic distribution coefficient and the change in the crystal frequency spectrum caused by defects with an arbitrary difference in mass from the atoms

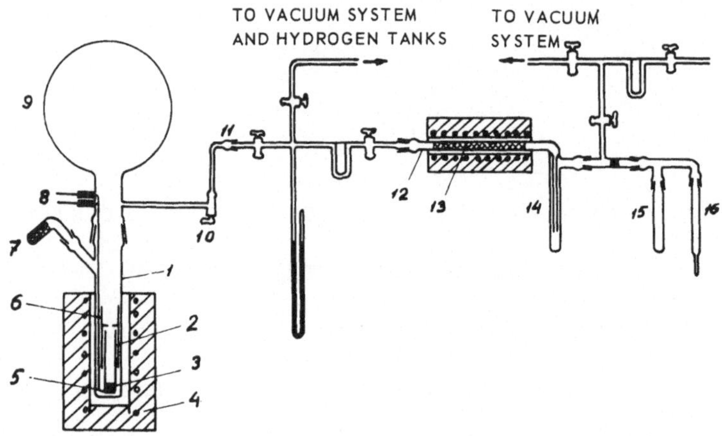

Fig.1. General view of the experimental apparatus. The explanation is given in the text.

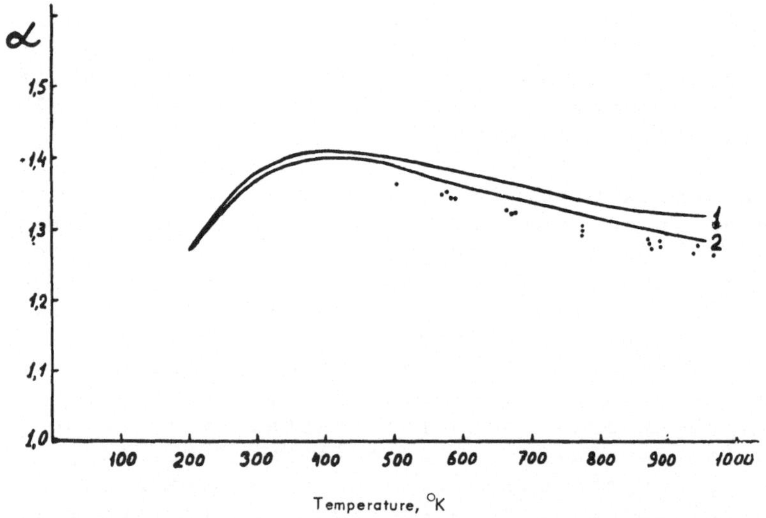

Fig. 2. Dependence of the distribution coefficient α on temperature. 1 and 2 are theoretical curves corresponding to $\zeta/e = 0.7$ and $\zeta/e = 0.85$, respectively.

of the host crystal. In connection with this, an interesting object for study should be a crystal containing isotopes of hydrogen, inasmuch as in this case the difference between the isotopic masses is a maximum. There exist data in the literature for the temperature dependence of the distribution coefficient for the isotopes H and D in the system gaseous hydrogen-uranium hydride[2]. However, uranium hydride is a very complicated crystal for calculations, which complicates comparisons between experimental and theoretical results. In reference 3 the value of the distribution coefficient for the isotopes H and D in the system gaseous hydrogen-lithium hydride was obtained at a temperature of 200°C. This work, as well as the investigation of the pressure for dissociation of hydrides and deuterides of sodium, potassium[4], rubidum, and cesium[5] indicate the possibility of an experimental investigation of the thermodynamic equilibrium for the system of alkali metal hydrides-gaseous hydrogen in a wide temperature interval.

In the present paper such an investigation of the thermodynamic equilibrium of gaseous hydrogen with crystalline lithium hydride in the presence of a small concentration of deuterium in the system, was undertaken.

I. Measurement of the equilibrium distribution of H and D isotopes between crystalline lithium hydride and gaseous hydrogen.

The distribution of isotopes between two phases in a state of thermodynamic equilibrium is conventionally characterized by the distribution coefficient, which is defined by the equation

$$\alpha = \frac{c_1}{1-c_1} \Big/ \frac{c_2}{1-c_2} \quad , \qquad (1)$$

where c_1 and c_2 are the relative atomic concentrations of one of the isotopes in phases 1 and 2 in an equilibrium state of the system.

The apparatus for carrying out measurements of the distribution coefficient of H and D between gaseous hydrogen and crystalline lithium hydride is shown in Figure 1.

Isotopic phase equilibrium was established in a reaction cell consisting of a test tube 1 and of a glass cylinder 9. The lithium hydride was placed in the crucible 2 made out of Armco steel. The hydrogen occupied the entire volume of the reaction cell to the spigot 10. The volume of the reaction cell was approximately 1800 cm^3. Since the reaction cell was heated by the heater 4 only at its lower end, then to ensure that the process of phase equilibration occurred isothermally the crucible was covered by a steel beaker 6 with an aperture in its bottom. The temperature of the hot zone of the cell was determined by a thermocouple 8, placed in the cavity 5 in the thick bottom of the crucible 2.

The isotopic composition of hydrogen in the gaseous and condensed phases was determined by the temperature float method from the density of water[6,7]. The transformation of the gaseous hydrogen into water took place in the tube 12 containing copper oxide 13, heated to 600° C. To obtain a correct result all of the hydrogen was allowed to transform. The water vapor resulting from the reaction of the hydrogen with copper oxide was frozen out in the test tube 14 which was cooled by liquid nitrogen. For purification, the resulting water was subsequently distilled from the test tube 14 and subsequently in test tubes 15 and 16. The determination of the density of the water was carried out immediately in test tube 16 with the aid of a quartz float calibrated in standard water.

For the extraction of hydrogen from lithium hydride for the determination of the isotopic composition of the solid phase, we used the reaction of lithium hydride with lead[8]. Granulated lead for this purpose was placed in the curved test tube 7. The decomposition of lithium hydride by the lead occurred at 600° C. The relative error in the determination of the concentration of deuterium in gaseous hydrogen and in lithium hydride by the temperature float method was about 0.3%.

The experiments were carried out in the following order. In a chamber filled with dry carbon dioxide gas approximately 1.5 grams of Li^7H in the form of a powder whose grain size was not larger than 0.26mm were placed in the reaction cell in crucible 2, and approximately 30 grams of lead were placed in test tube 7. After loading, the reaction cell was connected to the rest of the apparatus through a sleeve 11, and was evacuated to a pressure of 10^{-2}mm of mercury. Thereupon gaseous hydrogen was admitted into the reaction cell. The initial hydrogen used was obtained by electrolysis and contained deuterium in a concentration of between 3 to 4 atomic percent. The pressure of the hydrogen in the reaction cell ordinarily was between 630 to 650mm of mercury, which yielded an almost equivalent amount of hydrogen in the gaseous phase with that found in the lithium hydride. After the gas was introduced into the cell, it was heated to the temperature of the experiment. The time for isotopic equilibration of phases to occur depended on the temperature of the experiment and was determined in special experiments. At the conclusion of the equilibration of phases, an isotopic analysis was carried out of the composition of the gaseous hydrogen and of the lithium hydride. The gaseous phase was removed for analysis at the temperature of the experiment. The experimental conditions and the results of the experiments are presented in the table.

II. Discussion of the results.

The experimental values for the distribution coefficient for the H and D isotopes in the system Li^7H with D impurities (or Li^7H:D)plus a mixture of the gases H_2,HD, D_2, presented in the table indicate that in the entire temperature interval α is significantly greater than unity (by 30 - 40%). As is well known, in the classical limit of high temperatures the isotopic distribution coefficient is equal to unity. Its departure from this limiting value is due to quantum effects in one or both of the phases. The fact that in our case α differs significantly from unity indicates that the temperature range in which the experimental measurements were carried out is a quantum region. Quantum effects can arise from the gaseous phase, since the vibration frequencies of the molecules H_2, HD, D_2 are equal in energy units to 6327°K, 5486°K, and 4487°K, respectively. However, an attempt to explain the magnitude of α only through quantum effects arising from the gaseous phase does not lead to a quantitative agreement with experimental results.

In the case studied here, we must expect quantum effects associated with the crystalline phase as well, inasmuch as the melting temperature of LiH(T_M = 961°K[9]) is of the same order of magnitude as the maximum characteristic vibrational energy of the crystal[10].

The classical theory including quantum corrections [11] shows that the largest departure of the distribution coefficient from unity should occur in the case when quantum effects are significant only in one of the phases, at the same time that the second phase can be described classically. In the presence of quantum corrections from both phases α can be either larger or smaller than unity

Table 1. Distribution coefficient for H and D isotopes in the system $Li^7H:D$ plus gas.

Run	T^oC	Grain size (mm)	Duration of contact between phases- (hours)	D concentration c_1 in gaseous phase (at%)	D concentration c_2 in Li^7H (at %)	$\alpha=\frac{c_1/1-c_1}{c_2/1-c_2}$
1	230	≤0.05	7000	1.949	1.437	1.363
2	310	≤0.13	1037	2.233	1.672	1.343
3	314	0.13-0.26	1445	1.815	1.358	1.343
4	303	≤0.13	800	1.680	1.248	1.352
5	296	0.05-0.13	2100	2.016	1.501	1.350
6	390	0.13-0.26	185	1.717	1.298	1.328
7	398	-"-	340	1.687	1.280	1.323
8	404	-"-	570	1.774	1.346	1.324
9	500	-"-	170	1.616	1.242	1.306
10	500	0.05-0.13	70	2.004	1.560	1.290
11	501	0.13-0.26	110	1.752	1.354	1.299
12	617	≤0.13	91	2.098	1.652	1.276
13	615	-"-	115	2.098	1.640	1.285
14	598	0.13-0.26	84	1.813	1.415	1.286
15	600	-"-	24	1.779	1.392	1.283
16	602	-"-	45	1.683	1.327	1.273
17	665	-"-	46	1.872	1.482	1.268
18	670	-"-	48	1.890	1.486	1.277

depending on the relative magnitudes of these corrections.

We note that according to reference 1 α must be smaller than unity for a monatomic crystal, containing a heavy defect, in equilibrium with a classical monatomic gas, which in the present case would be D, if the phenomenon of separation is determined only by quantum corrections in the solid phase. In the present case, the crystal is in equilibrium with an essentially quantum gas. Quantum corrections from the gaseous phase act in the direction of increasing α and lead to a value of α greater than unity.

Recently, in reference 12, a theory of the thermodynamic equilibrium for systems consisting of a diatomic gas and a crystal was constructed which takes into account the details of the dynamics of a diatomic crystal containing defects. The isotopic distribution coefficient was calculated for the system (Li^7H:D) - (H_2, HD). The presence of D_2 molecules in the gas was neglected, since the theory was constructed on the basis of the assumption of a small concentration of impurities. The authors presented numerical results for two choices of input data, which differ from each other in the value of the ionic charges in a crystal of Li^7H, which were taken to be equal to $\zeta/e = 0.7$ or 0.85 (here ζ is the ionic charge, e is the electronic charge). In Figure 2 are presented numerical values for the distribution coefficient (curve 1 is for the case $\zeta/e = 0.7$ and curve 2 is for the case $\zeta/e = 0.85$), as well as the experimental data from the table. As can be seen from Figure 2, the agreement between theoretical and experimental results is entirely satisfactory. In this regard we note that curve 2 fits the experimental points somewhat better, corresponding to a value of $\zeta/e = 0.85$.

The authors thank I.P. Ipatova and A.A. Klochikhin for valuable advice during the discussion of the experimental results, and for acquainting them with the contents of reference 12 prior to its publication.

References

1. G. Benedek, I.P. Ipatova, A.A. Klochikhin, A.A. Maradudin, W. C. Overton, Jr., and R.F. Wallis, Proc. Int. Conf. on the Localized Excitations in Solids, Irvine (1967) (in print); A.F. Ioffe Physico-Technical Institute preprint, 1967; Zhur. Eksper. i Teor. Fiz. (in press).

2. J. Bigeleisen, A. Kant, J. Am. Chem. Soc. 76, 5957, (1954).

3. K.E. Wilzbach, L. Kaplan, J. Am. Chem. Soc. 72, 5795, (1950).

4. E.F. Sollers, J.L. Greenshaw, J. Am. Chem. Soc. 59, 2015, 2724, (1937).

5. A.L. Borocco, Compt. rend., 206, 1117, (1938).

6. T. Kirshenbaum, Physical Properties and Analysis of Heavy Water, (1951).

7. A.I. Shatenshteyn, E.A. Yakovleva, E.N. Zvyagintseva, Ya. M. Varshavskii, E.A. Izrailevich, N.M. Dykhno, Isotopic Analysis of Water (Academy of Sciences of the USSR, Moscow, 1957).

8. M.W. Mallet, A.F. Gerds, C.B. Griffith, Analyt. Chemistry 25, 116, (1953).

9. C.E. Messer, E.B. Damon, P. C. Maybury, J. Mellor, R.A. Seals, J. Phys. Chem. 62, 220, (1958).

10. A.S. Filler, E. Burstein, Bull. Am. Phys. Soc. 5, 198 (1963).

11. L. Landau and E. Lifshitz, Statistical Physics (Addison-Wesley Publishing Co., Inc., Reading, 1958).

12. G. Benedek, R.F. Wallis, I.P. Ipatova, A.A. Klochikhin, A.A. Maradudin, W.C. Overton, Jr., to be published.

LATTICE VIBRATIONAL IMPURITY MODES IN LiH:D AND LiD:H CRYSTALS

S. S. Jaswal and J. R. Hardy

Behlen Laboratory of Physics, University of Nebraska

Lincoln, Nebraska

INTRODUCTION

Some time ago, Misho[1] carried out an experimental study on the effects of isotopic composition on the fundamental infrared absorption frequency ($\omega_{t.o.}$) of Li(H,D) mixtures. His principal finding was that very small ($\sim 5\%$) concentrations of D^- were sufficient to shift this frequency from that characteristic of LiH to a value close to that of LiD.

Until the present, theoretical understanding of these results has been hampered by the absence of any reliable calculations of the normal modes of either pure LiH or pure LiD.

Recently, however, Verble, Warren and Yarnell[2] have measured most of the dispersion curves of Li^7D along the <100>, <110> and <111> directions. Thus it is now possible to attempt to construct a dynamical matrix which reproduces the measured curves and can then be used to derive eigenfrequencies and eigenvectors for a uniform sample of points throughout the reduced zone. With this information one can then proceed to make defect calculations. Moreover, it is entirely reasonable to take the dynamical matrix so constructed for LiD and to use it for LiH, simply substituting the H^- mass for the D^- mass.

Thus we have computed the normal modes of both crystals on the basis of the deformation dipole (D.D.) model[3], modified to include central second neighbor interactions between both types of ion and an angle bending force[4] which resists shearing of the 90° angle subtended by any pair of first neighbors at a given ion

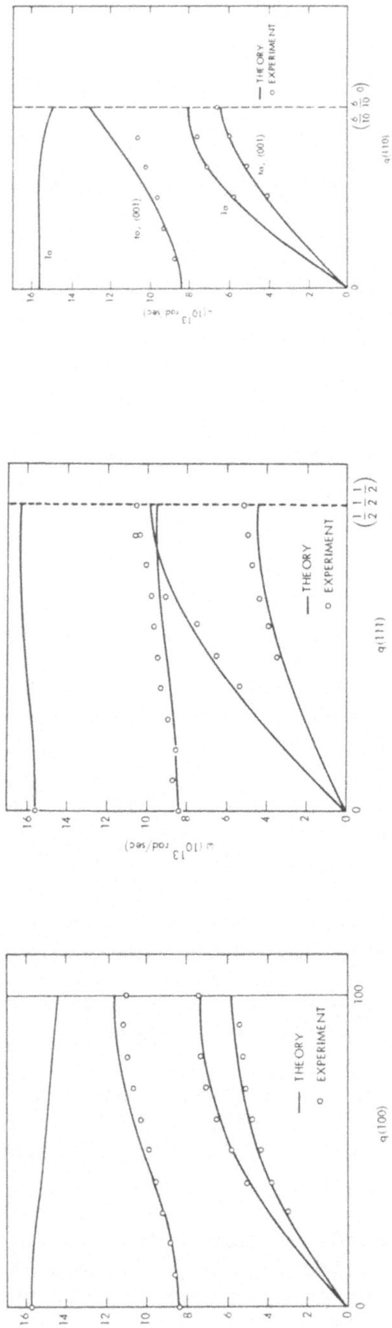

Fig. 1. Phonon Dispersion Curves in the Three Symmetry Directions for Li^7D.

(we assume the angle-bending force constant to be the same for both positive and negative ions). Also we have allowed the monopole charge Z to be disposable, an assumption to some extent justified by evidence for incomplete ionicity[5]. Furthermore, we do not treat the $Li^+ - Li^+$ and $H^- - H^-$ ($D^- - D^-$) interactions independently, but use their ratio as a second disposable constant. The various parameters are adjusted to obtain the best fit to the experimental dispersion curves, consistent with the observed elastic constants C_{11} and C_{44} (obtained from the slopes of the $\langle 100 \rangle$ acoustic branches at long wavelengths), the observed value of $\omega_{t.o.}$, and the static and high frequency dielectric constants. We also found it necessary to increase the positive ion polarizability α_+ at the expense of that of the negative ion α_-.

Thus with elastic constants:-

$$C_{11} = 6.63 \times 10^{11} \text{ dynes/cm}^2$$

$$C_{44} = 4.83 \times 10^{11} \text{ dynes/cm}^2$$

the best fit was given by:-

$$\text{Angle bending constant} = 0.841 \times 10^{11} \text{ c.g.s. units.}$$

$$Z = 0.7 \; |\text{electronic charge}|$$

$$\alpha_+ = 0.229 \times 10^{-24} \text{ cm}^3$$

$$\alpha_- = 1.664 \times 10^{-24} \text{ cm}^3$$

$$(\text{++ interaction}):(\text{-- interaction}) = 2/3$$

The measured and computed Li^7D dispersion curves are shown in fig. (1). The agreement is quite good, with maximum discrepancies $\sim$6% except for the transverse acoustic frequency at the symmetry point L and the transverse optic frequency at the symmetry point K where we have $\sim$13% misfit.

The resultant frequency spectrum for LiH appears in fig. (2) (sample density, 8000 wave vectors in the first zone); as can be seen there is a definite gap between $\omega = 9.83 \times 10^{13}$ and $\omega = 11.2 \times 10^{13}$ rad./sec.

We believe that our results are much more reliable than those of Benedek[6], who found no gap, since his calculations used input data which lead to imaginary frequencies within certain regions of the zone.

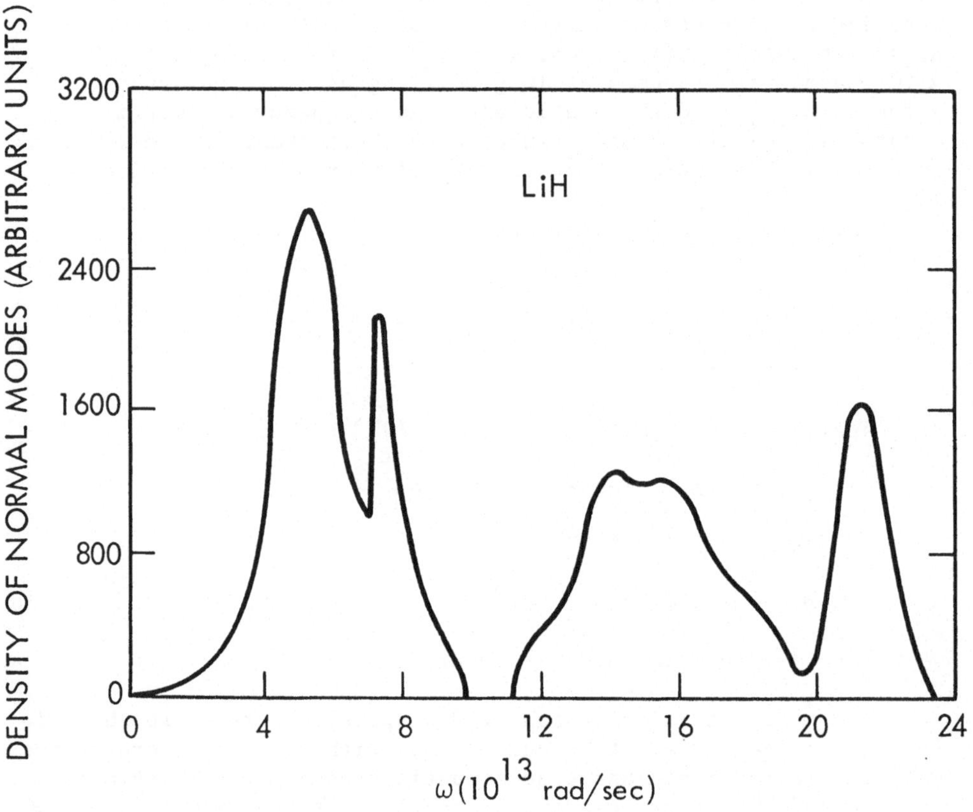

Fig. 2. Frequency Distribution for LiH.

IMPURITY MODES IN LiH:D

The presence of a gap in the LiH frequency spectrum means that a heavier impurity substituted for H^- can produce a local mode in this gap. For the particular case of a substitutional D^- ion, we have solved the standard integral equation of Dawber and Elliott(7), using the LiH eigenfrequencies and eigenvectors derived for the perfect lattice. This leads to a local mode frequency in the gap $\omega_\ell = 10.85 \times 10^{13}$ rad./sec., close to $\omega_{t.o.}$ for LiH which is 11.21×10^{13} rad./sec.

This result is exact in that we are dealing with a true isotopic impurity and it is this fact that makes the Li(H,D) system ideal for defect studies.

In order to understand Misho's(1) results we have to work out both the absorption frequency due to the local modes associated with finite numbers of D^- ions in an LiH host, and the integrated intensity of this absorption.

Maradudin(8) has given the following expression for this absorption frequency, valid in the limit of low defect concentration p.

$$\omega_{t.o.}^2 - \omega^2 + \epsilon \mu p \omega_{t.o.}^2 A^{-1}(\omega) = 0 \qquad 2.1$$

where $A(\omega) = 1 - \frac{\epsilon}{3N}\omega^2 \sum_{\underline{q}j} \frac{|\underline{W}(-|\frac{\underline{q}}{j})|^2}{\omega^2 - \omega_j^2(\underline{q})}$

where ϵ = the fractional change in mass at the defect site, ω is the absorption frequency to be determined, N is the number of unit cells in the crystal, the $\omega_j(q)$'s are the perfect lattice eigenfrequencies, the $\underline{W}(-|\frac{q}{j})$'s are the corresponding negative ion eigenvectors and $\mu = M_{Li}/(M_{Li}+M_H)$.

This equation has been solved for ω as a function of p and the results are shown in fig. (3). Thus, for example, we see that for p = 0.05, ω is ~6% less than ω_ℓ for an isolated defect.

The derivation of the total relative integrated local mode absorption (a_ℓ) is more difficult and we have only derived the appropriate expression for very low defect concentrations, such that the local modes centered on different defects do not interact.

In this limit:-

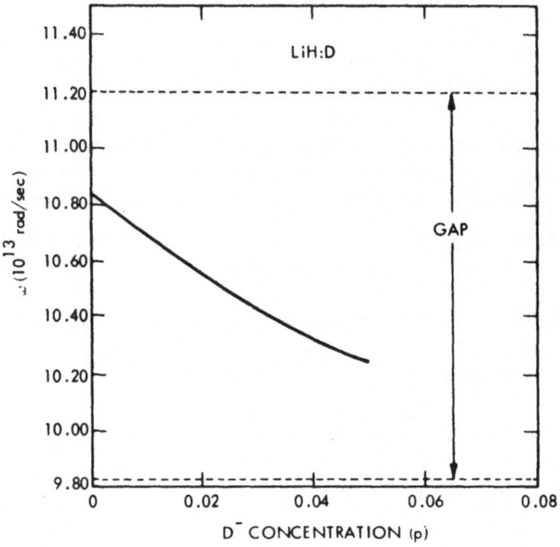

Fig. 3. Absorption Frequency as a Function of D^- Concentration in LiH

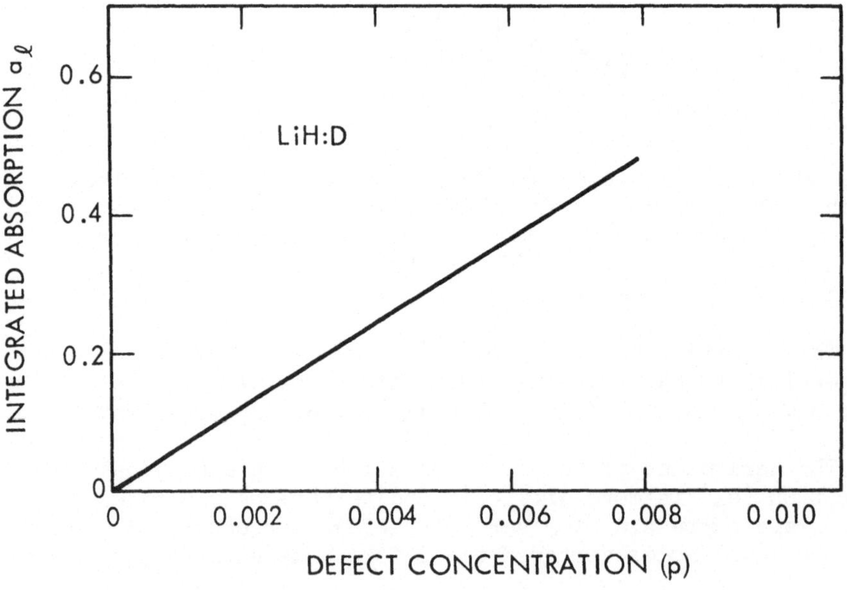

Fig. 4. Integrated Absorption in the Gap Mode as a Function of D^- Concentration in LiH

$$a_\ell = p\mu \frac{\epsilon^2 \omega_\ell^4}{(\omega_\ell^2 - \omega_{t.o.}^2)^2} M_H |\chi_\ell(0)|^2 \qquad 2.2$$

where M_H is the mass of the H^- ion and $\chi_\ell(0)$ is the modulus of the defect amplitude in the localized vibration.

In fig. (4) we show a_ℓ as a function of p and it is obvious that a_ℓ rises very rapidly with p:e.g. for p as low as 0.01, $a_\ell \sim 0.5$.

These results are in good qualitative agreement with Misho's[1] findings, even though eqn. (2.2) is unreliable when the predicted a_ℓ value is ~1 (obviously it must break down somewhere, since a_ℓ cannot be >1). Certainly one can reasonably infer that the local mode absorption is dominant for $p \sim 0.05$ or larger. However, even compared with the shifted frequencies predicted by eqn. (2.1), Misho's experimental absorption frequency is very much lower.

We have also investigated the in-band defect amplitude as a function of frequency and find a significant resonance at 19.8×10^{13} rad./sec. Whether this is genuine depends on how reliable the higher frequency part of our optical spectrum is, since the measured dispersion curves do not extend into this region and the high frequency maximum in our calculated spectrum lies above that measured by incoherent neutron scattering.[9]

We have also examined the vibrational properties of small amounts of H^- in LiD; these results are described more fully, together with the work outlined in this present paper, elsewhere[10].

Again we find a local mode, but this time it lies above the high frequency end of the whole spectrum at $\omega = 18.75 \times 10^{13}$ rad./sec. However its frequency is much less sensitive to defect concentration, although we find a slight rise with increasing p as opposed to the slight fall predicted by Elliott and Taylor[11] although in most other respects, our results are close to theirs, even though they were using an approximate frequency spectrum. Also the integrated absorption in the H^- local mode is about 100 times less than that in the D^- local mode for the same p value.

It seems to us that more experimental work on Li(H,D) infrared lattice absorption would be very fruitful; in particular it should be aimed at eliminating the discrepancy between the calculated and observed D^- local mode absorption frequencies and at observing the H^- local mode. Finally the realm of intermediate concentrations, i.e. comparable amounts of both H^- and D^- obviously warrants further study both theoretically and experimentally.

ACKNOWLEDGMENT

One of us (SSJ) is grateful to the University Research Council for the award of a faculty summer fellowship and other financial support.

REFERENCES

1. R. H. Misho, Ph.D. thesis, Michigan State University, 1961 (unpublished). See also D. J.Montgomery and J. R. Hardy, J. Phys. Chem.Solids, Supp. 1, 491 (1965).

2. J. L. Verble, J. L. Warren, and J. L. Yarnell, Bull. Amer. Phys. Soc. 12, 557 (1967). We should also like to thank these authors for providing us with a fuller account of their data.

3. J. R. Hardy, Phil. Mag. 7, 315 (1962).

4. A. A. Maradudin, (private communication). We should like to thank Prof. Maradudin for this information.

5. F. E. Pretzel, G. N. Rupert, C. L. Mader, E. K. Sturms, G. V. Gritton and C. C. Rushing, J. Phys. Chem. Solids 16, 10 (1960).

6. G. Benedek, Solid State Comm. 5, 101 (1967) (and private communication.

7. P.G. Dawber and R. J. Elliott, Proc. Roy. Soc. (London) A273, 222 (1963).

8. A. A. Maradudin, 1962, Brandeis University Summer Institute Lectures in Theoretical Physics (W. A. Benjamin, Inc., New York 1962), Vol. II.

9. A. D. B. Woods, B. N. Brockhouse, M. Sakomoto and R. N. Sinclair, Inelastic Scattering of Neutrons in Solids and Liquids (International Atomic Energy Agency, Vienna 1961), p. 487.

10. S. S. Jaswal and J. R. Hardy, Lawrence Radiation Laboratory Report, UCRL-70622.

11. R. J. Elliott and D. W. Taylor, Proc. Roy. Soc. (London) A296, 161 (1967).

FUNDAMENTAL ABSORPTION OF KI-NaI SOLID SOLUTION

KAIZO NAKAMURA and YOSHIO NAKAI

Department of Physics, Kyoto University, Kyoto, Japan

INTRODUCTION

The fundamental absorption of alkali iodides were measured early by Hilsch and Pohl,[1] and Fesefeld,[2] and later by Teegarden[3] who reported the fine structure of exciton absorption at low temperature, and also by Fischer and Hilsch[4] who observed firstly Wannier type exciton at the step absorption and investigated the change of exciton absorption with temperature. Eby, Teegarden, and Dutton[5] made a comprehensive measurement of fundamental absorption of alkali halides at room and liquid nitrogen temperature. Measurement at lower temperature (10°K) was carried out recently by Teegarden and Baldini.[6] Their results were interpreted by classifying the absorption peaks into p-s and p-d transition groups on the basis of the recent theoretical works.[7,8] In particular, their interpretation for the absorption spectrum of KI is different from that of NaI.

Several authors have discussed the structure of exciton absorption bands which are characteristic in the low energy region of the fundamental absorption of alkali halides. Knox and Inchauspe[9] predicted the minimum splitting values of spin-orbit doublets for the salts with NaCI structure. According to them, it holds for chlorides and bromides fairly well, but in iodides it no longer holds, perhaps owing to the presence of adjacent levels, except NaI.

It is, therefore, uncertain which pairs should be chosen as the spin-orbit partners, and so far some authors took the choice of 5.59 eV and 6.74 eV peaks for NaI, 5.82 eV and 6.71 eV for KI, and 5.72 eV and 6.64 eV for RbI. Recently, Phillips[10] has applied to alkali halides the band theory which is very successful in the study of semiconductors, and has given a new interpretation for them. Onodera et al[7] have recently made a relativistic band calculation for KI and have given a different proposal from Phillips'. Their remark on the importance of

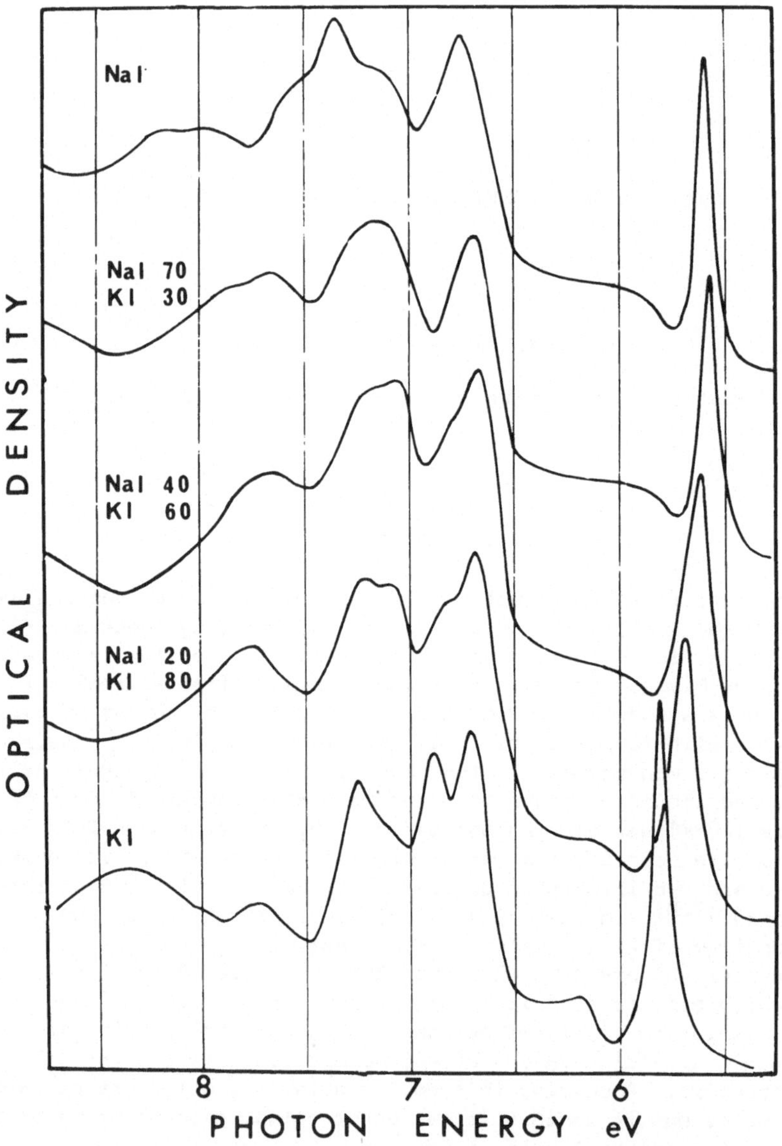

Fig. 1. Variation of the fundamental absorption in KI-NaI solid solutions measured at liquid nitrogen temperature. Compositions are given in mole percent for each curve. The ordinates are shifted appropriately for comparison.

d-orbitals in the conduction band calculation is consistent with the results of KCI calculated by Oyama and Miyakawa[8] and by DeCicco.[11]

To obtain the consistent interpretation, absorption measurement of mixed crystal system is useful. The absorption associated with the transition of the same origin will be found to change in a simple way when the concentration is changed in the solid solution of two kinds of materials. One of the authors[12] has recently measured the absorption spectra of KI-RbI mixed crystals. According to his measurements, each peak in KI corresponds to each in RbI, the order of peaks in the spectrum being kept unchanged over the whole range of concentration. Hence, it is expected that the mixed crystal of KI and NaI will give the decisive identification for the spin-orbit partner of the first exciton absorption of alkali iodide crystals. In this work the variation of the fundamental absorption spectrum in KI-NaI solid solution is studied at liquid nitrogen temperature with a primary purpose to clarify the assignment in the absorption spectrum in alkali iodide crystals.

EXPERIMENTAL

The measurments of optical absorption were made on films evaporated onto a LiF substrate. All the spectra were taken at liquid nitrogen temperature. Samples were prepared by melting the known fraction of mixture of KI and NaI commercial powders in a platinum crucible and quenching it down to room temperature, small pieces of which were evaporated from coiled platinum wire heater onto the substrate kept at room temperature. Evaporated films of mixed crystal are not stable and highly hygroscopic in open air. They can never exposed to air before absorption measurement. Good vacuum in cryostat is essential to be free from damage of samples. Reproducibility of the measurement can offer a reasonable check for the quality of every sample prepared.

The vacuum ultraviolet spectrometer used was a Seya type monochromator with a concave grating of one meter radius. The band-width of 1 or 2 A was employed in the experiment.

RESULTS

Absorption spectra obtained at liquid nitrogen temperature are shown in Fig.1. Fifteen compositions were measured and five results of which are given to show the variation of absorption spectrum with the change of composition. In Fig. 1 the ordinate for each spectrum is shifted appropriately to avoid complications. In NaI, the first exciton peak is found at 5.59 eV, the second at 6.74 eV which Teegarden and Baldini[6] and Phillips[10] have assigned as Γ-exciton. Beyond the second peak, a broad band composed of more than three components lies from 7.1 to 7.7 eV. Two broad bands are found round 8 eV. In KI, the first exciton peak is at 5.82 eV, the second at 6.71 eV, the third at 6.88 eV and the fourth prominent peak at 7.22 eV. A weak shoulder is at 7.1 eV. Two broad bands are at 7.77 and 8.35 eV. To be noted

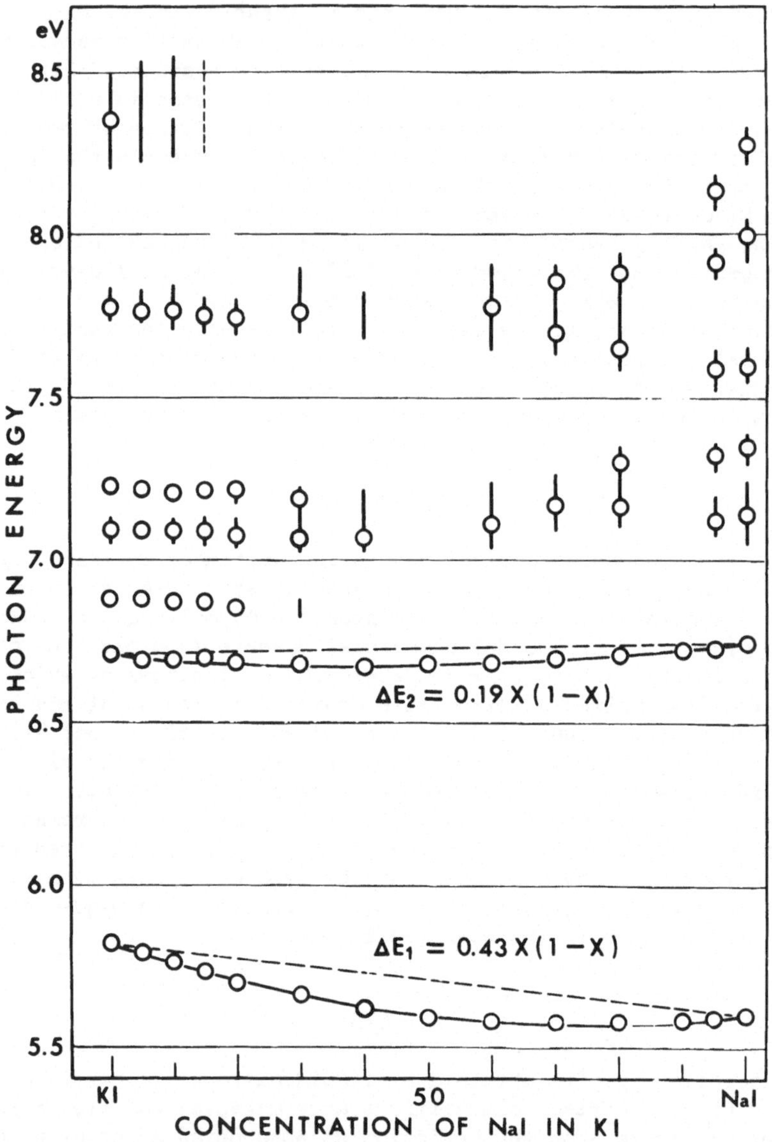

Fig. 2. Positions of dominant absorption peaks are plotted as a function of the concentration of NaI in KI. For two peaks of lowest energies, the changes of the positions are described well by the equation given in the figure in which ΔE (in eV) is the deviation of the positions from the straight line connecting the peak positions in pure crystals, and X is concentration of NaI in the solid solution.

here is the distinct shoulder in KI at 6.21 eV, the corresponding shoulder in NaI is not so distinct but is found at about 5.8 eV.

As increasing the content of NaI in KI, the first exciton is displaced quadratically with the change in concentration of NaI. The variation of the first peak is well described by the equation

$$\Delta E_1 = 0.43X(1-X)$$

where ΔE_1 is the energy shift of the first peak from the straight line connecting the peak energy of pure materials, and X stands for the concentration of NaI in solid solution. The variation of the second peak is more simple. It is almost unchanged. If its behavior is fitted to the equation above, the coefficient is about 0.19, but the agreement is not so good as the first. The behavior of the third peak is remarkable. It undergoes the radical change by adding a small amount of NaI, that is, it diminishes rapidly and vanishes entirely at about 40 percent of NaI. The fourth peak associated with 7.1 eV shoulder also suffers a drastic deformation as shown in the figure. The shoulder, which Phillips assigned as the spin-orbit partner of the 6.21 eV shoulder, grows rapidly into a peak almost to surpass or to be comparable with the fourth peak, and at last these two bands become unified into a broad band. The shoulder at 6.21 eV in KI becomes vague when doped with NaI, which is in contrast with the case of KI-RbI.[12]

In Fig. 2 the variation of the absorption spectrum with the change of composition is summarised. The circles show the positions of the absorption maxima and the bars represent the broadness and the uncertainty of the center position of the absorption bands. The complex behavior of the absorption spectrum in the solid solution with more than 70 percent of NaI was not shown in Fig. 1. But in Fig. 2, it can be followed for each peak with the increase in concentration up to 100 percent of NaI. The detailed measurement in this concentration region will be given elsewhere.

DISCUSSIONS

The second peak at 6.74 eV in NaI is generally believed to be the spin orbit partner of the first exciton band. Recentry Kunz[13] made valence band calculations for NaI. His result for the spin orbit separation, 1.25 eV, is consistent with the interpretation mentioned above. However the situation is a little more confusing in KI. As to three dominant peaks from the second to the forth, no definite choice is available in assigning the spin orbit partner.

Taking into account the persistent existence of the first and the second bands at any concentration, these two peaks are assigned to Γ-excitons associated with the transitions $\Gamma_{15}^{3/2} - \Gamma_1$ and $\Gamma_{15}^{1/2} - \Gamma_1$. This means that a tentative assignments proposed by Phillips for these bands in KI and NaI[2] are supported experimentally. The spin-orbit separation grows larger as the concentration of NaI is increased. The third peak in KI may be due to the transition of valence electron to a d orbit of metal ion, and the fact that Na lacks d orbitals[3] may explain the disappearance of this band at the concentration of NaI higher than 40 percent. The step at 7.1 eV in KI which Phillips explained as M_0 type edge with the transition $\Gamma_{15}^{1/2} - \Gamma_1$ may be some exciton transition at certain point of Brillouin zone which is sensitive to the deformation of the lattice. The step at 6.21 eV in KI becomes less distinct as NaI is increased, which may confirm the above statement.

REFERENCES

(1) R. Hilsch and R. W. Pohl, Z. Physik 57, 145 (1929); 59, 812 (1930).
(2) H. Fesefeld, Z. Physik 64, 623 (1930).
(3) K. Teegarden, Phys. Rev. 108, 660 (1957).
(4) F. Fischer and R. Hilsch, Nachr. Akad. Wiss. Gottingen, II. math-phys. Kl. No. 8, 241 (1959).
(5) J. E. Eby, K. J. Teegarden, and D. B. Dutton, Phys. Rev. 116, 1099 (1959).
(6) K. Teegarden and G. Baldini, Phys. Rev. 155, 896 (1967).
(7) Y. Onodera, M. Okazaki, and T. Inui, J. Phys. Soc. Japan 21, 2229 (1966).
(8) S. Oyama and T. Miyakawa, J. Phys. Soc. Japan 21, 868 (1966).
(9) R. S. Knox and N. Inchauspe, Phys. Rev. 116, 1093 (1959).
(10) J. C. Phillips, Phys. Rev. 136, A1705 (1964); 136, A1721 (1964).
(11) P. D. DeCicco, Phys. Rev. 153, 931 (1967).
(12) K. Nakamura, J. Phys. Soc. Japan 22, 511 (1967).
(13) A. B. Kunz, Phys. Rev. 151, 620 (1966).
(14) K. Nakamura and Y. Nakai, J. Phys. Soc. Japan 23, 455 (1967).

ENERGY FLOW IN DISORDERED LATTICES (DISCUSSION OF A FILM)*

D. N. Payton, III,† Marvin Rich, and William M. Visscher

University of California, Los Alamos Scientific Laboratory, Los Alamos, New Mexico

I. INTRODUCTION

In the course of an investigation of the thermal conductivity of disordered harmonic and anharmonic lattices,(1) it was decided that a better understanding of the nature of the energy flow could be obtained by following the propagation of an energy pulse through a lattice graphically. This paper describes a film which provides a technique for direct observation of the lattice motion.

The movie contains a number of examples of energy flowing in a two dimensional square lattice which is ten atoms wide and fifty atoms long. In every case the lattice is excited from rest by giving an identical impulse to each of the ten atoms at one end. The propagation of this energy pulse down the lattice has been followed by numerical integration of the equations of motion for the system. Results are displayed in 3D plots of the energy of the atoms of the array at successive times. The individual frames are the direct computer output to a SC-4020 plotter.

The sequences have been arranged to contrast the energy propagation through a region of randomly distributed light impurities with that of propagation through a heavy impurity region, and to show the changes which occur when anharmonic interatomic forces are added to the harmonic lattice. The first two examples show propagation in monatomic harmonic and anharmonic lattices. These are followed by situations in which a single light or heavy impurity has been placed one-third of the distance down the lattice. In the remaining sequences, the first third of the lattice is monatomic with the following two thirds containing low concentrations of randomly distributed light or heavy impurities. Results are shown for harmonic and anharmonic lattices with five and fifteen percent impurity concentrations.

The following section of this paper contains a brief discussion of the equations of motion of the square lattice used for the film, and of the method of calculation employed for their solution. Section III contains a detailed discussion of the individual sequences in the film.

II. THE LATTICE EQUATIONS

The lattice model used in making the film is illustrated in Fig. 1.

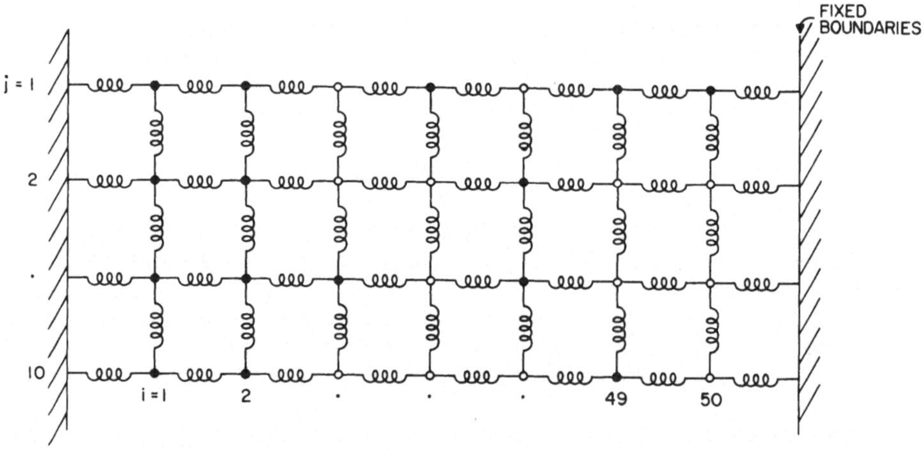

Fig. 1. Illustration of the binary disordered square lattice model used in making the film. Periodic boundary conditions are used in the j direction.

It consists of a ten by fifty array of atoms which interact through nearest neighbor central and non-central forces. Rigid boundary conditions were used at both ends of the long axis, and periodic boundary conditions in the short direction. Therefore, displacements of atoms from equilibrium satisfy the relations

$$x_{0,j} = y_{0,j} = 0 \qquad x_{51,j} = y_{51,j} = 0 \qquad j = 1, 2, \ldots, 10$$
$$x_{i,1} = x_{i,11} \qquad y_{i,1} = y_{i,11} \qquad i = 1, 2, \ldots, 50, \tag{1}$$

where x and y are coordinates relative to the long and short axes, respectively, and the subscripts specify the individual atoms.

The lattice can be considered cylindrical.

The equations of motion for the lattice atoms are

$$m_{ij} \frac{d^2 x_{ij}}{dt^2} = - \frac{d\Phi}{dx_{ij}} \qquad i = 1,\ldots,50 \qquad (2)$$

$$m_{ij} \frac{d^2 y_{ij}}{dt^2} = - \frac{d\Phi}{dy_{ij}} \qquad j = 1,\ldots,10.$$

In order to decouple the x motion from that in the y direction, it has been assumed that the potential Φ is the sum of two terms

$$\Phi = \Phi_x + \Phi_y \qquad (3)$$

where Φ_x is a function of the x_{ij} displacements only and Φ_y depends just on the y_{ij}. In the film, only the x motions of the lattice were excited. The y motions are identical and independent.

The potential Φ_x is a sum of nearest-neighbor interactions

$$\Phi_x = \sum_{i,j} V_c(x_{ij} - x_{i-1,j}) + \sum_{i,j} V_{nc}(x_{ij} - x_{i,j-1}). \qquad (4)$$

The central and non-central terms, V_c and V_{nc}, relate to interactions parallel and transverse to the line of centers between a pair of atoms. The functional forms of V_c and V_{nc} have been taken equal. They were obtained from an expansion of a Lennard-Jones potential about the equilibrium spacing r_o to fourth order

$$V(r) = \epsilon_o[(r_o/r)^{12} - 2(r_o/r)^6]$$

$$\qquad (5)$$

$$\cong - \epsilon_o + \tfrac{1}{2}\,\gamma(r-r_o)^2 - \tfrac{1}{3}\,\mu(r-r_o)^3 + \tfrac{1}{4}\,\nu(r-r_o)^4.$$

The magnitudes of μ and ν relative to γ were taken appropriate to noble gas solids for the anharmonic lattice films and were set equal to zero in the harmonic cases. With arbitrary units such that $\gamma = 1$, the potentials used are

$$V_c(x) = \tfrac{1}{2}\,x^2 - \tfrac{1}{3}\,\mu\,x^3 + \tfrac{1}{4}\,\nu\,x^4 \qquad (6)$$

$$\mu = 0.0 \qquad \nu = 0.0 \qquad \text{harmonic}$$
$$\mu = 0.35 \qquad \nu = 0.069 \qquad \text{anharmonic.} \tag{7}$$

The equations of motion take on the explicit form

$$m_{ij}\frac{d^2x_{ij}}{dt^2} = -\sum_{\lambda=-1}^{1}\{(x_{ij}-x_{i+\lambda,j}) - \lambda\mu(x_{ij}-x_{i+\lambda,j})^2 + \nu(x_{ij}-x_{i+\lambda,j})^3\}$$
$$-\sum_{\lambda=-1}^{1}\{(x_{ij}-x_{i,j+\lambda}) - \lambda\mu(x_{ij}-x_{i,j+\lambda})^2 + \nu(x_{ij}-x_{i,j-\lambda})^3\}. \tag{8}$$

Different methods of numerical integration of the equations of motion were used depending on whether or not the anharmonic forces were included. A simple iterative procedure was used in the harmonic case. This consisted of rewriting Eq. (8) as a pair of first order difference equations

$$m_{ij}\frac{dv_{ij}}{dt} \to m_{ij}\frac{v_{ij}(t+\Delta t)-v_{ij}(t)}{\Delta t} =$$
$$\left\{\sum_{\lambda=-1}^{1}[(x_{ij}-x_{i+\lambda,j}) - \lambda\mu(x_{ij}-x_{i+\lambda,j})^2 + \nu(x_{ij}-x_{i+\lambda,j})^3] + \sum_{\lambda=-1}^{1}[(x_{ij}-x_{i,j+\lambda}) - \lambda\mu(x_{ij}-x_{i,j+\lambda})^2 + \nu(x_{ij}-x_{i,j+\lambda})^3]\right\} \tag{9}$$

$$\frac{dx_{ij}}{dt} \to \frac{x_{ij}(t+\Delta t)-x_{ij}(t)}{\Delta t} = \tfrac{1}{2}[v_{ij}(t) + v_{ij}(t+\Delta t)]. \tag{10}$$

The displacements on the right hand side of Eq. (9) are assumed to refer to time $t + \frac{1}{2}\Delta t$, and to be defined by

$$x_{ij}(t + \tfrac{1}{2}\Delta t) \equiv \tfrac{1}{2}\,[x_{ij}(t) + x_{ij}(t + \Delta t)]. \quad (11)$$

Taking an initial guess of

$$x_{ij}(t + \Delta t) = x_{ij}(t) + \Delta t\; v_{ij}(t)$$

new values of the velocities at time t + Δt can be obtained from Eq. (9), followed by new displacements using Eq. (10). The procedure converges in two or three iterations for a sufficiently small time step. Convergence was determined by the requirement that no displacement shall change by more than 10^{-6} from one iteration to the next. This method required relatively small amounts of computer time and was very good in terms of energy conservation within the lattice, which was used as a check on the overall accuracy of the calculation. The total energy remained constant to one part in 10^6 throughout a calculation.

This iterative method was unstable when the anharmonic terms were included. A fourth order Runge Kutta integration scheme(2) was used in that case.

III. DISCUSSION OF THE FILM

In each example in the film, the atoms of the host lattice were assigned a mass of three. Light and heavy impurities were given masses of one and nine, respectively, and the impurity sites are the same in corresponding cases. The initial energy pulse has kinetic energy per atom corresponding to a temperature of about seven degrees Kelvin, and motion through the first third of the lattice is unperturbed in each case. The temperature of the wave front just before striking the impurity region is on the order of one degree. Even at this low temperature, the anharmonic forces are about ten percent of the harmonic forces for the interaction used.

The first sequence shows propagation of an energy pulse through a monatomic harmonic lattice. The initial pulse diminishes in amplitude as it moves and is followed by a trail of lesser waves as it deposits energy in the lattice. The pulse builds up again on reflection due to the pileup of energy at the rigid boundary during this process. The second example gives a comparison with propagation in a monatomic anharmonic lattice. The principal differences to be seen are that the anharmonic wave front is noticeably sharper than the harmonic front and also that its speed is greater. Figure 2 compares the harmonic and anharmonic energy pulses at equal times when the wave front is about two thirds of the way down the lattice. The figure consists of actual film frames.

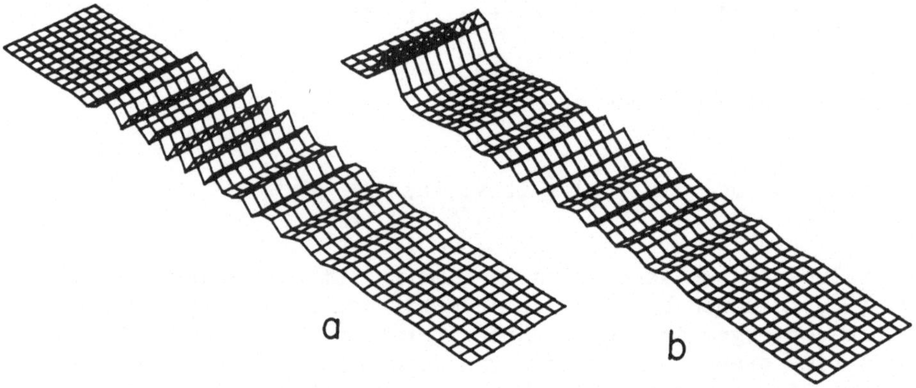

Fig. 2. Comparison of the energy waves at equal times in harmonic (a) and anharmonic (b) monatomic lattices.

The third and fourth examples show monatomic harmonic and anharmonic lattices with a single light impurity placed one-third of the way down the lattice. An indentation appears in the wave front as it passes over the light impurity, indicating that little energy has been transferred to this atom. This is due to the fact that the principal mode of excitation of the light impurity is a localized vibrational mode whose amplitude dies off exponentially with distance from the impurity. The initial impulse to the lattice leaves the localized mode practically unexcited, so that the light atom takes almost no part in the general oscillation of the lattice.

The opposite extreme of a single heavy impurity in an otherwise monatomic lattice is shown in examples five and six for harmonic and anharmonic forces, respectively. In this case, a sharp spike appears in the energy wave as it crosses the heavy atom. This has to do with the fact that the normal mode decomposition of the heavy atom motion is sharply peaked near the low energy end of the phonon spectrum, and within this narrow frequency range the modes act coherently to give the heavy impurity a large energy. The initial impulse populates all the available normal modes well, and since a band of these modes involves large motions of the heavy impurity, a substantial amount of energy can be transferred to this atom when the pulse arrives at its location.

The indentation, or spike, occurring when a wave front encounters a light or heavy impurity, respectively, is illustrated by movie frames in Figure 3.

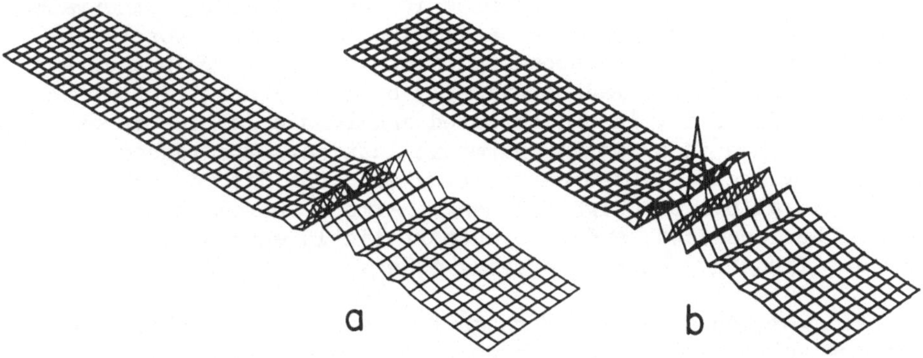

Fig. 3. Illustrations of the dip occurring in the energy wave as it passes over a light impurity (a) and the spike resulting from encounter with a heavy impurity (b).

The remaining film sequences show results with randomly distributed impurities in the second two-thirds of the lattice. The seventh and eighth examples illustrate energy penetration into a region containing five percent light impurities for harmonic and anharmonic interaction, respectively. Examples nine and ten give the corresponding results with five percent heavy impurities. Dips are visible in the successive waves as they cross the light impurity sites in the first pair of examples, while energy spikes occur at the heavy impurity sites in the cases nine and ten. The energy wave can be seen to penetrate more easily into a region of light impurities than into one of heavy defects, as is clear from Figure 4.

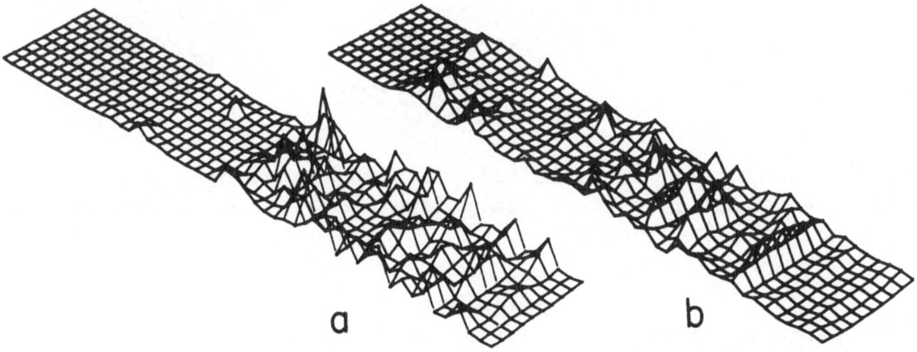

Fig. 4. Equal time comparison of energy penetration into a harmonic lattice containing 15 percent heavy impurities (a) with that into a 15 percent light impurity region (b). Impurity sites are identical in the two pictures.

The primary point of these examples is that the energy penetrates into the impurity region more easily when anharmonic forces are present than in the corresponding harmonic cases. This fact supports the observation made previously by the authors[1] that the thermal conductivity in a disordered anharmonic lattice is generally greater than in a similar harmonic lattice. The reason why anharmonicity increases the energy flow into the disordered region is that it allows the high frequency harmonic normal modes, which tend to be localized, and so transport little energy, to decay into low frequency energy-carrying modes.

Film sequences eleven through fourteen are equivalent to examples seven through ten, respectively, except that the impurity concentration has been raised to fifteen percent. The energy penetration into the region of disorder is much slower than at the lower impurity concentration in the pure harmonic cases. For heavy impurities, one sees little more than the energy spikes at the impurity sites themselves. The effect of anharmonicity in increasing the energy flow is more noticeable now than at the lower concentration. Otherwise, the comments made about the five percent impurity cases apply here as well. Figure 5 gives a comparison of the energy flow into the disordered heavy impurity region with and without anharmonic forces.

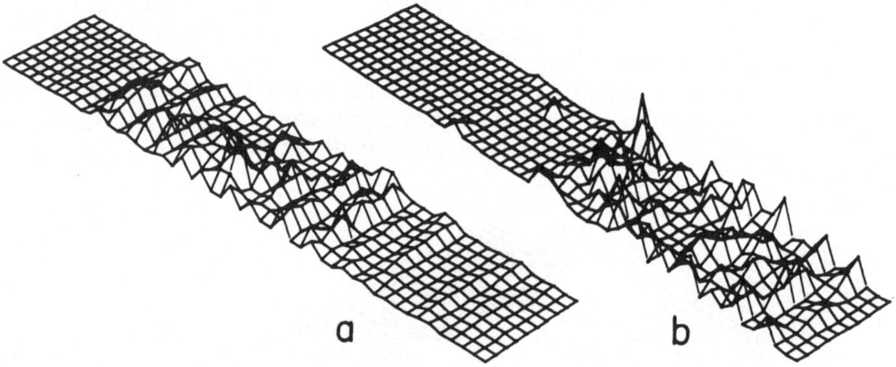

Fig. 5. Comparison at identical times of the energy penetration into a region of 15 percent heavy impurities for an anharmonic (a) and a harmonic (b) lattice.

REFERENCES

*Work performed under the auspices of the U. S. Atomic Energy Commission.

†Present address: Air Force Weapons Lab., Kirtland AFB, New Mex.

1. D. N. Payton, M. Rich, and W. M. Visscher, Phys. Rev. 160, 706 (1967).
2. Fr. A. Willers, Practical Analysis, p. 378 (Dover Publications, Inc., 1948).

PART J. SPECIAL TOPICS

THEORY OF LOCALIZED PLASMONS

L. J. Sham

Department of Physics, University of California,Irvine

1. Introduction

A vast amount of literature exists on the study of plasma oscillations in a homogeneous electron gas.[1] Recently there has been some interest in plasma oscillations in inhomogeneous electron systems. Treatments appropriate to finite systems, such as an atom, have been given by Brandt and Lundquist[2], and Wieder and Borowitz[3]. The case of a point impurity in an otherwise homogeneous electron gas has been considered by Sziklas[4].

Here we shall restrict our attention to the case Sziklas has investigated. In particular, we are interested in how a plasmon bound to the impurity may be formed. Sziklas has derived an equation governing the motion of a plasmon in this inhomogeneous system, treating the electron-electron interaction in the random phase approximation and the impurity potential as a first order perturbation. It appears impossible to satisfy simultaneously the conditions that, on the one hand, the impurity potential be weak enough for the first order perturbation to be valid and that, on the other hand, it be sufficiently strong to bind a plasmon. We shall remedy this situation by showing that Sziklas' equation, when written in terms of the electron density, is, in fact, valid to all orders of the impurity potential.

In Sec. 2, we shall first see how the equation for the plasmon motion can be derived from a classical model of a charged elastic continuum. In Sec. 3, almost the same equation is derived quantum mechanically within the time-dependent Hartree approximation. There will not be any restriction of treating the impurity potential as a weak perturbation. In Sec. 4, a rough estimate of the binding energy of the localized plasmon is made in an attempt to exhibit approximately its dependence on the impurity charge and the electron density of the host system.

2. The Elastic Continuum Model

Let the density distribution of the electrons in an inhomogeneous system be $n(\underset{\sim}{r})$. If a disturbance causes a displacement $\underset{\sim}{u}(\underset{\sim}{r})$ of the electrons from equilibrium at $\underset{\sim}{r}$, the restoring electric field is given by

$$\text{div}\ \underset{\sim}{E} = \text{div}\left\{4\pi\ en(\underset{\sim}{r})\underset{\sim}{u}(\underset{\sim}{r})\right\}, \qquad (1)$$

where $-e$ is the charge of an electron. The elastic force per particle is $(K/n)\,\nabla^2 u$, where K is the bulk modulus. Hence, the equation of motion is

$$m\ddot{\underset{\sim}{u}} = -e\underset{\sim}{E} + (K/n)\,\nabla^2\underset{\sim}{u}, \qquad (2)$$

m being the mass of an electron.

Analyzing the oscillation of $\underset{\sim}{u}$ into normal modes at frequency ω and using Eq.(1), we derive from (2),

$$\text{div}\left[m\omega^2\underset{\sim}{u} + (K/n)\,\nabla^2\underset{\sim}{u} - 4\pi ne^2\underset{\sim}{u}\right] = 0 \qquad (3)$$

For a homogeneous system with electron density n_o, Eq.(3) gives the plasma frequency at wave vector $\underset{\sim}{q}$,

$$\omega^2 = \omega_p^{\ 2} + Kq^2/mn_o\ , \qquad (4)$$

where

$$\omega_p^{\ 2} = 4\pi n_o e^2/m\ . \qquad (5)$$

In the random phase approximation[1],

$$K = (3/5)\ mn_o V_F^{\ 2}\ , \qquad (6)$$

where V_F is the Fermi velocity.

[1] One might be tempted to use the isothermal bulk modulus, $\frac{1}{3}mn_oV_F^{\ 2}$, for K. This measures the elastic force at low frequencies where the positive background has sufficient time to keep up with the electron cloud and is therefore not appropriate for the plasma oscillation which has a high frequency, in which case the positive background can be treated as stationary.

If an impurity is introduced into the homogeneous system and causes the electron density to become

$$n(\underset{\sim}{r}) = n_o + \delta n(\underset{\sim}{r}) \ , \tag{7}$$

the equation of motion for the plasmon is approximately

$$\mathrm{div}\left[(\omega^2 - \omega_p^2)\, \underset{\sim}{u} + (\kappa/n_o)\nabla^2 \underset{\sim}{u} - (\omega_p^2/n_o)\,\delta n \cdot \underset{\sim}{u}\right] = 0. \tag{8}$$

By Fourier transforming Eq.(8) and writing

$$\underset{\sim}{u}(\underset{\sim}{q}\omega) = -i\underset{\sim}{q}\varphi_e(\underset{\sim}{q},\omega), \tag{9}$$

and

$$\delta n(\underset{\sim}{q}) = \int \exp(-i\underset{\sim}{q}\cdot\underset{\sim}{r})\,\delta n(\underset{\sim}{r})\, d\underset{\sim}{r} \ , \tag{10}$$

we obtain the dispersion relation

$$(\omega^2 - \omega_p^2)\varphi_e(\underset{\sim}{q},\omega) - \frac{3}{5} v_F^2 q^2 \varphi_e(\underset{\sim}{q},\omega) - \frac{\omega_p^2}{\Omega}\sum_{\underset{\sim}{q}'} \frac{\underset{\sim}{q}\cdot\underset{\sim}{q}'}{q^2}\cdot\frac{\delta n(\underset{\sim}{q}-\underset{\sim}{q}')}{n_o}\varphi_e(\underset{\sim}{q}',\omega) = 0 \ , \tag{11}$$

Ω being the volume of the system.

Eq.(11) is in a form very similar to the effective mass equation of the shallow impurity donor in a semiconductor. The q^2 term of the dispersion of the square of free plasmon frequency provides the effective kinetic energy term of the plasmon in the impurity system and the attraction of the plasmon to the point impurity is provided by the lowering of plasmon energy due to a depletion of electron density near the impurity.

3. The Random Phase Approximation

Consider a homogeneous electron gas with density n_o occupying a large volume Ω, into which a point impurity of charge Ze, fixed in position, and Z electrons are introduced. A weak external potential $v_e(\underset{\sim}{r},\omega)$ oscillating at frequency ω will induce a potential $\varphi_e(\underset{\sim}{r},\omega)$ in this system, given by

$$\varphi_e(\underset{\sim}{r},\omega) = \int d\underset{\sim}{r}'\ \epsilon^{-1}(\underset{\sim}{r},\underset{\sim}{r}';\omega+i\eta)v_e(\underset{\sim}{r}',\omega), \tag{12}$$

where ϵ^{-1} is the inverse dielectric function and $\eta\to +0$. A plasma oscillation is a self-sustained oscillation of the system without external excitation. Hence, from Eq.(12), Fourier-transformed with respect to position, the plasma frequency is given by

$$\sum_{q'} \epsilon(\underset{\sim}{q},\underset{\sim}{q}' \,;\, \omega+i\eta) \quad \varphi_e(\underset{\sim}{q}',\omega) = 0. \tag{13}$$

The dielectric function is related to the proper polarization part $\tilde{S}(\underset{\sim}{q},\underset{\sim}{q}';\omega)$ by

$$\epsilon(\underset{\sim}{q},\underset{\sim}{q}';\omega) = \delta_{\underset{\sim}{q},\underset{\sim}{q}'} - \frac{4\pi e^2}{q^2}\tilde{S}(\underset{\sim}{q},\underset{\sim}{q}';\omega) \,. \tag{14}$$

In the homogeneous system, plasmons exist for wave vector q less than a certain cutoff, about a fraction of the Thomas-Fermi wave vector[1]. We are particularly interested in the onset of bound plasmon formation and, thus, important contributions to Eq.(13), when an impurity is present, come from interaction of free plasmons of small wave vectors q and q'. This is much the same as done in the derivation of the effective mass equation for the shallow impurity states in semiconductors[5]. In evaluating $\tilde{S}(q,q';\omega)$ we shall retain only the leading term in the power series of qv_F/ω and $q'V_F/\omega$.

In the time-dependent Hartree approximation, the proper polarization part is given by the analytic continuation of

$$\tilde{S}(\underset{\sim}{q},\underset{\sim}{q}' \,;\, i\omega_\ell) = \beta^{-1}\sum_n \Omega^{-1}\sum_{\underset{\sim}{k}\underset{\sim}{k}'} G_{\underset{\sim}{k}+\underset{\sim}{q},\underset{\sim}{k}'+\underset{\sim}{q}'}(iE_n+i\omega_\ell)\; G_{\underset{\sim}{k}'\underset{\sim}{k}}(iE_n), \tag{15}$$

where $\beta = 1/k_BT$, $E_n = (2n+1)\,\pi/\beta$, $\omega_\ell = 2\ell\pi/\beta$ and $G_{\underset{\sim}{k},\underset{\sim}{k}'}$ denotes the electron propagator in the Hartree potential of the system containing the impurity.

Sziklas expanded $G_{\underset{\sim}{k},\underset{\sim}{k}'}$ to first order in the impurity potential and found a very simple expression for $\tilde{S}(\underset{\sim}{q},\underset{\sim}{q}';\omega)$ in leading order of qv_F/ω, for $\underset{\sim}{q}\neq\underset{\sim}{q}'$,

$$\tilde{S}(\underset{\sim}{q},\underset{\sim}{q}' \,;\omega) = \frac{\underset{\sim}{q}\cdot\underset{\sim}{q}'}{m\omega^2\Omega}\,\delta n(\underset{\sim}{q}-\underset{\sim}{q}') \,, \tag{16}$$

in terms of $\delta n(q-q')$, the Fourier transform of the change of electron density $\tilde{\delta n}(\underset{\sim}{r})$ due to the presence of the impurity, as given in Eq.(10). Actually Eq.(16) holds to all orders in the impurity potential. It is not difficult to verify directly to second and third order. However, this manner of verification does not readily extend to all orders of the impurity potential.

There are two ways of proving Eq.(16). One is simply to note that $\tilde{S}(\underset{\sim}{q},\underset{\sim}{q}';\omega)$ as given by Eq.(15) is the density response function of an independent particle system under the influence of a one-particle potential equal to the Hartree potential of our impurity system. Thus, it has a spectral representation:

$$\tilde{S}(\underset{\sim}{q},\underset{\sim}{q}' ;\omega) = \int_{-\infty}^{\infty} d\zeta \; \sigma(\underset{\sim}{q},\underset{\sim}{q}' ;\zeta)/(\omega-\zeta)$$

$$= \omega^{-1} \int_{-\infty}^{\infty} d\zeta \; \sigma(\underset{\sim}{q},\underset{\sim}{q}';\zeta) + \omega^{-2} \int_{-\infty}^{\infty} d\zeta \; \sigma(\underset{\sim}{q},\underset{\sim}{q}';\zeta)\zeta + 0(\omega^{-3}) \qquad (17)$$

expanding in the inverse powers of ω. By the sum rules to be proved in the Appendix, the coefficients of ω^{-1} and ω^{-3} vanish and the ω^{-2} term yields Eq.(16).

The second method is equivalent to the first but does not refer to an auxiliary independent particle system. We simply expand the right hand side of Eq.(15) in the inverse powers of ω,

$$\tilde{S}(\underset{\sim}{q},\underset{\sim}{q}' ;\omega) = -$$

$$\omega^{-2}\Omega^{-1} \sum_{\underset{\sim}{k},\underset{\sim}{k}'} \int\int d\zeta d\zeta'\{f(\zeta)-f(\zeta')\}(\zeta-\zeta') \; g_{\underset{\sim}{k}+\underset{\sim}{q},\underset{\sim}{k}'+\underset{\sim}{q}'}(\zeta')g_{\underset{\sim}{k}'\underset{\sim}{k}}(\zeta) \qquad (18)$$

where g is the spectral density of the propagator G and f is the Fermi function. The integrals are evaluated in much the same way as the sum rules are proved in the Appendix, and simplified in the random phase approximation to yield Eq.(16).

The diagonal element of $\tilde{S}$ is approximately

$$\tilde{S}(\underset{\sim}{q},\underset{\sim}{q} ;\omega) = \frac{n_o q^2}{m\omega^2}\left(1 + \frac{3}{5} \cdot q^2 v_F^2/\omega^2\right) . \qquad (19)$$

Substituting Eqs.(14), (16), (19) into Eq.(13), we obtain the equation of motion for the plasmon,

$$\left(\frac{\omega^2}{\omega_p^2} -1 -\frac{3}{5}\frac{\cdot v_F^2}{\omega^2} q^2\right)\varphi_e(\underset{\sim}{q},\omega) - \Omega^{-1}\sum_{\underset{\sim}{q}'} \frac{\underset{\sim}{q}\cdot\underset{\sim}{q}'}{q^2} \cdot \frac{\delta n(\underset{\sim}{q}-\underset{\sim}{q}')}{n_o} \varphi_e(\underset{\sim}{q}',\omega) = 0, \qquad (20)$$

which would be identical with Eq.(11) if in the q^2 term ω^2 were replaced by ω_p^2.

If the impurity possesses spherical symmetry, φ_e can be analyzed in terms of spherical harmonics $Y_{\ell m}$,

$$\varphi_e(\underset{\sim}{q}) = q^{-2} f_\ell(q) \; Y_{\ell m}(\hat{\underset{\sim}{q}}) . \qquad (21)$$

Then, Eq.(20) reduces to the form

$$(q^2-E) \; f_\ell(q) + \Lambda \int_0^{\infty} dq' \; K_\ell \; (q,q') \; f_\ell(q') = 0 , \qquad (22)$$

where

$$E = \left(\frac{\omega^2}{\omega_p^2} - 1\right) / \frac{3}{5} \cdot \frac{v_F^2}{\omega^2} ,$$

$$\Lambda = \frac{1}{4\pi^2 n_o} / \frac{3}{5} \cdot \frac{v_F^2}{\omega^2} ,$$

and

$$K_\ell(q,q') = qq' \int_{-1}^{1} d\mu \, P_\ell(\mu) \, \mu . \delta n(\sqrt{q^2-2qq'\mu+q'^2}) . \tag{23}$$

Alternatively,

$$K_\ell(q,q') = \frac{8\pi qq'}{2\ell+1} \left[(\ell+1) \int_0^\infty dr \; j_{\ell+1}(qr) \; r^2 \; \delta n(r) \; j_{\ell+1}(q'r) + \ell \int_0^\infty dr \; j_{\ell-1}(qr) r^2 \delta n(r) \; j_{\ell-1}(q'r) \right] \tag{24}$$

We note that Eq.(22) is the form Sziklas first obtained, the only difference being that his equation was derived in terms of first order density change. In actuality, his equation has a wider validity.

4. Estimates of Binding Energy of the Localized Plasmon.

From the last section we learn that to calculate the frequency of a bound plasmon, all we need is the electron density distribution due to the presence of the impurity. The simplest approximation is the linear Thomas-Fermi expression due to Mott[6],

$$\delta n(\underset{\sim}{q}) = Z / (1+q^2/ k_{TF}^2) \tag{25}$$

where k_{TF} is the Thomas-Fermi screening wave vector. This was used by Sziklas to evaluate numerically the s-state bound plasmon. Here we make a further approximation on the kernel $K_\ell(q,q')$ of the integral equation (22) so as to obtain a more explicit, though rather qualitative, condition for the formation of bound plasmons.

It is convenient to follow Sziklas' usage of dimensionless quantities defined by,

$$\underset{\sim}{x} = \underset{\sim}{q}/k_{TF} ,$$

$$E = -\gamma^2 k_{TF}^2 ,$$

and

$$\lambda = \Lambda k_{TF} . \tag{26}$$

Then, Eq.(22) becomes

$$(x^2+\gamma^2)\ f_\ell(x)\ +\lambda \int_0^\infty dx'\ K_\ell(x,x')f_\ell(x') = 0\ . \tag{27}$$

We approximate the kernel by keeping only the leading term in a power series expansion of the wave vectors in either Eq.(23) or Eq.(24) and by providing a cutoff at k_{TF}. Thus,

$$\begin{aligned} K_o(x,x') &\simeq \frac{4}{3} Z\ x^2x'^2\ \theta(1-x)\ \theta(1-x')\ , \\ K_1(x,x') &\simeq \frac{2}{3} Z\ xx'\ \ \theta(1-x)\ \theta(1-x')\ , \\ K_2(x,x') &\simeq \frac{8}{15} Z\ x^2x'^2\ \theta(1-x)\ \theta(1-x')\ . \end{aligned} \tag{28}$$

$\theta(1-x)$ being the step function, vanishing for $x>1$. The cutoff is not unreasonable since not only the density δn diminishes rapidly for wave vector beyond the cutoff but also this is roughly the cutoff for the free plasmons. (See Sec.3).

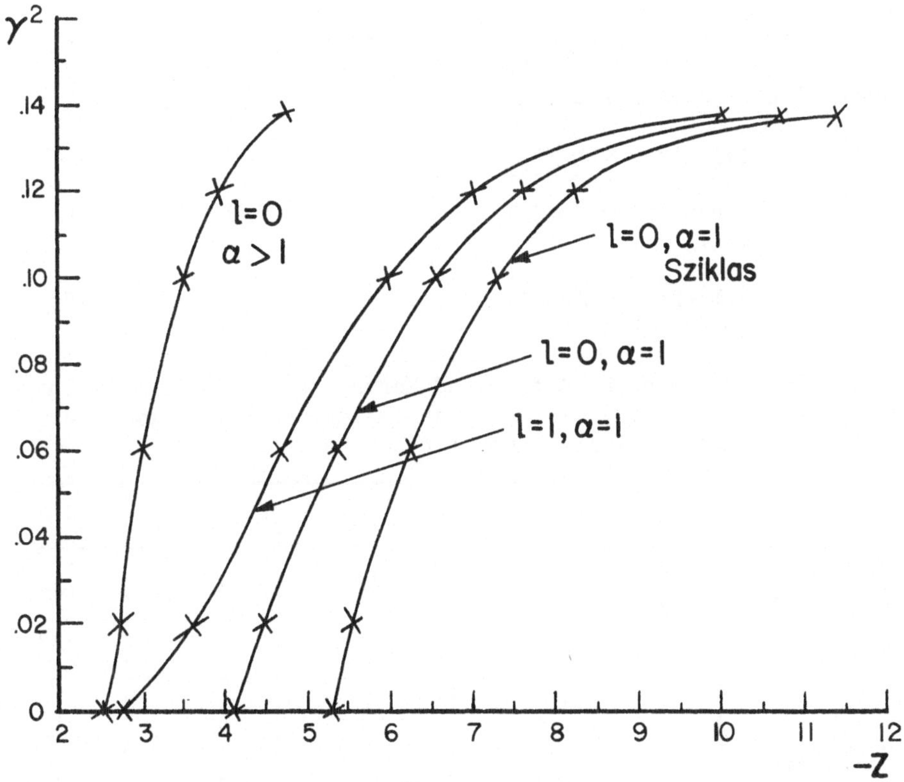

Fig. 1. Binding energy γ^2 as a function of impurity charge Z for s and p states at r_s=1.8.

With the separable kernels (28), the integral equation(27) is easily solved, giving the dispersion relation of γ^2 as a function of $Zr_s^{3/2}$ where r_s is the mean electron distance in the homogeneous host system. A plot of the binding energy, as measured by γ^2, against Z for s and p plasmons is given in Fig.1 for r_s=1.8. Sziklas result for the s-plasmon is included for comparison. The conditions for the formation of various plasmons are:

$$\ell = 0, \ -Zr_s^{3/2} > 81\pi^2/80 \simeq 10 \ ;$$

$$\ell = 1, \ -Zr_s^{3/2} > 27\pi^2/40 \simeq 20/3 \ ; \tag{29}$$

$$\ell = 2, \ -Zr_s^{3/2} > 81\pi^2/32 \simeq 25 \ .$$

It seems physically reasonable that the p-state lies lowest since the effective potential term in Eq.(20) is of p-type.

So far we have used the linear approximation (25) for the density in the estimate of the binding energy. In general, in powers of q^2, we can write

$$\delta n(\underset{\sim}{q}) = Z - \alpha Zq^2/k_{TF}^2 + \ldots \ . \tag{30}$$

In the linear approximation, α=1. To the next order in the impurity potential,

$$\alpha = 1 - Zr_s^{3/2}/3\pi \ . \tag{31}$$

α has been calculated by Alfred and March[7] self-consistently in the Thomas-Fermi approximation. The following table shows that Eq.(31) represents a fairly good fit to their values of α.

r_s	-Z		1	2	3
2.67 (Cu)	α	Ref. 7	1.40	1.85	2.34
		Eq.(31)	1.46	1.92	2.38
3.01 (Ag)	α	Ref. 7	1.49	2.04	2.65
		Eq.(31)	1.56	2.11	2.67

When the density (30) including higher order effects of the impurity is used in place of (25), the kernel $K_1(X,X)$ remains unchanged but K_0, K_2, etc are α times the expressions given in (29). Hence, p-states are unaltered while s and d states are lowered. A plot of the new binding energy for the s-state is is now seen to rival the p-state for the lowest energy level. We are presently computing the density due to point impurity self-consistently in Hartree approximation and plan to evaluate the dispersion relation with sufficient numerical accuracy to decide whether the qualitative features obtained in this section are valid. However, it seems unlikely from what we have done here that bound plasmons form for reasonable values of Z in an interesting class of materials, the degenerate semi-conductors, which generally have a value of r_s less than unity.

APPENDIX.

We now establish the sum rules for the density response function. Although only the sum rules for the non-interacting system are needed in Sec. 3, they are just as easy to establish for an interacting system.

Let $\rho(\underset{\sim}{r})$ denote the density operator as a function of position. For an infinite system, it is convenient to define the Fourier transform of the density operator by

$$\rho(\underset{\sim}{q}) = \Omega^{-1} \int d\underset{\sim}{r}\, e^{-i\underset{\sim}{q}\cdot\underset{\sim}{r}} \rho(\underset{\sim}{r}). \tag{32}$$

Note the difference of the factor of volume of the system between Eq. (10) and (32).

The spectral density, $\sigma(\underset{\sim}{q},\underset{\sim}{q}';\omega)$, as defined in (17), is the Fourier transform with respect to time of the correlation function

$$\sigma(\underset{\sim}{q},\underset{\sim}{q}';t) = (1/2\pi) < \left[\rho(\underset{\sim}{q},t),\ \rho^{\dagger}(\underset{\sim}{q}',0)\right] >, \tag{33}$$

where $\rho(\underset{\sim}{q},t)$ is the Heisenberg representation of $\rho(\underset{\sim}{q})$ and the square brackets are commutator brackets. Hence,

$$\sigma(\underset{\sim}{q},\underset{\sim}{q}';\omega) = \sum_{nm} e^{\beta(F-\varepsilon_n)} \{1 - e^{\beta(\varepsilon_n - \varepsilon_m)}\}\,. \\ <n|\rho(\underset{\sim}{q})|m> <m|\rho^{\dagger}(\underset{\sim}{q}')|n> \quad \delta(\omega + \varepsilon_n - \varepsilon_m) \tag{34}$$

where ε_n is the energy of the eigenstate $|n>$ of the system and F is the grand potential.

It is then easy to verify the first sum rule,

$$\int_{-\infty}^{\infty} d\omega\, \sigma(\underset{\sim}{q},\underset{\sim}{q}';\omega) = \Omega < \left[\rho(\underset{\sim}{q}),\ \rho^{\dagger}(\underset{\sim}{q}')\right] > = 0. \tag{35}$$

If J(q) denotes the current density operator, then from the equation of continuity,

$$\left[\rho(\tilde{q}),\, H\right] = \tilde{q}\cdot\tilde{J}(\tilde{q}), \tag{36}$$

H being the Hamiltonian of the system. From Eqs.(34) and (36), we obtain

$$\int_{-\infty}^{\infty} d\omega\ \sigma(\tilde{q},\tilde{q}\ ;\omega)\,\omega = \Omega < \left[\tilde{q}\cdot\tilde{J}(\tilde{q}),\rho^{\dagger}(\tilde{q}')\right] > = \tilde{q}\cdot\tilde{q}'\ n(q-q')/m, \tag{37}$$

where the electron density n(q) is the ensemble average of $\rho(\tilde{q})$. Eq.(37) is a slight extension of the usual f-sum rule.[1]

Similarly, one can verify

$$\int_{-\infty}^{\infty} d\omega\ \sigma(\tilde{q},\tilde{q}';\omega)\ \omega^2 = \Omega < \left[\tilde{q}\cdot\tilde{J}(\tilde{q})\ ,\tilde{q}'\cdot\tilde{J}^{\dagger}(\tilde{q}')\right] >$$
$$= (\tilde{q}\cdot\tilde{q}')\ (\tilde{q}+\tilde{q}')\cdot < \tilde{J}(\tilde{q}-\tilde{q}')> /m = 0. \tag{38}$$

REFERENCES.

*Supported in part by the U.S. Air Force Office, AFOSR Grant No. 1080-66.
†Present address: Department of Mathematics, Queen Mary College, University of London.

1. For an excellent treatment of the subject, see D. Pines and P. Nozieres. "The Theory of Quantum Fluids, Volume I: Normal Fermi Liquids", (W.A. Benjamin, Inc., New York 1966).

2. W. Brandt and S. Lundqvist, Phys. Rev. 132, 2135 (1963).

3. S. Wieder and S. Borowitz, Phys. Rev. Letters, 16, 724 (1966).

4. E. A. Sziklas, Phys. Rev. 138, A1070 (1965).

5. W. Kohn, Solid State Physics, 5, 257 (1957).

6. N.F. Mott, Proc. Camb. Phil. Soc., 32, 281 (1936).

7. N.H. March, Adv. in Physics, 6, 1 (1957).

THE THEORY OF ELECTROCONDUCTIVITY OF FERROMAGNETIC METALS CONTAINING IMPURITIES

Yu. Kagan, A. P. Zhernov, H. Pashaev

I. V. Kurchatov Institute of Atomic Energy

Moscow, USSR

1. INTRODUCTION

In a previous paper [1] a theory has been developed concerned with the electroconductivity of metals containing nonmagnetic impurities. The presence of impurities has been shown to be responsible for the appearance of not only static but also of dynamic inhomogeneity, which in turn leads to the existence of inelastic incoherent scattering of electrons. As a result, the impurity part of the electrical resistance becomes dependent on temperature, the dependence displaying quite anomalous features.

A similar situation must take place in the case of a ferromagnetic metal, where the electrical resistance is essentially connected with the scattering by the spin system. Indeed, the presence of impurity atoms leads to appearance of randomly distributed regions in which the spin system is dynamically perturbed. The scattering of electrons by these perturbed regions will have both elastic as well as inelastic character. This gives rise to a temperature dependent impurity part of the electrical resistance, the momentum nonconservation in such a process of electron scattering providing for a different temperature dependence than that valid for the pure host metal. It is natural, that in this case the Mathiessen's rule is violated.

In this paper we consider the electroconductivity of a ferromagnetic metal containing magnetic and nonmagnetic impurities of substitution. To reveal all qualitative aspects of the problem in a most clear form, we confine ourselves to some simple approximations for description of the spin and electron subsystems. For the electron system we adapt the free-electron approximation and neglect possible distortions of the spectrum upon introduction of impurity atoms. For description of the spin system a model of localized spins is used. In the analysis to follow we confine our treatment to the case of a simple cubic lattice and take account of the interaction between nearest neighbours only. This model has the advantage that within its limits the spectral properties of a single impurity problem have been previously analyzed in much detail [2], [3], [4]. The same assumptions as those made in paper [1] will be used here with respect to the phonon system and its interaction with the conduction electrons. The scattering of electrons will be as usual treated in Born approximation.

2.

Let us neglect the conduction electron polarization as well as the drag-effect. Then we have the following general expression for the electrical resistance:

$$\rho = \frac{1}{2T} \frac{\iint d\vec{k} d\vec{k}_1 (\varphi_{\vec{k}} - \varphi_{\vec{k}_1})^2 \, f^o_{\vec{k}}(1-f^o_{\vec{k}_1}) W_{\vec{k}\vec{k}_1}(\vec{q},\omega)}{\left| \int d\vec{k} \; e \, \vec{V}_{\vec{k}} \, \varphi_{\vec{k}} \frac{\partial f^o_{\vec{k}}}{\partial \epsilon_{\vec{k}}} \right|^2} \tag{2.1}$$

Here $\varphi_{\vec{k}}$ is a nonequilibrium correction to the distribution function of conduction electrons $f_{\vec{k}}$

$$f_{\vec{k}} = f^o_{\vec{k}} - \frac{\partial f^o_k}{\partial \epsilon_{\vec{k}}} \varphi_{\vec{k}}$$

$W_{\vec{k}\vec{k}_1}$ is the electron transition probability during scattering, defined as follows

$$W_{\vec{k}\vec{k}_1} = \frac{1}{4} \sum_{\sigma,\sigma_1} W_{\vec{k}\sigma,\vec{k}_1\sigma_1} \tag{2.2}$$

where σ is the spin index of a conduction electron. (In Eq. (2.1) and in our further treatment we omit everywhere the spin index when describing an electron state, bearing in mind that in the absence of polarization this cannot result in a misunderstanding).

$$\vec{q} = \vec{k}-\vec{k}_1 \quad , \qquad \omega = \epsilon_{\vec{k}} - \epsilon_{\vec{k}_1} \tag{2.3}$$

The amplitude of electron scattering by the n-th site occupied by the spin S_n is expressed in the form:

$$g_n(\vec{q}) = a_n(\vec{q}) + 2d_n'(\vec{q})\,\vec{s}\,\vec{S}_n \tag{2.4}$$

We calculate $W_{\vec{k}\vec{k}_1}$ neglecting therein the processes accompanied by simultaneous excitation of the spin and phonon systems. Making use of the relation between $W_{\vec{k}\vec{k}_1}$ and the corresponding scattering correlation functions $\Omega(\vec{q},\omega)$ we have

$$W_{\vec{k}\vec{k}_1}(\vec{q},\omega) = W^{(1)}_{\vec{k}\vec{k}_1}(\vec{q},\omega) + W^{(2)}_{\vec{k}\vec{k}_1}(\vec{q},\omega) + W^{(3)}_{\vec{k}\vec{k}_1}(\vec{q},\omega) \tag{2.5}$$

$$W^{(i)}_{\vec{k}\vec{k}_1} = \frac{(2\pi)^3}{m_*^2 V_o}\,\Omega^{(i)}(\vec{q},\omega) \tag{2.6}$$

where the second and the third terms describe inelastic scattering by the phonon system, $W_{\vec{k}\vec{k}_1}{}^{(2)}$ being identical with an expression found in [1].

$$\Omega^{(3)}(\vec{q},\omega) = \frac{1}{2N}\sum_{n,m} e^{i\vec{q}(\vec{R}^o_n - \vec{R}^o_m)} d_n d_m \times$$

$$\times \left[\sum_i \rho_i (\vec{S}_n)_{ii} (\vec{S}_m)_{ii}\right] \left\langle \left(\vec{q}\vec{U}_n(t)\right)\left(\vec{q}\vec{U}_m(0)\right)\right\rangle_\omega \tag{2.7}$$

(ρ is the density matrix of the spin system, index i labels the levels of the full spin Hamiltonian), and the first term in (2.5) describes elastic and inelastic scattering from the spin system.

$$\Omega^{(1)}(\vec{q},\omega) = \frac{1}{2N}\sum_{n,m} e^{i\vec{q}(\vec{R}^o_n - \vec{R}^o_m)} d_n(\vec{q})\, d_m(\vec{q}) \langle \vec{S}_n(t)\vec{S}_m(0)\rangle_\omega \;,$$

$$d_n(\vec{q}) = d_n'(\vec{q})\, e^{-W_n(\vec{q})/2} \tag{2.8}$$

(The other notations in Eqs. (2.5) - (2.8) are the same as those in [1]).

We write down $d_n(\vec{q})$ in the following form (in the approximation linear in concentration, W_n for sites of the host metal has the same value W_o).

$$d_n(\vec{q}) = d_o(\vec{q}) + C_n\,\Delta d(\vec{q}) \;,$$

$$\Delta d(\vec{q}) = d_1(\vec{q}) - d_o(\vec{q}) \tag{2.9}$$

where symbols 0 and 1 refer to atoms of the host metal and impurity, respectively, and C_n is equal to 1 in the impurity sites and 0 in the other sites.

The expression (2.1) for the electrical resistance must be averaged over all possible configurations of the impurity atoms (the corresponding process will be denoted by $\langle\ldots\rangle_c$). We choose $\varphi_{\vec{k}}$ in the form,

$$\varphi_{\vec{k}} = \vec{\mathcal{K}}\,\vec{k} \quad (\vec{\mathcal{K}} \text{ a unit vector}) \tag{2.10}$$

Substituting Eq. (2.9) into Eq. (2.8) and carrying out the averaging process, now over $\Omega(\vec{q},\omega)$, we get:

$$\begin{aligned}\langle\Omega^{(1)}(\vec{q},\omega)\rangle_c &= \frac{1}{2N} d_o^2 \sum_{n,m} e^{i\vec{q}(\vec{R}_n^o-\vec{R}_m^o)} \langle\langle\vec{S}_n(t)\vec{S}_m(0)\rangle_\omega\rangle_c + \\ &+ \frac{1}{2N} d_o \Delta d \sum_{n,m} e^{i\vec{q}(\vec{R}_n^o-\vec{R}_m^o)} \langle(c_n+c_m)\langle\vec{S}_n(t)\vec{S}_m(0)\rangle_\omega\rangle_c + \\ &+ \frac{1}{2N}(\Delta d)^2 \sum_n \langle c_n\langle\vec{S}_n(t)S_n(0)\rangle_\omega\rangle_c\end{aligned} \tag{2.11}$$

The correlation function of spin in Eq. (2.11) is

$$\begin{aligned}\langle\vec{S}_n(t)\vec{S}_m(0)\rangle_\omega &= \tfrac{1}{2}\left[\langle S_n^+(t)S_m^-(0)\rangle_\omega + \langle S_n^-(t)S_m^+(0)\rangle_\omega\right] + \\ &+ \langle S_n^z(t)S_m^z(0)\rangle_\omega \equiv \tilde{k}_{nm}(\omega) + \tilde{k}^z_{nm}(\omega)\quad .\end{aligned} \tag{2.12}$$

Let us consider first the correlation function $\tilde{k}_{nm}(\omega)$. This function is directly connected with the imaginary part of the Fourier component of the Green's function

$$\tilde{G}_{nm}(t) = -i\theta(t)\ \langle\left[S_n^+(t),S_m^-(0)\right]\rangle \tag{2.13}$$

Indeed, if one takes advantage of the well known relation

$$\tilde{k}_{nm}(\omega) = \langle S_n^+(t)S_m^-(0)\rangle_\omega = -\frac{2\,\mathrm{Im}\tilde{G}_{nm}(\omega+i\epsilon)}{1-e^{-\omega/T}} \tag{2.14}$$

then one gets

$$\tilde{k}_{nm}(\omega) = \tfrac{1}{2}\left[\tilde{k}_{nm}(\omega) + \tilde{k}_{nm}(-\omega)\, e^{\omega/T}\right] \tag{2.15}$$

Let us find the Dyson's equation for the Green's function (2.13). Taking the usual spin Hamiltonian

$$H = -\sum_{n,m} J_{nm}\,\vec{S}_n\vec{S}_m \tag{2.16}$$

and using the simplest form of decoupling (cf. for example [5]),

$$\langle [S_n^+(t)S_{n'}^z(t), S_m^-(0)] \rangle = \langle S_{n'}^z \rangle \langle [S_n^+(t), S_m^-(0)] \rangle$$

we get the following Dyson's equation

$$G_{nn_1}(\omega) = G^o_{nn_1}(\omega) + \sum_{n'n''} G^o_{nn'}(\omega)\, L_{n'n''}\, G_{n''n_1}(\omega) \tag{2.17}$$

where

$$\widetilde{G}_{nm}(t) = 2\sqrt{\sigma_n \sigma_m}\; G_{nm}(t) \quad , \qquad \sigma_n = \langle S_n^z \rangle \tag{2.18}$$

$$L_{nn_1} = 2\sum_{n'} \left(J_{nn'}\sigma_{n'} - J^o_{nn'}\sigma^o\right)\delta_{nn_1} - 2\left(J_{nn_1}\sqrt{\sigma_n \sigma_{n_1}} - J^o_{nn_1}\sigma^o\right)$$

$$G^o_{nm}(\omega) = \frac{1}{N}\sum_{\vec{q}} \frac{e^{i\vec{q}(\vec{R}^o_n - \vec{R}^o_m)}}{\omega - \omega_{\vec{q}}} \tag{2.19}$$

(The values characteristic for the ideal host metal are labeled with the symbol "0").

Being interested only in results that are linear in C, we neglect overlapping of perturbed regions, where the matrix L differs from zero. In this case

$$L_{nn_1} = \sum_p C_p V^p_{nn_1} \quad , \qquad V^p_{nn_1} \equiv V_{n-p,n_1-p} \tag{2.20}$$

For an impurity atom occupying the site $p = \ell$ we get from Eq. (2.17)

$$\hat{G}(\omega) = \hat{\mu}^{\vec{\ell}}\hat{G}^o + \sum_{p\neq\ell} \hat{\mu}^{\vec{\ell}}(\omega)\hat{G}^o(\omega)C_p\hat{V}^p\hat{G}(\omega) \quad , \qquad \hat{\mu}^{\vec{\ell}}(\omega) = [1 - \hat{G}^o(\omega)\hat{V}^{\vec{\ell}}]^{-1} \tag{2.21}$$

Using this expression for calculation of the second term in the right-hand side of the Eq. (2.17), we arrive at an iteration series that determines the true Green's function (cf., for example, [6])

$$\hat{G}(\omega) = \hat{G}^o(\omega) + \sum_p \hat{G}^o(\omega)C_p\hat{V}^p\hat{\mu}^p(\omega)\hat{G}(\omega) + \dots \tag{2.22}$$

To calculate the correlation functions in Eq. (2.11) to first order in C, one can confine oneself to the first term in Eq. (2.21) in the case that one or both indices coincide with the impurity site. To calculate the general correlation function when the indices are arbitrary one must take into account the first two terms of the series (2.22).

Introducing the notation

$$\tilde{k}_{nm}(\omega) = 2\sqrt{\sigma_n\sigma_m}\; k_{nm}(\omega) \tag{2.22'}$$

we get

$$\langle c_n k_{nm}(\omega)\rangle_c = ck_{om} = -c\,\frac{2\,\mathrm{Im}(\hat{\mu}^o\hat{G}^o)_{om}}{1-e^{-\omega/T}}\,, \tag{2.23}$$

$$\langle k_{nm}(\omega)\rangle_c = -\frac{2}{1-e^{-\omega/T}}\left[\mathrm{Im}G^o_{nm} + c\,\mathrm{Im}\,(\hat{G}^o\hat{V}^o\hat{\mu}^o\hat{G}^o)_{nm}\right]$$

To determine the correlation function for z-components, corresponding to the last term in Eq. (2.12), we consider separately low and high temperatures. In the first case we utilize the spin wave approximation which is practically equivalent to the decoupling used above. We find

$$\tilde{k}^z_{nm}(\omega) = \sigma_n\sigma_m\delta(\omega) +$$

$$+\frac{1}{4S_nS_m}\left[\langle S^-_n(t)S^+_n(t)S^-_m(0)S^+_m(0)\rangle_\omega - \langle S^-_nS^+_n\rangle\langle S^-_mS^+_m\rangle\delta(\omega)\right]\;. \tag{2.24}$$

or, approximately

$$\tilde{k}^z_{nm}(\omega) \approx \sigma_n\sigma_m\delta(\omega) +$$

$$+\frac{1}{4S_nS_m}\left[\langle S^-_n(t)S^+_m(0)\rangle\cdot\langle S^+_n(t)S^-_m(0)\rangle\right]_\omega \tag{2.24'}$$

In the high temperature region one can use a molecular-field approximation

$$\tilde{k}^z_{nm}(\omega) = \left[\sigma_n\sigma_m + \left(\langle (S^z_n)^2\rangle - \sigma^2_n\right)\delta_{nm}\right]\delta(\omega) \tag{2.25}$$

At sufficiently low temperatures the second term in Eq. (2.24) is small as compared with the first one and both approximations yield the same expression in the limit $T \to 0$

3. ELECTRICAL RESISTANCE AT LOW TEMPERATURES

An analysis shows that at low temperatures the second term in Eq. (2.24) leads only to small corrections to the electrical resistance. Therefore the contribution from this term can be neglected. Furthermore, we assume that σ_n has the same value for all atoms of the host metal.

We transform Eqs. (2.11) and (2.12), and get

$$\langle\Omega^{(1)}(\vec{q},\omega)\rangle_c=\sigma_o d_o^2\Big[\langle k_3(\vec{q},\omega)\rangle_c+2c\Big(\sqrt{\frac{\sigma_1}{\sigma_o}}-1\Big)k_2(\vec{q},\omega)+c\Big(\sqrt{\frac{\sigma_1}{\sigma_o}}-1\Big)^2 k_1(\omega)\Big]+$$

$$+2c\sqrt{\sigma_o\sigma_1}\,d_o\Delta d\Big[k_2(\vec{q},\omega)+\Big(\sqrt{\frac{\sigma_1}{\sigma_o}}-1\Big)k_1(\vec{q},\omega)\Big]+$$

$$+c\sigma_1(\Delta d)^2 k_1(\omega)+\tfrac{1}{2}c(d_o\sigma_o-d_1\sigma_1)^2\,\delta(\omega) \tag{3.1}$$

Here σ_o and σ_1 are mean values of the spin for atoms of the host metal and those of impurity, respectively. Besides, the following notations have been adopted:

$$k_1(\omega)=k_{oo}(\omega),\quad k_2(\vec{q},\omega)=\sum_m e^{-i\vec{q}\vec{R}_m^o}k_{om}(\omega),$$

$$k_3(\vec{q},\omega)=\frac{1}{N}\sum_{n,m}e^{i\vec{q}(\vec{R}_n^o-\vec{R}_m^o)}k_{nm}(\omega)\quad. \tag{3.2}$$

Now we transform the general expression for electrical resistance (2.1), using Eqs. (2.10) and (2.6)

$$\rho=\frac{\eta_1}{T}\int\frac{d\Omega_q}{4\pi}\int d\vec{q}q^3\int_{-\infty}^{+\infty}d\omega\cdot\omega\cdot n(\omega)\ \langle\Omega(\vec{q},\omega)\rangle_c\ ,\qquad \eta_1=\frac{\eta M}{2} \tag{3.3}$$

(η_1 has the same value as in [1]).

Substituting (3.1) into (3.3) one gets the part of the resistance which is due to the scattering of the spin system.

To calculate the correlation (3.2) we consider a simple cubic lattice and take into account the change of interaction between spins for nearest neighbours only. Applying the results obtained in the works [2], [3] we get the expressions for the correlation functions given in the Appendix.

To determine the electrical resistance at low temperatures one must know the low-frequency value for the k_i. One gets

$$k_1(\omega)\approx\frac{\sigma_1}{\sigma_o}\,g(\omega)\ \xi(\omega) \tag{3.4'}$$

$$k_2(\vec{q},\omega)\approx\sqrt{\frac{\sigma_1}{\sigma_o}}\Big[\Big(\sqrt{\frac{\sigma_1}{\sigma_o}}-1\Big)g(\omega)+\delta(\omega-\omega_{\vec{q}})\Big]\xi(\omega)\ , \tag{3.4''}$$

$$\langle k_3(\vec{q},\omega)\rangle_c = \delta(\omega-\omega_{\vec{q}})\xi(\omega) + c\Delta k_3(\vec{q},\omega) \quad , \tag{3.4'''}$$

$$\Delta k_3(\vec{q},\omega) = \left\{\left(1-\sqrt{\frac{\sigma_1}{\sigma_o}}\right)^2 g_o(\omega) - 2\left(1-\sqrt{\frac{\sigma_1}{\sigma_o}}\right)\delta(\omega-\omega_{\vec{q}}) - \right.$$

$$\left. -\left[\left(1-\frac{\sigma_1}{\sigma_o}\right)\omega - \frac{2}{D_p(0)}\left(1-\frac{J_1S_1}{J_oS_o}\right)\omega_{\vec{q}}\right]\frac{d}{d\omega}\delta(\omega-\omega_{\vec{q}})\right\}\xi(\omega) \quad ,$$

$$\xi(\omega) = \tfrac{1}{2}[(\bar{n}(\omega)+1)\theta(\omega)+\bar{n}(|\omega|)\theta(-\omega)] \quad .$$

(here $\bar{n}(\omega)$ is the Plank's distribution function).

Using the usual spin-wave spectrum of the perfect host lattice at small $\vec{q}$

$$\omega_{\vec{q}} = \beta\vec{q}^{\,2} \tag{3.5}$$

(here the gap Δ has been taken to be zero, because we bear in mind a region of temperatures much higher than Δ), retaining only the first two terms of the expansion in temperature, and substituting (3.4) into (3.1), we find for the impurity part of resistance which is due to inelastic scattering processes (the first three terms in Eq. (3.1)):

$$\Delta\rho_i^{(1)} = 2c\chi\chi\gamma P_{\frac{3}{2}}\left(\frac{T}{\theta_c}\right)^{\frac{3}{2}} +$$

$$+ c\chi P_2\left[3\left(\frac{S_1}{S_o}-1\right) + \frac{4}{D_p(0)}\left(1-\frac{J_1S_1}{J_oS_o}\right) + \frac{2\Delta d(0)}{d_o(0)}\cdot\frac{S_1}{S_o}\right]\left(\frac{T}{\theta_c}\right)^2 \tag{3.6}$$

For the impurity part of resistance due to elastic scattering of electrons and determined by the last term in Eq. (3.1), we have

$$\Delta\rho_e^{(1)} = c\chi P_{\frac{3}{2}}\frac{\int d\vec{q}\,\vec{q}^{\,3}[d_1(\vec{q})\sigma_1 - d_o(\vec{q})\sigma_o]^2}{d_o^2(0)S_o^2\left(\frac{\theta_c}{\beta}\right)^2} \quad . \tag{3.7}$$

In this case the general expression for the electrical resistance arising from the scattering by the spin system can be written as follows

$$\rho^{(1)} = \rho_o^{(1)} + \Delta\rho^{(1)} \quad , \quad \Delta\rho^{(1)} = \Delta\rho_i^{(1)} + \Delta\rho_e^{(1)} \quad . \tag{3.8}$$

where $\rho_o^{(1)}$ is the resistance of the pure host metal which corresponds to the first term in (3.4'''),

$$\rho_o^{(1)} = \chi P_2\left(\frac{T}{\theta_c}\right)^2 \tag{3.8}$$

and $\Delta\rho_o^{(1)}$ is the residual resistance.

In Eqs. (3.6) - (3.8) the following notations have been used:

$$X = \frac{\int dq q^3 [d_1(\vec{q}) S_1 - d_o(\vec{q}) S_o]^2}{d_o^2(0)\; S_o^2(\theta_c/\beta)^2} \quad ,$$

$$\chi = \tfrac{1}{2}\eta_1\left(\frac{\theta_c}{\beta}\right)^2 d_o^2(0) S_o \quad , \qquad P_m = \Gamma(m+1)\zeta(m) \tag{3.9}$$

here Γ and ζ are the gamma function and the Riemann zeta-function, respectively. The distribution function of the spin-wave frequency spectrum for the host metal at low frequencies has been taken in the form

$$g_o(\omega) = \gamma \frac{\omega^{\frac{1}{2}}}{\theta_c^{\frac{3}{2}}}$$

(see Appendix for the value of $D_p(0)$).

The elastic part of the resistance depends on temperature through σ_1 and σ_o. At low temperatures one has

$$\sigma_1(T) = S_1 - \int_{-\infty}^{\infty} d\omega k_1'(\omega) \quad , \tag{3.10}$$

where k_1' differs from k_1 in that a substitution $\xi(\omega) \to n(\omega)\theta(\omega)$ has been made. Considering Eq. (3.9), one finds

$$\sigma_1(T) = S_1 - \frac{2S_1}{3S_o}\gamma P_{\frac{3}{2}}\left(\frac{T}{\theta_c}\right)^{\frac{3}{2}} \quad . \tag{3.11}$$

(σ_o is obtained by substituting $S_1 = S_o$ in (3.11)). Introducing Eq. (3.11) into Eq. (3.7), we have for the residual resistance

$$\Delta\rho_o^{(1)} = cX\chi \tag{3.12}$$

and for the temperature dependent part

$$\Delta\rho_e^{(1)}(T) = - cX\chi\gamma P_{\frac{3}{2}}\left(\frac{T}{\theta_c}\right)^{\frac{3}{2}} \quad .$$

It is obvious that this term simply influences the magnitude of the numerical coefficient in the first term of Eq. (3.6).

To get the low temperature form for the impurity part of the total resistance $\Delta\rho(T)$, one must consider the second and the third terms in the general expression (2.5) for the probability of scattering. The correlation function in Eq.(2.7) has been calculated in [1]. It is possible to apply the results of that work for the determination of both terms. Retaining only the terms that are quadratic in temperature, one finally finds

$$\Delta\rho(T) = \tfrac{2}{3}cX\chi\gamma P_{\frac{3}{2}}\left(\frac{T}{\theta_c}\right)^{\frac{3}{2}} + cXP_2\left[3\left(\frac{S_1}{S_o}-1\right) + \right.$$

$$\left. + \frac{4}{D_p(0)}\left(1-\frac{J_1S_1}{J_oS_o}\right) + 2\frac{\Delta d(0)}{d_o(0)}\frac{S_1}{S_o}\right]\left(\frac{T}{\theta_c}\right)^2 + 2cdP_2\rho_\theta g_1'\left(\frac{T}{\theta}\right)^2 \quad . \qquad (3.13)$$

where in the last term the notations are identical with those used in [1], and g_1' differs from g_1 by a substitution

$$[\Lambda\, b(\vec{q})]^2 \rightarrow [\Delta\, b(\vec{q})]^2 + \tfrac{1}{2}[d_1(\vec{q})S_1 - d_o(\vec{q})S_o] \quad . \qquad (3.14)$$

From Eq. (3.13) one sees that at sufficiently low temperatures the main contribution comes from the term proportional to $T^{\frac{3}{2}}$. This term has a definite sign - it always causes an increase of resistance with respect to the magnitude of ρ at $T = 0$. The existence of this anomalous temperature dependence is due to the violation of momentum conservation for electron scattering from spin excitations in an irregular system. The magnitude of this term depends on the ratio of $d_i(\vec{q})S_i$ for the impurity atom to that for the atom of the host lattice.

The second term in Eq. (3.13) has a temperature dependence similar to that of the resistance (3.8) for the pure host metal, but this term may have either sign. The first two terms in the square brackets are connected with the deformation of the spin-wave spectrum. They cause the impurity part of the resistance even when the scattering amplitudes on the impurity atom and on the atom of the host lattice have equal magnitudes. The third term in the square brackets describes coherent scattering connected both with a change of the scattering amplitude and with a modification of the spin-wave spectrum (the sign of this term is simply determined by the sign of $\Delta d(0)$.

The way in which the deformation of the spin-wave spectrum influences the problem is connected with the values of the parameters J_1S_1/J_oS_o and J_1/J_o. If $J_1S_1/J_oS_o < 1$ and $J_1/J_o < 1$, and if we note that $1 > D_p(0) > 0$, we see that the resistance increases with increasing spectral density at low frequencies. When $J_1S_1/J_oS_o > 1$ and $J_1/J_o > 1$, then the spectral density decreases and results in a negative contribution to the electrical resistance. For intermediate values of these parameters, the sign may be either positive or negative.

The sign of the last term in Eq. (3.13) (representing the scattering by the phonon system) is always positive. We note that in many cases this term must be markedly smaller than the second one in Eq. (3.13). If this is the case and if in addition X, defined by Eq. (3.9), is small and the lowest temperatures are excluded from consideration, then $\Delta\rho(T)$ may have arbitrary sign.

4. ON THE ROLE OF THE QUASILOCAL LEVEL

The characteristic feature of the spin system under consideration is the appearance of a quasilocal level when exchange interaction between atoms of impurity and those of the host lattice decreases. The location of this resonant level can be derived from the equation (cf., for example [2]):

$$\mathrm{Re}\, D_S(\omega) = 0 \quad .$$

If one utilizes the expression for D_S given in the Appendix, then one easily finds in the limit $J_1/J_o \ll 1$

$$\omega_* \approx \frac{1}{2}\,\frac{J_1}{J_o}\,\omega_o \quad . \tag{4.1}$$

Here ω_o is the limiting frequency of the ideal spectrum which in the model under consideration has the magnitude $\omega_o = 24J_o\sigma_o$.

The existence of the quasilocal level gives rise to strong fluctuations of the impurity spin upon increasing temperature in the interval from zero to ω_*, that is in the interval in which the spin fluctuations of the host metal are yet small. This circumstance leads to a rapid increase of the electrical resistance in this temperature interval. It is natural that this picture has many common features with the strong influence of the quasilocal level in the phonon system on the electrical resistance [1]. It is interesting to note that in this sense a nonmagnetic impurity in a ferromagnetic host metal is essentially distinguished being able to shift ω_* practically to zero. Indeed, such impurity does not contribute by itself to the magnetic scattering of electrons and its influence on neighbouring spins doesn't give rise to anomalous behaviour of the spin system surrounding it.

The character of the electrical resistance temperature dependence with the quasilocal level being present can be directly derived, if one makes use of the general expressions for the correlation functions (3.1), given in the Appendix.

Here we confine ourselves to consideration of only that temperature interval which adjoins the resonant level from below, that is $T \gtrsim \omega_*/2$. In this case we can take advantage of the resonant behaviour characteristic of the correlation functions in the range of low frequencies near ω_*. A direct analysis indicates that approximately

$$k_1 = k_2 \approx \frac{1}{\pi} \frac{\Gamma}{(\omega-\omega_*)^2+\Gamma^2} \xi(\omega) \quad ,$$

$$\Delta k_3 \approx \frac{1}{\pi} \frac{\Gamma}{(\omega-\omega_*)^2+\Gamma^2} \xi(\omega) + \frac{2}{D_p(0)}\left(1-\frac{J_1\sigma_1}{J_o\sigma_o}\right)\omega_{\vec{q}} \frac{d}{d\omega} \delta(\omega-\omega_{\vec{q}})\xi(\omega) . \tag{4.2}$$

$$\Gamma = \pi g_o(\omega_*)\omega_*^2 \frac{\sigma_1}{\sigma_o} \quad . \tag{4.3}$$

We substitute this expression into (3.1) and (3.3). Taking into account that $\Gamma/\omega_* \ll 1$ and integrating over frequencies in explicit form, we find in the resonant approximation the following relation for inelastic part of the impurity resistance (again the first three terms in Eq. (3.1))

$$\Delta\rho_i^{(1)} \approx 2cXX_1\sigma_1 \frac{T}{\omega_*} \tag{4.4}$$

where

$$X_p = \frac{\int dq q^3 d_p^2(q)}{d_o(0) S_o^2\left(\frac{\theta_c}{\beta}\right)^2} \quad , \qquad p = 0,\ 1 \tag{4.5}$$

In order to calculate the elastic part of the electrical resistance we shall use not the last term in Eq. (3.1), but rather an expression that can be obtained after the substitution of Eq. (2.25) into Eqs. (2.11) and (2.12). Taking account of the impurity spin fluctuations only, we find

$$\Delta\rho_e^{(1)}(T) \approx cXX_1\left[\left\langle\left(S_1^z\right)^2\right\rangle - S_1^2\right] + cXX^1(S_1-\sigma_1) \tag{4.6}$$

where X^1 differs from X_1, by a substitution

$$d_1^2(\vec{q}) \rightarrow 2d_o(\vec{q})d_1(\vec{q})S_o \quad .$$

When determining an average spin on the impurity site, one can make use of the general expression relating $\sigma_1(T)$ with the integral $\int_{-\infty}^{\infty} d\omega\, k_1'(\omega)$ (cf., for example [5]). For the temperature interval of interest here one obtains, having used again the δ-functional behaviour of the correlation function [7]

$$\sigma_1 = B_{S_1}\left(\frac{\omega_*}{T}\right) \tag{4.7}$$

where B_{S_1} is the Brillouin function.

Comparing Eq. (4.4) (as well as Eq. (4.6)) with Eq. (3.3), one concludes that there takes place a very sharp increase of the impurity part of resistance ($\omega_* \ll \theta_c$). Upon increasing temperature the behaviour of the impurity resistance is close to linear with a characteristic frequency ω_* instead of θ_c, but later on at $\theta_c \gg T > \omega_*$ $(S_1+\frac{1}{2})$ (if $J_1S_1/J_oS_o \ll 1$), since $\sigma_1 \to S_1(S_1+1)\ \omega_*/3T$, $\Delta\rho_e^{(1)}(T)$ approaches practically a constant value. It should be noted, that all results derived in this section are valid if d_1 is of the same order of magnitude as or greater than d_o. In the case, when $d_1 \ll d_o$, the resonant approximation is insufficient and the general expressions for the correlation functions k_i must be used.

In order to show the influence of the quasilocal level on the impurity resistance $\Delta\rho(T)$, we carried out a numerical calculation. The results are given in Fig. 1.

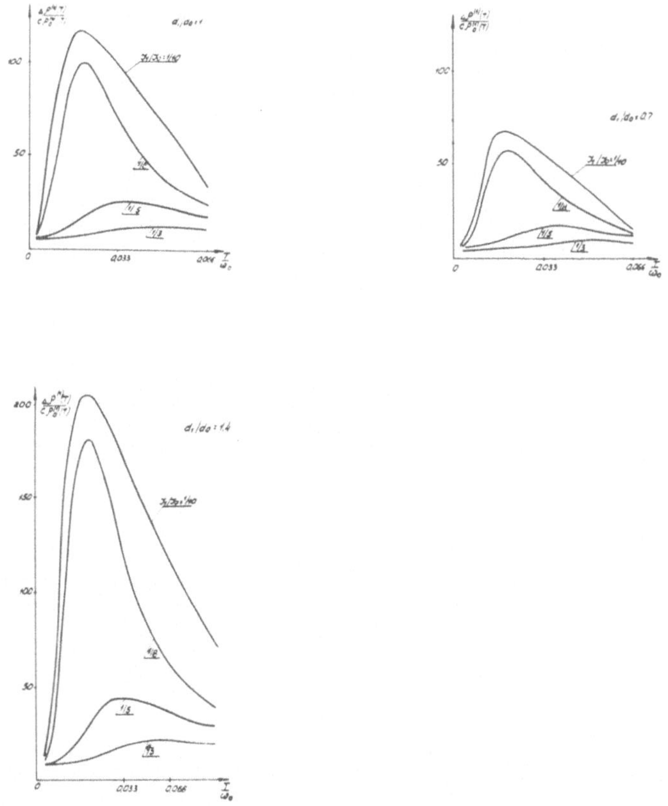

Fig. 1. The ratio $\Delta\rho^{(1)}(T)/c\rho_\delta^{(1)}(T)$ plotted against T/ω_o for $S_1 = S_o = \frac{1}{2}$.

Here on the ordinate axis the ratio of $\Delta\rho^{(1)}(T)$ to $c\rho_0^{(1)}(T)$ is plotted as a function of temperature. These results correspond to the case $S_1 = S_0 = \frac{1}{2}$; all other values of parameters are given in Fig. 1.

Figure 2 shows the same dependence but with $S_1 = 1$, $S_0 = \frac{1}{2}$. In both cases one can see that the effect is very pronounced. For example, if the ratio of exchange integrals is about 1/10, the increase of resistance is approximately one hundred times larger than concentration.

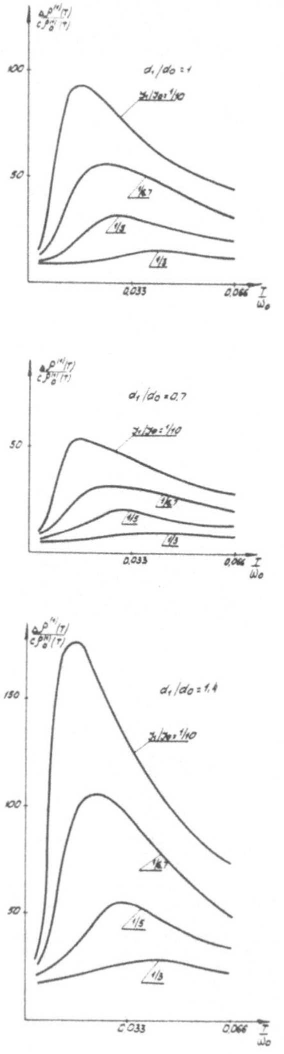

Fig. 2. The ratio $\Delta\rho^{(1)}(T)/c\rho_0^{(1)}(T)$ plotted against T/ω_0 for $S_1 = S_0 = \frac{1}{2}$.

The decrease of the exchange interaction seems to be a rather characteristic property for alloys of the type under consideration. Therefore the appearance of the quasilocal level and, as a result, of anomalous temperature dependence of the electrical resistance at low temperature has to be expected in many cases.

We note that in a number of the experimental works there were indications that the electrical resistance of ferromagnetic metals at low temperatures contained a term linear in T side by side with the term proportional to T^2. We imagine that one possible explanation of this fact is the influence of impurities. Indeed, the existance of the term $\sim T^{\frac{3}{2}}$ and the appearance of relations of the type (4.4), (4.6) if there exists the quasilocal level will in both cases imitate a temperature dependence close to a superposition of the linear and quadratic terms.

5. HIGH TEMPERATURES. MOLECULAR FIELD APPROXIMATION

In this section we treat the problem in the molecular field approximation. Although this approximation is rather crude, it gives on the one hand a possibility to obtain results at high temperatures and on the other hand - to find a simple approximate description suitable for any temperature.

Just as it has been done in the preceding sections, we assume here the average value of spin to be the same for all atoms of the host metal. This condition is equivalent to the following choice of the Hamiltonian for the model in question:

$$H = -\sum_n (1-c_n)\,\omega_o \hat{S}_o^z - \sum_n C_n \omega_1 \hat{S}_1^z \quad . \tag{5.1}$$

The usual variational procedure yields

$$\begin{aligned} \omega_o &= 2J^{00}\sigma_o + 2c\left(J^{10}\sigma_1 - J^{00}\sigma_o\right) \quad , \\ \omega_1 &= 2J^{10}\sigma_o \quad , \end{aligned} \tag{5.2}$$

where

$$J^{00} = \sum_m J_{om}^{00} \quad , \qquad J^{10} = \sum_m J_{om}^{1}$$

corresponds to the interaction between spins of the host metal (0,0) and between an impurity spin and those of the host metal (0,1), respectively.

We calculate the correlation function (2.11), using (2.12), (2.25) and the explicit form of the Hamiltonian (5.1).

$$\langle\Omega^{(1)}(\vec{q},\omega)\rangle_c=\tfrac{1}{2}(1-c)d_o^2(\vec{q})R_o(\omega)+\tfrac{1}{2}c[d_1(\vec{q})\sigma_1-d_o(\vec{q})\sigma_o]^2+\tfrac{1}{2}cd_1^2(\vec{q})R_1(\omega)$$

$$R_p(\omega)=\tfrac{1}{2}[\langle S_p^-S_p^+\rangle\delta(\omega+\omega_p)+\langle S_p^+S_p^-\rangle\delta(\omega-\omega_p)]+[\langle(S_p^z)^2\rangle-\sigma_p^2]\delta(\omega)\ , \tag{5.3}$$

$$p = 0,1\ .$$

Now we substitute this expression into Eq. (3.3). Taking into account the dependence (5.2) of ω_o on concentration, we find after a number of simple transformations the following expression, valid in linear approximation in C:

$$\Delta\rho^{(1)}=c\chi\left[X_1R_1(Z_1)-X_oR_o(Z_o)+2(J^{10}\sigma_1-J^{00}\sigma_o)X_o\frac{\partial R_o(Z_o)}{\partial Z_o}\right]+\Delta\rho_e^{(1)}, \tag{5.4}$$

where the last term is identical with (3.7), and

$$R_p(Z_p)=\tfrac{1}{2}Z_p n(Z_p)\langle S_p^+S_p^-\rangle+\tfrac{1}{2}Z_p[n(Z_p)+1]\langle S_p^-S_p^+\rangle + \\ + \langle(S_p^z)^2\rangle-\sigma_p^2\ , \qquad Z_p = \frac{\omega_p}{T}\ . \tag{5.5}$$

The average values in Eq. (5.5) can be represented in the form

$$\langle S_p^{\pm}S_p^{\mp}\rangle = S_p(S_p+1) - \langle(S_p^z)^2\rangle \pm \sigma_p$$

and in the molecular field approximation

$$\langle(S_p^z)^2\rangle = S_p(S_p+1) - \sigma_p\,\mathrm{cth}\,\frac{Z_p}{2}\ , \qquad \sigma_p = B_{s_p}$$

where B is the Brillouin function.

When $T \to 0$, $\Delta\rho^{(1)}$ approaches the value given by Eq. (3.12). At $T \geq \theta_c$ the impurity part of the resistance is determined by a simple expression:

$$\Delta\rho^{(1)} = c\chi[X_1S_1(S_1+1)-X_oS_o(S_o+1)]\ . \tag{5.6}$$

Of interest is the fact that $\Delta\rho^{(1)}$ is always positive when $T \to 0$, but it may have an arbitrary sign in the region $T \geq \theta_c$. When the value of (5.6) is negative in this region, $\Delta\rho^{(1)}$ passes through a maximum in an intermediate temperature region and then changes its sign at some $T < \theta_c$. A detailed analysis of the expression (5.4) shows that in general the temperature dependence of $\Delta\rho^{(1)}$ may be of a rather complicated character.

For purpose of illustration of the general behaviour of the impurity resistance in a wide temperature range Fig. 3 shows a plot of the ratio $\Delta\rho^{(1)}$ (Eq. (5.4)) to $c\rho_o^{(1)}(\theta_c)$ against T/θ_c.

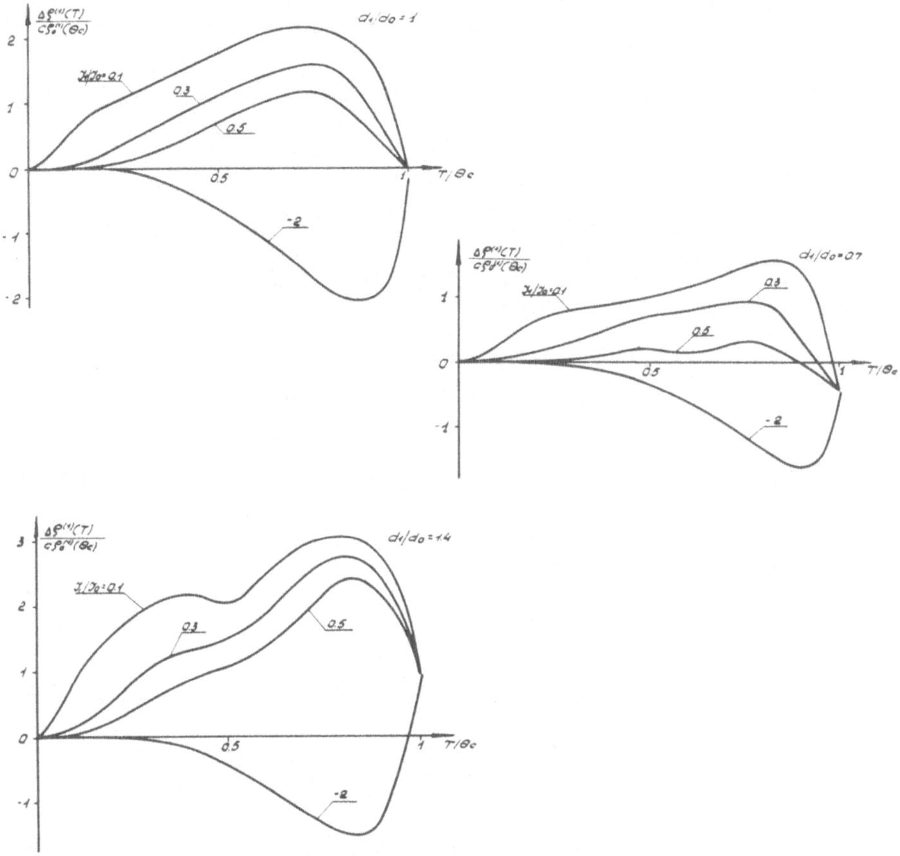

Fig. 3. The ratio $\Delta\rho^{(1)}(T)/c\rho_o^{(1)}(\theta_c)$ plotted against T/θ_c.

APPENDIX

Here we give values of the correlation functions k_i. In the approximations under consideration and taking into account the results obtained by Izjumov and Medvedev [3], [4], we have directly

$$k_1(\omega) = \frac{(J_1/J_o)^2(\sigma_1/\sigma_o)g_o(\omega)}{[\mathrm{ReD}_s(\omega)]^2[\mathrm{ImD}_s(\omega)]^2} \quad . \tag{A.1}$$

In this expression J_o exchanges integral for ideal and J_1 exchanges integral between impurity and six nearest neighbours. $\mathrm{ReD}_s(\omega)$ and $\mathrm{ImD}_s(\omega)$ are respectively real and imaginary parts of the function.

$$D_s(\omega) = \frac{J_1}{J_o} + 2\,\frac{\rho\omega}{\omega_o} - \left[\frac{J_1}{J_o}\left(1 - \frac{\sigma_1}{\sigma_o}\right)\omega + 2\,\frac{\rho\omega^2}{\omega_o}\right] G_o^o(\omega) \quad , \tag{A.2}$$

$$\rho = \frac{J_1\sigma_1}{J_o\sigma_o} - 1 \quad . \tag{A.3}$$

For function k_2 and k_3 we have

$$k_2(\vec{q},\omega) = \frac{(J_1/J_o)\sqrt{\sigma_1/\sigma_o}}{[\mathrm{Re}D_s(\omega)]^2 + [\mathrm{Im}D_s(\omega)]^2}\Bigg\{\mathrm{Re}D_s(\omega)\,\delta(\omega - \omega_{\vec{q}}) + {}$$
$$+ g_o(\omega)\left[\frac{(J_1/J_o)(1 - \sigma_1/\sigma_o) + \frac{2\rho\omega^2}{\omega_o}}{\omega - \omega_{\vec{q}}} + \frac{J_1}{J_o}\left(\sqrt{\frac{\sigma_1}{\sigma_o}} - 1\right) - \frac{2\rho\omega}{\omega_o}\right]\Bigg\}. \tag{A.4}$$

where $\omega_{\vec{q}} = \sigma_o \omega_{\vec{q}}^{0}/S_o$ is the normalization energy of a spin wave with the momentum $\vec{q}$.

$$k_3(\vec{q},\omega) = \frac{1}{\pi}\,\frac{c\,\mathrm{Im}\mathcal{P}_{\vec{q}}(\omega)}{[\omega - \omega_{\vec{q}} - c\mathrm{Re}\,\mathcal{P}_{\vec{q}}(\omega)]^2 + [c\;\mathrm{Im}\mathcal{P}_{\vec{q}}(\omega)]^2} \quad . \tag{A.5}$$

Here $\mathrm{Re}\,\mathcal{P}_{\vec{q}}(\omega)$ and $\mathrm{Im}\,\mathcal{P}_{\vec{q}}(\omega)$ are the real and imaginary parts of the polarization operator

$$\mathcal{P}_{\vec{q}}(\omega) = \frac{Z_s(\vec{q},\omega)}{D_s(\omega)} + 3\,\frac{Z_p(\vec{q})}{D_p(\omega)} + 2\,\frac{Z_d(\vec{q})}{D_d(\omega)} \quad . \tag{A.6}$$

We shall group atoms situated in the first coordination sphere near the impurity in pairs. Let us label pairs of atoms which are on the same straight line with the impurity by the index (12). The other pairs are labeled by the index (13). Then we have

$$D_p(\omega) = 1 - 2J_o\sigma_o\rho\,[G_o^o(\omega) - G_{12}^o(\omega)] \quad . \tag{A.7}$$

$$D_p(\omega) = 1 - 2J_o\sigma_o\rho\,[G_o^o(\omega) + G_{12}^o(\omega) - 2G_{13}^o(\omega)] \tag{A.8}$$

The following notations are introduced in (A.6)

$$Z_s(\vec{q},\omega) = \frac{J_1}{J_o}\left(1 - \frac{\sigma_1}{\sigma_o}\right)\omega + \frac{2\rho\omega}{\omega_o} + {}$$
$$+2\left[\frac{J_1}{J_o}\left(\sqrt{\frac{\sigma_1}{\sigma_o}} - 1\right) - \frac{2\rho\omega}{\omega_o}\right](\omega - \omega_{\vec{q}}) + \left[\frac{2\rho}{\omega} + \frac{J_1}{J_o}\left(\sqrt{\frac{\sigma_1}{\sigma_o}} - 1\right)^2\right](\omega - \omega_{\vec{q}})^2. \tag{A.9}$$

$$Z_p(\vec{q}) = 2J_o\sigma_o\rho\cdot\left[1 - \sum_{(\vec{1}-\vec{2})} \cos \vec{q}(\vec{1}-\vec{2})\right] \quad . \tag{A.10}$$

$$Z_d(\vec{q}) = 2J_o\sigma_o\rho \left[1 + \sum_{(\vec{1}-\vec{2})} \cos \vec{q}(\vec{1}-\vec{2}) - 2 \sum_{(\vec{1}-\vec{3})} \cos \vec{q}(\vec{1}-\vec{3}) \right] . \qquad (A.11)$$

In formulas (A.10) and (A.11) $\sum_{(\vec{1}-\vec{2})}$ means summing over all pairs of the type (12) with the subsequent division by the number of these pairs. $\sum_{(\vec{1}-\vec{3})}$ is defined in the same manner.

REFERENCES

1. Yu.Kagan, A.P.Zhernov, Zh. eksper. teor. Fiz., 50, 1107 (1966).
2. T.Wolfram, J.Callaway, Phys. Rev. 130, 2207 (1963).
3. Yu.A.Izjumov, M.V.Medvedev, Fiz. metallow i metallovedenie 22, 641 (1966).
4. Yu.Izjumov, Proc. Phys. Soc. 87, 505 (1966).
5. S.V.Tyablikov, Metody kvantovoy teorii magnetizma, Nauka, 1965.
6. Yu.Kagan, Zbornik "Fizika kristallov s defektami," Vol. II, 93, Tbilisi, 1966.
7. Yu.A.Izjumov, M.V.Medvedev, Fiz. tverd. Tela 8, 2117 (1966).

OBSERVATION OF LOCALIZED PHONON MODES BY SUPERCONDUCTIVE TUNNELING

B. S. Chandrasekhar and J. G. Adler

Department of Physics and Condensed State Center

Case Western Reserve University, Cleveland, Ohio

We present here a brief outline of the way in which the density of phonon states in a superconductor may be obtained from tunneling data, with illustrative examples from the lead-indium alloy system. A detailed paper will appear elsewhere.[1]

The method[2] is based on an inversion of the Eliashberg gap equation in terms of the normal and pairing self-energies, $\xi(\omega)$ and $\phi(\omega)$ respectively, for a strong-coupling superconductor. The relevant equations are

$$\xi(\omega) = [1 - Z(\omega)]\omega = \int_{\Delta_o}^{\infty} d\omega' \mathrm{Re} \left[\frac{\omega'}{(\omega'^2 - \Delta'^2)^{1/2}}\right] \times \int d\omega_q \alpha^2(\omega_q) F(\omega_q) [D_q(\omega'+\omega) - D_q(\omega'-\omega)] \quad (1)$$

$$\phi(\omega) = \int_{\Delta_o}^{\omega_c} d\omega' \mathrm{Re} \left[\frac{\Delta'}{(\omega'^2 - \Delta'^2)^{1/2}}\right] \times \left\{ \int d\omega_q \alpha^2(\omega_q) F(\omega_q) [D_q(\omega'+\omega) + D_q(\omega'-\omega)] - U_c \right\} \quad (2)$$

where $D_q(\omega) = (\omega+\omega_q-i0^+)^{-1}$, $\Delta(\omega) = \phi(\omega)/Z(\omega)$, and $\Delta_o = \Delta(\Delta_o)$. $F(\omega)$ is the phonon density of states, and $\alpha^2(\omega)$ is an effective electron-phonon coupling function for phonons of energy ω.

The tunneling density of states in a superconductor is given by[3]

$$\frac{N_s(\omega)}{N_n(0)} = \text{Re}\left\{\frac{|\omega|}{[\omega^2-\Delta^2(\omega)]^{1/2}}\right\} \tag{3}$$

and is directly measured by the ratio of the superconducting and normal conductances of a superconductor-normal metal junction at T = 0°K. It is also possible to compute the tunneling density of states from the characteristics of a superconductor-superconductor junction at finite temperatures, as in fact was the case in our experiments. One can then, by an iterative procedure, obtain an $\alpha^2(\omega)F(\omega)$ which gives the best fit between the density of states computed from equations (1)-(3) and that experimentally measured.

The above program was first carried out by McMillan and Rowell[2] for lead. Similar work on a lead + 3% indium alloy was reported by Rowell, McMillan and Anderson,[4] showing the existence of new phonon states over a narrow range of energies beyond the cut-off of the lead spectrum, which were identified as localized phonon modes due to the light indium impurity.

We have carried out tunneling measurements on the lead-indium alloy system over its entire range of primary solid solubility. The results have been analyzed to yield a variety of information on the superconductivity of this system. We report here briefly only on one aspect which is pertinent to the theme of this Conference. Figure 1 shows the computed $\alpha^2(\omega)F(\omega)$ for two alloys, as well as that for pure lead. The new structure above 9 MeV for the alloys is clearly associated with the indium impurity. It is of interest to note[5] that this structure is resolved even in concentrated alloys. Unpublished calculations by Taylor[6] of $F(\omega)$ for the lead-indium system are in qualitative agreement with our results, and in particular show the same structure beyond the cut-off of the lead spectrum.

The use of superconductive tunneling to obtain information about phonon spectra is relatively recent, and a few remarks may be appropriate about its potential. The structure which appears in the conductance of the tunnel junction as a function of voltage may physically be seen as follows: The lifetime of the tunneling quasiparticles is influenced by the density of phonon states of energy equal to the quasiparticle energy (referred to the gap edge), since they decay to the condensate by emission of phonons. This explains the appearance of $\alpha^2(\omega)F(\omega)$ in the equations above, emphasizing the involvement of both the electron-phonon coupling and the phonon density of states in the tunneling process. The analysis of the tunneling data gives only the product $\alpha^2(\omega)F(\omega)$, and does not permit a separation of the two factors. In the case of pure lead, a comparison of this quantity with the $F(\omega)$ derived from inelastic neutron scattering data suggests that the energy dependence of $\alpha^2(\omega)$ is weak. If we

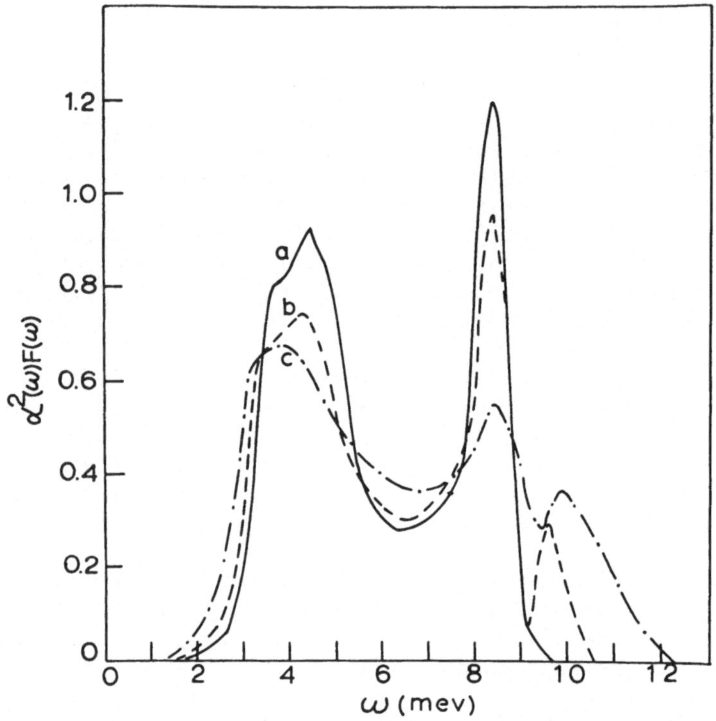

Figure 1. The dimensionless parameter $\alpha^2(\omega)F(\omega)$ as a function of phonon energy ω, as computed from tunneling data for (a) pure Pb, (b) $Pb_{0.25}\ In_{0.25}$, (c) $Pb_{0.6}\ In_{0.4}$.

assume that the same is true for the alloys, then the curves of Figure 1 will represent exactly the phonon density of states for the alloys. In general, however, one should bear in mind that it may not be always possible to separate out the two terms α and F.

The tunneling technique has the advantage (with the qualification mentioned in the previous paragraph) of directly giving $F(\omega)$, while most other techniques give the dispersion curves from which $F(\omega)$ has to be computed. It is applicable to a very wide range of materials, since superconductivity is a pervasive phenomenon in conducting solids. Present experimental techniques make it possible to see the phonon effects in strong-coupling (e.g. Pb, Hg) and intermediate-coupling (e.g. Sn, In) superconductors. There is no doubt that in the near future the techniques will be extended to include weak-coupling superconductors also.

The work reviewed here has been done in collaboration with J. E. Jackson, W. L. McMillan and T. M. Wu, with the support of the Air Force Office of Scientific Research, Office of Aerospace Research, U. S. Air Force, under AFOSR grant number 565-66.

REFERENCES

1. J. G. Adler, B. S. Chandrasekhar, J. E. Jackson, W. L. McMillan and T. M. Wu, to be published.

2. W. L. McMillan and J. M. Rowell, Phys. Rev. Letters 14, 108 (1965).

3. D. J. Scalapino, J. R. Schrieffer and J. W. Wilkins, Phys. Rev. 148, 263 (1966).

4. J. M. Rowell, W. L. McMillan and P. W. Anderson, Phys. Rev. Letters 14, 633 (1965).

5. J. G. Adler, J. E. Jackson and B. S. Chandrasekhar, Phys. Rev. Letters 16, 53 (1966).

6. D. W. Taylor, private communication.

EFFECT OF LOCALIZED AND RESONANT MODES ON ENERGY-GAP FUNCTION AND TRANSITION TEMPERATURE OF ISOTROPIC SUPERCONDUCTORS

Joachim Appel

General Dynamics, General Atomic Division

San Diego, California

In the contemporary theory of superconductivity two important parameters, namely the pair-breaking energy Δ and the transition temperature T_c depend on the electron-phonon interaction $\alpha^2(\omega)g(\omega)$.[1] The function $g(\omega)$ is the phonon density of states and $\alpha^2(\omega)$ is the interaction parameter depending on the energy ω of a phonon participating in an el-ph interaction. Furthermore, the energy gap function $\Delta(\omega)$ and T_c depend on the direct Coulomb interaction between two electrons. The question arises: "How do light or heavy impurity atoms, giving rise to localized or resonant modes,[2] respectively, affect $\Delta(\omega)$ and T_c of an isotropic superconductor?" Whereas impurities affect the shape of $\Delta(\omega)$ according to the modified phonon density of states and the electron-impurity mode interaction, the effect on T_c is gross property of the impure superconductor depending on a well-defined frequency average over the gap function of the host lattice and the frequency dependent electron-impurity mode interaction. The change of $\Delta(\omega)$ and T_c caused by this interaction and by the modification of the electronic structure, is denoted as valence effect. It is linear in the impurity concentration and has its origin in a small but basic change of the electronic and dynamic properties of the lattice. The mean-free-path effect, arising from elastic impurity scattering, will not be considered here.[3] It has been shown by Tsuneto[4] that elastic impurity scattering in itself does not affect the energy-gap function found from the Éliashberg gap-equation for an isotropic superconductor.

To study the valence effect on the energy gap function at $T = 0^\circ K$, a straightforward perturbation procedure is applied to the Éliashberg integral equation for the gap function and for the renormalization parameter. The perturbation consists of a small change of the interaction kernel,

$$K_{\pm 1}(\omega,\omega') = \frac{\lambda_1\omega_{11}}{2}\left(\frac{1}{\omega_{11}+\omega'+\omega-i0^+} \pm \frac{1}{\omega_{11}+\omega'-\omega-i0^+}\right)$$

$$- U_1 - \frac{N_1}{N_0} K_{\pm 0}(\omega,\omega') , \qquad (1)$$

where $K_{\pm 0}$ is the interaction kernel of the pure metal;[5] and where $\lambda_1 = 2\alpha^2(\omega_{11})/\omega_{11}$ = electron-impurity mode coupling constant, ω_{11} = impurity-mode frequency, $N_{1(0)}$ = number of impurity (host) atoms per cm^3. The perturbation of the Coulomb pseudopotential is for a simple metal (spherical energy band) given by

$$U_1 = A\frac{\delta N(0)}{N(0)} + B\frac{\delta k_F}{k_F} , \qquad (2)$$

where A and B are constants depending on $\pi e^2 N(0)/k_F^2$ (k_F = Fermi momentum, N(0) = density of states). Equation (1) presumes that the impurity-mode distribution is given by an Einstein distribution, $g(\omega_{11}) = \delta(\omega - \omega_{11})$. This presumption is sufficient for the calculation of the change of the BCS gap parameter $\Delta_{10} = \Delta_0 - \Delta_{00}$ where Δ_0 and Δ_{00} are respectively the gap parameters of the impure and the pure metal. However, for the calculation of the energy dependence of the impurity-gap function, $\Delta_1(\omega)$, the "spreading" of the impurity mode distribution must be taken into account. To this end the impurity kernel $K_{\pm 1}(\omega,\omega')$ is derived from a Lorentzian distribution of impurity modes $g(\omega_{11},\omega_{12})$, where ω_{11} characterizes the center and ω_{12} the halfwidth of the distribution. The changes $\Delta_1(\omega)$ and $Z_1(\omega)$ of energy-gap function and renormalization parameter are found by writing the interaction kernel of the impure metal in the form

$$K_{\pm}(\omega,\omega') = K_{\pm 0}(\omega,\omega') + \epsilon K_{\pm 1}(\omega,\omega') , \qquad (3)$$

where $\epsilon \ll 1$. The result of the perturbation calculation is a linear integral equation for $\Delta_1(\omega)$ given by

$$\Delta_1(\omega) = F(\omega) + \int_{\Delta_{00}}^{\omega_c} d\omega' \,\mathrm{Re}\left\{\frac{\Delta_1'}{(\omega'^2 - \Delta_0'^2)^{\frac{1}{2}}}\right\} K_{+0}(\omega,\omega') , \qquad (4)$$

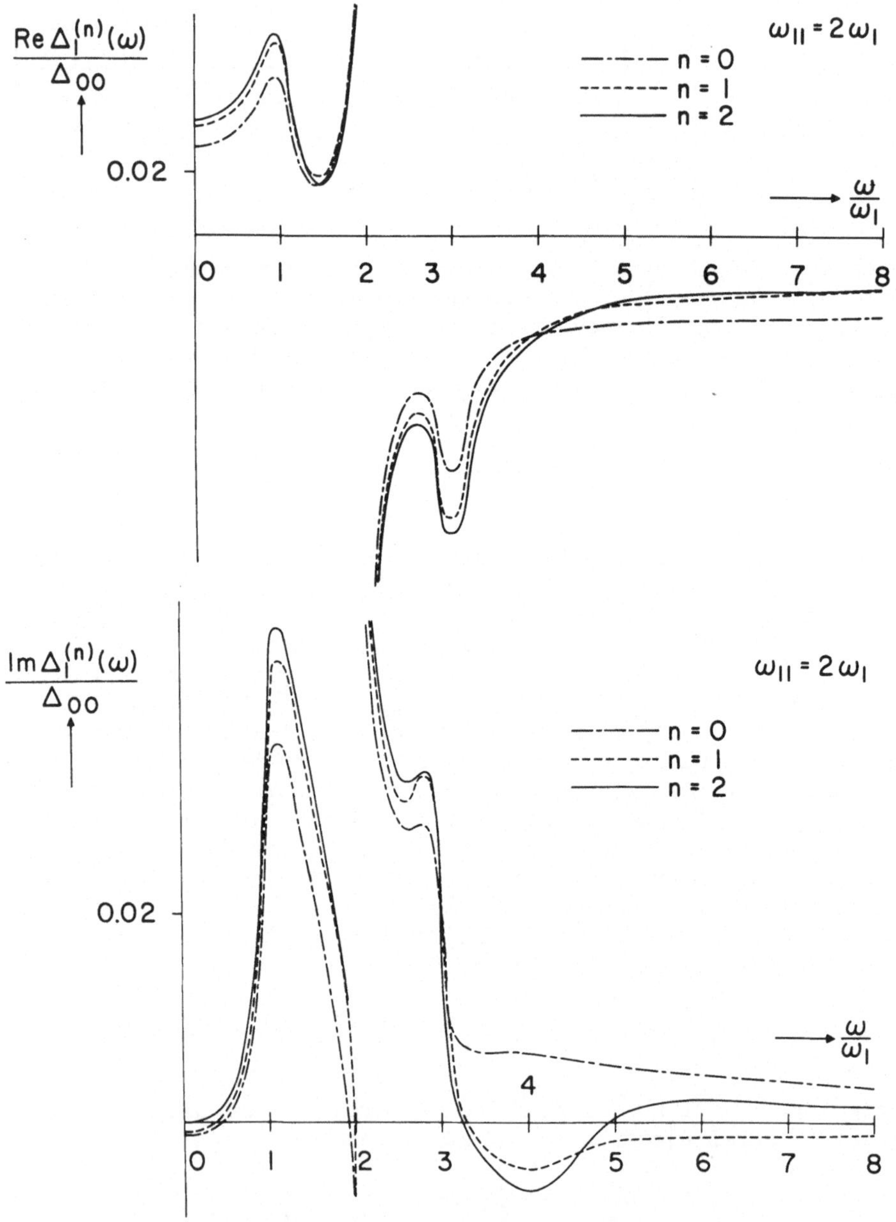

Fig. 1. Real and imaginary parts of the impurity-gap function for localized modes with Einstein frequency $\omega_{11} = 2\omega_1$. The phonon spectrum of the host lattice is characterized by a Lorentzian centered at ω_1.

where the inhomogeneous part $F(\omega)$ is given by

$$F(\omega) = \frac{1}{Z_0(\omega)}\left[-Z_1(\omega)\,\Delta_0(\omega) + \int_{\Delta_{00}}^{\omega_c} d\omega'\,\mathrm{Re}\left\{\frac{\Delta_0'}{(\omega'^2 - \Delta_0'^2)^{\frac{1}{2}}}\right\} K_{+1}(\omega,\omega') + \int_{\Delta_{00}}^{\omega_c} d\omega' \frac{\Delta_{10}\Delta_{00}}{(\omega'^2 - \Delta_{00}^2)^{3/2}}\left[\Delta_{00} K_{+0}(\omega,\omega') - \omega' K_{+0}(\omega,\Delta_{00})\right]\right], \tag{5}$$

and where $\Delta_1' = \Delta_1(\omega')$ etc. Equation (4) is correct for small impurity concentrations, independent of the strength of the el-ph interaction in the host metal. $\Delta_0(\omega)$ is the gap function of the pure metal, $\Delta_{00} = \Delta_0(\omega = \Delta_{00})$, $\Delta_{10} = \Delta_1(\omega = \Delta_{00})$; the change in the renormalization parameter, $Z_1(\omega)$, depends merely on Δ_{10}, not on $\Delta_1(\omega)$. The inhomogeneous part $F(\omega)$ can be calculated in closed form for a Lorentzian phonon spectrum of the host lattice and for an Einstein distribution of impurity modes.(5) If the inhomogeneous part $F(\omega)$ of Eq. (4) is known, the integral equation can be solved by iteration and the Neumann series is readily calculated. A few of the results are shown in Figs. 1 and 2. The impurity gap functions for a case of localized and a case of resonant impurity modes have been calculated for a host lattice characterized by a single-peaked phonon distribution centered at ω_1, having halfwidth $\omega_2 = 0.2\,\omega_1$; $\lambda_0 = 2\alpha^2(\omega_0)/\omega_0 = 0.35$ (= el-ph coupling constant), $U = 0.1$ (= Coulomb pseudopotential).

The possibility of observing localized modes in dilute lead alloys by measuring the I-V tunneling characteristics was suggested by Maradudin.(6) The phonon density-of-states peaks lead to structures in the tunneling density of states $N_T(\omega)$ that is given by(7)

$$\frac{N_T(\omega)}{N(0)} = 1 + \frac{(\mathrm{Re}\Delta)^2 - (\mathrm{Im}\Delta)^2}{2\omega^2}\,. \tag{6}$$

Structure due to localized modes has first been observed in dilute Pb-In alloys by Rowell, McMillan, and Anderson.(8) By using their tunneling data as input they invert the gap equation and determine the $\alpha^2(\omega)g(\omega)$ curves for pure lead and for the dilute alloy $Pb_{0.97}In_{0.03}$. We have calculated the change in the tunneling density of states by solving the impurity gap equation(4) taking the parameter values of Rowell, McMillan, and Anderson. Since the authors do not quote a value for the change in the Coulomb pseudopotential, U_1, we have determined U_1 by fitting the calculated value of the transition temperature change, δT_c, to the experimental value of Gamari-Seale and Coles.(9) The result is

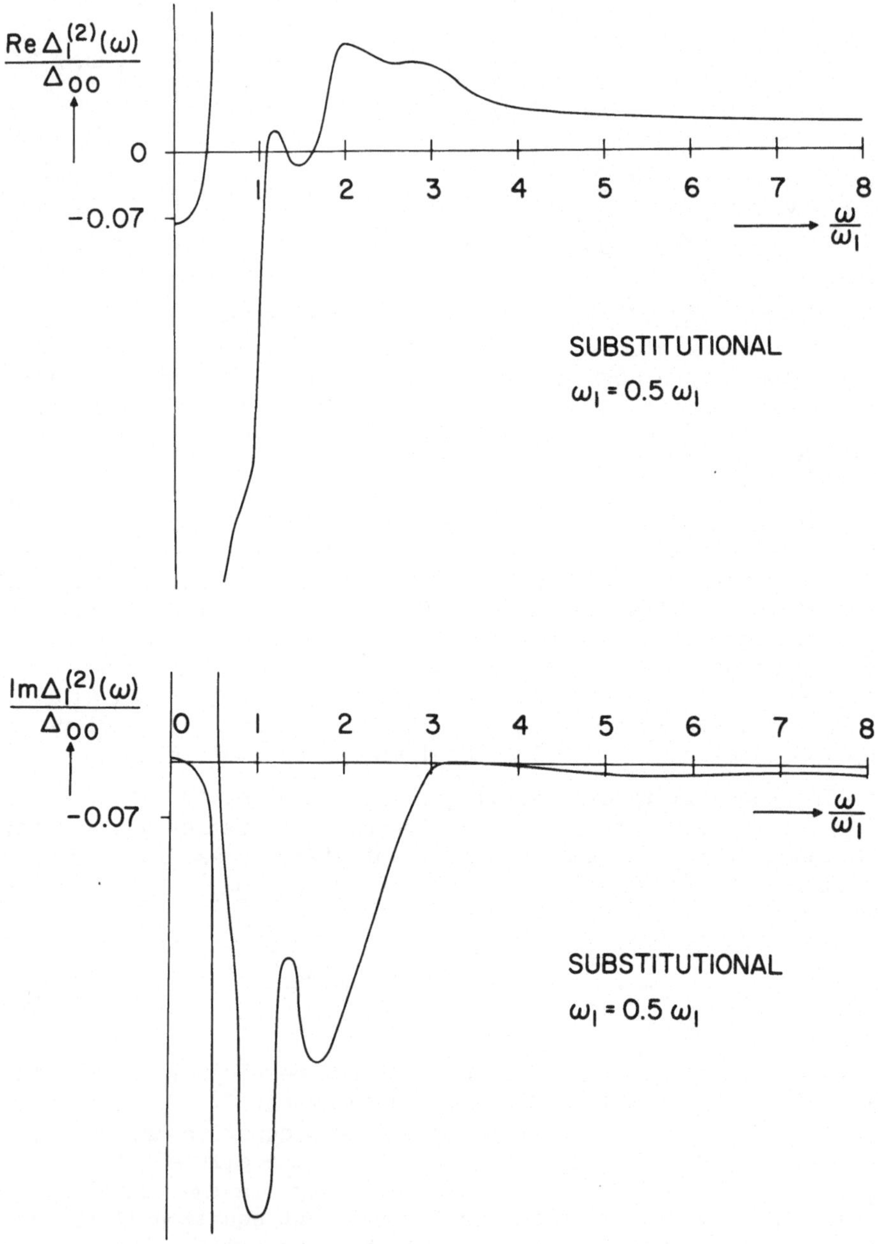

Fig. 2. Real and imaginary parts of the impurity-gap function for resonant modes with frequency $\omega_{11} = \frac{1}{2}\omega_1$.

$U_1 = 0.007$. The change in the tunneling density of states due to the localized modes of the In atoms is shown in Fig. 3. A sharp maximum is found at the impurity mode frequency $\omega_{11} = 9.5$ MeV and a shoulder at a somewhat higher frequency.

To study the valence effect on the transition temperature, we start from the energy-gap equation for finite temperatures which has solutions for $T < T_c$. Near the transition temperature, $(T_c - T)/T_c \ll 1$, the integral equation can be linearized,

$$\mathrm{Re}\{\Delta(\omega)\} = \int_0^{\omega_c} d\omega' \, \mathrm{Re}\{\Delta(\omega')\} \, \mathfrak{K}(\omega, \omega'; \beta) \tag{7}$$

Here, $\mathfrak{K}(\omega, \omega'; \beta)$ is the real part of the kernel,[5] $\beta = 1/kT$. A corresponding equation [independent of $\Delta(\omega')$] can be written down for the renormalization parameter. Near T_c the kernel of the impure metal is given by

$$\mathfrak{K}(\omega, \omega'; \beta) = \mathfrak{K}_0(\omega, \omega'; \beta) + (\beta - \beta_c)\left(\frac{d\mathfrak{K}_0(\omega, \omega'; \beta)}{d\beta}\right)_{\beta=\beta_c} + \mathfrak{K}_1(\omega, \omega'; \beta_c), \tag{8}$$

where $\mathfrak{K}_0(\omega, \omega'; \beta)$ is the kernel of the pure metal and where $\mathfrak{K}_1(\omega, \omega'; \beta)$ is the perturbation caused by impurities. Equation (8) is inserted into Eq. (7) to find the impurity-gap equation,

$$\mathrm{Re}\{\Delta_1(\omega)\} - \int_0^{\omega_c} \mathrm{Re}\{\Delta_1(\omega')\} \, \mathfrak{K}_0(\omega, \omega'; \beta_c) = \delta\beta_c \int_0^{\omega_c} \mathrm{Re}\{\Delta_0(\omega')\}\left(\frac{d\mathfrak{K}_0(\omega, \omega'; \beta)}{d\beta}\right)_{\beta=\beta_c}$$
$$+ \int_0^{\omega_c} \mathrm{Re}\{\Delta_0(\omega')\} \, \mathfrak{K}_1(\omega, \omega'; \beta_c) \, d\omega', \tag{9}$$

where $k\delta\beta_c = (T_c + \delta T_c)^{-1} - T_c^{-1}$. The left-hand side of this equation is of the same form as the real part of the homogeneous Eq. (7). Therefore, according to a well-known theorem of inhomogeneous Fredholm equations of the second kind, the inhomogeneous integral Eq. (7) has a solution only if the right side of this equation is orthogonal to the solution $\mathrm{Re}\{\tilde{\Delta}_0(\omega)\}$ of the transposed equation. The orthogonality condition gives the following expression for the change of the transition temperature:

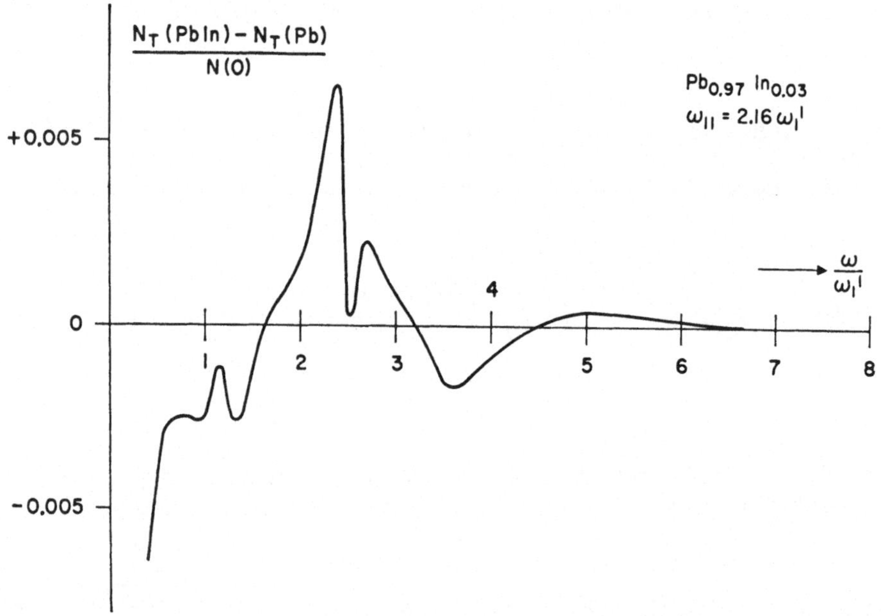

Fig. 3. The difference between the tunneling density of states for a lead-indium alloy and that of pure lead. For the interaction kernels of the impure and pure metal see Reference 5.

$$\delta\beta_c = - \frac{\int_0^{\omega_c} d\omega' \, \mathrm{Re}\{\widetilde{\Delta}_0(\omega')\} \int_0^{\omega_c} d\omega'' \, \mathrm{Re}\{\Delta_0(\omega'')\} \, \mathfrak{K}_1(\omega', \omega''; \beta_c)}{\int_0^{\omega_c} d\omega'' \, \mathrm{Re}\{\widetilde{\Delta}_0(\omega')\} \int_0^{\omega_c} d\omega'' \, \mathrm{Re}\{\Delta_0(\omega'')\} \left(\frac{d\mathfrak{K}_0(\omega', \omega''; \beta)}{d\beta}\right)_{\beta=\beta_c}} . \tag{10}$$

This result is exact. With Eq. (10) δT_c has been calculated as a function of the impurity-mode frequency for a host lattice characterized by a single-peaked phonon distribution ($\omega_2 = 0.2\ \omega_1$, $\lambda_0 = 0.35$, $U_0 = 0.1$). The result is shown in Fig. 4. Equation (10) has also been applied to a number of dilute lead alloys. Gamari-Seale and Coles[9] have measured δT_c for dilute alloys of lead with

In	Sn	Sb
Tl		Bi .

They account for the gap anisotropy of lead and for the corresponding mean-free-path effect on T_c with the help of the formula derived by Markowitz and Kadanoff.[10] The valence effect is then found from the relation $\delta T_c = (\delta T_c)_{exp} - (\delta T_c)_{anis}$. We have evaluated δT_c for substitution lead alloys as a function of the impurity-mode frequency ω_{11}, the associated coupling parameter $\alpha^2(\omega_{11})$, and the change of the Coulomb pseudopotential U_1. The results are shown in Table I. In the first case of dilute lead-indium alloys, the observed tunneling density of states has led, via the inversion of the gap equation,[8] to the determination of the phonon distribution in the impure metal and, thereby, to the following numbers for the impurity parameters: $\omega_{11} = 9.5$ MeV, $\alpha^2(\omega_{11}) = 1.34$ MeV. The third parameter U_1 which depends on the change of the electronic structure, is obtained by fitting the theoretical value for δT_c to the experimental value of Gamari-Seale and Coles. This procedure may be applied to a metal such as lead which has a nearly-free-electron Fermi surface. In the vicinity of the Fermi energy, the value of the density of states is not much different from that for free electrons. The slope of the density of states curve is, however, negative and, below the Fermi energy, Anderson and Gold[11] find a peak in this curve. This irregular dependence of the density of states is unimportant for small impurity concentrations. Each indium atom contributes only three conduction electrons for accommodation inside the Fermi surface, instead of four of a Pb atom. Therefore, δk_F is negative and, since B in Eq. (2) is positive, the second term in the expression for U_1 is positive ($a^2 = 0.38$ for lead). The first term of U_1 [in Eq. (2)] is positive if $\delta N(0) > 0$; this is the case if the free-electron Fermi surface shrinks. For Bloch electrons, the

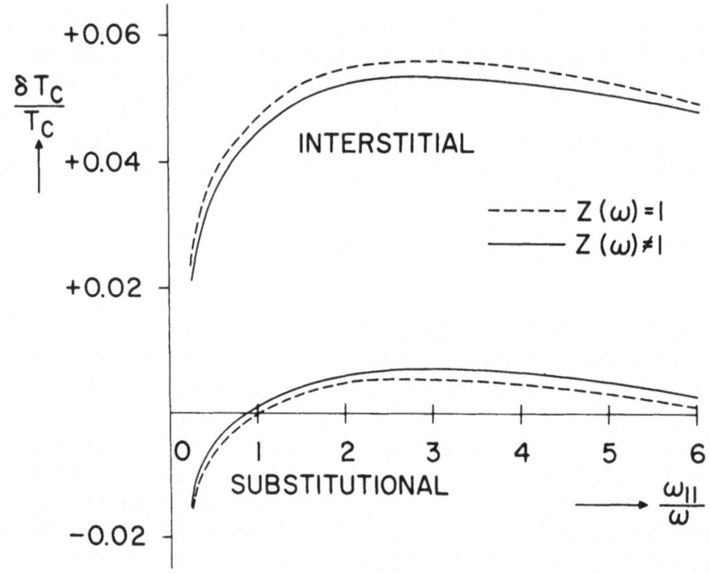

Fig. 4. Relative change of the transition temperature as a function of the mode frequency ω_{11}, taking into account and ignoring renormalization, respectively. The interaction kernel of the host lattice is characterized by a phonon distribution centered at ω_1 and having halfwidth $\omega_2 = 0.2\ \omega_1$, $\lambda_0 = 0.35$, and $U_0 = 0.1$. The parameters of the interaction kernel for the impurities are ω_{11}, $\lambda_1/\lambda_0 = N_1/N_0 = U_1/U_0 = 0.01$.

change of $N(0)$ with alloying is not merely determined by the change of k_F, or of the conduction electron density, but also by the effect of the impurity potential on the energy of a Bloch state, $E(\underset{\sim}{k}) = E_0(\underset{\sim}{k}) + N_1(\underset{\sim}{k}|U(\underset{\sim}{r})|\underset{\sim}{k})$, where $U(\underset{\sim}{r})$ is the perturbing potential caused by the impurity atoms. For free electrons, the matrix element is independent of $\underset{\sim}{k}$ (wave vector) and the density of states remains unchanged. Here it is assumed that the change in the electron concentration, or k_F, is determined by the valency of the impurity atoms; $\delta N(0)$ can then be considered as an adjustable parameter to fit experimental and theoretical values of δT_c.

The ratio $2\Delta_0/kT_c$ for a strong coupling superconductor such as lead decreases when high frequency impurity modes are introduced whose coupling constant is the same as that for pure lead at the frequency $\omega = \omega_{11}$. The gross effect of the impurity modes is to increase the characteristic temperature θ without a corresponding increase of the electron-phonon interaction. Therefore, the effective coupling measured by T_c/θ will decrease and hence the ratio $2\Delta_0/kT_c$ will decrease towards the weak-coupling ratio 3.5. A quantitative discussion of the ratio $2\Delta_0/kT_c$ for Pb-In and Pb-Tl alloys has been presented by Wu[12] using empirical values for $\alpha^2(\omega)g(\omega)$ for the pure and the impure metal that are found by inversion of the gap equations.

Table I. δT_c (valence effect) for substitutional lead alloys; ω_{11} = impurity-mode frequency, $\alpha^2(\omega_{11})$ = electron-impurity-mode coupling parameter, U_1 = change of the pseudo-Coulomb potential.

Impurity	ω_{11}/ω_1^1	$\alpha^2(\omega_{11})$ in MeV	U_1	δT_c in m°K/at.%
In	2.16	1.34	+0.0023	-0.010
Sn	2.16	1.34	+0.0011	+0.016
Sb	2.16	1.34	-0.0046	+0.140
Hg	....	....	0	0
Tl	....	....	+0.0040	-0.012
Bi	....	....	-0.0176	+0.053
TlBi	....	....	-0.0057	+0.017

ACKNOWLEDGEMENT

It is a pleasure to thank Professor W. Kohn for many stimulating discussions.

REFERENCES

1. J. R. Schrieffer, D. J. Scalapino, and J. W. Wilkins, Phys. Rev. Letters 10, 336 (1963); G. M. Éliashberg, J. Exptl. Phys. (U.S.S.R.) 38, 966 (1960); J.E.T.P. 11, 696 (1960).

2. I. M. Lifshitz, J. Phys., U.S.S.R. 7, 211, 249 (1943); 8, 89 (1944).

3. P. W. Anderson, J. Phys. Chem. Solids 11, 26 (1959).

4. T. Tsuneto, Prog. Theoret. Phys. (Kyoto) 28, 857 (1962). I am indebted to Professor Suhl for a discussion of the effect of elastic impurity scattering on the Éliashberg gap equation and its solution.

5. J. Appel, Phys. Rev. 156, 421 (1967).

6. A. A. Maradudin, Proceedings of the International Conference on Lattice Dynamics, Copenhagen, 1963, edited by R. F. Wallis (Pergamon Press, Inc., New York, 1965), p. 726.

7. J. R. Schrieffer, Theory of Superconductivity (W. A. Benjamin, Inc., New York, 1964), p. 148.

8. J. M. Rowell, W. L. McMillan, and P. W. Anderson, Phys. Rev. Letters 14, 633 (1965).

9. H. Gamari-Seale and B. R. Coles, Proc. Phys. Soc. 86, 1199 (1965).

10. D. Markowitz and L. Kadanoff, Phys. Rev. 121, 563 (1963).

11. J. R. Anderson and A. V. Gold, Phys. Rev. 139, A1459 (1965).

12. T. M. Wu, (to be published).

ELASTIC STRAINS IN IMPERFECT CRYSTALS

R. J. Elliott, J. A. Krumhansl*, T. H. Merrett

Theoretical Physics Department, Oxford, England

* LASSP, Cornell University, Ithaca, New York

1. INTRODUCTION

There have recently been a number of experimental measurements of the effect of elastic strains and electric fields on the properties of point imperfections in crystals. These properties normally depend on the local environment of the impurity. A prerequisite of their interpretation is therefore a knowledge of the strains induced locally by the applied stress. In this paper we show how these may be related to the elastic strains in a perfect crystal.

A finite concentration of such defects leads to a modification of the bulk elastic constants of the crystal. This will be considered by performing an average over the local strains and alternatively by a study of the velocity of sound in the long wavelength phonons.

2. LOCAL STRAINS

Let $\underline{\Phi}$ be the force constant matrix of a perfect crystal and let $\underline{\varphi}$ be the change induced by the defect. The former has, of course, the full periodic symmetry of the lattice. The latter is zero except for small blocks covering sites affected by the defect. Let $\underline{u}$ by a column matrix giving the displacements of the atoms in the perfect lattice. Let $\underline{w}$ give the displacements in the imperfect lattice and let the difference be $\underline{v}$, i.e. $\underline{w}=\underline{u}+\underline{v}$. Then if $\underline{F}$ is a column vector describing a general force, in our case the external force causing the elastic strain, the displacements are related by the equations

$$\underline{F} = \underline{\Phi}\,\underline{u} = (\underline{\Phi}+\underline{\varphi})\underline{w} \tag{1}$$

and

$$\underline{w} = (1+\underline{\Phi}^{-1}\underline{\varphi})^{-1}\underline{u} \tag{2}$$

$$\underline{v} = -(\underline{\Phi}+\underline{\varphi})^{-1}\underline{\varphi}\,\underline{u} \tag{3}$$

Thus if the elastic displacements in a perfect lattice are known those of the imperfect lattice can be obtained. In a non-Bravais lattice it is necessary to know the relative displacements in the unit cell as well as the elastic strains. In particular $\underline{v}$ is related to $\underline{\varphi}\,\underline{u}$ which because of the form of $\underline{\varphi}$ depends only on the displacements $\underline{u}$ in the vicinity of the defect where $\underline{\varphi}$ has non-zero elements.

The inverse matrices $-\underline{\Phi}^{-1}$ and $-(\underline{\Phi}+\underline{\varphi})^{-1}$ are the usual displacement Green's functions which are used to determine the lattice dynamics of defects [1]

$$\underline{g} = (\underline{M}\omega^2-\underline{\Phi})^{-1} \tag{4}$$

$$\underline{G} = (\underline{M}'\omega^2-\underline{\Phi}-\underline{\varphi})^{-1} \tag{5}$$

in the limit $\omega = 0$. $\underline{g}$ may be readily computed by transforming to the normal mode representation when $\underline{\Phi}$ becomes diagonal.

If attention is restricted to $\underline{w}$ on the sites where φ is changed it is only necessary to take the inverse of a finite matrix $(1-g\varphi)$. Using the local point symmetry this may be block diagonalised for displacements $\underline{u}$ and $\underline{w}$ belonging to particular representations of the point group. In simple cases some of these occur once and the equation becomes a scalar one

$$w' = (1-g'\varphi')^{-1}u' \tag{6}$$

The ratio goes to zero for large φ' since the strong forces prevent displacement in this case. It increases to unity as $\varphi \to 0$ and then to infinity when φ' becomes sufficiently negative to destroy the resistance of the perfect lattice to this type of displacement.

Detailed calculations are in progress for a face-centered cubic lattice and for CaF_2 where detailed experiments have been made on the local modes due to hydrogen [2].

3. ELASTIC CONSTANTS

In the perfect crystal the periodic symmetry ensures a uniform strain and the elastic constants are determined by $\underline{\Phi}$. In the imperfect crystal, as shown in §2, the strain is not uniform and the bulk elastic constants require an average over the lattice containing a finite concentration c of defects which introduce local changes $\underline{\varphi}$ in the force constants. Then taking a configuration average of eqn (1)

$$\underline{F} = \langle(\underline{\Phi}+\underline{\varphi})\underline{w}\rangle = \underline{\Phi}\langle\underline{w}\rangle + \langle\underline{\varphi}\,\underline{w}\rangle \qquad (7)$$

since $\underline{\Phi}$ is periodic but $\underline{w}$ and $\underline{\varphi}$ are not. We artificially separate the last term so that

$$\underline{F} = (\underline{\Phi}+\delta\underline{\Phi})\,\langle\underline{w}\rangle \qquad (8)$$

and the new elastic constants are given by $\underline{\Phi}+\delta\underline{\Phi}$ in place of $\underline{\Phi}$. Using (2), for a single defect

$$\underline{\varphi}\,\underline{w} = \underline{T}\,\underline{u} \qquad (9)$$

where the T matrix

$$\underline{T} = \underline{\varphi}(1-\underline{g}\,\underline{\varphi})^{-1} \qquad (10)$$

This only has elements for sites changed by the defects. To first order in c[3] $\langle\underline{T}\rangle$ is obtained by taking the result for a single site and multiplying by c. Writing this $c\underline{t}$ after some manipulation

$$\delta\underline{\Phi} = c\underline{t}\,(1+c\,\underline{g}\,\underline{t})^{-1} \qquad (11)$$

To first order in c the modified elastic constants are therefore to be calculated from $\underline{\Phi} + c\underline{t}$.

The form of (11) is equivalent to that found by Elliott and Taylor[3] in studying the dynamics of a random alloy at low c. In fact the elastic constants are most readily obtained by calculating the frequencies of the modes of low $\underline{k}$. The transformation of $\underline{\Phi}$ to normal co-ordinates leads to $\omega_j^2(\underline{k})$ of the perfect lattice. The modification of this is obtained by applying the same transformation to $c\underline{t}$. It is interesting to note that the static and dynamic approaches lead to the same results.

Explicit calculations of these effects have been made for a face-centered cubic crystal with nearest neighbour central forces. It is interesting to note that the Cauchy relations are no longer obeyed in the imperfect crystal because some atoms around the defect are no longer at centres of symmetry. Detailed calculations and results will be published elsewhere.

REFERENCES

1. R. J. Elliott, "Phonons" Aberdeen Summer School Lectures 1965 p. 377 (Oliver and Boyd 1966)

2. W. Hayes and H. F. Macdonald, Proc. Roy. Soc. 297, 503 (1967)

3. R. J. Elliott and D. W. Taylor, Proc. Roy. Soc. 296, 161 (1967)

LOCALIZED EXCITATIONS IN SOLID HYDROGEN

J. Van Kranendonk

Department of Physics, University of Toronto

Toronto, Canada

INTRODUCTION

Solid hydrogen is an ideal solid for studying various properties of localized excitations. Apart from the electronic excitations, which are also of great interest but which will not be considered here, three different types of excitations are possible in solid H_2, HD and D_2 : lattice vibrations (phonons), rotational excitations (rotons), and vibrational excitations (vibrons). The interaction between these various excitations gives rise to a number of interesting scattering and bound-state problems which have been studied by means of infrared and Raman techniques. Since the various interactions are known in detail, in contrast to, e.g., the interaction between an electron and an impurity in a metal, the study of the excitations in solid hydrogen provides a most useful testing ground for the theory of solid state scattering, bound-state and self-energy processes. The infrared and Raman spectra of solid H_2, HD and D_2 have been investigated extensively at the University of Toronto over a number of years, both experimentally[1,...,5] and theoretically [6,...,12]. The infrared activity in solid H_2 is due to the dipole moments induced in pairs of molecules by the intermolecular forces (free H_2 molecules have no electric dipole moments), and is analogous to the pressure-induced activity observed in the gaseous state[13,14,15]. The spectra of the solids are characterized by the appearance of a number of sharp zero-phonon lines and more extended zero-phonon bands, which are accompanied by broad phonon branches. The observed zero-phonon absorption features together with the corresponding Raman spectra, which contain no phonon branches, have yielded detailed information about the rotational and vibrational energy levels in solid H_2. A theoretical analysis of the experimental data has recently been completed[12] and the reader is referred to this paper for a detailed account of this work.

A characteristic property of the hydrogen solids is that the rotational motion of the molecules is free in the sense that the quantum number $J = \Sigma_i J_i$, where J_i is the rotational quantum number of molecule i and the sum extends over all the molecules, is a good quantum number. This free rotation persists down to the absolute zero, and is quite different from the free rotation observed in many molecular crystals at sufficiently high temperatures. In the hydrogen solids the anisotropic intermolecular interaction is so small relative to the spacing of the rotational levels that states corresponding to different values of J are not mixed appreciably. Thus, in the ground state of solid parahydrogen all the molecules are in the ground rotational state J = 0. In the first excited rotational state one of the N molecules is in the state J = 2 and the resulting level is 5N-fold degenerate. This degeneracy is removed by the anisotropic intermolecular interaction (mainly the quadrupole-quadrupole interaction), and the J = 2 level is broadened into a band about 20 cm^{-1} wide, whereas the separation between the J = 2 and J = 0 levels amounts to 356 cm^{-1}. The states in this band correspond to travelling rotational excitations, or rotons. We remark that these rotons are not analogous to magnons but rather to the rotons, or quantized vortices, which can exist in liquid helium[12].

The internal vibrational motion of the molecules exhibits similar properties. The intermolecular forces do not mix appreciably states corresponding to different values of $v = \Sigma_i v_i$, where v_i is the vibrational quantum number of molecule i, and v is a good quantum number also in the solid. The purity of v, in contrast to that of J, is a common property of most molecular crystals. The part of the interaction between two neighbouring molecules which depends on the vibrational coordinates in both molecules produces a coupling between the vibrations in the two molecules, which removes the N-fold degeneracy of the v = 1 level in the crystal. The resulting v = 1 band in solid H_2 is 4.0 cm^{-1} wide. The states in the v = 1 band correspond to travelling vibrational excitations, or vibrons. The coupling between the vibrons, rotons and phonons arising from the intra- and intermolecular forces produces a number of different bound-state complexes. In the next three sections we discuss in succession some properties of the resulting partially localized vibrons, rotons and phonons.

LOCALIZED VIBRATIONAL EXCITATIONS

The vibrational Raman cross section, $\sigma(J)$, per molecule in the state J, corresponding to the transitions $\Delta v = 1$, $\Delta J = 0$, in gaseous H_2 and D_2 is independent of J. This property is due to the fact that the isotropic part of the polarizability is practically independent of J. It has recently been observed[4,5] that in solid H_2, on the other hand, the ratio $\xi = \sigma(1)/\sigma(0)$ is by no means equal to unity, but varies from about 3 at small ortho (J=1) concentrations to about 2 at high ortho concentrations. In solid D_2 this ratio is even larger, varying from 30-50 at 3.7% para (J=1) concentration to about 9 at 33%. These large changes in the Raman cross sections in the condensed phases have been explained by James and Van Kranendonk[11] as due to the vibrational coupling between neighbouring molecules. A brief account of the main features of this work will be given here.

An arbitrary state in the manifold $v = \Sigma_i v_i = 1$ is given by

$$|\tau\rangle = \Sigma_i \, U_\tau(\vec{R}_i)|1_i\rangle \; , \tag{1}$$

where $|1_i\rangle$ is the state of the crystal in which the molecule at $\vec{R}_i$ is in the state $v_i = 1$. In a crystal of pure para- or orthohydrogen, the stationary states are Bloch states with energy

$$E(\vec{k}) = -\tfrac{1}{2}\,\varepsilon' \, \Sigma_{\vec{d}} \cos(\vec{k}\cdot\vec{d}) \; , \tag{2}$$

where the sum extends over the 12 nearest neighbours, and ε' is the vibrational coupling constant. Empirical values of ε' for H_2 and D_2 can be obtained from the observed frequency shifts in the vibrational infrared and Raman spectra, and are given by[11,12] : $\varepsilon'(H_2) = 0.49\text{cm}^{-1}$ and $\varepsilon'(D_2) = 0.37\text{cm}^{-1}$. The energies (2) extend from $-6\varepsilon'$ at $\vec{k} = 0$ to $+2\varepsilon'$ obtained at a set of values of $\vec{k}$. Let us now consider the impurity problem in which there is one J = 1 molecule in a matrix of J = 0 molecules. The wave function of a v = 1 vibron then satisfies the set of difference equations,

$$-\tfrac{1}{2}\,\varepsilon' \, \Sigma_{\vec{d}} \, U_\tau(\vec{R}_i+\vec{d}) + W\,\delta(\vec{R}_i;0)\,U_\tau(0) = E_\tau \, U_\tau(\vec{R}_i), \tag{3}$$

where $|W|$ is equal to the difference in the $v = 1$ excitation energy in a $J = 0$ and a $J = 2$ molecule. This difference arises from the intramolecular rotation-vibration interaction[6]. For H_2 we have $W = -5.9\ cm^{-1}$ for a $J = 1$ impurity in a $J = 0$ matrix, and $W = +5.9\ cm^{-1}$ for a $J = 0$ impurity in a $J = 1$ matrix. In D_2 we have $|W| = 2.1\ cm^{-1}$. Eq. (3) has one bound-state solution and N-1 scattering solutions. In H_2 the $v = 1$ excitation in the bound state due to a $J = 1$ impurity spends about 98% of its time on the impurity and only about 20% on the neighbouring $J = 0$ molecules. This imperfect localization leads nonetheless to an increase in the Raman cross section by a factor 3. This rather surprising result comes about as follows.

For small values of the parameter $|\varepsilon'/W|$ one can carry out a perturbation expansion[9]. For simplicity we assume here that the excitation will not be found beyond the nearest neighbour shell. This is not a very good approximation for H_2 and is quite inadequate for D_2, but is a useful model for showing the main effect in a simple way. For a more accurate treatment we refer to reference (11). If p denotes the probability of finding the excitation on one of the twelve neighbours, we get

$$\begin{cases} U_\tau(0) = (1-12p)^{\frac{1}{2}} , \\ U_\tau(\vec{d}) = \pm\ p^{\frac{1}{2}} . \end{cases} \qquad (4)$$

The +(-) sign holds for a $J = 1$ ($J=0$) impurity in a $J = -$ ($J=1$) matrix, for which the bound level lies below (above) the $v = 1$ band. In either case we get in first approximation,

$$p = (\varepsilon'/2W)^2 . \qquad (5)$$

The Raman cross section for transitions to the state $|\tau\rangle$ is proportional to the square of the matrix element of the total polarizability of the crystal,

$$\langle\tau|\Sigma_i\ \alpha_i|0\rangle = \alpha_{01}\ \Sigma_i\ U_\tau(\vec{R}_i) , \qquad (6)$$

where α_{01} is the matrix element for an isolated molecule. From Eqs. (4) and (6) we see that the cross section for a $J = 1$ molecule in a $J = 0$ matrix is enhanced relative to that of an isolated molecule by the factor

$$\eta_1{}^2 = |(1-12p)^{\frac{1}{2}} + 12p^{\frac{1}{2}}|^2 , \qquad (7)$$

whereas the cross section for a $J = 0$ molecule in a $J = 1$

matrix is decreased by a factor

$$\eta_0^2 = |(1-12p)^{\frac{1}{2}} - 12p^{\frac{1}{2}}|^2 \ . \qquad (8)$$

For H_2 we obtain $p = 0.0017$ and $12p^{\frac{1}{2}} = 0.20$, giving $\eta_1^2 = 2.2$ and $\eta_0^2 = 0.25$. The cross sections for the host molecules are not affected by a very small concentration of impurities. For the ratio, $\xi(c) = \sigma(1)/\sigma(0)$, of the cross sections (only the ratios are measured experimentally), where c is the ortho concentration, we therefore obtain $\xi(0) = 2.2$ and $\xi(1) = 4.0$. It is clear that these large values of ξ result from the fact that, although p is small, $12p^{\frac{1}{2}}$ is not particularly small. The accurate treatment, in which the contributions from all shells of neighbours are taken into account by means of the walk-counting method developed by James[16], yields the result[11] $\xi(0) = 3.42$ and $\xi(1) = 2.29$, in excellent agreement with the observed values $\xi(0) = 3.5 \pm 0.4$ and $\xi(1) = 2.4 \pm 0.4$[5]. We see that in spite of the fact that the excitation is well localized it is essential to take more distant shells of neighbours into account to obtain agreement with the observed concentration dependence of $\xi(c)$. It is clear that the Raman cross sections depend very sensitively on the degree of localization of the excitons, and that they provide an accurate test for the approximation methods used in the calculations.

In D_2 the bound states lie much closer to the exciton band, and the enhancement factors in the Raman cross section ratios are correspondingly much larger. Taking all shells of neighbours into account we obtain for the ratio $\xi(c)$, where c is the para (J=1) concentration, $\xi(0) = 55$ and $\xi(1) = 4.9$, whereas the observed[5] values are $\xi(3.7\%) = 50 \pm 10$ and $\xi(50\%) = 9.2 \pm 0.5$. The good agreement at low concentrations is not significant since the excitations are spread out so much that even at $c = 3.7\%$ the different impurity states are overlapping appreciably, so that a calculation based on isolated impurities is not adequate. A calculation carried out for a regular superlattice of J = 1 impurities in a J = 0 matrix yields the value $\xi(3.7\%) = 49$ in good agreement with the observed value. There is therefore no doubt that also in solid D_2 the anomalous Raman intensity ratios are due to the vibrational coupling and the resulting imperfect localization of the vibrational excitations on the impurity molecules.

LOCALIZED ROTATIONAL EXCITATIONS

In a crystal of pure parahydrogen a single $J = 2$ excitation is not localized but travels through the crystal under the influence of the quadrupole-quadrupole interaction, the $J = 2$ roton band being about 20 cm^{-1} wide. However, in the presence of a $J = 2$ roton and a $v = 1$ vibron a bound complex can exist in which the $J = 2$ roton spends most of its time on the $v = 1$ molecule. Transitions from the ground state to these bound states give rise to the $S_1(0)$ features in the infrared[1] and Raman[3,4] spectra. The states in which the $v = 1$ and $J = 2$ excitations are not bound form a mixed band which gives rise to the $Q_1(0) + S_0(0)$ infrared band[1] whose total width is equal to the sum of the $v = 1$ and $J = 2$ bands. The theory of the states in the $v = 1$, $J = 2$ manifold has been worked out in considerable detail[12] and we here discuss some effects arising from the localized nature of the bound complexes.

We first neglect the vibrational coupling and assume that the $v = 1$ excitation is localized on the molecule at the origin. An arbitrary state of the crystal in which there is an addition to the $v = 1$ excitation one $J = 2$ roton is given by

$$|\tau\rangle = \Sigma_{i,m}\, U_\tau(\vec{R}_i,m)\,|2_i,m\rangle \ , \qquad (9)$$

where $|2_i,m\rangle$ is the state in which the molecule at $\vec{R}_i$ is in the state $J_i = 2$, $J_{iz} = m$. The wave function $U_\tau^{\,i}(\vec{R}_{i,m})$ of the roton satisfies an equation similar to Eq. (4) but more complicated because of the spin of the roton. The quantity W arises from the intramolecular rotation-vibration interaction and is equal to -18.0 cm^{-1}. It represents the depression in the energy of the $J = 2$ roton when it is located on the $v = 1$ molecule and is due to the fact that the moment of inertia of a $v = 1$ molecule is larger than that of a $v = 0$ molecule. The impurity potential, $W\,\delta(\vec{R}_i;0)$, produces a bound state[6] but is not strong enough to localize the $J = 2$ roton completely, and the roton spends about 20% of its time away from the $v = 1$ molecule. This imperfect localization produces a lowering in the energy of the bound state which has been calculated[5,11] using second and third order perturbation theory and which amounts to -3.5 cm^{-1}. This value cannot be compared directly with the observed frequency of the $S_1(0)$ line since a number of other perturbations are also involved. However, a detailed analysis of all the infrared and Raman data shows[11] that the calculated shift is in agreement with the experimental data within the accuracy of about ± 0.1 cm^{-1}. The depression in the energy is not exactly equal for the m = 0, ±1, ±2 substates, and a small splitting arises which has been observed in the $S_1(0)$ Raman line[4]. We have so far neglected the vibrational coupling and the vibrational dependence of the quadrupolar interaction. When these interactions are introduced[12], the v = 1, J = 2 bound complexes are no longer completely stationary, and the resulting $S_1(0)$ band is about 1 cm^{-1} wide. The infrared and Raman active levels in this band correspond to $\vec{k} = 0$, and these levels lie very close to the band origin. The hopping of the $S_1(0)$ complexes therefore does not give rise to any observable effects.

Independent evidence concerning the structure of the v = 1, J = 2 bound complexes is obtained from the fine structure of the $S_1(0)$ infrared line[2] appearing in the presence of small concentrations of orthohydrogen, and from the splitting of the $S_1(0) + S_1(0)$ line in the infrared overtone spectrum arising from the creation of pairs of v = 1, J = 2 complexes. In both cases excellent agreement between theory and experiment is obtained when the imperfect localization of the rotons is taken into account[12]. In all these calculations we have assumed that the matrix elements of the molecular quadrupole moments between the various rotation-vibration states have the theoretical values calculated[17] for the isolated molecules. The successful analysis of the infrared and Raman data indicates the correctness of this assumption.

LOCALIZED LATTICE VIBRATIONAL EXCITATIONS

Because of the dependence of the anisotropic intermolecular forces on the intermolecular separations there is a coupling between the rotational motion of the molecules in solid H_2 to the lattice vibrations. This coupling gives rise to virtual phonon processes, and the resulting self-energy, or lattice polarization effects, give rise to observable shifts of the rotational levels[9][12]. Consider a localized J = 2 excitation in a matrix of J = 0 molecules. The classical rotational frequency corresponding to the transition J = 2 → J = 0 is larger than the Debye frequency and real one-phonon processes therefore cannot occur. Classically, the rotating molecule drives the lattice oscillators above their resonance frequencies producing a non-propagating, local distortion of the lattice in which the molecules vibrate about distorted equilibrium positions which are determined by the average charge density of the rotating molecule. In the quantum theory the process can be described by second-order perturbation theory. The molecule in the J = 2 state can emit a virtual phonon changing at most its orientation but not the value of J and then reabsorb the phonon. The existence of this effect can be inferred from the complete analysis of the spectroscopic data[12], and the lowering of the energy of the J = 2 state amounts to about 0.5 cm^{-1} in agreement with the theoretical estimate[9]. The corresponding effect for the vibrons is far too small to be observable.

REFERENCES

1. H.P. Gush, W.F.J. Hare, E.J. Allin and H.L. Welsh, Can. J. Phys. 38, 176 (1960).
2. H.P. Gush, J. Phys. et Radium 22, 149 (1961).
3. S.S. Bhatnagar, E.J. Allin and H.L. Welsh, Can. J. Phys. 40, 9 (1962).
4. V. Soots, E.J. Allin and H.L. Welsh, Can. J. Phys. 43, 1985 (1965).
5. A.H. McKague Rosevaer, G. Whiting and E.J. Allin, Can. J. Phys. (1967) to be published.
6. J. Van Kranendonk, Physica 25, 1080 (1959).
7. J. Van Kranendonk, Can. J. Phys. 38, 240 (1960).
8. H.P. Gush and J. Van Kranendonk, Can. J. Phys. 40, 1461 (1962).
9. V.F. Sears and J. Van Kranendonk, Can. J. Phys. 42, 980 (1964).
10. J. Van Kranendonk and V.F. Sears, Can. J. Phys. 44, 313 (1966).
11. H.M. James and J. Van Kranendonk, Phys. Rev. (submitted for publication).
12. J. Van Kranendonk and G. Karl, Phys. Rev. (submitted for publication).
13. D.A. Chisholm and H.L. Welsh, Can. J. Phys. 32, 291 (1954).
14. W.F.J. Hare and H.L. Welsh, Can. J. Phys. 36, 88 (1958).
15. J. Van Kranendonk, Physica 24, 347 (1958).
16. H.M. James, Phys. Rev. (submitted for publication).
17. G. Karl and J.D. Poll, J. Chem. Phys. 46, 2944 (1967).

INVESTIGATION OF LOCALIZED EXCITATIONS BY INELASTIC NEUTRON SCATTERING

A. R. Mackintosh and H. Bjerrum-Møller

Technical University, Lyngby, Denmark and

A.E.K. Research Establishment, Risø, Denmark

INTRODUCTION

In recent years, the inelastic scattering of neutrons has been extensively used in the study of the lattice vibrations and magnetic excitations in pure crystals, and the unique power of this technique has become widely recognized. It can also be applied to the study of the local modes in crystals in which the periodicity has been destroyed by the addition of impurities. In this paper we shall discuss briefly the theory of inelastic neutron scattering by such defect modes, and review the relatively small amount of experimental data which has so far been obtained.

An inelastic neutron scattering experiment consists essentially of scattering a monochromatic beam of neutrons of initial wavevector $\underline{k}_o$ and energy $\hbar^2 k_o{}^2/2m_o$ from a crystal, and of measuring the scattered intensity as a function of the final wavevector $\underline{k}$ and energy $\hbar^2 k^2/2m_o$. The energy and momentum transfer to the lattice are then

$$\hbar\omega = \hbar^2(k_o{}^2 - k^2)/2m_o \quad \text{and} \quad \hbar\underline{\kappa} = \hbar(\underline{k}_o - \underline{k}) \qquad (1)$$

respectively. The scattering cross section per unit solid angle and energy transfer may be written in the form

$$\frac{d^2\sigma}{d\Omega dE} = \frac{k}{hk_o} S(\underline{\kappa},\omega) \qquad (2)$$

The scattering function $S(\underline{\kappa},\omega)$ is closely related to the Fourier transform in space and time of the dynamical pair-correlation function[1], and it is this which makes neutron scattering such a powerful tool for the study of the dynamical behaviour of the local modes. The momentum and energy transfer to the lattice can both be readily measured, so that the spatial, as well as the temporal, variation of the local excitation can be studied. This is in contrast to optical studies of defects, where the momentum transfer to the crystal is normally too small to be observed. Optical measurements have provided a large amount of detailed information on local modes in insulators and semiconductors, but they are difficult to extend to metals, because of the high absorption. This limitation again does not generally apply to the neutron experiments, and much of the existing information on defect modes in metals has been obtained from measurements of inelastic neutron scattering.

In the following section, we shall outline the theory of neutron scattering by localized phonon and magnon modes, and discuss the information which can be obtained from the coherent and incoherent scattering cross sections. A comparison will be made between the theoretical predictions and those experimental results which have so far been obtained. Finally, the present status of these experiments will be summarized, and the special features, advantages, and difficulties of neutron scattering as a method of studying localized excitations will be briefly discussed.

LOCALIZED PHONONS

The scattering of neutrons by local phonon modes was first considered by Krivoglaz[2] and by Kagan and Iosilevskii[3]. These calculations were extended by Elliott and Maradudin[4], who made numerical computations based on specific models, and emphasized the power of coherent neutron scattering experiments for studying defect modes. Further aspects of the problem have been discussed by Dzyub[5], Privorotskii[6], Maradudin[7] and Elliott and Taylor[8], whose results we shall quote extensively below.

For a perfect Bravais lattice of identical atoms of mass M, the coherent scattering function in the one-phonon approximation is

$$S(\underline{\kappa},\omega) = a^2 D^2(\underline{\kappa})\left(\frac{\hbar}{2M\omega}\right)\sum_j \{n(\omega)\delta(\omega+\omega_j(\underline{\kappa})) +$$
$$(n(\omega)+1)\delta(\omega-\omega_j(\underline{\kappa}))\}(\underline{\kappa}\cdot\underline{\sigma}_j(\underline{\kappa}))^2 \qquad (3)$$

where a is the coherent scattering length, $D(\underline{\kappa})$ is the Debye-Waller factor, $\omega_j(\underline{\kappa})$ is the phonon dispersion relation in the periodic zone scheme, $\underline{\sigma}_j(\underline{\kappa})$ is the phonon polarization vector, and $n(\omega)$ is the Bose-Einstein population factor

$$n(\omega) = (e^{\hbar\omega/kT} - 1)^{-1} \qquad (4)$$

The two terms in the summation correspond to destruction and creation of a phonon, respectively. In future, we shall quote only the term corresponding to neutron energy gain. As may be seen from eqn. (3) therefore, the phonon dispersion relation can be determined by seeking peaks in the scattered intensity as a function of $\underline{\kappa}$ and ω.

The incoherent one-phonon scattering function is

$$S(\underline{\kappa},\omega) = \alpha^2 \frac{\hbar n(\omega)}{2M\omega} \sum_{j,\underline{q}} D^2(\underline{\kappa})(\underline{\kappa}\cdot\underline{\sigma}_j(\underline{q}))^2\delta(\omega+\omega_j(\underline{q})) \qquad (5)$$

where α is the incoherent scattering length, and the phonon wavevector in the primitive zone is denoted by $\underline{q}$. The intensity of incoherently scattered neutrons as a function of ω, for fixed $\underline{\kappa}$, is therefore closely related to the phonon density of states $g(\omega)$.

If a small concentration c of impurities of mass M' is substituted in the lattice, the appearance of local states modifies the scattering cross sections. If the impurity mass is greater than that of the host, a quasi-localized resonant mode in the continuum of band modes generally results and, assuming no change in the force constants, the coherent scattering function for a cubic Bravais lattice has the form

$$S(\underline{\kappa},\omega) = D^2(\underline{\kappa}) \frac{\hbar n(\omega)}{2M} \mathrm{Im}\{(a+c(a'-a)X(\omega))^2$$
$$\sum_j (\underline{\kappa}\cdot\sigma_j(\underline{\kappa}))^2 \frac{1}{\omega^2-\omega_j^2(\underline{\kappa})-c\varepsilon\omega^2 X(\omega)}\} \qquad (6)$$

where a' is the coherent scattering length of the impurity, $\varepsilon = (M-M')/M$ and $D(\underline{\kappa})$ has been assumed independent of the lattice site. $X(\omega)$ is completely determined by c, ε and the host density of states, and is given by

$$X_1(\omega)+iX_2(\omega) = \{1-\varepsilon\omega^2 \int \frac{g(\omega')d\omega'}{\omega^2-\omega'^2} - i\tfrac{1}{2}\pi\varepsilon\omega g(\omega)\}^{-1} \tag{7}$$

where the principal value of the integral is to be taken. For small concentrations, the cross section is very similar to that of the perfect lattice, except that the δ-function is replaced by a Lorentzian, with a cut-off in the tails. The peak is shifted from $\omega_j(\underline{\kappa})$ by an amount

$$\Delta_j(\underline{\kappa}) = \tfrac{1}{2}c\varepsilon\omega_j(\underline{\kappa})X_1(\omega_j(\underline{\kappa})) \tag{8}$$

and the width is

$$\Gamma_j(\underline{\kappa}) = c\varepsilon\omega_j(\underline{\kappa})X_2(\omega_j(\underline{\kappa})) \tag{9}$$

The scattering cross sections have been evaluated as a function of $\underline{\kappa}$ in this approximation for an alloy of 3% Au in Cu, by Svensson[9], and his results are shown in Fig. 1. At the resonance frequency, there is a rapid change from a negative to a positive frequency shift, with an associated broadening of the neutron groups.

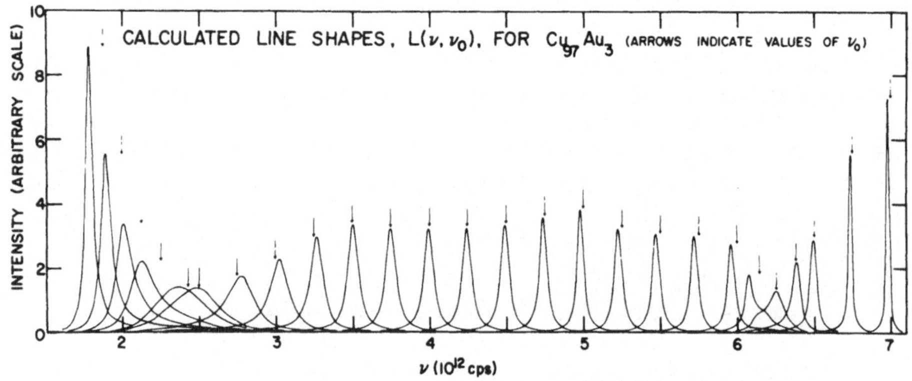

Fig. 1. Calculated frequency shifts and line shapes for Cu-3% Au, after Svensson (Ref. 9).

The first studies of crystals with impurities of very different mass by coherent neutron scattering were made on Cu-9% Au by Svensson, Brockhouse and Rowe[10] and on Cr-3% W by Møller and Mackintosh[11]. We have recently extended the latter measurements, and the dispersion curves for transverse waves propagating in the [100] and [110] directions are shown in Fig. 2. The characteristically rapid change of energy with wavevector about the resonance can clearly be observed. The half-value contour for the Gaussian resolution function, derived from measurements on the (220) Bragg reflection[12], is also shown. A constant q scan corresponds to a vertical movement of this function through the dispersion curve, and it is extremely important to know the resolution accurately when measuring the positions and natural widths of the neutron groups observed near the resonance. It is not, for example, generally adequate to subtract the widths of the pure crystal neutron groups from those

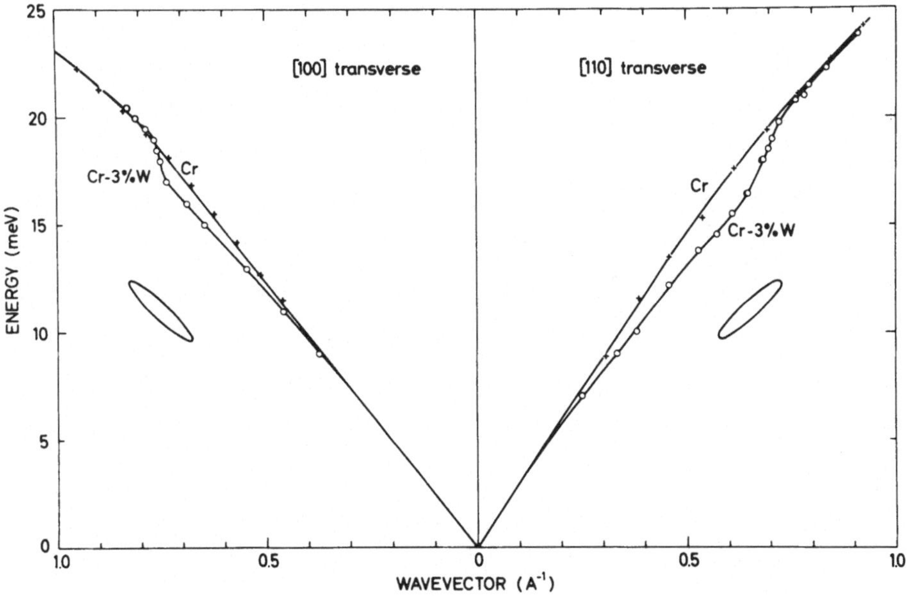

Fig. 2. Dispersion curves for transverse waves propagating in the [100] and [110] directions in Cr and Cr-3% W. The half-value contours for the Gaussian resolution functions are also shown.

for the defect crystal when attempting to determine the natural width of the latter, because the change in the shape of the dispersion curve alters the instrumental resolution width. This procedure is, however, satisfactory when the total shift is small compared with the energy resolution[13]. In this experiment, we found it generally more satisfactory to use constant E scans, corresponding to a horizontal motion of the resolution function.

The energy shift and natural width of the excitations in Cr-3% W are shown in Fig. 3. The shifts in the [110] direction are, within experimental error, identical with those previously reported[11]. In addition, the theoretical shifts and widths are shown. These were computed from eqns. (7), (8) and (9) by Svensson[9], using an approximate Cr density of states obtained by scaling that for W. In these calculations, the difference in coherent scattering amplitudes between Cr and W was not taken into account, but Elliott and Taylor[8] have shown that this leads to only a small change in the results.

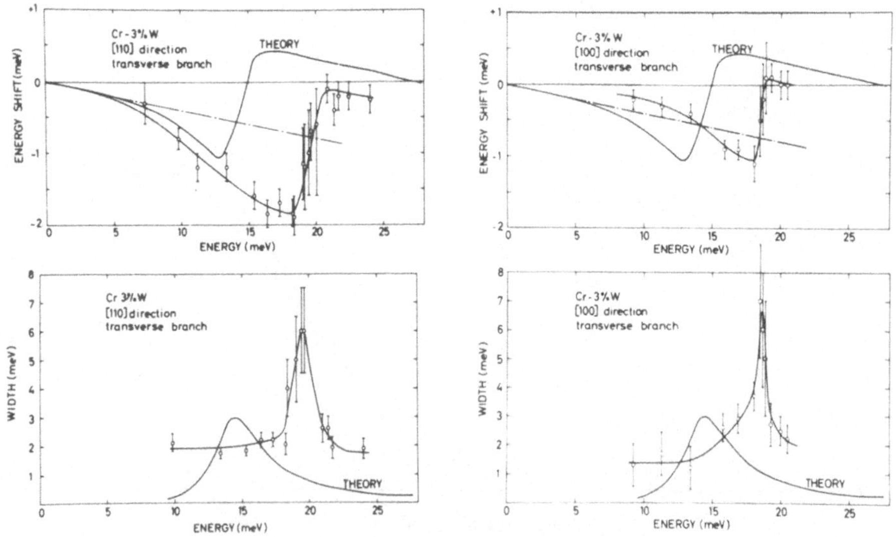

Fig. 3. The energy shifts, relative to pure Cr, and the natural widths for transverse phonons in Cr-3% W. The theoretical curves were calculated by Svensson (Ref. 9).

In the [110] direction, the resonance occurs at a substantially higher energy, and the energy shift across the resonance is somewhat greater than predicted by the mass defect theory. Furthermore, no positive energy shift is observed. The latter effect is probably due to the finite concentration of defects, but the other discrepancies indicate that the force constants in the vicinity of the defects are different from those in the pure metal. This conclusion is supported by the difference in the results between the [100] and [110] branches. Compared with the [110] branch, the magnitude of the energy shift in the [100] direction is smaller and the resonance energy is slightly different, and these features cannot be explained by the mass change at the impurity site. Behera and Deo[14] have shown that, as expected, the resonance energy can be made to agree with that observed experimentally by an appropriate change in the nearest neighbour force constants, but that the calculated energy shift is then substantially too small. Since the force constants in both Cr[15] and W[16] are known to be long range, however, the force distribution in the vicinity of the impurity may be extensively altered, and the effects of force changes to distant atoms are difficult to calculate.

The experimentally observed shift and width for the [110] transverse branch of Cu-3% Au[13] are in good agreement with the mass defect theory, as shown in Fig. 4, and this indicates that force constant changes are small in this system. However, the results for a 9% alloy do not agree with the theory. In the [100] transverse branch, shown in Fig. 4, the resonant behaviour is not very pronounced, and in the [100] longitudinal branch it is completely absent[9]. The drastic change between these two alloys may be due to nonlinear effects in the concentration, or to ordering of the Au impurities. It would clearly be of interest to study intermediate concentrations, and to measure different branches in the Cu-3% Au alloy.

The resonant modes may also be observed by incoherent scattering, which is also modified by the heavy impurities. The mass defect theory for a cubic Bravais lattice gives

$$S'(\underline{\kappa},\omega) = \frac{\hbar n(\omega)\kappa^2}{6M\omega} D^2(\underline{\kappa})\{\alpha^2(1-c)g(\omega) +$$

$$c((a'-a)^2+\alpha'^2)\ \frac{2X_2(\omega)}{\pi\varepsilon\omega}\} \qquad (10)$$

where α' is the incoherent scattering length of the impurity. The peak in $X_2(\omega)$ at the resonance is reflected in that part of the incoherent cross section due to the impurities, and such a peak was first observed in a Mg-2.8% Pb alloy by Chernoplekov and Zemlyanov[17]. They found that the position of the resonance was in good agreement with the predictions of the mass defect theory, which can also account semi-quantitatively for their results on a Ti-5% U alloy[18], where the resonance is split because of the hcp lattice symmetry. On the other hand, Mozer, Otnes and Thaper[19] observed no resonance in a V-5% Pt alloy, and they ascribe this to drastic changes in the force constants.

The substitution of a light impurity in a crystal generally results in a localized mode above the host band modes, with a frequency Ω which, in the mass defect theory, is determined by the equation

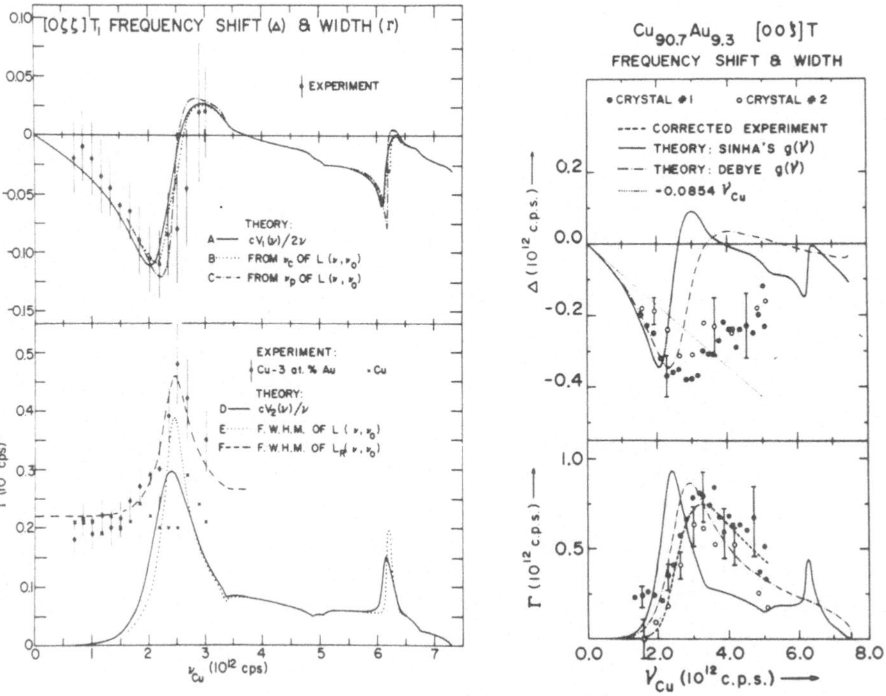

Fig. 4. The frequency shifts, relative to pure Cu, and widths for transverse phonons in Cu-3% Au and Cu-9% Au, after Refs. 13 and 10.

$$1 - \varepsilon\Omega^2 \int \frac{g(\omega')d\omega'}{\Omega^2-\omega'^2} = 0 \tag{11}$$

The total cross section from the local mode is given by

$$\tilde{S}(\underline{\kappa},\Omega) = \frac{c\hbar D^2(\underline{\kappa})}{2M\Omega} n(\omega)\delta(\omega+\Omega) \sum_j (\underline{\kappa}\cdot\underline{\sigma}_j(\underline{\kappa}))^2$$

$$\times\{\frac{1}{\Omega^2 B(\Omega)} \left(\left(\frac{a-a'}{\varepsilon} + \frac{a}{1-(\omega_j(\underline{\kappa})/\Omega)^2}\right)^2 + \frac{\alpha'^2-\alpha^2}{\varepsilon^2}\right) + \alpha^2\} \tag{12}$$

where

$$B(\Omega) = \int \frac{\omega'^2 g(\omega')d\omega'}{(\Omega^2-\omega'^2)^2} \tag{13}$$

The total cross section for a polycrystalline sample therefore has a peak at the local mode frequency. Such peaks were originally detected by Mozer, Otnes and Meyers[20] for alloys of Ni in Pd, and more recent measurements have been made on alloys with no coherent scattering[19]. The localized vibrations of interstitial H impurities in V have also been studied by this technique[21].

The observation of the $\underline{\kappa}$ dependence of the coherent scattering intensity at the local mode frequency allows the spatial variation of the atomic displacements around the impurity to be determined. In the mass defect theory this is contained in the term $a/(1-(\omega_j(\underline{\kappa})/\Omega)^2)$ in eqn. (12), which has been evaluated for an fcc lattice with nearest neighbour interactions by Elliott and Maradudin[4]. In fig. 5 is plotted the function

$$F(x) = \left(\frac{x}{1-(\omega_L \sin x/\Omega)^2}\right)^2$$

for different values of ε. In this expression, ω_L is the maximum lattice frequency and $x = R_o\kappa/4$, where R_o is the lattice parameter. For small values of ε, the local mode frequency is close to the maximum frequency of the host lattice, which occurs at the zone boundary in this model, so that the neighbours are out of phase with the impurity. The local mode is extended in space and the intensity is therefore sharply peaked in $\underline{\kappa}$. As ε is increased, the mode becomes more localized as its frequency increases, so that the intensity distribution becomes more uniform.

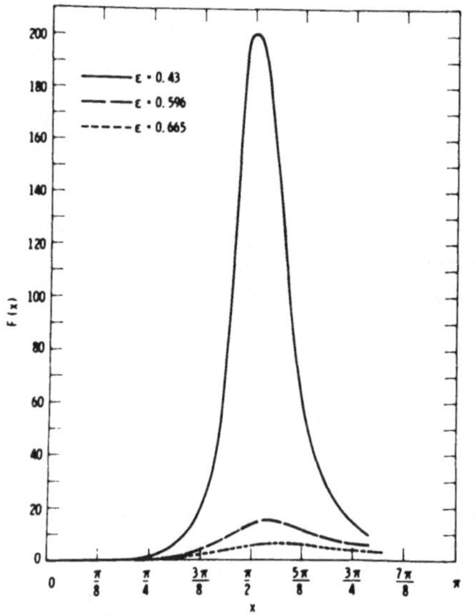

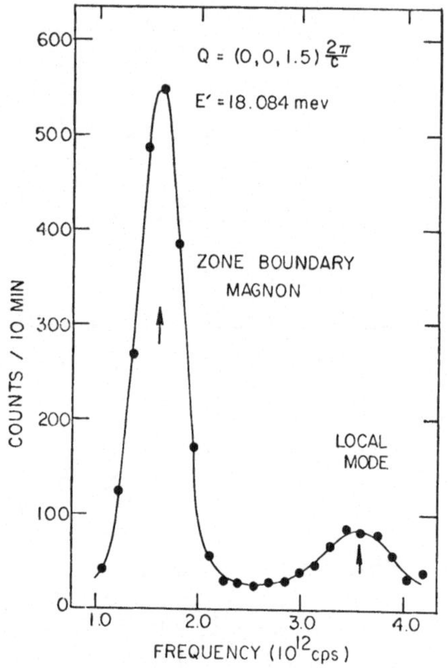

Fig. 5. The $\underline{\kappa}$ dependence of the scattered intensity at the localized mode frequency, in the [100] direction of an fcc crystal, after Elliott and Maradudin (Ref. 4).

Fig. 6. The scattered intensity as a function of frequency for fixed $\underline{\kappa}$ in MnF_2-5% Co, after Buyers et al. (Ref. 28). Both a band magnon and a localized magnon may be observed in this scan.

The coherent scattering from a Cu-10% Al crystal has recently been observed by Nicklow and co-workers[22]. They find a local mode frequency in good agreement with the prediction of the mass defect theory, and the scattered intensity at this frequency varies with $\underline{\kappa}$ qualitatively as expected, being greatest at those values corresponding to maxima in the dispersion relation for pure Cu.

LOCALIZED MAGNONS

The magnetic moment of the neutron interacts with the unpaired moments in a crystal, and the magnetic excitations may therefore be studied by inelastic neutron scattering. Below the magnetic ordering temperature, the scattering from a pure crystal is coherent and, in the one-magnon approximation, the scattering function for a ferromagnetic Bravais lattice is

$$S(\underline{\kappa},\omega) = \left(\frac{\gamma e^2}{mc^2}\right)^2 \frac{S}{2} D^2(\underline{\kappa}) f^2(\underline{\kappa})(1+\beta^2) n(\omega)\delta(\omega+\omega(\underline{\kappa})) \tag{15}$$

where γ is the magnetic moment of the neutron, S is the localized spin, $f(\underline{\kappa})$ is the magnetic form factor, β is the cosine of the angle between the magnetization and $\underline{\kappa}$, and $\omega(\underline{\kappa})$ is the magnon dispersion relation in the periodic zone scheme.

The substitution in the lattice of impurities of different spin S' and form factor $f'(\underline{\kappa})$ may have the effect of producing localized and quasi-localized magnon modes[23], analogous to local phonon modes. Since the substitution of a different spin necessarily changes the exchange coupling to the neighbours, there is no equivalent to the simple mass defect phonon problem and the cross section formulae tend to be rather cumbersome. They may be found in the papers of Izyumov[24] and Lovesey[25]

The introduction of a different magnetic species into the host lattice causes incoherent neutron scattering, which has a peak at the frequency of the localized or resonant mode. For the latter case, the incoherent cross section has a contribution of the form[26]

$$S'(\underline{\kappa},\omega) = \left(\frac{\gamma e^2}{mc^2}\right)^2 (1+\beta^2)(f'(\underline{\kappa})\sqrt{S'}-f(\underline{\kappa})\sqrt{S})^2 \frac{\hbar\Gamma}{\Gamma^2+(\omega-\omega_o)^2} \tag{16}$$

where $\hbar\omega_o$ is the energy of the quasi-localized state and $\hbar\Gamma$ its half-width. An additional scattering due to the substitution of 3% Mn in Fe has been observed by Kroo and Bata[27]. It has a peak at about 20 meV and a half-width of about 12 meV, and is ascribed by them to the occurrence of an anomalously broad resonance.

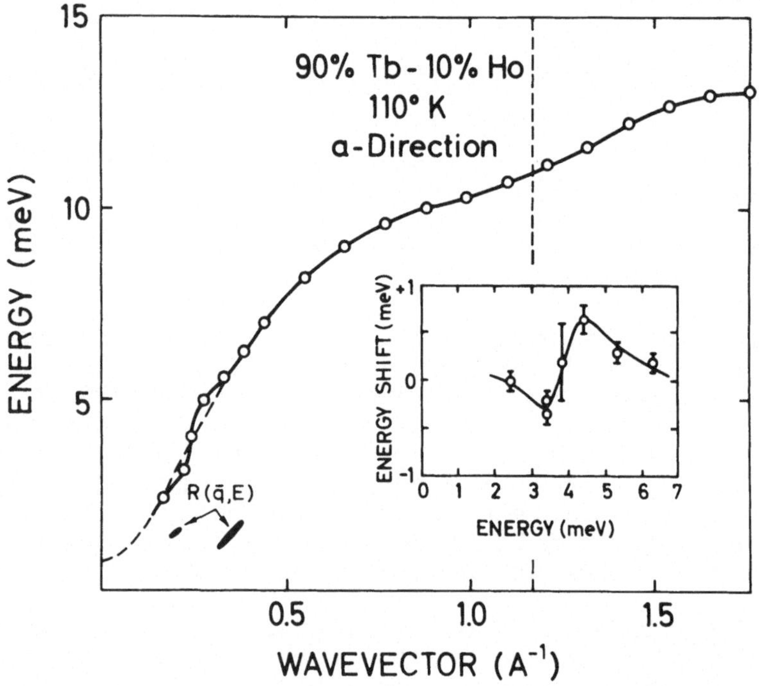

Fig. 7. The dispersion curve for magnons propagating in the a-direction of Tb-10% Ho at 110°K. The resolution functions used in obtaining these results are also shown. The insert shows the departure of the experimental points from the dashed straight line around 4 meV.

The most detailed information about local magnon modes may again be obtained from the dependence of the coherent scattering intensity on $\underline{\kappa}$ and ω. The localized magnon states in MnF_2 doped with 5% Co have been studied in this manner by Buyers et al.[28]. They observed an additional scattering above the magnon frequencies of pure MnF_2, as shown in Fig. 6. By measuring the intensity of this peak as a function of $\underline{\kappa}$, they were able to deduce that approximately 1.2% of the spin deviation associated with the local magnon is on the Mn^{++} neighbours rather than the Co^{++} impurity. Both the observed spatial distribution of the spin deviation and the local mode frequency are in good agreement with a calculation based on a nearest-neighbour cluster model.

We have recently studied the magnetic excitations in a Tb-10% Ho alloy by coherent neutron scattering and, as shown in Fig. 7, the magnon dispersion curve in the a-direction is strongly perturbed around 4 meV. Since the corresponding branch in pure Tb is quite linear over this region[29], we ascribe this perturbation to a resonance centred at 4 meV, with a width of about 1 meV. Since the Ho spin is $2\mu_B$, compared with $3\mu_B$ for Tb, its precessional frequency in the Tb lattice should lie in the band of magnon frequencies, so that a magnon resonance would be expected. Although the density of magnon states in Tb is known[29], a quantitative estimate of the resonance frequency is difficult to obtain because of the long range of the exchange constants.

CONCLUSION

The incoherent scattering of neutrons provides, in principle, a powerful method for studying the energies of localized and quasi-localized modes in solids. It may be applied to any material for which the neutron absorption cross section is not too high, and so is a valuable technique for the study of local states in metals. When optical methods can be applied, however, they are generally superior. The intensity of inelastically scattered neutrons is very low, so that the experiment takes a long time to perform and the energy resolution is poor. An uncertain background of coherently scattered neutrons must usually be subtracted from the results, and the scattered intensity from a localized mode is so weak that inconveniently high concentrations of impurities must be used.

The same limitations apply to the study of defect modes by coherent neutron scattering. They are, however, counterbalanced by the fact that such experiments are uniquely powerful for studying the dynamical behaviour of the defect and its surroundings, since the dependence of the scattered intensity on $\underline{\kappa}$ and ω measures directly the variation in space and time of the deviations of the ions or spins from their equilibrium configurations.

Of the four types of mode which can be observed by this technique, detailed results have only been published on resonant phonons and localized magnons. The limited results on Cu-3% Au are in quantitative agreement with the mass defect theory, but Cu-9% Au does not show clearly even the predicted qualitative effects. Cr-3% W behaves qualitatively as expected, but there are large quantitative discrepancies with the mass defect theory, probably due to force constant changes. There is good agreement between theory and experiment for the behaviour of the localized magnon mode in MnF_2-5% Co. Extensive preliminary measurements have been made on localized phonons in Cu-10% Al, and a resonant magnon mode has been observed in Tb-10% Ho. With the advent of reactors of higher flux, such experiments will become progressively easier, and we may expect them to make a large contribution to the study of the detailed behaviour of localized excitations in solids.

ACKNOWLEDGEMENTS

We have benefitted from discussions with R. J. Elliott, W. M. Hartmann, Yu. M. Kagan, D. W. Taylor and J. M. Rowe. E. C. Svensson kindly provided us with his unpublished experimental and theoretical results. We are also grateful to R. M. Nicklow and R. A. Cowley for communicating unpublished results to us. The assistance of J. C. G. Houmann, M. Nielsen, D. Heidebo and W. Kofoed in the experiments reported here is gratefully acknowledged.

REFERENCES

1. L. Van Hove, Phys. Rev. 95, 1374 (1954).

2. M. A. Krivoglaz, Zh. Eksperim. i Teor. Fiz. 40, 567 (1961); V. N. Kashcheev and M. A. Krivoglaz, Fiz. Tverd. Tela 3, 3167 (1961); M. A. Ivanov and M. A. Krivoglaz, ibid. 6, 200 (1964).

3. Yu. M. Kagan and Ya. A. Iosilevskii, Zh. Eksperim. i Teor. Fiz. 42, 259 (1962); ibid. 44, 1375 (1963).

4. R. J. Elliott and A. A. Maradudin, in Inelastic Scattering of Neutrons, Vol. 1 (I.A.E.A., Vienna, 1965).

5. I. P. Dzyub, Fiz. Tverd. Tela 6, 3691 (1964).

6. I. A. Privorotskii, Zh. Experim. i Teor. Fiz. 47, 1544 (1964).

7. A. A. Maradudin, in Solid State Physics, Vol. 18 (Academic Press Inc., New York, 1966).

8. R. J. Elliott and D. W. Taylor, Proc. Roy. Soc. A296, 161 (1967).

9. E. C. Svensson, Ph.D. Thesis, McMaster University (1967).

10. E. C. Svensson, B. N. Brockhouse and J. M. Rowe, Solid State Commun. 3, 245 (1965).

11. H. Bjerrum Møller and A. R. Mackintosh, Phys. Rev. Letters 15, 623 (1965).

12. H. Bjerrum Møller, Thesis submitted to the University of Copenhagen (1967).

13. E. C. Svensson and B. N. Brockhouse, Phys. Rev. Letters 18, 858 (1967).

14. S. N. Behera and B. Deo, Phys. Rev. 153, 728 (1967)

15. H. Bjerrum Møller and A. R. Mackintosh, in Inelastic Scattering of Neutrons, Vol. 1, (I.A.E.A., Vienna, 1965).

16. S. H. Chen and B. N. Brockhouse, Solid State Commun 2, 73 (1964).

17. N. A. Chernoplekov and M. G. Zemlyanov, Zh. Eksperim. i Teor. Fiz. 49, 449 (1965).

18. N. A. Chernoplekov, G. Kh. Panova, M. G. Zemlyanov, B. N. Samoilov and V. I. Kutaitsev, Phys. Stat. Sol. 20, 767 (1967).

19. B. Mozer, K. Otnes and C. Thaper, Phys. Rev. 152, 535 (1966).

20. B. Mozer, K. Otnes and V. W. Meyers, Phys. Rev. Letters, 8, 278 (1962).

21. R. Rubin, J. Peretti, G. Verdon, and W. Kley, Phys. Letters 14, 100 (1965).

22. R. M. Nicklow, private communication.

23. T. Wolfram and J. Callaway, Phys. Rev. 130, 2207 (1963).

24. Yu. A. Izyumov, Proc. Phys. Soc. 87, 521 (1966).

25. S. W. Lovesey, Proc. Phys. Soc. 91, 658 (1967).

26. Yu. A. Izyumov and M. V. Medvedev, Zh. Eksperim. i Teor. Fiz. 48, 574 (1966).

27. N. Kroo and L. Bata, Phys. Letters 24, A22 (1967); see also N. Kroo and L. Pal, Proceedings of the International Congress on Magnetism, Boston, 1967 (to be published).

28. W. J. L. Buyers, R. A. Cowley, T. M. Holden and R. W. Stevenson, Proceedings of the International Congress on Magnetism, Boston, 1967 (to be published).

29. H. Bjerrum Møller, J. C. Gylden Houmann and A. R. Mackintosh, Phys. Rev. Letters 19, 312 (1967) and Proceedings of the International Congress on Magnetism, Boston, 1967 (to be published).

IMPURITY MODES OF VIBRATION IN VANADIUM BASED ALLOYS*

Bernard Mozer

Brookhaven National Laboratory

(Now at the National Bureau of Standards)

Incoherent inelastic neutron scattering measurements on disordered substitutional alloys of vanadium yield information on the frequency dependence of the "self" part of the displacement-displacement correlation function for the sum of the atoms in the alloy. The frequency response of the square of the displacement of each atom in these alloys is measured and thus one has direct evidence for any defect modes of vibration that arise upon alloying. All alloys were examined to see that the solute atoms were in solution; the small (111) Bragg peak observed in vanadium via neutron diffraction showed smaller integrated intensity as vanadium was alloyed. The number of atoms in solution in the vanadium matrix determined by neutron diffraction agreed with the number determined by chemical analysis to within the experimental error. The inelastic neutron scattering data were taken on the slow-chopper facility at the Brookhaven Graphite Research Reactor. The resolution of this instrument and the beryllium filtered incoming beam is such that vibrational excitations less than 0.025 eV are broadened by the spread in the incoming beam, 0.0025 eV, and excitations between 0.025 to 0.050 eV are broadened by a combination of time resolution and incoming beam resolution of 0.004 eV. Measurements were taken at room temperature to insure sufficient population of the vibrational modes and at a number of scattering angles, but only the 90° scattering angle data are reported here.

Neutron scattering data taken on the time-of-flight instrument are shown in Figs. 1-4 for several $Be_x V_{1-x}$ alloys. Figure 3

*Work performed under the auspices of the U.S. Atomic Energy Commission.

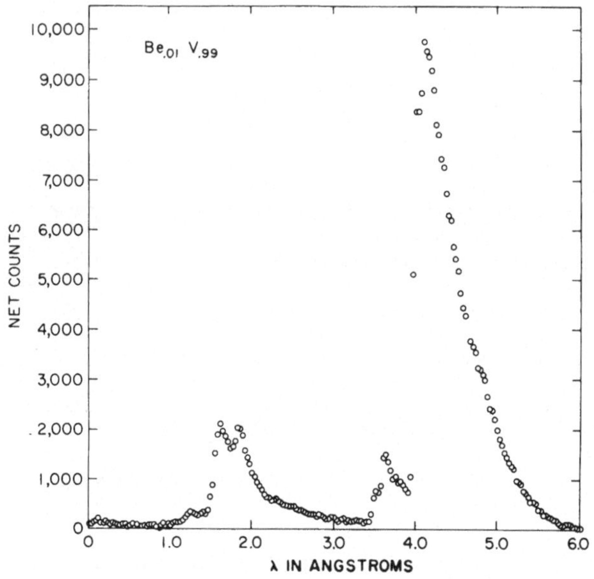

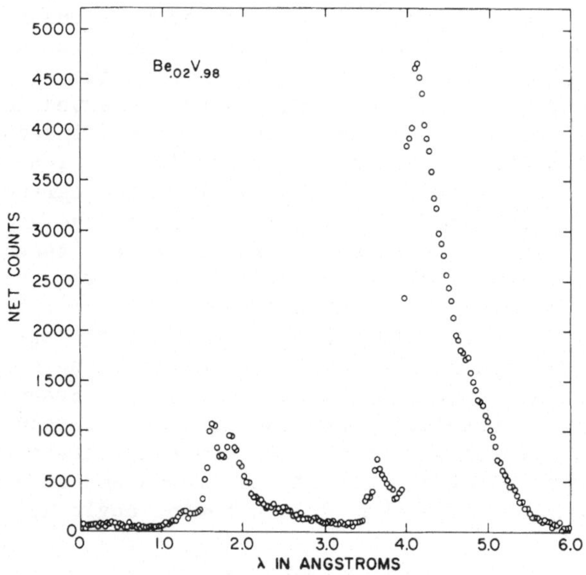

Figs. 1 and 2. Time-of-flight spectrum for two Be-V alloys consisting of raw data with a flat background subtracted. The incoming beam comes at 3.5 angstroms with the main part at 4 angstroms.

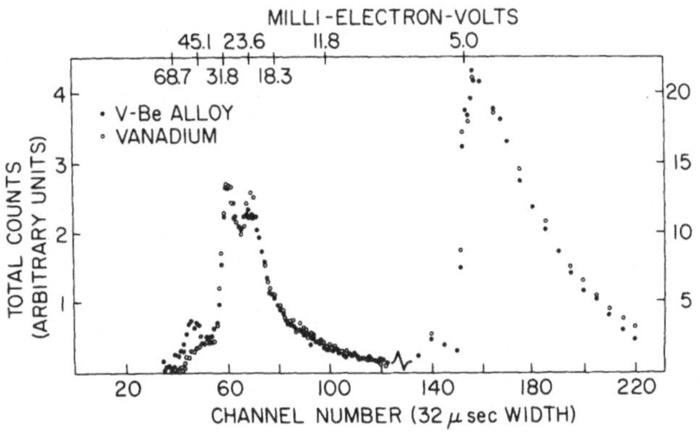

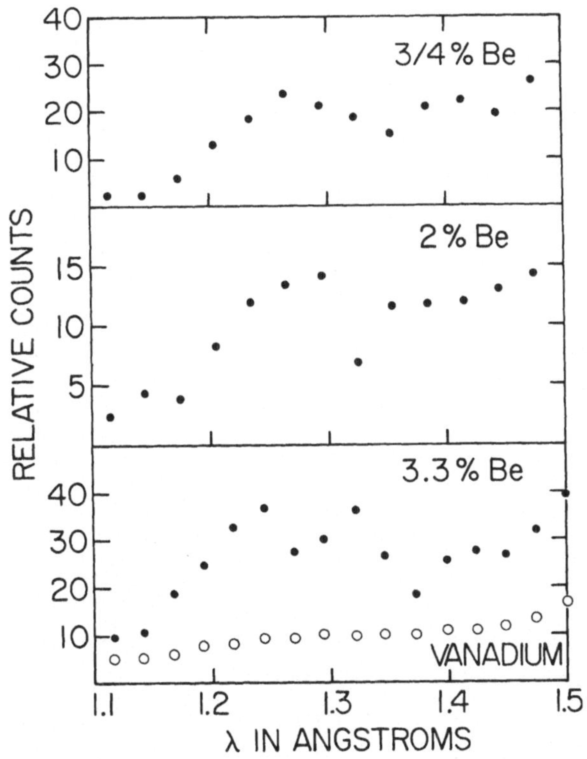

Figs. 3 and 4. Time-of-flight spectrum for $Be_{.033}V_{.967}$ and vanadium, and an expansion of the spectrum for the three alloys and vanadium. The figures are raw data and flat background subtracted.

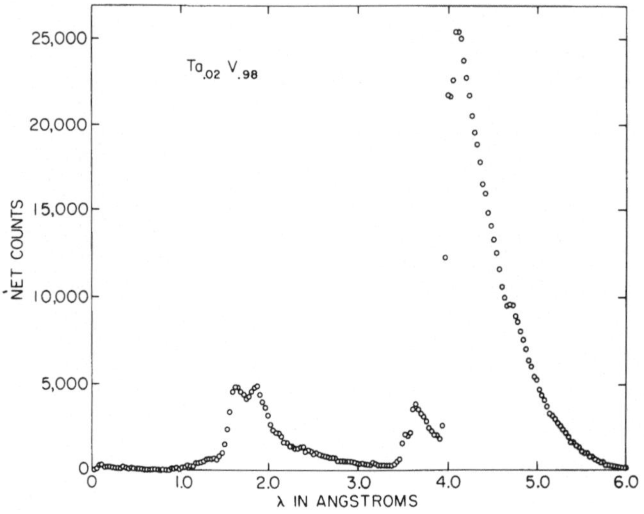

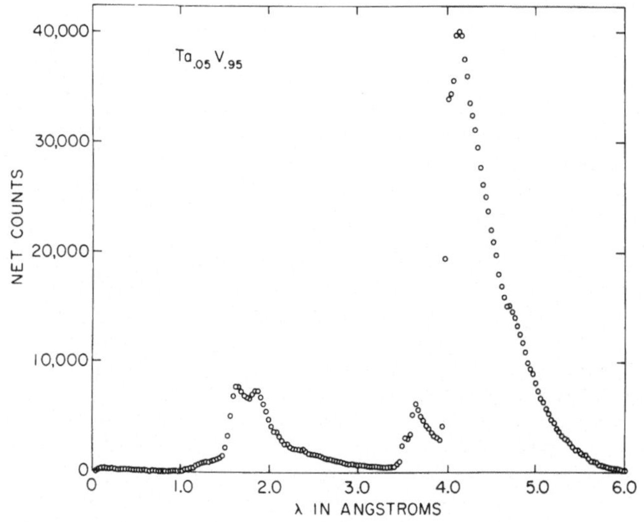

Figs. 5 and 6. Time-of-flight spectrum for Ta-V alloys with flat background subtracted from raw data.

shows one of the runs on the $Be_{.033}V_{.967}$ alloy and vanadium for comparison.[1] Figure 4 is an expanded scale of the data in the region of neutron wavelengths corresponding to an annihilation of a localized mode of vibration. Vibrational energies can be obtained from the data by subtracting 0.0035 eV, the average incoming energy, from the final neutron energy evaluated from $E = 0.082\ \lambda^{-2}$, where λ is in angstroms. The 3.3% data in Fig. 4 is a composite of many runs all of which exhibit the split peak. One notes at the lowest concentration only a single peak is observed; this peak is characteristic of localized vibrational modes of the beryllium atoms. The frequency of the localized mode is 0.048 eV and is well predicted by considering a single isotopic mass defect in vanadium. Increasing the concentration causes the peak to broaden, and at 3.3% the peak has split into two resolved peaks. The impurity band contribution to the density of states or displacement-displacement Green's function can be obtained from the data by correction factors including thermal occupation number, chopper transmission, etc.[1] The impurity band would be asymmetrical with the strongest peak at higher frequencies. Several theoretical explanations for this behavior are possible. Lifshitz[2] calculated the behavior of an isolated impurity band and noted that, when pair interactions were taken into account, an asymmetrical split peak was produced. Montroll calculated the splitting of the local mode peak into two symmetrical peaks for two impurities close together and suggests an asymmetry appears when his calculation is refined.[3] Evidence for asymmetrical splitting attributable to pairs of impurities was noted in calculations of Dean[4] and Payton and Visscher,[5] but more refined calculations are necessary for direct comparison. On the other hand, the Green's function technique of Langer,[6] Maradudin,[7] and Taylor[8] also indicate an asymmetric splitting of the impurity band. In these latter calculations any structure in the impurity band is only a reflection of the structure in the host lattice frequency distribution. Taylor's calculation is closest to reproducing the location and shape of the impurity band, but his attempt to add improvements by a self-consistent calculation causes a smearing of the spectra and a loss of structure. Here again no interaction between pairs is taken into account and no structure of the impurity band would occur for a structureless host spectrum. Thus our experiment and the other calculations emphasize the importance of pair interactions. One should also note the tendency for the alloy to vibrate with higher frequencies as indicated by the strengthening of the high frequency peak in the spectrum common to vanadium and these alloys.

Figures 5 and 6 show neutron scattering data for $Ta_{.02}V_{.98}$ and $Ta_{.05}V_{.95}$, respectively, in an additional search for resonant modes of vibration associated with the heavy mass metals. In contrast to the peak observed for light mass atoms, we see a

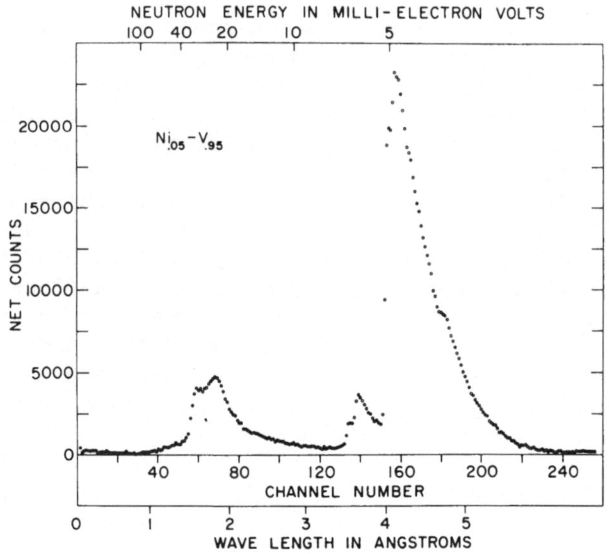

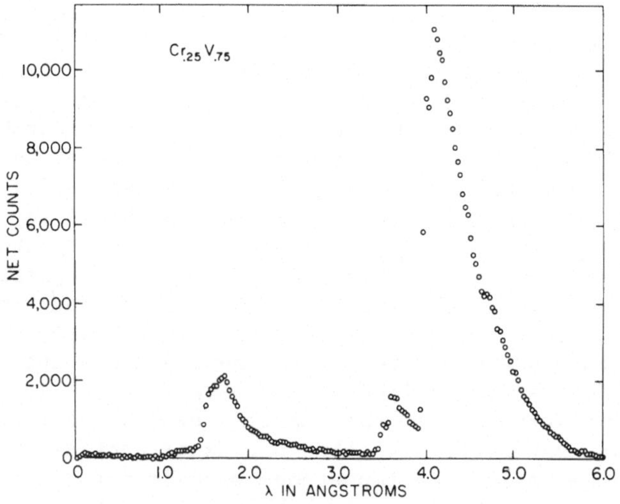

Figs. 7 and 8. Time-of-flight spectrum for $Ni_{.05}$ $V_{.95}$ and $Cr_{.25}$ $V_{.75}$ alloys with flat background subtracted from raw data.

spectrum almost identical to that of vanadium for these heavy mass alloys. Manipulation of the data by taking ratios of the alloy data to vanadium data or subtracting off a fraction of the vanadium data shows no large effects but only a tendency for the alloy to vibrate at lower frequencies by enhancing the peak near 2 angstroms. Even a broad peak, such as that observed in $Pt_{.05}V_{.95}$,[1] is not indicated in these alloys. Theoretical estimates for the effect to be observed for either a disordered alloy[5],[7] or a single isotope and its nearest neighbors for these concentrations indicate at least a 20% and 50% effect for the two concentrations even when one includes the additional broadening by the instrument. Our accuracy in the region of frequencies expected for the resonant mode is 4%, and we should have seen something at least one-half to one-fifth the theoretical estimates. It is possible that differences between tantalum ion cores and vanadium ion cores produce changes in the interatomic potentials which would lead to further broadening. Strong anharmonic effects and possibly pair interactions between impurities could also reduce the effect by broadening and splitting the "impurity band" but no calculations of such effects exist. Recently Chernoplekov and others[9] have reported similar type inelastic neutron scattering measurements on powder samples of Pb-Mg and U-Ti. Evidence for resonant modes of vibration can be seen in their data, but the effect is much smaller than predicted by theory and there is no indication of unusual broadening. Single crystal measurements of the width and shift of the perfect lattice phonons on alloying with heavy mass atoms of the same mass ratio we employed have been reported by Brockhouse et al.[10] and Möller and Mackintosh[10] with some agreement between theory and measurement and with no unusual broadening observed. Hence we are somewhat concerned that theories that predict width and shifts of the perfect lattice phonons cannot account for the observations in powder coherent samples or incoherent scatterers such as we measure.

Finally, we wish to present additional evidence for the type of long range interaction potential in alloys of solute atoms with different charge than solvent atoms. Previously[1] we showed that alloying nickel or platinum with vanadium produced shifts in the frequencies to higher values when the electron concentration was changed by 0.25 electron (5% alloy). Our conjecture was that the 5 electrons added to the system by the nickel or platinum would resemble the vibrational behavior of a Cr-V alloy of the same electron concentration. In Figs. 7 and 8 we present the time-of-flight data for $Ni_{.05}V_{.95}$ and $Cr_{.25}V_{.75}$ alloys. The vibrational spectra of the two alloys are essentially the same -- the peak positions are the same, as well as the overall structure. Thus our conjecture about the behavior of the nickel vanadium alloy is upheld. The behavior of the vibrational properties of alloys with such differing electronic configurations suggests that the electron states in this alloy are more like those expected from a rigid band model where nickel contributes 5 additional electrons per atom to the band. It appears then that the potential given by the difference in the nickel and vanadium cores is not sufficiently strong to bind the 5 additional electrons in bound states. The screening of this charge difference takes place via the band electrons and hence the interatomic potentials produced by these electrons are of long range.

References.

(1) B. Mozer, K. Otnes, and C. Thaper, Phys. Rev. 152, 535 (1966); see also M. G. Zemlyanov, V. A. Somenkov and N. A. Chernoplekov, Zh. Eksperim. i. Teor. Fiz. 52, 665 (1967).
(2) I. M. Lifshitz, Advan. Phys. 13, 483 (1964).
(3) E. Montroll, private communication.
(4) P. Dean and M. D. Bacon, Proc. Roy. Soc. (London) A283, 64 (1965).
(5) D. N. Payton and W. M. Visscher, Phys. Rev. 156, 1032 (1967).
(6) J. S. Langer, J. Math. Phys. 2, 548 (1961).
(7) A. A. Maradudin, Solid State Phys. 18, 316 (1966).
(8) D. W. Taylor, Phys. Rev. 156, 1017 (1967).
(9) N, A. Chernoplekov and M. G. Zemlyanov, Zh. Eksperim. i. Teor. Fiz. 49, 449 (1965), [English transl.: Soviet Phys. -- JETP 22, 315 (1966)]; N. A. Chernoplekov et al., Phys. Stat. Sol. 20, 767 (1967).
(10) See A. R. Mackintosh, this conference.

ANALYSIS OF ASSEMBLIES WITH LARGE DEFECT CONCENTRATIONS WITH SPECIAL APPLICATION TO THEORY OF DENATURATION OF COPOLYMERIC DNA

Elliott W. Montroll, University of Rochester*, and

Lee-Po Yu, University of Maryland*

I. GENERAL FORMALISM.

The aim of this paper is to present a formalism for the investigation of certain properties of periodic lattices with finite defect concentrations and to apply the general technique to the theory of denaturation of the linear chain Ising-like model of DNA.[1]

Let us consider a periodic lattice of any number of dimensions and suppose that defects exist at points $r_1, r_2, r_3, \ldots, r_m$. Our aim is to find the effect of these defects on some quantity which we identify by $F(r_1,r_2,\ldots,r_m) \equiv F(m)$ when m defects are located at the prescribed positions. For convenience we will employ periodic boundary conditions and, finally, we will be concerned with the limit of an infinite lattice in which as $m \to \infty$, the defects appear in concentration c. The function F will be chosen to be "thermodynamic" in character in that it will be proportional to the number of lattice points in a large lattice. In the specific example which will interest us in the application of our general method, F will correspond to the number of hydrogen bonds broken under specified conditions in a DNA molecule. Another case to which the formulae might be applied is the magnetization of a magnetic material with impurities.

The formalism begins as though the defects are of low concentration so that they are independent of each other. Then the correlation of pairs of defects, triples of defects, etc., are taken into account. When correlations are weak, the process converges rapidly; when they are strong, it converges slowly. The formalism is in the spirit of the Mayer cluster integral formalism and is not unlike a number of other proposals which lead to power series in defect concentrations. Sometimes it is possible to find F(m) when the defects are periodically distributed. In this case we will show how the convergence of our expansion in terms of increasing order of correlations can be improved by successively approximating arbitrary defect distributions by periodic ones.

We start with the sequence $F(0)$, $F(r_1)$, $F(r_1r_2),..,$ $F(r_1,..,r_m)$ and define a set of differences

$$\Delta(r_1) \equiv F(r_1)-F(0) \tag{1a}$$

$$\Delta(r_1r_2) \equiv F(r_1r_2)-F(r_1)-F(r_2)+F(0) \tag{1b}$$

$$\Delta(r_1r_2r_3)=F(r_1r_2r_3)-F(r_1r_2)-F(r_2r_3)-F(r_3r_1)+F(r_1)+F(r_2)+F(r_3)-F(0) \tag{1c}$$

$$\begin{aligned}\Delta(r_1r_2r_3r_4)=&F(r_1r_2r_3r_4)-F(r_1r_2r_3)-F(r_1r_2r_4)-F(r_1r_3r_4)-F(r_2r_3r_4)+\\&+F(r_1r_2)+F(r_1r_3)+F(r_1r_4)+F(r_2r_4)+F(r_3r_4)+F(r_2r_3)-\\&-F(r_1)-F(r_2)-F(r_3)-F(r_4)+F(0)\end{aligned} \tag{1d}$$

The Δ's have the property that in a dilute system, each Δ vanishes. These equations can be inverted so that the F's can be expressed in terms of the Δ's. Hence

$$F(1) = F(0) + \Delta(r_1) \tag{2a}$$

$$\begin{aligned}F(2) &= F(0) + [F(r_1)-F(0)] + [F(r_2)-F(0)] + \Delta(r_1r_2)\\&= F(0) + \Delta(r_1) + \Delta(r_2) + \Delta(r_1r_2)\end{aligned} \tag{2b}$$

$$F(3) = F(0) + \sum_{j=1}^{3} \Delta(r_j) + \sum_{j=1}^{3} \Delta(r_jr_{j+1}) + \Delta(r_1r_2r_3) \tag{2c}$$

$$\begin{aligned}F(4) = F(0) + \sum_{j=1}^{4} \Delta(r_j) + \sum_{j=1}^{3}\sum_{k=j+1}^{4} \Delta(r_jr_k) + \sum_{j=1}^{4} \Delta(r_jr_{j+1}r_{j+2}) +\\ + \Delta(r_1r_2r_3r_4)\end{aligned} \tag{2d}$$

Generally

$$F(m) = F(0) + S_1+S_2+... \tag{3a}$$

where

$$S_1 = \sum_{j=1}^{m} \Delta(r_j) \quad , \qquad S_2 = \sum_{j=1}^{m-1}\sum_{k=j+1}^{m} \Delta(r_jr_k) \tag{3b}$$

$$S_3 = \sum_{j=1}^{m-2}\sum_{k=j+1}^{m-1}\sum_{\ell=j+2}^{m} \Delta(r_jr_kr_\ell) \quad \text{etc.} \tag{3c}$$

Now let us suppose that the function F(m) can be expressed as a linear operation on the logarithm of partition function of an assembly of m defects; i.e.,

$$F(r_1..r_m) = \mathcal{O}\{\log Z(r_1..r_m)\} \quad , \tag{4a}$$

$\mathcal{O}$ representing the appropriate linear operator. Then if

$Z(0) \equiv$ partition function of perfect lattice,

$Z(r_j) \equiv$ partition function with single defect at r_j,

$Z(r_jr_k) \equiv$ partition function with one defect at r_j and another at r_k, etc.

$$F(0) = \mathcal{O}\{\log Z(0)\} \tag{4b}$$

$$F(r_j) = \mathcal{O}\{\log Z(r_j)\}, \text{ etc.} \tag{4c}$$

With these definitions

$$\begin{aligned}\Delta(r_j) &= F(r_j)-F(0)\\ &= \mathcal{O}[\{\log Z(r_j)\} - \{\log Z(0)\}]\\ &= \mathcal{O}\{\log[Z(r_j/Z(0)]\}\end{aligned} \tag{5a}$$

$$\begin{aligned}\Delta(r_jr_k) &= F(r_jr_k)-F(r_j)-F(r_k)+F(0)\\ &= \mathcal{O}\{\log[Z(r_jr_k)Z(0)/Z(r_j)Z(r_k)]\}\end{aligned} \tag{5b}$$

$$\Delta(r_jr_kr_\ell)=\mathcal{O}\{\log[Z(jk\ell)Z(j)Z(k)Z(\ell)/Z(jk)Z(k\ell)Z(j\ell)Z(0)]\} \tag{5c}$$

where

$$Z(r_j) \equiv Z(j);Z(r_j,r_k) \equiv Z(j,k) \text{ etc.} \tag{6}$$

Also

$$\Delta(r_jr_kr_\ell r_m)=\mathcal{O}\left\{\log \frac{Z(jk\ell m)Z(jk)Z(j\ell)Z(jm)Z(k\ell)Z(km)Z(\ell m)Z(0)}{Z(jk\ell)Z(k\ell m)Z(jkm)Z(j\ell m)Z(j)Z(k)Z(\ell)Z(m)}\right\} \text{ etc.} \tag{7}$$

If equations (3) and (5) are combined, we find

$$S_1 = \sum_{j=1}^{m} \mathcal{O}\{\log[Z(r_j)/Z(0)]\} \tag{8a}$$

$$S_2 = \sum_{j=1}^{m-1} \sum_{k=j+1}^{m} \mathcal{O}\{\log[Z(r_jr_k)Z(0)/Z(r_j)Z(r_k)]\} \text{ etc.} \tag{8b}$$

Hence

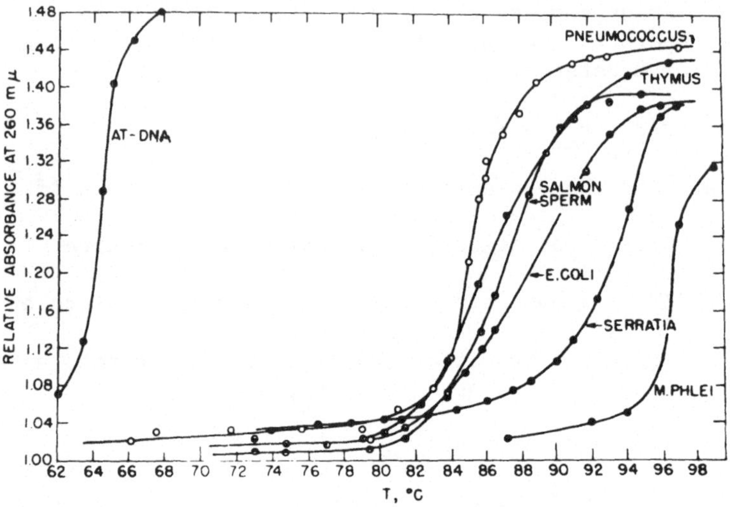

Fig. 1. Absorbence (260 mμ) vs. temperature curves for various samples of DNA in 0.15 NaCl+ 0.015M $Na_3C_6H_5O_7$.

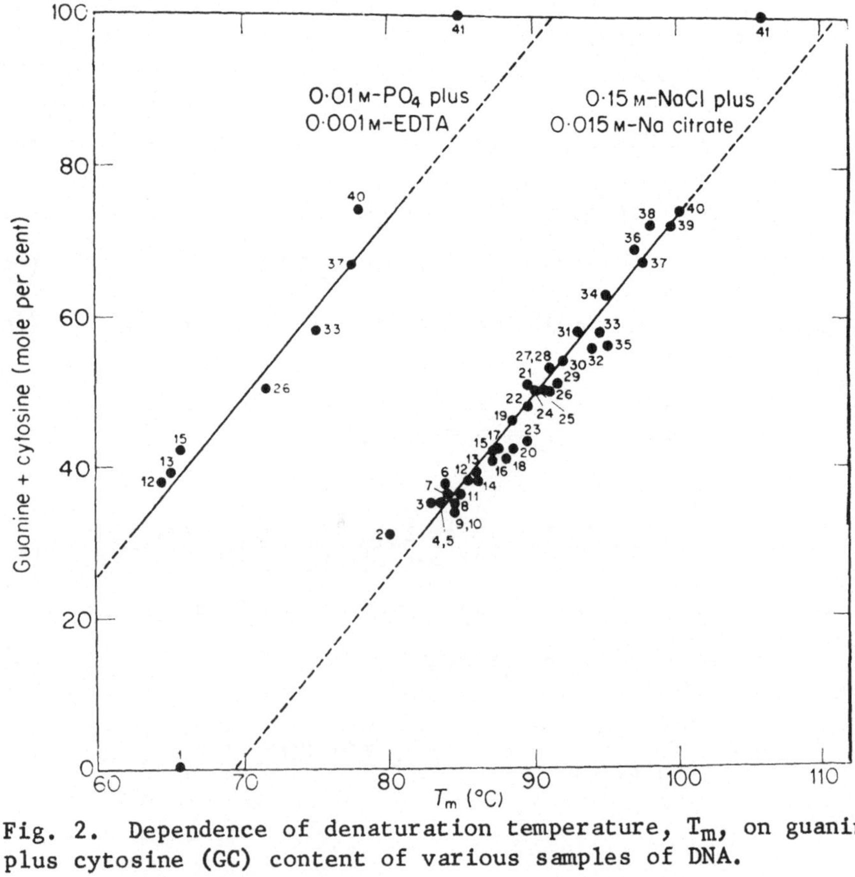

Fig. 2. Dependence of denaturation temperature, T_m, on guanine plus cytosine (GC) content of various samples of DNA.

$$F(m) = F(0) + \sum_{j=1}^{m} \mathcal{O}\{\log[Z(r_j)/Z(0)]\} + \ldots \tag{9}$$

Now consider two distributions of m defects, one whose F(m) value is $F^{(1)}(m)$ and whose defects are at $r_1^{(1)}, r_2^{(1)}, \ldots, r_m^{(1)}$, and the other whose F(m) value is $F^{(2)}(m)$ and whose defects are at $r_1^{(2)}, \ldots, r_m^{(2)}$. Then

$$F^{(1)}(m) - F^{(2)}(m) = \sum_{j=1}^{m-1} \sum_{k=j+1}^{m} \{\Delta(r_j^{(1)}, r_k^{(1)}) - \Delta(r_j^{(2)}, r_k^{(2)})\} + \ldots \tag{10}$$

since the zero and first order terms are independent of the defect distribution. If one scans the possible (j,k) pairs, a certain number will be nearest neighbor lattice points, some the next larger possible distance, etc. In many problems $\Delta(r_i, r_j) \to 0$ rapidly as $|r_i - r_j| \to \infty$. Hence if the defect concentration is small, the sum on the right of (10) converges rapidly when defect distributions (1) and (2) have the same number of defect pairs which are at nearest neighbor lattice points, the same number of second neighbor pairs and, perhaps, the same number of third order pairs. Since in some applications the function F(m) can be found easily for periodic two-component assemblies, one can then choose distribution (2) to be periodic so that (10) becomes a rapidly convergent expression for $F^{(1)}(m)$ for a more random distribution.

We now apply the scheme discussed above to a theory of denaturation of DNA molecules with a random distribution of base pairs.

2. NEAREST NEIGHBOR CORRELATION MODEL OF DNA

The DNA molecule has a double helix structure. Each strand of the helix contains a sequence of bases chosen from the set of four, adenine (A), thymine (T), cytosine (C), and guanine (G). Several thousand of these bases are on each strand of a naturally occurring DNA molecule in some more or less random order. A and T bases appear bonded together (as do the G and C bases) on the double helix with an A on one helix being bonded to a T on the other by two hydrogen bonds. A G on one helix is always bonded to a C on the other with three hydrogen bonds.

Long DNA molecules can be extracted from the chromosomes of the organism of which they are a part and can be investigated as solutes in solutions at various temperatures. If the temperature is increased sufficiently above room temperature, the hydrogen bonds which hold the strands of the molecule together start to break. This is detected by observing the magnitude of certain peaks in the absorption spectrum. When the absorbence at 260 mμ is measured as a function of temperature for various samples of DNA, one obtains curves similar to those plotted in Fig. 1.

The fraction of the way one has progressed from the low temperature to the high temperature level of the absorbence curve is identified with the average fraction of base pairs whose hydrogen bonds have been broken. The "melting point" of the molecule is that temperature at which half the hydrogen bonds have been broken. As shown in Fig. 2, the melting point varies linearly with the fraction of GC pairs in the molecule. A chain containing only AT pairs under certain pH and ionic strength conditions melts at 67°C while one containing only GC pairs would, under the same conditions, melt at 106°C.

There seems no doubt that the breaking of the hydrogen bonds in base pairs is a cooperative effect.[2] If a single base pair becomes unbonded, it is easier for the neighboring pairs to become unbonded, etc. We examine this cooperative effect first by considering a synthetic DNA chain which is either of the pure AT or pure GC type, and then generalize our ideas to the copolymer case in which both types of base pairs exist on the chain. We introduce a sequence of parameters $\sigma_1, \sigma_2, \ldots, \sigma_N$ such that $\sigma_j = 1$ if the bonding complex which holds the jth base pair together is intact and $\sigma_j = -1$ if it is broken. A separated "bonding complex" will be called one whose "bond" is broken even though several bonds are involved. It will be assumed that the hydrogen bonds associated with a given pair are all broken or all intact.

Under any specified condition, the number of "bonds" intact is

$$N(\sigma_1..\sigma_N) = \tfrac{1}{2}\sum_{i=1}^{N}(1+\sigma_i) \tag{1}$$

so that, statistically, the fraction of bonds intact is

$$f = \tfrac{1}{2}\sum_{i\sigma}\sum(1+\sigma_i)F(\sigma_1..\sigma_N)/N \tag{2}$$

where $F(\sigma_1..\sigma_N)$ is the statistical weight assigned to the sequence $\sigma_1,..,\sigma_N$. The simplest model which leads to a cooperative chain breaking process is one in which only nearest neighbor bond pairs influence each other directly. The most general nearest weight function with nearest neighbor correlations only, is (since $\sigma^2=1$)

$$F(\sigma_1..\sigma_N) = Z^{-1}\prod_{i=1}^{N} f(\sigma_i\sigma_{i+1}) \tag{3}$$

where

$$f(\sigma,\sigma') = \exp\left[U\sigma\sigma' - \tfrac{1}{2}J(\sigma+\sigma')\right] \tag{3a}$$

and

$$Z = \sum_{\sigma_1=\pm1}\ldots\sum_{\sigma_N=\pm1}\prod_{i=1}^{N} f(\sigma_i\sigma_{i+1}) \quad , \tag{3b}$$

a normalizing factor which we call the partition function of our chain. We have, for simplicity, assumed that the ends of our chain are bound together to form a ring $\sigma_1 \equiv \sigma_{N+1}$. The function Z has the same mathematical form as the standard statistical mechanical partition function of the Ising model of a one-dimensional ferromagnet and can be so analyzed.[3]

If we let F be a 2X2 matrix with matrix elements $f(\sigma,\sigma')$, then

$$Z = \text{trace } F^N = \lambda_1^N + \lambda_2^N \tag{4}$$

where λ_1 and λ_2 are the characteristic values of the matrix F. As $N \to \infty$,

$$Z \sim \lambda_{max}^N \quad \text{where } \lambda_{max} = \max(\lambda_1,\lambda_2). \tag{5}$$

By definition λ_1 and λ_2 are roots of the equation

$$\begin{vmatrix} e^{U-J} - \lambda & e^{-U} \\ e^{-U} & e^{U+J} - \lambda \end{vmatrix} = 0 \tag{6}$$

the larger being

$$\lambda_{max} = e^U \cosh J + e^U |\sinh J| (1+e^{-4U} \operatorname{csch}^2 J)^{\frac{1}{2}} \quad . \tag{7}$$

As $N \to \infty$,

$$\frac{\partial \log Z}{\partial J} = -\tfrac{1}{2} \sum_{j=1}^{\infty} \langle \sigma_j + \sigma_{j+1} \rangle$$

$$= -\langle \Sigma \sigma_j \rangle = N \frac{\partial \log \lambda_{max}}{\partial J} \tag{8}$$

Hence for a typical σ_j (since all σ_j are equivalent because of our circular arrangement of bonded pairs)

$$\langle \sigma \rangle = -\frac{1}{\lambda_{max}} \frac{\partial \lambda_{max}}{\partial J} \tag{9}$$

$$= -\frac{\sinh J}{(\sinh^2 J + e^{-4U})^{\frac{1}{2}}}$$

From (2) the fraction of bonds intact is[1]

$$f = \tfrac{1}{2}(1 + \langle \sigma \rangle)$$

$$= \tfrac{1}{2}\left\{1 - \frac{\sinh J}{(\sinh^2 J + e^{-4U})^{\frac{1}{2}}}\right\} \tag{10}$$

and the fraction of bonds broken is

$$1\text{-}f = \tfrac{1}{2}\left\{1 + \frac{\sinh J}{(\sinh^2 J + e^{-4U})^{\frac{1}{2}}}\right\} \quad . \tag{11}$$

If one sets

$$J = a(T - T_o) \quad , \tag{12}$$

where T_o is the melting point of our model and a is a constant, then the variation of (1-f) with temperature is that observed experimentally. The slope of our theoretical melting curve at $T = T_o$ is

$$-\partial f/\partial T = -\tfrac{1}{2}a \exp 2U. \tag{13}$$

As discussed in reference 1, experimental melting curves can be fitted by choosing a to be about 0.018°C and U in the range 1.5 - 2.0. Note that as $U \rightarrow 0$ (bonds breaking in an uncorrelated manner)

$$1 - f \rightarrow \tfrac{1}{2}\,[1 + \tanh a(T - T_o)] \tag{14a}$$

while as $U \rightarrow \infty$ (bonds are strongly correlated)

$$1 - f = \tfrac{1}{2}\,[1 + \text{sign } a(T - T_o)] \tag{14b}$$

which is a step function.

The slope of the melting curve of copolymeric DNA (which contains both AT and GC pairs) depends on the base sequence as well as on the coupling parameters U and a. To see this, consider the two sequences

```
    A  A        A  G  G  . .  G
a)  |  |  . . . |  |  |  . .  |
    T  T        T  C  C       C

    A  G  A  G        A  G
b)  |  |  |  |  . . . |  |
    T  C  T  C        T  C
```

which contains 50% AT and 50% GC pairs. The melting curve of (a) would be flat near the melting point because the A-T pairs would tend to break at the melting temperature of pure AT-DNA; then few other bonds would break until one reached the temperature at which GC pairs break. Hence a long flat range would exist in the melting curve. On the other hand, in case (b), the AT pairs would try to break but would be clamped by neighboring GC pairs until a sufficiently high temperature was reached so that all the AT's would work together to break the GC bonds. Then all bonds would break in a very narrow temperature range so that the melting curve would be sharp.

In order to unravel the effect of the base sequence from the correlation contribution to the melting curves of naturally occurring DNA, one can employ recently synthesized DNA with known periodic base pair sequences to determine the correlation parameters. Then with these parameters, one might deduce some statistical properties of natural DNA sequences from the melting curves of the natural DNA. For this purpose, techniques must be developed for the calculation of a melting curve for a given sequence of base pairs.

3. MELTING CURVES OF COPOLYMER DNA

i. A General Formalism

Naturally occurring copolymer DNA is composed of both AT and GC pairs. For empirical reasons discussed in references 1 and 4, the parameter U does not seem to vary much between (AT,AT), (GC,GC) and (AT,GC) nearest neighbor correlations. A discussion of how one might further check this point experimentally by employing synthetic periodic copolymers is given in reference 5. Partly for simplicity and partly because it may not be too unrealistic, we assume that only the parameter J varies between AT and GC pairs and, indeed, that we can write

$$J_1 = a(T-T_1) \tag{15a}$$

$$J_2 = a(T-T_2) \tag{15b}$$

where J_1 refers to the AT J value and T_1 is the melting temperature of pure AT-DNA, while J_2 refers to the GC J value and T_2 is the melting temperature of pure GC-DNA. Another simplifying assumption is that the pair correlation between AT and GC is the same as that between AT and CG, etc.

With these assumptions we consider a DNA molecule which can be pictured schematically as a ladder as is done in Fig. 3 where we have "unwound" the double helix for ease in drawing, and, second, to identify successive bond pairs and rungs on the ladder. We will refer to the GC pairs as defects or abnormal pairs introduced into a matrix of normal AT pairs.

A typical copolymer is exhibited in Fig. 3 with an abnormal bond at rung 1, a second n_1 rungs above, a third n_2 rungs above the second, etc. The partition function of this ladder is

$$Z = \sum_{\sigma_1=\pm1,\ldots\sigma_N=\pm1} \cdots \sum \exp-\{J_2\sigma_1+J_1\sigma_2+J_1\sigma_3+J_1\sigma_4+J_1\sigma_5+J_1\sigma_6+\ldots\}$$

$$\times \exp U(\sigma_1\sigma_2+\sigma_2\sigma_3+\ldots+\sigma_N\sigma_1) \tag{16}$$

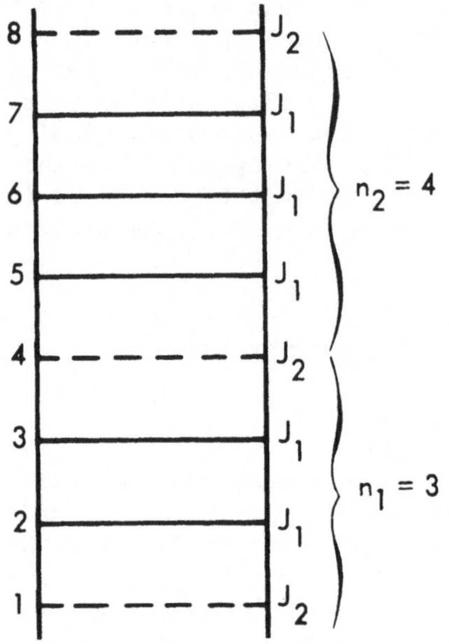

Fig. 3. Ladder representation of double-stranded molecule.

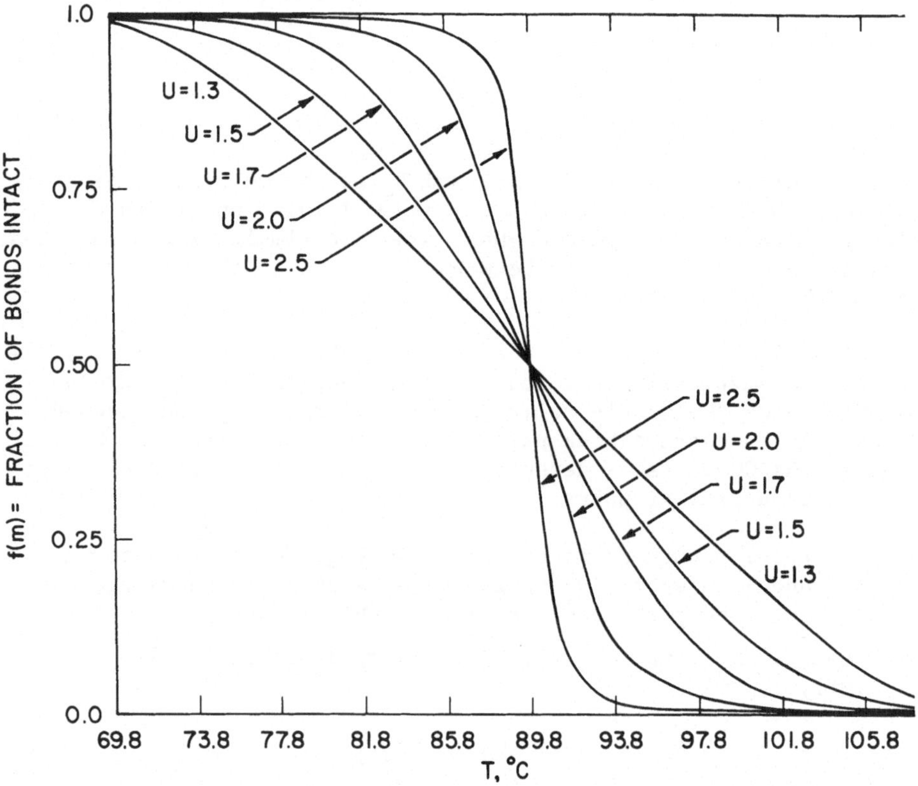

Fig. 4. Fraction of bonds intact versus temperature for random distribution of bonds in a chain with half AT and half GC pairs.

Using the well-known relation between elements of a matrix and its characteristic vectors and characteristic values, we find

$$\exp[U\sigma_1\sigma_2-\tfrac{1}{2}J_1(\sigma_1+\sigma_2)] = \sum_{j=1}^{2}\lambda_j\psi_j(\sigma_1)\psi_j(\sigma_2) \tag{17a}$$

$$\sum_{\sigma_2}\exp\{U\sigma_1\sigma_2-\tfrac{1}{2}J_1(\sigma_1+\sigma_2) + U\sigma_2\sigma_3-\tfrac{1}{2}J_1(\sigma_2+\sigma_3)\} =$$

$$= \sum_{j=1}^{2}\lambda_j^2\psi_j(\sigma_1)\psi_j(\sigma_3) \text{ etc.} \tag{17b}$$

Hence if we let[1]

$$F(\sigma,\sigma';n) = \Sigma\lambda_j^n\psi_j(\sigma)\psi_j(\sigma') \tag{18}$$

and remember that there are m defect bonds with successive separation distances $n_1,n_2,\ldots,n_m$,

$$Z = \sum_{\sigma_1=\pm1}\ldots\sum_{\sigma_m=\pm1} e^{-\epsilon(\sigma_1+..+\sigma_m)}F(\sigma_1\sigma_2;n_1)F(\sigma_2\sigma_3;n_2)\ldots F(\sigma_m\sigma_1;n_m) \tag{19}$$

where

$$\epsilon = J_2-J_1 \tag{20}$$

In deriving (19) we summed over all σ in (16) whose coefficients in the exponent were J and we employed (18) to eliminate these variables. Then we renumbered the bonds of type 2 by letting the first, (which is 1 in Fig. 3) be 1, the second (which is 4 in Fig. 3) be 2, the third (which is 8 in Fig. 3) be 3, etc. Finally, if the total number of bonds is N

$$n_1 + n_2 + \ldots + n_m = N \tag{21}$$

Equation (19) can also be rewritten as

$$Z = \sum_{j_1,j_2..=1,2}\lambda_{j_1}^{n_1}\lambda_{j_2}^{n_2}\lambda_{j_3}^{n_3}\ldots(j_1,j_2)(j_2,j_3)\ldots(j_m,j_1) \tag{22}$$

where

$$(j,k) = \sum_{\sigma}\psi_j(\sigma)e^{-\epsilon\sigma}\psi_k(\sigma) \tag{23}$$

It can be shown that (see reference 1 where the detailed properties of the characteristic vectors of f are discussed)

$$(1,2)^2 = e^{-4U}(\sinh^2\epsilon)/(s_1^2 + e^{-4U}) \tag{24a}$$

where

$$s_\alpha \equiv \sinh J_\alpha \tag{24b}$$

We also use the following abbreviations

$$\omega = \lambda_2/\lambda_1 \tag{25}$$

$$a_{ij} \equiv (i,j)/(1,1) \text{ and } b_{ij} = (i,j)/(2,2) \tag{26}$$

In the one defect case $n_1 = N$ (remembering that we are concerned with a "circular" molecule)

$$Z(r) = \sum_{j=1}^{2} \lambda_j^N (j,j) = (1,1)\lambda_1^N + (2,2)\lambda_2^N \tag{27a}$$

$$\sim (1,1)\lambda_1^N \quad \text{as } N \to \infty. \tag{27b}$$

We put this in a form that will come naturally for higher order Z's by setting

$$Z(r) \sim (1,1)\lambda_1^N h_1 \text{ with } h_1 \equiv 1. \tag{28}$$

In the case of two defects, the jth and k^{th} chosen from the set 1,2,...,m

$$Z(r_j,r_k) = \sum_{j_1=1}^{2} \sum_{j_2=1}^{2} \lambda_{j_1}^{n_{jk}} \lambda_{j_2}^{N-n_{jk}} (j_1,j_2)(j_2,j_1)$$

$$= (1,1)^2 \lambda_1^N f_2(n_{jk}) \tag{29a}$$

where

$$n_{jk} = n_j + n_{j+1} + \ldots + n_{k-1} \tag{29b}$$

so that

$$n_{jk} + n_{k\ell} = n_{j\ell} \tag{30}$$

and

$$f_2(t_1 t_2) = \left[1 + a_{12}a_{21}\left(\omega^{t_1} + \omega^{t_2}\right) + \omega^N a_{22}^2\right] = f_2(t_1, N-t_1) = f_2(N-t_2, t_2) \tag{31}$$

If

$$0 < t_1 < \alpha N \text{ with } 0 < \alpha < 1, \tag{32a}$$

$$f_2(t_1 t_2) \sim \left(1 + a_{12}^2 \omega^{t_1}\right) \equiv h_2(t_1) \quad \text{as } N \to \infty \tag{32b}$$

so that

$$Z(r_j r_k) \sim (1,1)^2 \lambda_1^N h_2(t_1) \tag{33}$$

The 3-defect partition function is

$$Z(r_j r_k r_\ell) = \sum_{j_1=1}^{2} \sum_{j_2=1}^{2} \sum_{j_3=1}^{2} \lambda_{j_1}^{t_1} \lambda_{j_2}^{t_2} \lambda_{j_3}^{t_3} (j_1 j_2)(j_2 j_3)(j_3 j_1)$$

$$= (1,1)^3 \lambda_1^N f_3(t_1 t_2 t_3) \tag{34}$$

where

$$t_1 = n_{jk},\ t_2 = n_{k\ell},\ t_3 = n_{\ell j}$$

$$t_1+t_2+t_3 = n_{jk} + n_{k\ell} + n_{\ell k} = N \tag{35}$$

and

$$f(t_1 t_2 t_3) = 1 + a_{12}a_{21}\left(\omega^{t_1}+\omega^{t_2}+\omega^{t_3}\right) + a_{12}a_{22}a_{21}\left(\omega^{t_1+t_2}+\omega^{t_2+t_3}+\omega^{t_3+t_1}\right)+a_{22}^3\omega^N \tag{36}$$

Again, if

$$0<t_1<N\alpha_1\ ,\quad 0<t_2<N\alpha_2 \quad \text{with } 0 < \alpha_j < 1,\ \ 1 = 1,2 \tag{37a}$$

then as $N \to \infty$, $t_3 = N(1-\alpha_1-\alpha_2) \to \infty$ and

$$f_3(t_1 t_2 t_3) \sim 1 + a_{12}a_{21}(\omega^{t_1}+\omega^{t_2}) + a_{12}a_{22}a_{21}\omega^{t_1+t_2} \equiv h_3(t_1 t_2) \tag{37b}$$

so that

$$h_3(t_1 t_2) = \frac{h_2(t_1)h_2(t_2)}{h_1}\left\{1 + \frac{a_{12}a_{21}(a_{22}-a_{21}a_{12})\omega^{t_1+t_2}}{h_2(t_1)h_2(t_2)}\right\} \tag{38}$$

and

$$Z(r_j r_k r_\ell) \sim (1,1)^3\lambda_1^N h_3(t_1,t_2) \tag{39}$$

Proceeding in the same way, we find that

$$Z(r_j r_k r_\ell r_m) = (1,1)^4\lambda_1^N f_4(t_1 t_2 t_3 t_4) \tag{40}$$

where

$$t_1 \equiv n_{jk}, t_2 \equiv n_{k\ell}, t_3 \equiv n_{\ell m}, t_4 \equiv n_{mj},\ \Sigma_1^4 t_j = N \tag{41}$$

and

$$f_4(t_1 t_2 t_3 t_4) = 1+a_{12}a_{21}\sum_{\nu=1}^{4}\omega^{t_\nu}+a_{12}^2 a_{21}^2\left(\omega^{t_1+t_3}+\omega^{t_2+t_4}\right) + + a_{12}a_{22}a_{21}\Sigma_1^4\omega^{t_j+t_{j+1}}+a_{12}a_{22}^2a_{21}\Sigma_1^4\omega^{t_j+t_{j+1}+t_{j+2}}+\omega^N a_{22}^4 \tag{42}$$

Again, if

$$0 < t_j < N\alpha_j\ ,\quad j = 1,2,3;\ \text{and } 0 < \Sigma_1^3\alpha_j < 1 \tag{43}$$

then, as $N \to \infty$, $t_4 = N(1 - \Sigma_1^3\alpha_j) \to \infty$ and

$$f_4(t_1t_2t_3t_4) \sim 1 + a_{12}a_{21}\left(\omega^{t_1} + \omega^{t_2} + \omega^{t_3}\right) + a_{21}^2 a_{12}^2\, \omega^{t_1+t_3} \quad (44)$$

$$+ a_{12}a_{22}a_{21}\left(\omega^{t_1+t_2} + \omega^{t_2+t_3}\right) + a_{12}a_{22}^2 a_{21}\, \omega^{t_1+t_2+t_3} \equiv h_4(t_1t_2t_3)$$

so that

$$h_4(t_1t_2t_3) = \frac{h_3(t_1t_2)h_3(t_2t_3)}{h_2(t_2)}\left\{1 + \frac{a_{21}a_{12}(a_{22}-a_{12}^2)^2\omega^{t_1+t_2+t_3}}{h_3(t_1t_2)h_3(t_2t_3)}\right\} \quad (45)$$

and

$$Z(r_jr_kr_\ell r_m) \sim (1,1)^4\lambda_1^N h_4(t_1t_2t_3) \quad (46)$$

It can be proven by induction that

$$Z\left(r_{k_1}, r_{k_2}, \ldots, r_{k_n}\right) \sim (1,1)^n\lambda_1^N h_n(t_1t_2\ldots t_{n-1}) \quad (47)$$

where

$$h_n(t_1t_2\ldots t_{n-1}) = \frac{h_{n-1}(t_1..t_{n-2})h_{n-1}(t_2..t_{n-1})}{h_{n-2}(t_2\ldots t_{n-2})} \times$$

$$\times \left\{1 + \frac{a_{21}a_{12}(a_{22}-a_{12}^2)^{n-2}\omega^{t_1+..+t_{n-1}}}{h_{n-1}(t_1..t_{n-2})h_{n-1}(t_2..t_{n-1})}\right\}. \quad (48)$$

Incidentally,

$$a_{12}a_{21} - a_{22} = -1/(1,1)^2 \quad (49)$$

The fraction of intact bonds in a copolymer with the abnormal bonds at $(r_1, r_2, .., r_m)$ is

$$f(m) = \tfrac{1}{2}\{1 - \partial \log Z\} \quad (50a)$$

where

$$\partial \equiv \partial/\partial J_1 + \partial/\partial J_2 \quad . \quad (50b)$$

Following the general formalism of section I, the function f(m) can be expanded as

$$f(m) = \tfrac{1}{2}\left\{1 - \frac{1}{N}\partial\log Z(0) - \frac{1}{N}\sum_{j=1}^{m}\partial\log\frac{Z(j)}{Z(0)} - \frac{1}{N}\sum_{j=1}^{m-1}\sum_{k=j+1}^{m}\partial\log\frac{Z(j,k)Z(0)}{Z(j)Z(k)}\right.$$

$$\left. - \frac{1}{N}\sum_{j=1}^{m-2}\sum_{k=j+1}^{m-1}\sum_{\ell=j+2}^{m}\partial\log\frac{Z(j,k,\ell)Z(j)Z(k)Z(\ell)}{Z(j,k)Z(k,\ell)Z(j,\ell)Z(0)}\ \ldots\right\}$$

$$= \tfrac{1}{2}[f(0) - s_1 - s_2 \ \ldots\ ] \quad (51)$$

With these definitions one finds

$$f(0) = 1 - \partial\log\lambda_{max}; \quad s_1 = (m/N)\partial\log(1,1) \quad (52a)$$

$$s_2 = N^{-1}\sum_{j=1}^{m-1}\sum_{k=j+1}^{m} \partial\log h_2(t)/h_1 = N^{-1}\sum_{t=1}^{N-1} g(t)\partial\log h_2(t)/h_1 \tag{52b}$$

$$s_3 = \sum_{0<t_1,t_2<N}\sum g(t_1,t_2)\partial\log h_3(t_1,t_2)/h_2(t_1)h_2(t_2)h_2(t_1+t_2) \tag{52c}$$

where $g(t)$ is the number of defect pairs that are separated by t base pairs of either type; $g(t_1,t_2)$ is the number of three defect pairs such that t_1 base pairs (of either type) lie between the first two defect pairs and t_2 between the second and third defect pairs.

To simplify our notation, we define

$$U_o = h_1 \quad ; \quad U_1(t) = h_2(t) \tag{53a}$$

$$U_2(t_1,t_2) = h_3(t_1,t_2)/h_2(t_1)h_2(t_2) \tag{53b}$$

$$U_3(t_1,t_2,t_3) = h_4(t_1,t_2,t_3)h_2(t_2)/h_3(t_1,t_2)h_3(t_2,t_3) \tag{53c}$$

Generally

$$U_n(t_1\ldots t_n) = \frac{h_{n+1}(t_1\ldots t_n)h_{n-1}(t_2\ldots t_{n-1})}{h_n(t_1\ldots t_{n-1})h_n(t_2\ldots t_n)} \tag{54}$$

Then

$$s_2 = N^{-1}\sum_{t=1}^{N-1} g(t)\partial\log[U_1(t)/U_o] \tag{55a}$$

$$s_3 = N^{-1}\sum_{0<t_1,t_2<N}\ldots\sum g(t_1,t_2)\partial\log[U_2(t_1,t_2)/U_1(t_1+t_2)] \tag{55b}$$

$$s_4 = N^{-1}\sum_{0<t_1,t_2,t_3<N}\ldots\sum g(t_1t_2t_3)\partial\log\left\{\frac{U_3(t_1,t_2,t_3)U_1(t_1+t_2+t_3)}{U_2(t_1+t_2,t_3)U_2(t_1,t_2+t_3)}\right\} \tag{55c}$$

$$s_5 = N^{-1}\sum_{0<t_1,\ldots,t_4<N}\ldots\sum g(t_1,\ldots,t_4)\partial\log\left\{\frac{U_4(t_1t_2t_3t_4)U_2(t_1+t_2+t_3,t_4)}{U_3(t_1,t_2+t_3,t_4)U_3(t_1+t_2,t_3,t_4)}\right.$$

$$\left.\times\frac{U_2(t_1+t_2,t_3+t_4)U_2(t_1,t_2+t_3+t_4)}{U_3(t_1,t_2,t_3+t_4)U_1(t_1+t_2+t_3+t_4)}\right\} \tag{55d}$$

By now the reader has probably observed the pattern. The numerator of (55d) contains all the partitions of $t_1+t_2+t_3+t_4$ into an even number of partitions

$$4: \ (t_1,t_2,t_3,t_4) \qquad 2: \ (t_1+t_2+t_3,t_4),\ (t_1+t_2,t_3+t_4),\ (t_1,t_2+t_3+t_4) \tag{56a}$$

while the denominator is a product of the U's over all partitions of $t_1+t_2+t_3+t_4$ into an odd number of partitions

$$3: \ (t_1,t_2,t_3+t_4),\ (t_1,t_2+t_3,t_4),\ (t_1+t_2,t_3,t_4) \qquad 1: \ (t_1+t_2+t_3+t_4) \tag{56b}$$

Similarly, the next order term has all the odd partitions of $t_1+..+t_5$ in the numerator and the even partitions in the denominator. These are, respectively,

$$5: \ (t_1,t_2,t_3,t_4,t_5)$$
$$3: \ (t_1,t_2,t_3+t_4+t_5),\ (t_1,t_2+t_3,t_4+t_5),\ (t_1,t_2+t_3+t_4,t_5),\ (t_1+t_2,t_3,t_4+t_5),\ (t_1+t_2,t_3+t_4,t_5),\ (t_1+t_2+t_3,t_4,t_5)$$
$$1: \ (t_1+t_2+t_3+t_4+t_5)$$
$$4: \ (t_1,t_2,t_3,t_4+t_5),\ (t_1,t_2,t_3+t_4,t_5),\ (t_1,t_2+t_3,t_4,t_5),\ (t_1+t_2,t_3,t_4,t_5)$$
$$2: \ (t_1+t_2+t_3+t_4,t_5),\ (t_1+t_2+t_3,t_4+t_5),\ (t_1+t_2,t_3+t_4+t_5),\ (t_1,t_2+t_3+t_4+t_5) \quad \text{etc.} \tag{56c}$$

ii. Random Distribution

We will now apply these general formulae to a random sequence of AT and GC pairs.

Let the defect base pairs be of concentration c and distributed at random. The quantity we need in s_2 (see eq. 52b) is $g(t)$, the number of defect pairs separated by t units. Given a defect pair, the probability that another one is t units away is c. Since the total number of defects is cN, the expected number of defect pairs separated by t intervening pairs of either type is in a very large ring Nc^2 independently of t. Hence

$$s_2 = c^2 \sum_{t=1}^{\infty} \partial \log U_1(t) \quad (57)$$

To find the number of three defect pairs with successive separations t_1 and t_2 we note that given the location of the first defect pair, the probability that another is t_1 units away is c and that another is t_1+t_2 units away from the first is also c. Since there are Nc "first" defect pairs, our required number of three defect pairs with the appropriate separation is Nc^3 independently of t_1 and t_2. Hence

$$s_3 = c^3 \sum_{t_1=1}^{\infty} \sum_{t_2=1}^{\infty} \partial \log U_2(t_1,t_2)/U_1(t_1+t_2) \quad (58)$$

Generally

$$s_n = c^n \sum^{\infty} \ldots \sum^{\infty}_{t_1 \ldots t_n=1} \partial \log \left\{ \frac{U_n(t_1,\ldots,t_n)}{U_{n-1}(t_1 \ldots)} \ldots \right\} \quad (59)$$

where the combination of U's in the bracket is the appropriate one required from the discussion above. The expression (51) for f(m) with the s_j defined above can be simplified by some rearrangements. Let us first collect $\partial \log U_1(t)$ from all the terms. Then from the term proportional to c^2 we find

$$c^2 \sum_{t=1}^{\infty} \partial \log U(t)$$

From the term proportional to c^3, the smallest value of (t_1+t_2) is 2. Then we get:

$$c^3 \left[\partial \log U_1(2) + 2\partial \log U_1(3) + 3\partial \log U_1(4) + \ldots + (k-1)\partial \log U_1(k) + \ldots \right].$$

he coefficient (k-1) results from the fact that there are (k-1) ways of writing $t_1+t_2=k$. From the term proportional to c^4 we obtain (note that the smallest value of $(t_1+t_2+t_3)$ is 3)

$$c^4 \left\{ \partial \log U_1(3) + 3\partial \log U_1(4) + 6\partial \log U_1(5) + \ldots + \left[(k-1)(k-2)/2! \right] \partial \log U_1(k) + \ldots \right\}.$$

The coefficients $(k-1)(k-2)/2!$ result from the number of ways of partitioning k into the sum $t_1+t_2+t_3$. If we collect all the coefficients of $\partial \log U_1(k)$, we find

$$c^2 \partial \log U_1(1) + c^2(1-c)\partial \log U_1(2) + c^2(1-c)^2 \partial \log U_1(3) + \ldots$$
$$= c^2 \sum_{t=1}^{\infty} (1-c)^{t-1} \partial \log U_1(t) \quad .$$

Let us now collect all terms of the form $\partial \log U_2(t_1,t_2)$. Those proportional to c^3 add up to

$$c^3 \sum_{t_1,t_2=1}^{\infty} \partial \log U_2(t_1,t_2)$$

Those proportional to c^4 are

$$c^4 \partial \sum_{t_1 t_2 t_3} \left\{ \log U_2(t_1,t_2+t_3) + \log U_2(t_1+t_2,t_3) \right\}$$

Since t_2+t_3 can equal an integer t_2' in $(t_2'-1)$ ways when t_1 and t_2 are positive integers, this becomes

$$c^4 \sum_{t_1,t_2=1}^{\infty} (t_1+t_2-2)\partial \log U_2(t_1,t_2)$$

The term proportional to c^5 is

$$c^5 \partial \sum_{t_1}..\sum_{t_4} \left\{ \log U_2(t_1,t_2+t_3+t_4)+\log U_2(t_1+t_2,t_3+t_4)+\log U_2(t_1+t_2+t_3,t_4) \right\}$$

Since $t_2+t_3+t_4$ can equal an integer t_2' in $\frac{1}{2}(t_2'-1)(t_2'-2)$ ways and t_1+t_2 can equal the integer t_1' in $(t_1'-1)$ ways, this term becomes

$$c^5 \Sigma \left\{ \tfrac{1}{2}(t_2-1)(t_2-2)+(t_1-1)(t_2-1)+\tfrac{1}{2}(t_1-1)(t_1-2 \right\} \partial \log U_2(t_1,t_2)$$

$$= c^5 \Sigma \tfrac{1}{2}(t_1+t_2-2)(t_1+t_2-3)\partial \log U_2(t_1,t_2) \quad \text{etc.}$$

If all the terms proportional to $\partial \log U_2(t_1,t_2)$ are added together, one finds the sum to be

$$c^3 \sum_{t_1,t_2=1}^{\infty} (1-c)^{t_1+t_2-2} \partial \log U_2(t_1,t_2) \quad .$$

The terms of the form $\partial \log U_3(t_1,t_2,t_3)$ can be analyzed in the same way. The sum of all terms of this type is

$$c^3 \partial \Big\{ \Sigma \log U_3(t_1 t_2 t_3) - c\Sigma \Big[\log U_3(t_1+t_2,t_3,t_4)+\log U_3(t_1,t_2+t_3,t_4)+$$
$$+\log U_3(t_1,t_2,t_3+t_4) \Big] + c^2 \Sigma \Big[\log U_3(t_1,t_2,t_3+t_4+t_5)+$$
$$+\log U_3(t_1,t_2+t_3,t_4+t_5)+\log U_3(t_1,t_2+t_3+t_4,t_5)+$$
$$+\log U_3(t_1+t_2,t_3,t_4+t_5)+\log U_3(t_1+t_2,t_3+t_4,t_5)+$$
$$+\log U_3(t_1+t_2+t_3,t_4,t_5) \Big] + \ldots \Big\}$$
$$= c^3 \Sigma \Big\{ 1-c\Big[(t_1-1)+(t_2-1)+(t_3-1) \Big]$$
$$+c^2 \Big[\tfrac{1}{2}(t_3-1)(t_3-2)+(t_2-1)(t_3-1)+\tfrac{1}{2}(t_2-1)(t_2-2)$$

$$+(t_1-1)(t_3-1)+(t_1-1)(t_2-1)+\tfrac{1}{2}(t_1-1)(t_1-2\Big]-$$

$$-\ ..\Big\}U_3(t_1,t_2,t_3)$$

$$=c^3\Sigma\Big\{1-c(t_1+t_2+t_3-3)+$$

$$+c^2(t_1+t_2+t_3-3)(t_1+t_2+t_3-4)/2!\ldots\Big\}U_3(t_1,t_2,t_3)$$

$$=c^3\Sigma(1-c)^{t_1+t_2+t_3-3}U_3(t_1,t_2,t_3) \qquad \text{etc.}$$

The final expression for the fraction of bonds intact as a function of concentration is

$$f(c)=\tfrac{1}{2}\Big\{1-\partial\log\lambda_{max}-c\partial\log(1,1)$$

$$-\sum_{j=1}^{\infty}c^{j+1}\sum_{t_1..t_j=1}^{\infty}\sum^{\infty}(1-c)^{t_1+t_2+..+t_j-j}\partial\log U_j(t_1..t_j)\Big\} \qquad (60)$$

The approximate result which corresponds to stopping this series with the j-1 term has been given by N. S. Goel and one of the authors.[1] The algorithm for the calculation of f(c) is now clear. The quantities h_1,h_2...are derived using the recurrence formula (48) and the U_j's which are to be inserted into (60) are derived from the defining relation for the U 's, eq. (54).

Another algorithm for the f(c) has recently been derived by G. W. Lehman.[6]

The variation of the fraction of bonds intact is plotted in Fig. 4 as a function of temperature for random chains of equal concentration of AT and GC pairs. The calculations were made on a high speed computer using the algorithm eq. (60). The correlation parameter U in each case is related to the parameter a through the relationship a = exp 2U as derived from eq. (13) to fit experimental data on the slope of one component pure AT-DNA (a discussion of the choice of parameters is given in references 1 and 5). When U>2 the convergence of (60) becomes rather slow.

One can consider the unperturbed chain to be either a pure AT or a pure GC chain. It turns out that the more rapid convergence of (60) is achieved by considering GC pairs as defects in an AT lattice at temperatures above T_c and AT pairs as defects in the GC lattice when $T<T_c$. The reason for this is that certain denominators appear in the calculation of derivatives of log $U_2(t)$ which are small when $T\simeq T_1$, the critical temperature of a pure AT chain when it is considered to be the unperturbed system. This small denominator problem is avoided by using pure GC as the unperturbed system when $T\simeq T_1$. The same difficulty must be avoided when $T\simeq T_2$,

the critical temperature of pure GC-DNA by using an AT lattice as the unperturbed system when $T>T_c$.

iii. Melting Curves for Large U and Approximation of Arbitrary Distributions by Periodic Ones.

It was shown in reference 1 that if $U\to\infty$ the melting curve becomes step function with the melting point at $T_c = c_1T_1+c_2T_2$ (c_j being the concentration of the jth component) independently of the details of the sequence. At $T<T_c$ all bonds are intact and at $T>T_c$ all are broken. Hence there should be a U' such that for all $U>U'$, the melting curves are practically the same for two sequences with the same number of GC pairs which are nearest neighbors and second nearest neighbors. Two sequences which agree up to, say 4th neighbors, should give similar melting curves for $U>U''$ where $U''<U'$, etc. We now discuss various periodic approximations to the random distribution and, at the end of this section, give an estimate of the range of validity of these approximations.

Let us seek a sequence of periodic chains which approximate a random sequence composed of half AT pairs and half GC pairs. If the period is small, only nearest and, perhaps, next nearest neighbors occur in the right number but, as the period becomes longer, one can expect more distant neighbor pairs to appear in the proper proportion. In a random chain of length N the number of pairs of B's separated by an integral number of lattice spacings is N/4.

In the chain of period 4, (letting A represent an AT pair and B a GC pair), and length N

$$\text{AABBAABBAABB} \ldots \tag{61}$$

the number of pairs of B's which are nearest neighbors is N/4, second neighbors none, third neighbors N/4, or, as indicated below

1. N/4	5. N/4	9. N/4
2. 0	6. 0	10. 0
3. N/4	7. N/4	11. N/4
4. 2N/4	8. 2N/4	12. 2N/4 ....

Generally, for separations,

4j-3 with j = 1,2,...	the number is N/4
4j-2 with j = 1,2,...	the number is 0
4j-1 with j = 1,2,...	the number is N/4
4j with j = 1,2,...	the number is 2N/4

The absence of neighbors separated by 4j-2 spaces is compensated by those of 4j. While this sequence of period 4 agrees with a random chain in number of nearest neighbor B pairs, it differs in the number of second neighbor pairs.

The chain of period 8

$$|AAABABBB|AAABABBB| \; ... \tag{62}$$

has the required number, N/4, of pairs of B's separated by

8j-7;8j-6;8j-4;8j-2;8j-1 with j = 1,2,...

only half the required number, N/8, occur for separations 8j-5 and 8j-3. These are compensated by separations 8j which appear in twice the required number $\frac{1}{2}N$. In doubling the length of the period, we obtain the correct number of nearest and next nearest pairs and the deficiency in third neighbor pairs is less than the deficiency of second neighbor pairs in the chains of period 4.

A chain of period 12 which agrees with a random chain in the number of first, second, third and fourth neighbors is

$$BBBBABAAABAA|BBBBABAAABAA \tag{63}$$

Another is obtained by interchanging the A's and B's.

One period of an example of a case of period 16 which agrees with a random chain up to sixth neighbors is

$$BBBAAAABAABABBBABB \tag{64}$$

A second such chain can be obtained by interchanging the A's and B's.

When U=3 the chains (62), (63), and (64) yield melting curves which differ from each other through the entire temperature range by, at most, 0.001, an amount that cannot be detected in graphs on the scale given in Fig. 4. When U=2.5 there is still essentially no difference between the melting curves of sequences (63) and (64), (see Table I), which differ in the number of pairs which are fifth and sixth neighbors. Since (64) differs from a random distribution by even more remote neighbors, those at least 7-base spacings away, we can conclude that (64) gives essentially the same melting curves as a random base distribution (50% AT, 50% GC) as long as U≥2.5.

In conclusion the authors wish to thank Dr. Narendra S. Goel for a number of useful discussions.

Table I

Comparison of Fraction of Bonds Intact with Regular Distributions of Period = 12 and Period = 16 for U = 2.5. See (63) and (64) of Text.

FRACTION OF BONDS INTACT f(m)

Temp., °C	Period=12	Period=16
79.80	0.9972	0.9971
80.80	0.9966	0.9964
81.80	0.9957	0.9955
82.80	0.9944	0.9941
83.80	0.9924	0.9921
84.80	0.9892	0.9887
85.80	0.9834	0.9827
86.80	0.9716	0.9705
87.80	0.9420	0.9401
88.80	0.8436	0.8400
89.80	0.5000	0.5000
90.80	0.1564	0.1600
91.80	0.0580	0.0599
92.80	0.0284	0.0295
93.80	0.0166	0.0173
94.80	0.0108	0.0113
95.80	0.0076	0.0079
96.80	0.0056	0.0059
97.80	0.0043	0.0045
98.80	0.0034	0.0036
99.80	0.0028	0.0029

BIBLIOGRAPHY

* This work was partially supported by The Office of Naval Research.

1. E. W. Montroll and N. S. Goel, Biopolymers 4, 855 (1966).
2. J. Marmur, R. Rownd, and C. L. Schildkraut, in PROGRESS IN NUCLEIC ACID RESEARCH, Vol. 1, Academic Press, New York, 1963, pp. 231-300.
   G. Felsenfeld and H. Todd Miles, in Ann. Rev. of Biochem., 1967, pp. 407-448.
3. G. Newell and E. W. Montroll, Rev. Mod. Phys., 25, 353 (1953).
4. N. S. Goel, N. Fukuda and R. Rein, preprint (1967).
   R. Rein, N. S. Goel, N. Fukuda, M. Pollak and P. Claverie, Annals N. Y. Acad. Sci., to be published.
5. N. S. Goel and E. W. Montroll, preprint (1967).
6. G. W. Lehman; to be published in PROC. OF IUPAP CONF. ON STATISTICAL MECHANICS, Copenhagen, July 11-16, 1966.

Anharmonic Correlations of the U-Center Local Mode with Phonons in the U-Center Sidebands

R.Zeyher and H.Bilz

Institut für Theoretische Physik der Universität

Frankfurt/Main, Germany

Introduction

In a recent paper[1] we discussed the influence of the anharmonic broadening of the U-Center local mode on the sideband absorption. The main results are shifts of the maxima positions and a drastic reduction of the absorption in the gap region between the acoustic and optic modes. In addition the ratio of intensities of the first to the second maximum is increased, improving the agreement in the acoustic region between the experiment and a simple theory which neglects the effect under consideration[2,3,4]. Results have been given for summation bands in KI and KBr containing U-Centers. In this paper we extend the investigation to KCl, NaBr and NaI.

1. Harmonic version of sidebands[1].

Usually, the following formula has been used in calculating sidebands of U-Centers[2,3,4]:

$$\varepsilon''^{(0)}(\omega) = 4\pi M_{1L}^2 \frac{4\omega_L^2 \Gamma_L^{(0)}(\omega)}{[\omega_L^2 - \omega^2]^2} \tag{1}$$

Here M_{1L} means the linear dipole moment of the localized mode with the experimental frequency ω_L and the damping function $\Gamma_L^{(0)}(\omega)$ which describes the anharmonic decay of the local mode in the sideband region. This should be compared with a more correct formula[1]:

FIG. 1 [6)]

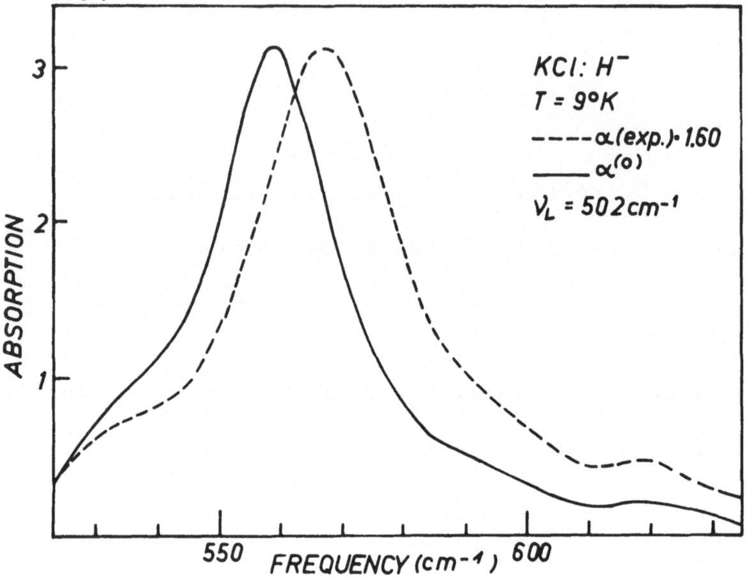
KCl: H⁻
T = 9°K
----α(exp.)·1.60
——α(o)
νL = 502cm-1
ABSORPTION
FREQUENCY (cm-1)
3
2
1
550
600

FIG. 2 [6)]

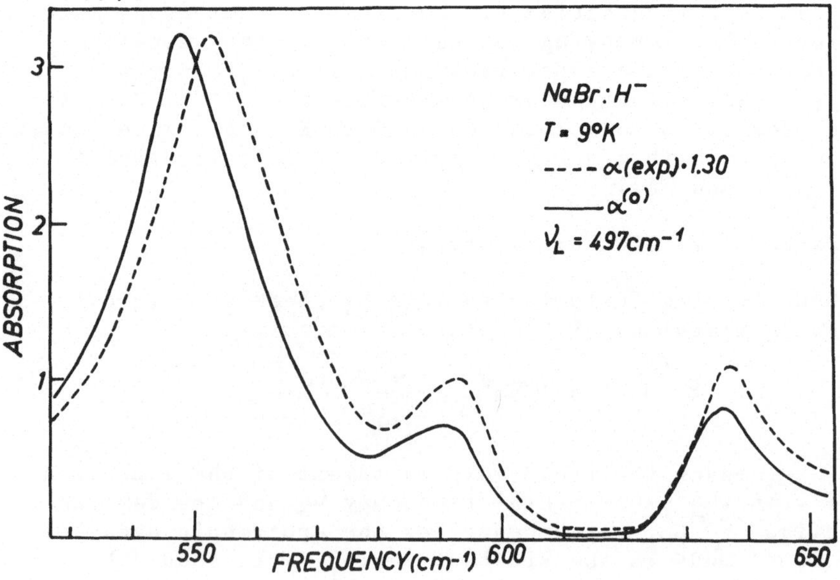
NaBr: H⁻
T = 9°K
----α(exp.)·1.30
——α(o)
νL = 497cm-1
ABSORPTION
FREQUENCY(cm-1)
3
2
1
550
600
650

$$\varepsilon''(\omega) = 4\pi M_{1L}^2 \frac{4\,\omega_L^{(0)2}\Gamma_L(\omega)}{[\tilde{\omega}_L^2(\omega) - \omega^2]^2 + 4\,\omega_L^{(0)2}\Gamma_L^2(\omega)} \qquad (2)$$

with the frequency-dependent eigenfrequency

$$\tilde{\omega}_L^2(\omega) = \omega_L^{(0)2} + 2\omega_L^{(0)}\Delta_L(\omega) \quad ; \quad \omega_L = \tilde{\omega}_L(\tilde{\omega}_L) \qquad (3)$$

Here Δ_L is the real part of the self-energy

$$\Pi_L(\omega) = \Delta_L(\omega) - i\Gamma_L(\omega) \qquad (4)$$

$\Gamma_L(\omega)$ contains all types of decay processes including correlations between the local mode and band phonons in the sideband region while $\Gamma_L^{(0)}(\omega)$ in eq. (1) describes those types of final states only where no correlations between the local mode and band phonons are considered (harmonic version of the damping function).

Formula (2) is the correct formula for the local mode dipole absoption in so far as contributions of the non-linear dipole moment and interference processes with band phonons - especially the infrared active dispersion oscillator - are negligable. This seems to be a reasonable assumption for most of the alkali halides i.e. corrections should be smaller than 20%.(see the discussion in [5]).

As shown in ref. 1), there exists a straightforward iteration procedure which allows one to calculate $\varepsilon''^{(0)}(\omega)$ (eq. (1)) from the experimentally determined $\varepsilon''(\omega)$ assuming that eq. (2) describes the measured data within the abovementioned limits. An essential point of the calculation is a self-consistent determination of $\Delta_L(\omega)$ and $\Gamma_L(\omega)$ via a Kramers-Kronig transformation.

The calculations are made in the limit of zero tempera - ture and neglecting the self-energy of the band phonons. In ref. [1] results are given for KBr and KI. In figs. 1.,2., and 3. corresponding results are shown for KCl, NaBr and NaI. The calculated harmonic versions of the absorption constants $\alpha^{(0)}$ are always normalized so as to have the same value of the first sideband maximum as the experimental absorption α. This shows more clearly the shift of the first maximum to lower frequencies in the harmonic version as compared with the experimental curve and a decreased halfwidth of this part of the side-band spectrum. The re-sonance character of the first maximum is more pronounced for $\alpha^{(0)}$.

FIG.3 7)

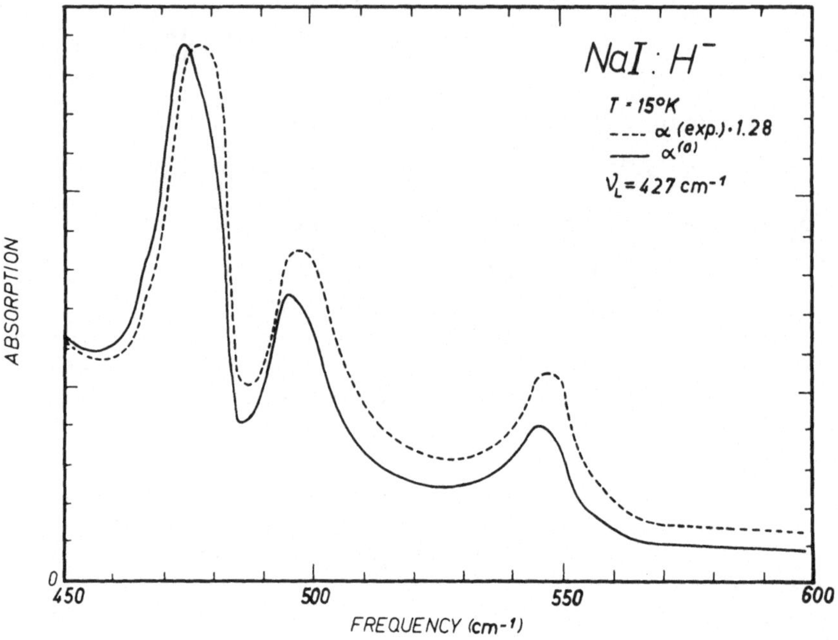
NaI : H⁻
T = 15°K
---- α (exp.)·1.28
—— α^(0)
ν_L = 427 cm⁻¹
ABSORPTION
0
450
500
550
600
FREQUENCY (cm⁻¹)

FIG. 4 6)

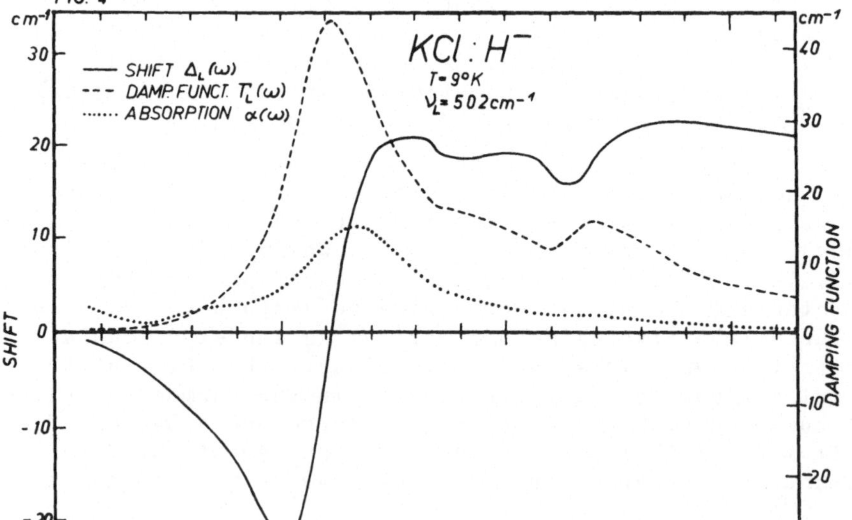
KCl : H⁻
T = 9°K
ν_L = 502cm⁻¹
SHIFT Δ_L(ω)
DAMP. FUNCT. Γ_L(ω)
ABSORPTION α(ω)
cm⁻¹
30
20
10
0
-10
-20
SHIFT
cm⁻¹
40
30
20
10
0
-10
-20
-30
DAMPING FUNCTION
500
550
600
650
FREQUENCY (cm⁻¹)

2. Effect of the shift function $\Delta_L(\omega)$.

At low temperatures $(T \ll \Theta)$ the shift function $\Delta_L(\omega)$ in the low-frequency region of the sideband is quite insensitive to contributions of difference processes. This means that the first sideband absorption maximum in this region is not shifted going from α to $\alpha^{(0)}$.

For the high-frequency region fig. 4 shows the shift and the damping functions $\Delta_L(\omega)$ and $\Gamma_L(\omega)$ for KCl:H^-. There exists a remarkable jump of about 40 cm^{-1} in the region of the first maximum. This causes a shift of about 6 cm^{-1} to lower frequencies in the harmonic version as compared with the experimental α . Table 1 shows the distances Δ_+ and Δ_- of the first maxima in the high and low frequency region from the local mode position.

Tab. 1

	Δ_+	Δ_-	$\Delta_+^{(0)}$	$\Delta_-^{(0)}$	
KCl : H^-[6]	64	59	58	59	cm^{-1}
NaBr : H^-[6]	54	50	51	50	cm^{-1}

In the harmonic version, Δ_+ and Δ_- are nearly equal as implicitely required by eq. (1)

3. Discussion

The results for KCl:H^- and NaBr:H^- are similar to those for KBr and KI. In the case of NaI, the gap absorption is nearly unchanged in the harmonic version (keeping in mind the factor 1.28 of fig. 3.) This means that the observed gap absorption in NaI is not due to the correlation of the local mode and band phonons in contrast to KBr[6] and KI. (In NaBr the observed gap absorption is very small[6]). It is not clear whether or not two-phonon processes caused by quartic anharmonicity are responsible for this effect.

Recently , H.Dötsch measured the sidebands of NaF and LiF[8]. At least in the case of LiF, observed effects of interaction of U-Centers make the usefulness of a harmonic version doubtful.

Acknowledgement: The authors are indebted to the Deutsche Forschungsgemeinschaft for financial support of numerical calculations.

References

1) H.Bilz , R.Zeyher and R.K.Wehner, phys. stat. sol. 20, K167 (1967)
2) Th.Timusk and M.V.Klein , Phys. Rev. 141,664 (1966)
3) X.X.Nguyen, Solid State Commun. 4, 9 (1966)
4) J.B.Page and B.G.Dick, to be published
5) B.Fritz, Proc. Int. Conf. Loc. Excit., Irvine, California (1967)
6) B.Fritz, private communication
7) D.Bäuerle, Diplomarbeit Stuttgart, 1967
8) H.Dötsch, Thesis, Frankfurt, 1967

CONFERENCE SUMMARY*

R. J. Elliott

Theoretical Physics Department, Oxford University

Oxford, England

It is essentially impossible to provide a suitable summary for a conference which lasts for five days and covers the sort of area which this one did. In any case we had a very masterful summary in the very first paper by Alex Maradudin of the experimental and theoretical position up to date. The conference has, I think, extended that situation and indicated that progress is going forward on a very broad front into a wide variety of phenomena associated with the topic of localized excitations. It is rather difficult to pick out particular contributions, so I hope that you will take what I have to say now as rather personal reflections on what happened here this week.

The title of the conference was Localized Excitations in Solids, but you realize that we have really only talked about half that topic, namely localized bosons and have deliberately, I think as the policy of the conference, excluded the properties of localized electrons in materials. Had we not done that, it would certainly have doubled at least the size of the conference and probably made it very unwieldly, but it is perhaps worth bearing in mind as we go away that there is an enormous literature of related topics, the localized states in semiconductors, the localized states in metals associated with the kind of effects that Friedel has often discussed, and nowadays if you came from the Magnetism Conference last week as I did, the Kondo effect, all of which are very important phenomena associated with localized excitations.

*This is a transcript of Dr. Elliott's talk and is not a prepared manuscript.

To return to our own meeting, three-quarters of the papers have in fact been concerned with localized vibrations. As has been pointed out in the opening talk, it is now seven years since the first detailed experiments on localized vibrations in solids began to appear in the literature. During that seven years there has been a very strong coupling between theory and experiment which has produced a large amount of work and a great deal of progress. In the case of lattice vibtations I think we have now passed beyond the use of the very simple models and are passing into a much more sophisticated period when the detailed experimental results are going to require a much more detailed discussion in theoretical interpretation. If we look a little at the sort of experiments which have been reported here, the most direct experiments are those which measure one phonon — one-phonon absorption, for example, one-phonon scattering by neutrons. Here I think the theorist can now tell you the form of the modes for isolated impurities — localized, gap or resonance band modes — and as long as one is doing an experiment where one knows the coupling constant between the probe and these modes, the confrontation between theory and experiment is usually in good shape. This perhaps points up the view that the neutron scattering technique, which is anyway a very powerful one, is one where you really do know the interaction between the thing you are using, the neutron, and the nucleus, because it is really the same as the single nuclear scattering length which you can determine by other experiments. This perhaps indicates that this powerful but rather little used technique, because of the difficulty of getting hold of a pile and neutrons from it, should perhaps be pushed a little more than is being done at the moment.

When it comes to the much commoner, simpler from the experimental view, optical experiments, it has been clear during the course of the meeting that there has been growing dissatisfaction with the simplest pictures of the way in which the light wave couples to the lattice vibrations. I think that the very simple picture, where you just consider an atom to be an entity with an electrical charge whose oscillation provides the coupling to the light, has been given a death blow by Szigeti when he wrote on the blackboard his calculation of what the charge is on the defect atom. In the very simplest case of a semiconductor where you might have expected it to be one, he wrote down approximately zero. And in fact it appears that in semiconductors and in ionic crystals it is the polarization of the atoms, the polarization of the charge cloud caused by short range forces and by long range electronic forces, which gives the dominant coupling of the lattice vibrations to the light wave. This calls for a great extension, I think, of the theoretical model, and I hope it will lead to much more extensive work on detailed theories of the way in which the polarization of atoms is affected by crystal vibrations. We have seen the beginnings of many more sophisticated models of this sort, in particular, for example, the breathing shell model which was discussed today, but even these models have a serious fundamental drawback in that one knows that atoms are really not like that — they don't consist of a point in the middle and a shell on the outside even if you divide the shell up into six parts. We must in the future, I think, look much more closely theoretically into the way in which the electrons respond when the nuclei are displaced. In particular, I think the phenomenon of looking at localized vibrations — that is, looking at perturbed lattice vibrations — will give us much more information about these polarization effects, when we come to understand them some more, than will just the perfect lattice by itself, so that here again the interaction between theory and experiment should prove very fruitful. On the one hand, better models will help us to interpret these more detailed experiments, but on the other hand, the detailed experiments might in the end give us a much better fundemantal insight into the way in which the polarization of atoms occurs in a solid.

Perhaps one should add a word of encouragement at this time for those brave souls who attempt first principles calculations of the way in which the energy varies with displacement of atoms in a solid. The results they get always seem quite a long way from experiment, but eventually, if we wish to understand what solids are about, we shall have to do calculations of this kind.

Passing from one phonon effects to two phonon effects, it is clear that if you cannot understand the intensity of the one phonon effect, it is going to be a lot worse to try to understand the two phonon effect. Here, the second order dipole moment, the polarization caused by the atomic distortions, is essential, and the kind of extentions to the theory which I've just been hinting at are again essential here for a proper interpretation of this data.

There was comparatively little at this conference about vibronic sidebands, about the vibrational fine structure on the sides of electronic transitions. I find it a little surprising because it is a subject which has been studied extensively experimentally and to some extent theoretically over the past few years. I don't think that we can read into this any particular trend which indicates a slowing off of interest in this important branch of the subject, but just a random fluctuation that didn't bring here the people who are predominantly interested in it.

Another aspect of the vibrations of defects which has received considerable attention here and must continue to do so are the anharmonic effects. Having gotten the harmonic model straight, it is essential to see the way in which a realistic lattice, which is always anharmonic, produces modifications in those simple predictions. The intensity and the width of the lines seen by some probe such as neutrons or photons are related to the anharmonic effects, and one of the most striking things to me as a theoretician was the way in which narrow resonance modes look from this point of view, from the way they broaden and vary with temperature, exactly like localized modes. While there is clearly some similarity in a simple physical picture, the normal method of treating these theoretically is quite different. Here again there are a number of calculations, Maradudin and the German groups, but I think much more effort in this field is going to be needed before we really feel that we understand all the anharmonic effects. When we do, we shall have used localized excitations to tell us a great deal more about crystals — namely, more about the anharmonic properties which are rather difficult to get at in a microscopic way from the bulk properties of perfect crystals.

An aspect of localized vibrations in this conference which appears to be having a vogue at this moment is the study of pairs of impurities. There were a considerable number of papers on this, particularly in semiconductors, but also in ionic crystals. It is clear that this technique promises much more detailed information about the coupling between defect atoms in crystals, about their agglommeration, about the tendency for them to form pairs or not, and I think here again we should expect an extension in the use of this method to try and tell us more about the way in which impurities behave in crystals.

From pairs of impurities we pass to the high concentration random lattices which we have heard about mainly this afternoon. Here again there are a number of very interesting experimental papers which have been presented here. This represents an aspect of a very important basic theoretical physics problem which is essentially unsolved. You really can do only as much as you do about the theory of crystals because of the very high degree of symmetry which they have, the high translational symmetry, which really simplifies the calculation by orders of magnitude. For single isolated impurities the breakdown of this symmetry is somewhat compensated by the short range of the perturbation, and this is the property which is used to deal with the theory of localized impurities. When one has completely broken down the translational symmetry, none of the standard theoretical techniques really works. We have seen attempts today to deal with the one-dimensional model, and there have been machine calculations to attempt to deal with more realistic models. It may be that the problem is just too hard for the theorist, and we shall be forced back to this semi-empirical derivation of the properties — namely, by putting the models on machines and seeing how they come out. In a sense this is more an experiment than a theory.

Perhaps it might be worth pointing out that there are two sorts of experiments which have been done on these high concentration random lattices. The first is the sort of experiment which is able to look at the response of the system for all values of the wavevector k or some distribution of it. Neutron scattering would do this, although it has not been extensively applied to this particular problem. The sort of work that has applied it here was the work on superconductive tunnelling. This also suffers a bit from the difficulty of knowing the interaction between the probe you are using, the electrons in this case, and the phonons. There is no detailed theory of the electron-phonon interaction, but we do know enough about it to be able to make some interpretation of the results in terms of the density of states of the modes. It is interesting that in the case that was described here one did get what one would generally expect. In low concentration at one end of the series it looked like the host lattice with the dominant local mode or resonance behavior of the defect superimposed as a small effect; if you went to the other end, you got the same result — another host lattice now and another sort of defect — but when you got in the middle it was really a mess with some features of both lattices still there, but all the features very broad.

Now the other kind of experiment that is predominantly done is the $k = 0$ mode experiment using optical properties which we heard about this afternoon. Here, it appears that there are two types of behavior in which you either see a sort of virtual crystal result where a single strong mode couples to the light and this moves continuously as you vary the concentration, or you see something which looks like the property of each constituent lattice with one intensity going down and the other rising as you vary the concentration. This suggests all kinds of analogies. In the electron problem it suggests the analogy between the mean lattice band theory, for example, where you apply to an alloy the band theory appropriate to a perfect lattice half way inbetween, or the sort of localized electron theory where you apply to each atom the properties that it would have by itself. It is clear, however, that much more theoretical work will have to be done in this range if we are to understand either of these two types of experiment with any degree of completeness.

Finally, under the topic of lattice vibrations one should draw attention to the discussion of surface modes which took place here. The defect in this case, a large surface, is of course much bigger than the defects that one normally talks about in this respect; however, the high degree of symmetry that remains in the rest of the crystal after just cutting across at one end allows one to get a considerable way with the theory, though probably not as far as one can get with the single defect case. The lack of coupling between probes and the surface modes seems to be a drawback, but I think the most striking impression that one was left with here is that we really need much more information, and I hope that the experimentalists will take this to heart and provide us with some before very long.

Besides lattice vibrations, one of the other two main topics of discussion was that of magnons. Here the basic theory is more difficult than in lattice vibrations because one is always in a rather anharmonic situation except possibly at very low temperatures. Moreover, if there is any kind of antiferromagnetic coupling in the problem, there is some difficulty in deciding what the ground state is. Most of these problems are overcome by the kind of theoretical discussion which we heard here. Experimentally the study of magnons is also difficult simply because the coupling to all the standard probes is rather weak. Light interacting with a magnetic transition normally has only a magnetic dipole character and this of course is much weaker than an electric dipole transition which one is concerned with in the vibrational case. Even neutrons have a weaker cross section with magnetic ions than they do through the nucleus as a general rule. It does seem that a very good idea to pursue from the experimental point of view is the use of electrons. Two quite separate things were suggested here — one, that the electrical conductivity, where you use the conduction electrons as your probe to study localized magnetic spins on the lattice, would tell us a great deal about defect magnons. The other was that one might use electrons to study surface spin waves. I think that the theory of defect magnons is still in a fairly elementary stage. Most of the theoretical calculations are still done on the very simplest models and not on the kind of realistic magnon spectra which we have now begun to know something about. The effort involved will clearly be rather considerable, and I am sure that nobody wants to make it until there is a greater variety of experimental evidence available in this field.

In contrast to the fact that we heard nothing about vibronic sidebands on electronic transitions, we did hear a considerable amount about magnon sidebands on electronic transitions (optical transitions) in crystals. This technique promises to tell us a great deal about excited electronic states in magnetic crystals, and this kind of information is not easy to come by. However, if one wants, for example, to understand about the exchange interaction which is responsible for ordered magnetism and if one has some technique for finding out what the exchange interaction is like in the excited state, it should be a great help with the ground state.

Finally, the other topic which occupied us here was the study of excitons, mainly, as it turned out, in semiconductor-like materials. There is, however, the possibility of getting localized excitons in the kind of tightly bound ionic materials where the excitons are of the Frenkel type. Although these were not discussed here, it does represent another aspect of this subject which has had some attention and will certainly have some more. We heard about some extremely beautiful experiments on a wide variety of materials. Here the basic theory is considerably more difficult than in the case of either magnons or lattice vibrations simply because these excitons are large objects and also fairly complex ones. They certainly contain an electron and a hole and when bound to the impurity they often have another electron there as well. The beginning of a fundamental theory of these things was discussed here and in addition to that there has been considerable phenomenological interpretation of the detailed results which have been obtained, and this has been used to tell us much more about the band structure of these materials. We heard a little bit about the strong interaction which might take place between electrons or holes or their combination as excitons and the lattice — namely, the formation of polaron-like effects and the self-trapping of excitations when the interaction between the particles and the lattice is important. There is no doubt that this will be a fruitful field for investigation in the future, and for the moment for my own part I would say that the situation as regards the significance of these effects is far from clear.

So, what can we expect over the next few years if we should have a conference of this kind again? Clearly, the basic ideas of localized excitations in solids are now pretty clear. One knows for a large part what sort of theory one should be doing in order to explain the results, but to make a detailed comparison between theory and experiment much work needs to be done. We need a much deeper insight into the polarization effects in vibrations. We need a much deeper insight into the random lattice problem. So I hope that you will accept that as a personal review of what I thought about the meeting. You will see that it was severely prejudiced toward the theoretician's cause, but I hope that you will accept that as a normal prejudice in a theoretician.

It simply remains for me not as a reviewer but as a member of the conference to say a few words of thanks to the people who made this conference possible. The first of these, of course, is Alex Maradudin who really seemed to take almost the whole burden of conceiving, organizing, and running the conference. He has as you know made a great many personal contributions to this subject, he has added to these the service of providing us with an excellent and authoritative review of the situation in localized vibrations, and now he has added to that the social service of providing us with a conference in which we can come together and discuss all these things of mutual interest. As well as Alex, the people I think mainly involved have been Eli Burstein — I am never quite sure what Eli does in these meetings, he is a kind of benevolent "eminence gris" in the background, but what I do know is that they would not run so smoothly and they would not occur with such frequency were it not for Eli's particular gift. Dick Wallis perhaps should be mentioned. Having edited the proceedings of the Copenhagen Conference, I am surprised that he would ever take on a job like that again. It must be a task more unrewarding than the one

I have here, but he too deserves our thanks and our admiration if the timetable which has been suggested is even approached. And finally the staff of the Physics Department here at the University of California, Irvine, particularly Rod Rose and his colleagues, have been so helpful in getting us from the airport, looking after us while we are here, and taking us off to enjoy ourselves when we could get a little rest from physics. So I hope you will join me in saying thanks to all these people.

AUTHOR INDEX